CONVERSION FACTORS AND SELECTED CONSTANTS

$\ln x = 2.303 \log_{10} x$
$\ln 10 = 2.303 \ldots$
$e = 2.718 \ldots$
$\pi = 3.1416 \ldots$
1 in. = 2.54 cm = 0.0254 m
1 lb mass = 454 g = 0.454 kg
1 lb force = 4.44 newtons
1 Å = 10^{-8} cm = 0.1 nm
°C = (°F − 32)/1.8
K = °C + 273
°R = °F + 460
1 poise = 0.1 Pa-sec
1 lb force/in.^{2} = 6.9×10^3 N/m^2 = 6.9×10^{-3} MPa = 7.03×10^{-4} kg/mm^2
1 cal = 4.186 J
1 eV = 1.6×10^{-19} J
1 erg = 1×10^{-7} J
R = gas constant = 1.987 cal/mole·K
k = Boltzmann's constant = 1.3×10^{-16} erg/K = 1.3×10^{-23} J/K
h = Planck's constant = 6.62×10^{-27} erg-sec = 6.62×10^{-34} J·sec
Avogadro's number = 6.02×10^{23} atoms/at. wt.
= 6.02×10^{23} molecules/mol. wt.
Density of water = 1 g/cm^3 = 62.4 lb/ft^3 = 0.0361 lb/in.^{3}
1 gal water weighs 8.33 lb
Electron charge = electron hole charge = 1.6×10^{-19} coul
1 Bohr magneton = 9.27×10^{-24} amp-m^2 = 0.927 erg/gauss
c = velocity of light = 3×10^{10} cm/sec

Prefixes

giga	G	10^9
mega	M	10^6
kilo	k	10^3
milli	m	10^{-3}
micro	μ	10^{-6}
nano	n	10^{-9}

ENGINEERING MATERIALS AND THEIR APPLICATIONS
Second Edition

ENGINEERING MATERIALS AND THEIR APPLICATIONS
Second Edition

Richard A. Flinn
University of Michigan, Ann Arbor

Paul K. Trojan
University of Michigan, Dearborn

HOUGHTON MIFFLIN COMPANY / BOSTON
Dallas Geneva, Illinois Hopewell, New Jersey Palo Alto London

To Edwina and Barbara

Credits for chapter-opening photographs: Chapter 3, Pratt and Whitney Aircraft Group; Chapter 4, John Mardinly; Chapter 5, LeRoy F. Grannis; Chapter 6, International Nickel Company; Chapters 7 and 8, Corning Glass Works; Chapter 9, G. S. Yeh/University of Michigan; Chapter 13, Wide World Photos; Chapter 14, Gregory Head; Chapter 15, C. C. Wu and B. Roessler

On the cover: The scanning electron micrograph was done by Stephen Krause

Printed in the U.S.A.

Library of Congress Catalog Card Number: 80-82840

ISBN: 0-395-29645-5

CONTENTS

14 Electrical Properties of Materials 586

15 Magnetic Properties of Materials 628

16 Optical and Thermal Properties of Materials 656

PREFACE

A good preface should accomplish two things: First, it should be a heart-to-heart talk with students about what is in the book and, specifically, about how the material covered can be useful to them in their engineering careers. It should point out the rough spots in the course and explain why some purely theoretical concepts are needed as a background for understanding the important applications discussed later.

Second, the Preface, along with the Instructor's Manual, provides an opportunity for candid conversation with our colleagues as to the ways we have changed the coverage of topics to increase student interest. With proper attention, many students who may take the course grudgingly because it is required can be motivated to explore the field of materials further.

TO THE STUDENT

Now let's examine what the course is all about and why it's important to an engineer. It would be easy to put together some grandiose statements about the importance of metallurgy, ceramics, and plastics in society. But such statements might later fall flat during the dreary exam days. Instead, let's take a hardheaded engineering approach. Let's extract from each chapter one important idea or principle and see how each one puts the engineer in a position to understand, specify, and even develop new materials.

In the first chapter we find that all engineering materials may be divided into three classes: *metals, ceramics,* and *plastics*. The different properties we encounter in these groups depend basically on the structure of the particular material, that is, the way the atoms are arranged and bonded.

Chapters 2 through 6 are devoted to *metallic* structures and their properties. In Chapter 2 the punch line is that a grain (or crystal) of metal is built up of just simple molecules (unit cells) that give rise to planes of atoms. These planes of atoms form the structural framework of the metal, just as planes of steel beams form the skeleton of a skyscraper. This may seem a little academic at first, but in Chapter 3 we find that from this model we can explain the effects of stress and temperature on a metal. A metal is ductile because *slip* instead of rupture occurs between the planes of atoms. When you heat a metal that has been deformed, this causes recrystallization into new grains so that you can continue working and deforming the metal to a desired shape.

Chapter 4 takes up phase diagrams. There is no quicker way for a metallurgist to satisfy or disperse the average engineering audience than to say, "We will now have a few slides of the phase diagrams on which our new alloy is based." However, in this text we use the phase diagram very simply as a map to show the effects of alloying elements on the basic structure. For example, we find that the addition of 8% nickel and 18% chromium to iron changes the usual structure of iron at room temperature to a totally new structure—that of austenitic stainless steel. This example shows that the alloys nickel and chromium are added not because of their individual chemical properties, but because of their effect on the structure of the iron alloy, as indicated by the phase diagram.

Chapters 5 and 6 examine the structures and properties of the important nonferrous and ferrous alloys from this point of view.

At the end of these six chapters you should understand the structures of more than 95% of metallic materials and should be aware of how properties depend on structures. An added bonus comes in the form of numerous examples of selection of alloys for different applications.

Chapters 7 and 8 then go on to take up ceramics. Here we have to start with a negative punch line that ceramics are not just the brittle stuff dishes are made of. Because of the covalent or ionic bonds, ceramics have the greatest hardness and strength and the highest melting points of all materials, but usually the lowest toughness or ductility. For a long time we have used ceramics alone in vital applications, such as in diamond-cutting tools or concrete. However, only recently have ceramics been used in combination with other materials, in everything from cemented carbide tools to delicate graphite-plastic fishing rods and optical fibers.

Chapters 9 and 10 investigate the third group of materials—the plastics. The important point here is that we need to understand the *molecules* of the polymer instead of the unit cell. There are two great families of plastics. In the thermosetting class, such as is used to make a bowling ball, we have one hard, strong giant molecule made up of carbon, hydrogen, and oxygen atoms. In the other class, such as is used to make polyethylene bags, we have a mass of many separate large molecules that can be readily formed and liquefied by heating.

Now we have reached the three-quarter point of the text. We understand the structures, properties of metals, ceramics, and polymers—and some applications. But prospective engineers are never told (or should not be), "Find the best *plastic* for this application." They should always think, "What is the best *material*?" Therefore we need to compare, contrast, and combine the different materials. This is done in the last quarter of the book. Chapter 11 discusses composite materials. Then Chapter 12 examines the effects of corrosion on materials. Then we go on in Chapter 13 to analyze the ways in which materials fail and how failure may be prevented. Finally, Chapters 14, 15, and 16 take up electrical, magnetic, thermal, and optical properties of materials. We find that these properties, like tensile strength and hardness, depend on the structure.

And now we come to the bottom line of this section: How are students who take this course better equipped to deal with professional problems as a result of this course? First, they will be better able to specify a material for a given application. By understanding the structure, they can go far beyond the tensile and hardness data of the handbook in predicting what will happen if the material is welded, pounded, twisted, corroded, or heated. If failure occurs, they will be able to analyze the reason and suggest improvements in materials or processing. Finally, they will have a firmer basis for understanding new materials as they are developed.

TO OUR COLLEAGUES

Let us begin this section on a note of gratitude to our colleagues at other institutions who loyally supported our first effort to provide a book on materials with a proper balance between theory and application. We have adhered to the same principle in this text and included in this first course only those theoretical principles that we believe will benefit the professional engineer in the use of materials.

As a first step in planning this second edition, we sent out a "report card" to those who had used the first edition. We asked them to grade the merits of individual chapters and to suggest topics they wished to see added. As a result of their responses, we have made the following changes.

We have added two completely new chapters—Composite Materials and Analysis and Prevention of Failure. The latter chapter contains a good deal of new material on fatigue as well.

We added much new material to Chapter 14 on Semiconductors and Devices, to Chapter 5 on Metal Processing, and to Chapter 11 on Concrete and Wood.

We included many new example problems on how to choose materials. We also doubled the number of problems. The entire text has been reworked for greater clarity.

As we say in the instructor's manual, the added material need not lengthen the course. The new material simply provides a greater variety of optional topics and makes this a better reference book.

Throughout the book, a vertical line in the right-hand margin indicates that the material thus set off is optional material.

A final word on examinations. Students often complain that the course covers too much material and that it is hard to know what to study for exams. To alleviate this problem, we begin with the premise that students remember material given in an examination best of all. Two or three days before each of four tests we give the students a general outline of the exam. This concentrates study on the most important topics. Then, after students take the exam and go over corrections, they have reviewed the critical material enough so that it becomes a permanent part of their professional background. This technique does wonders for student–faculty relations, and in addition really benefits the student.

ACKNOWLEDGMENTS

We wish to acknowledge the many suggestions we have received from the staff of Rensselaer Polytechnic Institute, and from our colleagues at The University of Michigan. We also wish to thank those who helped in the planning and reviewing of this edition: K. L. Wiggins, Memphis State University; S. J. Hruska, Purdue University; William H. Talbott, Clemson University; J. E. Ritter, Jr., University of Massachusetts; Ernest L. Multhaup, College of San Mateo; Khalid H. Khan, University of Portland; James C. O'Hara; Charles F. Heuer, Valparaiso University; Paul G. Shewmon, Ohio State University; Victor Strautman, University of Bridgeport; R. J. Brungraber, Bucknell University; James F. Shackelford, University of California; John Morral, University of Connecticut; and Christine Frasier. Mrs. A. G. Crowe gave valuable help in preparing the manuscript. In addition, we appreciate the contributions of Paul A. Flinn to Chapter 14 and J. Wayne Jones to Chapter 13.

R.A.F.
P.K.T.

ENGINEERING MATERIALS AND THEIR APPLICATIONS

Second Edition

1

THE PROBLEM OF MATERIALS SELECTION AND DEVELOPMENT

THIS photograph illustrates the three important families of materials available to the engineer—the metallic, such as steel and silver; the ceramic, such as glass and china; and the plastics or polymers, such as polyethylene and wood.

In this chapter we shall discuss how it is necessary to attain a basic understanding of the structure *of each of these groups in order to anticipate how they will perform under service conditions. We shall define the word "structure" as indicating (1) the nature of the atoms of which a material is composed, (2) the arrangement of the atoms into structural units called "unit cells" (also "molecules" in the case of the polymers), and (3) the grouping of unit cells to form grains or crystals.*

1.1 General plan of action

Before setting out on a trip into new territory, the seasoned backpacker or canoeist studies the ground to be covered and plans the time and resources needed. Later, when the person encounters a few boggy spots or black flies, he or she can be helped by the prospect of the overall goal.

In a similar way we shall use this first chapter to chart the path that we'll follow to an understanding of engineering materials. Also—just as no hike has the sole objective of climbing Mt. X, but is enriched by sights along the way—we can assure you that you will be enthralled by the structures you'll see through the microscope, and by the intricacies of crystal growth. In charting our path, we have three important points to establish:

1. There are thousands of materials available. But it helps to divide them into three simple classes: metallic, ceramic, and polymeric (plastics, wood, and so forth).
2. These divisions are made because the bonding forces that depend on fundamental atomic arrangements are similar within a family. This makes it easier to understand and classify the properties of a material. Therefore we'll discuss the differences among these bonding forces.
3. Bearing in mind points 1 and 2, we can discuss our path through this text. We shall begin with the metals, progress to ceramics and plastics, and finally take up special topics common to all materials: Failure Analysis, Corrosion, and Electrical and Magnetic Properties.

1.2 The three families of materials: metals, ceramics, polymers

The chapter-opening photograph illustrates the three families of materials, and also their competition. Engineers who are specifying materials soon find out that there is no ideal material for any application. As a corollary, they soon realize that as soon as they believe that they have a satisfactory material, other materials compete as replacements.

As an example, consider the advantages and disadvantages of the six cups in the chapter-opening photograph.

Practically everyone is familiar with the Sierra cup. Above all it is durable. The bowl is stamped as described in the chapters on metals. A tough steel wire is anchored in place by bending the lip of the cup over it. A loop of the wire forms the handle. The open design of the handle permits handling under most conditions. A disadvantage of this cup is the high thermal conductivity of metal. One can burn one's lips without even tasting the liquid.

The second cup illustrates other options in processing metal. The bowl is hand-hammered of silver. Handles are cast separately of silver, and the silver is brazed in place. The silver cup has the same basic features as the Sierra cup, but a higher resistance to corrosion. A disadvantage is that a silver cup is hardly suitable for a backpacker.

The English bone china cup is more comfortable to use and is esthetically more pleasing. Disadvantages: The colored decorations will fade if frequently washed in a dishwasher. It's more expensive to make this ceramic cup than a metal one because it must be formed from a clay-water mixture, then fired at high temperature, cooled, painted with glaze, and refired. And again, it is not a reliable companion for backpacking.

The glass can withstand a dishwasher, has the advantage of transparency, and comes in many beautiful shapes. Glass can be heated to a temperature at which it is plastic and then blown to any desired shape. Disadvantage: Glass has the same type of bonding as ceramic structures and suffers from the same brittleness.

Now let us turn to the two polymeric materials. Both plastic and wood are made up of large molecules with a carbon backbone and hydrogen plus other added elements.

The plastic cup is made of polystyrene, the least costly material to process. It belongs to the class called thermoplastics, meaning that it can be melted at relatively low temperatures and squirted (injected) into a mold made of metal. A small projection on the bottom of the cup is the only clue to the method. Disadvantages: The same property that makes it cheap to produce restricts its use for hot materials. Also plastic is not as durable as metal, and is scratched more easily than ceramics.

The wood cup is cut (turned on a lathe) from existing materials. Wood is durable. However, two disadvantages caused a longitudinal crack to form in the wood while the cup was simply stored on a shelf. (1) Wood is strongly *an*isotropic, meaning that it has directional properties. (By contrast, glass is isotropic, meaning that it has similar properties in all directions.) Therefore wood has properties across the grain that are different from its properties with the grain. (2) Wood can increase in volume by 1% when the humidity of the atmosphere increases. This expansion varies with direction. These properties caused the cup to crack. On the other hand, by laminating the wood or by using it in an environment with controlled humidity, one can prevent such cracking.

1.3 The structure of materials

We were able to divide the materials used for cups into three principal families. In a similar way we can classify all materials into these groups, and combinations of them. And in order to better understand all materials, we must develop a deeper knowledge of the materials and the reasons for properties such as strength, toughness, conductivity, and resistance to corrosion.

To understand the structure of a material, we need to know (1) which atoms are present and (2) how they are bonded (arranged). We can show the arrangement by means of a simple unit cell, such as a cubic or hexagonal structure, consisting of just a few atoms. Millions of these cells are arranged together to form the grains of the material. The grains are visible with a

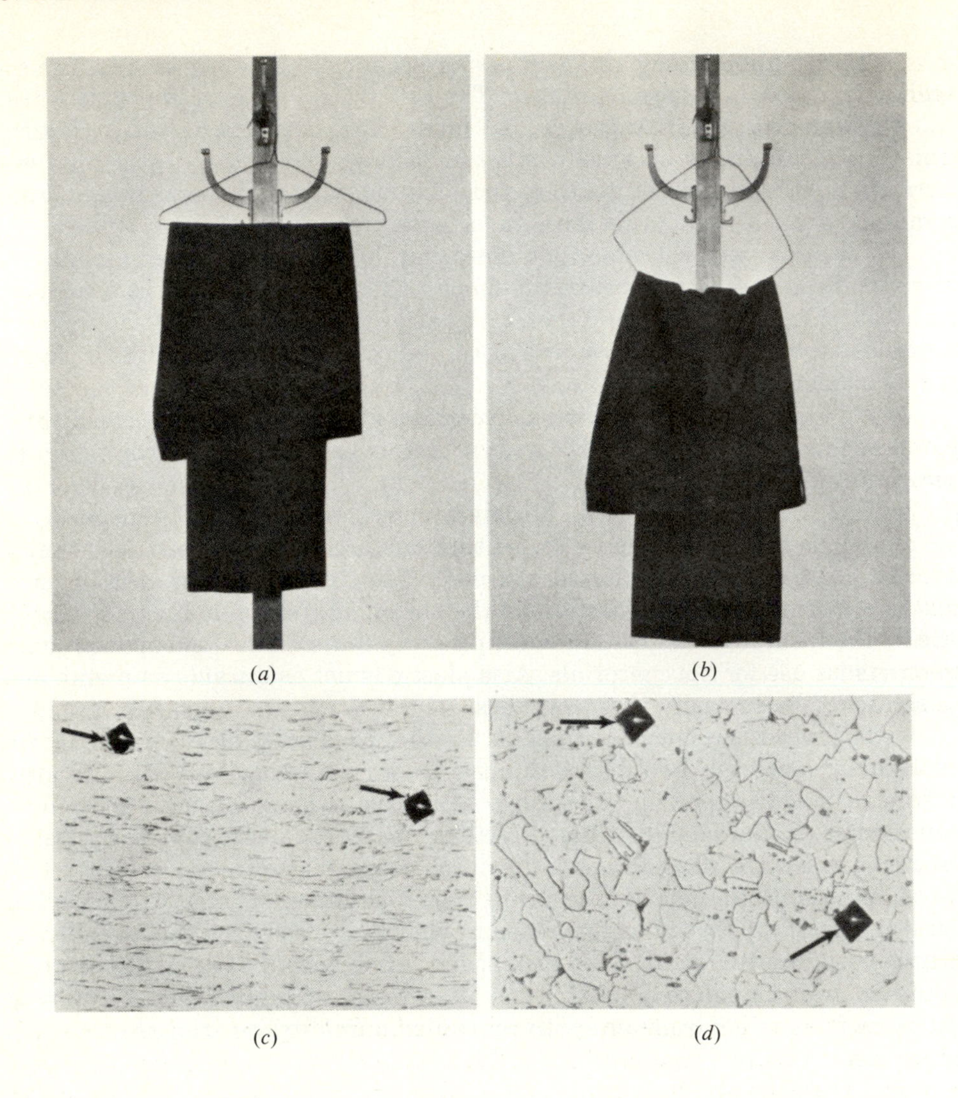

Fig. 1.1 *Effect of annealing and cold work on microstructure and service performance of a clothes hanger. (a) Cold-worked hanger, normal form. Grains are elongated in direction in which wire was drawn through a die. (b) Annealed clothes hanger. The larger equiaxed grains make hanger soft and weak. (c) Normal hanger, Vickers hardness number (VHN) 173 to 183; 500×, 2% nital etch. (d) Hanger annealed at 1700°F (910°C) $\frac{1}{2}$ hr, VHN 108–118; 500×, 2% nital etch.*

Grain structures shown are from samples that were longitudinally cut, then polished and etched with 2% nitric acid in alcohol (nital). The "normal" hanger was work-hardened by drawing through a die. When it was heated, new softer grains were formed. The VHN was obtained by pressing a light 25-g load into individual grains with a pyramid-shaped diamond indenter (arrows).

microscope. Often, such as in a brass doorknob, they are large enough to be seen with the naked eye.

When we understand these elements of structure—atomic structure, unit cell structure, and microstructure—we can classify and simplify the hundreds of thousands of materials into a few typical structures and their combinations.

For simplicity, when we use the term "structure," we'll be referring to any or all of these aspects. For example, if we talk about the effect of nickel on the structure of iron, we mean the effect of atomic interaction, the effect on the dimensions of the unit cell, and the changes in microstructure and macrostructure. Naturally we'll discuss each effect in detail where necessary.

After describing these structures and how they respond to stress, temperature, electrical potential, and magnetic fields, we will be able not only to comprehend the specific properties of these materials, but also to predict how they will behave under unusual circumstances.

Let us digress a moment to illustrate the basic differences between understanding the structure and simply knowing the chemical analysis of a material. Consider a coat hanger of steel wire (Fig. 1.1), which adequately supports a heavy pair of trousers. Now suppose we heat the hanger to a red color (about 1700°F or 910°C) and let it cool slowly. The chemical analysis is unchanged. But the change in structure, as seen in the photomicrograph, leads to the hanger's failure to perform. In Chapter 3 we shall take up in detail the structural changes that occur when a material is heated. In Chapter 6 we shall see that one can understand the properties of thousands of different steels, processed in many ways, by following the changes in only seven simple microstructures.

The greater simplicity of this new structural approach is only one of its advantages. With the old approach, one can look up mechanical properties in a handbook. However, if the steel will be processed in fabricating a component—by welding or rolling, for example—it is far easier to anticipate how the properties will be affected if you have a basic knowledge of how the microstructures will be changed rather than using data from a handbook, if you can find them at all. In addition, in analyzing the success or failure of the component under complex service conditions involving wear, impact, fatigue, and other problems, you can find valuable information by observing the changes in the structure in service. Now let us see how the structures of engineering materials vary.

1.4 Bonding forces in different materials

With the concepts of structure in mind, let us examine the differences in bonding in metallic, ceramic, and polymeric materials. How do these differences lead to different properties?

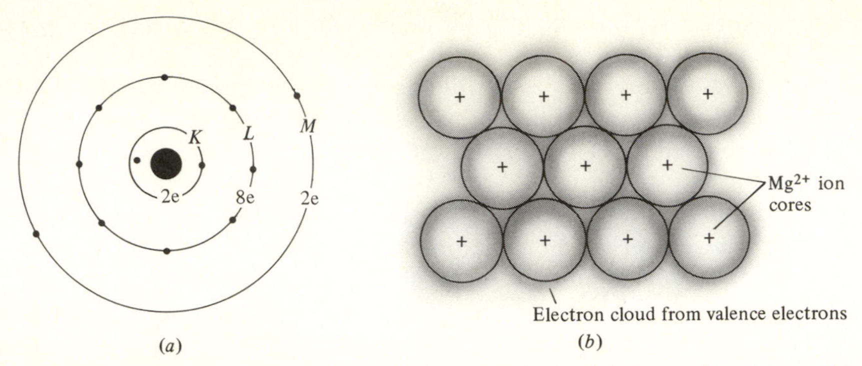

Fig. 1.2 *(a) A magnesium atom with its electron shells. (b) Atoms (ions) in solid magnesium surrounded by "electron gas."*

The Metallic Bond We know from elementary chemistry that the atomic structure of any element is made up of a positively charged nucleus surrounded by electrons revolving around it at different distances. The atomic number is equal to the number of these "planetary" electrons. We also know that it is helpful to consider these electrons as present at different energy levels, called "shells" or "rings." For example, magnesium, with an atomic number of 12, has two electrons in the first shell (nearest the nucleus), eight electrons in the second shell, and two electrons in the outer shell (Fig. 1.2*a*).

A common characteristic of the metallic elements is that they contain only one, two, or three electrons in the outer shell. This is the key to the formation of the metallic bond. When an element has only one, two, or three electrons in the outer shell, these electrons are bonded relatively loosely to the nucleus. Therefore, when we put a number of magnesium atoms together in a block of magnesium, the outer electrons leave individual atoms to enter a common "electron gas" (Fig. 1.2*b*). The atoms are therefore changed to Mg^{2+} ions. These repel each other, but they are still held together in the block because the negative electrons are attracted to the positive ions. The end result is that the Mg^{2+} ions arrange themselves in the regular pattern shown. This is the arrangement in one plane cut through a crystal of magnesium. It represents the closest possible packing for a group of equally sized spheres. It is the arrangement we would get if we shook a box of tennis balls. Many other metals show this same structure. Let us stop our discussion of metallic structure at this point and see how even this very simple model helps us to understand a few metallic properties.

First, it explains the high electrical conductivity of metals. If we apply a voltage across the crystal, the electrons in the electron gas (which are loosely bonded) will move readily, giving a current. Next, we can understand the ductility or ability to deform without fracturing. We note that the atoms are closely packed in planes. If we apply a shearing stress, as in Figure 1.3, one

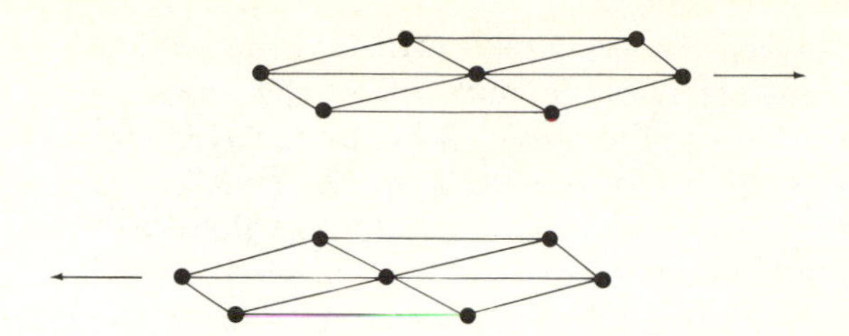

Fig. 1.3 *Slip between close-packed atom planes in magnesium*

plane will slide over the other and there will be no fracture, because after a movement of one atomic diameter the same forces between atoms are still operative. In the chapters that follow, we shall discuss many other features of the metallic structure. But this brief illustration will serve for the present.

The ceramics and plastics contain by contrast ionic, covalent, and van der Waals bonds.

The Ionic Bond This bond occurs between metallic and nonmetallic elements (Fig. 1.4*a*). We recall from chemistry that the outer shell in a nonmetallic element contains a larger number of electrons than the outer shell in a metal. For example, the outer shell of a chlorine atom contains seven electrons, and there is a strong driving force to attract an electron to make a stable group of eight. When sodium (one valence electron) reacts with chlorine (seven), the sodium gives up its electron to form a stable outer shell for the chlorine. This is called an *ionic bond* because Na^+ and Cl^- ions are formed and are naturally attracted because of their opposite charges. If the sodium chloride is dissolved in water, the ions exist independently in solution. In a *crystal* of sodium chloride, however, the ions arrange themselves in an electrically balanced structure, as shown by a section through the crystal (Fig. 1.4*b*).

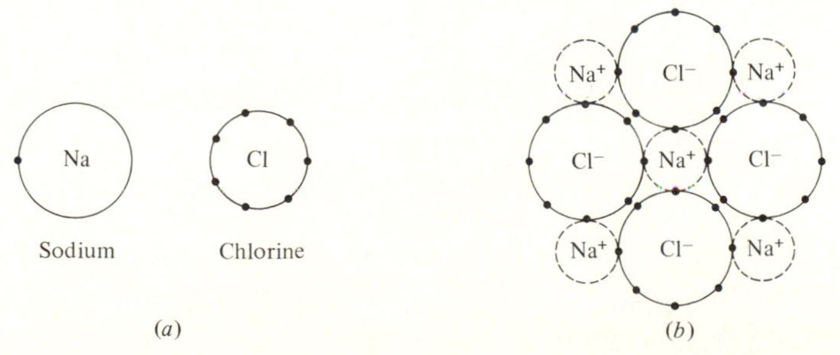

Fig. 1.4 *Formation of the ionic bond in sodium chloride. (a) Un-ionized sodium and chlorine atoms. (b) Ionized sodium and chloride ions. (The eight outer electrons in the chloride ion are at different energy levels and not at fixed distances.) The valence electron of sodium is bonded to the chloride ion. For simplicity, only the outer electron shells are shown.*

Let us compare the properties of this structure with those of the metallic structure. The electrical conductivity of the solid is many orders of magnitude lower than that of the metals because the electrons are tightly held in place instead of being free in the electron gas. The crystal fractures in brittle fashion because, when we attempt to slide one plane of ions across another, the electrical fields of different ions are opposed. Furthermore, the actual fracture follows cleavage planes which have certain ionic arrangements, such as that shown in Fig. 1.4*b*.

The Covalent Bond Many ceramic materials also have covalent bonds. The feature of this bond is that electrons are tightly held and equally shared by the participating atoms. An outstanding example is the diamond crystal (Fig. 1.5*a*), in which each carbon atom is in the center of a tetrahedron, sharing each of its four electrons with the adjoining atoms. In many cases, such as silica (SiO_2), only a portion of the bonding is covalent. This takes place when there is a compromise between perfectly balanced attraction for the shared electrons, as in diamond, and strong ionic attraction, as in sodium chloride. In the case of silica, the four outer electrons of silicon can be shared with one of the outer-shell electrons from each of four oxygen atoms, to make a stable shell of eight. Thus each silicon atom is surrounded by a tetrahedron of four oxygen atoms. Each oxygen atom is between two silicon atoms and shares an electron on each side (Fig. 1.5*b*). The bond is also partly ionic because the oxygen is somewhat more electronegative than silicon and tends to attract electrons more strongly. In one form of silica, quartz, this bonding results in high hardness and low electrical conductivity. Other forms also have high hardness and low conductivity, but the values depend on which form is present.

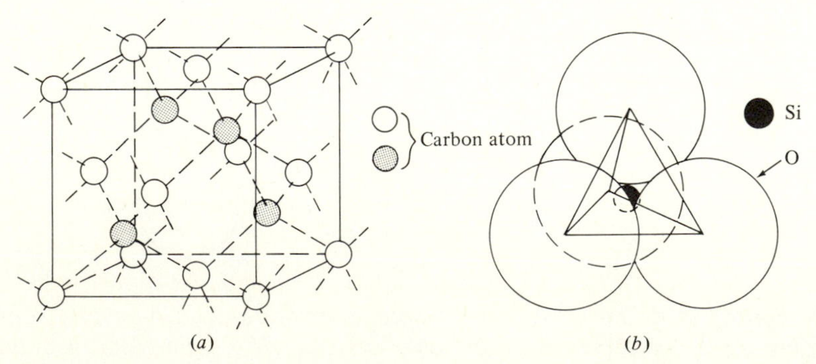

Fig. 1.5 *(a) Covalent bonding of carbon atoms in the unit cell of diamond. Atoms at centers of tetrahedra are shaded. (b) Tetrahedral structure of an* SiO_4^{4-} *ion. Compare the Si—O bonds with those of diamond.*

> EXAMPLE 1.1 Sketch the valence electron configuration in an SO_4^{2-} ion.
>
> ANSWER Assume that the valence electrons of sulfur and oxygen interact and are shared. In the figure the electrons for atoms are denoted by • or × to indicate the parent element, although it must be appreciated that the identity is lost in the electron-sharing process. (Sulfur and oxygen in the un-ionized state both have six electrons in their outer shells.)
>
> O
> O S O
> O
>
> Two electrons from some outside source to satisfy the desired configuration of eight. (Such a source might be one each from Na atoms, which then become Na^+ and form the compound Na_2SO_4.)

The van der Waals Bond This type of bond occurs to some extent in all materials, but is particularly important in plastics or high polymers. The molecules are made up of a backbone or skeleton of covalently bonded carbon atoms with other atoms—such as hydrogen, nitrogen, fluorine, chlorine, sulfur, and oxygen—attached to the other carbon bonds at the sides (Fig. 1.6).

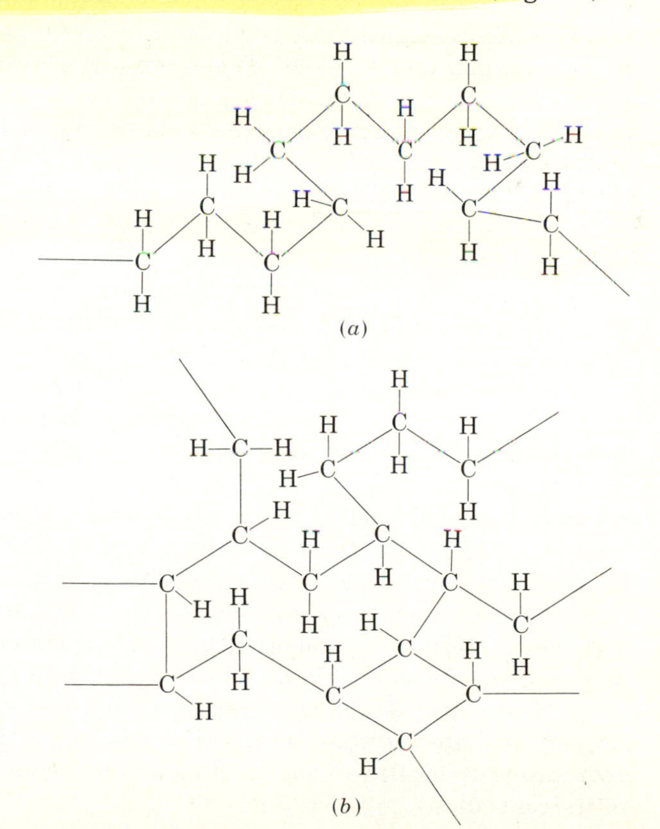

Fig. 1.6 *(a) Carbon backbone in a linear polymer (thermoplastic). (b) Carbon network in a thermosetting polymer.*

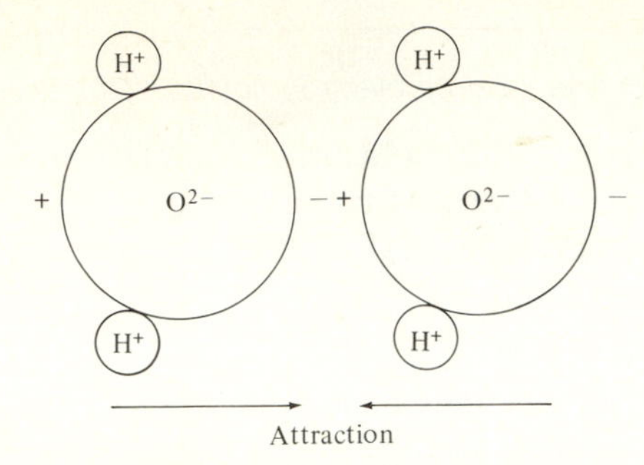

Fig. 1.7 *Van der Waals forces between water molecules*

The covalent bonds within the molecule are very strong and rupture only under extreme conditions. It is the bonds between the molecules that allow the sliding (and finally rupture) that are important. These bonds, called "van der Waals forces," require separate discussion.

Even when ionic and covalent bonds are present, there is still some imbalance in the electrical charge in a molecule. For example, it can be shown that in a simple water molecule hydrogen atoms are connected to oxygens by bonds at an angle of 104.5°. This gives a positive polarity at the hydrogen-rich end of the molecule and a negative polarity at the other end (Fig. 1.7). As a result, the molecules attract one another. This is the force that holds water molecules together in the liquid state. Let us see how these concepts explain some properties of plastics.

As molecules become larger, the total of the van der Waals forces between molecules increases. In polyethylene molecules, for example, we can have the same ratio of hydrogen to carbon atoms as in simple ethylene gas. But as the number of atoms in the molecule increases, we pass from a gas to a liquid and finally to a solid plastic.

Later we shall discuss two important families of high polymers, the thermoplastic and thermosetting materials. We can mention here, however, that thermoplastic materials are converted to liquids by heating. They harden on cooling. The effect of the heat is to overcome the van der Waals forces between the molecules by thermal agitation, making flow possible. By contrast, in the thermosetting materials, instead of the long chainlike molecules of the thermoplastics, we have three-dimensional networks of covalent bonds. These are not easily broken by heating, and we therefore have harder, stronger materials. On the other hand, these materials cannot be formed into shapes as easily as the thermoplastics (Fig. 1.6).

These are just a few examples of the correlation between structure and properties. Later we shall see another example of this, when we show how the extraordinary elastic extension of rubber is due to the shape and distribution of its molecules.

EXAMPLE 1.2 The boiling points of CH_4, NH_3, and CCl_4 are quite different, as shown in the table. Explain why this might be anticipated.

	Boiling Temperature (°C)
CH_4	−161.4
NH_3	− 33.4
CCl_4	+ 76.8

ANSWER First we find that carbon shares four pairs of electrons, nitrogen three pairs, and hydrogen one pair.

Let us first consider CH_4 and CCl_4. We can see from the sketches of the valence-electron interaction that they both have a high degree of symmetry and primary covalent bonding.

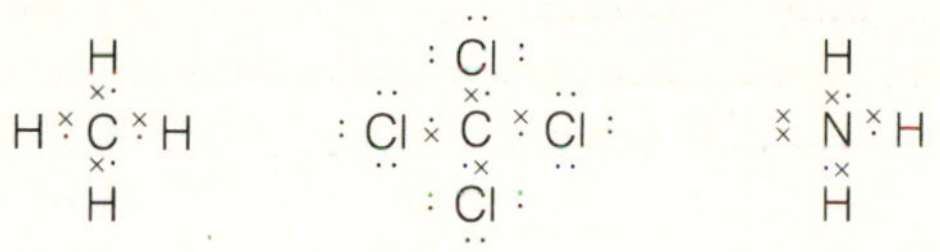

```
                        ..
   H                  : Cl :                 H
   x.            ..    x.    ..              x.
H x. C .x H    : Cl x. C .x Cl :         x x N x. H
   x.            ..    .x    ..              .x
   H                  : Cl :                 H
                        ..
```

Even though the sketches are planar representations, you might still expect the symmetry to be maintained in three dimensions (the carbon-hydrogen bond has an angle of 109°). The basic difference between CH_4 and CCl_4 is the molecular weight. For the *same* relative symmetry, the bonding is stronger for substances with higher molecular weight.

On the other hand, NH_3 is asymmetric, though not to the extent shown (hydrogen-nitrogen bond angles are 107°). Therefore, NH_3 has a much higher boiling point than CH_4, due to development of a dipole (dipoles are oppositely charged ends on a molecule), even though the molecular weights are approximately equivalent.

1.5 Outline of the text

From the preceding discussion we see that we have three types of materials to understand and that these materials exhibit a wide range of properties. In the chapters ahead we shall follow this program:

1. *The structures of metals, ceramics, and plastics* In Chapters 1 to 10, in each case we shall discuss the structures, then take up the analyses and some mechanical properties and applications of each important to commercial materials.
2. *Composite materials* This very important group of materials includes both older materials, such as wood and concrete, and more recent additions,

such as fiberglass-resin composites. Chapter 11 catalogs these materials on the basis of whether they are completely man-made (polymer-reinforced fibers), completely natural (wood), or combinations of man-made and natural materials, such as are found in concrete.

3. *Corrosion* To this point, the emphasis has been on the correlation of mechanical properties with structure. In many cases the most important characteristic of the service application is resistance to corrosion or to combined stress and corrosion effects. Chapter 12 covers the nature of corrosion and the resistance of materials to different environments.
4. *Analysis and prevention of failure* Increasing consumer awareness and lawsuits have made failure analysis an important research and development activity. After you develop an understanding of the factors affecting the structure of materials, you are equipped to understand the use of microstructural methods as an investigative tool and as a means of quality control. Chapter 13 covers these topics.
5. *Electrical and magnetic properties* A number of vital new industries are based on the application of the principles of structural analysis and control. Chapters 14 and 15 discuss the application of physical metallurgy and ceramics to the development of electrical materials, such as semiconductors and superconductors, and of magnetic materials, such as metallic and ceramic magnets and magnetic bubbles.
6. *Optical and thermal properties* Materials exhibit interesting properties related to the emission, absorption, transmission, reflection, and refraction of electromagnetic energy. Examples range from solar cells and lasers to photochromatic glass. Materials can also be placed in a more excited state by the application of thermal energy, as evidenced by their differences in conductivity, heat capacity, and coefficient of expansion. Chapter 16 treats these topics.

EXAMPLE 1.3 What types of bonding are present in iron, clay, and polyethylene?

ANSWER

1. *Iron* Metallic bonding is present. *Pure* iron has good thermal and electrical conductivity due to the freedom of motion of the valence electrons that form the electron cloud bonding together the iron atoms. Pure iron is not to be confused with steel or cast irons, which are alloys of iron and carbon with lower thermal and electrical conductivities (to be explained in later chapters).
2. *Clay* The greasy feeling of dry clay particles suggests a sheet type of structure. Bonding is both covalent and ionic within a sheet, but there are van der Waals bonds between sheets. The effect of addition of water is to plasticize the clay through attachment by van der Waals attraction. The addition of too much water gives a water-to-water bond, which is not as strong as the clay–water bond. Hence the consistency becomes soupy. Chapters 7 and 8 discuss these points.

3. *Polyethylene* Literally polyethylene means "many ethylenes." The C_2H_4 ethylene molecules are bonded together to form a long-chain polymer. The chain is bonded covalently to form the carbon backbone. Bonding between chains is by mechanical entanglement and van der Waals bonds. Due to the symmetry, the van der Waals bonds are weak. That is, the surfaces of the chains are not polar. This makes polyethylene components difficult to "glue," since wetting of the surface by van der Waals attraction is one method used in adhesives. (See Chapters 9 and 10.)

SUMMARY

At the end of each chapter we shall summarize important ideas introduced in that chapter and then list important definitions.

In this chapter there are only a few new basic ideas. We first discussed the problems of selecting materials for containers. However, it was apparent that at this point we could treat the three classes of materials—metals, ceramics, and polymers—only in a superficial way. To understand and predict properties, one needs to understand the structures of the different types of materials. The illustration of cold-worked and then heat-treated coat hangers showed that structure concerns not just which atoms are present, but how they are organized in the microstructure of the material. Next we discussed the type of bonding in each group of materials and how it affects the properties. In metals the metallic bond leads to an electron gas that provides excellent conductivity. In ceramics the combination of ionic and covalent bonds leads to high strength, brittle fracture, and cleavage. In polymers the covalent bonds within the molecules and the van der Waals forces between molecules lead to phenomenal elongation under stress in linear (thermoplastic) polymers. On the other hand, in thermosetting polymers the network structure leads to high strength and low elongation.

The contents of the text were then outlined.

DEFINITIONS

Bonding

Covalent Bonding in which atoms of the same element (such as carbon) or different elements share electrons, resulting in a strong bond.

Ionic Bonding in which metal and nonmetal ions bond as follows: The metal gives up an electron or electrons, which are taken up by the outer shell of the nonmetal. Therefore positive ions of the metal and negative ions of the nonmetal are produced, and attract each other.

Metallic Bonding in which metal atoms give up their electrons to an electron gas and take up a regular arrangement.

Van der Waals A secondary bonding between molecules, primarily due to charge attraction. The attraction results from unsymmetrical charge distribution rather than from primary bonds. Note that molecules rather than ions (as in the ionic bond) are attracted.

Ceramic materials Materials exhibiting ionic or covalent bonds or both.

Metallic materials Materials characterized by metallic bonds.

Plastics (high polymers) Materials exhibiting principally covalent bonding. The residual bonding forces, van der Waals bonds, are also important.

Structure The organization of a material, defined by specifying which atoms are present and their amounts, how they are arranged in unit cells, and the grains resulting from the unit cells. Also the presence of several different unit cells leads to grains or phases with different properties.

PROBLEMS

1.1 List a number of important components in a car and categorize each as metallic, ceramic, plastic, or composite. Examples: engine block, spark plug, body, windows. (Sections 1.1 through 1.4)

1.2 Graphite and diamond are both pure carbon. The bonding in one structure is perfectly covalent, while that of the other is partly metallic. One material is a fairly good conductor of electric current, and the other is an insulator. Identify each type of bonding with its properties. (Sections 1.1 through 1.4)

1.3 Despite the strength of the ionic bond, many materials with ionic bonds are not considered "good engineering materials." Why not? (*Hint:* Consider how ionics react to different environments.) (Sections 1.1 through 1.4)

1.4 Sketch the valence-electron configuration for a CO_3^{2-} ion (see Example 1.1). (Sections 1.1 through 1.4)

1.5 Sketch the valence-electron configuration for water (H_2O) and formaldehyde (CH_2O). (Sections 1.1 through 1.4)

1.6 Ethylene (C_2H_4) has the following structure and a boiling point of $-103.9°C$. Another compound of ethylene, trichlorethylene (C_2HCl_3), which is used as a dry-cleaning agent or solvent, has a boiling point of $+87.2°C$. Sketch the valence-electron structure of trichlorethylene; account for the higher boiling point. Data are as follows. (Sections 1.1 through 1.4)

Ethylene

```
H   H
·×  ·×
C ×× : C
·×  ·×
H   H
```

	C	*H*	*Cl*
Atomic number	6	1	17
Atomic weight	12.01	1	35.46
Valence	4+	1+	1−
Boiling point (°C)	4830	−253	−34.7

1.7 In a given type of bonding the different bond strengths manifest themselves macroscopically as variations in melting or boiling points. The materials listed below are covalently bonded with secondary van der Waals bonds. Arrange them in the anticipated order of lowest to highest boiling points. (If necessary, use sketches to support your conclusions.) (Section 1.5)

a. C_2H_4, a single ethylene molecule
b. CH_4, a single methane molecule
c. C_2H_3Cl, a single vinylchloride molecule
d. $(C_2H_4)_n$, multiple joined ethylene molecules (polyethylene)

1.8 What type(s) of bond(s) are found in the following materials? *a.* bromine *b.* sulfuric acid *c.* glass *d.* graphite (Section 1.5)

1.9 Look at various objects around you and state the type of bonding present. Determine which of the following processes was used in manufacturing each. (1) Casting, (2) forging, (3) machining, (4) powder-pressing, (5) rolling, (6) other. (Section 1.5)

1.10 When aluminum is added to liquid steel (an iron-carbon alloy) to remove oxygen dissolved in the steel, the compound Al_2O_3 is formed. An excess of aluminum is usually added to ensure removal of the oxygen.

a. What type of bond is present in the Al_2O_3?
b. What type of bond is present between the unreacted aluminum atoms and the iron atoms in the solid steel?
c. What weight of Al does one need to react with 1 kg of steel containing 0.1 wt.% oxygen and still leave 0.05 wt.% Al dissolved in the steel? (Section 1.5)

PROBLEMS COVERING OPTIONAL SECTIONS

1.11 Review the Periodic Table of the Elements. Classify the elements as

a. Metals
b. Strongly nonmetallic
c. Between metals and nonmetals

(Requires chemistry, or Section 2.2)

1.12 Draw a circle around each of the following elements, if any, that you would consider as material for a car body (ignore cost). (Requires chemistry, or Section 2.2)

Atomic Number	
3	$1s^2\ 2s^1$
7	$1s^2\ 2s^2\ 2p^3$
20	$1s^2\ 2s^2\ 2p^6\ 3s^2\ 3p^6\ 4s^2$
26	$1s^2\ 2s^2\ 2p^6\ 3s^2\ 3p^6\ 3d^6\ 4s^2$
29	$1s^2\ 2s^2\ 2p^6\ 3s^2\ 3p^6\ 3d^{10}\ 4s^1$
37	$1s^2\ 2s^2\ 2p^6\ 3s^2\ 3p^6\ 3d^{10}\ 4s^2\ 4p^6\ 5s^1$

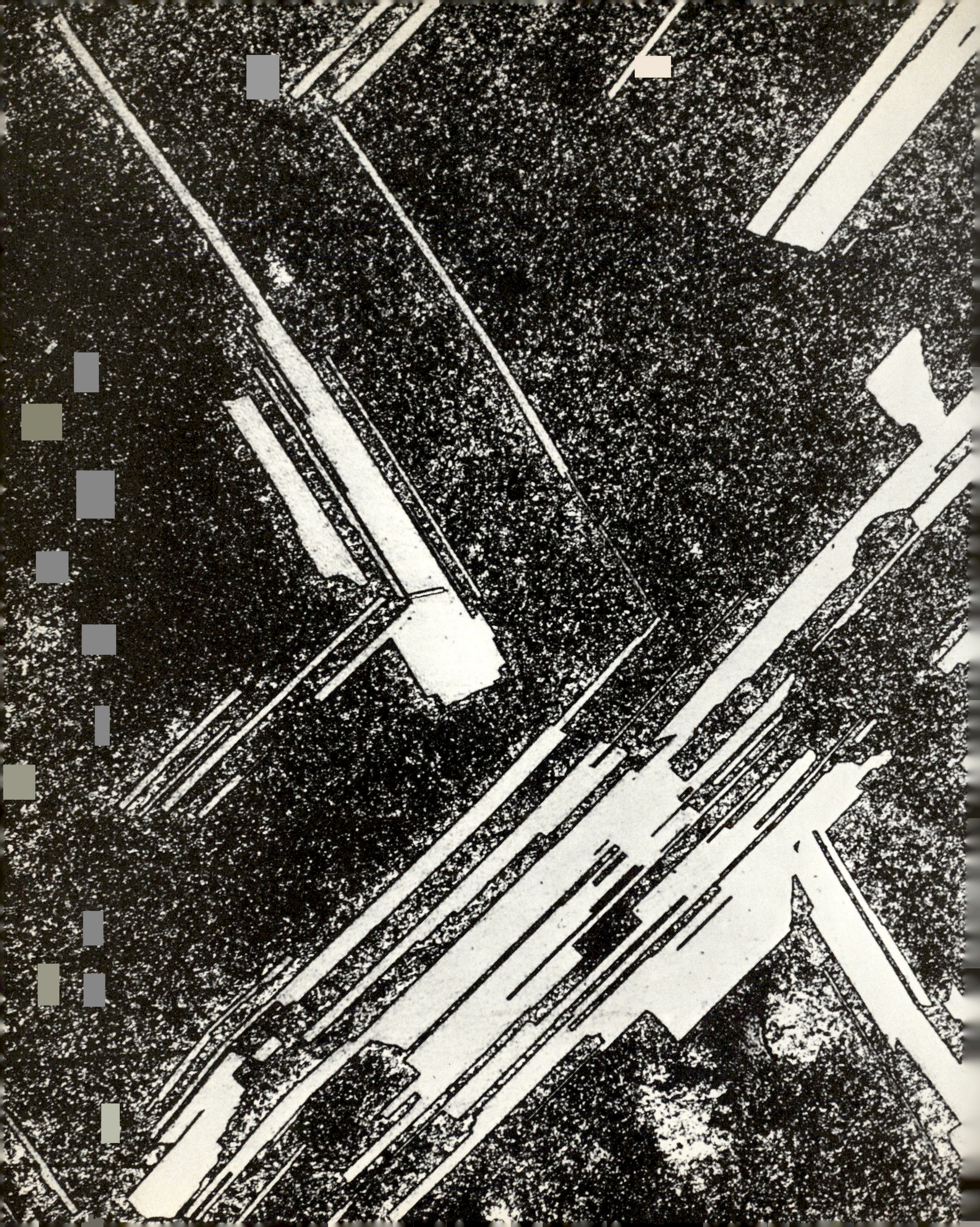

2

METALLIC STRUCTURES; THE UNIT CELL

THIS is a photograph of a specimen of native copper collected in the upper peninsula of Michigan. The sample was sectioned, polished, and etched with a weak acid solution to disclose the crystalline structure. The picture is called a photomicrograph because it is taken through a microscope (magnification 100×).

The crystals or grains show a cubic structure. In this chapter we shall discuss how this structure is due to the arrangement of atoms in cubic unit cells. *From an understanding of these elementary building blocks, we shall be able to show how metals—and materials in general—behave under stress and other service conditions.*

2.1 General overview

In this chapter we begin our study of metallic materials. The structures of these materials are simpler than those of the ceramics or plastics, but what we learn about the basic features of metallic structures will be valuable in understanding the other materials. To obtain a basic knowledge of any material, it is not enough merely to look at it under the microscope. After all, the fundamental building blocks are the *atoms* themselves, so we need to start with some atomic properties. Next we find that the atoms are grouped in *unit cells*. By assembling millions of unit cells, we obtain the *grains* we see under the microscope. It is important to understand the grain structure because a metal part does not stretch uniformly, like a piece of taffy. It stretches by deformation of individual grains. The properties of a metal are different in different directions in the grain. We shall begin, therefore, with a short review of the properties of metallic atoms. Then, after discussing the geometry of unit cells, we'll examine metallic unit cells and their role in the structure of grains. We shall also begin our study of alloys, which are simply materials produced when one or more elements are added to a metallic element (without forming a covalent or ionic compound); for example, copper plus 30% zinc gives an alloy we know as brass.

2.2 Metal atoms

The properties of a metal atom are largely determined by the number of electrons around the nucleus and their arrangement. Recall that the atomic number tells us the number of electrons in orbit around the nucleus and that the nucleus has a positive charge balancing the negative charge of the electrons.

Even early scientists realized that electrons had different energies and rotated at different distances from the nucleus. They made simple sketches of the atom, showing the electrons in rings or shells of different energy levels, as illustrated for magnesium (atomic number = 12) in Fig. 2.1. It is important to review this early concept because some of the nomenclature is still used. For example, we still call the shell or ring closest to the nucleus the K shell, the next the L shell, and so forth (Fig. 1.2*a*).

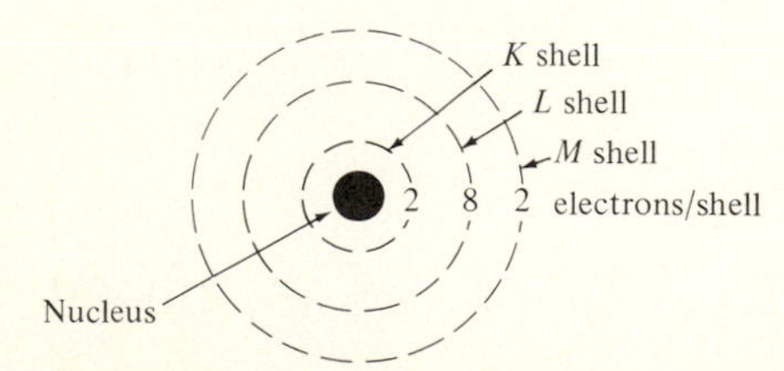

Fig. 2.1 *Early concept of the atomic structure of magnesium (atomic number = 12). Numerals on shells indicate the number of electrons per shell.*

Later investigators, however, found differences in energy and in spin direction among electrons in the same shell. It became necessary to assign four quantum numbers to each electron rather than merely to state the shell in which it was found.†

In simple terms we can say that if we are to locate an object, such as an electron, in space, we need three coordinates. We need a fourth if we want to define the direction of spin. The basic explanation is found in quantum theory and is much more complex. The symbols for these four numbers are:

- n The principal quantum number gives the shell or ring. The numbering begins with the inner shell. Thus for the electrons in the K shell $n = 1$, for the L shell $n = 2$, and so forth.
- l The second quantum number is limited by the value of n and must be a positive integer equal to or less than $n - 1$. Therefore, in the second shell of magnesium, $n = 2$ and l can have values of 1 or 0.
- m_l The third quantum number is equal to or less than l, but can be positive or negative. Therefore, in the second shell of magnesium, m_l can have values of $+1$, 0, or -1.
- m_s The fourth quantum number can have values of only $+\frac{1}{2}$ or $-\frac{1}{2}$. This number tells the direction of spin of the electron. We shall see in Chapter 15 that magnetic properties are basically related to spin.

We need one other concept to develop the quantum numbers of the electrons in magnesium. The *Pauli postulate* states that no two electrons can have the same four quantum numbers. With this in mind, and starting with the inner ring $n = 1$, we can write the quantum numbers for the 12 electrons of magnesium as follows:

Electron Number	*n*	*l*	m_l	m_s	*Abbreviated Designation*
1	1	0	0	$+\frac{1}{2}$	$1s^2$
2	1	0	0	$-\frac{1}{2}$	
3	2	0	0	$+\frac{1}{2}$	$2s^2$
4	2	0	0	$-\frac{1}{2}$	
5	2	1	-1	$+\frac{1}{2}$	$2p^6$
6	2	1	0	$+\frac{1}{2}$	
7	2	1	$+1$	$+\frac{1}{2}$	
8	2	1	-1	$-\frac{1}{2}$	
9	2	1	0	$-\frac{1}{2}$	
10	2	1	$+1$	$-\frac{1}{2}$	
11	3	0	0	$+\frac{1}{2}$	$3s^2$
12	3	0	0	$-\frac{1}{2}$	

†The vertical line at the right indicates optional material, as explained in the Preface.

Note that we have also given an abbreviated designation, which is often used as a shorthand notation for the more cumbersome quantum numbers. The first number is the first or principal quantum number, while the letters s, p, etc., designate the value of l as follows.

$l =$	0	1	2	3	4	5
Letter =	s	p	d	f	g	h
Maximum electrons =	2	6	10	14	18	22

The exponent tells how many electrons of this type are present.

Thus we use the abbreviations $1s^2$, $2s^2$, $2p^6$, $3s^2$ to mean two electrons $n = 1, l = 0$; two electrons $n = 2, l = 0$; six electrons $n = 2, l = 1$; two electrons $n = 3, l = 0$.

One other fact is that in p, d, and f electrons, we find first the maximum number of positive spins, then the negative spins, instead of alternating directions. For example, electrons 5, 6, and 7 all have $s = +\frac{1}{2}$. (This is called *Hund's rule.*) Table 2.1 gives electronic structures for all the elements.

Table 2.1 ELECTRONIC CONFIGURATION OF THE ELEMENTS*

Atomic No.	Element	K	L		M			N				O				P				Q
		1	2		3			4				5				6				7
		s	s	p	s	p	d	s	p	d	f	s	p	d	f	s	p	d	f	s
1	H	1																		
2	He	2																		
3	Li	2	1																	
4	Be	2	2																	
5	B	2	2	1																
6	C	2	2	2																
7	N	2	2	3																
8	O	2	2	4																
9	F	2	2	5																
10	Ne	2	2	6																
11	Na	2	2	6	1															
12	Mg	2	2	6	2															
13	Al	2	2	6	2	1														
14	Si	2	2	6	2	2														
15	P	2	2	6	2	3														
16	S	2	2	6	2	4														
17	Cl	2	2	6	2	5														
18	Ar	2	2	6	2	6														
19	K	2	2	6	2	6		1												
20	Ca	2	2	6	2	6		2												
21	Sc	2	2	6	2	6	1	2												
22	Ti	2	2	6	2	6	2	2												

Table 2.1 ELECTRONIC CONFIGURATION OF THE ELEMENTS* (*Continued*)

		K	*L*		*M*			*N*				*O*				*P*				*Q*
		1	*2*		*3*			*4*				*5*				*6*				*7*
Atomic No.	*Element*	*s*	*s*	*p*	*s*	*p*	*d*	*s*	*p*	*d*	*f*	*s*	*p*	*d*	*f*	*s*	*p*	*d*	*f*	*s*
23	V	2	2	6	2	6	3	2												
24	Cr	2	2	6	2	6	5†	1												
25	Mn	2	2	6	2	6	5	2												
26	Fe	2	2	6	2	6	6	2												
27	Co	2	2	6	2	6	7	2												
28	Ni	2	2	6	2	6	8	2												
29	Cu	2	2	6	2	6	10†	1												
30	Zn	2	2	6	2	6	10	2												
31	Ga	2	2	6	2	6	10	2	1											
32	Ge	2	2	6	2	6	10	2	2											
33	As	2	2	6	2	6	10	2	3											
34	Se	2	2	6	2	6	10	2	4											
35	Br	2	2	6	2	6	10	2	5											
36	Kr	2	2	6	2	6	10	2	6											
37	Rb	2	2	6	2	6	10	2	6	..		1								
38	Sr	2	2	6	2	6	10	2	6	..		2								
39	Y	2	2	6	2	6	10	2	6	1		2								
40	Zr	2	2	6	2	6	10	2	6	2	..	2								
41	Nb	2	2	6	2	6	10	2	6	4†	..	1								
42	Mo	2	2	6	2	6	10	2	6	5	..	1								
43	Tc	2	2	6	2	6	10	2	6	6	..	1								
44	Ru	2	2	6	2	6	10	2	6	7	..	1								
45	Rh	2	2	6	2	6	10	2	6	8	..	1								
46	Pd	2	2	6	2	6	10	2	6	10†	..	..								
47	Ag	2	2	6	2	6	10	2	6	10	..	1								
48	Cd	2	2	6	2	6	10	2	6	10	..	2								
49	In	2	2	6	2	6	10	2	6	10	..	2	1							
50	Sn	2	2	6	2	6	10	2	6	10	..	2	2							
51	Sb	2	2	6	2	6	10	2	6	10	..	2	3							
52	Te	2	2	6	2	6	10	2	6	10	..	2	4							
53	I	2	2	6	2	6	10	2	6	10	..	2	5							
54	Xe	2	2	6	2	6	10	2	6	10		2	6							
55	Cs	2	2	6	2	6	10	2	6	10	..	2	6	..	..	1				
56	Ba	2	2	6	2	6	10	2	6	10	..	2	6	..	..	2				
57	La	2	2	6	2	6	10	2	6	10	..	2	6	1	..	2				
58	Ce	2	2	6	2	6	10	2	6	10	2†	2	6	..	..	2				
59	Pr	2	2	6	2	6	10	2	6	10	3	2	6	..	..	2				
60	Nd	2	2	6	2	6	10	2	6	10	4	2	6	..	..	2				
61	Pm	2	2	6	2	6	10	2	6	10	5	2	6	..	..	2				
62	Sm	2	2	6	2	6	10	2	6	10	6	2	6	..	..	2				
63	Eu	2	2	6	2	6	10	2	6	10	7	2	6	..	..	2				
64	Gd	2	2	6	2	6	10	2	6	10	7	2	6	1	..	2				
65	Tb	2	2	6	2	6	10	2	6	10	9†	2	6	..	..	2				
66	Dy	2	2	6	2	6	10	2	6	10	10	2	6	..	..	2				

(*Continued*)

Table 2.1 ELECTRONIC CONFIGURATION OF THE ELEMENTS* (*Continued*)

Atomic No.	Element	K	L	M	N	O	P	Q
		1	2	3	4	5	6	7
		s	s p	s p d	s p d f	s p d f	s p d f	s
67	Ho	2	2 6	2 6 10	2 6 10 11	2 6	2	
68	Er	2	2 6	2 6 10	2 6 10 12	2 6	2	
69	Tm	2	2 6	2 6 10	2 6 10 13	2 6	2	
70	Yb	2	2 6	2 6 10	2 6 10 14	2 6	2	
71	Lu	2	2 6	2 6 10	2 6 10 14	2 6 1 ..	2	
72	Hf	2	2 6	2 6 10	2 6 10 14	2 6 2 ..	2	
73	Ta	2	2 6	2 6 10	2 6 10 14	2 6 3 ..	2	
74	W	2	2 6	2 6 10	2 6 10 14	2 6 4 ..	2	
75	Re	2	2 6	2 6 10	2 6 10 14	2 6 5 ..	2	
76	Os	2	2 6	2 6 10	2 6 10 14	2 6 6 ..	2	
77	Ir	2	2 6	2 6 10	2 6 10 14	2 6 9† ..	0	
78	Pt	2	2 6	2 6 10	2 6 10 14	2 6 9 ..	1	
79	Au	2	2 6	2 6 10	2 6 10 14	2 6 10 ..	1	
80	Hg	2	2 6	2 6 10	2 6 10 14	2 6 10 ..	2	
81	Tl	2	2 6	2 6 10	2 6 10 14	2 6 10 ..	2 1	
82	Pb	2	2 6	2 6 10	2 6 10 14	2 6 10 ..	2 2	
83	Bi	2	2 6	2 6 10	2 6 10 14	2 6 10 ..	2 3	
84	Po	2	2 6	2 6 10	2 6 10 14	2 6 10 ..	2 4	
85	At	2	2 6	2 6 10	2 6 10 14	2 6 10 ..	2 5	
86	Rn	2	2 6	2 6 10	2 6 10 14	2 6 10 ..	2 6	
87	Fr	2	2 6	2 6 10	2 6 10 14	2 6 10 ..	2 6	1
88	Ra	2	2 6	2 6 10	2 6 10 14	2 6 10 ..	2 6	2
89	Ac	2	2 6	2 6 10	2 6 10 14	2 6 10 ..	2 6 1 ..	2
90	Th	2	2 6	2 6 10	2 6 10 14	2 6 10 ..	2 6 2 ..	2
91	Pa	2	2 6	2 6 10	2 6 10 14	2 6 10 2†	2 6 1 ..	2
92	U	2	2 6	2 6 10	2 6 10 14	2 6 10 3	2 6 1 ..	2
93	Np	2	2 6	2 6 10	2 6 10 14	2 6 10 4	2 6 1 ..	2
94	Pu	2	2 6	2 6 10	2 6 10 14	2 6 10 6	2 6	2
95	Am	2	2 6	2 6 10	2 6 10 14	2 6 10 7	2 6	2
96	Cm	2	2 6	2 6 10	2 6 10 14	2 6 10 7	2 6 1 ..	2
97	Bk	2	2 6	2 6 10	2 6 10 14	2 6 10 9	2 6	2
98	Cf	2	2 6	2 6 10	2 6 10 14	2 6 10 10	2 6	2
99	Es	2	2 6	2 6 10	2 6 10 14	2 6 10 11	2 6	2
100	Fm	2	2 6	2 6 10	2 6 10 14	2 6 10 12	2 6	2
101	Md	2	2 6	2 6 10	2 6 10 14	2 6 10 13	2 6	2
102	No	2	2 6	2 6 10	2 6 10 14	2 6 10 14	2 6	2
103	Lr	2	2 6	2 6 10	2 6 10 14	2 6 10 14	2 6 1 ..	2
104	Rf	2	2 6	2 6 10	2 6 10 14	2 6 10 14	2 6 2 ..	2
105	Ha	2	2 6	2 6 10	2 6 10 14	2 6 10 14	2 6 3 ..	2

*Devised by Laurence S. Foster. †Note irregularity.

References: Therald Moeller, "Inorganic Chemistry," John Wiley & Sons, New York, 1952, pp. 98–101. Joseph J. Katz and Glenn T. Seaborg, "The Chemistry of the Actinide Elements," Methuen & Co., Ltd., London, 1957; John Wiley & Sons, New York, 1957, p. 464.

2.3 The periodic table

This review of quantum numbers of electrons is important here because it is part of the background of the Periodic Table, Table 2.2 (page 27). This table is obtained by placing the elements with $n = 1$ (hydrogen and helium) in the top row, those with $n = 2$ in the next row, and so forth. The table is valuable in the development of metallic materials because of the interrelation of elements in the same column, and in some cases in the same row. Let us consider some examples.

The first vertical column, group IA (omitting hydrogen, which is a special case) is composed of the very active metals lithium through francium. This reactivity is due to the fact that all have one loosely held s electron in the outer shell. All react rapidly with oxygen (which has a high affinity for electrons to attain an outer shell of eight). For this reason they are used to eliminate undesired oxygen from liquid melts of other metals. Lithium, for example, is added to liquid copper to remove oxygen as an impurity and thereby produce a pure copper structure with high electrical conductivity.

The relative tendency of an element to gain an electron is expressed by a factor called *electronegativity*. As we would expect, the active metals have the lowest values (see Table 2.3, page 29). We can also say that these metals are strongly electropositive. In general, electronegativity *decreases* as the total number of electrons in a vertical column increases because the outer electrons are farther from the positive nucleus. Also, electronegativity *increases* as we go across the table because the higher the number of electrons in the outer shell, the greater the attraction for other electrons, until a stable group of eight is reached (two for helium).

The second column, group IIA, contains elements with two loosely held electrons in the outer shell. These are only slightly less active than the elements in group IA. However, magnesium can be protected by covering it with a suitable coating or by adding other elements to it. It is used extensively, even though it may react in pure form. For example, the engine block of a Volkswagen automobile is principally magnesium.

From group IIIB to group VIII, we have the first group of transition metals (atomic numbers 21 through 28). In this range the changes in electrons are in the inner rings. In the usual case (potassium and calcium) the $4s$ level fills before the $3d$ level does.

EXAMPLE 2.1 What is the electron configuration (shorthand notation) for the three transition elements vanadium (atomic number = 23), iron (atomic number = 26), and nickel (atomic number = 28)? What is the ionized electron configuration for nickel?

ANSWER By definition, the transition elements are those with partially filled $3d$ shells, but the $3d$ shells are not the outer shells. These elements have similar properties.

V	$1s^2, 2s^2, 2p^6, 3s^2, 3p^6, 3d^3, 4s^2$
Fe	$1s^2, 2s^2, 2p^6, 3s^2, 3p^6, 3d^6, 4s^2$
Ni	$1s^2, 2s^2, 2p^6, 3s^2, 3p^6, 3d^8, 4s^2$

When a metal is ionized, it loses its outer-shell electrons, which for these transition elements would be the 4*s* electrons. Hence

$$Ni^{2+} \quad 1s^2, 2s^2, 2p^6, 3s^2, 3p^6, 3d^8$$

The so-called *lanthanide series* of rare earths, from cerium (atomic number = 58) to lutetium (atomic number = 71), are even more alike than the transition metals, because the only changes in electrons are in the second shell below the outer shell. That is, the 5*p* and 6*s* levels are filled first, then the 4*f* levels. Since the 4*f* level can contain 14 electrons, there are 14 rare earths. All exhibit typical metallic properties and are finding increasing use. Cerium, for instance, is used in lighter "flints" and in producing a new casting alloy called "ductile iron." Europium has unique emission properties which lead to its use in television tubes. Samarium is alloyed with cobalt to form the strongest known permanent magnets.

Group IB contains the noble metals, copper, silver, and gold. Although the outer shell of these metals contains only one electron, this electron is tightly held because it is close to the energy level of the 3*d* electrons and tends to associate with the other electrons. For example, the electron structure of copper is $1s^2, 2s^2, 2p^6, 3s^2, 3p^6, 3d^{10}, 4s^1$, and the 4*s* electron is close to the 3*d* electron. By contrast, in the alkali metals, in group IA, the shell beneath the outer *s* electron is a very stable group of either two or eight electrons, and the outer electron is loosely held.

For the same reason the elements of group IIB—zinc, cadmium, and mercury—are stabler than those of IIA, such as magnesium and calcium, which have a shell of eight beneath the outer electrons. Zinc, for example, is used widely in die castings for automobiles.

The most important element in group IIIA is aluminum. Despite the fact that its three outer electrons are loosely held (there is a shell of eight beneath), making it quite reactive, it is widely used as a building material. We shall see in Chapter 12 that successful use of aluminum depends on the formation of an adherent film of aluminum oxide, which protects it from further reaction with the air.

Finally in group IVA we reach the end of the metallic elements. The lower members of the group, such as lead and tin, have metallic characteristics, while the upper members, such as germanium and silicon, are semiconductors, and the diamond form of carbon shows only covalent bonding and is an insulator. We shall discuss this behavior in detail in Chapter 14 when we consider transistor materials.

Table 2.2 PERIODIC TABLE OF THE ELEMENTS*

KEY TO CHART

Atomic Number →	50 +2	← Oxidation States
Symbol →	Sn +4	
Atomic Weight →	118.69	
	18 18 4	← Electron Configuration

1a	2a	3b	4b	5b	6b	7b	8	8	8	1b	2b	3a	4a	5a	6a	7a	0	Orbit
1 H +1 −1 1.0079 1																	2 He 0 4.00260 2	K
3 Li +1 6.94 2-1	4 Be +2 9.01218 2-2											5 B +3 10.81 2-3	6 C +2 +4 −4 12.011 2-4	7 N +1 +2 +3 +4 +5 −1 −2 −3 14.0067 2-5	8 O −2 15.9994 2-6	9 F −1 18.998403 2-7	10 Ne 0 20.17_9 2-8	K-L
11 Na +1 22.98977 2-8-1	12 Mg +2 24.305 2-8-2	Transition Elements					Group 8			Transition Elements		13 Al +3 26.98154 2-8-3	14 Si +2 +4 −4 28.0855 2-8-4	15 P +3 +5 −3 30.97376 2-8-5	16 S +4 +6 −2 32.06 2-8-6	17 Cl +1 +5 +7 −1 35.453 2-8-7	18 Ar 0 39.948 2-8-8	K-L-M
19 K +1 39.0983 -8-8-1	20 Ca +2 40.08 -8-8-2	21 Sc +3 44.9559 -8-9-2	22 Ti +2 +3 +4 47.90 -8-10-2	23 V +2 +3 +4 +5 50.9415 -8-11-2	24 Cr +2 +3 +6 51.996 -8-13-1	25 Mn +2 +3 +4 +7 54.9380 -8-13-2	26 Fe +2 +3 55.847 -8-14-2	27 Co +2 +3 58.9332 -8-15-2	28 Ni +2 +3 58.71 -8-16-2	29 Cu +1 +2 63.546 -8-18-1	30 Zn +2 65.38 -8-18-2	31 Ga +3 69.735 -8-18-3	32 Ge +2 +4 72.59 -8-18-4	33 As +3 +5 −3 74.9216 -8-18-5	34 Se +4 +6 −2 78.96 -8-18-6	35 Br +1 +5 −1 79.904 -8-18-7	36 Kr 0 83.80 -8-18-8	-L-M-N
37 Rb +1 85.467_8 -18-8-1	38 Sr +2 87.62 -18-8-2	39 Y +3 88.9059 -18-9-2	40 Zr +4 91.22 -18-10-2	41 Nb +3 +5 92.9064 -18-12-1	42 Mo +6 95.94 -18-13-1	43 Tc +4 +6 +7 98.9062 -18-13-2	44 Ru +3 101.07 -18-15-1	45 Rh +3 102.9055 -18-16-1	46 Pd +2 +4 106.4 -18-18-0	47 Ag +1 107.868 -18-18-1	48 Cd +2 112.41 -18-18-2	49 In +3 114.82 -18-18-3	50 Sn +2 +4 118.69 -18-18-4	51 Sb +3 +5 −3 121.75 -18-18-5	52 Te +4 +6 −2 127.60 -18-18-6	53 I +1 +5 +7 −1 126.9045 -18-18-7	54 Xe 0 131.30 -18-18-8	-M-N-O
55 Cs +1 132.9054 -18-8-1	56 Ba +2 137.33 -18-8-2	57* La +3 138.9055 -18-9-2	72 Hf +4 178.49 -32-10-2	73 Ta +5 180.947_9 -32-11-2	74 W +6 183.85 -32-12-2	75 Re +4 +6 +7 186.207 -32-13-2	76 Os +3 +4 190.2 -32-14-2	77 Ir +3 +4 192.22 -32-15-2	78 Pt +2 +4 195.09 -32-16-2	79 Au +1 +3 196.9665 -32-18-1	80 Hg +1 +2 200.59 -32-18-2	81 Tl +1 +3 204.37 -32-18-3	82 Pb +2 +4 207.2 -32-18-4	83 Bi +3 +5 208.9804 -32-18-5	84 Po +2 +4 (209) -32-18-6	85 At (210) -32-18-7	86 Rn 0 (222) -32-18-8	-N-O-P
87 Fr +1 (223) -18-8-1	88 Ra +2 226.0254 -18-8-2	89** Ac +3 (227) -18-9-2	104 — +4 (260) -32-10-2	105 — (260) -32-11-2	106 (263) -32-12-2													O P Q

*Lanthanides	58 Ce +3 +4 140.12 -20-8-2	59 Pr +3 140.9077 -21-8-2	60 Nd +3 144.24 -22-8-2	61 Pm +3 (145) -23-8-2	62 Sm +2 +3 150.4 -24-8-2	63 Eu +2 +3 151.96 -25-8-2	64 Gd +3 157.25 -25-9-2	65 Tb +3 158.9254 -27-8-2	66 Dy +3 162.50 -28-8-2	67 Ho +3 164.9304 -29-8-2	68 Er +3 167.26 -30-8-2	69 Tm +3 168.9342 -31-8-2	70 Yb +2 +3 173.04 -32-8-2	71 Lu +3 174.967 ± 0.003 -32-9-2	N O P
**Actinides	90 Th +4 232.0381 -18-10-2	91 Pa +5 +4 231.0359 -20-9-2	92 U +3 +4 +5 +6 238.029 -21-9-2	93 Np +3 +4 +5 +6 237.0482 -22-9-2	94 Pu +3 +4 +5 +6 (244) -24-8-2	95 Am +3 +4 +5 +6 (243) -25-8-2	96 Cm +3 (247) -25-9-2	97 Bk +3 +4 (247) -27-8-2	98 Cf +3 (251) -28-8-2	99 Es +3 (254) -29-8-2	100 Fm +3 (257) -30-8-2	101 Md +2 +3 (258) -31-8-2	102 No +2 +3 (259) -32-8-2	103 Lr +3 (260) -32-9-2	O P Q

Numbers in parentheses are mass numbers of most stable isotope of that element

*Reprinted with permission from the CRC *Handbook of Chemistry and Physics*, 1980, The Chemical Rubber Co., CRC Press, Inc.

The rest of the elements in the Periodic Table are generally considered to be nonmetallic, although a few, such as antimony and bismuth, have semimetallic characteristics. We shall return to these nonmetallic elements in our discussion of polymers, where carbon, nitrogen, and oxygen are important, and in ceramics, which are defined as combinations of metallic and nonmetallic elements.

Now let us go on to discuss the unit cell and grains.

EXAMPLE 2.2 By analyzing their electronic structure, classify the elements with the atomic numbers 8, 14, 19, and 29 as very active metals, corrosion-resistant metals, semiconductors, or nonmetals.

ANSWER *Element No. 8:* $1s^2$, $2s^22p^4$ This element has 6 electrons in the shell beyond the stable helium shell of 2. It therefore tends to accept 2 additional electrons to form a stable outer shell of 8. Since the element is an electron acceptor, it is a nonmetal. The element is oxygen. It reacts readily with metals to form oxides.

Element No. 14: $1s^2$, $2s^22p^6$, $3s^23p^2$ This element has 4 electrons in the 3 quantum shell, which is halfway to the stable number of 8. It is a semiconductor, not a conductor of electricity like the metals. It may either lose or gain 4 electrons to achieve the stable configuration of 8 electrons. Therefore, it tends to share electrons. This element is silicon. Its electrical properties are discussed in detail in Chapter 14.

Element No. 19: $1s^2$, $2s^22p^6$, $3s^23p^6$, $4s^1$ This element has a stable shell of 8 electrons at the 3 quantum level just beneath the outer electron. Therefore the outer electron is only lightly held, and the metal is very active. This element is potassium. If exposed to air or water, it readily forms potassium oxide.

Element No. 29: $1s^2$, $2s^22p^6$, $3s^23p^63d^{10}$, $4s^1$ Here the 4*s* electron has a shell of 18 beneath it, which is not particularly stable. The bonding energy of the 4*s* electron is quite close to that of the 3*d* electrons beneath. Thus it is held quite closely. The element—copper—is quite corrosion resistant (unreactive). Copper may give up a 3*d* electron in addition to the 4*s* electron to form the familiar Cu^{2+} ion, which imparts a blue color to copper salts. This is further evidence that the binding energies of the 3*d* and 4*s* electrons are similar.

2.4 Crystals and grains

Instead of describing the arrangement of many millions of atoms (10^{10} millions) in a grain, it is much easier to describe the unit cell. Just as a building is made up of modules or units, so a grain is made up of identical unit cells. This concept of grain structure is basic to later problems.

In cavities in rocks we encounter cubic crystals of copper. The relation between a *crystal* of copper and a *grain* is quite simple. The grain is merely

Table 2.3 ELECTRONEGATIVITIES OF THE ELEMENTS*

IA	IIA	IIIB	IVB	VB	VIB	VIIB	VIIIB			IB	IIB	IIIA	IVA	VA	VIA	VIIA	O
1 H 2.1																	2 He –
3 Li 1.0	4 Be 1.5											5 B 2.0	6 C 2.5	7 N 3.0	8 O 3.5	9 F 4.0	10 Ne –
11 Na 0.9	12 Mg 1.2											13 Al 1.5	14 Si 1.8	15 P 2.1	16 S 2.5	17 Cl 3.0	18 Ar –
19 K 0.8	20 Ca 1.0	21 Sc 1.3	22 Ti 1.5	23 V 1.6	24 Cr 1.6	25 Mn 1.5	26 Fe 1.8	27 Co 1.8	28 Ni 1.8	29 Cu 1.9	30 Zn 1.6	31 Ga 1.6	32 Ge 1.8	33 As 2.0	34 Se 2.4	35 Br 2.8	36 Kr –
37 Rb 0.8	38 Sr 1.0	39 Y 1.2	40 Zr 1.4	41 Nb 1.6	42 Mo 1.8	43 Tc 1.9	44 Ru 2.2	45 Rh 2.2	46 Pd 2.2	47 Ag 1.9	48 Cd 1.7	49 In 1.7	50 Sn 1.8	51 Sb 1.9	52 Te 2.1	53 I 2.5	54 Xe –
55 Cs 0.7	56 Ba 0.9	57–71 La–Lu 1.1–1.2	72 Hf 1.3	73 Ta 1.5	74 W 1.7	75 Re 1.9	76 Os 2.2	77 Ir 2.2	78 Pt 2.2	79 Au 2.4	80 Hg 1.9	81 Tl 1.8	82 Pb 1.8	83 Bi 1.9	84 Po 2.0	85 At 2.2	86 Rn –
87 Fr 0.7	88 Ra 0.9	89-103 Ac–Lr 1.1–1.7	104 (Rf)	105 (Ha)													

*From Linus Pauling, *The Nature of the Chemical Bond,* 3d ed. Copyright 1939 and 1940, 3d ed. © 1960, by Cornell University Press.

a crystal without smooth faces because its growth was impeded by contact with another grain or a restraining surface. Within the grain the arrangement of unit cells is just as perfect as it is within a crystal with smooth faces.

To indicate the planes formed by alignment of the unit cells of copper, let us perform a simple experiment. First we cut a small bar of copper 1 in. long and 0.030 by 0.030 in. (0.76 by 0.76 mm) square. We polish one surface and find that we can observe grains (Fig. 2.2*a*).

Now let us bend the specimens, as shown in Fig. 2.2*b*. Straight dark lines appear in the grains near both edges of the sample. If we watch carefully while the sample is being bent, we see that the metal is being displaced on either side of the black lines by a shearing, slipping type of action. Chapter 3 will show that the black lines or slip bands are simply the intersections of slip planes with the surface. Furthermore, we shall see that the slip plane is made up of a very closely packed mass of atoms on a common plane. Although the grains at the upper edge of the sample are in tension and those at the lower edge are in compression, the mechanism of deformation is by shearing in both cases.

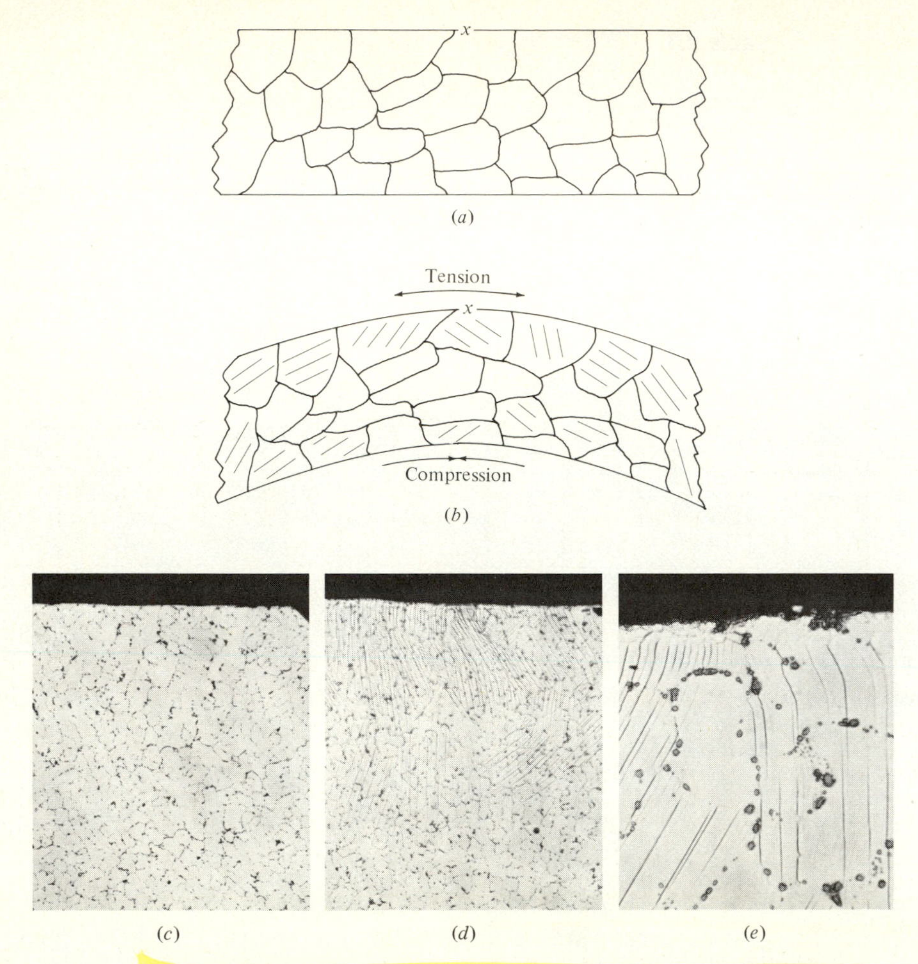

Fig. 2.2 *(a) Grains in a polished bar of copper. (b) Slip occurring during bending. (c) Photomicrograph of region x in (a), 100×. (d) Photomicrograph of region x in (b), 100×. (e) Photomicrograph of region x in (b) 500×.*

2.5 The unit cell

A description of the unit cell will enable us to describe these movements more accurately. To describe the unit cell and later the movement of an atom in the cell, we need a system for specifying: (1) Atom positions or coordinates, (2) Directions in the cell, and (3) Planes in the cell.

Position The position of an atom is described with reference to the axes of the unit cell and the unit dimensions of the cell Suppose that a grain

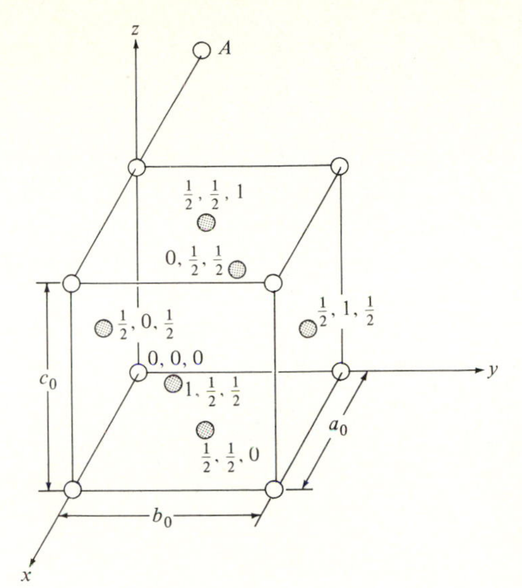

Fig. 2.3 *Coordinates of atoms in the face-centered positions in a unit cell. The atom A shown in the next unit cell would have coordinates $\bar{1}$, 0, 1. Note: $\bar{1}$ is equivalent to -1.*

or crystal is built of unit cells of dimensions a_0,† b_0, c_0 angstroms*, as shown in Fig. 2.3. In this case the axes are at right angles to each other. To construct the cell, we merely lay out a_0, the lattice parameter, in the x direction, b_0 in the y direction, and c_0 in the z direction. The figure shows the coordinates of the corners of the cell. An atom at the center would have coordinates $\frac{1}{2}$, $\frac{1}{2}$, $\frac{1}{2}$, while an atom in the center of the face in the xy plane would have coordinates $\frac{1}{2}$, $\frac{1}{2}$, 0. It is important to note that commas after coordinates in space refer to *points* in that space. These coordinates are not enclosed in parentheses; we did not want you to confuse them with planes, which we shall discuss in a later section.

Up to this point we have used relatively simple cells as examples. In nature 14 different types of crystal lattices are found; see Fig. 2.4 on p. 32. These lattices cover the variations in the lengths of a, b, and c and in the angles between the axes.

Fortunately, in metals we find mostly the three simple types of cells shown in Fig. 2.5: body-centered cubic (BCC), face-centered cubic (FCC), and hexagonal close-packed (HCP). (See p. 33.) Some of the other types are encountered occasionally in metals and ceramics.

---

†The subscript 0 specifies that this dimension is measured at a standard temperature, usually 68°F (20°C). When the unit cell expands with temperature, a, b, and c change.
*The angstrom (Å) = 10^{-8} cm. In the SI system of units, the linear unit of measure is the meter. Therefore 1 Å = 10^{-10} meter = 0.1 nanometer (nm), where 1 nm = 10^{-9} meter. Both Å and nm will be used in the following discussion of unit cell dimensions.

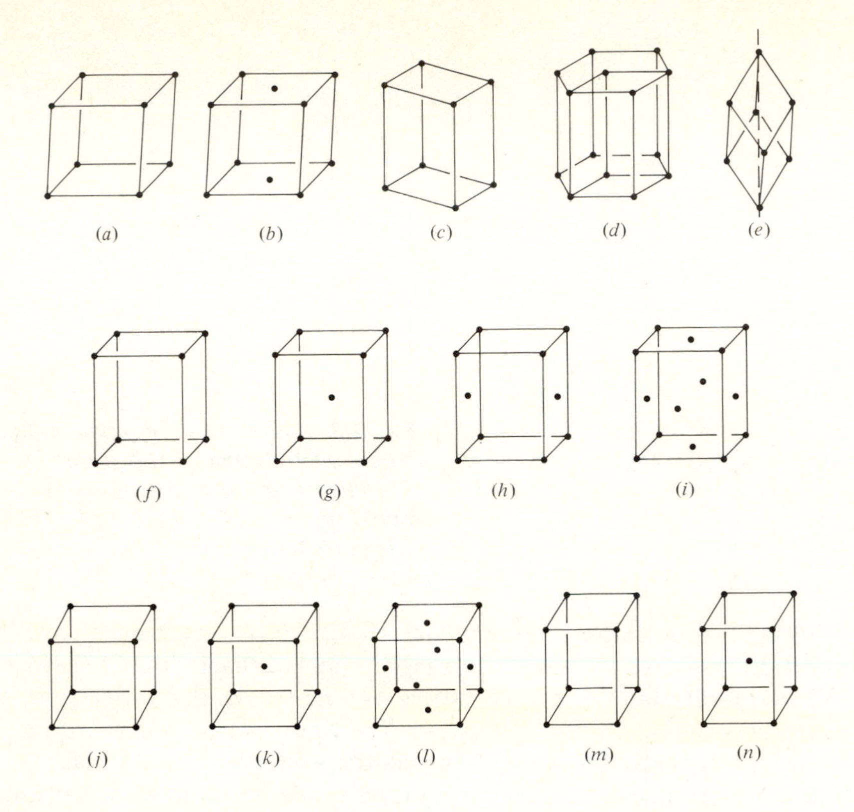

System	*Axes*	*Axial Angles*
Cubic	$a_1 = a_2 = a_3$	All angles = 90°
Tetragonal	$a_1 = a_2 \neq c$	All angles = 90°
Orthorhombic	$a \neq b \neq c$	All angles = 90°
Monoclinic	$a \neq b \neq c$	Two angles = 90°; one angle ≠ 90°
Triclinic	$a \neq b \neq c$	All angles different; none equals 90°
Hexagonal	$a_1 = a_2 = a_3 \neq c$	Angles = 90° and 120°
Rhombohedral	$a_1 = a_2 = a_3$	All angles equal, but not 90°

Fig. 2.4 *The 14 crystal lattices and their geometric relationships. The lattices continue in three dimensions. (a) Simple monoclinic. (b) End-centered monoclinic. (c) Triclinic. (d) Hexagonal. (e) Rhombohedral. (f) Simple orthorhombic. (g) Body-centered orthorhombic. (h) End-centered orthorhombic. (i) Face-centered orthorhombic. (j) Simple cubic. (k) Body-centered cubic. (l) Face-centered cubic. (m) Simple tetragonal. (n) Body-centered tetragonal.*

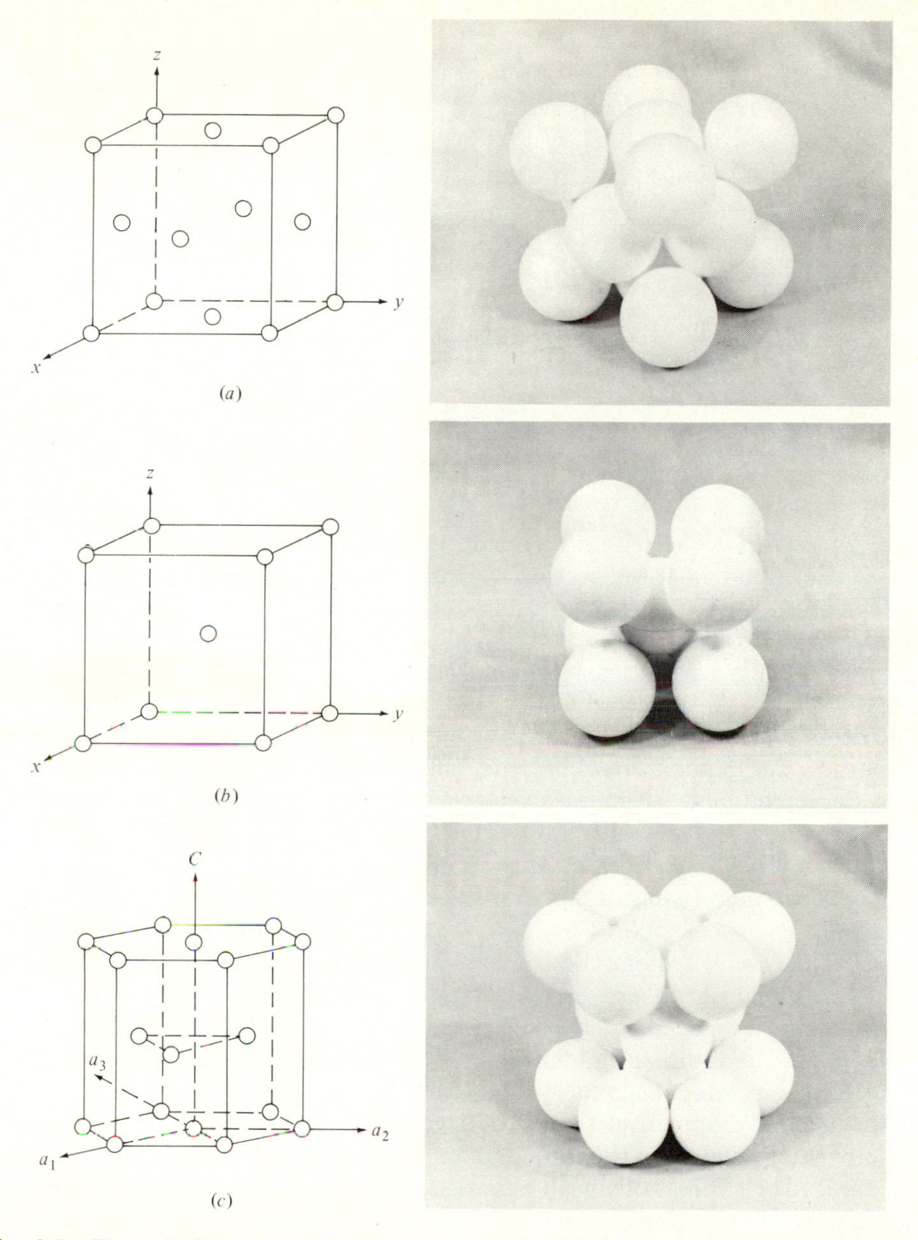

Fig. 2.5 *The principal crystal structures of metals. (a) Face-centered cubic. (b) Body-centered cubic. (c) Close-packed hexagonal. For each case, a sketch and a photograph of the hard-sphere model are shown.*

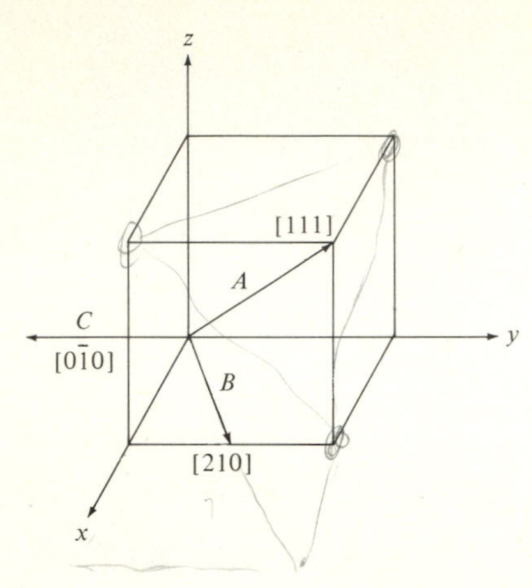

Fig. 2.6 *Specification of directions*

Direction To specify a direction in the unit cell, we merely place the base of the arrow of the direction vector at the origin and follow the shaft until we encounter integral coordinates (Fig. 2.6, above). Instead of constructing other cells, we can use a point that has fractional intercepts in the unit cell and multiply by the least common denominator. Thus direction A is obviously [111], but B has coordinates 1, $\frac{1}{2}$, 0 at the edge of the cell. These become [210]. Note that in specifying a direction we put square brackets around the numbers to distinguish direction from the notation for coordinates (and for a plane, described later). Note also that we can have negative indices, as shown by C. We indicate these with an overbar.

If we wish to find the indices of a direction that does not pass through the origin, we merely pass a parallel direction through the origin and proceed as before.

In the cubic system the dimensions of the unit cell are the same: $a_0 = b_0 = c_0$. The order in a given set of indices, such as [110], depends on which directions we happened to choose for x, y, and z in the crystal, since in the cubic system the indices are identical. Often we wish to signify a set of directions that are related, such as [110], [101], and [011]. These are all directions across the face diagonal of the cube. To specify these similar directions, we use pointed brackets and the indices of only one direction. Thus $\langle 110 \rangle$ = [110], [101], and [011]. Directions with negative indices, such as [$\bar{1}$10], are also considered part of the same set.

It is important to realize that unit distances in the three directions x, y, and z are a_0, b_0, and c_0, respectively. Therefore, the [111] direction in a noncubic system [orthorhombic (three side lengths unequal); Fig. 2.4] means a direction starting at coordinates 0, 0, 0 with the tip at a_0 distance in the x direction, b_0 distance in the y direction, and c_0 distance in the z direction.

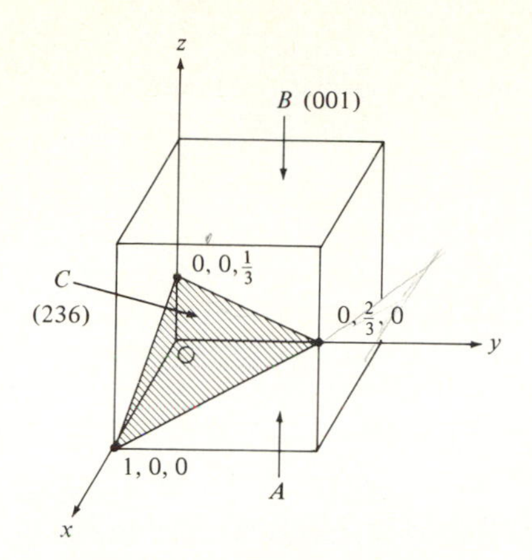

Fig. 2.7 *Calculation of Miller indices of a plane*

Plane The method for finding the Miller indices of a plane differs from that for describing coordinates and directions (Fig. 2.7). Follow this sequence carefully.

1. Select a plane in the unit cell that does not pass through the origin. For example, to describe the plane at face A that passes through 0,0,0, we use plane B, which is in the same set of planes because it is a unit distance away from A. (That is, B would have the same number of atoms and would be parallel to the original plane.)
2. Record the intercepts of the plane as multiples of a_0, b_0, and c_0 on the x, y, and z axes in that order.
3. Take reciprocals and clear fractions. This gives (001) for plane B.† Note that we use parentheses in this case to show that we are describing a plane rather than a direction. Also commas do not separate the indices, as they do in atom positions. The plane should be read as the "zero-zero-one" plane.

Following the same rules, we find the indices for plane C of intercepts 1, $\frac{2}{3}$, $\frac{1}{3}$. We take reciprocals: 1, $\frac{3}{2}$, 3. Then we clear fractions: 2, 3, 6 or (236).

As in the case of directions, we often want to specify a family of planes, such as the cube faces. Here we use braces; thus

$$\{100\} = (100), (010), (001)$$

We can also have the corresponding negative values, such as $(\bar{1}00)$, depending on our choice of position for the origin 0, 0, 0.

---

†$\frac{1}{\infty}, \frac{1}{\infty}, \frac{1}{1}$

EXAMPLE 2.3 Determine the Miller indices of the plane shown in the figure.

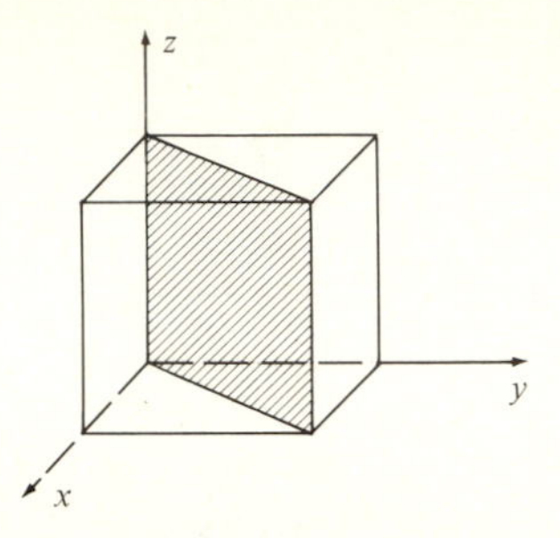

ANSWER Since the plane passes through the origin, we must go through a unit translation of the axes to obtain the intersection of the plane with the axes. This unit translation does not influence the outcome of our problem because of the symmetry of the unit cells. There are two possible translations:

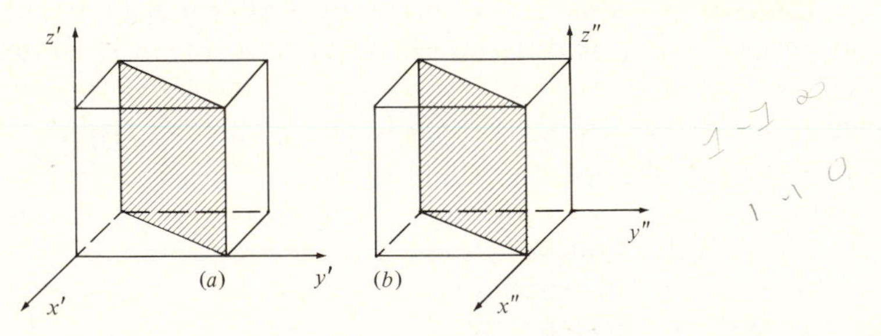

In (*a*) x', y', z', the intercepts are -1, 1, ∞, which give the Miller indices $(\bar{1}10)$. In (*b*) x'', y'', z'', the intercepts are 1, -1, ∞ or $(1\bar{1}0)$. Since these indices are equivalent, we can see the advantage of referring to families of planes, such as $\{110\}$.

2.6 Correlation of data on unit cells with measurements of density

Let us see whether we can check the dimensions of the unit cell with engineering data, such as density. After all, we are not going to use the metals in unit-cell-size pieces. We should satisfy ourselves that the characteristics of the unit cell relate to engineering-size components.

For example, the density of copper is 8.96 g/cm^3 (8.96×10^3 kg/m^3 = 8.96 megagrams/m^3) at 20°C (68°F). If our data are correct, we should find the same value using the mass of copper in the volume of a unit cell.

We can find the mass (M) and volume (V) from previous definitions. We know the weight of an atom of copper, since each gram atomic weight of copper

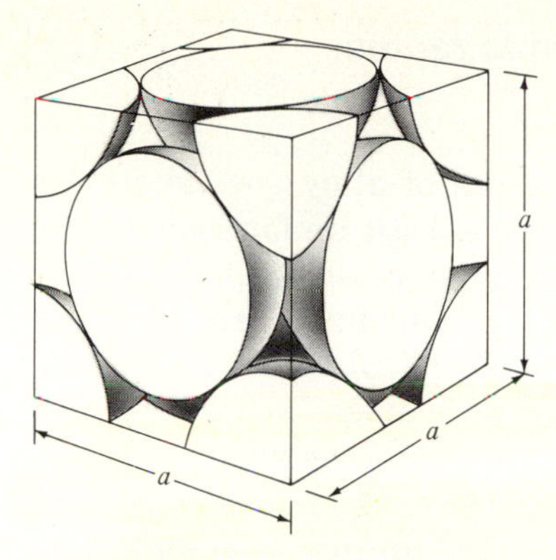

Fig. 2.8 *Illustration of the number of atoms (four) in an FCC unit cell*

(63.5 g/at. wt.) has 6.02×10^{23} atoms (Avogadro's number). The volume of a unit cell is merely the lattice dimension cubed. Before we can complete the calculation, however, we must determine the number of atoms per unit cell.

Figure 2.8 shows how to determine the number of atoms in a single cell. If we were to set up thin glass walls as faces of the FCC, these would pass through the centers of the atoms in the faces and extend only to the centers of the corner atoms. Counting the portions of the atoms that are within the cell, we have

6 face atoms of which $\frac{1}{2}$ of each is inside the glass = 3 atoms
8 corner atoms of which $\frac{1}{8}$ of each is inside the glass = 1 atom
Total atoms per unit cell = 4 atoms

In a similar manner we find that the number of atoms per unit cell in a BCC is two, while that in an HCP is six.

EXAMPLE 2.4 We know from x-ray data discussed later in this chapter that a_0 = 3.61 Å (0.361 nm) at 20°C (68°F) for copper. Calculate the theoretical density of copper, using the knowledge that the cell of this element is an FCC.

ANSWER

$$\text{Density} = \frac{M}{V} = \frac{4 \text{ atoms/unit cell} \dfrac{63.5 \text{ g/at. wt.}}{6.02 \times 10^{23} \text{ atoms/at. wt.}}}{(3.61)^3 \times 10^{-24} \text{ cm}^3\text{/unit cell}}$$

$$= 8.98 \frac{\text{g}}{\text{cm}^3} \left(8.98 \times 10^3 \frac{\text{kg}}{\text{m}^3} = 8.98 \text{ mg/m}^3 \right)$$

This is quite close to the density of a block of copper (8.96 g/cm³). In Chapter 3 we shall see that there are voids that explain the difference.

2.7 Other unit cell calculations: atomic radius, planar density, and linear density

Atomic Radius When we discuss the design of alloys, we shall see that when a second element is added, its atoms can substitute for some of the atoms of the major element. The extent to which this substitution can take place is governed by the similarity of the two metals. One important measure of similarity is the atomic radius, which we can easily calculate, once we know the dimensions of the unit cell. First we consider the atoms as spheres and find some dimension of the unit cell along which the spheres are in contact. We can calculate any dimension of the cube, such as a body or face diagonal, by geometry. Then we merely divide the figure by the number of atomic radii present (Fig. 2.9).

Planar Density We will see that when slip occurs under stress (plastic deformation), it takes place on the planes on which the atoms are most densely packed. To calculate planar density, we use the following convention.

If an atom belongs entirely to a given area, such as the atom in the center of a face in an FCC structure, we note that the trace of the atom on

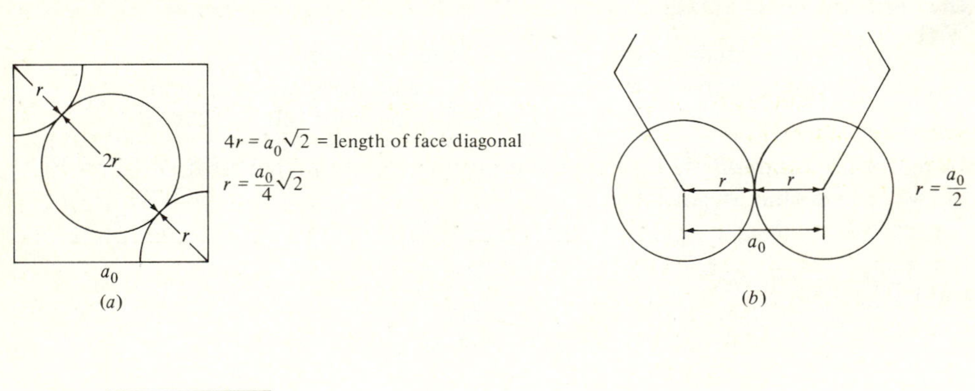

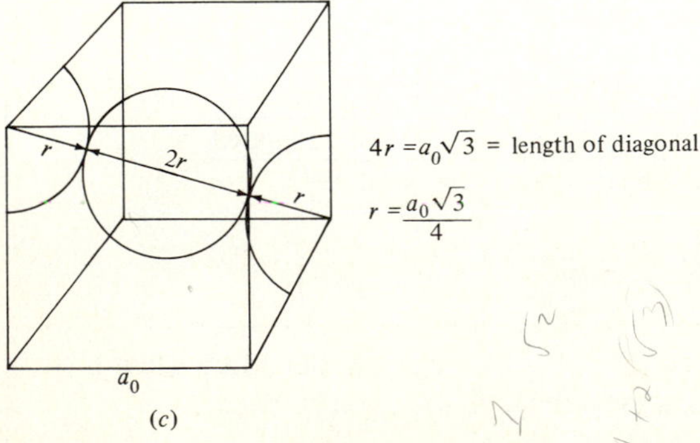

Fig. 2.9 *Calculation of atomic radius. (a) FCC unit cell: spheres (atoms) in contact along the face diagonal. (b) HCP unit cell: spheres (atoms) in contact on the edge. (c) BCC unit cell: spheres (atoms) in contact along the body diagonal.*

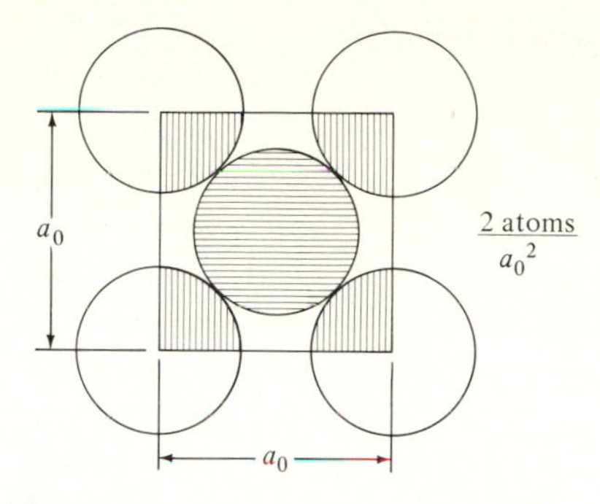

Fig. 2.10 *Calculation of planar density on the face plane of an FCC unit cell*

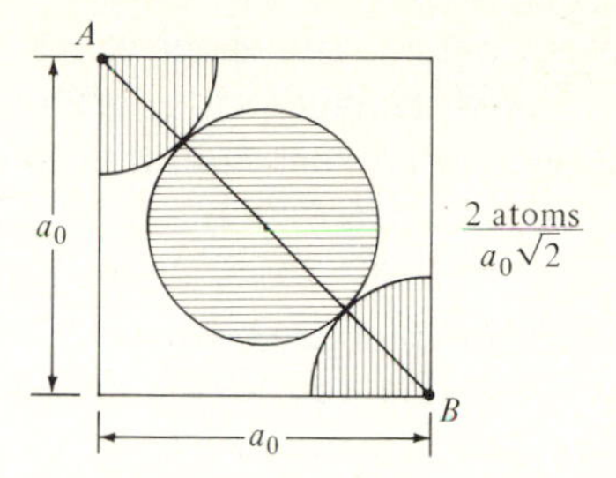

Fig. 2.11 *Calculation of linear density on the face diagonal of an FCC unit cell*

the plane is a circle (Fig. 2.10). Therefore, in the area a_0^2 we count one atom for the center but one-quarter atom for each of the corners because each has a trace of only one-quarter circle on the area a_0^2. The planar density is $2/a_0^2$ (atoms/Å^2). It should be added that in all these density calculations one of the ground rules is that a plane or a line must pass through the center of an atom or the atom is not counted in the calculations.

Linear Density This is an important concept because when planes slip over each other, the slip takes place in the direction of the closest packing of atoms on the planes. We calculate this by the following convention.

If a line passes completely through an atom, the trace of the atom on the line is one diameter (Fig. 2.11). In an FCC face the center atom counts for one atom on the face diagonal. The corner atoms make traces equal to only one-half diameter each on the line of length AB. Therefore, the linear density in the AB direction [110] in an FCC structure is

$$\frac{1 + \frac{1}{2} + \frac{1}{2}}{a_0\sqrt{2}}\frac{\text{atoms}}{\text{Å}} = \frac{2}{a_0\sqrt{2}}\frac{\text{atoms}}{\text{Å}} \quad \text{or} \quad \frac{\text{atoms}}{\text{nm}}$$

By the same reasoning the linear density in a BCC in the [111] direction is

$$\frac{2}{a_0\sqrt{3}}\frac{\text{atoms}}{\text{Å}} \quad \text{or} \quad \frac{\text{atoms}}{\text{nm}}$$

(Recall that the atoms touch along the body diagonal of a BCC.)

EXAMPLE 2.5 Calculate linear and planar atomic density for a face-centered-cubic structure in the [112] direction and the (111) plane, respectively.

ANSWER

When determining linear atomic density, we must remain in the reference cell. Therefore the centers of the atoms cut in an FCC are at 0, 0, 0 and $\frac{1}{2}$, $\frac{1}{2}$, 1. There are 2 radii or 1 atom. We can determine the length of the ray in several ways. One is to use the geometry of the large right triangle.

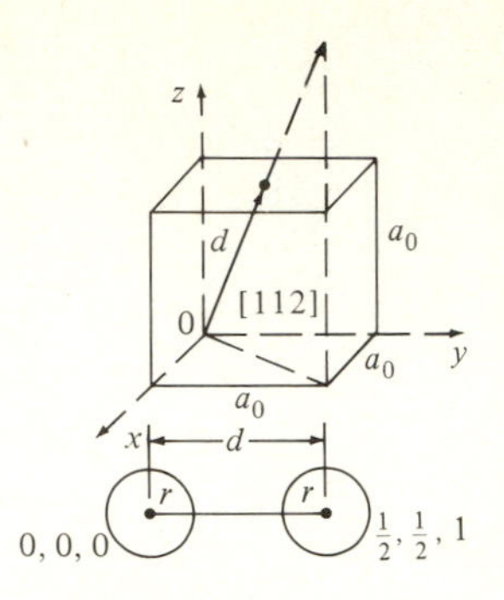

$$(2d)^2 = (a_0\sqrt{2})^2 + (2a_0)^2 \quad \text{from which } d = a_0\sqrt{6}/2$$

Therefore $\quad$ Linear atomic density $= \dfrac{1 \text{ atom}}{a_0\sqrt{6}/2} = \dfrac{2}{a_0\sqrt{6}} \dfrac{\text{atoms}}{\text{Å}}$

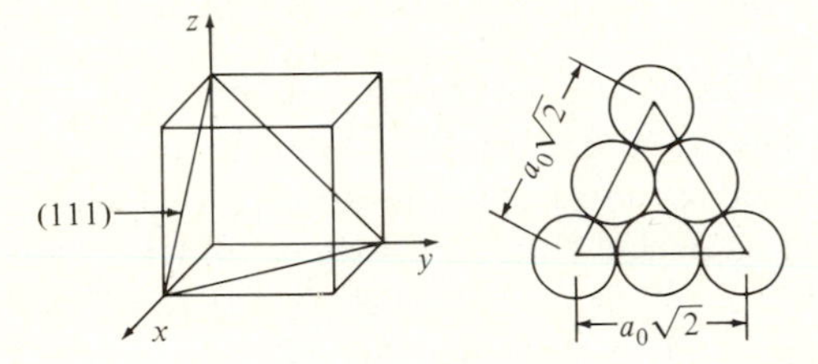

Number of atoms $= 3 \times \frac{1}{6} + 3 \times \frac{1}{2} = 2$ atoms

Area of equilateral triangle $= \dfrac{\sqrt{3}\,a_0^2}{2}$ (see Example 2.7 for derivation)

Therefore $\quad$ Planar atomic density $= \dfrac{2}{\sqrt{3}\,a_0^2/2} = \dfrac{4}{a_0^2\sqrt{3}} \dfrac{\text{atoms}}{\text{Å}^2}$

2.8 Close-packed hexagonal metals

To describe hexagonal structures, we need a few simple modifications of the Miller indices of directions and planes. These are called *Miller-Bravais indices* (Fig. 2.12). Instead of three axes x, y, z, we use four axes—three in the horizontal xy plane, called a_1, a_2, a_3, at 120° to each other, and the fourth, c, in the z direction. Using the extra axis makes it easier to see the relations between similar planes in the hexagonal structure. Let us locate some planes using this reference frame (Fig. 2.12*a*).

The most important planes in an HCP cell are the basal planes. Finding the intercepts of one of these ($K'N'LPMO'$), we obtain $a_1 = \infty$, $a_2 = \infty$, $a_3 = \infty$, $c = 1$. Taking reciprocals as before, we find that this is a (0001) plane. Now taking the intercepts of K, K', N', N, we get 1, ∞, $\bar{1}$, ∞ or $(10\bar{1}0)$. Similarly, for

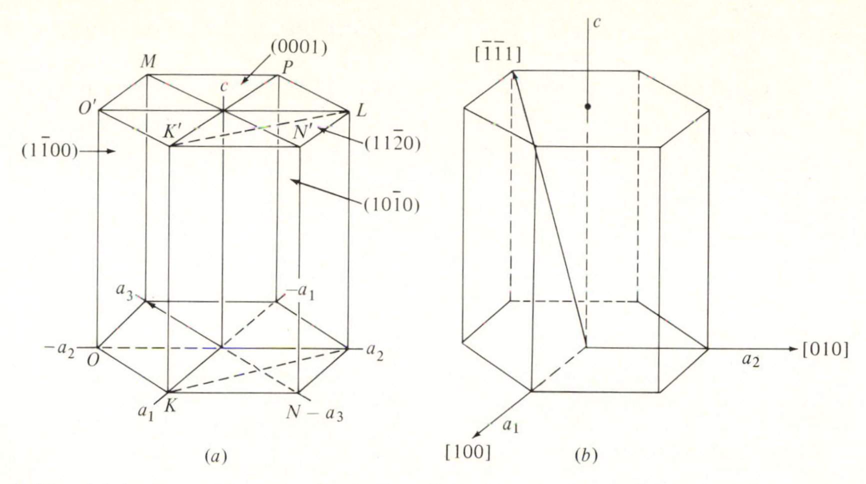

Fig. 2.12 *(a) Indices of some planes in a hexagonal crystal. (b) Indices of some directions in a hexagonal crystal.* [Part (a) from C.S. Barrett and T.B. Massalski, *The Structure of Metals*, 3d ed., McGraw-Hill, New York, 1973, Fig. 1.7, p. 12. Used with permission of McGraw-Hill Book Company.]

K, O, O', K', we find intercepts 1, $\bar{1}$, ∞, ∞ or (1$\bar{1}$00). For the other faces we find other combinations of 1$\bar{1}$00. Thus these planes are of the family {10$\bar{1}$0}. We call these Miller-Bravais indices h, k, i, and l. If we select h, k, and l, then i cannot be independent and by vector geometry always equals $-(h + k)$.

The notation for a direction is developed by using either only three axes (a_1, a_2, and c) or four axes. Note again that a_1 and a_2 are at 120°. We shall use the simple three-axes notation. Figure 2.12*b* shows that this is similar to the method explained for the cubic system.

In a hexagonal *close-packed* system, additional atoms are present in the unit cell (Fig. 2.5). The packing on the additional plane is the same as on the planes above and below. However, the stacking sequence changes.

Note: An HCP structure and an FCC structure exhibit similar packing. The figures for the (111) plane and the (0001) plane of an HCP show the same atomic arrangement. Similarly, using the hard, perfect-sphere model for atoms, the c/a_0 ratio for an HCP is 1.63, which is derivable from the geometry of a regular tetrahedron. (The height of a regular tetrahedron with a side length a is 0.866a. So this is the distance from the basal plane to the center of the close-packed atoms. The total HCP height becomes twice this value, or 1.63a.) The *atomic packing factor,* defined as the volume of atoms divided by the volume of the unit cell, is 0.74 for both the FCC and an ideal HCP. This is further proof of their similarity. The similarity is important to our later discussion of why FCC and HCP structures form extensive solid solutions (Section 2.13).

Most HCP metals do *not* exhibit perfect packing with a c/a_0 of 1.63. The atoms in this arrangement do not behave as perfect spheres. Examples of experimental c/a_0 ratios are Be (1.57), Ti (1.58), Mg (1.62), Zn (1.86), and Cd (1.89).

The basic difference between FCC and HCP structures lies in the way the close-packed planes are stacked above each other. (See Problem 2.23.)

Here is a summary of the various crystal structures that are important in metals.

Structure	*Atoms/unit cell*	*Unit cell–radius relationships*
Body-centered cubic	2	$4r = a_0\sqrt{3}$
Face-centered cubic	4	$4r = a_0\sqrt{2}$
Hexagonal close-packed	6	$2r = a_0$; $c/a_0 = 1.63$†

†See text and Example 2.7 for explanation of why this ratio is seldom achieved.

2.9 Determination of crystal structure; x-ray diffraction

Although our principal interest is in the use of the unit cell dimensions rather than in the technique of their determination, it is helpful to know a few basic features of x-ray methods. For example, x rays are used in the microprobe to determine the analysis of different microstructures as well as in structure determination.

The x-ray diffraction method does not measure the positions of the individual atoms directly. It measures instead the distances between planes of atoms. To illustrate this in two dimensions, consider a simple cubic unit cell *ABCD* (Fig. 2.13). If we determine the distance between planes containing *AB* and *CD* and between planes containing *AC* and *BD*, we have determined the cell.

We measure the interplanar distances by x-ray diffraction in the manner shown in Fig. 2.14, where the planes are similar to *AB* and *CD* in Fig. 2.13. The differences between diffraction and reflection are important. In reflection (as with a mirror) we can shine a light beam at a surface at any angle and it will be reflected at the same angle. By contrast, in diffraction the x-ray beam penetrates beneath the surface. When it strikes the atoms, they re-emit the radiation at the same wavelength. However, if

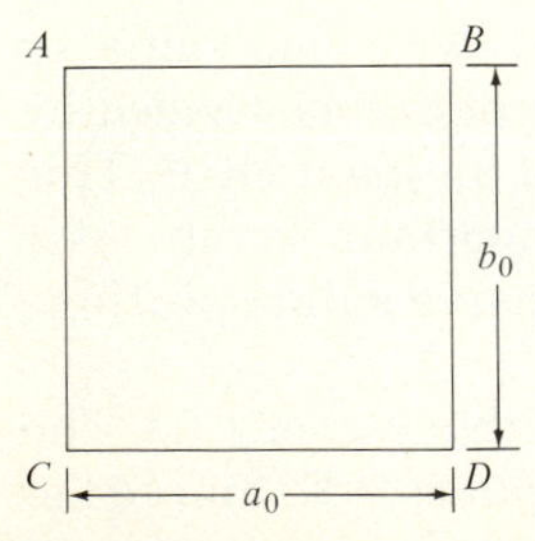

Fig. 2.13 *Calculation of a_0b_0. By finding distances between planes AC and BD, one can determine a_0.*

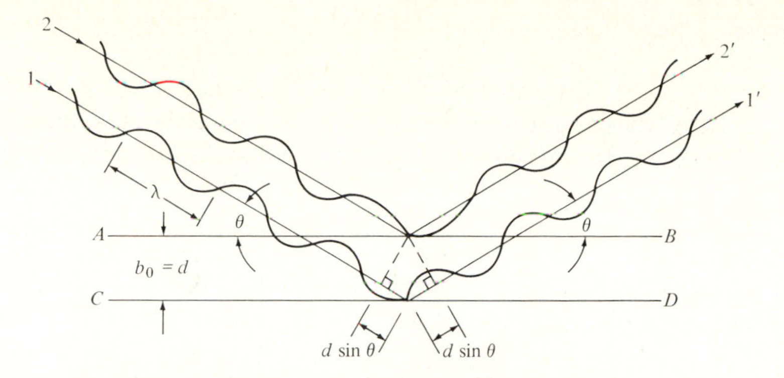

Fig. 2.14 *Diffraction of x rays from planes of atoms*

the radiations being emitted are not in phase, they will cancel out and there will be no diffracted beam.

If the radiations are in phase, there will be a beam, as shown in Fig. 2.14. To obtain this beam we must have a certain relationship between the angle of incidence, the wavelength of the radiation λ, and the interplanar distance d. For example, the path of the x-ray beam (1–1′) that goes one planar distance beneath the surface is longer than that of the beam diffracted by the surface atoms (2–2′) by the distance $2d \sin \theta$. For the radiations to be in phase, $2d \sin \theta$ must equal the wavelength λ, or some multiple $n\lambda$. This is known as *Bragg's law* and is written

$$n\lambda = 2d \sin \theta$$

Therefore, if we know θ and λ, we can find d. We obtain a beam of x rays of constant wavelength λ from an x-ray tube and observe the angle θ at which diffraction is found.

EXAMPLE 2.6 A sample of BCC chromium was placed in an x-ray beam of $\lambda = 1.54$ Å. Diffraction from 110 planes was obtained at $\theta = 22.2°$. Calculate a_0. Assume n (the order of diffraction) = 1.

ANSWER

$$\lambda = 2d \sin \theta \qquad d = \frac{1.54 \text{ Å}}{2(0.378)} = 2.04 \text{ Å}$$

If we inspect the BCC unit cell, we see that the distance between 110 planes is one-half the face diagonal, or

$$d_{110} = \frac{a_0 \sqrt{2}}{2}$$

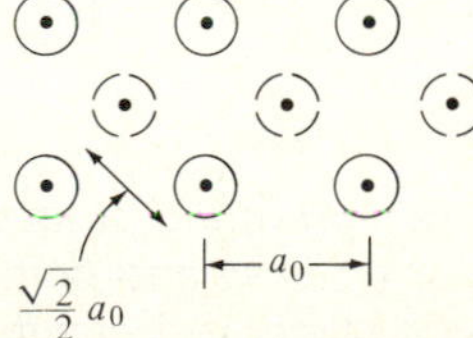

Therefore

$$a_0 = 2.04\sqrt{2} = 2.88 \text{ Å}$$

(This is a very much simplified case. Determining crystal structure by diffraction methods is an interesting and important field covered in advanced courses.)

This example can be generalized for the determination of distances between planes in cubic structures with the relationship

$$d_{hkl} = \frac{a_0}{\sqrt{h^2 + k^2 + l^2}} \quad \text{(for cubics only)}$$

In practice only specific planes will diffract, since Bragg's law is derived for the mutual support of the x radiation. The existence or extinction of diffraction from specific sets of planes lets us determine what type of cubic structure is diffracting the x rays. For example, in BCC structures $h + k + l$ must be even if diffraction is to occur, while for FCC, h, k, and l must be either all odd or all even if diffraction is to occur.

Finally, x radiation of a fixed wavelength must be available for the diffraction study. Such radiation is called *characteristic* x radiation. It should not be confused with *white* radiation or multiple-wavelength x radiation, commonly used in medical diagnosis.

To produce characteristic x radiation, we generate electrons by heating a filament, then accelerate them in an electromagnetic field. The high-energy electrons strike an anode with sufficient velocity to penetrate the outer electron shells of the anode material. When an electron strikes a K-shell electron, the incident electron removes the K-shell electron from orbit (Fig. 2.15a).

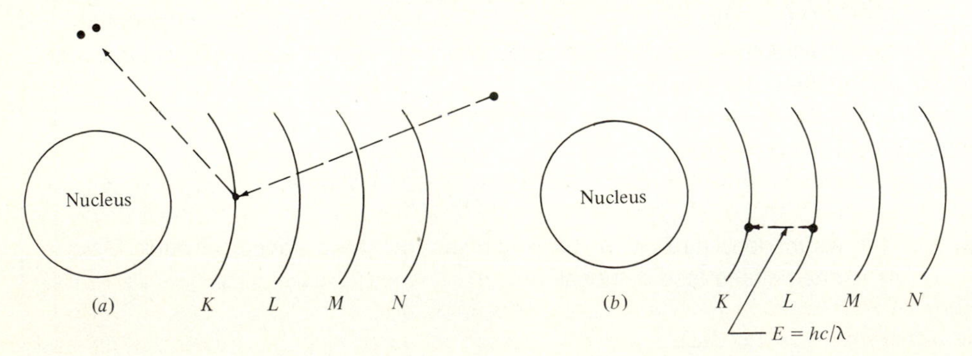

Fig. 2.15 *(a) In the production of a characteristic x ray, an incident electron strikes a K-shell electron and both leave the atom. (b) The movement of an L-shell electron into the K shell produces the characteristic x ray known as* K_α.

The atom that has lost one of its K-shell electrons is in a metastable condition. Therefore, an L-shell electron drops in to fill up the K shell (Fig. 2.15b). The energy E associated with this is

$$E = \frac{hc}{\lambda}$$

where h = Planck's constant (6.62×10^{-27} erg · sec or 0.662×10^{-33} J · sec); $c = 0.3 \times 10^9$ m/sec (speed of light); λ = wavelength.

For a given atom species, such as iron, the energy E is constant, and therefore the wavelength λ is also constant. This gives the *monochromatic* x radiation required for diffraction.

We might expect that this characteristic x radiation, referred to as K_α, would be of higher energy and thus shorter wavelength for those elements with more tightly bound K-shell electrons. The following short table shows this to be the case.

Element	*Atomic Number*	*K_α Wavelength (Å)*
Cr	24	2.29
Fe	26	1.94
Co	27	1.79
Cu	29	1.54
Mo	42	0.709
W	74	0.209

Chapter 16 gives a further discussion of this principle and its application to identification of the chemical composition of metals.

2.10 Definition of a phase; solid phase transformations in metals

At this point let us explain what we mean by a *phase* and a *phase transformation*. Everyone knows that H_2O can exist as a gas, a liquid, and a solid. These are three different phases. The general definition of a phase is a homogeneous aggregation of matter.

Under different conditions of temperature and pressure, different crystal structures, that is, different unit cells, for ice are formed, called ice I, II, III, etc. These are different solid phases with the same chemical composition.

Similarly, we encounter different crystal structures in the same metal. A bar of iron at room temperature has a BCC structure. However, if we heat the bar above 1670°F (910°C), the structure changes to FCC. Instead of calling these iron I and iron II, as we did with ice, we call the BCC iron α (alpha) and the FCC γ (gamma). In general, Greek letters are used to designate the different phases in solid metals.

Table 2.4 PERIODIC TABLE OF THE ELEMENTS AND THEIR CRYSTAL STRUCTURES*

	IA	IIA	IIIA	IVA	VA	VIA	VIIA	← VIII	VIII	VIII →	IB	IIB	IIIB	IVB	VB	VIB	VIIB	O		
1st							1 H A3 A1	2 He (A3) (A2)												
2nd	3 Li A2 A1 A3	4 Be (A2) A3											5 B H T R	6 C R H A4	7 N H C	8 O C (R)	9 F	10 Ne A1		
3rd	11 Na A2 A3	12 Mg A3											13 Al A1	14 Si A4	15 P C O C	16 S O M R	17 Cl A1 A3 A2	18 A A1		
4th	19 K A2	20 Ca A2 A1	21 Sc (A2) A3	22 Ti A2 A3	23 V A2	24 Cr (A1) A2	25 Mn A2 A1 C C	26 Fe A2 A1 A2	27 Co A1 A3	28 Ni A1	29 Cu A1	30 Zn A3	31 Ga O	32 Ge A4	33 As A7	34 Se A8 M	35 Br O	36 Kr A1		
5th	37 Rb A2	38 Sr A2 A3 A1	39 Y A2 A3	40 Zr A2 A3	41 Nb A2	42 Mo A2	43 Tc A3	44 Ru A3	45 Rh A1	46 Pd A1	47 Ag A1	48 Cd A3	49 In A6	50 Sn A5 A4	51 Sb A7	52 Te A8	53 I O	54 Xe A1		
6th	55 Cs A2	56 Ba A2 (T) (H)	57 La A2 A1 H	72 Hf A2 A3	73 Ta A2	74 W A2	65 Re A3	76 Os A3	77 Ir A1	78 Pt A1	79 Au A1	80 Hg R T	81 Tl A2 A3	82 Pb A1	83 Bi A7	84 Po R C	85 At	86 Rn		
7th	87 Fr	88 Ra	89 Ac A1																	
6th (continued) →				58 Ce A2 A1 H	59 Pr A2 H	60 Nd A2 H	61 Pm	62 Sm (A2) R	63 Eu A2	64 Gd (A2) A3	65 Tb (A2) A3	66 Dy (?) A3	67 Ho (?) A3	68 Er A3	69 Tm A3	70 Yb A2 A1	72 Lu (?) A3			
7th (continued) →				90 Th A2 A1	91 Pa T	92 U A2 T O	93 Np A2 T O	94 Pu A2 T A1 O M M	95 Am H	96 Cm	97 Bk	98 Cf	99 Es	100 Fm	101 Md	102 No	103 Lr	104 (Rf)	105 (Ha)	106

*These other designations are used to denote the structures of the allotropic forms which are given in the order of their appearance. *A*1 = FCC; *A*2 = BCC; *A*3 = HCP; *A*4 = diamond cubic; *A*5 = body-centered tetragonal; *A*6 = face-centered tetragonal; *A*7 = rhombohedral; *A*8 = trigonal; H = hexagonal (usually *ABAC* . . . close-packed); R = rhombohedral; O = orthorhombic; C = complex cubic; T = tetragonal; M = monoclinic; ( ) = uncertain.

†Note that the authors of this table have used a different convention for the A and B subgroups which has not been followed in Tables 2.2 and 2.3.

(From C. S. Barrett and T. B. Massalski, "The Structure of Metals," 3d ed., McGraw-Hill Book Company, New York, 1973, Table 10.1, p. 227. Used with permission of McGraw-Hill Book Company.)

Table 2.4 summarizes the crystal structures of the elements. We see that many metallic elements can have different crystal structures. We also note that most of the metallic structures are BCC, FCC, or HCP. When a material has more than one crystal structure, we say it exhibits *allotropy* and has different *allotropic* forms.

Now let us use some lattice parameter data to estimate volume changes resulting from transformation of a metal from one solid phase to another.

EXAMPLE 2.7 On cooling through 880°C (1615°F), titanium goes through a phase change analogous to iron, except that in this case the crystal structure changes from BCC to HCP:

BCC: $a = 3.32$ Å
to HCP: $a = 2.956$ Å $c = 4.683$ Å

What is the volume change?

ANSWER The volume of the BCC is $(3.32 \text{ Å})^3 = 36.6 \text{ Å}^3$ for the two atoms per unit cell.

To determine the volume of the HCP, observe that there are six equilateral triangles in the basal plane:

$$\text{Area} = \tfrac{1}{2}ah$$

$$\sin 60° = \frac{\sqrt{3}}{2} = \frac{h}{a}$$

or $$\text{Area of the triangle} = \frac{\sqrt{3}}{4}a^2$$

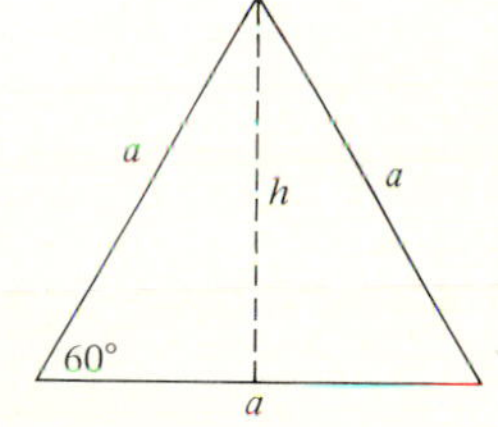

Therefore the area of the hexagonal base is $(6\sqrt{3}/4)a^2$, and the volume of the HCP equals

$$\left(6\frac{\sqrt{3}}{4}a^2\right)(c) = (\tfrac{3}{2}\sqrt{3})\,(2.956)^2(4.683) = 106.3 \text{ Å}^3$$

However, this HCP structure has six atoms per unit cell. To compare the volumes, we must consider the same number of atoms in each structure. We therefore compare one-third of the HCP volume (two atoms) with the two atoms in the BCC, or

$$36.6 \text{ Å}^3 \longrightarrow \frac{106.3}{3} = 35.4 \text{ Å}^3$$

Therefore there is a 3.3% contraction when BCC changes to HCP. Note that the c/a ratio of the HCP is not 1.63. We may not calculate the volume on the basis of a hard-sphere model.

2.11 Effects of the addition of other elements on the structure of pure metals

Only a few elements are widely used commercially in their pure form; pure copper in electrical conductors is one example. Generally, as in the case of iron, other elements are present to produce greater strength, to improve corrosion resistance, or simply as impurities because of the cost of refining. Whatever the reason, we need to know the effect these elements have on structure, because structure determines properties.

When a second element is added, two basically different structural changes are possible:

1. The atoms of the new element form a solid solution with the original element, but there is still only one phase, such as an FCC or liquid.
2. The atoms of the new element form a new second phase, usually containing some atoms of the original element. The entire microstructure may change to this new phase or two phases may both be present.

Let us examine each possibility separately.

2.12 Solid solutions

If we take a crucible containing 70 g of liquid copper and add 30 g of nickel while heating to maintain a completely liquid melt, we obtain a liquid solution of copper and nickel. If we cool the solution slowly and examine the grains, we find that we have an FCC structure and that a_0, the edge of the unit cell, is between the values for copper and nickel. This is called a *substitutional* solid solution because nickel atoms have substituted for copper atoms in the FCC structure (Fig. 2.16). This structure is stronger than either pure copper or pure nickel because the interaction between the atoms gives greater resistance to slip.

Another type of solid solution is called *interstitial*. The new atoms are located in the interstices or spaces between the atoms in the matrix. Naturally, the atoms that form interstitial solid solutions are those with small radii, such as hydrogen, carbon, boron, and nitrogen. Carbon, for example, forms an interstitial solid solution in iron, giving steel. Position C $\frac{1}{2}, \frac{1}{2}, \frac{1}{2}$ in Fig. 2.17 is an interstitial position in the FCC allotropic form of iron that can be filled with carbon.

When a substitutional or interstitial solid solution is formed, the Greek letter that was used for the pure metal is retained.

2.13 New phases, intermediate phases

If instead of nickel we added 30 wt.% lead to the copper melt, we would again obtain a liquid solution. However, when we examined the frozen melt, we would find two different phases. There would be grains of copper with a very

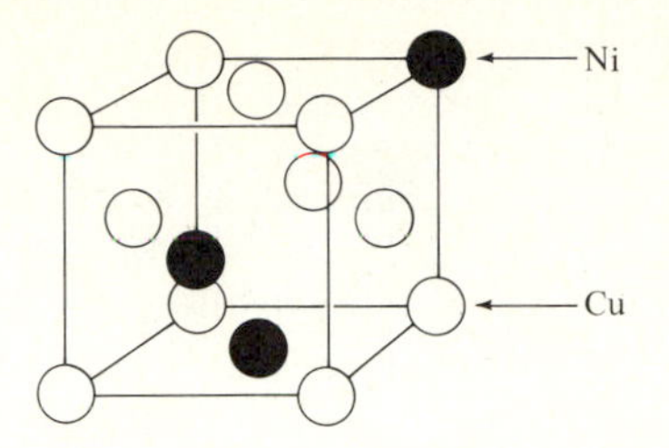

Fig. 2.16 *Substitutional solid solution of nickel in an FCC copper unit cell*

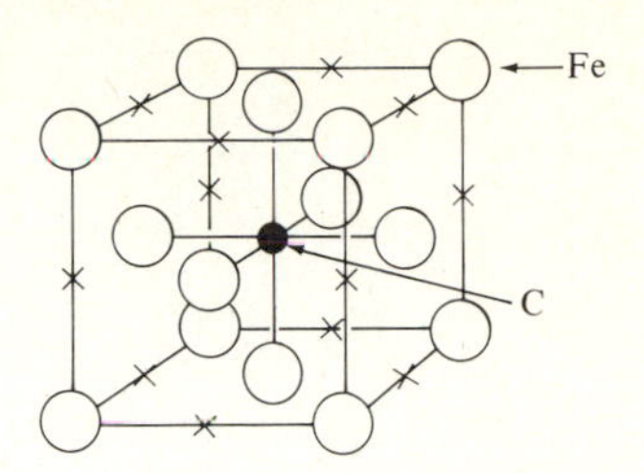

Fig. 2.17 *Interstitial solid solution of carbon in the FCC form of iron. Positions marked × are positions in the FCC equivalent to* $\frac{1}{2}, \frac{1}{2}, \frac{1}{2}$ position.

small amount of lead in solid solution, and the balance of the lead would have the form of small spherical grains of practically pure lead. Both structures are FCC, but the unit cell of lead is much larger, which means that the atomic radius is greater. Therefore, we find two phases, α and β. Although the added lead lowers the strength, a material of this type is quite useful in bearings. When a load is applied, the lead from the spherical grains tends to flow out and coat the entire surface of the bearing with a film that has a low coefficient of friction.

In this case, the new phase has the same structure as the element being introduced. Frequently the new phase has a structure different from either element. This is called an *intermediate phase.* For example, if we add 15 wt.% tin to copper, some of the tin will go into solution in the FCC copper phase, but we will find another phase with the approximate formula $Cu_{31}Sn_8$. This is a hard phase completely different from copper or tin.

We may ask if there are any general principles that determine whether a given element B will form a separate phase or a solid solution when added to another element A. This is important in alloy development because, in general, a solid solution will be more ductile and more easily shaped, while a two-phase material will usually exhibit greater hardness and strength.

The most important considerations in forming solid solutions are related to the atomic radii and chemical properties. To form an extensive solid solution (greater than 10 atomic percent soluble),† we should be aware of the following general rules of W. Hume-Rothery:

1. The difference in atomic radii should be less than 15 percent.
2. Proximity within the Periodic Table is important (Similar electronegativity).
3. For a complete series of solid solutions, the metals must have the same crystal structure.

---

†To find atomic percent, calculate the percentage atoms of element B for the total number of atoms present.

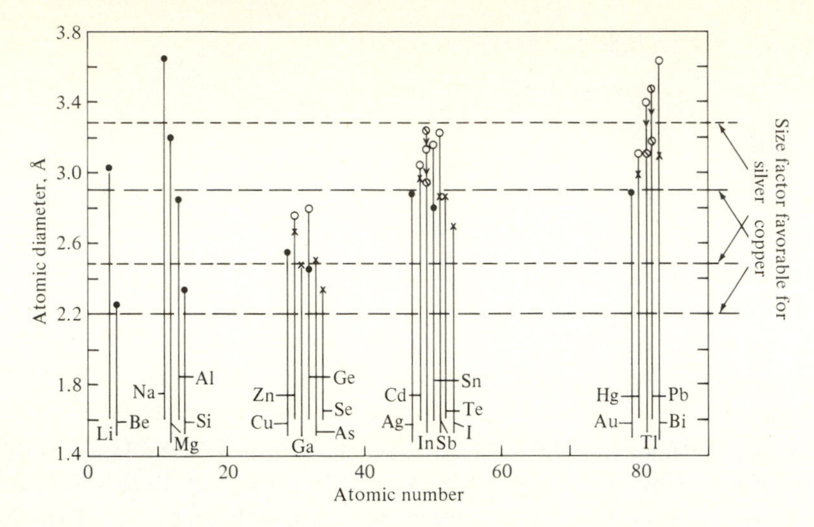

Fig. 2.18 *Favorable atomic diameter for elements soluble in copper and silver. When the size is within ± 15%, there is extensive solid solution.* (Courtesy of W. Hume-Rothery)

Elements close to one another in the Periodic Table have similar electronegativities (Table 2.3) and are more compatible.

As an example, in Fig. 2.18 we have plotted the atomic diameters ($2r$) of a number of elements. The atomic diameter of copper ±15 percent is shown by the dashed heavy lines, that of silver ±15 percent by the dashed light lines. If we include elements with an FCC or HCP structure, we find that zinc, which is in the favorable zone for both copper and silver, forms extensive solid solutions with both elements. Cadmium, however, forms extensive solid solutions with silver but not with copper because it is outside the favorable zone.

EXAMPLE 2.8 Predict whether extensive substitutional solid solubility would occur between copper as the solvent and aluminum, nickel, and chromium as the solute elements.

Element	*Atom Radius (Å)*	*Crystal Structure*	*Electronegativity*
Cu	1.28	FCC	1.9
Al	1.43	FCC	1.5
Ni	1.25	FCC	1.8
Cr	1.25	BCC	1.6

ANSWER The radius difference is as follows:

$$\left(\frac{\text{Final} - \text{initial}}{\text{Initial}} \times 100\right) \quad \text{or} \quad \frac{r_x - r_{Cu}}{r_{Cu}} \times 100$$

Cu-Al	+11.7%
Cu-Ni	− 2.3%
Cu-Cr	− 2.3%

All the elements fall within the required ±15% radius requirement. The electronegativity values are not vastly different. This indicates somewhat similar chemical activity for the metals.

However, the atom packing within the BCC Cr is not the same as in the FCC Cu. Therefore, we would not expect extensive solid solubility between the two elements. The order of expected solid solubility in copper based on radius difference and crystal structure, along with the observed experimental values, is as follows:

Element	*Observed At. % Soluble*
Ni (best)	100
Al	17
Cr (poorest)	<1

The importance of considering atomic percentage rather than weight percentage can be seen if 17 at. % Al is converted to wt. %.

Basis: 100 atoms

17 atoms Al 83 atoms Cu

$$\frac{17 \text{ atoms} \times 26.98 \text{ g/at. wt.}}{6.02 \times 10^{23} \text{ atoms/at. wt.}} = \text{weight of 17 Al atoms}$$

$$\frac{83 \text{ atoms} \times 63.54 \text{ g/at. wt.}}{6.02 \times 10^{23} \text{ atoms/at. wt.}} = \text{weight of 83 Cu atoms}$$

Therefore

$$\text{Wt. \% Al} = \frac{\dfrac{17 \times 26.98}{6.02 \times 10^{23}}}{\dfrac{17 \times 26.98}{6.02 \times 10^{23}} + \dfrac{83 \times 63.54}{6.02 \times 10^{23}}} \times 100 = 8.0$$

Since aluminum has a low atomic weight compared to copper, the weight percentage is much less than the atomic percentage. Nickel, however, is close to copper in atomic weight, and so values for the atomic and weight percentages would be very close.

SUMMARY

We began this chapter by studying the nature of metal atoms arising from their outer electrons and relative position in the Periodic Table. Next we considered the arrangement of atoms in the unit cell and methods for describing their location. We also discussed directions and planes in the unit cell. We found that we can calculate density by dividing the mass of the atoms in the unit cell by the cell volume. Other important calculations included linear and planar density of atoms and atomic radius. We then described the determination of the unit cell by x-ray diffraction and the calculation of volume change resulting from phase transformation. The concepts of solid solution of alloying elements in a phase and the formation of a new phase such as an intermetallic compound were considered.

DEFINITIONS

Allotropy The occurrence of two or more crystal structures in the same chemical composition.

Atomic radius A measure calculated from the unit cell dimensions by finding a direction in which atoms are in contact, as for example the diagonal of the cube face in an FCC. Since the diagonal length can be calculated from a_0, the edge of the unit cell, the atomic radius can be found.

Coordinates Location of an atom in the unit cell, found by moving specified distances from the origin in the x, y, and z directions.

Density Mass divided by volume, usually expressed as grams per cubic centimeter or kilograms per cubic meter.

Indices of direction Descriptions of direction in a unit cell found by translating the direction passing through the origin, finding the coordinates of the point at which the arrow leaves the unit cell, and clearing fractions. These indices are enclosed in brackets.

Interstitial solid solution A solution formed when the dissolved element fills holes (interstices) in the solvent element lattice.

Linear density The number of atoms that have their centers located on a given line of direction for a given length.

Miller indices Descriptions of a plane found by taking reciprocals of the intercepts of the plane on the x, y, and z axes and clearing fractions. Miller indices are enclosed in parentheses.

Number of atoms per unit cell A number found by connecting the corners of the cell to form an imaginary box. Then determine the fraction of each atom contained within the box and add the fractions. As an example, an atom at a corner position of a cube is counted as one-eighth because it is shared equally by eight unit cells meeting at the center of the atom.

Periodic Table A tabular grouping of the elements in order of atomic number. The metallic elements are on the left-hand side, since in general these elements contain one, two, or three electrons in the outer shell.

Phase A homogeneous aggregation of matter. In solid metals grains composed of atoms of the same unit cells are the same phase. Alloying atoms may also be present in a one-phase structure.

Planar density The number of atoms with centers located within a given area of the plane. The planar area selected should be representative of the repeating groups of atoms.

Quantum number A number describing the average position of an electron in the electronic structure.

Shorthand notation A system for describing the number of electrons in each shell of an atom and their subgrouping.

Solid solution A solution formed when the addition of one or more new elements still results in a single-phase structure.

Substitutional solid solution A solution formed when the dissolved element substitutes, i.e., replaces, an atom or atoms of the solvent element in its unit cell.

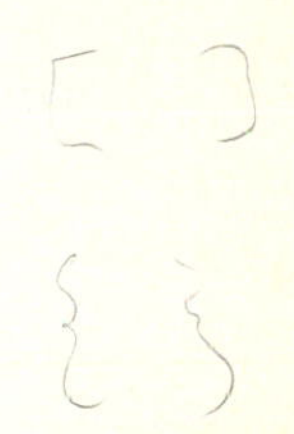

Unit cell A geometric figure illustrating the grouping of atoms in a solid. This group or module is repeated many times in space within a grain or crystal.

PROBLEMS

2.1 A foundry used selenium to control a certain property (the inclusion shape) in its castings. During a national emergency selenium became scarce, and tellurium was tried and found satisfactory. Give a basic reason for the similarity. (A humorous side issue: Exposure to selenium gave the workers garlicky breath. Their spouses did not appreciate the fascinating scientific fact that tellurium reproduced the effects of selenium even to the breath.) (Sections 2.1 and 2.3)

2.2 Identify the following elements as very active metals, less active metals, semiconductors, or nonmetals by examining their electronic structures. Then check the Periodic Table to find the actual elements and their positions. (Sections 2.1 and 2.3)

Element Number	*Structure*
3	$1s^2, 2s^1$
6	$1s^2, 2s^2, 2p^2$
9	$1s^2, 2s^2, 2p^5$
28	$1s^2, 2s^2, 2p^6, 3s^2\ 3p^6, 3d^8, 4s^2$
29	$1s^2, 2s^2, 2p^6, 3s^2, 3p^6, 3d^{10}, 4s^1$
32	$1s^2, 2s^2, 2p^6, 3s^2, 3p^6, 3d^{10}, 4s^2, 4p^2$

2.3 Indicate the un-ionized and ionized shorthand electron notation for the following elements. (Sections 2.1 and 2.3)

Cu	$\longrightarrow$	Cu^{+} and Cu^{2+}	atomic number = 29
Fe	$\longrightarrow$	Fe^{2+} and Fe^{3+}	atomic number = 26
K	$\longrightarrow$	K^{+}	atomic number = 19
V	$\longrightarrow$	V^{3+} and V^{5+}	atomic number = 23
Ga	$\longrightarrow$	Ga^{3+}	atomic number = 31

2.4 Use one of the following phrases to describe the element: active metal, inert gas, nonmetal, semiconductor, transition element. (The same phrase may be used several times.) (Sections 2.1 and 2.3)

a. $1s^2\ 2s^1$ ___________
b. $1s^2\ 2s^2\ 2p^6\ 3s^1$ ___________
c. $1s^2\ 2s^2\ 2p^6\ 3s^2\ 3p^2$ ___________
d. $1s^2\ 2s^2\ 2p^6\ 3s^2\ 3p^5$ ___________
e. $1s^2\ 2s^2\ 2p^6\ 3s^2\ 3p^6\ 3d^5\ 4s^2$ ___________

2.5 Sketch the following directions and planes in a BCC unit cell, and give the coordinates of the atoms they intersect. (We always deal only with intersections through atom centers.) (Sections 2.4 through 2.7)

[100]	(100)
[110]	(110)
[111]	(111)

2.6 Repeat Prob. 2.5 for an FCC structure. (Sections 2.4 through 2.7)

2.7 Calculate the linear density (atoms/Å) of atoms in the [100], [110], and [111] directions in BCC iron (a_0 = 2.86 Å). (Sections 2.4 through 2.7)

2.8 Calculate the linear density of atoms in the [100], [110], and [111] directions in FCC copper (a_0 = 3.62 Å). (Sections 2.4 through 2.7)

2.9 Calculate the planar density of atoms (atoms/Å^2) in BCC iron (a_0 = 2.86 Å) in the (100), (110), and (111) planes. (Sections 2.4 through 2.7)

2.10 Calculate the density of BCC iron at room temperature from the atomic radius of 1.24 Å. Compare your results with the experimental value of 7.87 g/cm^3 (7.87 × 10^3 kg/m^3, 7.87 mg/m^3). (Sections 2.4 through 2.7)

2.11 One of your associates has calculated the density of body-centered cubic iron as given here (atomic radius = 1.24 Å and atomic weight = 55.85):

$$\text{Density} = \frac{\dfrac{4\ \text{atoms/u.c.} \times 55.85\ \text{g/at. wt}}{6.02 \times 10^{23}\ \text{atoms/at. wt}}}{\left[\dfrac{2(1.24 \times 10^{-8}\ \text{cm})}{\sqrt{2}}\right]^3} = 68.8\ \text{g/cm}^3$$

Since materials do not have densities this high, there must be errors. What are these errors? (Assume that the error is not a result of having pushed the wrong buttons on the calculator.) (Sections 2.4 through 2.7)

2.12 We have a hypothetical element Q of atomic weight 64.09 which is FCC. An experimental density has been determined as 8.41 g/cm^3. (Sections 2.4 through 2.7)

a. Calculate the approximate atomic radius of Q.
b. Why is the calculation only approximate?

2.13 The lattice parameter of BCC iron is 2.86 Å and the density is 7.87 g/cm^3. Calculate the *atomic weight.* (Sections 2.4 through 2.7)

2.14 Label the directions shown (use integers) in the figure. Label the planes shown in the figure. Which of these directions are directions of *maximum* linear density for a BCC crystal? For an FCC crystal? Which of these planes are planes of maximum atomic density for an FCC crystal? (Sections 2.4 through 2.7)

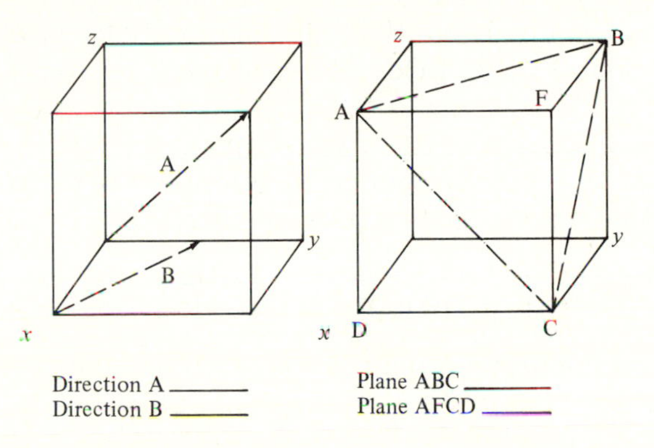

2.15 Another national emergency involved a shortage of tungsten, used in producing high-speed steel. A typical analysis of this steel was 18% W, 4% Cr, 1% V, balance Fe. It was found that molybdenum could be substituted for more than half the tungsten. What basic relationships would lead you to believe that these elements can be substituted for one another? The new steel is still widely used today. (Sections 2.10 through 2.13)

2.16 During the recent accelerated development of titanium alloys for aircraft and underwater research, one group concentrated on alloys that would be single phase. Decide which of the following metals would be expected to form extensive solid solutions with titanium. (Sections 2.10 through 2.13)

Ti is HCP	$a_0 = 2.95$ Å
Be is HCP	$a_0 = 2.28$ Å
Al is FCC	$a_0 = 4.04$ Å
V is BCC	$a_0 = 3.04$ Å
Cr is BCC	$a_0 = 2.88$ Å

Defining extensive solid solutions as 10 atomic percent, calculate the value in weight percent that would correspond to 10 atomic percent in titanium for the elements that form extensive solid solutions.

2.17 As we mentioned, pure iron undergoes an allotropic transformation at 910°C (1670°F). The BCC form is stable at temperatures below 910°C, while the FCC form is stable above 910°C. Calculate the volume change for the transformation BCC → FCC, if at 910°C, a = 3.63 Å for FCC and a = 2.93 Å for BCC. (Sections 2.10 through 2.13)

2.18 Solid solutions occur to some extent in all alloys. Given the FCC unit cell of iron, a_0 = 3.60 Å: Calculate the radius of an atom for which you would expect a high probability of *interstitial* solid solution. Calculate the range in radius for an atom for which you would expect extensive *substitutional* solid solution. (Sections 2.10 through 2.13)

2.19 There is another important transformation in iron, in addition to the $\alpha + \gamma$ change discussed. When heated to 2534°F (1390°C), the structure changes from FCC to BCC. The a_0 of the FCC structure is 3.680 Å; the a_0 of the BCC structure is 2.926 Å. Does a given mass of iron contract or expand in volume as the transformation from FCC to BCC takes place? Calculate the percent change in volume that takes place during this transformation. (Sections 2.10 through 2.13)

PROBLEMS COVERING OPTIONAL SECTIONS

2.20 Write out all four quantum numbers for the following common elements. (Section 2.2)

K (19)	$1s^2, 2s^2, 2p^6, 3s^2, 3p^6, 4s^1$
Fe (26)	$1s^2, 2s^2, 2p^6, 3s^2, 3p^6, 3d^6, 4s^2$
Cu (29)	$1s^2, 2s^2, 2p^6, 3s^2, 3p^6, 3d^{10}, 4s^1$
Mo (42)	$1s^2, 2s^2, 2p^6, 3s^2, 3p^6, 3d^{10}, 4s^2, 4p^6, 4d^5, 5s^1$

2.21 Calculate the atomic radius of zinc (HCP, a_0 = 2.66 Å, c_0 = 4.95 Å). Do the same for copper (FCC) and iron (BCC), using the data of Probs. 2.7 and 2.8 (Section 2.8)

2.22 Calculate the planar density of atoms in the (0001) plane of zinc. Why is the answer the same as for the (111) plane of copper? Will this be true for any HCP and FCC structures? (Section 2.8)

2.23 Although packing in the (0001) plane in the hexagonal close-packed structure is the same as that in the {111} planes of the face-centered-cubic structure, the unit cells are quite different, due to the way the layers of atoms are stacked. In both HCP and FCC, the first two planes are stacked as shown in the top-view sketch opposite. The positions of the atoms in the lower plane are *A* positions and the positions of the atoms in the second plane are *B* positions. HCP and FCC differ in the location of the atoms in the third plane. In case 1, the atoms are placed

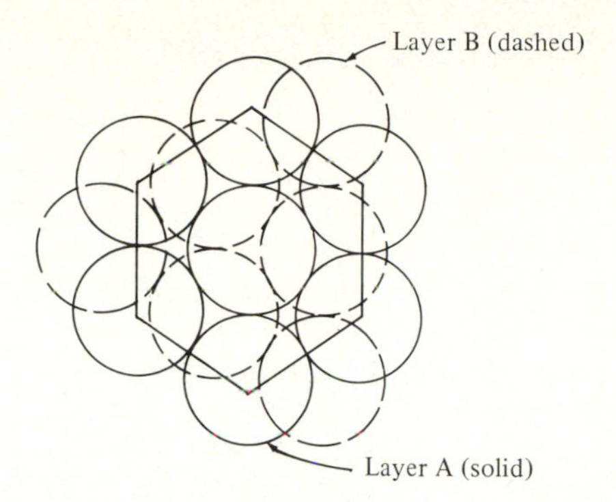

directly above the atoms in the first plane, giving an *ABA* stacking. In case 2, the atoms are placed in positions that are not over *A* or *B* atoms. This is called the *C* position. (After the *C* position, in the fourth layer, the atoms revert to *A* position.) Sketch top views of case 1 and case 2 as overlays on the sketch given. Which is HCP and which is FCC? (Section 2.8)

2.24 A specimen of silver (FCC, r_0 = 1.44 Å) is placed in an x-ray camera and irradiated with molybdenum-characteristic radiation (0.709 Å). It is observed that θ for 111 planes decreases by 0.11° as the silver is heated from room temperature to 800°C (1470°F). Given that the crystal structure remains the same, find the change in a due to the heating. (*Hint:* $d_{111} = a/\sqrt{3}$.) (Section 2.9)

2.25 An investigator is trying to determine whether a structure in an 18% Cr, 18% Ni stainless steel is FCC or BCC. (In this composition both structures are encountered at room temperature.) X-ray diffraction shows that the distance between 111 planes is 2.1 Å. For BCC iron a_0 = 2.86 Å, and for FCC iron a_0 = 3.63 Å. Which structure is present? (Section 2.9)

2.26 X rays with a wavelength λ of 2.29 Å are diffracted from BCC iron (lattice constant = 2.866 Å) at a Bragg angle θ of 34.5° (sin 34.5° = 0.566). Was this diffraction from 110 planes? (Section 2.9)

2.27 We are given an x-ray tube with anode material unlabeled. We obtain an x-ray diffraction pattern for FCC aluminum and find the diffraction angle to be 24.5° for the 111 planes. Determine the anode material from the elements given in Section 2.9. (Section 2.9)

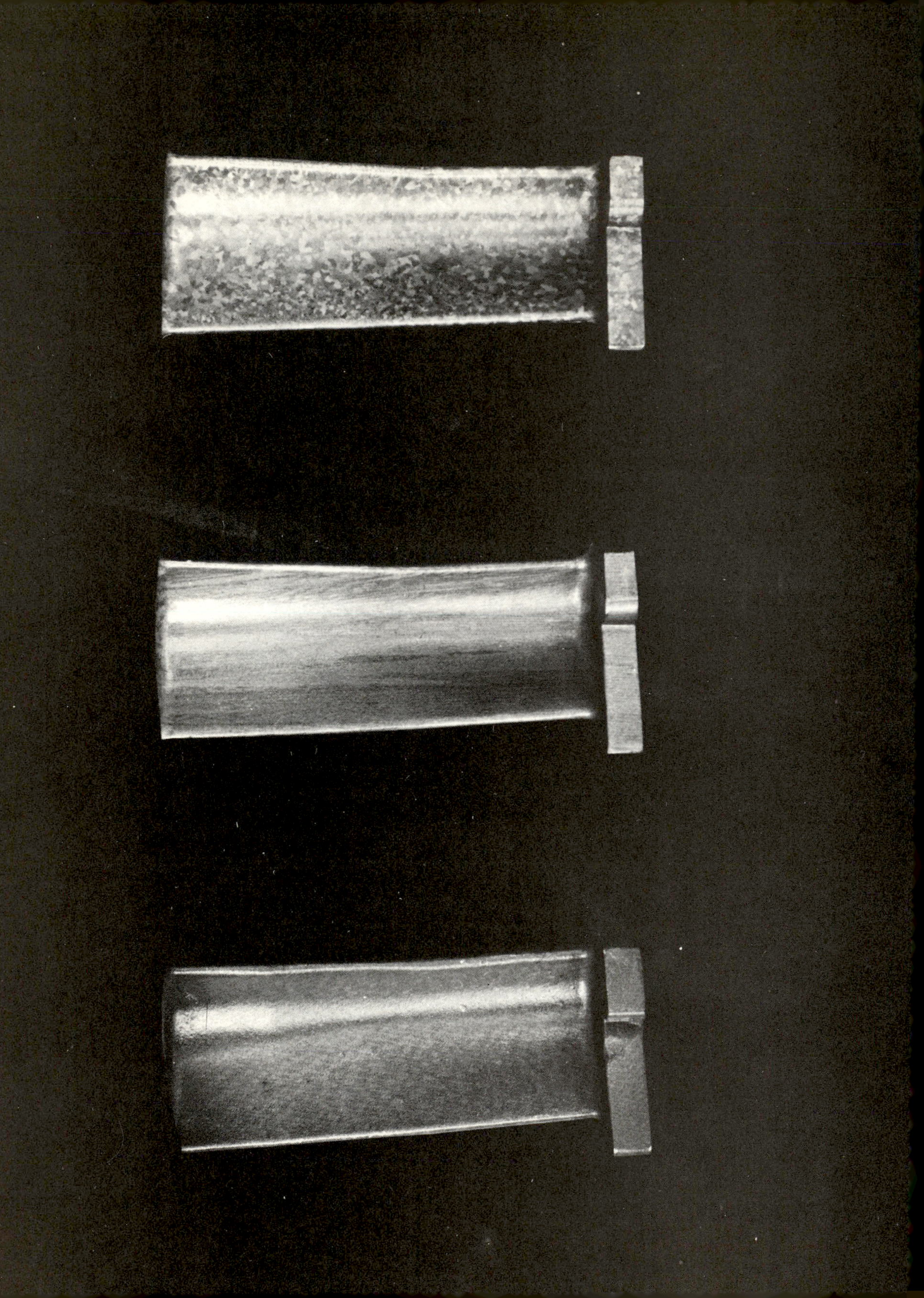

3

EFFECTS OF STRESS AND TEMPERATURE ON SIMPLE METAL STRUCTURES

THE blades shown in the illustration are typical of those used in the most severe application in modern jet aircraft, the combustion section of the turbine. These castings illustrate the rewards of careful development of the proper structure to resist fracture. The lowest-strength blade (top) shows a randomly oriented grain structure. The first step in improvement (middle) was to grow the grains by controlled solidification so that a columnar structure formed and no grain boundaries were formed at right angles to the axis of the blade, since rupture in this case initiates at grain boundaries. Finally it was found possible to eliminate the grain boundaries by growing a single grain (bottom) with the direction of greatest strength oriented to resist the condition of maximum stress. This represents a great advance in the design of jet engines.

In this chapter we will study the effect of stress on the structures described in Chapter 2. We will see that both elastic and plastic strains are produced. Finally we will consider the combined effects of stress and temperature.

3.1 General

Now that we have described the nature of metal structures, let us note the effects of stress and temperature on them. These effects are important for two reasons.

1. To form parts into the required shape and to obtain the desired properties, we use combinations of stress and heat, as in rolling, forming, annealing, heat treating.
2. In service all parts are subjected to stress, and in many cases to elevated temperature as well, as in an automobile engine.

Let us consider first the effects of stress, then temperature, and finally the combined effects in the phenomena called "creep" and "stress relief." We shall deal with one-phase structures in this chapter, and with more complex two-phase structures in Chapter 4.

3.2 Effects of stress on metal structures

If we examine the objects about us in the street or in our homes we see that the most useful property of metals is their ability to resist or transmit stress. There are dozens of illustrations of this in an automobile. The frame and fenders resist imposed stresses, while the power train transmits stress from the pistons to the wheels. In this case all the parts deform a little and then spring back when the stress is removed. This deformation is *elastic* strain.

Another important effect of stress on metals is permanent deformation or *plastic* strain. If we exceed the proper load in a car, the springs may sag permanently. If we hit a tree, the fender deforms. This ability to deform and not shatter is a vital feature of metals, not merely in case of automobile accidents, but so that we can form metal sheets and bars into shapes such as car bodies, structural columns, beams, and aircraft wings.

We shall now take up these vital characteristics of metal deformation under stress: elastic strain and plastic strain.

Elastic Strain Let us first review the engineering aspects of elastic strain and then discuss the relation of the structure to the observations.

A common experiment in elementary physics is to suspend a long steel wire from the ceiling and then hang increasingly heavy weights from the end. The length of the wire increases proportionately with the load. Upon removal of the load, the wire returns to its original length. In general terms, we define

$$\text{"Engineering" stress} = \sigma = \frac{\text{load}}{\text{original area of wire cross section}}$$

$$= \frac{P}{A_0} \qquad \text{units: } \begin{cases} \text{pounds per square inch (psi) or} \\ \text{newtons per square meter (N/m}^2\text{) or} \\ \text{pascals; } 1\ \text{N/m}^2 = 1\ \text{Pa} \end{cases}$$

$$\text{"Engineering" strain} = \epsilon = \frac{\text{change in length}}{\text{original length}} = \begin{cases} \dfrac{\text{in.}}{\text{in.}} \text{ (units dimensionless)} \\ \\ \dfrac{\text{m}}{\text{m}} \end{cases}$$

We use the terms "engineering" stress and "engineering" strain because later we shall define *true* stress and strain. In this example of elastic deformation, the difference is very small. If we calculate the stress in the wire and the strain produced, we find the relation

$$E = \frac{\sigma}{\epsilon} \qquad \text{(units: psi or N/m}^2 \text{ or Pa)}$$

where E is called the *modulus of elasticity*. As long as we are careful not to reach too high a stress, the deformation is principally elastic and E is a constant. For each group of materials E has a characteristic value. For example, $E = 30 \times 10^6$ psi (2.07×10^5 MPa)† for all steels and 10×10^6 psi (0.69×10^5 MPa) for aluminum alloys. The modulus is basically related to the bonding between atoms.

This is the macroscopic picture of elastic strain, and it can be deceivingly simple. Let us go to the other extreme and test a single crystal of iron rather than a wire, which contains thousands of grains or crystals.

If we stress the single crystal along different crystal directions, we get values quite different from 30 million psi (2.07×10^5 MPa):

Crystal direction:	[111]	[100]
E (10^6 psi):	41	18
E (10^5 MPa)	2.83	1.24

Although this seems astonishing at first, we recall that in BCC structures (iron at room temperature), the atomic packing is densest in [111], the direction of highest E. We would expect that the interatomic forces would be greatest along this direction and therefore that the stress required to produce a given strain would be highest.

How do we explain the practically constant value of 30×10^6 psi (2.07×10^5 MPa) for steel? We get this value when there are many crystals of different orientations and our measurement gives the average value. There are occasional important exceptions for which we must remember the properties of the single crystal. Let us consider some examples.

A number of identical dentures for teeth were cast of Vitallium, a cobalt alloy, and tested for deflection under constant stress. The deflections differed although the cross sections were the same. It was found that because of the thin cross section, only one or two grains were present at the highly stressed

---

†1 psi = 6.9×10^3 N/m^2 = 6.9×10^{-3} MN/m^2 = 7.03×10^{-4} kg/mm^2 (MN = meganewton). 1 MPa = 1 MN/m^2 (MPa = megapascal).

region. Since the modulus E varies with direction in a grain, the orientation of these grains determined the amount of deflection.

In some cases we want to produce a part with the crystals oriented in one direction. This is called development of *preferred orientation*. An outstanding example is the production of transformer steel sheet with what is termed a cubic texture. By special processing the ⟨100⟩ directions of BCC iron are aligned in the plane of the sheet. In this case the ease of magnetization, like the modulus, varies with direction in the crystal. Thus when transformer laminations are stamped from sheet with this preferred orientation, the hysteresis losses (magnetic energy losses manifested as heat) in the finished part are lower.

A more recent example is a process that casts blades with controlled directional properties for aircraft gas turbines, so that the best value of strength is in the direction in which the operating stress is highest (see the frontispiece for Chapter 3).

3.3 Plastic strain, permanent deformation, slip

We described elastic strain using the physics experiment with a wire suspended from the ceiling and loaded with weights at the bottom. Occasionally an inventive student adds a really massive weight from some other equipment and finds (1) that the extension in length is more than expected from the equation $\epsilon = \sigma/E$, and (2) that when the load is removed, the wire does not return to its original length. The portion of the total deformation under load that does not disappear when the load is removed is called *plastic* or *permanent* deformation. Let us now discuss what occurs within the structure during plastic deformation.

3.4 Critical resolved shear stress for plastic deformation

To illustrate plastic deformation, let us first grow a number of single crystals of zinc in the form of rods. To do this, we melt the zinc in a test tube in a vertical furnace and then lower the tube very slowly out of the bottom of the furnace. Freezing will start, and if we are careful, a single crystal will form and the rest of the metal will freeze with this as a nucleus, giving us a single crystal in the form of a rod. After breaking away the glass, let us grip the ends of the rods in a tensile machine and pull until there is appreciable permanent deformation. There will be a great difference among the specimens in the axial stress required to cause this permanent deformation or strain. In all cases, however, *the flow occurs by shearlike movement (called "slip") on the {0001} planes*. The specimens that require the least stress for slip (plastic flow) are those with both the normal to the {0001} planes and the ⟨110⟩ directions at 45° to the axis of the specimen.

To explain this we need an important but simple derivation.

Consider the rod in Fig. 3.1*a*, which is a single crystal of zinc. We locate the orientation of the {0001} planes, such as A_2, by x rays. The (0001) planes are the basal planes of the HCP structure in the rod crystal. Slip takes place on these planes and in the [110] direction.

We first find the component of the *force* in the slip plane. This is $F \cos \lambda$. The shear stress in the slip plane is this force divided by the area A_2. Next A_2 may be related to the known area A_1 by trigonometry: $A_2 = A_1/\cos \varphi$. Therefore, the shear stress τ is

$$\tau = \frac{F}{A_1} \cos \lambda \cos \varphi = \sigma \cos \lambda \cos \varphi$$

If we take the different axial stresses required to cause slip and the angular measurements and calculate the resolved shear stress in each case, we get the same value (within experimental error). This is the *critical resolved shear stress* τ_c. When $\varphi = \lambda = 45°$, the maximum shear stress is obtained from a given axial stress.

We finally have the answer to the unexpected behavior of the grains of copper in the bar in Fig. 2.2. Although slip occurred in some regions away from the region of maximum tensile stress, these were regions of maximum resolved shear stress. That is, because of the favorable *orientation* of these grains, τ_c was reached sooner than in the other grains.

In hexagonal close-packed metals the slip occurred in the {0001} planes and in the ⟨110⟩ directions. As illustrated in Fig. 3.1*b*, three directions and one plane are involved. We define a plane and a slip direction in the plane as a *slip system;* therefore, at room temperature magnesium has three slip systems. The slip system is usually made up of the planes with highest atomic density (closest packing) and the directions with highest linear density. In

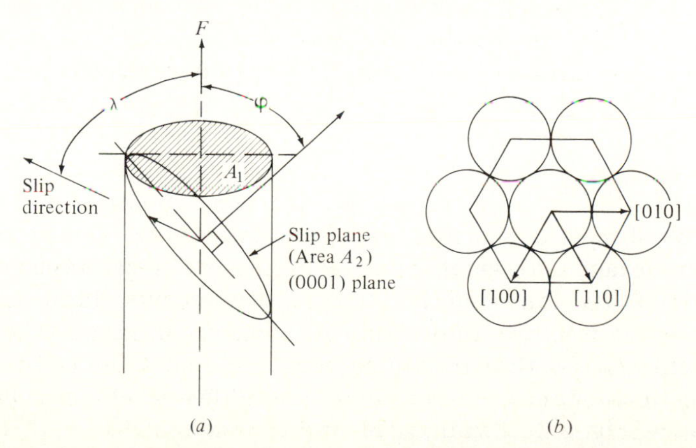

Fig. 3.1 *(a) Critical resolved shear stress model. (b) Slip directions in HCP.*

EXAMPLE 3.1 An investigator prepares four single crystals of magnesium in cylinders of the same cross-sectional area and finds that different axial stresses are required for permanent deformation (0.2% plastic strain). Furthermore, by x rays the investigator finds the orientation of the slip plane and of the slip direction relative to the axis.

Are the differences in yield strength due to imperfections in the crystals? What are the critical resolved shear stresses? The data are as follows.

Crystal	*Yield stress, g/mm²*	φ	λ
1	200	45°	54°
2	230	30°	66°
3	400	60°	66°
4	1,000	70°	76°

ANSWER

Crystal	F/A	×	*(cos φ)*	×	*(cos λ)*	=	τ_c, *g/mm²*
1	200		0.707		0.587		83
2	230		0.865		0.407		81
3	400		0.500		0.407		81
4	1,000		0.342		0.242		83

The critical resolved shear stresses are the same (within experimental error). Difference in orientation, not imperfections, causes variation in yield strength.

the FCC metals the {111} planes and the ⟨110⟩ directions reach maximum density. There are no exceptions to this rule. In BCC the ⟨111⟩ directions are closest packed and slip always occurs in these directions. There is no plane of maximum packing density, and therefore a number of planes such as (110) are involved. In HCP metals the *c/a* ratio (see Fig. 2.12) is never the ideal of 1.633. When this value is lower than 1.633, planes other than the {0001} planes may slip.

3.5 Twinning

Another type of plastic deformation is called *twinning*. This is particularly important in hexagonal crystals because normally slip can occur on only one plane, (0001). If this plane is normal to the specimen axis, there is no shearing stress, and brittle fracture would occur if twinning could not take place. The essential difference is that in slip each atom on one side of the slip plane moves a constant distance, whereas in twinning the movement is proportional to the distance (Fig. 3.2). Twinning is most common in BCC and HCP metals. It can take place much more rapidly than slip. Therefore, we often find me-

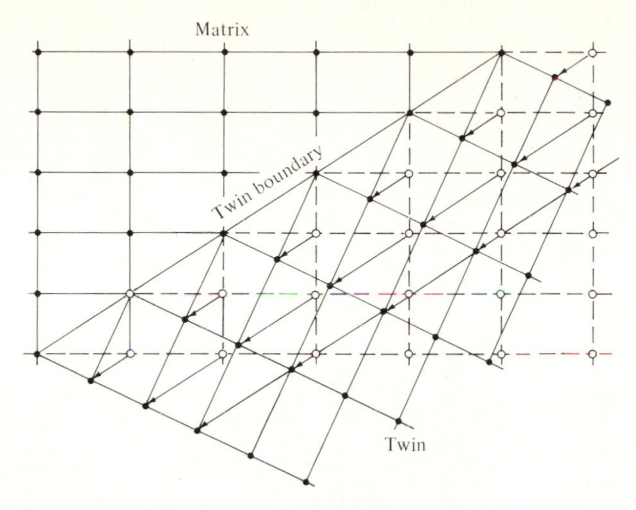

Fig. 3.2 *Formation of a twin in a tetragonal lattice by a uniform shearing of atoms parallel to the twin boundary. The dashed lines represent the lattice before twinning, the solid lines after twinning.* (H.W. Hayden, W.G. Moffatt, and John Wulff, *The Structure and Properties of Materials*, Vol. 3: Mechanical Behavior, John Wiley, New York, 1965, Fig. 5.10, p. 111. By permission of John Wiley & Sons, Inc.)

chanically formed twins rather than slip bands after shock loading. In FCC structures twins are usually formed only on heating (annealing) of cold-worked structures (Fig. 3.24 on page 91).

The differences between slip and mechanical twinning are sometimes hard to understand. Figure 3.3 shows two single crystals after plastic deformation. In Fig. 3.3*a* the deformation is by slip. Each plane, approximately 1000 atoms thick, moves an integral amount relative to an adjacent plane. Putting your hands on the top and bottom of a squared deck of cards and then moving them in opposite directions would be analogous.

Figure 3.3*b* shows deformation by twinning. The atoms move a distance proportional to the distance from the twin boundaries. The layer of atoms between the boundaries is called the mechanical twin.

3.6 Engineering stress-strain curves

Now, let us look at the way the strength of commercial materials is tested and specified. In the specifications for materials of the principal engineering groups, such as the American Society for Testing Materials, The American Iron and Steel Institute, or the Society of Automotive Engineers, the most common basis for testing is the 0.505-in.- (1.28-cm-) diameter tensile specimen (Fig. 3.4). In other words, whether we are buying material for a bridge or a crankshaft, a common tensile test is used to evaluate the mechanical properties. Usually modulus of elasticity, tensile strength, yield strength, percent

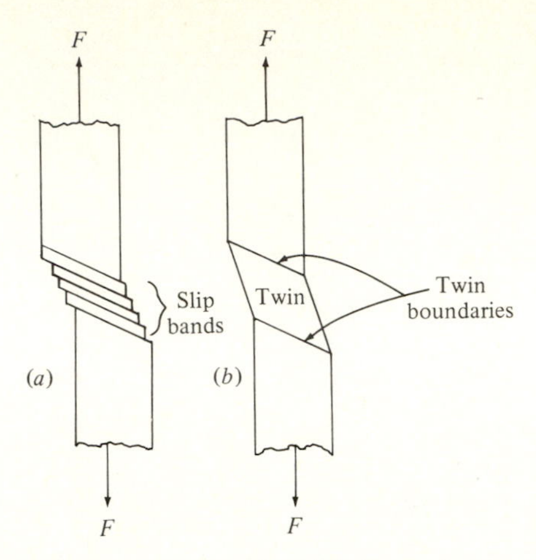

Fig. 3.3 *(a) Single crystal that exhibits slip. (b) A mechanical twin shown in a single crystal.*

elongation, and reduction of area are specified. Let us discuss how these are determined.

The "0.505 bar," as it is called from its diameter, is machined from the stock being bought or from separately cast specimens from the same ladle of metal. The bar is screwed into a pair of grips, which in turn are part of a tensile testing machine (Fig. 3.5).

Next the grips are pulled apart by mechanical means. The load on the specimen is recorded continuously. The load can be converted to engineering stress by multiplying by 5, since the area of the 0.505-in.-diameter specimen is 0.2 in.^{2} and engineering stress is load divided by the original area.

It is important to measure the extension or strain of the bar at the same time. A variety of devices, from mechanically operated extensometers to electric strain gages, are used. In the usual mechanical extensometer the gage is anchored to a 2-in. gage length at the start of the test. Note that the raw data are in terms of *load* and *extension*. These are converted as follows:

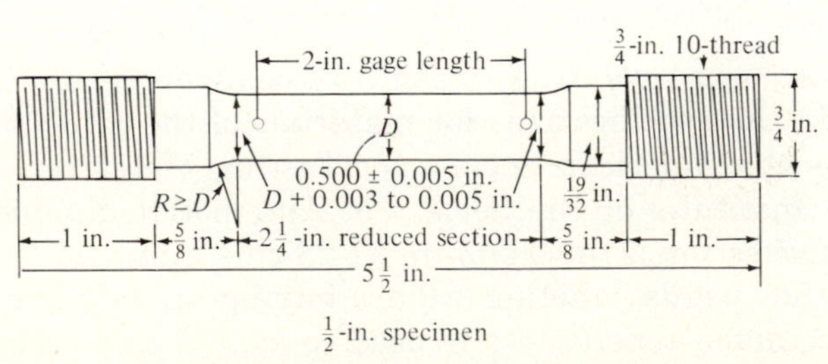

Fig. 3.4 *Design of a tensile test specimen. (To convert to millimeters, multiply by 25.4.)*

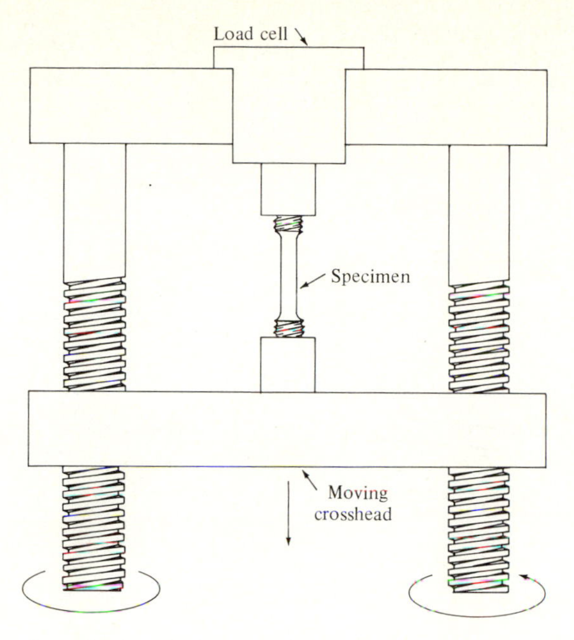

Fig. 3.5 *Cross section of a tensile test machine.* (H.W. Hayden, W.G. Moffatt, and John Wulff, *The Structure and Properties of Materials,* Vol. 3: Mechanical Behavior, John Wiley, New York, 1965, Fig. 1.1, p. 2. By permission of John Wiley & Sons, Inc.)

$$\text{Engineering stress} = \sigma = \frac{\text{load}}{\text{original area}}$$

$$\text{Engineering strain} = \epsilon = \frac{\text{change in length}}{\text{original length}}$$

We then plot the stress vs. strain and obtain curves, as shown in Fig. 3.6*a* and *b*.

The following data, obtained from a tensile test, are used in specifications.

Modulus of elasticity (psi or MPa)
= (stress/strain) in elastic range (slope of stress-strain curve)
Tensile strength (psi or MPa)
= maximum stress on the stress-strain curve
Yield strength at 0.2% offset (psi or MPa)
= stress at which 0.2% permanent or plastic strain is present
Percent elongation at fracture
= $(l_f - l_o)/l_o \times 100$, where l_f = final length and l_o = original length
Percent reduction of area
= $(A_o - A_f)/A_o \times 100$, where A_o = original area and A_f = final area

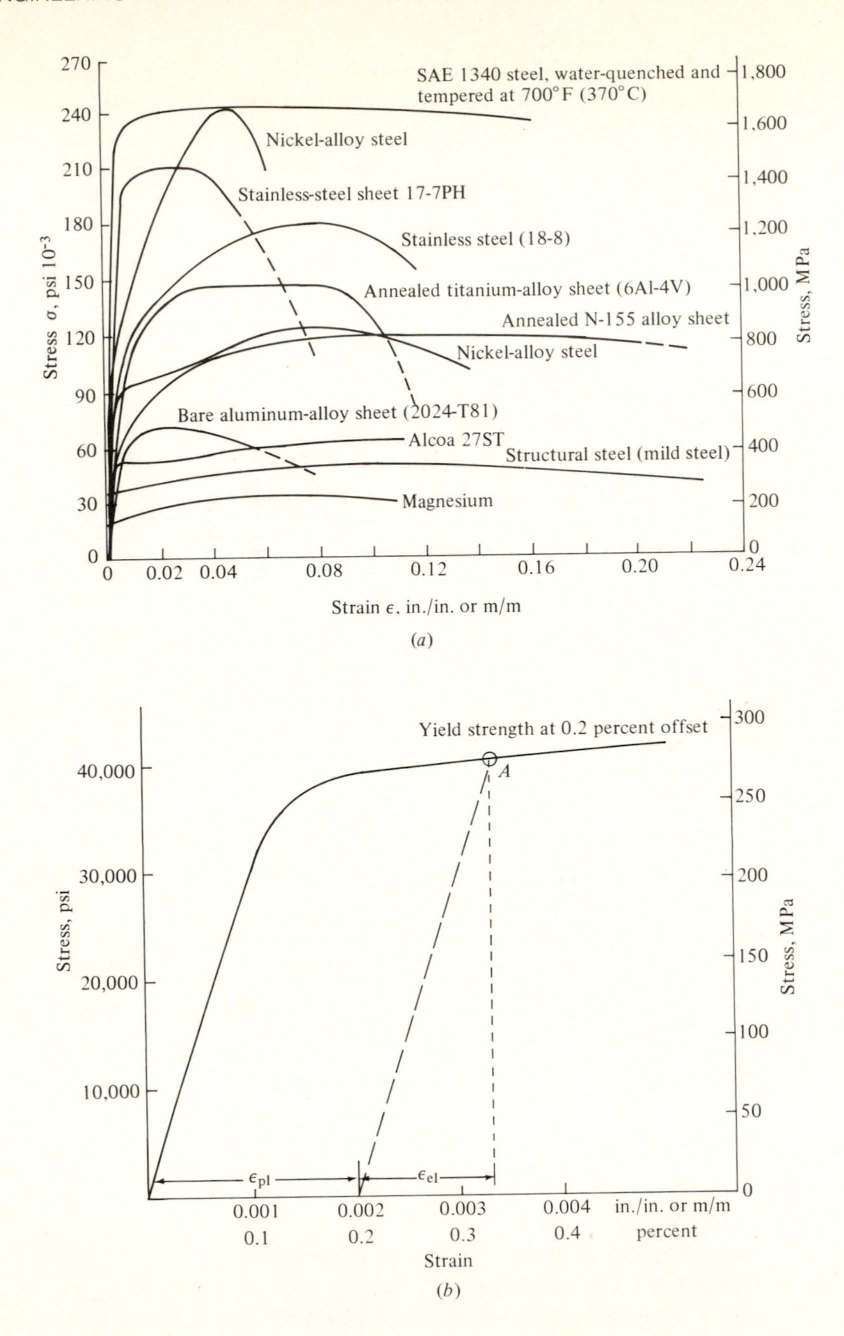

Fig. 3.6 *(a) Stress–strain curves for various alloys. (b) Details of a stress–strain curve for mild steel (0.3% carbon).* [Part (a) from Joseph Mann, *Mechanical Behavior of Engineering Materials*, © 1962, Fig. 1.10, p. 24. *Reprinted by permission of Prentice-Hall, Inc., Englewood Cliffs, N.J.)*

Breaking stress, the engineering stress at fracture, is also noted but not included in specifications for ductile materials. For brittle materials it is hard to distinguish from the tensile strength.

The *modulus of elasticity* is used to calculate the deflection of a given part under load. For example, a crankshaft will bend a certain amount between bearings, and clearance must be allowed for this. Also, a material may not be substituted for another of the same strength without considering the modulus. For instance, an aluminum crankshaft will deflect three times as much as a steel one because the modulus is 10×10^6 psi rather than 30×10^6 psi. Note also that the modulus does not change with strength. Characteristic moduli are given in Table 3.1.

The *tensile strength* is an index of the quality of the material. It is not used much in design for ductile materials because there has been a great deal of plastic strain by the time this load-carrying capacity has been reached. However, it is a good indication of defects. If the bar has flaws or harmful inclusions, it will not reach the same maximum stress.

The *yield strength* is the most important value for design. The significance of the phrase "0.2% offset" needs to be explained. As the tensile specimen is loaded, two types of strain develop: elastic and plastic.

As we mentioned earlier, the elastic strain will disappear upon unloading, whereas the plastic strain resulting from permanent deformation by slip will remain. It was once thought that the bar behaved elastically up to a certain point, called the *yield point*, then plastic strain began. However, we know now from precise strain measurements and from microscopic observation of slip that plastic strain begins at low stresses. The question then becomes: How much set (plastic elongation) can the designer tolerate? The percentage of set that can be tolerated in the spring of a delicate balance is different from that in the boom of a steam shovel. However, for most engi-

Table 3.1 MODULI OF ELASTICITY FOR SEVERAL METALLIC MATERIALS

Material	*Modulus of Elasticity,* psi $\times 10^6$*
All steels, alloyed and unalloyed	30
Nickel alloys	26 to 30
Copper alloys	15 to 18
Aluminum alloys	10 to 11
Magnesium alloys	6.5
Cast iron, depending on amount and type of graphite	15 to 22
Ductile iron, depending on amount of graphite	22 to 25
Malleable iron, depending on amount of graphite	26 to 27
Molybdenum	47

*Multiply psi by 6.9×10^{-3} to obtain MPa or by 7.03×10^{-4} to obtain kg/mm^2.

neering uses a plastic strain of 0.2% can be tolerated. The stress at which this occurs is called the *yield strength*.

We calculate this value using the following construction (Fig. 3.6*b*). First 0.002 strain (0.2% strain) is laid off from the origin on the x axis. Next a line is drawn through this point parallel to the straight line portion of the stress-strain curve until it intersects the curve. This intersection gives the value of the yield strength, 40,000 psi (276 MPa). If we follow normal testing procedures with a tensile specimen to this point and then remove the load, we will have 0.2% set, which is tolerable in most cases. In general, to find the amount of permanent and elastic strain at any point on a stress-strain curve, we draw a line parallel to the straight-line portion from the point to the strain axis, as shown in Fig. 3.6*b*. The total strain at A is approximately 0.0033 in./in. (m/m), the elastic component ϵ_{el} is 0.0013 in./in. (m/m), and the plastic component ϵ_{pl} is 0.002 in./in. (m/m). If less plastic strain can be tolerated, the yield strength is specified at 0.1% offset or lower.

In other alloys, such as copper, aluminum, and magnesium, the stress-strain graph begins to curve at low stresses. In this case the yield strength is usually specified as the stress at 0.5% *total* strain. This is read directly from the graph as the sum of the elastic and plastic components.

The *percent elongation* at fracture serves several purposes. It is possibly a better index of quality than the tensile strength because if inclusions or porosity are present, the elongation is drastically lowered. Second, the elongation multiplied by the tensile strength is an index of toughness at low rates of strain. The toughest steel available, Hadfield's 12% manganese steel, is used for railroad crossings, safe parts, and in ore crushers, as discussed in Chapter 6. It has elongation of over 40% and tensile strength of over 100,000 psi (690 MPa). The gage length over which the percent elongation is calculated must be noted.

EXAMPLE 3.2 The following data were obtained for a high-strength aluminum alloy. Plot the engineering stress-strain curve. A 0.505-in.-diameter tensile specimen with a 2-in. gage length was used.

Load, lb	*Stress, psi*	*Gage length, in.*	*Strain*
0	0	2.0000	0
4,000	20,000	2.0041	0.002
8,000	40,000	2.0079	0.004
10,000	50,000	2.0103	0.005
12,000	60,000	2.0114	0.006
13,000	65,000	2.0142	0.007
14,000	70,000	2.0202	0.010
16,000	80,000	2.0503	0.025
16,000 (maximum)	80,000	2.0990	0.050
15,600 (fracture)	78,000	2.1340	0.067

Calculate the modulus of elasticity, yield strength at 0.2% offset, percent elongation, and percent reduction of area. The diameters at maximum load and fracture are 0.485 in. and 0.468 in., respectively.

ANSWER The data for load are converted to stress by multiplying by 5, since the area of a 0.505-in.-diameter tensile test bar is 0.2 in.^{2}. The data are then plotted, as shown in Fig. 3.7. Note that to find the modulus of elasticity and yield strength a magnified scale is used on the x axis.

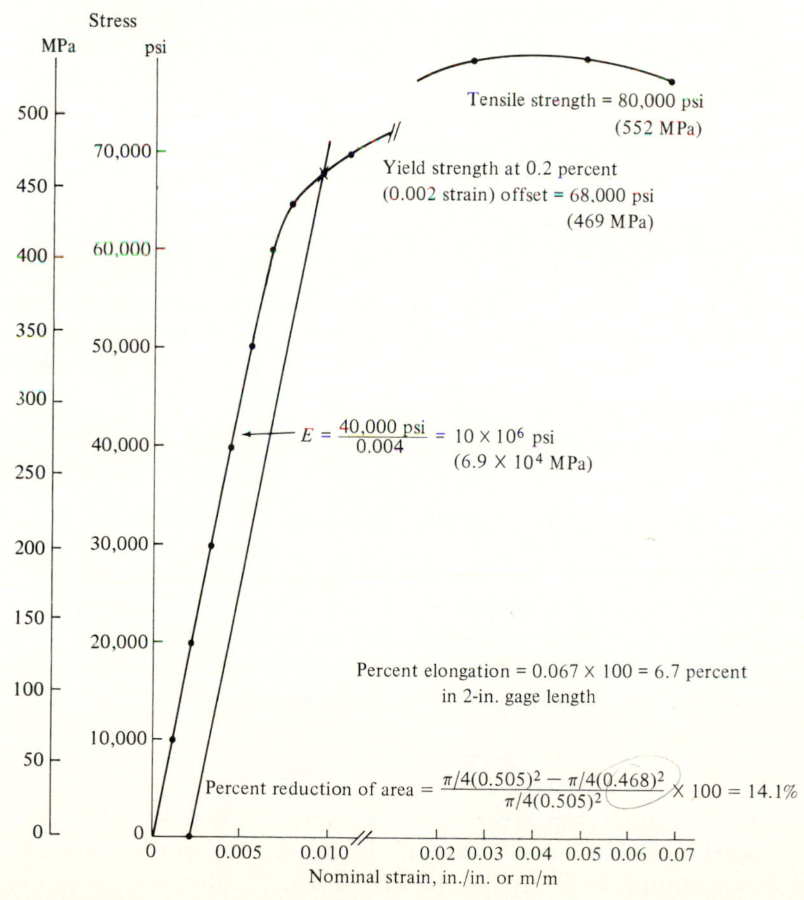

Fig. 3.7 *Stress-strain curve. (See Example 3.2.)*

EXAMPLE 3.3 Assume that we have a rectangular bar 1 in. × 0.5 in. × 10 in. (25.4 mm × 12.7 mm × 254 mm) long of the aluminum alloy in Example 3.2. The bar is hung at one end and a 35,000-lb (156,250 newtons) load is placed on the other end. The bar undergoes an increase in length. Determine the stress, the elastic and plastic components of strain, and the bar length when under load.

ANSWER

$$\text{Stress} = \frac{\text{force}}{\text{area}} = \frac{35{,}000 \text{ lb}}{1 \text{ in.} \times 0.5 \text{ in.}} = 70{,}000 \text{ psi (483 MPa)}$$

At this stress the total strain is 0.010 in./in. (Example 3.2). Therefore the strained bar length is

$$10.0 \text{ in.} + 10.0 \text{ in.} \times 0.010 \text{ in./in.} = 10.10 \text{ in. (256.5 mm)}$$

When the bar is unloaded the stress follows a straight line that depends on the modulus of elasticity E. Therefore we can calculate the elastic component of strain.

$$\epsilon_{el} = \frac{\text{stress}}{E} = \frac{70{,}000 \text{ psi}}{10 \times 10^6 \text{ psi}} = 0.007$$

Since

$$\epsilon_{total} = \epsilon_{el} + \epsilon_{pl}, \qquad \epsilon_{pl} = 0.010 - 0.007 = 0.003$$

(Note that the elastic component of strain is not fixed. It is dependent on the stress value even after plastic deformation begins. This is an important consideration in plastic deformation processes such as forming a fender, where we must know "spring back" if we are to control the final dimensions.)

3.7 True stress–true strain relations

True stress–true strain curves are used in development and research more than in routine testing because the data are more difficult to obtain and plot than engineering stress and strain. We shall discuss the basic features here and add further material in the problem section. True stress is simple to understand because it follows the real definition of stress,

$$\sigma_t = \frac{\text{load}}{\text{true area at the time}}$$

In calculating engineering stress we used the original area throughout. The area decreases only a small amount during testing in the elastic range but decreases significantly in the plastic range. To obtain true stress we must measure the diameter at several intervals during testing.

True strain ϵ_t is a slightly more difficult concept. Suppose we stretch the gage length from 2 to 3 in. Now let us further stretch the bar 0.1 in.

The added engineering strain would be $\Delta l/l_o$ or 0.1/2. However, if we consider strain as the change in length divided by the length *at the time*, the added strain would be 0.1/3.05. In this case 3.05 is the average length during the additional straining.

We obtain true strain, therefore, by summing a succession of Δl's divided by the length at the time the Δl is produced. The solution is found by calculus to be

$$\epsilon_t = \ln \frac{l}{l_o}$$

Due to volume conservation, this expression is equivalent to

$$\epsilon_t = \ln \frac{A_o}{A} = 2.3 \log_{10} \frac{A_o}{A}$$

Note that we use the symbols σ = engineering stress, σ_t = true stress, ϵ = engineering strain, ϵ_t = true strain. Figure 3.8 shows a typical curve.

The true stress–true strain data are most useful in helping us to

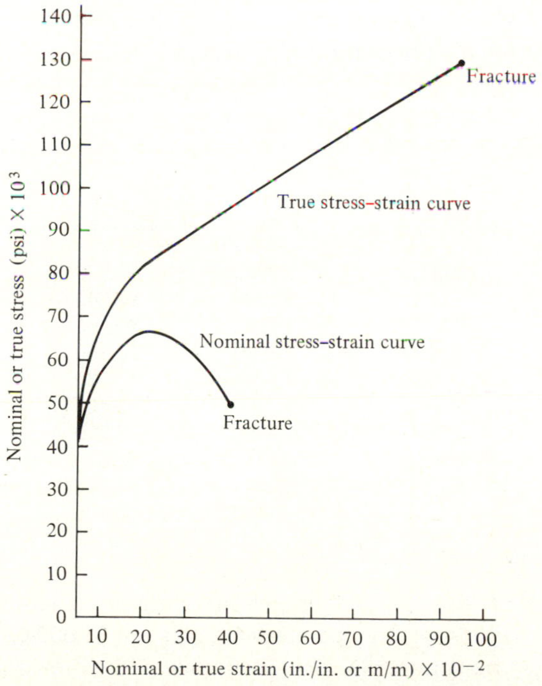

Fig. 3.8 *Comparison of true and nominal (engineering) stress–strain curves based on measurements of diameter. The material is a low carbon steel (0.20C).*

understand forming operations involving high plastic deformation. If we plot log stress vs. log strain, we obtain a straight line (Fig. 3.9). From the graph we can then read the amount of stress needed to produce a given plastic deformation.

As a practical illustration, the public often clamors for automobile bodies of stainless steel to prevent corrosion. The added cost is not merely a question of alloy expense. The stainless steel is more difficult to work (higher stress for a given strain), as shown by the true stress curve (Fig. 3.9), and causes greater die wear.

EXAMPLE 3.4 Using the data in Example 3.2, determine the true stress–true strain curve for the aluminum alloy.

ANSWER We can calculate the true areas from the conservation of volume relationship

$$l_oA_o = lA \quad \text{or} \quad A = \frac{2.000\ \text{in.} \times \pi/4(0.505\ \text{in.})^2}{l} = \frac{0.400\ \text{in.}^3}{l}$$

However, we must use the actual diameters at the tensile strength and at higher strains, since the material is necking and the gage length is well beyond this localized area. Therefore actual areas are necessary. We must also use the area relationship for the true strain calculation, i.e., $\epsilon_t = \ln l/l_o$ for strains up to necking (tensile strength) and $\epsilon_t = \ln A_o/A$ for higher strains or deformation. Figure 3.10 shows the required curve.

Load, lb	*Gage Length, in.*	*Area, in²*	*True Stress, psi*	*True Strain*
0	2.0000	0.2000	0	0
4,000	2.0041	0.1996	20,040	0.0020
8,000	2.0079	0.1992	40,160	0.0039
10,000	2.0103	0.1990	50,250	0.0051
12,000	2.0114	0.1989	60,330	0.0057
13,000	2.0142	0.1986	65,460	0.0071
14,000	2.0202	0.1980	70,710	0.0100
16,000	2.0503	0.1951	82,010	0.0248
16,000 (max)	2.0990	0.1847*	86,640*	0.0796*
15,600 (break)	2.1340	0.1720*	90,700*	0.1508*

*Values based on measured diameters

Note: If we assumed conservation of volume, the area at fracture would be $A = 0.400\ \text{in.}^3/2.1340 = 0.1874\ \text{in.}^2$ and the breaking stress = 15,600/0.1874 = 83,240 psi. This value would be impossible, as it would give a true breaking strength less than the true tensile strength.

3.8 Dislocations

There used to be an afternoon colloquium in Cambridge, Massachusetts, in which metallurgists and physicists of leading institutions would meet after tea to present new data. Often a budding metallurgist, beaming with success, would present data for a new steel with greatly improved tensile strength, such as 300,000 psi (2.07×10^3 MPa). Invariably, at the end of the presentation a pipe-puffing physicist would puncture the young metallurgist with a statement such as: "Interesting. But when are you going to reach at least the order of a million psi predicted by simple theoretical calculations of strength?"

Finally one day some scientists at the Bell Telephone Laboratories received samples of tiny tin crystals called "whiskers," which were shorting out capacitors. Out of curiosity they built a microtesting device and found that these whiskers did indeed have the theoretical strength level of over 1 million psi (6.9×10^3 MPa).

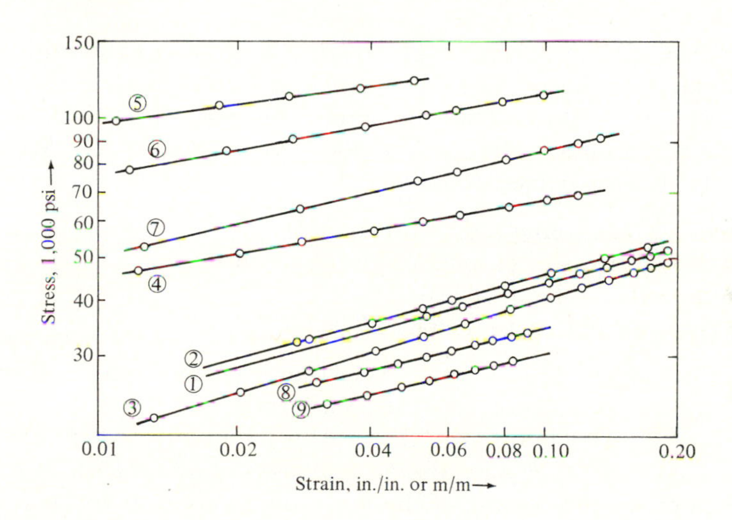

Material	Treatment
1. 0.05 percent carbon rimmed steel	Annealed
2. 0.05 percent carbon killed steel	Annealed and temper-rolled
3. Same as 2, completely decarburized	Annealed in wet hydrogen
4. 0.05 to 0.07 percent phosphorus low-carbon steel	Annealed
5. SAE 4130 steel	Annealed
6. SAE 4130 steel	Normalized and temper-rolled
7. Type 430 stainless steel (17 percent chromium)	Annealed
8. Alcoa 24-S aluminum	Annealed
9. Reynolds R-301 aluminum	Annealed

Fig. 3.9 *Logarithmic true-stress–true-strain tensile relations for various materials. Tests are by J.R. Low and F. Garafalo. (To obtain MPa, multiply psi by 6.9×10^{-3}. To obtain kg/mm^2, multiply by 7.03×10^{-4}.)*

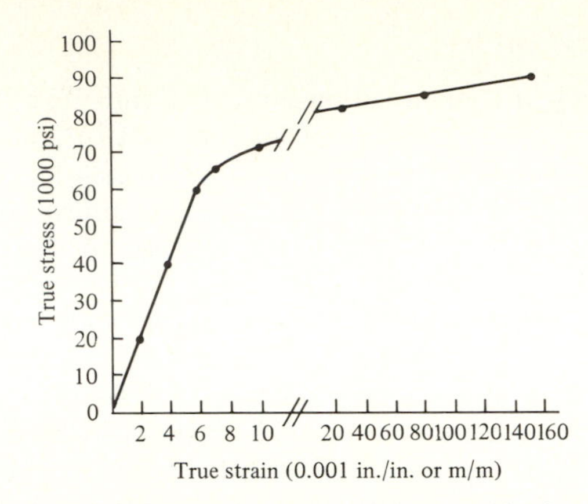

Fig. 3.10 *True-stress–true-strain curve for the high-strength aluminum alloy whose engineering or nominal stress-strain curve is shown in Fig. 3.7*

For a long time it was suspected that actual and theoretical values differed because of the presence of microdefects. Indeed Griffith had shown that by testing glass fibers of fine diameters (where the probability of a defect was low) he could obtain strengths of 500,000 psi (3.45×10^3 MPa). It was postulated that these defects could be:

1. Point defects or missing atoms (vacancies)
2. Line defects or rows of missing atoms (dislocations)
3. Area defects (grain boundaries)
4. Volume defects as in actual cavities

The first type is not of great importance where strength is concerned, but is significant in diffusion, as discussed later.

We are already familiar with the third type at the face of the grain boundary. The fourth type may also be dismissed at present because it results only from improper processing of the material, leading to voids, or from uneven diffusion conditions.

The second type, the line defect or dislocation, is of primary importance in understanding the reason for the gap between commercial and theoretical strengths. To illustrate this, suppose a block of metal has an extra plane of atoms extending halfway through it (Fig. 3.11).

This will result in a core of unstable material going back into the block, as shown by the five circled atoms in the face section. This core is a line defect called an *edge dislocation*. An important characteristic is that the atoms in the upper side are in compression, those in the lower side in tension. Clearly the bonding forces are not as strong as in a perfect lattice.

Now let us apply a shearing stress (Fig. 3.12). In the perfect lattice we need to attain a uniformly high stress level to displace the *A* atoms; this is the theoretical yield strength obtained in whiskers. However, where we have

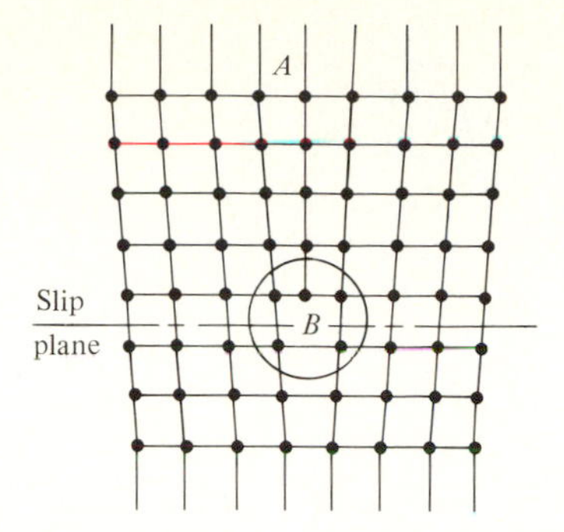

Fig. 3.11 *Atomic packing near edge dislocation. The dislocation B is circled. The atoms at A are in a normal configuration.*

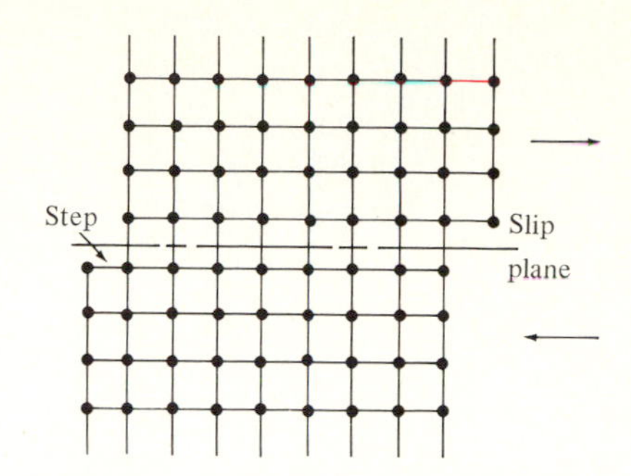

Fig. 3.12 *Correlation of slip with a line dislocation*

a dislocation, the atoms nearby are not so firmly held and the dislocation moves easily to the right. In other words, the yield strength value is lower.

In an actual case millions of dislocations result from casting, rolling, etc. Although they are on an atomic scale, evidence of their presence and movement can be seen with an electron microscope. There is a great amount of literature on the subject, but we shall use the concept to understand rather simple phenomena.

By now it should be clear that the high strength of the whiskers is due to the absence of dislocations in the direction in which they were tested. It is questionable whether material can be prepared that is completely free of dislocations. However, whiskers of many materials have been tested and found to be exceptionally strong. As a matter of fact, sapphire whiskers are now being used as strengtheners in composite materials.

We shall see that the concept of dislocation and dislocation movement is valuable as a "thinking tool" in explaining many phenomena.

3.9 Cold work or work hardening

We have already seen this phenomenon in the stress–strain curve. To illustrate, let us take a bar of metal and apply stress until we reach point *A* on the curve, far above the yield strength of 50,000 psi (345 MPa), but below the tensile strength of 100,000 psi (690 MPa) (Fig. 3.13*a*).

Now let us unload the sample, take it from the machine, and give it to another technician for testing. This technician will obtain a yield strength of over 80,000 psi (483 MPa) (point *B* in Fig. 3.13*b*) instead of 50,000 psi (345 MPa). To check this effect, recall that on unloading (Fig. 3.13*a*) the points would follow the line *AA* and on restressing would follow the same line.†

†There are small "aftereffects" of a lesser order of magnitude that we need not consider at this time.

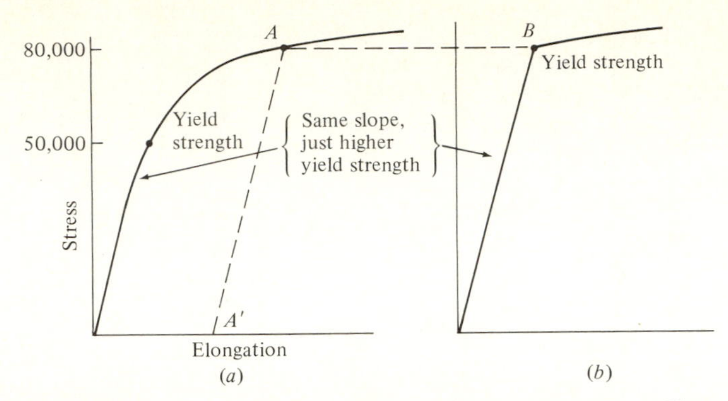

Fig. 3.13 *Correlation of work-hardening with the stress–strain curve. The units of stress are psi.*

Now visualize what is happening. As we stress into the region of plastic strain, slip takes place on the favorably oriented planes, producing dislocations and their movement. However, as more and more slip occurs, dislocations interact and pile up, and dislocation tangles form. This makes it more and more difficult for further slip to take place. The rising stress–strain curve shows this: To produce more strain, more stress is needed.

We come now to the key point of the argument. When we reached point A in Fig. 3.13*a*, all the planes and dislocation sites for easy slip had been used up. When the load was removed, this situation did not change. Therefore, when the load was reapplied, no plastic strain could take place until the stress level of point A was reached. Thus, we encountered only elastic strain to a much higher level than in Fig. 3.13*a* and the yield strength was correspondingly higher.

This phenomenon is called *work hardening, strain hardening,* or *cold work*. The term "cold" is relative. It means working at a temperature that does not alter the structural changes produced by the work. In other words, cold work causes atoms to move and dislocation tangles to form. As we shall see in Sec. 3.16, this effect can be removed by working at higher temperatures.

3.10 Methods of work hardening

Naturally if we wish to raise the yield strength by work hardening, we do not need to put the part in a tensile machine. We only need to produce slip. From our experiences with the copper bar, we know that slip is produced in the same way (by shearing) under compressive stress. So various methods are available, as shown in Fig. 3.14. Note also that at the same time we can be producing the final shape we desire.

Other processes include *spinning*, in which a rotating sheet is forced into a rotating backing form with a tool; *swaging*, in which small die-shaped

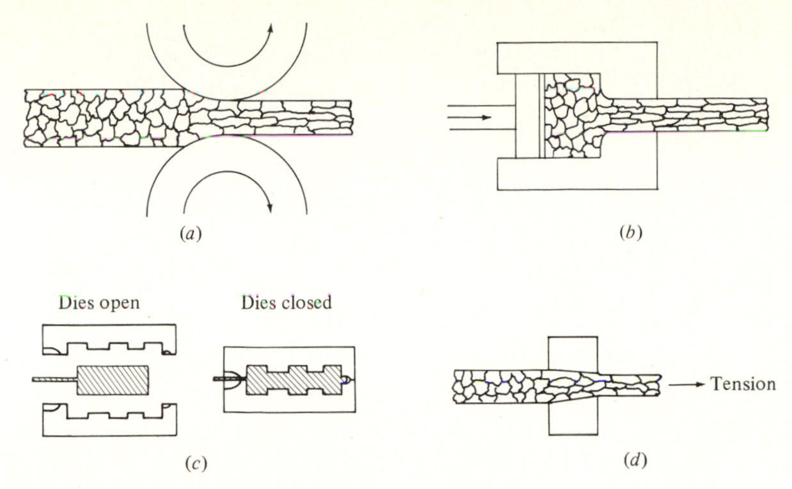

Fig. 3.14 *Methods for work-hardening during processing. (a) Cold-rolling bar or sheet. (b) Cold extrusion. (c) Cold forming, stamping, forging. (d) Cold drawing.*

hammers beat on a rod to reduce it in diameter; and *shot peening*, in which metal shot is thrown at the surface.

In all processes the percent cold work is defined as $(A_o - A_f)/A_o \times 100$, which is the same as percent reduction of area in the tensile test.

In all these cases the part is strengthened by the cold working. Another important effect is that the part is hardened, making it more resistant to wear and galling. Since actual hardness testing involves cold work, this is a good chance to analyze hardness measurements. Hardness tests are as important as tensile tests in many specifications.

3.11 Hardness testing

Hardness is usually defined as resistance to penetration. Let us review a few of the most common tests and see how closely they fit this definition.

Brinell Hardness Number (BHN) This is one of the oldest tests. It is still the most common standard (Fig. 3.15).

A specimen with a flat upper surface is placed on the anvil. A steel or tungsten carbide ball is pressed into the sample with a load of either 500 or 3,000 kg. The lighter load is used for the softer nonferrous metals such as copper and aluminum, and the heavier load is used for iron, steel, and hard alloys. The load is left in place for 30 sec and then removed. The diameter of the impression in millimeters is then read with a low-power microscope with a filar (measuring) eyepiece. Next the observer reads the Brinell hardness number (BHN) that corresponds to the impression diameter from a table of values for the load used. We will not analyze the method used to derive the

Test	Indenter	Shape of Indentation: Side View	Shape of Indentation: Top View	Load	Formula for Hardness Number
Brinell	10-mm sphere of steel or tungsten carbide	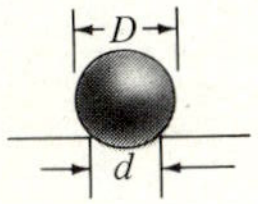 D, d	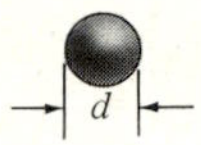 d	P	$BHN = \frac{2P}{\pi D(D - \sqrt{D^2 - d^2})}$
Vickers	Diamond pyramid	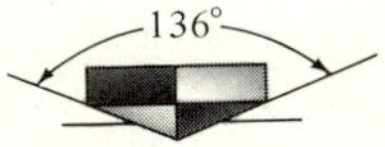 136°	 d_1, d_1	P	$VHN = 1.72\ P/d_1^2$
Knoop microhardness	Diamond pyramid	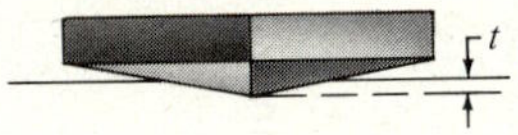 t; $l/b = 7.11$, $b/t = 4.00$	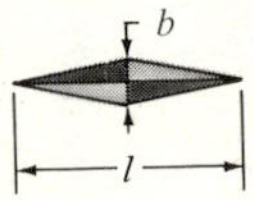 b, l	P	$KHN = 14.2\ P/l^2$
Rockwell A, C, D	Diamond cone	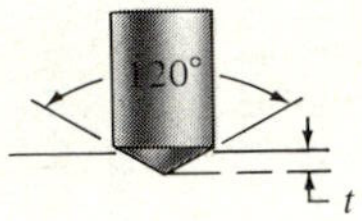 120°, t		60 kg, 150 kg, 100 kg	$R_A = R_C = R_D = 100 - 500t$
B, F, G	$\frac{1}{16}$-in.-diameter steel sphere	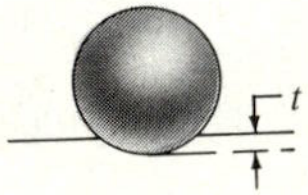 t		100 kg, 60 kg, 150 kg	$R_B = R_F = R_G = 130 - 500t$
E	$\frac{1}{8}$-in.-diameter steel sphere			100 kg	$R_E = 130 - 500t$

Fig. 3.15 *Hardness testing methods.* (H.W. Hayden, W.G. Moffatt, and John Wulff, *The Structure and Properties of Materials*, Vol. 3: Mechanical Behavior, John Wiley, New York, 1965. By permission of John Wiley & Sons, Inc.)

numbers. We merely note that the more difficult the penetration, the higher the BHN. The table is developed so that the BHN is about the same whether the 500- or 3,000-kg load is used, although obviously the impression diameter is different. The lighter load is used because in very soft materials the 3,000-kg load will continue to penetrate until the ball is deeply sunken.

Vickers Hardness Number (VHN) This is an improvement on the Brinell test. Here a diamond pyramid is pressed into the sample under loads much lighter than those used in the Brinell test. The diagonal of the square impression is read, and the Vickers hardness number (VHN) is read from a chart. As shown in Fig. 3.16, the VHN is close to the BHN from 250 to 600. The figure does not show that the VHN climbs steadily with strength at higher values, whereas the BHN is not used above 750. The advantages of the Vickers test are in obtaining hardness measurements at high levels and in measuring the hardness of a small region. On the other hand, the BHN gives a better averaging effect because of the larger impression.

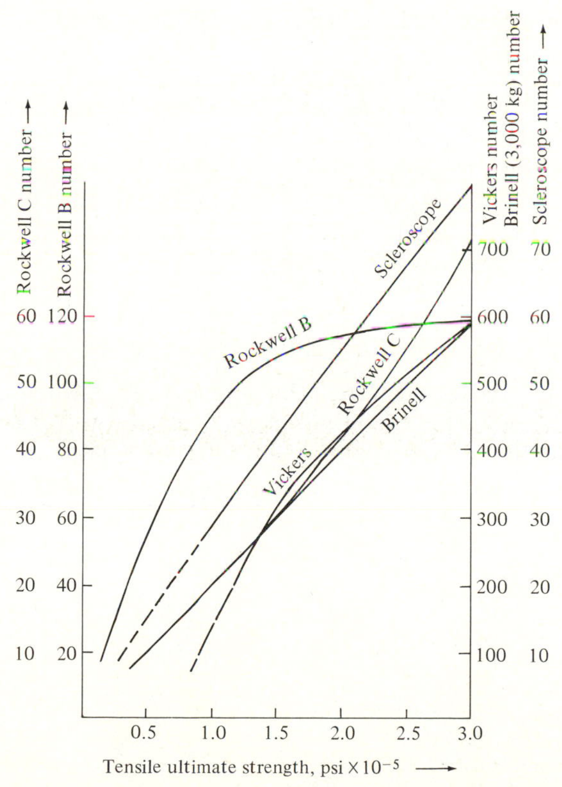

Fig. 3.16 *Conversion values for Brinell, Vickers, and Rockwell tests.* (From Joseph Marin, *Mechanical Behavior of Engineering Materials*, © 1962, Table 10.2, p. 450. Reprinted by permission of Prentice-Hall, Inc., Englewood Cliffs, N.J.)

Rockwell Hardness Testing (R_A, R_B, etc.) The chief advantage of the Rockwell test is that the hardness is read directly from a dial. The indenter for the R_C test is a suitably supported diamond cone or "brale." The observer first turns a handle which presses the diamond cone a slight standard amount into the sample. This is called the "preload." Next the standard R_C load of 150 kg (Fig. 3.15) is released. This forces the diamond farther into the sample. The same lever is used to remove the load. At this point the observer reads the R_C hardness from the dial and then unloads the specimen. The principle of this test is that the dial, through a lever system, records the depth of penetration between the preload and the 150-kg load and reads directly in R_C. The R_C is approximately $\frac{1}{10}$ BHN (Fig. 3.16). The R_B scale is used for softer materials. It employs a $\frac{1}{16}$-in.-diameter ball and a 100-kg load. It is also direct reading. Some Brinell test machines are also direct reading.

The scleroscope hardness test is used chiefly for checking large rolls on which it is difficult to use the other tests. The value is obtained by measuring the height of rebound of a small weight under standard conditions.

3.12 Correlation of hardness, tensile strength, and cold work

All the hardness tests depend on resistance to plastic deformation. It is no surprise, therefore, to find that the hardness of a bar is higher after cold working, because all the sites for easy slip have been used up. Thus there is good correlation between hardness and strength. For steel a simple relation to remember is that the tensile strength in pounds per square inch is 500 times the BHN (derived from Fig. 3.16).

3.13 Solid-solution strengthening

To this point we have emphasized the effect of cold working in increasing strength and hardness. However, the strength may be further improved by alloying the material with one or more additional elements to provide solid-solution strengthening. The combined effects of solid-solution strengthening and cold working can be seen in two common aluminum alloys.

	Yield Strength, psi	*Percent Elongation*	*Brinell Hardness*
Commercially pure aluminum, annealed	4,000	43	19
Cold-worked aluminum, 50% reduction	18,000	6	35
Aluminum with 1.2% Mn in solid solution, annealed	6,000	30	28
Same alloy cold-worked, 50% reduction	27,000	4	55

These data show that adding manganese in solid solution raised the yield strength in the annealed condition from 4000 to 6000 psi. When cold working was added to the solid-solution effect, a yield strength of 27,000 psi was attained, compared with 18,000 psi in the absence of manganese. We shall encounter many single-phase alloys of this type in which solid-solution strengthening and cold working are combined to attain high strength.

3.14 Fatigue

A survey of the broken parts in any automobile scrap yard reveals that the majority failed at stresses below the yield strength. This is due not to imperfections in the material but to the phenomenon called *fatigue*. If a bar of steel is loaded a number of times to a stress, 80% of the yield strength, for example, it will ultimately fail if it is stressed through enough cycles. Furthermore, even though the steel would show 30% elongation in a normal tensile test, no elongation is evident in the appearance of the fatigue fracture. A typical crankshaft failure is shown in Fig. 3.17.

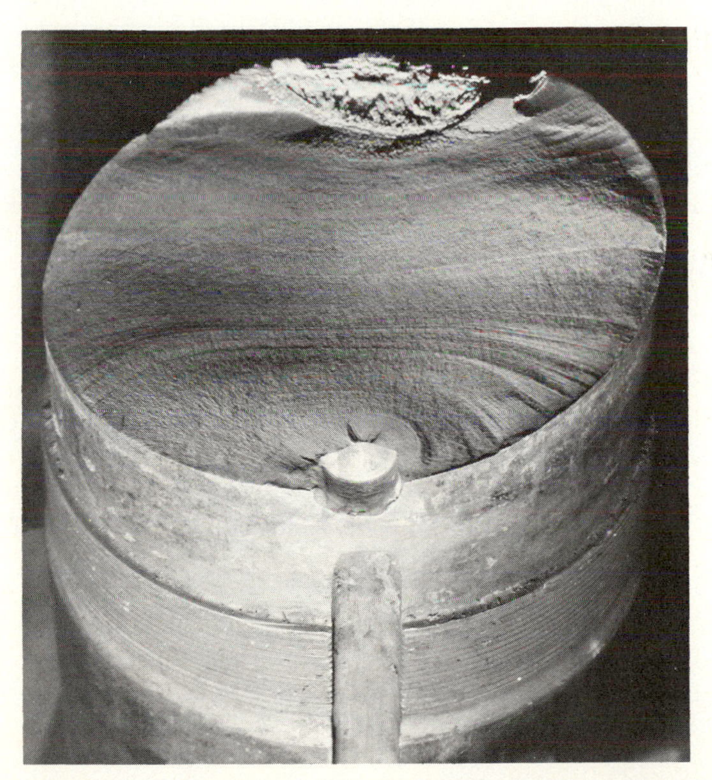

Fig. 3.17 *Typical fatigue failure of a crankshaft. The crack began at the stress concentration at the hole (front of picture). Steps in the fracture caused by successive cycles of loading led to an oyster-shell effect. The final failure took place suddenly, as shown by the small fracture surface at the rear.* (Courtesy of H. Mindlin, Battelle)

Fatigue testing in its simplest form involves preparing test specimens with carefully polished surfaces and testing them at different stresses to obtain an *S-N* curve relating *S* (stress required for failure) to *N* (number of cycles) (Fig. 3.18). As we would expect, the lower the stress, the greater the number of cycles to failure.

The curve for ferrous materials exhibits what is called the *fatigue strength* or *endurance limit* (Fig. 3.19). In other words, beyond 10^6 cycles, further cycling does not cause failure. On the other hand, for nonferrous materials such as aluminum, the curve continues to decline. The material must be tested for the number of cycles it will encounter in service.

Since fatigue tests are time-consuming and the fatigue data require statistical treatment because of reproducibility difficulties, attempts have been made to relate the tensile test to fatigue data. The *fatigue ratio* or *endurance ratio* is defined as the ratio of fatigue strength to tensile strength. It has values of 0.45 to 0.25, depending on the material.

The fatigue strength is also greatly affected by the following variables:

1. Stress concentrations, such as radii at fillets and possible notches
2. Surface roughness, which indicates that results depend on the type of machining used
3. Surface residual stress
4. Environment, such as corrosion

Ceramics, particularly certain glasses and oxides, exhibit a phenome-

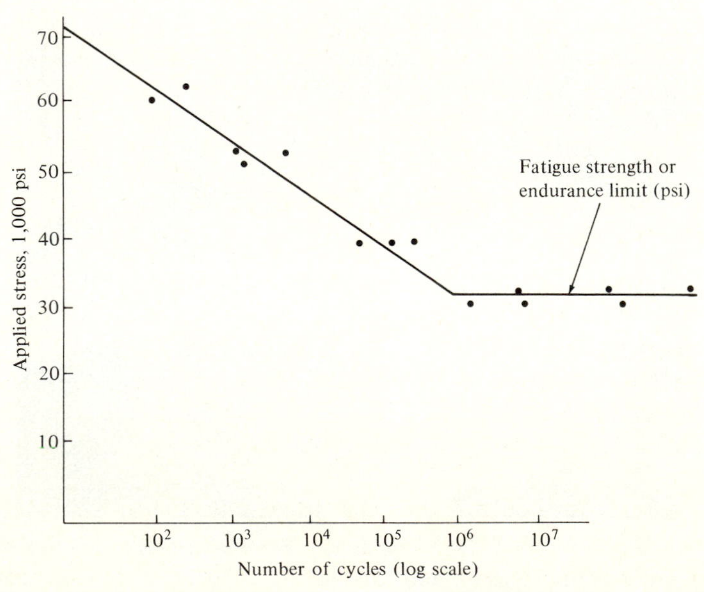

Fig. 3.18 *Typical S-N curve (fatigue) for a ferrous material, showing an endurance limit (schematic)*

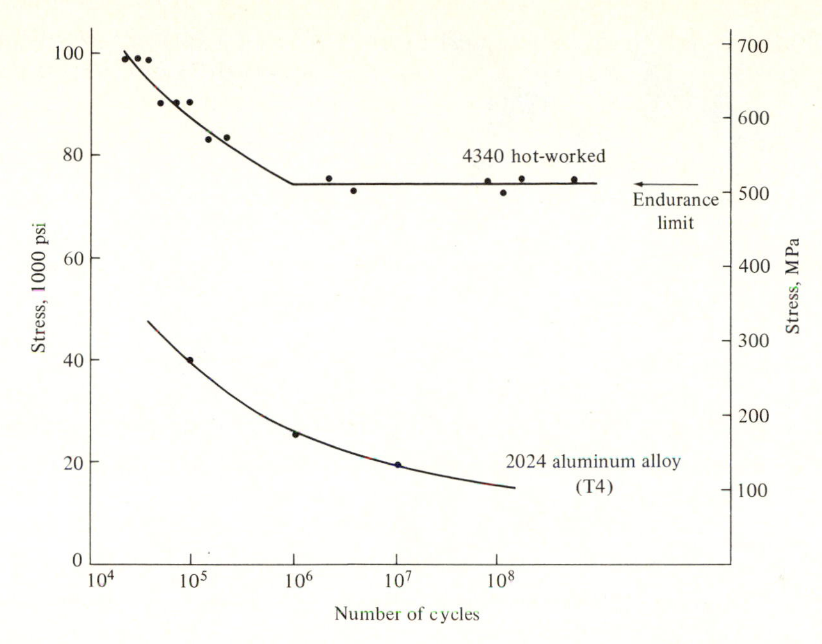

Fig. 3.19 *Typical S-N curves for ferrous and nonferrous alloys (after Garwood and Alcoa). 2024 aluminum: 4*Cu, *0.6*Mn, *1.5*Mg; *4340 steel: 0.4*C, *0.7*Mn, *0.25*Si, *1.85*Ni, *0.3*Cr, *0.25*Mo.

non known as *static fatigue*. These materials and even some ultra-high-strength alloys will withstand a high static load for a period of time, then fail suddenly. This type of failure does not take place in dry air or *in vacuo*, so it is related to a chemical reaction between the water in the atmosphere and the highly stressed surface. (See Chapter 13)

3.15 Impact testing and effects of low temperatures

Let us take up now the effects of the temperature of testing on the fracture strength. One of the most common tests is the so-called "Charpy impact test," Fig. 3.20*a*. A square bar with a V notch is struck by a calibrated swinging arm, and the energy absorbed is measured. It is relatively simple to examine the effects of temperature by immersing several specimens in advance in liquids at different temperatures and then transferring them quickly to the test fixture.

The types of data obtained are shown schematically in Fig. 3.20*b*. FCC metals show high impact values and no important change with temperature. However, BCC metals, polymers, and ceramics show a transition temperature below which brittle behavior is found. It should be emphasized that the actual transition temperatures for different materials varies greatly. For metals and

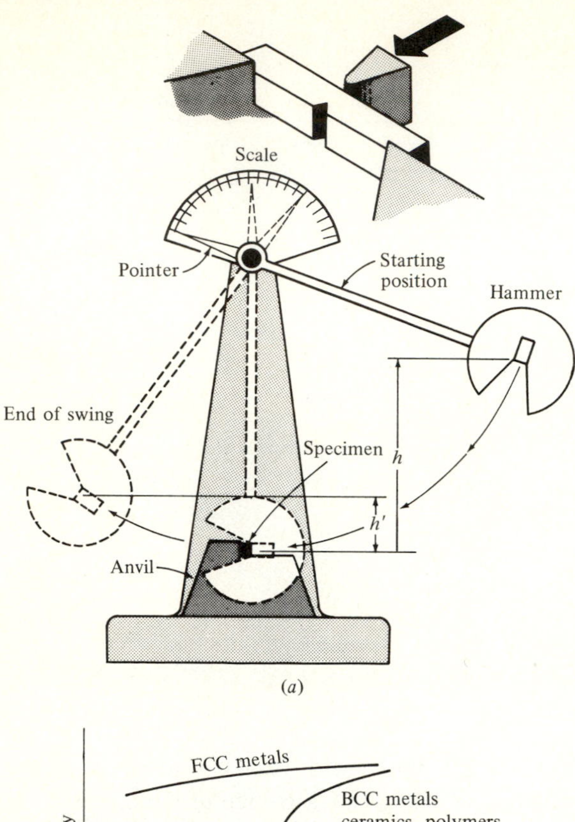

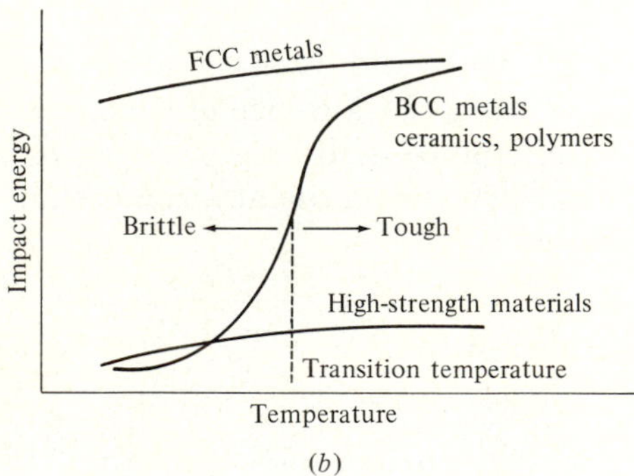

Fig. 3.20 *(a) Operation of a Charpy impact test. (b) Effect of temperature on the impact strength of various materials (schematic).* [Part (a) from H.W. Hayden, W.G. Moffatt, and John Wulff, *The Structure and Properties of Materials*, Vol. 3: Mechanical Behavior, John Wiley & Sons, Inc. New York, 1965]

Fig. 3.21 *(a) Charpy impact vs. temperature for SAE 1020 steel, hot-rolled. (b) Charpy impact fractures after testing at (right to left) −196, 0, 25, 50, 93°C. (c), (d), and (e) are scanning electron micrographs of Charpy specimens, showing shear failure above transition temperature, cleavage below transition temperature, and mixed shear and cleavage at intermediate temperature; 1020 steel, hot-rolled, 1500×. (c) Tested at 93°C, ductile (shear) fracture, 170 ft-lb. (d) Tested at 25°C, mixed fracture, 40 ft-lb. (e) Tested at −196°C, cleavage fracture, <1 ft-lb.*

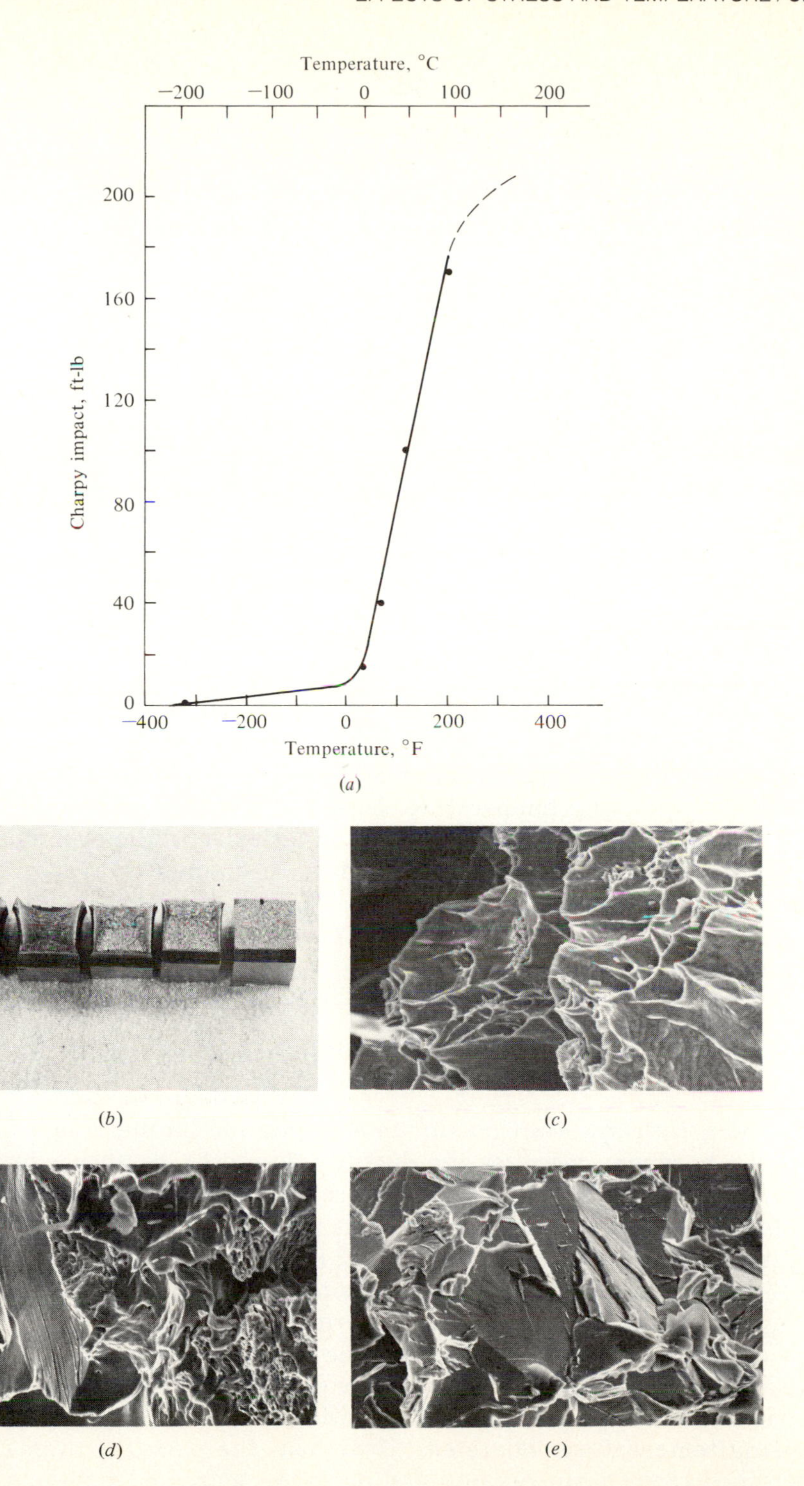
Temperature, °C
−200
−100
0
100
200
Charpy impact, ft-lb
200
160
120
80
40
0
−400
−200
0
200
400
Temperature, °F
(a)
(b)
(c)
(d)
(e)

polymers it is between −200 and 200°F (−129 and 93°C), while for ceramics it is above 1000°F (538°C).

There is a distinct difference in the appearance of the fractures of low-carbon steels, depending on whether the specimen was tested and broken below or above the transition temperature. As indicated in Fig. 3.21*b*, the appearance of Charpy V notch specimens varies from ductile to brittle as the specimen temperature is reduced from 200 to −321°F (93 to −196°C). Careful observation shows that a shear type fracture, as shown by the presence of a shear lip, is characteristic of the specimens tested at higher temperatures, while shear is absent in the specimens tested at the lowest temperatures. Photomicrographs taken with the scanning electron microscope provide additional evidence (Fig. 3.21*c*, *d*, and *e*). The specimen with highest impact strength shows a dimpled surface with myriads of cuplike projections of deformed metal. In contrast, the brittle sample shows mainly cleavage fractures, almost like split platelets of mica. The specimen at 77°F (25°C) shows a mixture of both types of fracture, as expected.

One of the most striking examples of the importance of transition temperature is the failure of the Liberty ships produced during World War II. These ships were made of low-carbon steel which showed good ductility in the usual tensile test. However, 25% of them developed severe cracks, often while anchored in harbor, and many broke in two. The failures were analyzed as the result of constraint (stress concentration) caused by square hatches in the deck, coupled with the use of steel with a transition temperature (Charpy) near the operating temperature. However, the Charpy test does not give a wholly satisfactory answer, as shown by the recent failures of oil tankers, and the Naval Research Laboratory has developed a new test called the *dynamic tear test.* In this test notched samples of varying sections can be tested with a drop weight. The transition temperatures this test indicates are more reliable indicators of steel quality.

3.16 Effects of elevated temperature on work-hardened structures

We do not always want maximum strength and hardness in a part, because as the hardness increases the ductility decreases, as shown by the percent elongation (Fig. 3.22). Suppose we are fabricating a simple metal cup 4 in. (10.16 cm) high. When we try to press the cup to this depth, the sheet cracks at the corners when formed only to a 2-in. (5.08-cm) depth. We have "used up" the plastic elongation. We encounter similar situations in the other cold-forming operations. What can we do to permit deeper drawing?

Basically we wish to restore the original structure by eliminating the extensive slip and dislocation tangles. Remember that atoms are not rigidly fixed but can diffuse from their positions. Diffusion increases rapidly with rising temperature. Therefore, if we heat the part, the atoms in severely strained regions can regroup and move to unstrained positions. This is ad-

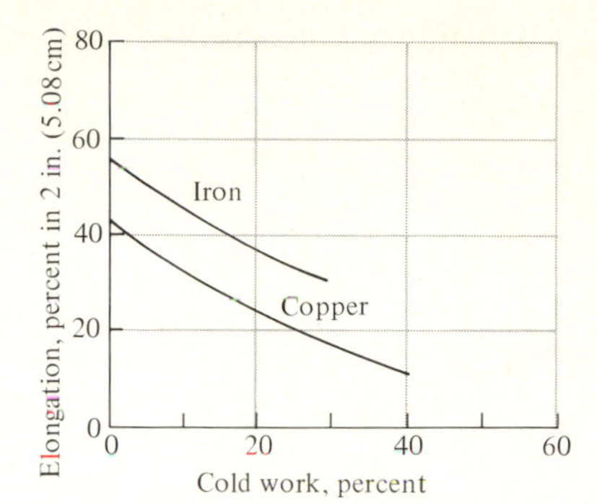

Fig. 3.22 *Correlation of cold work and elongation in tensile testing.* (L.H. Van Vlack, *Elements of Materials Science*, 2d ed., Addison-Wesley Publishing Company, Reading, Mass., 1964, Fig. 6.24, p. 150)

justment to strain on a microscopic scale. The partly formed shape does not change dimensions.

Therefore, we take the cold-worked part and heat-treat it in a furnace in the process called *annealing*. The metal softens as a function of heating temperature (Fig. 3.23) and time at temperature. We can see that profound changes take place in the microstructure (Fig. 3.24). At elevated temperatures new small grains of equal dimensions in all directions (*equiaxed* grains) grow inside the old distorted grains and at the old grain boundaries. At higher temperatures the grain size is larger. Hardness and strength decrease with grain size, but elongation increases. This can be explained from a microscopic view. Slip on a given plane stops when we reach a grain boundary, an area of high dislocation density and entanglements. Thus the more grain boundaries, the greater the limitation to slip and the higher the strength. Ductility (elongation) is lower because slip is limited by the larger grain-boundary area.

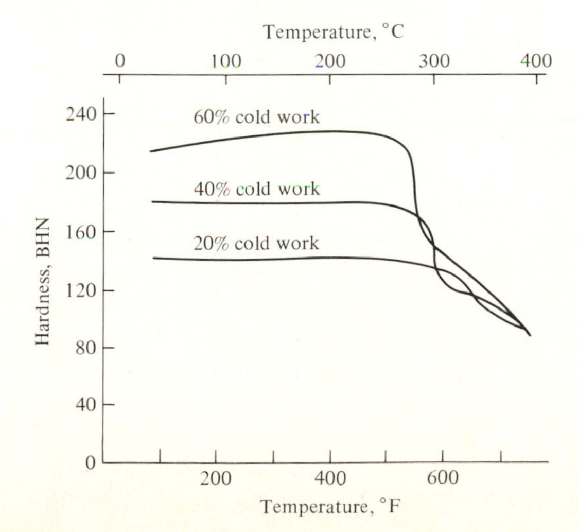

Fig. 3.23 *Effect of heating on hardness of cold-worked 65% Cu, 35% Zn brass, 1 hr.* (L.H. Van Vlack, *Elements of Materials Science*, 2d ed., Addison-Wesley Publishing Company, Reading, Mass., 1964, Fig. 6.28, p. 153)

3.17 Recovery, recrystallization, grain growth

It is useful to divide the effects of temperature on cold-worked material into three regions, in order of increasing temperature (Fig. 3.25, page 92).

Recovery This is the temperature range just below recrystallization. With the electron microscope we can see that stresses are relieved in the most severely slipped regions. Dislocations move to lower-energy positions, giving rise to subgrain boundaries in the old grains. This process is called *polygonization*. Hardness and strength do not change greatly, but corrosion resistance is improved.

Recrystallization In this temperature range the formation of new stress-free and equiaxed crystals leads to lower strength and higher ductility. There is an interesting relation between the amount of previous cold work and the grain size of the recrystallized material. With less cold work there are fewer nuclei for the new grains, and the resulting grain size is larger.

Grain Growth As the temperature is raised further, the grains continue to grow. This is because the large grains have less surface area per unit of volume. There are fewer atoms leaving a unit grain boundary area than with a small grain, and the larger grains grow at the expense of the smaller.

It should be added that all three effects are time dependent as well (see Prob. 3.19). These effects are summarized in Fig. 3.26 on page 92.

3.18 Selection of annealing temperature

The temperature required for recrystallization varies with the metal. It is approximately one-third to one-half the melting temperature of a pure metal, expressed on an absolute scale (Fig. 3.27). Therefore, steel is annealed at red heat (1600 to 1800°F) (870 to 980°C), while aluminum is still liquid at this temperature. On the other hand, lead can hardly be cold-worked at room temperature because the recrystallization temperature is so low. Furthermore, the recrystallization temperature is really a range, not a sharp point. Also, the greater the amount of cold work, the lower the recrystallization temperature because of the stored energy. The sudden drop in hardness for the 65% Cu, 35% Zn brass in Fig. 3.23 shows the order of magnitude of the drop in recrystallization temperature to be expected with increased cold work.

We may now summarize the variables that influence the recrystallization temperature. In effect it is not a fixed temperature. Rather, it depends on the alloy, percent cold work, original grain size, and time held at temperature. The recrystallization temperature increases with increasing alloy. However, it decreases for increased cold work, finer grain size, and longer holding times.

The cold work and grain-size effects can be explained by the realization

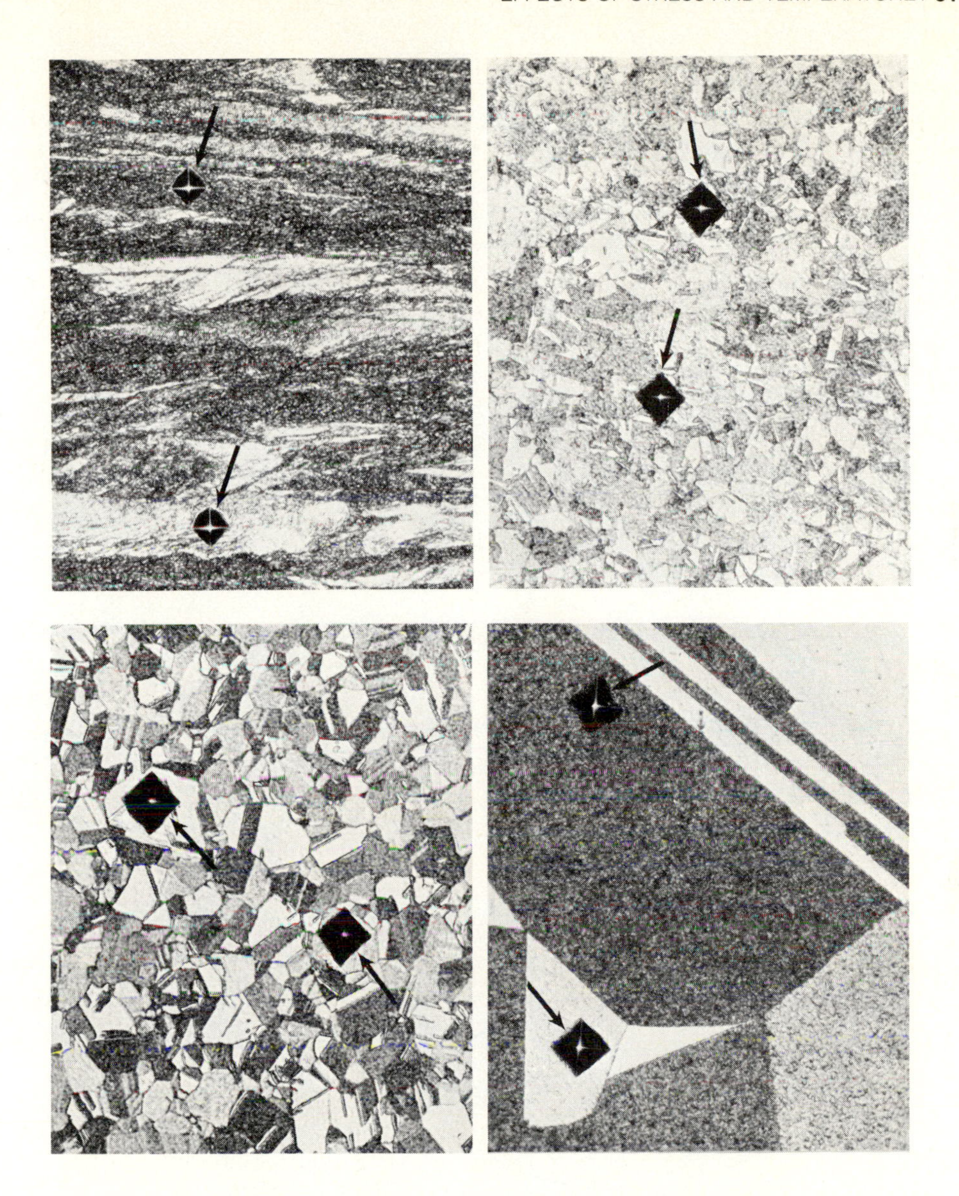

Fig. 3.24 *Upper left: cold-worked strip of single-phase 30% Zn brass, 50% reduction, microhardness 140 to 170 VHN, 25-g load. Upper right: partly recrystallized after 1 hr at 600°F (316°C), microhardness (arrows) 93 to 109 VHN. Lower left: recrystallized at 880°F (471°C) for 1 hr, small grains, microhardness 90 to 100 VHN. Lower right: recrystallized at 1400°F (760°C) for 1 hr, large grains, microhardness 65 to 87 VHN. Magnification 500× in all cases, peroxide etch.*

	Copper, OFHC			*70% Cu, 30% Zn Brass**				
	Prior Cold Work			*Prior Cold Work*		*Tensile Strength, psi × 10^3*	*Percent Elongation in 2 in.*	*Grain Size, mm*
	30%	*50%*	*80%*	*50% F.G.*	*50% C.G.*			
Initial	$86R_H$	$91R_H$	$95R_H$	$99R_X$	$97R_X$	80	8	
30 min								
150°C	85	90	94	101	98	81	8	
200°C	80	88	93	102	100	82	8	
250°C	74	75	65	103	101	82	8	
300°C	61	54	42	82	98	76	12	
350°C	46	40	34	66	80	60	28	0.02
450°C	24	22	27	50	58	46	51	0.03
600°C	15	17	22	38	34	44	66	0.06
750°C				20	14	42	70	0.12
Final grain size	0.15	0.12	0.10	0.08	0.12			

*F.G. = originally fine-grained; C.G. = originally coarse-grained; R_H = Rockwell scale, $\frac{1}{8}$-in. ball, 60-kg load; R_X = $\frac{1}{16}$-in. ball, 75-kg load.

Fig. 3.25 *Recovery, recrystallization, and grain growth as produced on heating cold-worked material.* (After Brick and Phillips)

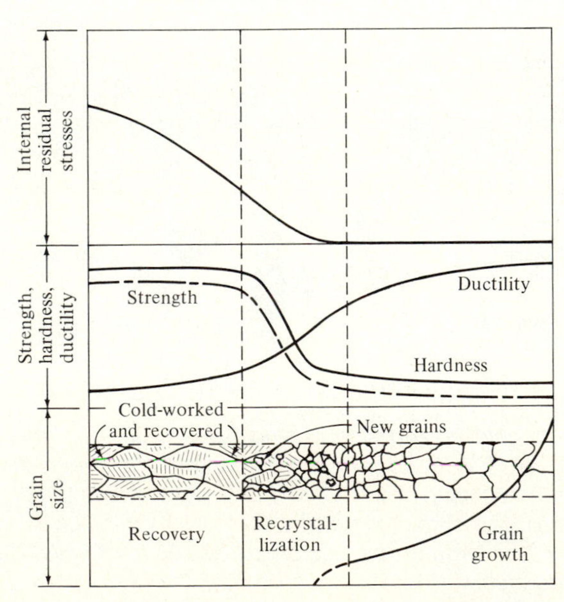

Fig. 3.26 *Summary of annealing effects, time and temperature dependent (schematic representation).* (After Sachs)

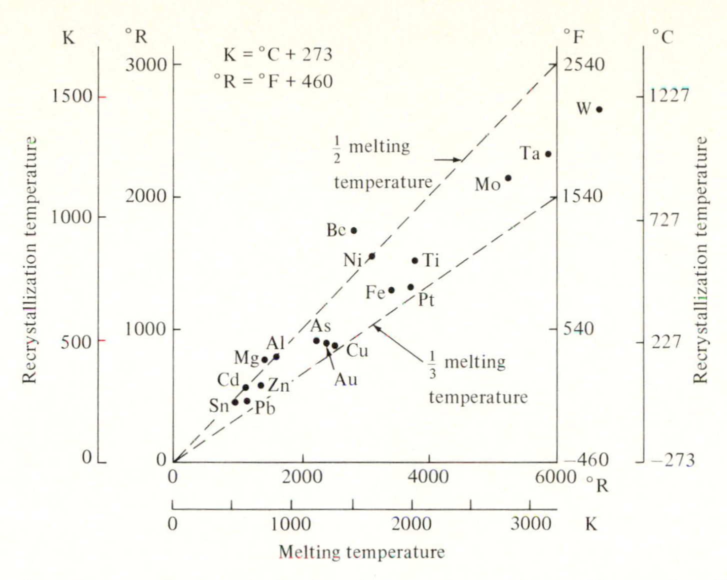

Fig. 3.27 *Correlation of recrystallization temperature with melting point. See text for reasons why the recrystallization temperature is not fixed.* (L.H. Van Vlack, *Elements of Materials Science*, 2d ed., Addison-Wesley Publishing Company, Reading, Mass., 1964, Fig. 6.29, p. 154)

that the new grains can be more readily nucleated at sites of high dislocation density (grain boundaries and slip planes). The alloy and time effects depend on diffusion, to be discussed in Chapter 4. However, it is reasonable to expect that atom motion is required in recrystallization and that the presence of foreign atoms (alloying elements) makes atom motion more difficult. Similarly longer holding times give atoms more time to move, and so recrystallization can occur at a lower temperature. Therefore, it is common to use 1 hr at temperature to determine the recrystallization temperature.

3.19 Effect of grain size on properties

The preceding discussion showed that we can control the grain size of the final part by the combination of cold work and annealing. With large amounts of cold work and rapid annealing the grain size will be small, because many nuclei for growth of new grains are present. Furthermore, grain boundaries are an impediment to slip. We would expect, then, that a fine-grained material would have higher strength. This is given in the Petch relation: $\sigma = \sigma_0 + kd^{-1/2}$, where σ is the yield strength, d is the grain size, and σ_0 and k are constants for a particular material. An added advantage to fine-grained material is that the plastic flow under stress is more even and the surface of stamped or formed parts is smoother than with coarse-grained material. Figure 3.28 gives an example of the effect of grain size on various alloys.

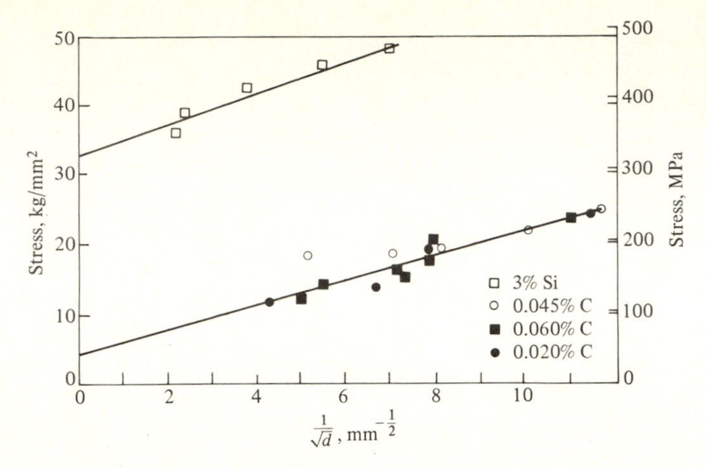

Fig. 3.28 *Relation between yield stress and grain diameter of iron alloys.* (W.J.M. Tegart, *Elements of Mechanical Metallurgy*, Macmillan Publishing Co., New York, 1966, Fig. 6.14, p. 175.)

3.20 Engineering application of cold work and annealing

The cold work–annealing sequence may be repeated as desired. We can start with a simple sheet of metal and roll or form it extensively. Each time the material becomes difficult to work or in danger of cracking, we can anneal it. Furthermore, we can ensure the right combination of strength and hardness on the one hand and ductility on the other by the final sequence of operations. There are two possibilities:

1. Finish the part with higher hardness and lower ductility than desired and then anneal it to obtain the desired combination.
2. Use the proper amount of cold work for the final operation to reach the required hardness level.

3.21 Hot working; hot rolling, forging, extrusion

At this time we may well ask: Instead of cold working plus annealing, what would happen if we worked the part *at* the annealing temperature? Would not recrystallization take place, so that we could continue to work the part without interruption? The answer is yes, and therefore we can conduct practically all the operation of working at the annealing temperature or above. Steel, for example, is passed through a set of rolls at high temperature. In some cases the heat produced by the mechanical work actually increases the temperature! The disadvantages of hot working are oxidation of the metal surface and shorter die life caused by heating from the part. Cold working can produce a better surface because the part can be annealed in a furnace with a protective

atmosphere after working. Also, if the part is finished by hot working, the strength and hardness will be lower.

In many cases a combination of hot and cold working is used. A large shape such as an ingot is cast, hot-rolled to bar stock, and then cleaned. The surface is then preserved by cold rolling and annealing in protective atmospheres.

EXAMPLE 3.5 In the manufacture of practically all stamped and cold-formed parts it is necessary to balance the properties desired in the final part with the economic and engineering problems of forming the part. Let us recall the following aspects of the problem.

Effect of:	*Strength*	*Hardness*	*Percent Elongation*	*Percent Reduction of Area*
Increased work before testing	Increased	Increased	Decreased	Decreased
Increased annealing temperature	Decreased	Decreased	Increased	Increased

Therefore, for maximum workability we want a well-annealed part, but for highest strength and hardness we want maximum cold work. Suppose we have some annealed 70% Cu, 30% Zn brass bar stock (cylindrical rod) of 0.35-in. (0.889-cm) diameter.

PROBLEM Produce bar stock of 0.21-in. (0.533-cm) diameter with a tensile strength of over 60,000 psi (414 MPa) and an elongation of over 20%.

ANSWER One way to solve the problem is first to determine what the final step in cold working should be. From Fig. 3.29, to obtain 60,000 psi (414 MPa) tensile strength we need more than 15% cold work. Also, to obtain the required elongation we should cold-work less than 23%. Let us choose to cold-work 20% as the final working step. This will meet the tensile strength and elongation specifications. Then

$$\text{Percent cold work} = \frac{\Delta A}{A} = \frac{\frac{1}{4}\pi d^2 - \frac{1}{4}\pi(0.21)^2}{\frac{1}{4}\pi d^2} \times 100 = 20$$

where d is the diameter at which we wish to start the final deformation (starting, of course, with an annealed structure). Solving, we have

$$d = 0.235 \text{ in. } (0.596 \text{ cm})$$

Therefore, we obtain 20% cold work going from a 0.235- to 0.21-in. (0.596- to 0.533-cm) diameter.

The 20% cold work required to meet the final specification is not to be confused with the 20% elongation specification. The cold work is a forming operation, $\Delta A/A$, whereas the elongation specification is derived from a tensile test of a specimen that may have been previously cold worked. It is defined as $\Delta l/l$.

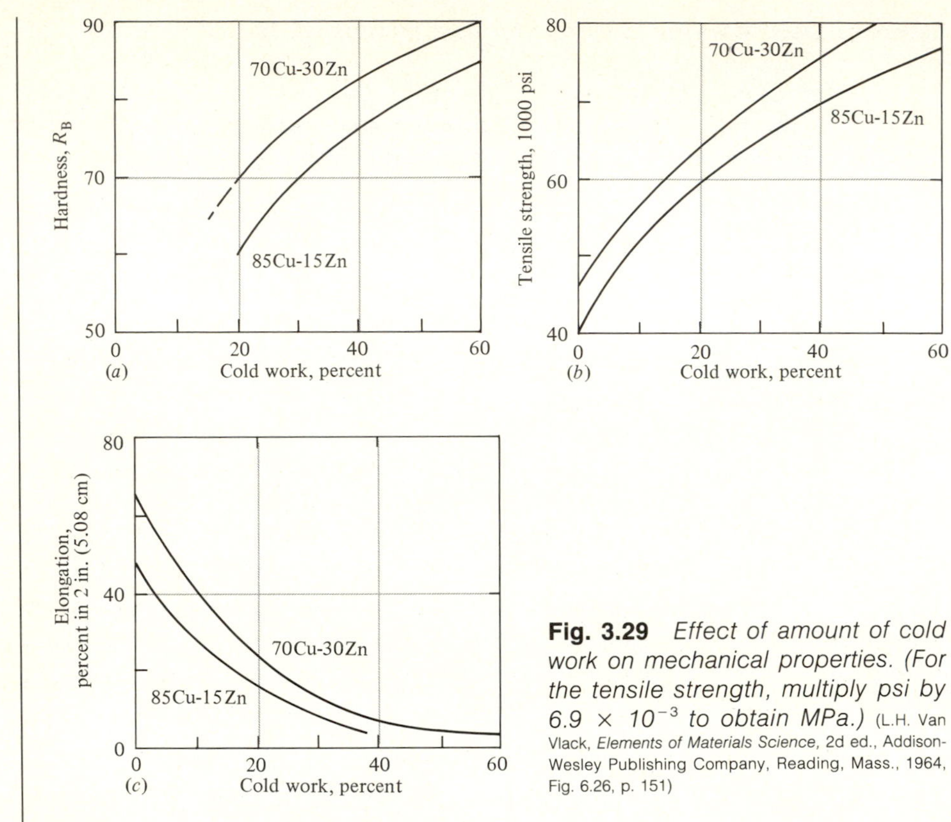

Fig. 3.29 *Effect of amount of cold work on mechanical properties. (For the tensile strength, multiply psi by 6.9 × 10^{-3} to obtain MPa.)* (L.H. Van Vlack, *Elements of Materials Science*, 2d ed., Addison-Wesley Publishing Company, Reading, Mass., 1964, Fig. 6.26, p. 151)

Now we have to reduce the diameter from 0.35 to 0.235 in. (0.889 to 0.596 cm). The cold work required in this step would be

$$\frac{\frac{1}{4}\pi(0.35)^2 - \frac{1}{4}\pi(0.235)^2}{\frac{1}{4}\pi(0.35)^2} \times 100 = 54.9\%$$

Since we received the material in the annealed state, we could cold-work from 0.35- to 0.235-in. (0.889- to 0.596-cm) diameter, then anneal, and finally cold-work the required 20% to achieve our tensile strength and elongation specifications, along with a final diameter of 0.21 in. (0.533 cm). Had the original cold-work step been greater than 60%, we would run the risk of fracture during cold work. We would have to use several cold-work and annealing steps. Remember, the effects of cold work are cumulative, but they can be removed by annealing.

An alternative would be to hot-work (at or above the recrystallization temperature) to a 0.235-in. (0.596-cm) diameter. After hot-working, the material would have a recrystallized structure that could then be cold-worked. The most economical solution would depend on the surface quality desired and the equipment available.

3.22 Properties in single-phase alloys

Let us review the methods of controlling properties in single-phase alloys. (In Chapter 4 we shall discuss methods of controlling properties when more than one phase is present.)

Alloying Figure 3.30 shows the effect of two substitutional solid-solution elements (Zn, Ni) on the mechanical properties of copper. In general the strength and hardness increase, whereas the ductility may increase or decrease depending on the substitutional element.

Cold Work As shown in Fig. 3.29, cold work also increases the tensile strength and hardness while significantly decreasing the ductility.

Annealing Annealing removes the influence of cold work, leading to a gain in ductility but a loss in strength and hardness.

As discussed in Sec. 3.13, a combination of alloying and cold working may produce the most desirable combination of mechanical properties.

EXAMPLE 3.6 We require a bar of copper-base alloy with a minimum tensile strength of 35,000 psi (241.5 MPa) and a minimum elongation of 30%. We have in stock an annealed 85% Cu, 15% Ni alloy, an annealed 85% Cu, 15% Zn alloy, and a 20% cold-worked 85% Cu, 15% Zn alloy. Which alloy would you select?

ANSWER Using Figs. 3.29 and 3.30, we find the following properties.

Alloy	*Tensile Strength, psi*	*Percent Elongation*
Annealed 85% Cu, 15% Ni	42,000 (290 MPa)	41
Annealed 85% Cu, 15% Zn	41,000 (283 MPa)	46
20% cold-worked 85% Cu, 15% Zn	59,000 (407 MPa)	15

The cold-worked alloy does not meet the minimum elongation specification of 30%. Therefore either annealed alloy would be a proper selection. Since nickel is much more expensive than zinc, the annealed 85% Cu, 15% Zn alloy would have the lowest initial cost. But if salt-water corrosion were involved, as in desalinization equipment, the copper-nickel alloy would give longer servce and lowest cost per hour of service.

3.23 Creep and stress relief

To conclude this chapter we should consider the effect of temperature on stress-strain relations. These data are important in designing parts for operation at higher temperatures and in the operation called *stress relief*.

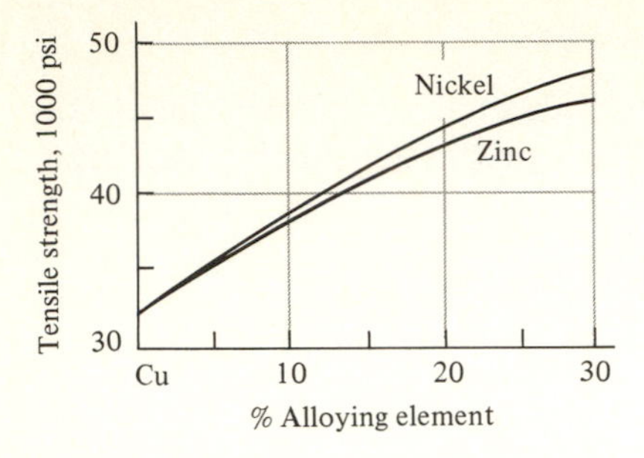

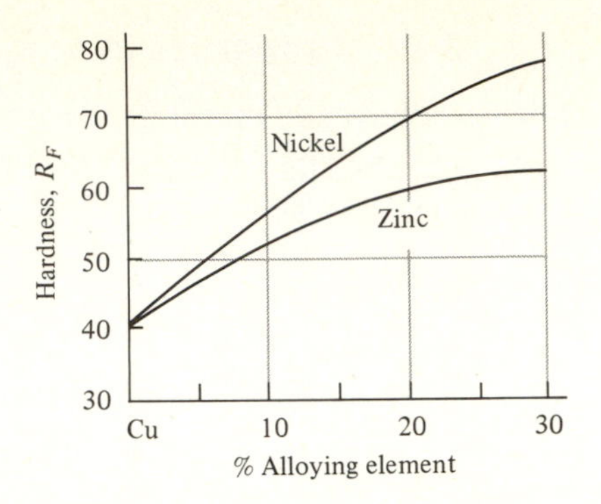

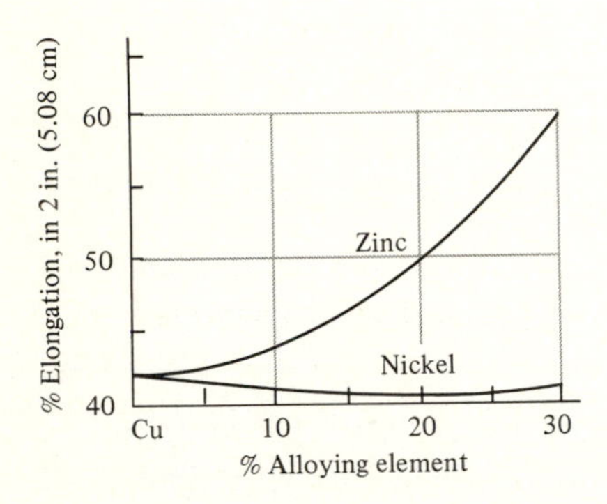

Fig. 3.30 *Influence of solid-solution elements zinc and nickel on annealed mechanical properties of copper-base alloys. Nickel forms a complete solid solution with copper, while the maximum solubility of zinc in copper is approximately 40 wt. %.*

If we apply a stress of 10,000 psi (69 MPa) to a typical bar of structural steel, measure the elongation, and then leave the specimen under load overnight, we will not find any appreciable additional length change in the morning. However, if we surround the specimen with a resistance furnace and conduct the same experiment with the specimens at 1000°F (540°C), we find that plastic elongation occurs as a function of time. This phenomenon is called *creep*. Measuring it is very important for parts subjected to elevated temperatures. For example, it would be useless to develop a turbine blade material of exceptionally high tensile strength if at operating temperature it would elongate and contact the housing after a few hours of operation.

Figure 3.31 shows the type of data obtained in creep testing. Bars are equalized at temperature in a test furnace, then a fixed load is applied through a lever system. The extension is plotted as a function of time. Note that we obtain increased creep rates (slopes) and shorter rupture times by increasing either the load or the temperature.

Three stages of creep are recognized: stage I, in which initial yielding is fairly rapid; stage II, in which a straight line is obtained; and stage III, in which yielding is again rapid and failure occurs. From these data it is possible to plot the stress required to produce various amounts of deformation and to rupture as a function of time.

The test temperature has a pronounced effect on the creep and rupture properties of all materials. This will be discussed again in Chapter 13.

Stress relief is an important concept related to creep. High internal or

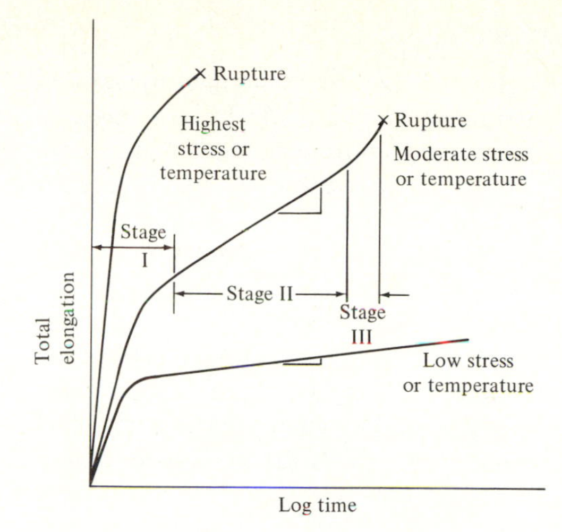

Fig. 3.31 *Typical creep curves*

residual stresses may develop in components as a result of uneven temperature distribution during processing. The residual stress is actually present as an elastic strain, just as if the section were held in a tensile test machine under load. If we heat the entire part to a temperature at which creep takes place, the elastic strain is converted to plastic strain. If the part has been cold-worked, recrystallization may also occur. The part is then cooled slowly to avoid setting up new elastic strains resulting from temperature gradients in the material during cooling.

EXAMPLE 3.7 Giving strength, ductility, and hardness specifications for engineering components is common practice. For the applications below, indicate whether creep, impact, or fatigue specifications should also be included: automotive exhaust manifold; fan blade; automotive brake drum; paneling nail.

ANSWER

Automotive exhaust manifold. Service is primarily at high temperatures, and so creep and rupture are important. After dark it is possible to see the manifold glowing red after expressway travel.

Fan blade. Since the fan rotates at high speed, it is subjected to cyclic stress, and fatigue resistance is important. In Chapter 13 we shall see that surface blemishes may lower the fatigue strength.

Automotive brake drum. Since the drum heats during braking, we might suspect that creep resistance should be specified. However, failure is often by thermal fatigue. The cyclic heating and cooling of the brake drum causes thermally induced cyclic stresses.

Paneling nail. Impact resistance is an obvious specification. The existence of notches in a serrated nail tends to lower the impact strength.

SUMMARY

When stress is applied to a metallic material, the strain at low stresses is principally elastic and varies with direction in a crystal. At higher stresses both elastic and plastic (permanent) strain are encountered. For design purposes the yield strength is defined as the stress at which the plastic strain reaches 0.2%.

The modulus of elasticity is an important property in design because it gives the elastic strain ϵ produced by a given stress σ: $\epsilon = \sigma/E$. Other design values are tensile strength, percent elongation, and percent reduction of area.

A basic understanding of plastic deformation is obtained from the critical resolved shear stress, which is a constant for a given material. While different values of tensile stress may be required to cause yielding, there is a constant critical resolved shear stress to be reached. A further basic consideration is the role of dislocations in promoting slip. In material that is free from dislocations, the yield strength is over 10^6 psi (6.9×10^3 MPa).

The hardness of a material is determined by the size of the cavity produced by a hardened indenter. Prominent scales of hardness are Brinell (BHN), Rockwell (R_A, R_B, etc.), and Vickers (VHN).

After a material has undergone plastic deformation, as by cold rolling, for example, the increase in hardness and decrease in ductility may be removed by annealing. As the specimen is heated, recovery, recrystallization, and grain growth take place at increasingly higher temperatures. By combining cold working and annealing we can attain the desired balance between strength and hardness on the one hand and ductility, as measured by percent plastic elongation.

Creep occurs when a metal is heated to a temperature at which it is relatively hot and plastic. Care must be exercised to avoid placing metals in service where substantial deformation will occur with time. On the other hand, if elastic strain is to be removed, as in residual stresses, the part is heated to allow creep to occur. The elastic strain is thus replaced by plastic.

DEFINITIONS

Annealing In general, a heat treatment in which a part is heated to soften the material. In this chapter the treatment leads to the recrystallization of cold-worked material.

Brinell hardness number, BHN The value obtained from a Brinell indentation hardness test.

Cold working The deformation of material below its recrystallization temperature.

Creep Permanent strain that increases as a function of time under load.

Critical resolved shear stress, τ_c The applied stress resolved in the slip direction in the slip plane, $(F/A) \cos \lambda \cos \varphi$.

Dislocation A line of missing atoms, leading to a region of easy slip.

Elastic strain The elastic displacement of atoms from their normal positions, as for example by applying a tensile or compressive stress. When the stress is removed, the atoms return to the normal spacing.

Engineering strain, ϵ Change in length divided by original gage length, $\Delta l / l_o$.

Engineering stress, σ Load divided by original area, P/A_o.

Engineering stress–strain curve The results, usually of a tensile test, plotted with σ as the y axis and ϵ as the x axis.

Equiaxed grains Grains that are of equivalent dimensions in all directions, i.e., not longer in one direction than in another, as is common in cold working.

Fatigue testing If a specimen is subjected to cycles of loading and unloading in tension, compression, or bending, the stress required for fracture is much lower than in a single stress application. The *fatigue strength* (also called the *endurance limit* in steels) is the stress required to produce failure in a specified number of cycles, usually 10^6.

Grain growth The development of larger grain size by preferential growth of larger crystals.

High-temperature testing For practically every material there is a range of temperature in which it softens and deforms in increasing amounts as a function of time. When this occurs, it is necessary to obtain test data of stress vs. time. These are obtained from high-temperature tests. The properties measured include creep and stress required for rupture.

Hot working The deformation of material at or above its recrystallization temperature.

Impact testing A bar, usually notched, is struck by a calibrated pendulum and the energy required to cause fracture is measured. Although the term *impact strength* is used, the results are expressed in terms of energy. The principal reason for the fracture is not the impact but the presence of the notch, which causes a brittle stress condition known as a triaxial stress condition. This is discussed in texts on strength of materials.

Modulus of elasticity, E Stress divided by strain, σ/ϵ in elastic range.

Percent cold work $\dfrac{\text{Change in cross-sectional area}}{\text{Original area}} \times 100$

Percent elongation at fracture Plastic (engineering) strain times 100,

$$\frac{l_f - l_o}{l_o} \times 100$$

Percent reduction of area $\dfrac{\text{Change in area}}{\text{Original area}} \times 100 = \dfrac{A_o - A_f}{A_o} \times 100$

Plastic strain The permanent displacement of atoms from a given starting position, as by slip or twinning.

Recovery The relief of elastic strain in the early stages of annealing.

Recrystallization The growth of new stress-free equiaxed crystals in cold-worked material.

Rockwell hardness number, R_A, R_B, R_C, etc. The number obtained from a Rockwell hardness test.

Slip system A combination of a slip direction and a slip plane containing the direction.

Stress relief The disappearance of elastic strain, especially during annealing.

Tensile strength The maximum engineering stress encountered during a tensile test.

Transition temperature If the samples for the impact test are tested at different temperatures, some materials such as steel show a rapid fall in impact strength at a low temperature, such as 0°C. This is called the transition temperature. The fracture changes from ductile to brittle at this point. (This brittleness with temperature usually does not exist in FCC metals.)

True strain, ϵ_t $= \ln (l/l_o) = 2.3 \log_{10} (l/l_o)$.

True stress, σ_t Load divided by area at the given load, P/A.

Twinning The shifting of the atoms on one side of a twinning plane by an amount proportional to the distance from the plane. The twin produces a mirror image of one part of the grain with the other.

Vickers hardness number, VHN The value obtained from a Vickers test.

Work hardening An increase in hardness and strength due to plastic deformation.

Yield strength The stress at which a specified amount of plastic strain, usually 0.2%, is produced.

PROBLEMS

3.1 Re-examine the bent microspecimen of copper (Fig. 2.2*b*), and note that some slip planes appear to make an angle of nearly 90° with the surface. Explain on the basis of critical shear stress. (Sections 3.1 through 3.5)

3.2 Slip planes and directions for three common metal structures are:

Structure	*Directions*	*Planes*
FCC	⟨110⟩	{111}
BCC	⟨111⟩	{110}
HCP	⟨110⟩	{0001}

Are these the planes and directions of densest packing (greatest atomic density)? (Sections 3.1 through 3.5)

3.3 A slip system is defined as a combination of a plane and a direction. Show that the number of slip systems in FCC is 12 and in HCP is 3, using only the most densely packed planes and directions. (Sections 3.1 through 3.5)

3.4 Two rods in the form of single crystals of pure zinc show widely different yield strengths at 0.2% offset. What index of yield strength should be the same? (Sections 3.1 through 3.5)

3.5 Given that the critical resolved shear stress is 82 psi (0.566 MPa), plot axial stress for slip as a function of $\cos \varphi \cos \lambda$. See Example 3.1. (Sections 3.1 through 3.5)

3.6 In BCC iron, slip occurs in the directions of closest packing of atoms. Sketch two of these in a BCC unit cell *and* label each with its indices of direction.

In FCC iron, slip occurs on the planes of greatest atomic density. Sketch one of these in a unit cell and give its Miller indices. (Sections 3.1 through 3.5)

3.7 Old specifications frequently refer to a *yield point,* defined as the stress required to cause the onset of permanent deformation in a specimen. Discuss the difficulty of measuring this quantity in a commercial metal. (Section 3.6)

3.8 From the following data, plot the engineering stress–strain curve and calculate the modulus of elasticity, yield strength at 0.2% offset, percent elongation, percent reduction of area, and tensile strength. (Section 3.6)

Load, lb	*Gage length, in.*
0	2.0000
5,000	2.00167
10,000	2.00340
12,000	2.00425
13,000	2.0047
14,000	2.0052
13,900	2.0065
15,000	2.0079
18,000	2.0118
20,000	2.0156
26,000	2.056
39,000 (maximum)	2.472
32,000 (fracture)	2.820 (after fracture)

Original diameter: 0.505 in.
Diameter at maximum load: 0.454 in.
Diameter at fracture region: 0.284 in.
(To convert stress to MPa, multiply psi by 6.9×10^{-3}.)

3.9 From the following data for a metal bar 6 in. long × $\frac{1}{2}$ in. wide × $\frac{1}{4}$ in. thick (152.4 mm × 12.7 mm × 6.35 mm), give the maximum load that can be applied *without* extensive permanent deformation: tensile strength

= 60,000 psi (414 MPa); yield strength = 40,000 psi (276 MPa); breaking strength = 55,000 psi (380 MPa); elongation = 25%; modulus = 25 × 10^6 psi (1.725 × 10^5 MPa). (Section 3.6)

3.10 From the data given, construct an approximate stress–strain curve: modulus = 10 × 10^6 psi (6.9 × 10^4 MPa); 0.2% offset yield strength = 40 × 10^3 psi (276 MPa); tensile strength = 60 × 10^3 psi (414 MPa); breaking strength = 50 × 10^3 psi (345 MPa); elongation = 6%. (Section 3.6)

3.11 Rather than machining a round tensile specimen out of a piece of sheet metal, we cut a flat tensile out of an 0.500-in.-thick plate as shown on the accompanying graph. The specimen is 0.500 in. wide in the gage length. From the data given (load versus change in 1-in. gage length), calculate the tensile strength, yield strength (0.2% offset), modulus of elasticity, and percent plastic elongation (fracture). (Section 3.6)

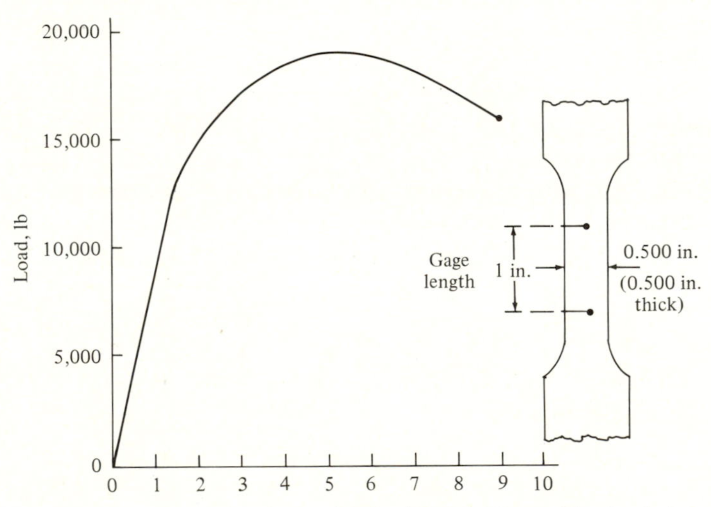

Change in one-inch gage length caused by load. Scale in *thousandths* of an inch.

3.12 Mark the following statements true or false *and* illustrate your answers with schematic graphs, axes marked with units, of impact energy vs. T.

a. It is possible to have two steels with about the same percent elongation in tension and the same Brinell hardness, but with impact properties that vary by 2:1 at 0°F (−18°C).

b. It is possible to have one steel with about the same percent elongation at 0°F (−18°C) and 70°F (21°C) but with impact properties that vary by 3:1 at these temperatures. (Sections 3.8 through 3.16)

3.13 The room-temperature properties of two steels being considered for lightweight cylinders that will contain liquid nitrogen are as follows.

	Tensile Strength, ksi	*Yield Strength, ksi*	*Percent Elongation*
Heat-treated 4340	330 (2277 MPa)	290 (2001 MPa)	8
18% Cr, 8% Ni (304)	85 (587 MPa)	35 (242 MPa)	60

The structure of the 4340 is a fine dispersion of iron carbide in BCC iron. The 18-8 is a FCC single-phase alloy. Which material would you specify? (Sections 3.8 through 3.16)

3.14 A designer is replacing a steel that has an endurance limit of 75,000 psi (518 MPa) (10^6 cycles) with an aluminum alloy that has an endurance limit of 24,000 psi (166 MPa). She calculates that the cross section should increase by a factor of 75/24. Do you agree? (Sections 3.8 through 3.16)

3.15 A 70% Cu, 30% Zn 0.1-in.- (0.254-cm-) diameter brass wire with a minimum tensile strength of 60,000 psi (414 MPa), a hardness of 75 R_B minimum, and an elongation of over 10% is needed. You have some 0.25-in.- (0.635-cm-) diameter stock available which has been cold-rolled 40%. Specify your procedure. Assume that 70% Cu, 30% Zn brass fails at cold work in excess of 60%. (Sections 3.17 through 3.23)

3.16 A bar of 85% Cu, 15% Zn alloy (red brass) that has a 0.5-in. (1.28-cm) diameter is to be cold-rolled to a bar of 0.125-in. (0.317-cm) diameter. Specify the procedure needed to obtain a final tensile strength of 60,000 psi (414 MPa) minimum and an elongation of 10% minimum. (Sections 3.17 through 3.23)

3.17 A rolled 70% Cu, 30% Zn brass plate 0.500 in. (1.28 cm) thick has 2% elongation as received from the supplier. The final thickness desired is 0.125 in. (0.317 cm), with a tensile strength of 70,000 psi (483 MPa) minimum and an elongation of 7% minimum. Assume that the rolling is conducted so that the width of the sheet is unchanged. (This means that in calculating the area, $A = wt$, with width w constant.) Specify all steps in the procedure, including heat treatments. (Sections 3.17 through 3.23)

3.18 Sketch the microstructures (longitudinal and transverse to the rolling direction) you would expect to see in the material of Prob. 3.17 as received, after intermediate annealing, and after final working. Show the differences in grain shape, slip bands, and annealing twins. (Sections 3.17 through 3.23)

3.19 The times for 50% recrystallization of pure copper are as follows:

Temperature:	203°C	162°C	97°C	72°C
Time:	6 sec	1 min	100 min	1000 min

Consider the value

$$\frac{1}{\text{Time for 50\% recrystallization}}$$

as representing the rate. Plot ln [1/time (50%)] against $1/T$, where T is degrees absolute (°C + 273). From the plot estimate the time for 50% recrystallization when the temperature is 20°C (68°F). Under these conditions would you worry about the loss of strength in a cold-worked copper cable over a period of 10 yr? (Sections 3.17 through 3.23)

3.20 Suppose that you are on a desert island and want to make a bar of copper with a single-phase FCC structure and maximum strength. You have a melting furnace, an ingot mold, a rolling mill, and a heat-treating furnace. You have a supply of copper (FCC, a_0 = 3.61 Å [0.361 nm]) and the following three metals. (Sections 3.17 through 3.23)

	Structure
Bolonium	FCC a_0 = 4.11 Å (0.411 nm)
Manurium	BCC a_0 = 2.13 Å (0.213 nm)
Zippium	HCP a_0 = 2.66 Å (0.266 nm)

a. Which one of the three metals would probably form the most extensive range of FCC solid solutions with copper using the Hume Rothery rules? Show calculations.

b. After casting an ingot 2 × 2 × 10 in. (0.051 m × 0.051 m × 0.254 m), how would you make a bar of 1 in. (0.0254 m) cross section and maximum strength? Obviously some experimentation is needed. What would be your first attempt to reach the maximum strength?

c. If you encountered cracks in your first attempt, what changes would you make in the procedure?

3.21 We have a piece of brass that we know to be cold worked, but we are uncertain whether the metal is a 70-30 or 85-15 alloy. However, someone has run a tensile test for us and found the tensile strength to be 60,000 psi (414 MPa) with an elongation of 15%. Determine the alloy composition, amount of cold work, and tensile strength in the annealed condition. (Sections 3.17 through 3.23)

3.22 Given the following data for 70% Cu, 30% Zn brass: Material annealed, then cold-rolled 50%, then heated ½ hr at the temperatures shown, cooled to room temperature, and tested.

		Temperature, °C							
Treatment	*As Cold-rolled*	*150*	*200*	*250*	*300*	*350*	*450*	*600*	*750*
Hardness R_H	97	98	100	101	98	80	58	34	14
Tensile strength (10^3 psi)	80	81	82	82	76	60	46	44	42
Percent elongation	8	8	8	8	12	28	51	66	70

Assume that you have a coil of 0.5-in. (12.7 mm)-diameter wire that you wish to process to 0.1 in. (2.54 mm) diameter. You want the following properties in the 0.1.-in.-(2.54 mm) diameter material.

Hardness R_H, min: 75 Tensile strength, min: 50,000 psi (345 MPa)
Percent elongation, min: 30%

Specify the procedure quantitatively. (Sections 3.17 through 3.23)

3.23 The following data were obtained by cold-rolling specimens of 70% Cu–30% Zn brass (50% cold-worked), then heating them $\frac{1}{2}$ hr at the indicated temperatures. After heating, the samples were cooled to 70°F (21°C) and tested with R_x [$\frac{1}{16}$-in. ball (1.6 mm), 75-kg load]. (Sections 3.17 through 3.23)

	R_x	*Grain Size (average diameter, in.)*
As rolled	97	Not determined
After 150°C $\frac{1}{2}$ hr	98	Not determined
200°C $\frac{1}{2}$ hr	100	Not determined
250°C $\frac{1}{2}$ hr	101	Not determined
300°C $\frac{1}{2}$ hr	98	Not determined
350°C $\frac{1}{2}$ hr	80	0.02
450°C $\frac{1}{2}$ hr	58	0.03
600°C $\frac{1}{2}$ hr	34	0.06
750°C $\frac{1}{2}$ hr	14	0.12

a. The recrystallization temperature for this material lies within the range (select one): 150–250, 300–350, 350–450, 450–700°C.

b. The recovery temperature lies within the range (select one): 150–300, 350–450, 600–750°C.

c. The slip bands will not be present in the specimens heated above (select one): 150, 250, 350°C.

d. The elongation in the as-rolled material is 8%. The most probable value for the specimen heated at 350°C is (select one): 7%, 8%, 28%.

e. The tensile strength of the as-rolled material is 80,000 psi (552 MPa). The most probable value for the specimen heated at 750°C is (select one): 85,000 psi (587 MPa), 80,000 psi (552 MPa), 42,000 psi (290 MPa).

3.24 Samples of a given metal are heated to different temperatures for 1 hr, then cooled to 70°F (21°C). The hardness is then measured as shown in the accompanying graph. The metal had been cold-worked 30% prior to

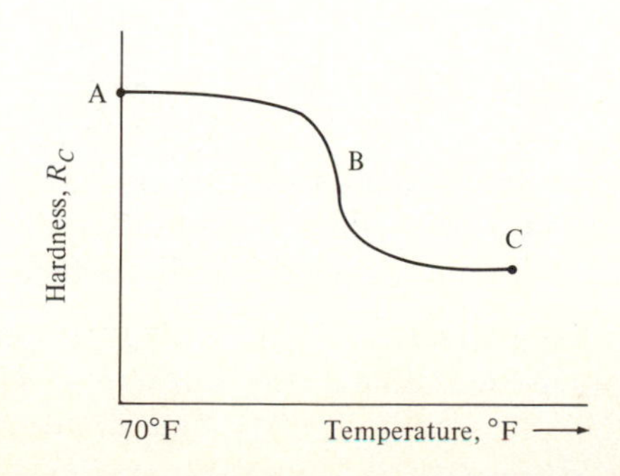

heating. Other samples of the same metal cold-worked 50% were heat-treated in the same way. Draw the type of curve you would expect on the same axes, paying particular attention to the location of the second curve relative to the first curve at regions *A*, *B*, and *C* (Sections 3.17 through 3.23).

3.25 The following cold-worked properties are obtained for a 90% Cu, 10% Ni (cupro-nickel) alloy.

Percent Cold Work	Tensile Strength, psi	Percent Elongation	Hardness	
			R_F	R_B
10	50,000 (345 MPa)	25	80	30
20	55,000 (380 MPa)	16	95	63
30	59,000 (407 MPa)	12	100	70
40	63,000 (435 MPa)	11	102	74
50	65,000 (449 MPa)	10	105	79

Graph these data. Select a cold work for both a 90% Cu, 10% Ni and a 85% Cu, 15% Zn alloy to achieve the following final properties:

Minimum tensile strength: 54,000 psi (373 MPa)
Maximum hardness: 65 R_B
Minimum elongation: 12%

When compared with annealed properties, do copper-zinc alloys or copper-nickel alloys appear to show the most work hardening? (Comparison can be made only for the low solute contents.) (Sections 3.17 through 3.23)

PROBLEMS COVERING OPTIONAL SECTIONS

3.26 Derive the relation between original area A_o and the area at σ_{true} and true strain. [*Hint:* Take a small volume at the region of fracture. With no load the volume would be l_oA_o. After fracture the volume would be l_fA_f and the volume would be equal. Now substitute for the ratio l_f/l_o in the formulas for true strain.] (Section 3.7)

3.27 Calculate the true stress–true strain curve using the data of Prob. 3.8. Use the relation

$$\text{True strain} = \ln \frac{A_o}{A_f} = 2.3 \log \frac{A_o}{A_f}$$

for the strain at maximum load and fracture. (Section 3.7)

3.28 A tensile specimen of a new alloy broke at 27,000 lb (120,536 newtons). The original and final gage lengths and diameters were (Section 3.7)

l_o = 2.00 in. (50.8 mm) d_o = 0.505 in. (12.83 mm)
l_f = 2.40 in. (61.0 mm) d_f = 0.460 in. (11.68 mm)

a. What were the engineering and true breaking strengths?
b. What was the final true strain?
c. What was the ductility? (% elongation and % reduction in area)

3.29 The accompanying graph is a recorder plot from a tensile test of a 0.505-in.- (12.83 mm) diameter specimen with a 2-in.- (50.8 mm) gage length, along with other data. Calculate E (modulus of elasticity), yield strength at 0.2% offset, ultimate tensile strength, percent reduction in area, and true stress at fracture. (Section 3.7)

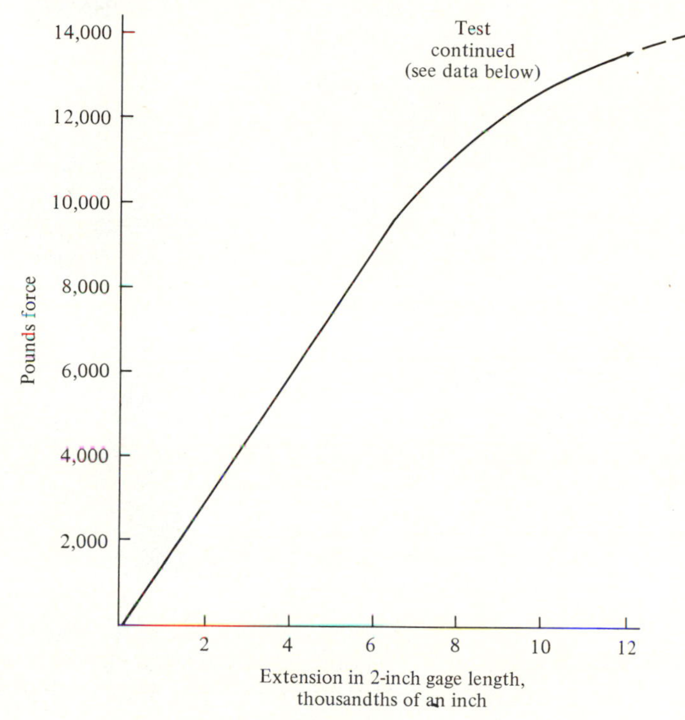

Additional data:		
	Maximum load,	16,000 pounds force
	Load at fracture,	12,000 pounds force
	Diameter of fracture,	0.400 inch
	Gage length at fracture,	2.50 inch

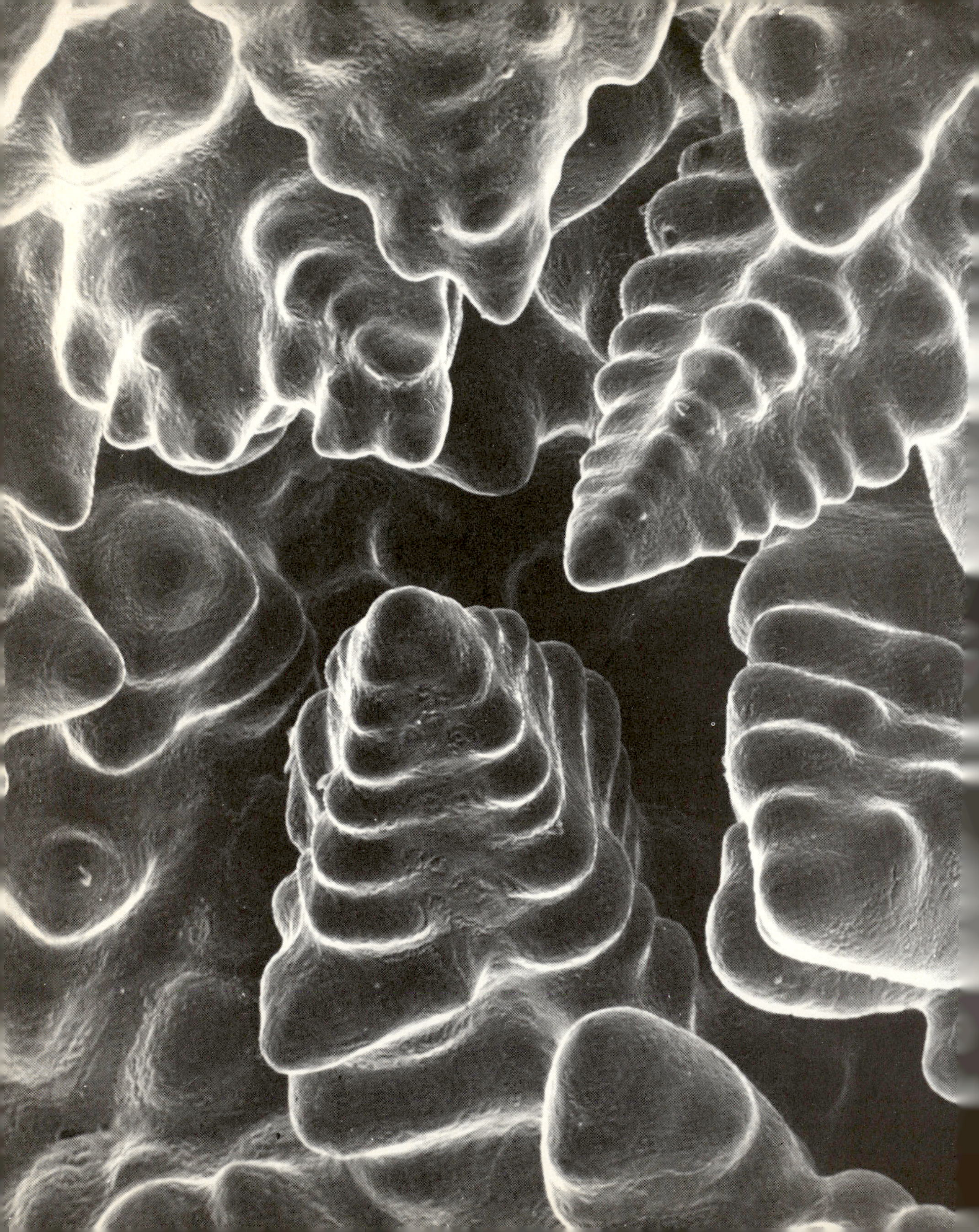

4

CONTROL OF POLYPHASE STRUCTURES IN METALS: UNDER EQUILIBRIUM CONDITIONS (PHASE DIAGRAMS); UNDER NONEQUILIBRIUM CONDITIONS (PRECIPITATION REACTIONS)

THE somewhat fernlike growths in the illustration are called "dendrites." They are actually crystals of a copper alloy growing into a cavity in a casting, photographed with a scanning electron microscope (100×). This is the most common type of growth in metal parts. Of course, it is not necessarily accompanied by cavity formation.

This chapter is devoted entirely to the formation of metal structures from the liquid, such as dendrites, and from changes taking place in the solid state.

4.1 General

So far we have discussed the properties of only *single-phase* materials, such as pure metals, solid solutions, and intermetallic compounds. These are useful materials, but the metallic structures with highest strength, hardness, and wear resistance are composed of two or more phases in a well-controlled dispersion. This microstructure is usually not obtained in the original casting or ingot. It is formed by carefully controlled processing that involves hot-working and heat treatment. For example, the properties of a common aluminum alloy, 2014, used for structural members in aircraft, are as follows.

	Yield Strength at 0.2% Offset, psi × 10³†	*Tensile Strength, psi × 10³*	*Percent Elongation*
Alloy 2014,‡ annealed	14	27	18
Alloy strengthened by heat treatment	60	70	13

If an engineer had to design on the basis of yield strength of the annealed material, an aircraft would have to be 4 times heavier. It might never get off the ground!

Here we introduce a general statement of the greatest importance for understanding the relations between structure and properties, not only for metals but for all materials. The properties of a material depend on the nature, amount, size, shape, distribution, and orientation of the phases. This point is illustrated in large-scale fashion in Fig. 4.1*a*, which shows how the properties of a reinforced concrete slab may vary. Let us review this figure and consider the equivalent effects in different microstructures.

1. *Nature* of the phases. In the concrete slab the individual properties of the concrete and the steel affect the overall strength of the slab. In the aluminum alloy 2014 two phases, a solid solution of copper in aluminum and an intermetallic compound $CuAl_2$, are involved.

†To obtain MN/m^2 (MPa), multiply by 6.9×10^{-3}.
‡4% Cu, 0.8% Si, 0.8% Mn, 0.6% Mg, balance Al.

Fig. 4.1 *(a) Schematic representation of how the nature, size, shape, amount, distribution, and orientation of phases control physical properties (as applied to steel and concrete). Photographs show microstructure of aluminum-silicon alloy used in an aluminum automobile engine block. The composition is 17% Si, 4% Cu, 0.5% Mg, balance Al. The large gray crystals are β (almost pure silicon) and are precipitated from the liquid. Balance of the structure is a mixture of fine α, β, and a trace of a copper-rich phase θ (light gray). (b) Silicon crystals 1070 VHN; matrix 130 VHN, 500×, etched. (c) Silicon gray, θ light gray, matrix white, 1500×, unetched.*

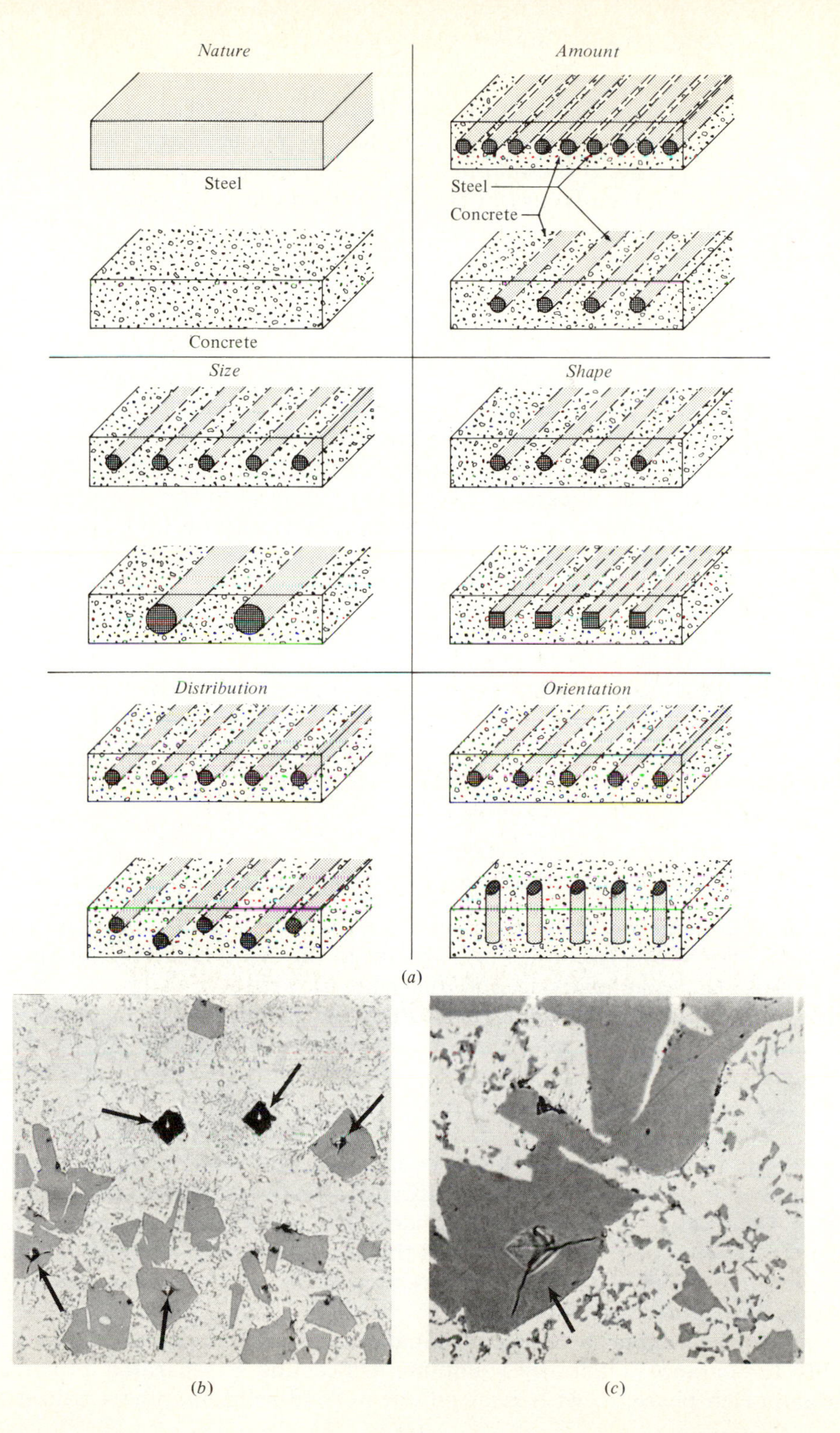
Nature
Amount
Steel
Steel
Concrete
Concrete
Size
Shape
Distribution
Orientation
(a)
(b)
(c)

2. *Amount* of the phases. Just as the relative amounts of steel and concrete are important, so are the amounts of the two phases in the alloy.
3. *Size* of the phases. If there were only a few large steel reinforcing rods, the structure would not be as strong as it is when the same amount of steel is used in thinner bars. Similarly, a fine dispersion of $CuAl_2$ is preferred.
4. *Shape* of the phases. If the steel is in square bars, stress concentration will occur at the corners, causing cracking of the concrete; hence round bars are preferred. In the case of a precipitate in the microstructure, the properties obtained depend on whether the shape is spheroidal or platelike.
5. *Distribution* of the phases. If the slab is to encounter bending stresses, the steel should be placed closer to the surface rather than at the center. Similarly, in a microstructure properties will differ if a precipitate phase is localized at grain boundaries instead of being uniformly distributed.
6. *Orientation* of the phases. If the slab is to be used to resist typical traffic, a horizontal orientation of the steel is preferred. Similarly, we will see later that for optimum magnetic properties a certain alignment of particles or precipitates in a magnetic tape is very important.

From these examples we see that it is essential to understand the methods of control of polyphase structures.

We shall begin with a discussion of the structural changes that take place under equilibrium (slow cooling), because these treatments are useful in softening a material for machining or forming and because the nonequilibrium treatments discussed next are usually based on the control or suppression of the equilibrium reactions.

4.2 Phase diagrams for polyphase alloys

Up to this point we have discussed only the simple but important case of single-phase alloys. In these, all the grains in the metal have the same crystal structure, the same analysis, and the same properties.

As an example of a two-phase alloy, let us look at the microstructure of an aluminum-silicon alloy used in many engine blocks (Fig. 4.1*b* and *c*). It is apparent that two different major phases are present. (Recall that a phase is a homogeneous and physically distinct region of matter.)

We can show that the properties of the phases are different by a *micro*hardness test. In this procedure we locate a diamond indenter (carefully ground to a pyramid-shaped point) above the phase we wish to test. A load of 25 g, for example, is applied to the indenter to press it into the sample, as in the Vickers test discussed earlier (but this is a much lower load). The diagonal of the impression is measured and related to the hardness; the smaller the value, the harder the phase.

In the sample shown, the original melt or liquid contained 17% silicon, with the balance essentially aluminum. On cooling, it separated into an aluminum-rich phase, α, with some silicon in solid solution, plus a silicon-rich phase, β, with very little aluminum. (Greek letters are used to distinguish the different solid phases in a given alloy. These solid phases are seldom pure, but rather are solid solutions or compounds. The symbol L is used for the liquid phase; L_1 and L_2 are used if there are two different liquid phases.) It is apparent from the hardness tests that the alloy contains a hard structure in the softer background or continuous phase, usually called the *matrix*. Further, some of the silicon phase is in fine needles, whereas other particles are large. The wear resistance of the cylinder walls of an engine block depends on the combination of these hard particles and the soft matrix.

These observations are a good introduction to the case in which two or more phases are present—the polyphase alloys. Most metallic materials are of this type.

4.3 Phase diagrams

The most useful tool in understanding and controlling polyphase structures is a knowledge of *phase diagrams*. The phase diagram is simply a map showing the structures or phases present as the temperature and overall composition of the alloy are varied. We shall thus discuss these in detail at this point.†

The Aluminum-Silicon Diagram

Let us construct the very important aluminum-silicon phase diagram by direct observations from several experiments. If we heat pure aluminum in a crucible, we find that it melts sharply at 660°C. Similarly, on cooling, crystallization from the liquid takes place at the same constant temperature.

Now let us remelt the aluminum and add 5% silicon to the bath. The silicon dissolves just like sugar in water. It is important to avoid saying that the silicon "melts," for the melting point of silicon is far above the temperature of the liquid aluminum. The silicon *diffuses* from the surface of the solid into the liquid. After the silicon is dissolved, we cool the melt. We find two important differences compared with the freezing of the pure aluminum.

1. The alloy starts to freeze or crystallize at a lower temperature, 624°C compared with 660°C for pure aluminum.
2. The alloy is in a mushy condition (liquid plus solid) over a range of temperatures, rather than freezing at a constant temperature. More and more solid precipitates from the liquid until the alloy is finally solid at 577°C. We find the same type of behavior at 10% silicon, but the start of freezing is delayed until 588°C.

---

†In the sections concerning nonequilibrium conditions (precipitation reactions) we shall discuss the driving force leading to solidification.

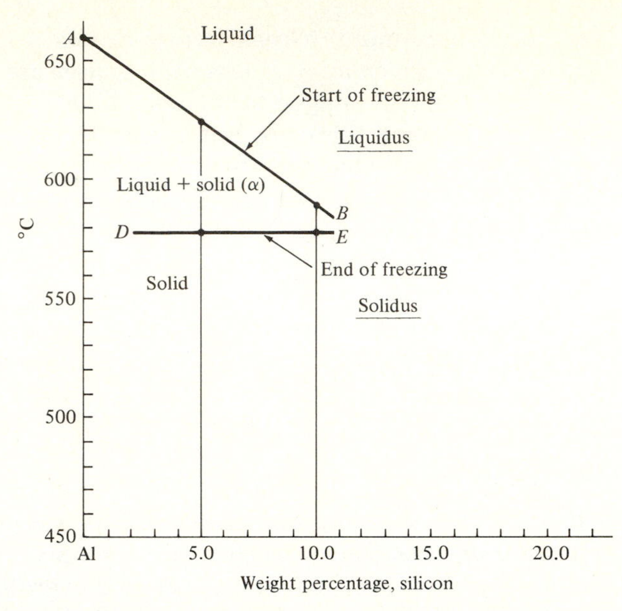

Fig. 4.2 *Construction of a portion of the aluminum–silicon phase diagram*

Now we have enough data to begin constructing the aluminum-silicon phase diagram (Fig. 4.2). As we said earlier, this is simply a map showing which phases are present at different compositions as the temperature is changed. We lay off percentage silicon by weight on the x axis and temperature (°C) on the y axis and plot the points we have obtained.

Above line AB the melt is entirely liquid. AB is called the *liquidus*. Below DE the melt is entirely solid. DE is called the *solidus*. Between AB and DE we label the region liquid plus solid (L + α).

If we continue our experiments, we can complete the diagram as shown in Fig. 4.3. This diagram contains many new lines, which we will now discuss. First, as the percentage of silicon increases, the liquidus temperature falls until 11.6% silicon is reached; then the temperature rises. This pattern is quite common in alloys. The composition at the minimum freezing point has a special name, *eutectic* (E), meaning "of low melting point." (In usage, the temperature at which a liquid of eutectic composition freezes is called the eutectic.) Note that although the melting point of silicon is twice that of aluminum, adding it *lowers* the melting range of aluminum until the alloy has 11.6% silicon. Note also that the phases present to the right of the eutectic and above the solidus are β + L instead of α + L. In other words, if we cool an alloy with 16% silicon, the first phase to crystallize is β (crystals of almost pure silicon). Only after crossing the solidus at 577°C do α crystals begin to form.

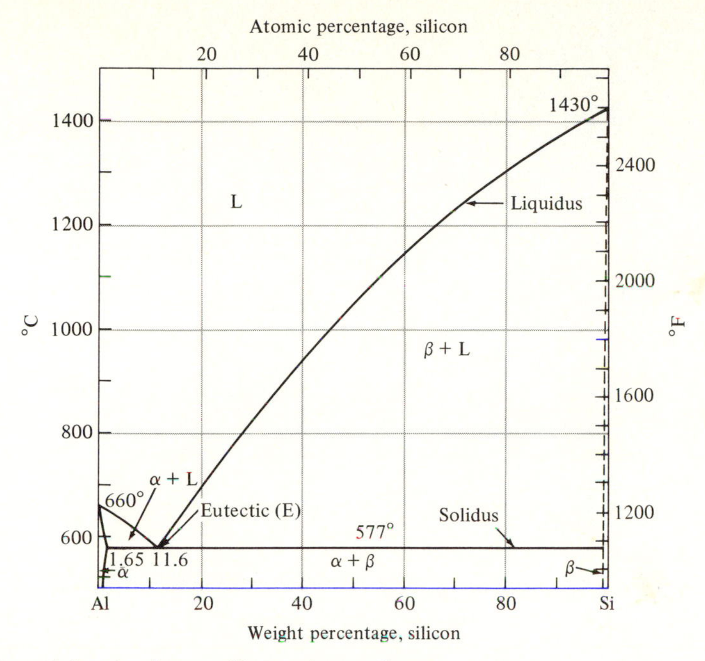

Fig. 4.3 *Complete aluminum–silicon phase diagram.* (American Society for Metals Handbook, 8th ed., Vol. 8: *Metallography, Structures, and Phase Diagrams,* Metals Park, Ohio, 1973)

Let us interrupt this discussion to point out the practical significance of what we have learned. First, the aluminum-silicon alloys are important engine-block materials. They have been substituted for cast iron in some cases. It is essential to pour these castings at the proper temperature. If the liquid is too cold, it will not fill the mold cavity completely; if it is too hot, it can react with the mold wall and give a poor surface. Usually the pouring temperature is set at about 110°C above the liquidus. We see, therefore, that the upper line of the diagram gives us a guide for establishing pouring temperature. At any silicon level we merely add 110°C to the liquidus to obtain the pouring temperature. Second, many alloys are heat-treated. The *solidus* gives us a guide for determining the maximum heating temperature, since above this line the parts begin to melt. (As a matter of fact, it is advisable to stay 20°C below the solidus because melting may start at lower temperatures if impurities, which lower the solidus, are present.)

We can get a good deal of additional information from this diagram. To illustrate, let us discuss the determination of line *AD* in Fig. 4.4.

If we make a melt with 0.2% silicon, we find that it starts to freeze at point 1 and is solid at point 2 in the enlarged diagram. Similarly, using 0.4% silicon, we obtain the liquidus and solidus points 3 and 4. In this way we determine *AD,* the line between the *single-phase field,* marked α, and the *two-phase field,* marked α + L.

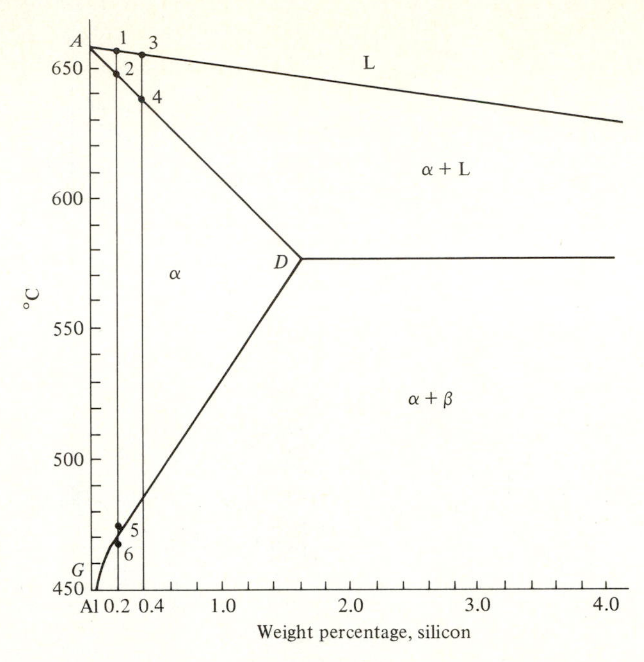

Fig. 4.4 *Enlarged region of the aluminum–silicon phase diagram (the high aluminum end)*

Now the question is how *DG* is determined. If we follow the cooling of our sample containing 0.2% silicon, we see that it is in a single-phase α field from 648 to 470°C. However, when we cross *DG* at 470°C, the specimen enters a two-phase field, meaning that β will begin to precipitate from solid α. This is the key to our problem.

To answer the question we must digress to explain an important tool of the metallurgist, the heat treatment cycle involving heating and quenching. If we observe a sample of our 0.4% silicon alloy that was slowly cooled to 20°C, we find two phases. Now let us heat the sample to 550°C, where only one phase is stable according to the diagram. We let the sample soak at temperature for about 1 hr to let the α + β change to α only. Now we quench the sample in cold water. We find only α under the microscope. The reason for this is that for β to precipitate, a certain time is required. Silicon atoms must diffuse from their dispersed solid-solution positions in the α phase to form the new 99% silicon β phase. Diffusion is so slow at 20°C that the alloy contains the same structure (α) it showed at 550°C. In other words, quenching a sample from a given temperature tends to fix the structure that was present at that temperature. There are important exceptions we shall discuss later, but the technique can be used successfully in many cases.

Now we are ready to consider the location of line *DG*. This time let us take five samples of the alloy, heat them all to 550°C, and soak them for 1 hr.

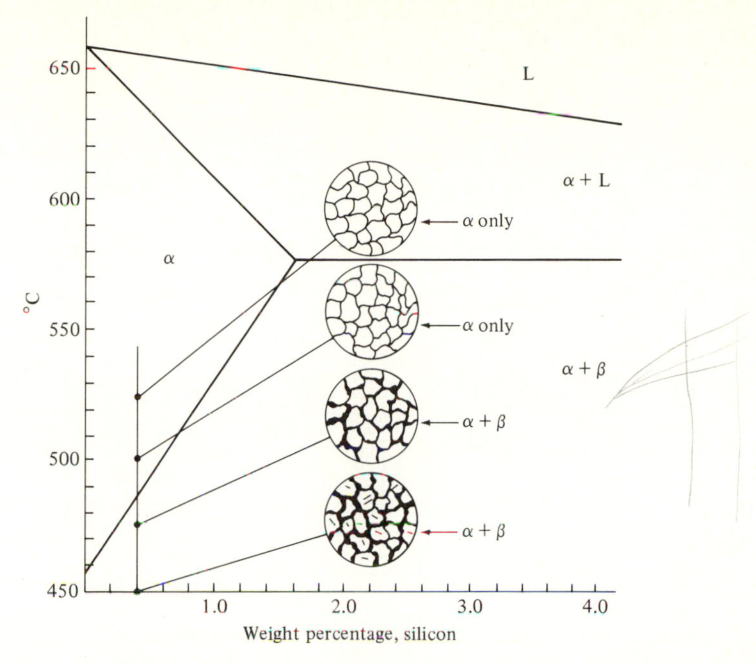

Fig. 4.5 *Microstructure of heat-treated 0.4% silicon–aluminum alloys (slowly cooled from 550°C to the indicated temperature and then quenched to room temperature)*

We then slow-cool a sample to 525°C and quench it. We slow-cool the other samples to 500, 475, 450, and 425°C, respectively, and quench them. We find the structures shown in Fig. 4.5.

Since no β was present under slow-cooling (equilibrium) conditions at 500°C but β appeared in the 450°C sample, the line *DG* is between 450 and 500°C. By further testing we could locate the line with greater accuracy.

4.4 Phase compositions (phase analyses)

If we have two phases present, we really have two different materials making up the structure, each with a different chemical analysis. For example, if we dissolve 15% silicon in liquid aluminum, we have a single liquid phase, but we find two solid phases after freezing, similar to the illustration in Fig. 4.1*b* and *c*. If we separate these phases at room temperature and analyze them, we find that α contains over 98% aluminum and β over 99% silicon. (With a new device called the *electron microprobe* we can analyze the individual phases in the sample. The procedure is related to the use of x rays, discussed in Chapter 2. A tiny beam of electrons is focused on the particular phase. These cause the atoms of the elements that are present to emit x rays. Since each element gives off x rays of its own characteristic wavelength, we need only record this x-ray spectrum, just as the visible spectrum is recorded in a spectrograph. The

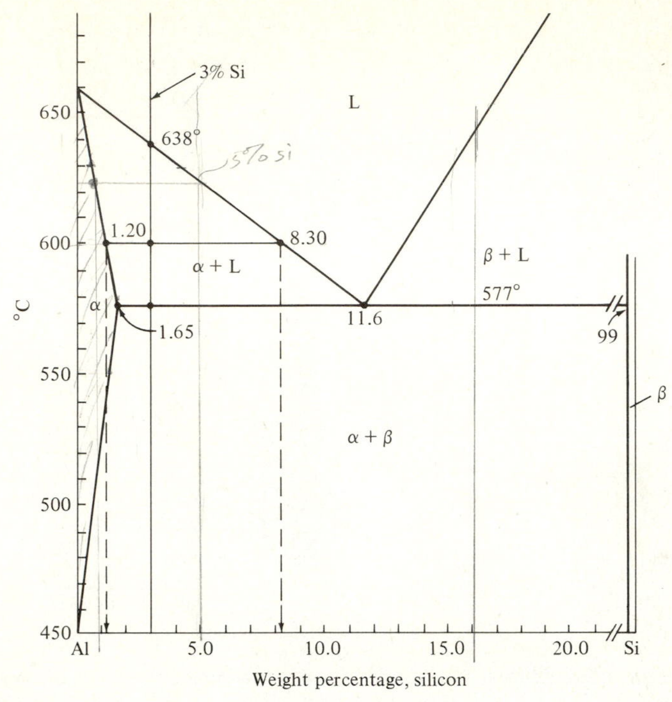

Fig. 4.6 *Determination of phase compositions from a phase diagram*

percentage of a given element in the phase is related to the intensity of radiation of its characteristic wavelength.)

Let us suppose we have an alloy of 97% aluminum and 3% silicon at 600°C. This is definitely in the two-phase field α + L. To find the phase compositions we simply draw a horizontal line across the two-phase field until it touches the single-phase fields (Fig. 4.6). Then we drop a perpendicular from each contact point to the x axis and read off the particular phase composition. Thus we see that the phase composition of α in the 3% silicon alloy at 600°C is 1.2% silicon. Using the same construction, we find that even if the overall composition of the alloy changes from 3 to 8% silicon, the α phase at this temperature still contains 1.2% silicon. Similarly, using the same horizontal, we see that the composition of the liquid in all these alloys is 8.3% silicon. There is nothing artificial about this procedure; it is merely a way of recovering the information that we found experimentally. This method of determining the phase compositions in a two-phase field is so important that we will state it as a simple rule.

RULE 1. FOR DETERMINING PHASE COMPOSITIONS (CHEMICAL ANALYSIS OF PHASES). In a two-phase field draw a horizontal line (called a "tie line") at the temperature desired and touching the single-phase fields. Drop perpendiculars from the points where the tie line meets these fields, and read the phase compositions on the x axis.

Note that in a phase diagram single-phase fields are always at the ends of the tie line in a two-phase field. This is a consequence of the phase rule, to be discussed later.

We have used several terms in this section that have exactly the same meaning, yet are used interchangeably. These terms are: *phase composition, phase analysis,* and *chemical analysis of the phases.*

EXAMPLE 4.1 For a 1.0% and 2.0% silicon alloy, what are the liquidus and solidus temperatures, the composition of the first solid to form, and the composition of the last liquid to disappear upon cooling?

ANSWER From Fig. 4.6 we can obtain the following information.

1% silicon. All liquid above 655°C and all solid below 610°C. These are therefore the liquidus and solidus temperatures. Between these temperatures, we have the two phases α + L.

Drawing a tie line at 655°C, we find liquid with 1% silicon in equilibrium with α of approximately 0.1% silicon (the composition of the first solid to form). A second tie line at 610°C gives α of 1% silicon in equilibrium with liquid of approximately 7% silicon (the composition of the last liquid to disappear).

2% silicon: The alloy is completely liquid above approximately 645°C and completely solid below 577°C (liquidus and solidus, respectively).

A tie lie at 645°C gives an α composition of 0.3% silicon and a liquid composition of 2.0% silicon. A tie line at 578°C (just within the α + L) region gives a solid α composition of 1.65% silicon and a liquid composition of 11.6% silicon. Note that any total composition (alloy composition) between 1.65 and 11.6% silicon will give the same liquid and solid phase compositions at 578°C.

4.5 Amounts of phases

It is important to know the *amounts* of phases because the properties of a two-phase mixture depend on these percentages. These amounts can be found by calculation or by a simple graphic treatment of the phase diagram. Although the graphic method is faster, the calculation will be presented first because it deepens our understanding of what is meant by *overall* composition compared with *phase* composition.

Let us consider again the two-phase region α + L of the aluminum-silicon diagram (Fig. 4.6).

Let us ask the question: What are the relative amounts of solid and liquid in the alloy containing 97% aluminum and 3% silicon overall composition when it is at equilibrium at 600°C? In other words, it is evident from the diagram that solid and liquid are present. If we pour off the liquid and weigh the liquid and solid separately, what percentage of the total weight will each be?

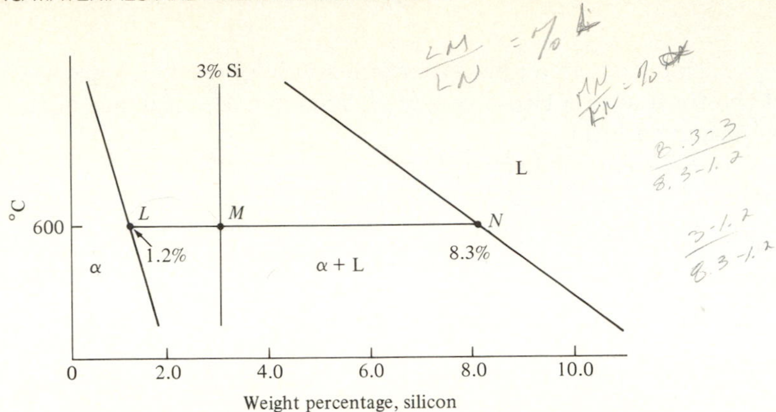

Fig. 4.7 *Calculation of phase amounts from a phase diagram*

Let us start by making a homogeneous melt at 700°C using 97 g of aluminum and 3 g of silicon. Now let us cool the all-liquid alloy to 600°C, where two phases are present. Even though there are two phases, 3 g of silicon must still be present in these phases, so that the amount of silicon in α plus the amount of silicon in the liquid equals 3 g. We add to this the information obtained from a horizontal at 600°C (Fig. 4.7), that from Rule 1 the compositions of the phases are

α: 98.8% Al, 1.2% Si L: 91.7% Al, 8.3% Si

Now let the grams of α phase equal x. Then the grams of the liquid phase equal $100 - x$. The total amount of silicon is

(grams of α)(fraction of silicon in α) + (grams of L)(fraction of silicon in L) = total grams of silicon, or

$$(x)(0.012) + (100 - x)(0.083) = 3$$

$$-0.071x = -5.3 \qquad x = 74.6 \text{ (grams of } \alpha)$$

Therefore we have 74.6% by weight α and 25.4% by weight liquid.

Now let us use a simpler graphic method called the *inverse lever rule*.

RULE 2. LEVER RULE TO DETERMINE AMOUNTS (PERCENTAGES) OF PHASES PRESENT. We again draw a tie line across the two-phase field at the desired temperature (600°C) and touching the one-phase fields (Fig. 4.7). Next we draw a vertical line at the overall composition of the melt (3% silicon). The inverse lever rule (which can be proved by geometry) states that the percentage of α is $(MN/LN) \times 100$ and the percentage of L is $(LM/LN) \times 100$. Note that the amount of α is found by taking the length of the tie line on the *far* side of the overall composition, hence the expression *inverse* lever rule.

Making the calculation, we have

$$\text{Percentage of } \alpha = \frac{8.3 - 3}{8.3 - 1.2} \times 100 = 74.6 \text{ g of } \alpha$$

$$\text{Percentage of L} = \frac{3 - 1.2}{8.3 - 1.2} \times 100 = 25.4 \text{ g of L}$$

These are the weight percentages of the phases. They are the same as those obtained by the algebraic method.

Note again that we have used several terms interchangeably in this section. The terms are: *phase amounts, quantity of phases,* and *percentage of phases present.*

EXAMPLE 4.2 Calculate the phase amounts present at 600°C for a 2% and a 6% silicon alloy.

ANSWER First we must determine *what* phases are present. They are α (a solid solution of silicon in the aluminum lattice) and liquid, Fig. 4.7.

Next the compositions of the phases in equilibrium are needed. These are found to be the same as discussed previously.

α: 98.8% Al, 1.2% Si L: 91.7% Al, 8.3% Si

In other words, the same two phases with the same phase compositions exist at 600°C for both the 3% and 6% alloys. Only the quantities change. These are determined by the inverse lever rule.

We could use a ruler to determine the lengths between the total composition and the ends of the tie lines; however, there is a "built in" ruler on each phase diagram, the composition scale. Therefore

2% silicon alloy:

$$\text{Percentage of } \alpha = \frac{8.3 - 2.0}{8.3 - 1.2} \times 100 = 88.7$$

$$\text{Percentage of L} = \frac{2 - 1.2}{8.3 - 1.2} \times 100 = 11.3$$

6% silicon alloy:

$$\text{Percentage of } \alpha = \frac{8.3 - 6.0}{8.3 - 1.2} \times 100 = 32.4$$

$$\text{Percentage of L} = \frac{6.0 - 1.2}{8.3 - 1.2} \times 100 = 67.6$$

Thus increasing the silicon of the alloy (total composition) increases the amount of liquid while decreasing the amount of solid. This is not totally unexpected, since increasing the silicon of the alloy at 600°C pushes us closer to the liquidus. A materials balance for the 6% silicon alloy gives us a check on our previous calculation.

Basis: 100 g of alloy; silicon balance

$$\begin{aligned}\text{Total silicon in alloy} &= \text{silicon in } \alpha + \text{silicon in liquid} \\ &= (32.4 \text{ g of } \alpha)(0.012 \text{ Si in } \alpha) + (67.6 \text{ g of L})(0.083 \text{ Si in L}) \\ &= 0.39 + 5.61 = 6 \text{ g Si in 100 g of alloy, or 6\% Si}\end{aligned}$$

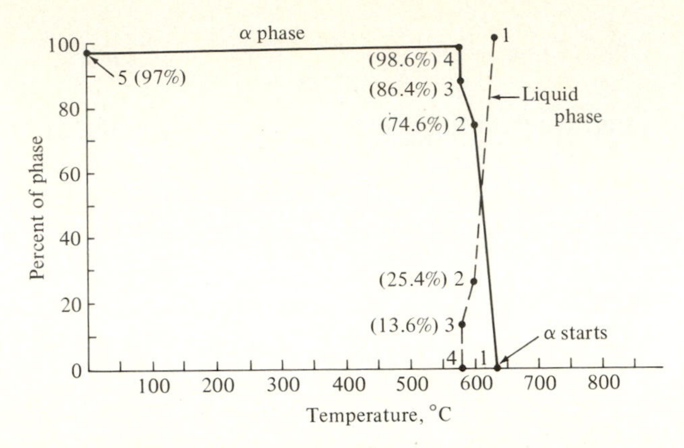

Fig. 4.8 *Phase fraction chart calculated from phase diagram for 3% silicon in aluminum. Only α and liquid are shown. The β phase is 1.4% at 576°C and increases to 3% at 20°C.*

4.6 Phase fraction chart of amounts of phases present as a function of temperature

To summarize our interpretation of a phase diagram, a phase fraction chart (Fig. 4.8) is quite useful. This chart shows how the percentages of phases present in an alloy of fixed overall composition change as the temperature changes. Let us use the 3% silicon alloy just discussed (Fig. 4.6) as an example. In this case the vertical axis represents the percentage of phases present and the horizontal axis the temperature. A separate line graph is used for each phase. Taking the data for α as an example, if we follow the vertical overall composition line, we can calculate and show the following:

1. α appears at 638°C (liquidus).
2. At 600°C we have 74.6% α and 25.4% liquid, as calculated before.
3. At just above the eutectic, say at 578°C, we have, by the inverse lever rule,

$$\text{Percentage of } \alpha = \left(\frac{11.6 - 3}{11.6 - 1.65}\right) 100 = 86.4$$

$$\text{Percentage of L} = 13.6$$

4. At 576°C (just below the eutectic) we have α + β present and no liquid.

$$\text{Percentage of } \alpha = \left(\frac{99 - 3}{99 - 1.65}\right) 100 = 98.6$$

$$\text{Percentage of } \beta = \left(\frac{3 - 1.65}{99 - 1.65}\right) 100 = 1.4$$

The change in α from 86.4% to 98.6% is abrupt because the 13.6% of liquid of eutectic composition that was present at 578°C formed $\alpha + \beta$ at the eutectic.

5. At 450°C there is more β.

$$\text{Percentage of } \alpha = \left(\frac{99 - 3}{99 - 0}\right) 100 = 97$$

$$\text{Percentage of } \beta = \left(\frac{3 - 0}{99 - 0}\right) 100 = 3$$

It is important to our later discussions dealing with the control of the shape of phases to note that the α precipitated under two different conditions. In the temperature interval 638 to 578°C, the α precipitated alone from a liquid. This is called the *primary* α. It forms large crystals or dendrites. The second period of importance was at the eutectic. Here α and β precipitated together, giving a mixture of fine α and β crystals. Therefore, 86.4% of the α will be in the form of large primary grains, and 12.2% will be small grains in the eutectic mixture. [Total α (98.6%) = primary α (86.4%) + eutectic α (12.2%).]

This concludes our simplified discussion of phase diagrams.

EXAMPLE 4.3 Calculate the volume percentages of phases present in an alloy of 16% by weight silicon and 84% by weight aluminum. [Density of silicon = 2.35 g/cm^3 (2.35 × 10^3 kg/m^3); density of aluminum = 2.70 g/cm^3 (2.70 × 10^3 kg/m^3).]

ANSWER From the phase diagram we find that we will have approximately 16 g of β and 84 g of α because at room temperature these phases are essentially pure silicon and aluminum.

$$\text{Volume of } \alpha = \frac{84 \text{ g}}{2.70 \text{ g/cm}^3} = 31.2 \text{ cm}^3$$

$$\text{Volume of } \beta = \frac{16 \text{ g}}{2.35 \text{ g/cm}^3} = 6.8 \text{ cm}^3$$

$$\text{Total volume} = 31.2 + 6.8 = 38.0 \text{ cm}^3$$

$$\text{Volume percentage of } \alpha = \frac{31.2 \text{ cm}^3}{38 \text{ cm}^3} \times 100 = 82$$

$$\text{Volume percentage of } \beta = \frac{6.8 \text{ cm}^3}{38 \text{ cm}^3} \times 100 = 18$$

Note: Physical properties are more easily related to the volume percentages of phases than to the weight percentages.

4.7 The phase rule

Now that we have a "feel" for the use of phase diagrams in calculating the composition and quantity of phases as functions of temperature and overall composition, we should take one further step to complete our understanding of this important tool of the materials engineer.

The essence of this powerful concept, the phase rule, is that nature imposes certain restrictions on the possible variations in a phase diagram. These restrictions are very helpful to the engineer in planning experiments to determine phase diagrams and in exposing false data. For example, we would now, from our limited experience, be ready to criticize an observer who reported the existence of liquid plus solid over a temperature range for a pure metal. But suppose the observer showed an area (or field) indicating three phases in equilibrium for a two-metal diagram? Furthermore, what type of diagram should we expect when we have more than two elements in an alloy?

The phase rule, developed by J. Willard Gibbs, states:

$$F = C - P + 2$$

where F = degree of freedom, C = number of components, and P = number of phases in equilibrium. (The degree of freedom is the number of variables not controlled by nature in a given situation, or, in other words, the number of variables that we can change, such as temperature, phase composition, etc. The components are the materials at the endpoints of the phase diagram, such as aluminum and silicon in the diagrams just discussed. However, one or both can be a compound, such as Fe and Fe_3C.)

The number "2" in the equation refers to the normal phase variables, temperature and pressure, which we would expect to change the nature of our phase diagrams.

In the usual case we are investigating a phase diagram at a fixed pressure of 1 atm. This means that we have reduced the variables F that we can change. We must rewrite the equation as

$$F = C - P + 1 \qquad \text{(pressure = 1 atm)}$$

In effect, the degree of freedom F tells us how many variables must be fixed if we are to describe the system completely. As an analogy, in mathematics if we have four independent equations and five independent unknowns, we need one more relation if we are to fix the values of the variables. We say, therefore, that we have one degree of freedom.

Now let us apply the phase rule to the aluminum-silicon diagram. First, consider the case of a one-phase field (liquid, for example). From the phase rule we have

$$F = C - P + 1 = 2 - 1 + 1 = 2$$

This means that to describe fully to a colleague the inherent characteristics of a liquid in the one-phase liquid field, we would need to fix two things. For example, if we specified the composition of the liquid and the temperature, our colleague could obtain a liquid of the same density, conductivity, or any other measurable property. If we only specified the composition, this would not be true, because without the temperature of the liquid fixed, the other properties would vary.

Now let us look at the two-phase field $\alpha + L$. The phase rule states:

$$F = C - P + 1 = 2 - 2 + 1 = 1$$

Therefore, in a two-phase field, we have only one degree of freedom. Let us test this by using up our one degree of freedom.

1. If we fix temperature, we also fix the *phase* compositions. These are given by the intersections of the horizontal tie line with the single-phase fields at this temperature. (We are interested in specifying the *nature* of the phases, *not* their amounts. The overall composition can be any value in the two-phase field.)
2. Instead of temperature, let us fix the phase composition of α in equilibrium with L. This fixes temperature and the composition of the L phase, again by the tie line relationship.

Suppose an investigator reports a *field* where α, β, and L are present.

$$F = C - P + 1 = 2 - 3 + 1 = 0$$

The phase rule states that if two components and three phases are present, there is no degree of freedom. This is illustrated by the eutectic. At 577°C, the temperature fixed by nature, not by us, we have three phases of fixed phase composition in equilibrium (α: 1.65% silicon, L: 11.6% silicon, β: 99% silicon). Thus if we say $P = 3$, then $F = 0$, and all conditions are already fixed. We may change the *overall* composition horizontally along the line, but the composition of the individual phases (the phase rule variables) does not change. Therefore $\alpha + \beta + L$ does not occur in a field, but rather along a unique three-phase tie line—the eutectic in our example.

4.8 Complex phase diagrams

The most complex binary diagram is made up of just two types of lines, horizontal and nonhorizontal. Let us take up the nonhorizontal lines first. All of them are very simple, since they merely give the boundary between a one-phase field and a two-phase field. Let us check, for example, the copper-zinc diagram (Fig. 4.9) to convince ourselves of this. Two two-phase fields never touch along a vertical line, say $\alpha + \beta$ and $\beta + \gamma$, because this would mean that three phases, α, β, and γ, could exist over a range of temperatures at the composition of the dividing line. Varying the temperature would leave one degree of freedom to be specified, which is contrary to our previous phase rule

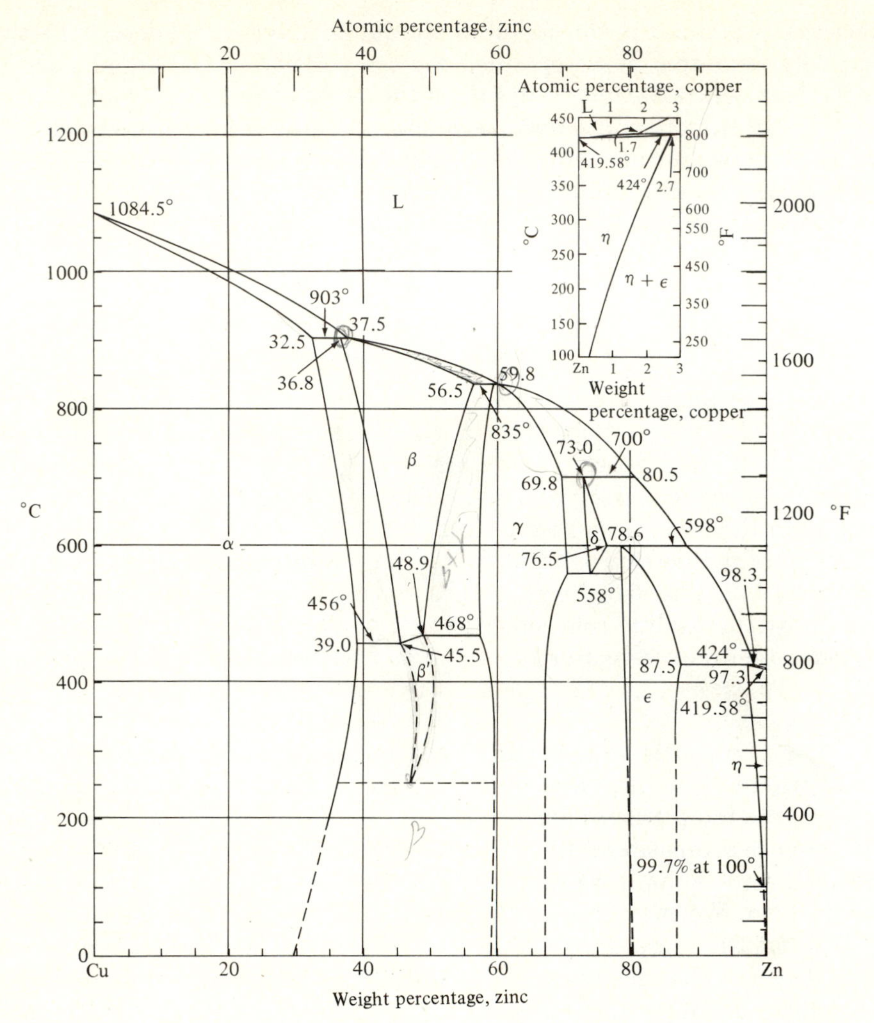

Fig. 4.9 *Copper–zinc phase diagram.* (American Society for Metals Handbook, 8th ed., Vol. 8: *Metallography, Structures and Phase Diagrams,* Metals Park, Ohio, 1973, p. 301)

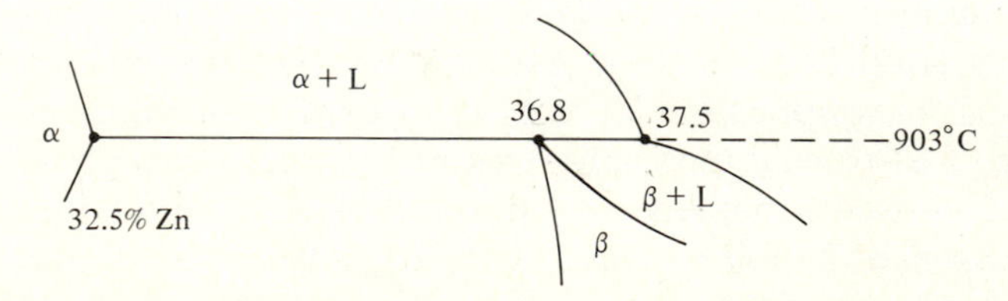

Fig. 4.10 *Peritectic transformation in copper-zinc diagram*

analysis for two components and three phases in equilibrium at 1 atm (F = 0). Note therefore that as we go across the phase diagram at a given temperature, we go from one- to two-phase fields alternately.

Now let us consider the horizontal lines in the copper-zinc diagram. At 903°C we have the situation shown in Fig. 4.10. Let us consider an alloy with 36.8% zinc at 902°C containing β. If this is heated to 904°C, we find α + L. Thus one solid phase has transformed to a new solid phase plus liquid. When we cool the same material, we have a solid α (S_1) that reacts with liquid to form one new solid S_2:

$$\underset{(\alpha)}{S_1} + L \xrightarrow{\text{cooling}} \underset{(\beta)}{S_2}$$

This is different from the *eutectic* reaction in the 11.6% silicon alloy in the aluminum-silicon system which reacts on cooling:

$$L \xrightarrow{\text{cooling}} \underset{(\alpha)}{S_1} + \underset{(\beta)}{S_2}$$

We therefore give the $S_1 + L \xrightarrow{\text{cooling}} S_2$ reaction a different name: *peritectic*.

At 835°C we encounter the same type of reaction, $\beta + L \rightarrow \gamma$, at 59.8% zinc, and at 598°C we have $\delta + L \rightarrow \epsilon$ at 78.6% zinc.

Another type of reaction, $\delta \rightarrow \gamma + \epsilon$, is encountered on cooling an alloy of 74% zinc at 558°C. This is analogous to the eutectic reaction, but since it involves an all-solid transformation, it has the special name *eutectoid*.

There are a few other types of horizontal lines involving three phases which we will cover in the problems.

Let us make a few final comments about the copper-zinc phase diagram. First, there are several dotted lines that may be present on phase diagrams for two reasons. One possibility is that the exact position of the solubility lines is not firmly established. A second consideration, especially at lower temperatures, is that equilibrium may require extended times to achieve, and so the phase relationships may not be observed in reasonable lengths of time. An example of this is the horizontal lines at 250°C in the copper-zinc system. A 40% zinc alloy at room temperature would be found to consist of α and a small amount of β′ rather than α and γ as indicated by the equilibrium diagram. In theory equilibrium can take an infinite amount of time.

A second source of confusion in the copper-zinc system is the occurrence of β and β′. The primed notation refers to an ordered solid solution or preferred positions for the solute atom in the solvent lattice. The unprimed β occurs at high temperatures, precisely where we might expect to find a random or more disordered structure.

EXAMPLE 4.4 In a two-component system, why are there tie lines for three-phase equilibria but not for two-phase equilibria?

ANSWER From the phase rule,

$$F = C - P + 1 \text{ (constant pressure)} = 2 - P + 1 = 3 - P$$
$$F = 0 \quad \text{for 3 phases in equilibrium} \quad \text{and} \quad F = 1 \quad \text{for 2 phases in equilibrium}$$

When two phases are present, the one degree of freedom allows us to fix another variable. We might do this by fixing the temperature, or in other words by providing a tie line. However, since two phases exist over a temperature range, there are an infinite number of tie lines. It is unreasonable to include all of these on a phase diagram.

When three phases are in equilibrium, the temperature is fixed because there are zero degrees of freedom. The tie line, tying together the three phase compositions, is no longer variable and therefore appears on the phase diagram.

4.9 Ternary diagrams

So far we have discussed only systems with two components, leading to binary diagrams. In many alloys we have three principal elements present, as in 18/8 stainless steel, which contains 18% chromium, 8% nickel, and about 74% iron. Let us compare the degrees of freedom of this system with the binary system:

Binary: $F = C - P + 2 = 4 - P$
Ternary: $F = C - P + 2 = 5 - P$

If we fix pressure at 1 atm in the binary system, $F = 3 - P$. Thus if we specify that three phases are in equilibrium, we have nothing else to specify; the temperature and phase compositions are fixed by nature.

However, in a ternary system under the same conditions (fixed pressure, $F = 4 - P$), we can encounter three phases over a range of temperatures and four phases at a given temperature.

Obviously we need some sort of three-dimensional map to represent this added variable. We use a triangular graph (Fig. 4.11).

Let us consider first the representation of composition at constant temperature, then the added dimension for varying temperature. Perhaps the quickest way to visualize the net shown is as a three-sided football field. The point at the iron corner is 100% iron. Any point on the 90% iron line contains that amount of iron. Next let us locate an 18% chromium, 8% nickel, and 74% iron stainless steel called 18/8. The 74% iron line is shown, and the composition must lie somewhere on this line.

Next we recognize that the chromium corner represents 100% chromium. We move away to find the 18% chromium line. The point where this intersects the 74% iron line is 18% chromium and 74% iron. We do not need

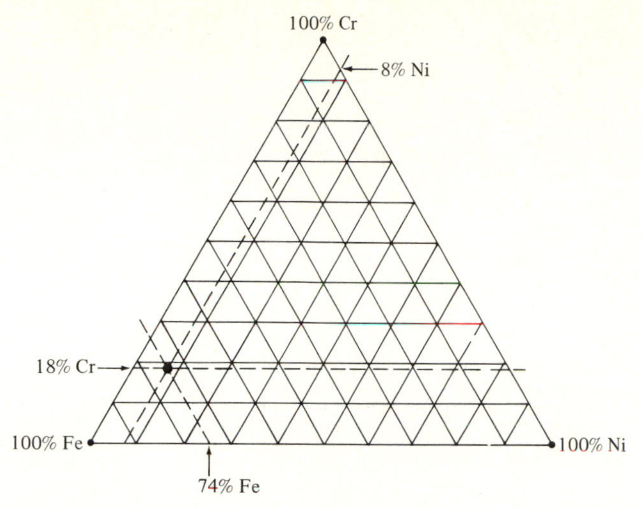

Fig. 4.11 *Location of a point in the iron–nickel–chromium ternary diagram.*

to draw the 8% nickel line except as a check, because this is determined by the difference from 100% (that is, % Ni = 100 − 74% Fe − 18% Cr).

RULE. To find a composition, locate the proper isocomposition lines for each element starting from that element's corner of the triangle.

4.10 Representation of temperature in ternaries

A complete three-component or ternary diagram at constant pressure but varying temperature is shown in Fig. 4.12 (next page).

In some cases the determination of the liquidus can be a million-dollar research problem. For example, during pig iron production in a blast furnace, the slag must be kept liquid, or a very expensive shutdown will occur. The principal components of the ternary diagram for the slag are three oxides, CaO, SiO_2, and Al_2O_3, instead of three metals. The million-dollar liquidus is shown in Fig. 4.13. The significance is that a low melting combination occurs in the range 49% CaO, 39% SiO_2, 12% Al_2O_3 at 1315°C. Knowing this, the metallurgist adds the right amounts of oxides to the ore to avoid a freeze-up and, equally important, to give a fluid slag which is active in removing impurities such as sulfur.

The ternary diagram is equally important in dealing with metals. A third element can be added to lower the melting point of an alloy. For example, an alloy of 51% bismuth, 40% lead, and 9% cadmium will melt in boiling water (212°F, 100°C), whereas the individual melting points are: bismuth: 520°F (271°C); lead: 621°F (327°C); cadmium: 609°F (320°C). This type of alloy is used in plugs in automatic sprinkler heads that melt when a fire occurs.

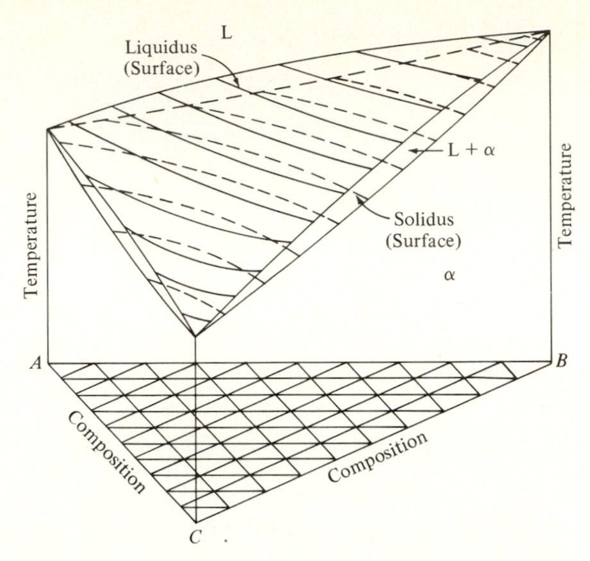

Fig. 4.12 *Temperature–composition space diagram of a ternary isomorphous system (complete solubility in the solid state).* (F. Rhines, *Phase Diagrams in Metallurgy*, McGraw-Hill, New York, 1956. Used with permission of McGraw-Hill Book Company)

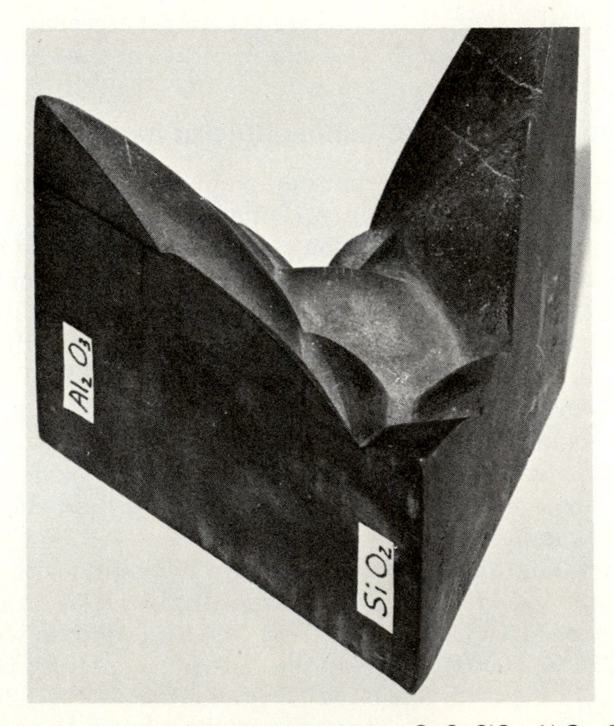

Fig. 4.13 *Liquidus surface of the ternary system* CaO–SiO_2–Al_2O_3. *The constituent with the highest melting point is CaO in the right rear of the photograph. The horizontal lines or planes would be constant temperature contours.*

4.11 Nonequilibrium conditions

All we have learned about reading phase diagrams to determine which phases are present at a given temperature applies to equilibrium conditions. We must also consider nonequilibrium conditions, as we see from the following example. One of the most important structures in metallurgy—the microstructure in hardened steel parts such as drills, saws, and roller bearings—is called "martensite." It cannot be found on any phase diagram. On the other hand, it cannot be produced under controlled conditions without a knowledge of the iron-carbon phase diagram, because to produce it one must first produce a stable phase called "austenite" (FCC iron with carbon in interstitial solid solution) at high temperature.

There are three fundamentals to be presented that bear directly on nonequilibrium conditions: (1) diffusion, (2) nucleation and growth, and (3) segregation effects.

4.12 Diffusion phenomena

Diffusion is the migration of atoms; it depends on time and temperature. Let us look at examples in the gaseous, liquid, and solid states.

If we open a partition between a glass chamber filled with nitrogen gas and another filled with bromine, we can see the brown color of the bromine become fainter as the gas molecules migrate and intermix. If we carefully pour a layer of water and a layer of alcohol in a vertical cylinder, the two liquids will become one after a time. Similarly, in the solid state, if we take a U.S. quarter (made up of layers of copper and a copper-nickel alloy) and heat it at an elevated temperature, the atoms will interdiffuse.

There are two principal types of diffusion: interstitial and vacancy or substitutional (Fig. 4.14). In interstitial diffusion small atoms such as hydrogen, carbon, and nitrogen are involved. These jump from one interstitial position to another. This is the most rapid type of diffusion, because most interstitial sites or positions are empty. Substitutional atoms diffuse more slowly, since the atoms have to wait for a vacancy to jump into. Even in the most carefully prepared single crystal there is an equilibrium number of va-

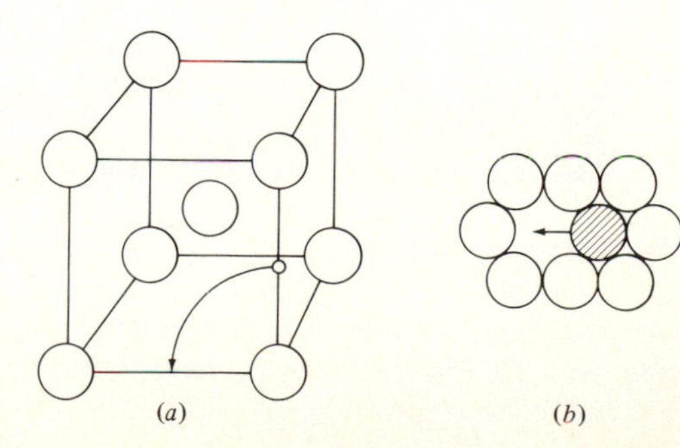

Fig. 4.14 *Mechanisms of (a) interstitial and (b) substitutional diffusion. In substitution, the atom moves into a vacancy normally considered to be a substitutional atom position and not an interstitial position.*

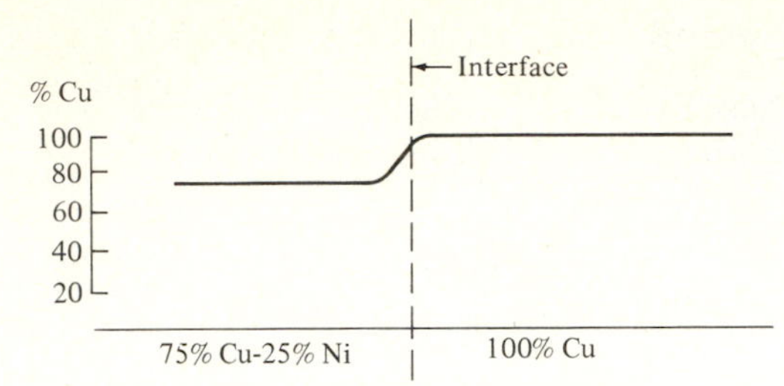

Fig. 4.15 *Analysis of the interface of a U.S. quarter after diffusion*

cancies. This number increases with temperature. In addition, grain boundaries and dislocations provide diffusion paths.

Let us review some of the many practical cases of diffusion, such as bimetals. Diffusion makes it possible to roll a metal sandwich. Layers of copper-nickel alloy over a copper core give us the U.S. quarter. Because of diffusion, the bond between the layers is not mechanical but a true metallurgical bond in which the composition changes gradually across the interface (Fig. 4.15). Later we shall discuss "aluminum-clad" aircraft materials, in which a layer of pure aluminum protects the less corrosion-resistant alloy beneath.

Another somewhat different example is a carburized gear (Fig. 4.16). In this case there is a gradual change from high carbon at the surface to low-carbon steel in the interior. This gives high hardness and good wear resistance at the surface, supported by the tough low-carbon steel beneath the surface. The carbon gradient is produced by heating the part in a carbon-rich atmosphere. The carbon atoms dissolve and build up in the surface layers, and a portion diffuses to the layers beneath.

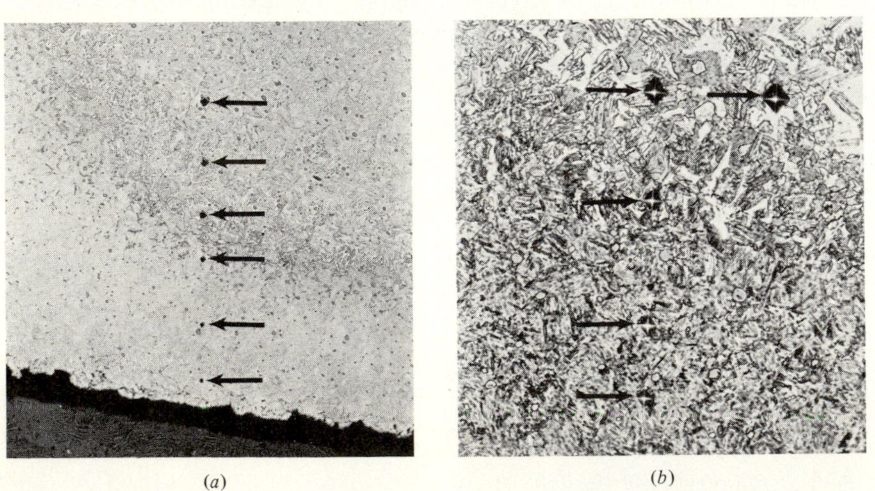

Fig. 4.16 *Microstructure and microhardness survey of a steel gear that was carburized and then quenched. Surface layers are martensite, but soft regions contain ferrite. (a) VHN (reading upward from bottom surface): 740, 710, 390, 220, 193, 171; 100×; 2% nital etch. (b) Transition region: martensite at bottom, martensite and ferrite at top; 500×; 2% nital etch.*

Diffusion processes are also important in many cases involving change in the internal structure of a part, not merely the surface layers. The most important effect of heat-treating steel is the change from a one-phase to a two-phase structure. The precipitation of the second phase, iron carbide, depends on the movement of atoms by diffusion. The controlled precipitation of hardening phases in nonferrous alloys also depends on movement by diffusion. In materials such as ceramics, the diffusion of ions is very important.

4.13 Diffusion equations; Fick's First Law

Three principal relationships govern diffusion rate: Fick's first and second laws, and the variation of the diffusion constant with temperature.

Fick's First Law Fick's first law describes the diffusion of an element under steady-state conditions. A simple case is the rate of loss of nitrogen under constant pressure through the side walls of a container (Fig. 4.17). This is analogous to a more familiar case: heat loss through the wall of a house in winter (Fig. 4.18). The rate of heat loss (or the thermal flux) through the wall is

$$\text{Heat flux} = J = K\frac{\Delta T}{\Delta X}$$

Heat flux is defined as energy per square foot (or square meter) of area per second. The driving force for heat flow is the thermal gradient $\Delta T/\Delta X$. K (thermal conductivity) is a constant that depends on the material of the wall. Similarly, the flow of nitrogen through a wall of thickness X is

$$N_2 \text{ flux} = -D\frac{\Delta C}{\Delta X}$$

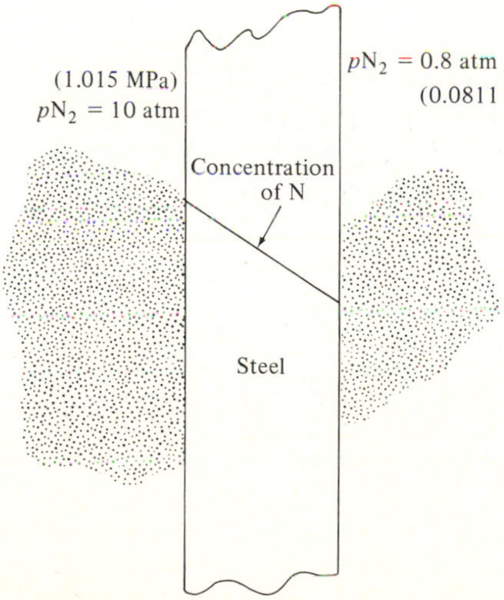

(1.015 MPa)
pN_2 = 10 atm
pN_2 = 0.8 atm
(0.0811 MPa)
Concentration of N
Steel

Fig. 4.17 *Diffusion of nitrogen through a steel wall*

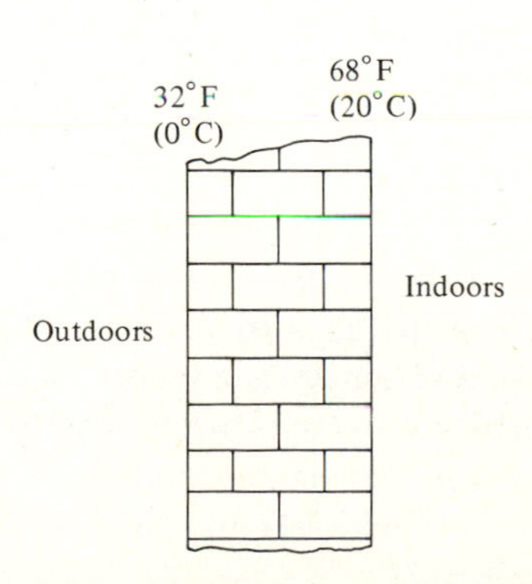

Fig. 4.18 *Differences in temperature between the indoors in a heated building and the outdoors*

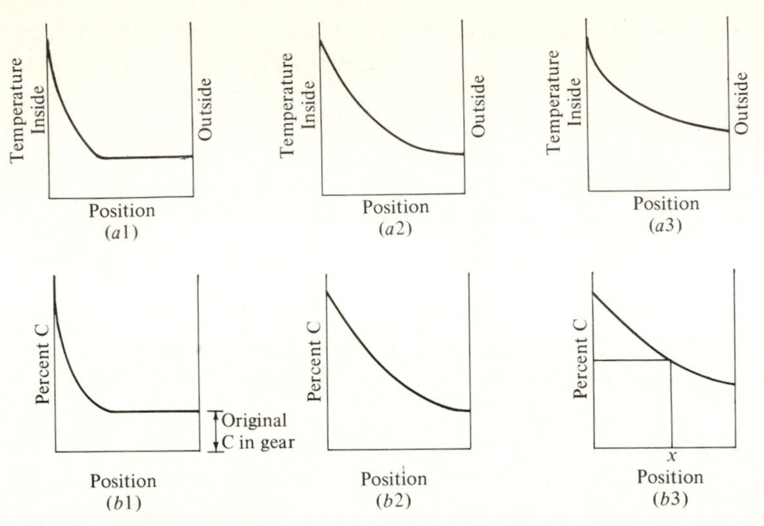

Fig. 4.19 *(a) Unsteady-state heat transfer in heating the wall of a house. (a1) After fire lit, t_0. (a2) after time t_1. (a3) After time t_2. (b) Unsteady-state diffusion in carburizing a gear. (b1) Time = t_0. (b2) Time = t_1. (b3) Time = t_2. (In diffusion, the curves may show different shapes, depending on placement of substitutional or interstitial atoms and on dependence of D on the chemical composition.)*

where $\Delta C/\Delta X$ is the concentration gradient instead of the temperature gradient and D, the diffusivity, is a constant that depends on the material. (We use a minus sign because the diffusion is in the direction opposite to the concentration gradient.)

4.14 Fick's second law

So far we have discussed only diffusion under steady-state conditions, which are rarely encountered. To return briefly to the analogy of a house, we have considered heat flow *after* the inside wall has been brought up to constant temperature (20°C, 68°F). Now suppose we light the fireplace on our return from a long vacation. The wall surface of the room will reach 20°C (68°F) rather quickly, but the interior of the wall will take some time to heat up (Fig. 4.19*a*). In the same way, if we place a steel gear in a carbon-rich furnace atmosphere, the surface reaches a high carbon level quickly, but the carbon diffuses inward as a function of time (Fig. 4.19*b*).

If we desire a carbon content C_x at a depth x in order to provide adequate wear resistance and strength, we need to wait a time t at the furnace temperature used. The relation governing these conditions is a solution of Fick's second law:

$$\frac{C_s - C_x}{C_s - C_0} = \operatorname{erf}\left(\frac{x}{2\sqrt{Dt}}\right)$$

where C_s = surface concentration of carbon produced immediately by the atmosphere (a constant), C_0 = initial uniform concentration of carbon throughout steel, C_x = concentration of carbon at distance x from surface at time t, x = distance from surface, D = diffusivity (described earlier), and t = time.

The quantity erf is called the *error function*. It can be found from standard tables (Table 4.1) or from a graph. In other words, we substitute numerical values for x and Dt and then find the error function of the result.

Table 4.1 TABLE OF THE ERROR FUNCTION

z	erf(z)	z	erf(z)	z	erf(z)	z	erf(z)
0	0	0.40	0.4284	0.85	0.7707	1.6	0.9763
0.025	0.0282	0.45	0.4755	0.90	0.7970	1.7	0.9838
0.05	0.0564	0.50	0.5205	0.95	0.8209	1.8	0.9891
0.10	0.1125	0.55	0.5633	1.0	0.8427	1.9	0.9928
0.15	0.1680	0.60	0.6039	1.1	0.8802	2.0	0.9953
0.20	0.2227	0.65	0.6420	1.2	0.9103	2.2	0.9981
0.25	0.2763	0.70	0.6778	1.3	0.9340	2.4	0.9993
0.30	0.3286	0.75	0.7112	1.4	0.9523	2.6	0.9998
0.35	0.3794	0.80	0.7421	1.5	0.9661	2.8	0.9999

EXAMPLE 4.5 Many sliding and rotating parts such as gears call for a hard structure in the surface layers, backed by a tough structure in the interior. The first step in the process is to diffuse carbon into the surface of a steel, raising the level from about 0.2% carbon, the original carbon level, to 0.5 to 0.9% carbon for 0.005 to 0.050 in. (0.0127 to 0.127 cm).

If we place the gear in a furnace at 1000°C with an atmosphere rich in hydrocarbon gas, the surface reaches a carbon content of about 0.9% very rapidly. The carbon content beneath the surface then rises gradually as a function of time. Calculate the carbon content at 0.010 in. (0.0254 cm) beneath the surface after 10 hr (36,000 sec) at 1000°C.

ANSWER

$$\frac{C_s - C_x}{C_s - C_0} = \text{erf}\left(\frac{x}{2\sqrt{Dt}}\right)$$

$C_s = 0.9 \quad C_x$ = desired value

$C_0 = 0.2$ (assume an SAE 1020 steel is used)

$D = 0.298 \times 10^{-6}$ cm²/sec (from Example 4.7)

Let x = 0.01 in. = 0.0254 cm, since D is in cm²/sec. Then

$$\frac{0.9 - C_x}{0.9 - 0.2} = \text{erf}\left(\frac{0.0254 \text{ cm}}{2\sqrt{(0.298 \times 10^{-6} \text{ cm}^2/\text{sec})(3.6 \times 10^4 \text{ sec})}}\right)$$

$$= \text{erf } 0.123$$

We need the error function of the number 0.123. Interpolate from Table 4.1.

z	erf z
0.150	0.1680
0.123	x
0.100	0.1125

$$\frac{0.150 - 0.123}{0.150 - 0.100} = \frac{0.1680 - x}{0.1680 - 0.1125} \quad \text{or } x = 0.1380$$

Therefore

$$\frac{0.9 - C_x}{0.7} = 0.138 \qquad C_x = 0.803$$

EXAMPLE 4.6 In Example 4.5, calculate the time necessary to raise the carbon level to 0.60% at 0.010 in. (0.0254 cm) beneath the surface. The diffusion coefficient is again 0.298×10^{-6} cm²/sec at 1000°C.

ANSWER

$$\frac{C_s - C_x}{C_s - C_0} = \text{erf}\left(\frac{x}{2\sqrt{Dt}}\right)$$

where $C_s = 0.9$, $C_0 = 0.2$, $C_x = 0.6$

$$\frac{0.9 - 0.6}{0.9 - 0.2} = \text{erf}\left(\frac{0.0254 \text{ cm}}{2\sqrt{(0.298 \times 10^{-6} \text{ cm}^2/\text{sec}) \times t}}\right)$$

$$0.4286 = \text{erf}\left(\frac{23.26}{\sqrt{t}}\right)$$

We need a number Z whose error function is 0.4286. From Table 4.1, we find that this number is 0.40. Therefore

$$0.40 = \frac{23.26}{\sqrt{t}} \quad \text{or} \quad t = 3{,}381 \text{ sec} = 56.35 \text{ min (approximately 1 hr)}$$

Lowering the carbon requirement from 0.80 to 0.60 at 0.010 in. (0.0254 cm) below the surface decreases the furnace time from 10 hr to 1 hr, a considerable saving in time. Chapter 6 will treat the carbon requirements for this diffusion process.

4.15 Effects of temperature

In Fick's first and second laws we see that the movement of the diffusing material is proportional to the diffusivity D. We would expect D to increase with temperature, since the atomic motion and number of vacancies both increase. The relationship is found to obey the equation

$$D = Ae^{-Q/(RT)}$$

where A = constant, Q = constant for the diffusing substance and solvent involved, R = gas constant (1.987 cal/mole-K), and T = absolute temperature (K). Therefore, taking $\log_e$ of both sides, we have

$$\ln D = \ln A - \frac{Q}{RT}$$

which is the equation of a straight line if we plot $\ln D$ vs. $1/T$.

This approach is important because if we determine D for only two temperatures, we can solve for A and Q and get a general relation for any temperature. In a practical case, if we wish to estimate the effect of changing a furnace temperature for carburizing a gear, we can calculate the time needed at the new temperature to obtain equivalent results.

EXAMPLE 4.7 Calculate D for 1000°C as used in the previous example, given $A = 0.25\ \text{cm}^2/\text{sec}$ and $Q = 34{,}500$ cal/mole for the diffusion of carbon in γ iron.

ANSWER

$$D = (0.25\ \text{cm}^2/\text{sec})e^{-\frac{34{,}500\ \text{cal/mole}}{(1.987\ \text{cal/mole-K})(1273\ \text{K})}} = \frac{0.25}{e^{13.64}}$$

since

$$\ln e^{13.64} = 13.64,$$

$$\log_{10} e^{13.64} = \frac{13.64}{2.3} = 5.92 \qquad \text{antilog } 5.92 = 8.38 \times 10^5$$

$$e^{13.64} = 8.38 \times 10^5$$

then

$$D = \frac{0.25}{8.38 \times 10^5} = 0.298 \times 10^{-6}\ \text{cm}^2/\text{sec}$$

4.16 Other diffusion phenomena

It is interesting to analyze the value of D for different combinations of materials (Fig. 4.20). In particular, elements of small diameters that form interstitial solid solutions diffuse very rapidly. The faster diffusion rate of carbon in BCC iron compared with FCC iron seems paradoxical until we recall that the BCC cell is a less dense structure. Therefore, although the FCC cell centers have more room for the carbon atoms, the passageways through the unit cell are tighter for carbon movement.

The role of dislocations and grain boundaries is also of interest. The looser packing leads to a more rapid movement of the diffusing substance, and this can lead to a factor as high as 10:1 in penetration along grain boundaries.

4.17 Nucleation and growth

It is important to re-emphasize that we frequently find phases that are not indicated by the phase diagram, i.e., phases that are out of equilibrium. In discussing the aluminum-silicon diagram, we mentioned that we could quench

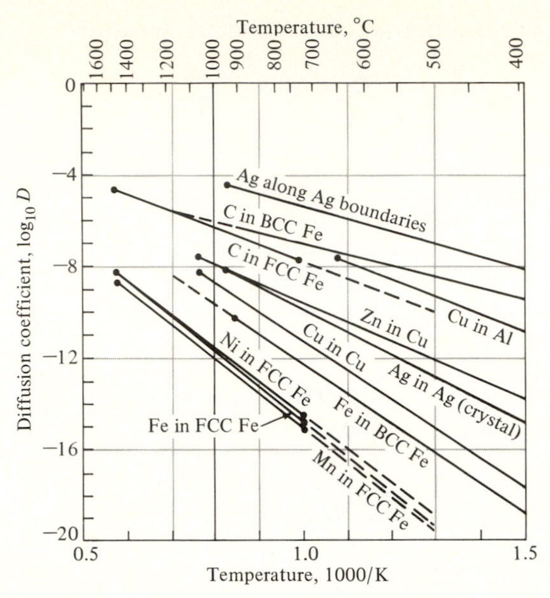

Fig. 4.20 *Variation in diffusivity D as a function of temperature for various materials (units of D are cm²/sec).* (From L.H. Van Vlack, *Elements of Materials Science,* 2d ed., Addison-Wesley Publishing Company, Reading, Mass., 1964)

an all-α alloy from a high temperature and retain the same structure at room temperature even though the equilibrium diagram called for α + β at 0.4% silicon at 20°C. The reason is that silicon atoms must diffuse from their random solid solution in the FCC lattice to form the silicon crystals with a new structure. In addition to requiring time for diffusion, the material must form a nucleus.

The problem of nucleation has wide application. For example, everyone is familiar with experiments in cloud seeding in which silver iodide crystals are supplied as nuclei. As another example, in a special container pure water can be cooled to −30°C without freezing. Also, common metals such as nickel can be cooled to 200°C below the equilibrium freezing point without solidifying. The usual impression is that this supercooling or undercooling is done by fast cooling, but it is important to understand that these liquids can be maintained for a period of time at these low temperatures.

The reason for these effects lies in the important phenomenon of nucleation. Once a liquid is cooled below its equilibrium freezing temperature, there is a driving force for solid to precipitate. This force is measured as the difference in bulk free energy between the liquid and the solid. For instance, for a *spherical* volume the difference is [(free energy of liquid/cm³) − (free energy of solid/cm³)] × (volume of sphere, cm³) or $\Delta F_V \, \frac{4}{3}\pi r^3$.

Bulk free energy is a quantity precisely defined in chemistry. For those unfamiliar with the concept, a mechanical analogy may be useful. Consider

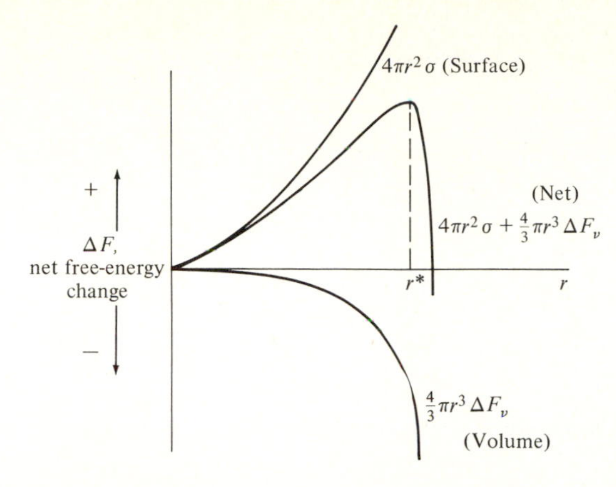

Fig. 4.21 *Change in free energy as a function of the radius of the nucleus*

an old-fashioned screen door that slams shut when released because of an overhead spring. As we open the door, tension builds up in the spring. The further the door is opened, the greater the energy in the spring for shutting. When a liquid is cooled below the equilibrium freezing temperature, the greater the undercooling, the greater the driving force for solidification (or the greater the bulk free energy).

However, there is a force to be overcome in growing a solid sphere of radius r in place of the liquid. A new surface has to be formed. We need something like the energy required to blow a soap bubble against surface-tension effects. The energy required to create a new surface for the sphere is equal to the surface area times the surface tension, or $4\pi r^2\sigma$, where σ is the surface tension in ergs per square centimeter of surface area.

If we plot these two energies and their sum, we obtain the curve shown in Fig. 4.21. Since the difference in free energy or bulk free energy is a cubic function, whereas the surface energy is a quadratic equation, their sum results in an inflection point, which we call r^*. We recall that in nature a system tends to minimize its energy. Therefore, for r values less than r^*, the easiest way to go to lower energy is to have r values become still smaller, or to have the nucleus dissolve. On the other hand, for values of r^* or greater, minimum energy is obtained by an increase in r, so that the nucleus will grow. These are two important cases.

CASE 1. If no nuclei are present in the melt, it will have to be supercooled greatly for nucleation to occur. The lower the temperature, the greater the difference in bulk free energy and the greater the driving force to transform. (The surface tension does not change appreciably with temperature.) With great undercooling, r^* becomes very small and *homogeneous* nucleation takes place.

CASE 2. If we introduce a solid nucleus of radius greater than r^*, it will grow. This nucleus may be another material that is "wet" by the liquid and hence acts like a nucleus of the metal. This is called *heterogeneous*

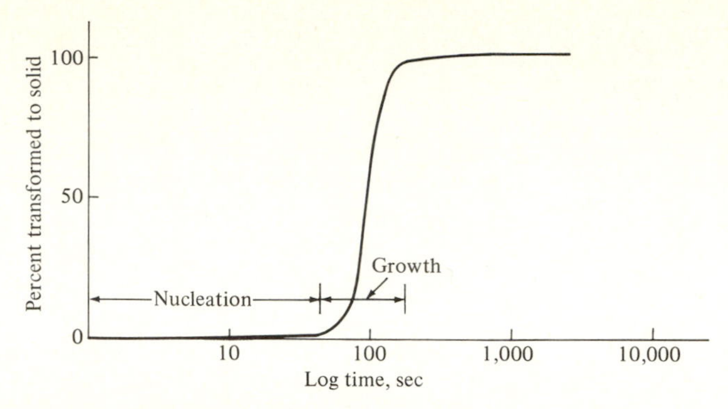

Fig. 4.22 *Representation of a liquid transforming to a solid as a function of time*

nucleation. Because this nucleus is usually relatively large, it produces freezing close to the equilibrium temperature.

Once a nucleus or nuclei are present, growth occurs. The typical growth curve is shown in Fig. 4.22. At first the rate of growth is slow because there is limited solid surface. Then, as the surface increases, the growth curve rises rapidly. Finally the rate of growth slows down because solid surfaces come in contact, reducing the area of the solid-liquid interface.

These nucleation and growth phenomena are also present in solid–solid reactions. One additional nucleation effect of importance also affects the structure of the solid precipitate. The solid often comes out in the form of needles or plates rather than spheres or cubes. The reason is that an additional effect operates against the growth of the nucleus. This effect is related to the volume change in going from one solid to another. If the second solid is more voluminous than the first, the nucleus builds up compressive stress. This effect is diminished if the nucleus is needlelike in shape instead of spherical. Therefore, in transformations that occur at low temperatures, where the structure is high strength, the precipitate is usually in the form of needles or plates.

A practical application is found in the control of the size and shape of the silicon crystals in the aluminum-silicon engine block. For many years hypereutectic (16% silicon) alloys could not be used because large, hard, bladelike silicon crystals formed from the liquid. These were difficult to machine and gave the block a rough surface. Finally it was found that if phosphorus was introduced, many small aluminum phosphide crystals formed. These in turn nucleated many small silicon crystals, giving the desired refinement.

4.18 Segregation

The key to understanding the phenomenon of segregation is, first, to realize that the crystals that freeze initially are different in composition from the liquid. Second, to satisfy equilibrium conditions, the composition of the solid that freezes originally must change by diffusion as freezing continues. Figure 4.23 illustrates these points.

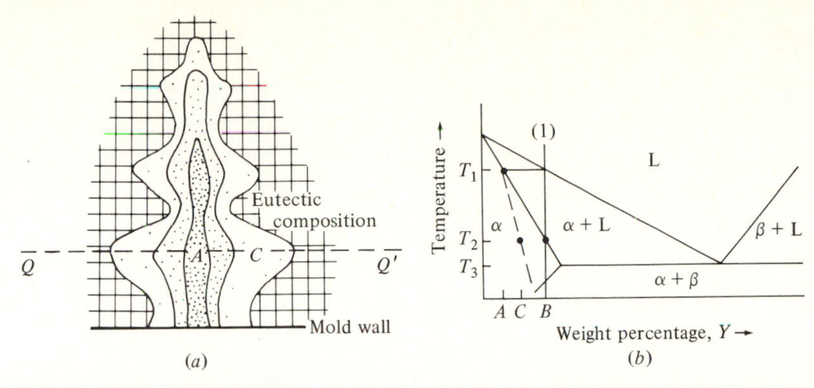

Fig. 4.23 *(a) Segregation in dendrite growth. (b) Phase diagram illustrating segregation. The dashed line shows the new solidus temperature.*

If we start freezing overall composition (1) at T_1 (Fig. 4.23*b*), the first crystals will have composition *A*. Under equilibrium at a lower temperature T_2 *all* the solid will have composition *B*. However, if there is not time for diffusion to occur throughout the solid, the average composition of the solid will be *C* at T_2. The solid will have a cored structure, as shown in Fig. 4.23*a*. One consequence is that freezing will not be complete when temperature T_2 is reached, and we can reach the eutectic temperature T_3 with eutectic liquid present. This phenomenon is called *segregation*. A survey of chemical analysis would show the features indicated in Fig. 4.24.

At times the last liquid to freeze will be forced between the crystals from the interior to the surface of the ingot or casting, causing *inverse segregation*. Often droplets of eutectic are exuded from the surface.

In cast ingots that are subsequently rolled, it is possible to improve on the as-cast structure. It is important first to avoid shrinkage cavities in the ingot. This is done in one of two ways. First, a "hot top," which is similar to a riser on a casting, may be used. (These are liquid metal reservoirs which freeze last and supply liquid metal to fill voids caused by liquid-solid shrinkage.) Second, the ingot is allowed to freeze with a deliberate evolution of gas (called *rimming*). The ingot then contains a number of sealed-in gas voids

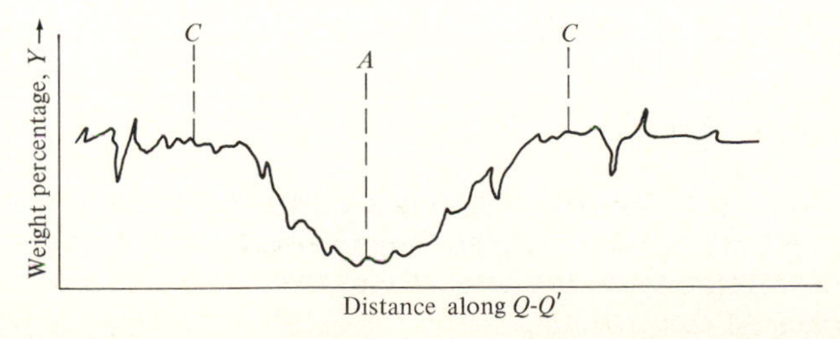

Fig. 4.24 *Variation in composition in Fig.4.23 due to nonequilibrium cooling*

instead of shrinkage cavities. These voids are later closed by the mechanical action of the rolls. Segregation is still present in the rolled material. For specialty steels several processes have been developed to minimize it. In one method an electrode of the desired analysis is prepared, then an ingot is cast by striking an arc between the electrode and the base of the ingot mold. The metal is cast almost drop by drop as the electrode melts and the metal falls into the water-cooled mold.

4.19 Strengthening by nonequilibrium reactions

From the preceding discussion we see that we need not accept the combinations of structures obtained under equilibrium conditions following the phase diagram. We can get improved structures by proper processing at a number of points in the manufacture of a component. Methods of doing so include:

1. Control of the liquid-to-solid reaction (nucleation and chilling effects)
2. Control of distribution and fineness of products in reactions involving precipitation of one or more phases from a *solid* matrix (age hardening, dispersion hardening, martensite-type reactions)

Let us consider these in order.

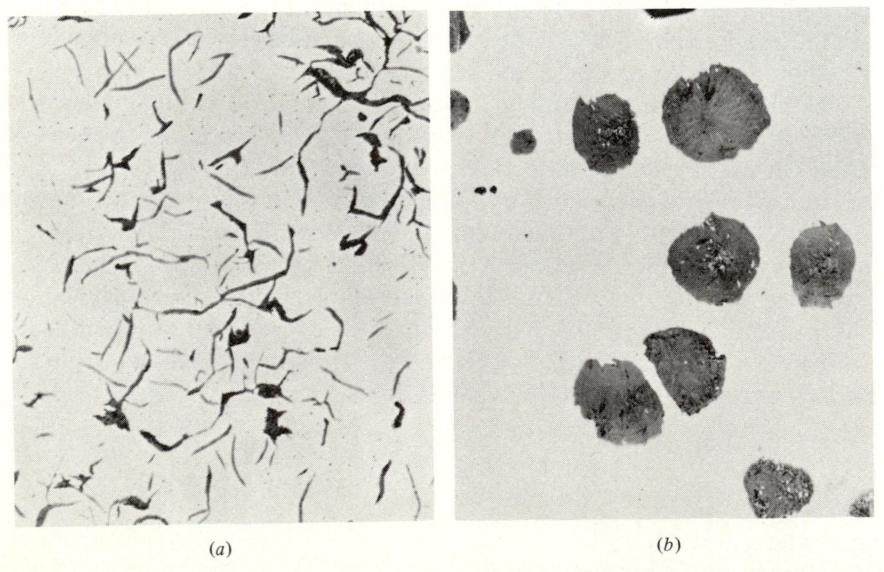

(*a*) (*b*)

Fig. 4.25 *(a) Flake graphite in gray cast iron; 100×, unetched. (b) Spheroidal graphite in ductile cast iron; 100×, unetched. Both irons have approximately the same analysis (3.5% carbon, 2.5% silicon), but the graphite shape is spheroidal in (b) because of the presence of 0.05% magnesium. As a result, the strength and ductility of this iron are more than twice as great as those of (a).*

4.20 Control of liquid-to-solid reactions

We have already referred to the nucleation of silicon in aluminum. Another change of great engineering importance is the change in the shape of graphite particles in iron alloys. In normal gray cast iron, the graphite precipitates in flakelike particles (Fig. 4.25*a*). The matrix of the material is similar to steel with good ductility, but the flakes act as notches and the ductility is below 1%. Adding a small amount of magnesium (0.05%) causes the graphite to crystallize in the form of spheres (Fig. 4.25*b*). The ductility is raised to as high as 20% and the strength is increased several times. This material, discussed in detail in Chapter 6, is known as "ductile" or "nodular" iron.

4.21 Control of solid-state precipitation reactions

As examples of this effect, let us discuss the very important control of precipitation of iron carbide in steel and the age hardening of aluminum alloys by the $CuAl_2$ phase.

There are two heat-treatment cycles for the hardening of steel. Both refine the dispersion of iron carbide, but the mechanisms are quite different. The first is a time-dependent nucleation and growth reaction; the second is a rapid shearlike change in structure. Let us consider both these reactions in a typical steel containing 0.8% carbon.

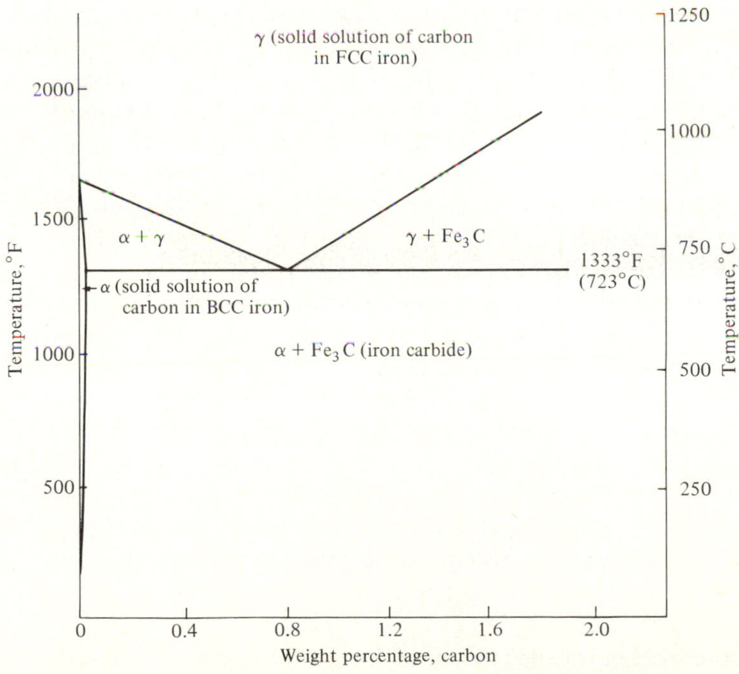

Fig. 4.26 *Section of the iron–iron carbide phase diagram*

Pearlite Formation Figure 4.26 shows the pertinent section of the iron–iron carbide phase diagram. Note that at 0.8% carbon a single phase, γ, is present above 1333°F (723°C). This consists of FCC iron with all the carbon in interstitial solid solution. Under equilibrium conditions, the diagram shows that on cooling below 1333°F (723°C), γ changes to two phases—α, which is essentially pure, soft BCC iron, and iron carbide, which is hard. This is called a *eutectoid reaction*. The two-phase mixture is called *pearlite*. Under slow-cooling conditions the dispersion of plates of Fe_3C in α is coarse (Fig. 4.27*a*), and the hardness low, VHN 230. If we heat a piece of the same steel into the γ range and then quench it in a bath of liquid lead or salt at 1000°F (538°C), the reaction will take place over a period of less than 1 min, and we will get a much finer dispersion (Fig. 4.27*b*) with a higher hardness, VHN 300. We can obtain still finer dispersion and further increases in hardness and strength by transforming the γ to α plus carbide at still lower temperatures. Figure 4.28 shows the effect of these changes on properties. The strength increases as the spacing† between carbide particles decreases.

Martensite Reaction The second heat-treatment cycle is more complex. If the 0.8% carbon steel, originally all γ, is quenched quickly below 400°F (205°C), a very rapid shearlike change in structure occurs. Instead of α plus carbide, a single new phase, called *martensite,* is formed. The structure is BCT (body-centered tetragonal), which can be considered an elongated BCC (Fig. 4.29). The carbon is still distributed in atomic form as in γ, but in a supersaturated solid solution. When the martensite is reheated at a relatively low temperature [400°F (205°C), for example], the carbon precipitates as very fine iron carbide particles and the tetragonal iron structure changes to the normal BCC structure of iron. By following this processing we can obtain the finest carbide dispersions. This reaction therefore involves two steps: quenching to low temperature to form martensite and reheating (called "tempering") to form fine iron carbide.

Age Hardening As an example of this process, consider the hardening of aluminum-copper alloys by a three-step process. This process involves a solution treatment at about 950°F (510°C) in which (1) the $CuAl_2$ phase is dissolved in the matrix, (2) the alloy is quenched to room temperature, retaining the solid solution formed at high temperature, and (3) following the quench, the alloy is either aged naturally (i.e., held at room temperature), or aged artificially by heating at a relatively low temperature (200 to 400°F, 95 to 205°C). During the aging step the $CuAl_2$ precipitates and hardens the alloy. The final dispersion of the precipitate is much finer than the dispersion in the

†The average spacing is called the *mean free path*. We get a straight-line relationship if we plot the log of the mean free path against yield strength.

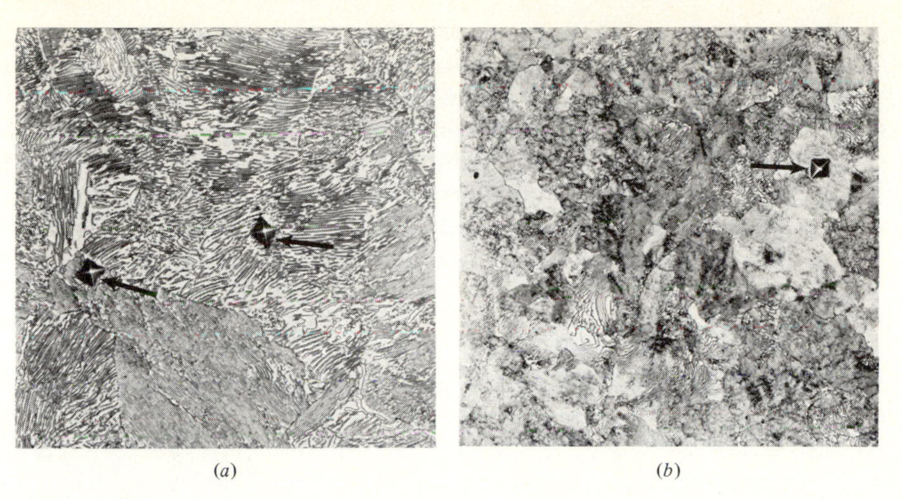

Fig. 4.27 *(a) Coarse iron–iron carbide distribution (coarse pearlite), VHN 196 to 266; 500× 2% nital etch. (b) Fine iron–iron carbide distribution (fine pearlite), VHN 270 to 320; 500×, 2% nital etch. Distribution of the finer carbide results in higher hardness.*

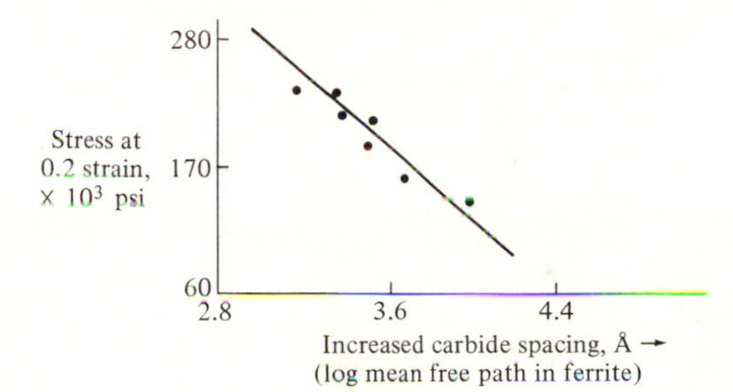

Fig. 4.28 *Effect of carbide spacing on yield strength.* (Gensamer et al.)

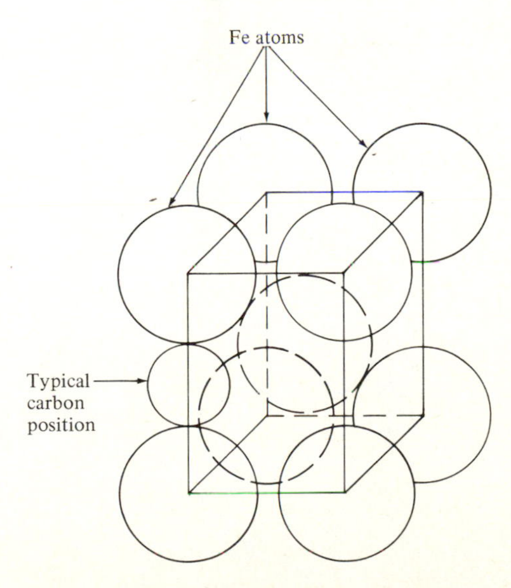

Fig. 4.29 *Body-centered tetragonal structure of martensite*

original ingot. Figure 4.30 shows the relation of these steps to the aluminum-copper phase diagram. The line between the α and α + θ fields (called the *solvus*) tells us that in the α phase a great deal more copper dissolves at elevated temperatures than at room temperature. After the alloy is quenched to room temperature, we can read the degree of supersaturation from the diagram. Since diffusion is slower at room temperature, the equilibrium amount of θ is slow to precipitate and the rate of hardening is slow. At an intermediate temperature we obtain more rapid hardening. However, if the

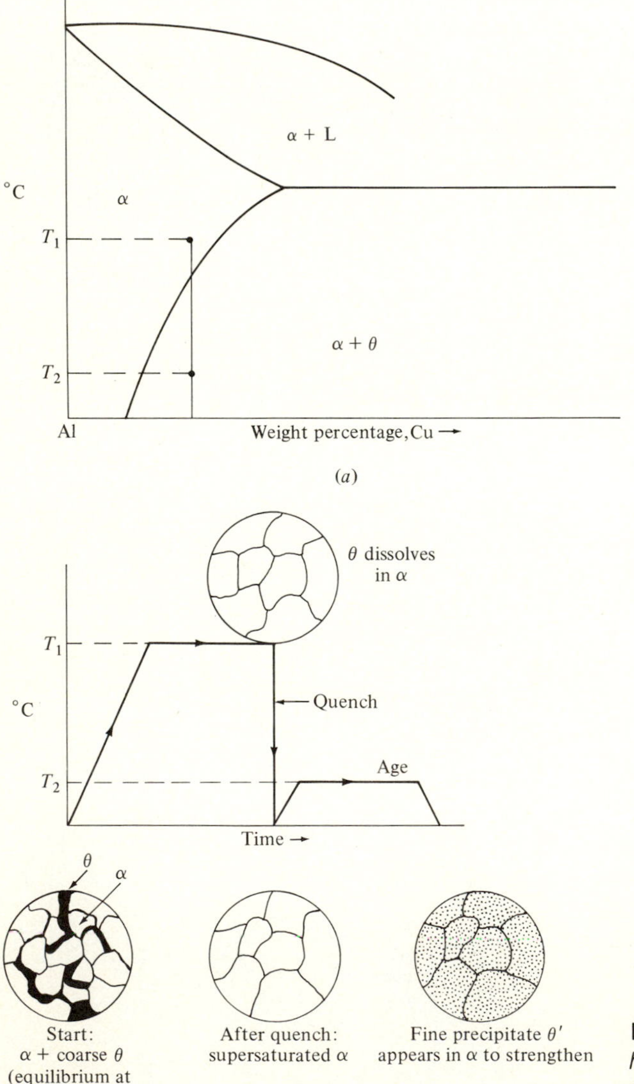

Fig. 4.30 *Age-hardening heat treatment. (a) Phase diagram. (b) Chart of heat treatment and microstructures.*

alloy is heated for extended periods or at too high a temperature, the precipitate coarsens and the hardening is diminished. This is called "over-aging" (Fig. 4.31). Note that not all precipitates produce this degree of hardening in the aged condition. A so-called "coherent" precipitate is obtained only when some atom spacing in the precipitate is close to a spacing in the matrix. A region of strained material then surrounds the precipitate particle, making it difficult for slip to occur in this region. Enhanced hardening results.

Also, the aluminum alloy in Fig. 4.31 (2014) contains not only 4.5% copper but also small quantities of magnesium and silicon (see Chapter 5). This produces two hardness peaks from different phases. The hardness at zero time (Fig. 4.31) reflects some early natural aging rather than merely the solution-treated hardness. The early aging hardness is partially lost when the alloy is artificially aged at the indicated temperatures.

Decreasing solid solubility on the phase diagram, as in Fig. 4.30, does not always guarantee coherent precipitation hardening. As mentioned above,

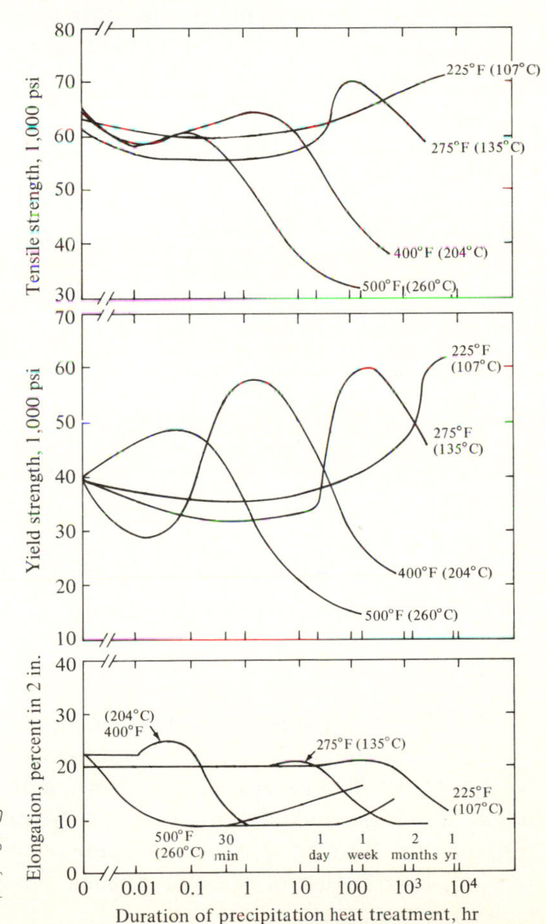

Fig. 4.31 *Effect of age hardening on the mechanical properties of a 4.5% copper–aluminum alloy (2014).* (American Society for Metals Handbook, 8th ed., Vol. 2: *Heat Treating, Cleaning and Finishing,* Metals Park, Ohio, 1964)

the strain developed between the matrix and the precipitate causes local resistance to slip, with correspondingly high strength and hardness when the precipitating phase is coherent. In overaging, the coherency strains are relieved, and therefore the hardness decreases.

In some systems the difference in atom spacing between the matrix and the second phase is too great, preventing the development of coherency. The hardness is therefore not as high and the process is referred to as *dispersion hardening*. Chapter 11 (Composite Materials) discusses the characteristics of dispersion strengthening further.

As might be expected, the most useful engineering alloys are those that produce coherency. However, even in the absence of coherency, the age-hardening–heat-treatment sequence may produce more desirable properties. For example, copper dissolves 1% chromium at high temperatures but less than 0.1% at room temperature. The chromium phase is not coherent; however, an age-hardening heat treatment provides a purer matrix and a more desirable distribution of chromium, giving higher electrical conductivity. This material is stronger than pure copper. One of its uses is for electrical resistance (spot welding) electrodes.

4.22 Summary of strengthening mechanisms in multiphase metals

As pointed out at the beginning of the chapter, the way to obtain desired mechanical properties is to control the nature, size, shape, amount, distribution, and orientation of the phases. This is accomplished by liquid-to-solid reactions or solid-to-solid reactions, which may be listed as follows:

A. Liquid to solid
   1. Variation in grain size
   2. Eutectic
   3. Change in phase shape (for example, graphite)
B. Solid to solid
   1. Eutectoid
   2. Order–disorder solid solutions
   3. Martensite
   4. Age hardening

(Phase shape and size may be superimposed on the above variables.)

The differences between age hardening and martensite reactions may be confusing. The reactions are shown schematically at the top of the opposite page. The important difference is the formation of a new crystal structure in martensite. Both processes use quenching and reheating to give a more desirable distribution of the two phases. Chapter 6 treats the martensite reaction in more detail.

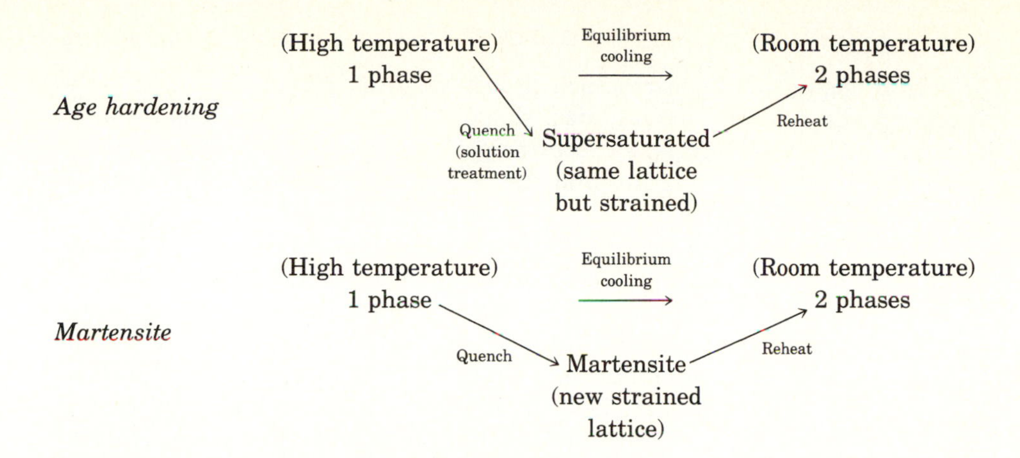

EXAMPLE 4.8 Four alloys have been identified on the accompanying phase diagram. Indicate the most probable strengthening mechanism(s) for each.

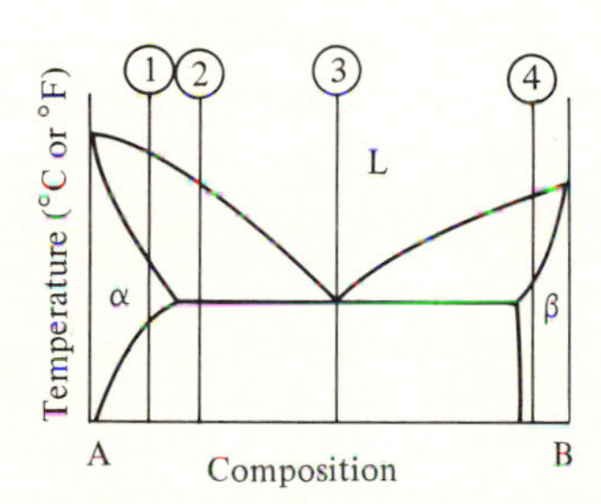

ANSWER Alloy ①: Age hardening likely because of decreasing solid solubility. If β is soft and ductile, a slow-cooled alloy may be work-hardened.

Alloy ②: At room temperature the microstructure would be composed of α plus a small amount of eutectic. The α *may still* be age-hardened after solution treatment just below the eutectic temperature.

Alloy ③: Strengthening is by the eutectic reaction. In general the ductility of eutectics is low. Therefore work hardening is not used. The α phase portion of the eutectic does not generally respond to age hardening because after solution treatment and aging, the β phase grows larger rather than forming new β within the α of the eutectic. In other words, diffusion distances between the α and β are very short, and growth takes less energy than nucleation.

Alloy ④: Since the alloy is single-phase and already solid solution–strengthened, we may only cold-work the alloy to increase its strength.

SUMMARY

We obtain the optimum mechanical properties from polyphase alloys when we control the nature, amount, size, shape, distribution, and orientation of the phases. The metallurgist's two principal tools for accomplishing this are the equilibrium phase diagram and the relations governing the kinetics of nonequilibrium conditions.

The phase diagram may be used as a map showing the phases present as functions of temperature and composition. For any overall composition we can read the order of separation of phases on cooling, as well as the *compositions* and *amounts* of the phases. Similarly, we can readily determine the phase compositions and amounts obtained on heating to a given temperature. This is a very powerful tool in itself. In addition, heating a structure such as γ iron to a known high temperature is often the prelude to a step involving nonequilibrium conditions, such as a rapid quench.

Before discussing the strengthening mechanisms involving nonequilibrium conditions, we described three important phenomena: diffusion, nucleation and growth, and segregation effects. Diffusion is important because it governs the time required for a strengthening precipitate to go into solution or to precipitate. Also, the hardening of a surface by diffusing carbon from the atmosphere, as in carburizing, is important.

Nucleation is important because in some cases the shape and size of a precipitate can be controlled by proper nucleation.

Segregation is significant because it may lead to the development of composition differences or even nonequilibrium phases.

The principal strengthening mechanisms described were the control of liquid-solid reactions, such as the production of spheroidal graphite in ductile iron, the pearlite reaction showing the effect of temperature on pearlite fineness, the martensite reaction, and age hardening. These are the reactions most often employed to strengthen the two-phase materials described in the chapters that follow.

DEFINITIONS

Age hardening A three-step process consisting of heating an alloy to dissolve all or part of a second phase in the matrix phase, quenching to retain the solute in a supersaturated solid solution, and finally allowing the second phase to precipitate in fine particles that are coherent with the matrix.

Bimetal A two-layer structure made by rolling or plating a layer of one metal or alloy on another.

Carburizing A process for adding carbon to the surface layers of an iron part.

Degree of freedom, F The number of variables that the experimenter can specify or control and still have the number of phases and components originally substituted in the phase-rule equation. For example, in the aluminum-silicon system, if two phases (α + L) are present, $F = 2 - 2 + 1 = 1$, and the temperature can be varied over a wide range without leaving the two-phase field (constant pressure).

Diffusion Migration of atoms in a solid, liquid, or gas. Even in a pure material atoms will change position. This is called *self-diffusion*.

Diffusivity, D The rate of diffusion of a substance at a given temperature. D varies exponentially with temperature. $D = Ae^{-Q/(RT)}$.

Eutectic composition The composition of a liquid that reacts to form two solids at the eutectic temperature. In the aluminum-silicon system this is 11.6% silicon, the balance aluminum. Note, however, that some liquid of eutectic composition will be obtained during the freezing of any alloy of overall composition between 1.65 and 99% silicon.

Eutectic temperature The temperature at which a liquid of eutectic composition freezes to form two solids simultaneously under equilibrium conditions. Also, the temperature at which a liquid and two solids are in equilibrium.

Heterogeneous nucleation The development of a new phase by the addition of foreign material (seeding).

Homogeneous nucleation The development of a new phase by the formation of nuclei of the new phase from the parent phase.

Interstitial diffusion Migration of atoms through a space lattice using interstitial positions.

Isomorphous Of the same structure. Note in Fig. 4.12 that there is only one solid phase α; hence only one structure is present. However, the lattice parameter will change continuously with composition.

Liquidus The temperature at which a liquid *begins* to freeze under equilibrium conditions (solid first forms).

Martensite A metastable, body-centered tetragonal phase formed by quenching γ iron containing carbon.

Matrix The continuous phase in a two-phase material. It usually forms the "background" of the microstructure.

Microhardness test A test for measuring the hardness of a grain of a particular phase by using a very light load on the indenter.

Nucleation The development of nuclei that act as centers of crystallization for a new phase.

Number of components, C The number of materials, elements or compounds, for which the phase diagram applies. For example, the iron-nickel and iron–iron carbide systems are both binary or two-component systems.

Number of phases, P The sum of all the solid, liquid, and gas phases. There can only be one gas phase, since all gases are intersoluble in all proportions. (See Chapter 2 for a detailed discussion of phases.)

Pearlite A two-phase mixture of α iron and iron carbide produced when γ iron transforms by a eutectoid reaction.

Peritectic reaction A reaction in which a solid goes to a new solid plus a liquid on heating, and the reverse occurs on cooling:

$$S_1 \underset{\text{cooling}}{\overset{\text{heating}}{\rightleftarrows}} S_2 + L$$

Phase diagram A graph showing the phase or phases present for a given composition as a function of temperature (a collection of solubility lines).

Phase rule $F = C - P + 2$, where F = degree of freedom, C = number of components, and P = number of phases in equilibrium. This formula is usually written $+1$ instead of $+2$ because we use up one degree of freedom in fixing the pressure at 1 atm.

Polyphase material A material in which two or more phases are present.

Segregation The development of a concentration gradient or nonequilibrium structure as a result of freezing under nonequilibrium conditions.

Shrinkage cavity A void produced within a casting or ingot because the solid, which is denser, cannot fill the mold volume originally occupied by the liquid.

Single-phase material A material in which only one phase is present, such as BCC iron. Any number of elements may be present, but these must be in solid solution.

Solidus The temperature at which the liquid phase disappears on cooling or at which melting begins on heating.

Steady-state condition A system in which diffusion or heat transfer is taking place and there is no change in composition or temperature at different points in the system with the passage of time.

Substitutional diffusion Migration using substitutional positions.

Ternary system A three-component system. The additional component gives an additional degree of freedom.

Two-phase material A material in which two different phases are present and grains or crystals of two different materials, with different unit cells, for example, can be found.

Unsteady-state condition A system in which temperature or composition is changing as a function of time.

PROBLEMS

4.1 Consider an alloy of 5% Si, 95% Al. (Sections 4.1 through 4.4)

a. What is the percentage of silicon in the α phase at 630, 600, 577, and 550°C?

b. What is the percentage of silicon in the liquid phase at 630, 600, and 577°C?

c. What is the percentage of silicon in the β phase at 550°C?

To answer this last part, draw the horizontal tie line until it touches the β field.

4.2 Consider the typical engine-block alloy, which is 16% Si, 84% Al. (Sections 4.1 through 4.4)

a. At what temperature will the first crystals of solid appear on slow-cooling the melt?
b. At what temperature will the alloy be completely solid?
c. Just before the alloy is all solid, at say 578°C, what will be the analyses of the β and the liquid, respectively?
d. At this temperature will the analysis of the liquid be greatly different from that of the liquid in the alloy of Prob. 4.1?
e. What will be the phase analyses of the α and β in this alloy at 550°C?

4.3 Suppose you are going to pour some castings of red brass (85% Cu, 15% Zn) and some of yellow brass (60% Cu, 40% Zn) (Fig. 4.9). (Sections 4.1 through 4.4)

a. What pouring temperature would you use in each case, assuming that a temperature 200°C above the liquidus is needed to give the metal sufficient fluidity to fill the molds?
b. At what temperature would each alloy be completely solid?

4.4 Draw a graph showing the maximum temperature to which copper-zinc alloys can be heated during heat treatment. [*Hint:* About 20°C below the solidus (Fig. 4.9).] (Sections 4.1 through 4.4)

4.5 In Fig. 4.32 on the next page, the single-phase regions are labeled. Label the others. [*Hint:* A horizontal tie line drawn in a two-phase field gives the type and analyses of the phases at the points where it intersects the single-phase fields. This is the important iron–iron carbide diagram; carbide is a solid-phase Fe_3C and is also the component of the right-hand side.] (Sections 4.1 through 4.4)

4.6 What are the percentages of α and liquid in a 5% Si, 95% Al alloy at 620, 600, and 578°C? What are the percentages of α and β in this alloy at 576 and 550°C? (Sections 4.5 through 4.6)

4.7 What are the percentages of α and liquid in a 1% Si, 99% Al alloy at 630, 600, and 578°C? (Sections 4.5 through 4.6)

4.8 In a *hyper*eutectic (more than eutectic) analysis, 16% Si, 84% Al, calculate the amounts of liquid and β at 578°C and of α and β at 576°C. (Sections 4.5 through 4.6)

4.9 Prepare a fraction chart for the hypereutectic 16% Si, 84% Al alloy. (Sections 4.5 through 4.6)

4.10 Prepare a fraction chart for a 60% Cu, 40% Zn alloy. (Sections 4.5 through 4.6)

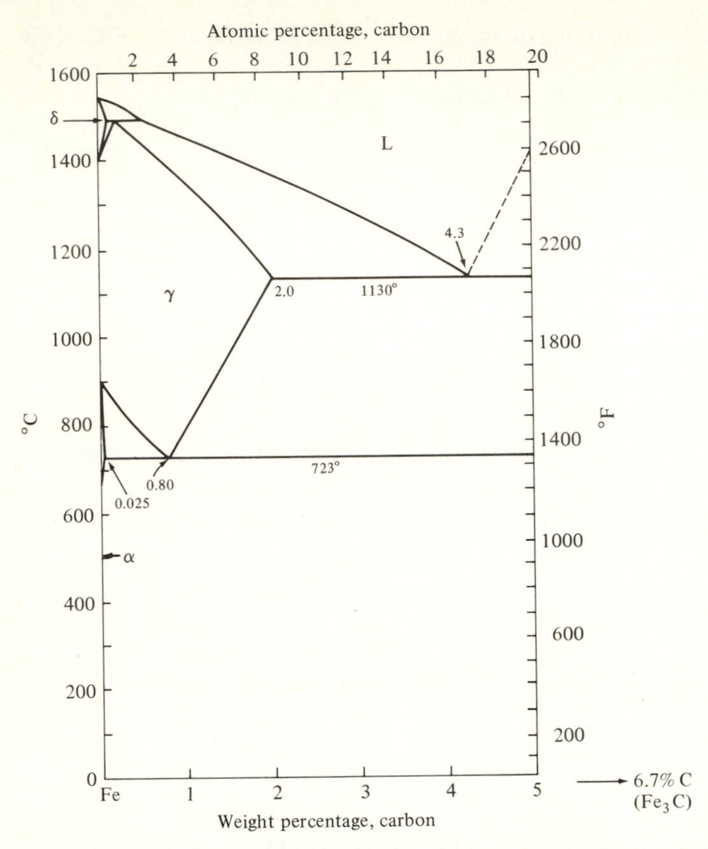

Fig. 4.32 *Unlabeled portion of iron-iron carbide phase diagram (Problem 4.5)*

4.11 Prepare a fraction chart for a 65% Cu, 35% Zn alloy. What is the chief difference in the freezing of this alloy compared with the 60% Cu, 40% Zn alloy of Prob. 4.10? (Sections 4.5 through 4.6)

4.12 Calculate the volume percent of graphite for a cast-iron engine block. Assume the weight percents of phases are 3% carbon as graphite and 97% α (essentially iron). Assume the density of graphite is 2.25 g/cm^3 (2.25×10^3 kg/m^3) and that of iron is 7.87 g/cm^3 (7.87×10^3 kg/m^3). (Sections 4.5 through 4.6)

4.13 Assume that you are the new metallurgist at the Chevrolet Foundry and you are asked the following questions regarding production of the new 18% silicon, 82% aluminum block: At what temperature will the alloy start to solidify? If the castings are stress-relieved by heating after casting, what is the maximum temperature to which they may be reheated? Draw a graph to explain when the hard β phase precipitates and the amount present as a function of temperature. (Sections 4.5 through 4.6)

4.14 A common brass, often called "Muntz metal," is 60% Cu, 40% Zn. Over what temperature ranges are two phases present? One phase? At what temperature is the analysis of one of the phases highest in zinc content (Fig. 4.9)? (Section 4.8)

4.15 The β phase is quite important and is given the formula CuZn. If the phase were exactly this formula, what would be the weight percent of copper and the weight percent of zinc? (Section 4.8)

4.16 In the copper-zinc diagram (Fig. 4.9), locate the temperatures at which three phases can exist at equilibrium and name the reactions (eutectic, peritectic, etc.). (Section 4.8)

4.17 Using the accompanying phase diagram, plot the weight fraction of the α phase as a function of the temperature that would be encountered under equilibrium conditions in an alloy containing 40% R, 60% K. (Section 4.8)

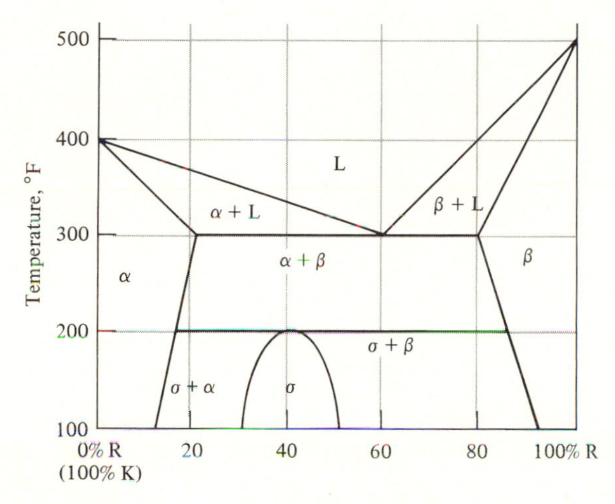

4.18 *a.* In the diagram at the top of the next page, label the blank fields in the parentheses in the phase diagram given.

b. Assuming all lines are straight, plot an accurate graph giving the percent of the β phase as a function of temperature for a 3% Z alloy. [*Hint:* Since lines are straight, it is necessary to calculate β at only a few carefully selected temperatures to determine the graph. Show all changes in slope carefully, as there are locations at which the graph is vertical.] (Section 4.8)

4.19 Given the following data, sketch a phase diagram. Be as accurate and complete as your knowledge permits. (Section 4.8)

Melting point of A = 715°F; melting point of B = 655°F;
501°F and 25% A and 75% B. All liquid

499°F and 25% A and 75% B; 25% of a BCC metal, 75% of a FCC metal

499°F and 35% A and 65% B; 50% of a BCC metal, 50% of a FCC metal

400°F and 40% A and 60% B; 60% of a BCC metal, 40% of a FCC metal

400°F and 20% A and 80% B; 20% of a BCC metal, 80% of a FCC metal

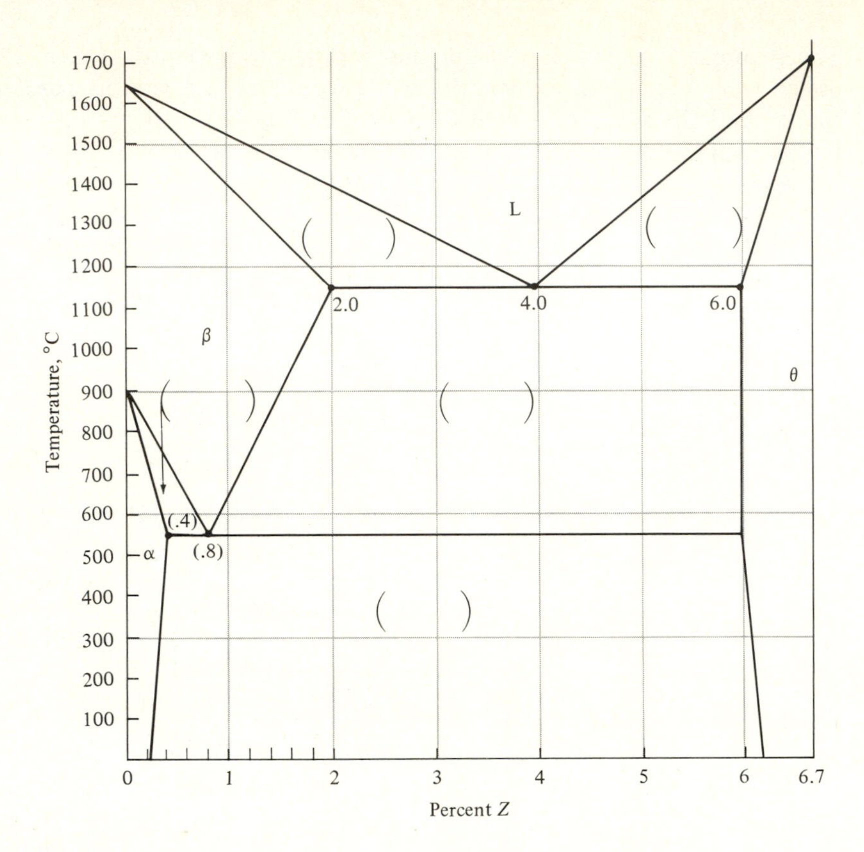

4.20 In 5% Si, 95% Al alloys the β phase is in a fine intimate mixture with α in localized regions. However, in 15% Si alloys the β is in two distinct forms: large blocky crystals and a fine mixture. Explain from the phase diagram. (Sections 4.17 through 4.21)

4.21 A contractor has a specification for soldering sections of copper water pipe. The recommendation is for 50% Pb, 50% Sb, and the contractor suspects that 50% Pb, 50% Sn is intended. Both phase diagrams are given in the diagram shown at the top of the opposite page. (Sections 4.17 through 4.21)

a. What are the liquidus and solidus temperatures for each of the alloys?
b. What is the primary phase for each alloy?
c. Calculate the percentage primary phase present for each alloy at 1°C below the eutectic temperature.
d. Why would a 50% lead, 50% tin alloy be easier to solder than a 50% lead, 50% antimony alloy?

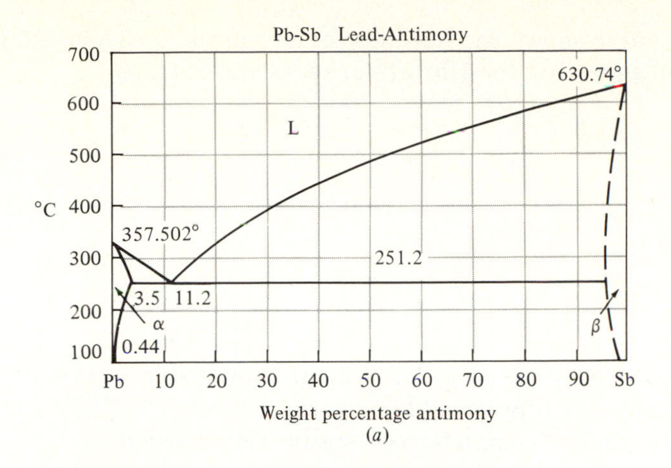

(*a*)

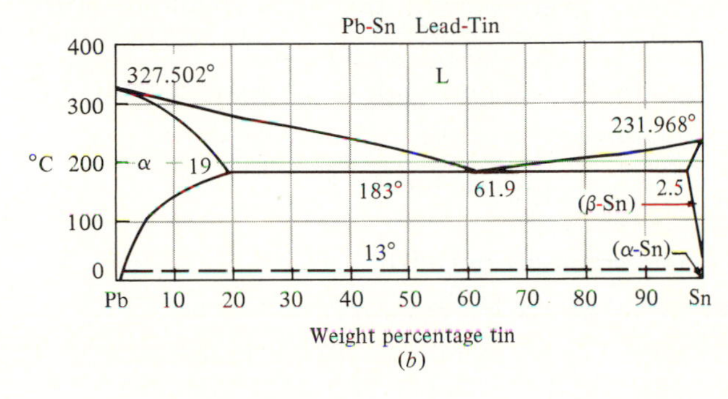

(*b*)

4.22 Answer the following with the appropriate words in the blanks indicated by (***); or, for blanks indicated by (IND), the *letter* I, N, or D for the most appropriate choice of I = increase, N = not change significantly, D = decrease. (Sections 4.17 through 4.21 and 3.16 through 3.22)

Example Water has a melting point of (***). When it is heated, its temperature will (IND). 0° / I

1. Cold-working a metal causes:

   a. its yield strength to (IND) ______
   b. its hardness to (IND) ______
   c. its ductility to (IND) ______
   d. the concentration of dislocations in it to (IND) ______
   e. the mobility of dislocation in it to (IND) ______
   f. its elastic modulus to (IND) ______

2. The curve shows the hardness of a cold-rolled copper alloy after annealing for 1 hr at various temperatures.

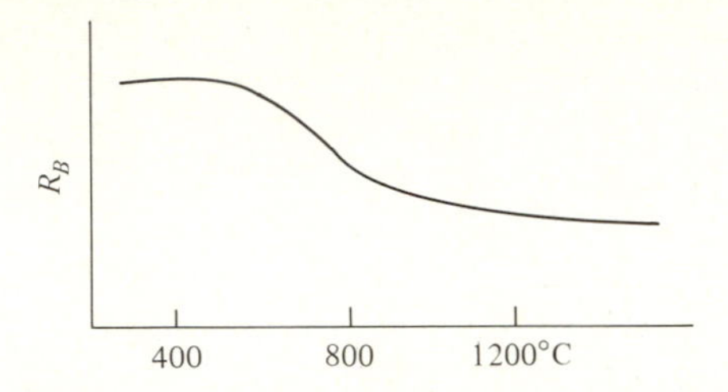

   a. To achieve stress relief without appreciable recrystallization, this alloy should be heated 1 hr at (***)°C. ______
   b. To achieve complete recrystallization without appreciable grain growth, this alloy should be annealed for 1 hr at (***)°C. ______

3. The addition of up to 30% zinc as an alloying element in copper causes:

   a. the melting point to (IND) ______
   b. the cost to (IND) ______
   c. the yield strength to (IND) ______
   d. the critical resolved shear stress to (IND) ______

4. Three elements that form interstitial solid solutions in iron are carbon, (***), and (***). ______ ______

4.23 In the copper-zinc system, commercial alloys with more than approximately 40 weight percent zinc are not available. Suggest reasons why this might be so. (Sections 4.17 through 4.21)

4.24 As shown in Fig. 4.31, why is a longer time required to reach the maximum hardness at lower aging temperatures? Why is the maximum hardness higher for lower aging temperatures? (Sections 4.17 through 4.21)

4.25 Assume that a *large* aluminum structure is to be made from an age-hardened alloy.

   a. At what stage in the aging process would cold working be done? Why? (Sections 4.17 through 4.21)
   b. Why would an aluminum alloy that exhibits natural aging be selected?
   c. Why might welding be difficult?

4.26 In Example 4.8 change the liquid region in the phase diagram to a solid solution δ. Discuss the strengthening mechanisms for alloys ① through ④ for the revised phase diagram. (Sections 4.17 through 4.21)

PROBLEMS COVERING OPTIONAL SECTIONS

4.27 A student attempts to apply the phase rule to a 5% Si, 95% Al alloy in the α + L field. He sees that he is in a two-phase field, and at a pressure of 1 atm the equation is $F = C - P + 1$, or $F = 2 - 2 + 1 = 1$. He then says that since this is an area where there are two variables, composition and temperature, the phase rule is incorrect in giving $F = 1$. Do you agree? (Section 4.7)

4.28 Another student applies the phase rule to the eutectic temperature for the aluminum-silicon system. She says quite correctly that there are three phases and two components; hence $F = 0$. She then complains that three phases will be in equilibrium from 1.6 to 99% silicon at 577°C, so there is still a variable to be fixed if she is to describe the system to another student. Can you help her? (Section 4.7)

4.29 The β phase is a random solid solution of copper and zinc, but β' is an ordered phase. The ordering temperature increases with the amount of zinc. Why is the line, which shows the temperature of β ordering, horizontal in both the $\alpha + \beta$ fields and the $\beta + \gamma$ fields but slanted in the pure β field? (Section 4.7)

4.30 Given the single-phase fields in the horizontal section of the ternary diagram (Fig. 4.33), label the others. This is the important iron-nickel-chromium diagram, the basis of many stainless steels and superalloys. Note that the fields containing three phases are triangles (straight sides), that there is a two-phase field at each side, that the corners of the triangle touch single-phase fields, and that all the phases found in the fields touching the triangle are also found in the triangle. (Sections 4.9 through 4.10)

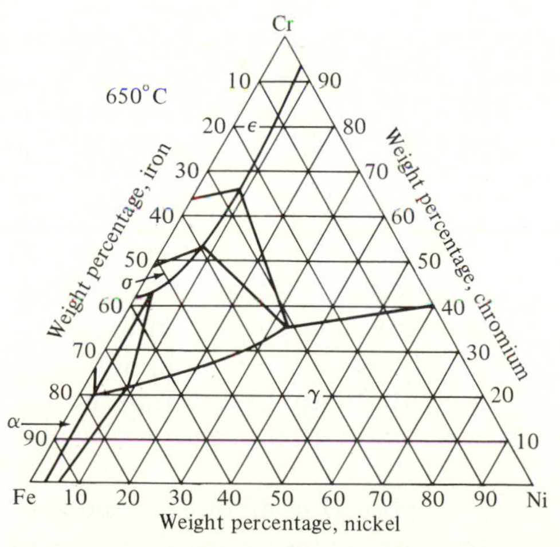

Fig. 4.33 *Unlabeled 650°C isothermal section of iron–chromium–nickel ternary phase diagram (Problem 4.30).* (American Society for Metals Handbook, 8th ed., Vol. 1: *Properties and Selection of Metals,* Metals Park, Ohio, 1961, p. 214)

4.31 Assume that the heat-treatment supervisor in charge of processing the gears described in Example 4.5 is on an economy program directed toward getting better life from the electric furnaces. The supervisor recommends reducing the carburizing temperature from 1000 to 900°C, saying that the furnace life is much longer at the lower temperature and that the carburizing time needed to get the same results will be only 1000/900 times or about 10% longer. Show that the supervisor is sadly in error by calculating D for 900°C and calculating the time required to reach the same percentage of carbon as in Example 4.5. (Sections 4.12 through 4.16)

4.32 When segregation exists from solidification conditions, we can use diffusion calculations to estimate the time required to homogenize the material. The simplest calculation is to obtain the time needed to reduce the difference in concentration to one-half. For example, suppose that we have a carbon concentration C_s at one point and zero carbon at another. We calculate the time for C_x to equal 0.5 C_s. (Sections 4.12 through 4.16)

a. Show that the homogenization time is approximately $t = x^2/D$.

b. Referring to Fig. 4.20, what would be the relative homogenization times for nickel in FCC iron compound with carbon in FCC iron at 1000°C?

4.33 A gear is placed in a carbon-rich furnace atmosphere at 1700°F, and the surface rapidly attains 1.3% carbon. The original carbon content of the gear is 0.20%. Given: $A = 0.25$ cm^2/sec; $Q = 34{,}500$ cal/mol for carbon in FCC iron.

a. What is the carbon content at 0.01 in. beneath the surface after 10 hr?

b. What furnace temperature would be needed to attain the same carbon content at 0.01 in. beneath the surface in 1 hr? (Sections 4.12 through 4.16).

4.34 You are asked to determine the time necessary to aluminize a piece of steel (dip steel into pure molten aluminum). For the aluminum to adhere, it is found by experiment that 10% aluminum is required at 0.002 in. below the surface of the steel. If the diffusion coefficient for aluminum is 7.2×10^{-9} cm^2/sec at the dipping temperature (1000°C), determine the time necessary to obtain 10% aluminum at 0.002 in. (Sections 4.12 through 4.16)

4.35 Given an activation energy Q of 40,000 cal/mole and an initial temperature of 1000 K, find the temperature that will increase the diffusion coefficient D by a factor of 10. (Sections 4.12 through 4.16)

4.36 The text (Prob. 4.32) states that $t = x^2/D$ to reduce the concentration difference to 50% when a material is homogenized. Recalculate the time (in terms of x and D) if the concentration difference is to be only 5% ($C_x = 0.95\ C_s$; $C_0 = 0$). From this simplification, suggest the amount of time it would take to completely homogenize a material. (Sections 4.12 through 4.16)

5

ENGINEERING ALLOYS IN GENERAL; NONFERROUS ALLOYS—ALUMINUM, MAGNESIUM, COPPER, NICKEL, AND ZINC

THE hang glider is a good illustration of the use of high-strength lightweight alloys. For centuries people have tried to make wings with which they could glide like a bird. The success of the present equipment is related to the use of high-strength aluminum alloy 6061-T6, discussed in this chapter, in the ribs, and the high-strength polymer fabrics described in Chapter 9.

In this chapter we will use the concepts of structure and structural control that we discussed in the earlier chapters. We will review the structures of the important nonferrous alloys to understand the interrelation of structure, mechanical properties, and processing.

5.1 General

We are now entering a new portion of the text, devoted to the actual engineering alloys used in structures and components. Let us spend a few moments reflecting on what we have studied and how it applies to this new material. In the first chapter we agreed that to reach a basic understanding of metals, we needed to know first the nature of the different atoms—how they were arranged in metallic structures and the effects of stress and temperature on these structures. As a result, we are now in the powerful position of being able to look over the entire field of thousands of metallic alloys and rapidly understand their properties. We can do this, despite the almost infinite number of combinations, because the alloys are made up of just a few different metallic structures with which we are already familiar.

The simplest way to describe these commercial alloys is to arrange them into families based on the principal element present, such as the aluminum alloys, the copper alloys, and so forth. We should mention the traditional division into nonferrous alloys and ferrous (iron-base) alloys. In some countries even the metallurgists are divided into these two specialized groups. This is really an unscientific point of view; for example, the nickel alloys, while "nonferrous," have more in common with iron-base alloys than with zinc alloys.

Therefore, let us take the more systematic approach of dividing the alloys into families. We find that over 95% of the tonnage is in the aluminum, magnesium, copper, nickel, and iron-base alloys (Table 5.1). In fact, over 90% is in the iron-base family alone. Although the percentages for magnesium and nickel alloys are small, these have special importance and can be discussed conveniently with the others.

Several other features of the data are important to our background. First we may ask: Since the iron-base alloys are low in price and highest in strength, why bother with the others? We must realize that strength is only one of the requirements of a component. Satisfactory service can depend on density, corrosion resistance, effects of temperature, and electrical and magnetic properties. As examples, let us consider some parts for which the different alloys are especially suited.

Aluminum alloys: aircraft parts (high strength per pound)
Magnesium alloys: aircraft castings (competitive with aluminum)
Copper alloys: electrical wiring (high conductivity)
Nickel alloys: gas turbine parts (high strength at elevated temperatures)

Another question we may ask is: Why did we bother to survey so many metals in Chapter 2 when only four or five are so important? The answer is that at least 20 others are of vital importance in controlling and modifying the structures of the principal metals. It is by this modification, combined with processing treatments, that we obtain the tremendously improved properties of the alloys. Table 5.1 shows that by experimenting with alloys and

Table 5.1 PROPERTIES OF IMPORTANT ELEMENTS AND THEIR ALLOYS

	Properties of Element (Annealed)					*Properties of Common Alloy of Highest Strength*			
Element	*Yield Strength,* psi × 10^3*	*Percent Elongation*	*BHN*	*Ingot,† dollars/lb*	*Alloy No.*	*Yield Strength,* psi × 10^3*	*Percent Elongation*	*BHN*	*Use as Engineering Materials (U.S.), tons/yr*
Aluminum	4	43	19	0.60	7178	78	10	160	2 million
Magnesium	Not used in pure form				AZ31B-H24	42	16	82	$<$100,000
Copper	10	50	25	1.00	172	140	7	380	1 million
Nickel	22	47	90	2.05	301	150	10	400	$<$100,000‡
Iron	20	48	70	0.05	4340	270	11	500	100 million

*To obtain MN/m^2 (MPa), multiply psi by 6.9×10^{-3}. To obtain kg/mm^2, multiply psi by 7.03×10^{-4}.
†1980 prices.
‡As nickel-base alloy.

heat treatment, metallurgists have been able to raise the strength of pure aluminum 20 times, copper 14 times, and iron 13 times, while retaining satisfactory ductility in all cases.

5.2 Processing methods

As we study the specifications for commercial alloys, we find that the processing method is given. For example, the properties are accompanied by a term such as cast, forged, wrought, rolled, or extruded. There are two major divisions—cast and wrought—and recently powder metallurgy has also become important.

We have already discussed hot and cold working and their effects on properties. Now let us consider the casting process. To obtain a "hands-on" feel for this process, let us follow the step-by-step production of a familiar casting—a frying pan—and then discuss other casting methods. In this elementary example, the casting itself is used as a pattern, although we will not obtain an exact duplicate because of the phenomenon of metal shrinkage. (This first method is crude, but it is valuable in emergency replacement of a failed part that can be glued together and used as a pattern.)

We begin by positioning the pan on a plate (Fig. 5.1*a*) and placing an open box called a *flask* (or molding box in Great Britain) over the pattern. Next we prepare a green* sand mixture of approximately 93% silica sand, 4% clay (bentonite) and 3% water. Due to the clay-water-silica bond, this mixture has adequate strength when rammed to form a relatively firm surface. After the flask is rammed full, Fig. 5.1*b*, the surface is leveled with a simple bar called a *strike,* then a *bottom plate* is positioned on the struck surface and the mold is rolled over. The sand does not fall out of the mold box because of its strength and because of the lip of the flask, which retains the sand.

Next, the *parting surface* is cut, Fig. 5.1*c*, with a trowel and finished with a smaller tool called a *slick*. It is important to design the parting surface so that the pattern can be *drawn* later, i.e., pulled cleanly from both halves of the mold without tearing the sand surface.

*Called green not because of color, but because of water content!

Fig. 5.1 *(a) Frying pan pattern positioned on a bottom plate with wood gating system attached. (b) Drag flask filled with sand and leveled (struck) ready for roll over. (c) Parting surface being cut with a slick in order to make possible easy removal of the pattern. (d) Cope flask in place over pattern. Vertical portion of gating (downsprue) is centered over gating well. (e) Cutting pouring basin in cope. Sprue has been removed. Loose sand is blown out of sprue after removal of cope. (f) Cast iron pattern being removed from drag with aid of a magnet. (g) Casting being poured into previously clamped flasks. (h) Casting shaken out of mold. Note that some metal ran up into one of the vent holes. (i) Casting after removal of gate, then grinding and sandblasting.*

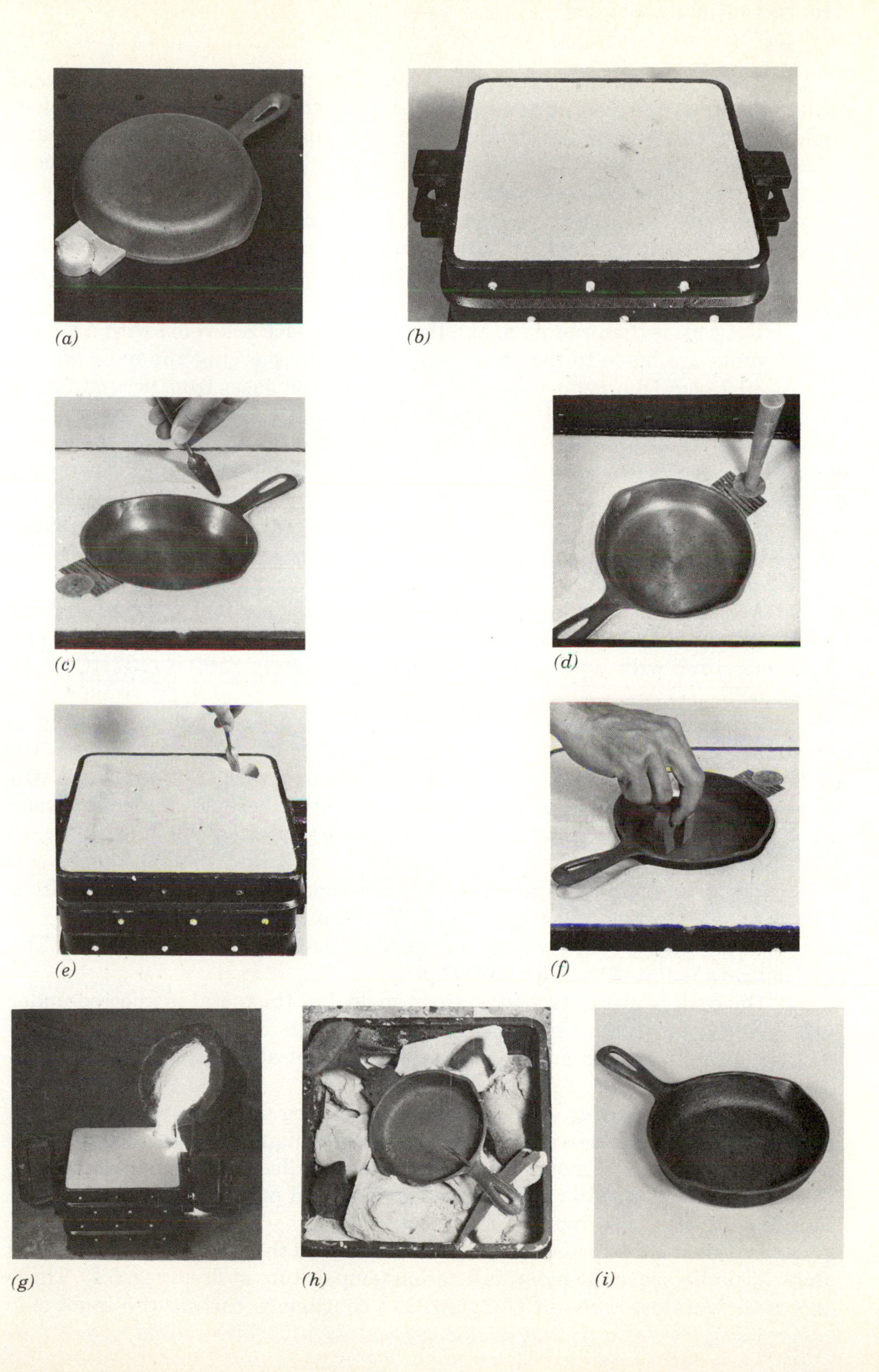

(a) (b)

(c) (d)

(e) (f)

(g) (h) (i)

A second flask called the *cope* is placed in position over the lower half, called the *drag,* Fig. 5.1*d*. Pins that project from the upper half fit into bushed holes to locate the cope in a fixed position over the drag. A channel is needed to introduce liquid metal into the mold. This is called the *gating*. Loose pieces of wood were already positioned in the drag to form this, as shown in Fig. 5.1*a*. However, the vertical portion of the gating (*downsprue*) is in the cope (Fig. 5.1*d*). At this point the surface of the drag is dusted lightly with a ceramic powder to prevent the sand of the drag surface from bonding to the sand of the cope.

The cope is then rammed, and the upper surface is leveled with a strike. Next, vents are made to the inner cope surface by piercing the mold with a 0.1-in.-(0.25-cm-) diameter wire. This prevents mold gases from accumulating in the top of the mold and retarding filling. Finally, a *pouring basin* is cut around the vertical portion of the gating (Fig. 5.1*e*). The pattern for the downsprue has been drawn upward because its draft is designed to taper downward for better metal flow.

The cope is then drawn carefully from the pattern and the drag. Next the pattern is drawn from the drag with the aid of a magnet (Fig. 5.1*f*). The gating is removed with a wood screw.

After the casting cavity is inspected, the flask sections are reassembled accurately with the aid of the pins and bushings and clamped together. The mold is poured with liquid cast iron at approximately 2500°F (1371°C) (Fig. 5.1*g*).

After 30 min the mold is shaken out (Fig. 5.1*h*). Then the casting is sandblasted, the gating is cut off, and the casting is ground smooth (Fig. 5.1*i*). If the dimensions of the final casting are compared with the pattern, it will be seen that the casting is approximately 1% smaller because of the shrinkage of the metal from the solidification temperature, about 2100°F (1149°C), to room temperature.

The process we have just described can be used to produce castings from less than 30 g to over 150 metric tons. Other processes are as follows.

In *permanent mold casting* the mold is made of metal or graphite. The cooling rate is faster than for sand.

Die casting also uses a permanent mold, but the metal is injected under pressure. This is particularly useful for metals with low melting points—such as zinc, aluminum, and magnesium—which do not attack the metal die rapidly.

Investment casting is a method for producing small precision castings. A wax replica of the desired part is made and a liquid, cementlike slurry is poured around it in a mold. After the cement (called an "investment") sets, the mold is heated. This eliminates the wax and preheats the mold. Then liquid metal is poured into the cavity.

In the *wrought* processes an ingot is cast and then worked to the desired shape, usually above the recrystallization temperature at first (Fig. 5.2). These processes were described in Chapter 3. In continuous casting the ingot step

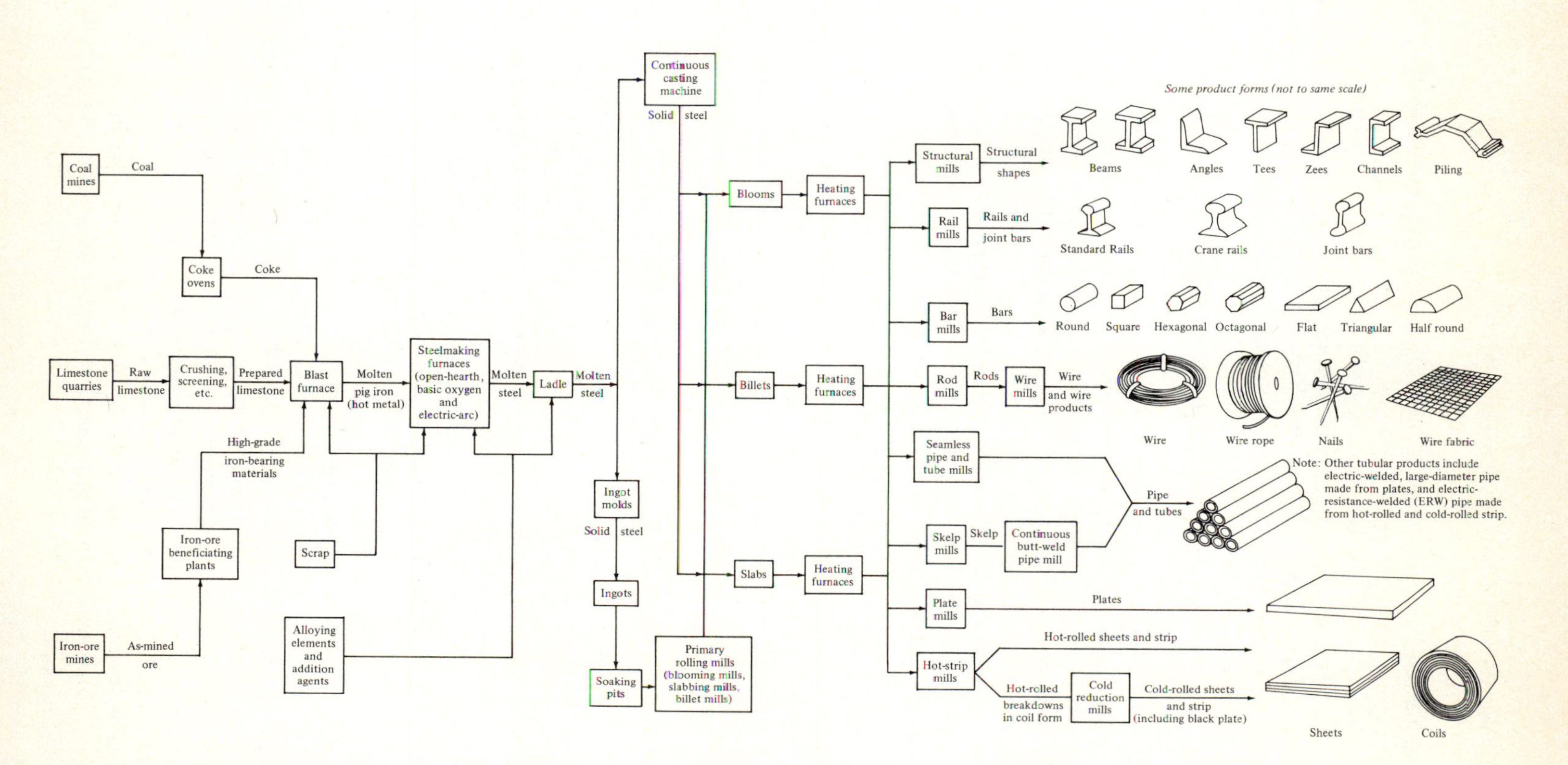

Fig. 5.2 *Conversion of raw materials into different steel shapes.* (United States Steel Corp., *The Making, Shaping, and Treating of Steel*, 9th ed., 1971, p. 2.)

is eliminated. A continuous strand of metal issues from the bottom of a water-cooled die.

All alloys can be cast, but wrought alloys are limited to those that can be hot-worked without cracking. Examples of alloys that cannot be worked are in the aluminum alloys the low-ductility–high-silicon analyses and in the ferrous alloys the cast irons. These distinctions will be made as we consider the different alloys.

In *welded* structures combinations of cast and wrought parts can be used.

In *powder metallurgy* powders of the desired materials are mixed and then compressed in a die to the desired shape. Next the mass is treated at a high temperature to promote bonding of the powder grains by sintering, a type of diffusion bonding. This method is especially important for producing carbide tools for machining.

One final word: In covering this material it is not necessary to memorize analyses. The important point is to recognize how analysis and heat treatment control the structure of a material and how the properties of the material respond to the changes. There are four important special cases of control of the structure that we discussed in the previous chapter. These are: (1) controlled crystallization from the melt, (2) controlled nucleation in solid, (3) martensite reaction, (4) age-hardened. These, of course, are all techniques for controlling the nature, size, shape, amount, distribution, and orientation of the phases, as pointed out previously.

Aluminum Alloys

5.3 General

In discussing these materials, our first group of alloys for actual engineering applications, let us take special care to examine how even complex specifications are based on the fundamental factors we have learned. For each alloy it is important to understand why certain alloying elements are used and why the processing results in the properties given.

Since this family has an aluminum base, let us review briefly what we know about pure aluminum. The atomic structure is $1s^2$, $2s^2$, $2p^6$, $3s^2$, $3p^1$ or, more simply, a tightly bonded core surrounded by three valence electrons that are readily given up. This results in a light, reactive metal with good electrical and thermal conductivity. The unit cell is FCC, which leads to a number of plane and direction combinations for easy slip, $\{111\}$ and $\langle 110 \rangle$, so that we expect excellent ductility and formability. All these predictions are borne out by our everyday experiences with aluminum foil and wire.

Unalloyed or relatively pure aluminum has many uses. One can raise yield and tensile strengths by cold working, discussed in Chapter 3. Table 5.2 (pages 174–175) gives two sets of properties for commercially pure aluminum

(1060), for conditions marked 0 and H18. We shall discuss these symbols in detail shortly, but for the present we may consider them as denoting annealed (recrystallized) and cold-worked conditions, respectively. The microstructures are similar to those of the cold-worked and annealed α brasses discussed in Chapter 3. As in the case of brass, we can obtain different combinations of strength, elongation, and hardness between the extremes listed in Table 5.2 by controlling the sequence and amount of annealing and cold working.

Now let us advance to the specifications involving the second method for improving the strength of a single-phase alloy, namely solid-solution strengthening. This is the mechanism operating in alloys 3003 and 5052, listed in Table 5.2. From the aluminum-magnesium phase diagram (Fig. 5.3), we would expect to find the 2.5% magnesium of alloy 5052 in solid solution in the α (FCC) phase. Similarly, the manganese in 3003 is essentially in solid solution.

In both these alloys in the annealed (0) condition, the tensile and yield strengths are much higher than for 1060 aluminum with somewhat lower elongation. However, the microstructures still show the simple grains of a single phase. Basically, the solute atoms have raised the stress required for slip. Note that in the cold-worked conditions (marked H18 and H38), the alloys are still proportionately higher in strength than the cold-worked pure aluminum. This means that the effects of solid-solution strengthening and work hardening can be used together; that is, the effects are *additive*. This method of alloy design is used in all the other systems as well. In other words, when we wish to retain the ductility and formability of the base metal, we add alloys that dissolve in solid solution to raise the strength in the annealed condition, then cold-work to the desired properties.

Now let us examine the *two-phase* wrought alloys. The most important element in 2014 is the 4.5% copper. Turning to the aluminum-copper phase diagram (Fig. 5.4), we have a single-phase alloy at 550°C but a two-phase material at room temperature. As discussed in Chapter 4, this alloy is a candidate for age hardening if the precipitate is coherent, a condition which is found to exist. To develop a fine coherent precipitate, we give the alloy a two-step treatment. First we heat the annealed material to 500 to 550°C to dissolve the θ phase, then quench. This gives supersaturated κ. We then produce a fine precipitate by aging at 170°C for 10 hr; this precipitate is so fine it is not visible by conventional microscopy. A tensile strength of 70,000 psi (483 MPa) develops (marked T6 condition), as compared with 27,000 psi (182 MPa) (annealed condition).

Casting alloys are usually two-phase. A notable exception is the casting of relatively pure aluminum cooling fins around iron laminations in electric rotors, where it is desired to capitalize on the high thermal conductivity of aluminum. In one group of heat-treatable casting alloys, such as 356.0, age hardening is produced as in the wrought alloys. In some cases the solution-treating step may be omitted if the cooling rate of the casting is fast enough

Table 5.2 TYPICAL PROPERTIES OF ALUMINUM ALLOYS

Alloy Number	*Chemical Analysis, percent**	*Condition*	*Tensile Strength,† psi × 10^3*	*Yield Strength,† psi × 10^3*	*Percent Elongation*	*BHN*	*Typical Use*
			Single-phase Wrought Alloys				
1060	99.6 minimum Al	0	10	4	42	19	Sheet, plate,
		Hard H18	19	18	6	35	tubing
3003	1.2 Mn	0	16	6	30	28	Truck panels,
		Hard H18	29	27	4	55	ductwork
5052	2.5 Mg, 0.2 Cr	0	28	13	25	47	Bus bodies,
		Hard H38	42	37	7	77	marine applications
5050	1.2 Mg	0	21	8	24	36	Sheet, trim,
		Hard H38	32	29	6	63	gas lines
			Two-phase Wrought Alloys				
2014	4.5 Cu 0.8 Si 0.8 Mn	Annealed	27	14	18	45	Airplane
	0.5 Mg	Heat-treated T6	70	60	13	135	structures
6061	1 Mg 0.6 Si 0.2 Cr	Annealed	18	8	30	30	Transportation
	0.3 Cu	Heat-treated T6	45	40	12	95	equipment, pipe
7178	7 Zn, 0.3 Mn, 3 Mg, 0.3 Cr,	Annealed	33	15	16	60	Structural
	2 Cu	Heat-treated T6	88	78	10	160	parts in aircraft

Two-phase Cast Alloys

296.0	4.5 Cu	Solution heat-treated (T4)	37	19	9	75	Aircraft fittings, pump bodies
		Aged (T6)	40	26	5	90	
356.0	7 Si, 0.4 Mg	Aged T5	25	20	2	60	Auto transmission casings, wheels
		Aged T6	38	27	5	80	
712.0	5.5 Zn 0.6 Mg 0.5 Cr 0.15 Ti	Aged T5	35	25	3	75	Machine parts
208.0	3 Si, 4 Cu	As cast (F)	21	14	2	55	General
380.0	8 Si, 3.5 Cu	As cast (F)	47	23	4	80	Die casting
390.0	17 Si, 1 Fe, 4.5 Cu, 0.5 Mg	As cast (F)	41	35	<3	120	Die casting

*Balance aluminum.
†Multiply psi by 6.9×10^{-3} to obtain MPa or by 7.03×10^{-4} to obtain kg/mm^2

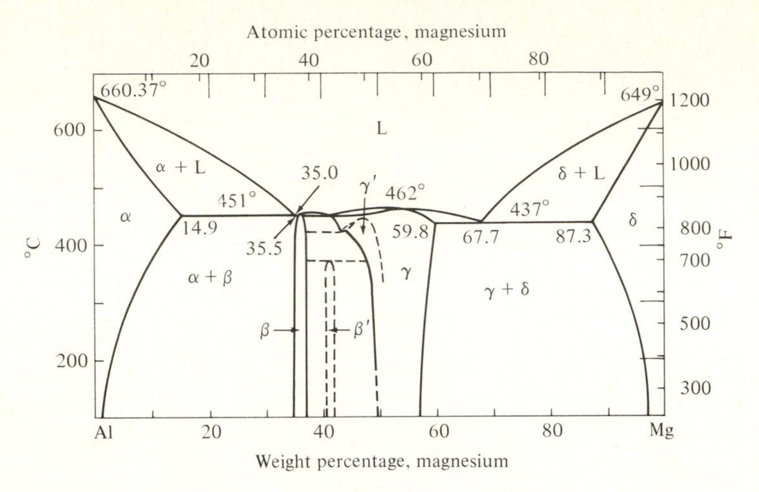

Fig. 5.3 *Aluminum–magnesium phase diagram.* (American Society for Metals Handbook, 8th ed., Vol. 8, *Metallography, Structures and Phase Diagrams,* Metals Park, Ohio, 1973, p. 261)

to produce a supersaturated solid solution. In this case we need only age the casting. This is denoted by T5. (See the code given at end of this section.) In other cases the second phase acts simply as a hard dispersion to improve hardness and wear resistance. This is the case with the silicon-rich phase in the 390.0 alloy for engine blocks. The size and shape of the dispersion are controlled by the cooling rate and by nucleation with phosphorus (forming AlP). Fast cooling rates and abundant nuclei give a dispersion of fine equiaxed crystals of silicon rather than long bladelike crystals, thereby improving strength and ease of machining. The alloy can be further hardened with a solution heat treatment and aging because of the presence of 4% copper, as discussed for 2014.

Now let us consider some details of the actual engineering specifications, which may seem complex after our simple discussions of working and heat treatment. For example, the specification for an ordinary aluminum frying pan would read:

Alloy 1100: 99.0% minimum Al, 1.0% maximum (Fe + Si), 0.20% maximum Cu, 0.05% maximum Mn, 0.10% maximum Zn, 0.05% maximum each of other elements, total of which shall be 0.15% maximum.

Mechanical properties: H16 temper, 20,000 psi (138 MPa) yield strength, 21,000 psi (145 MPa) tensile strength, 6% elongation.

To understand this we must learn to see through all the specification language and determine what structure we are talking about.

First, pure aluminum is an FCC structure and therefore has good ductility. If we consult the important phase diagrams for aluminum (Fig. 5.5), we see that the objective of specifying small maximum amounts of impurities is to avoid the formation of any quantities of hard second phases that would

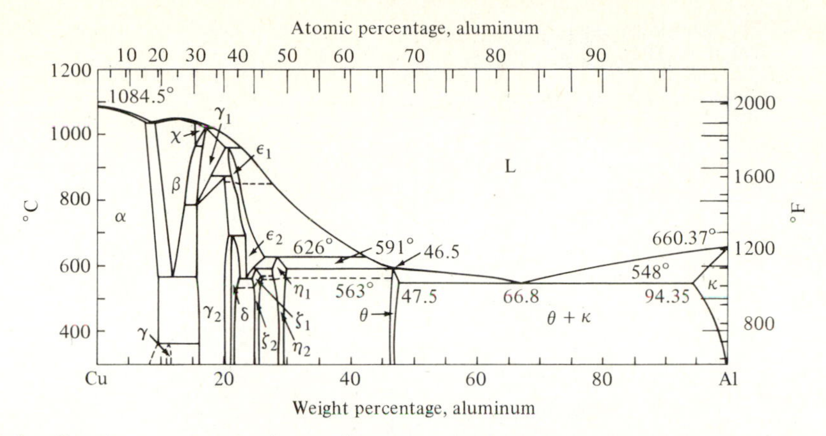

Fig. 5.4 *Aluminum–copper phase diagram.* (American Society for Metals Handbook, 8th ed., Vol. 8, *Metallography, Structures and Phase Diagrams*, Metals Park, Ohio, 1973, p. 259)

reduce ductility and increase corrosion, as discussed in Chapter 12. One could specify a purer material, but this would raise the cost.

Next we note the specification of "H16 temper." This denotes a certain amount of cold working, such as cold rolling, which raises the yield point and prevents denting of the frying pan. (The temperature encountered during normal heating of the pan would be too low for recrystallization.) In essence, then, the specification is written to give a cold-worked single-phase structure of commercially pure aluminum.

Let us now review the general specifications. We shall see that the alloys can be divided into two general classifications: single-phase and polyphase alloys. For single-phase alloys only working and annealing are used to control the properties of a given alloy. For the polyphase alloys combinations of working and precipitation hardening are employed.

The principal alloying elements are copper, manganese, silicon, magnesium, zinc, nickel, and tin. For the wrought alloys a code has been developed to make it easy to recognize the type of alloy.

Code	*Type of Alloy (Major Element)*	*Example*
1XXX	Essentially pure Al	1060 (99.6% minimum Al)
2XXX	Cu (two-phase)	2014 (4.5% Cu)
3XXX	Mn (one-phase)	3003 (1.3% Mn)
4XXX	Si (two-phase)	4032 (12.5% Si)
5XXX	Mg (one-phase)	5050 (1.2% Mg)
6XXX	Mg and Si (two phase)	6063 (0.4% Si, 0.72% Mg)
7XXX	Zn (two-phase)	7075 (5.6% Zn)

Unfortunately there is no simple difference in numbering, separating the single-phase and two-phase alloys, and so it is necessary to recognize that the 2XXX, 4XXX, 6XXX, and 7XXX series are two-phase alloys. The type of processing is covered by the following code.

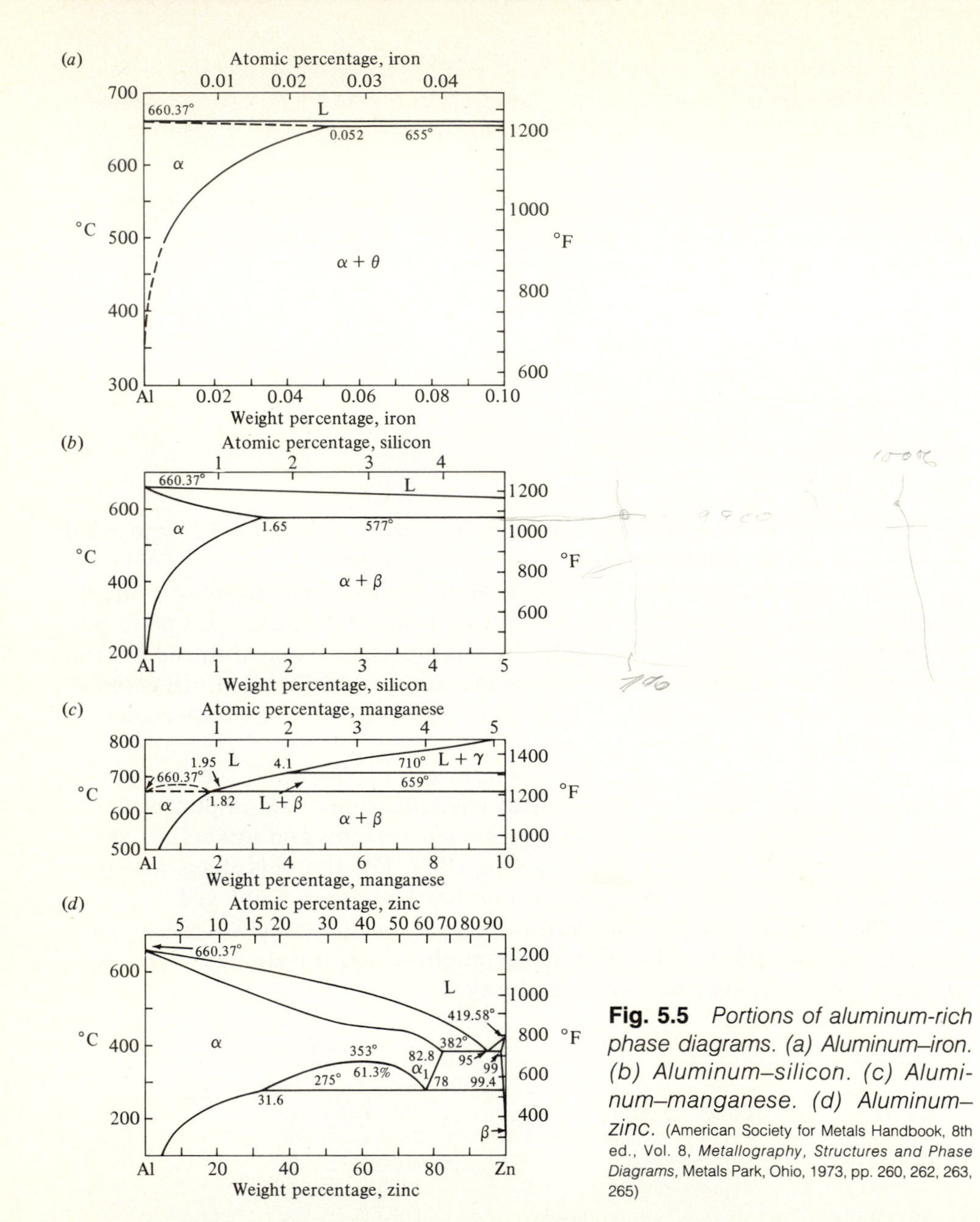

Fig. 5.5 *Portions of aluminum-rich phase diagrams. (a) Aluminum–iron. (b) Aluminum–silicon. (c) Aluminum–manganese. (d) Aluminum–zinc.* (American Society for Metals Handbook, 8th ed., Vol. 8, *Metallography, Structures and Phase Diagrams,* Metals Park, Ohio, 1973, pp. 260, 262, 263, 265)

H1X, cold-worked. The higher the number used for X, the greater the cold working. For example, 1060-H14 denotes a 1060 alloy cold-worked about half of the total possible, while 1060-H19 represents the maximum.

H2X, cold-worked and annealed. The X still represents the severity of cold work.

H3X, cold-worked and stabilized. The X still represents the amount of cold work. Stabilizing means heating to 50 to 100°F (30 to 55°C) above the maximum service temperature so that the material does not soften in service.

Specifications with the letter T involve heat treatment to produce age hardening. These are used only for alloys that develop coherent precipitates, the 2XXX, 6XXX, and 7XXX series. The code is:

T3 Solution treatment followed by strain hardening and then natural aging (i.e., held at room temperature).
T4 Solution treatment plus natural aging.
T5 Aged only. In special cases where the part is cooled quickly enough from the forging or casting temperature, the solution heat treatment is omitted.
T6 Solution heat treatment plus artificial aging.
T7 Solution heat treatment plus stabilization.
T8 Solution heat treatment plus strain hardening plus artificial aging.
T9 Solution heat treatment plus artificial aging plus strain hardening.
0 The annealed condition in wrought alloys.
T2 The as-cast condition in castings.
F The as-fabricated (as-rolled, etc.) condition in wrought alloys and as-cast in castings.

The code for casting alloys is quite different:

Numeral	1XX.X	2XX.X	3XX.X	4XX.X	5XX.X	6XX.X	7XX.X	8XX.X
Family	Al 99. min	Al-Cu	Al-Si Cu, Mg	Al-Si	Al-Mg	Unused series	Al-Zn	Al-Sn

The same codes for heat treatment are used where applicable. The T3, T8, and T9 designations, which call for strain hardening, apply only to wrought products, however.

EXAMPLE 5.1 The T5 designation means aged only. How can alloy 712.0 attain properties close to 356.0-T6 if no solution treatment step is used after casting? (See Table 5.2.)

ANSWER A thin section in a casting cools rapidly enough to provide a solution treatment; i.e., supersaturated α is obtained. Next, at room temperature, natural aging takes place and a coherent precipitate results in an age-hardened alloy. In a casting with grossly varying section sizes, the physical properties in large sections may not attain those of lighter sections unless the material is solution-treated.

EXAMPLE 5.2 How much $CuAl_2$ (θ phase, Fig. 5.4) is present as a function of temperature in a 4.5% copper-aluminum alloy cooled under equilibrium conditions? It is experimentally found that there is more $CuAl_2$ present in an annealed alloy than in a naturally aged alloy but that the yield strength of the latter is greater. Why?

ANSWER

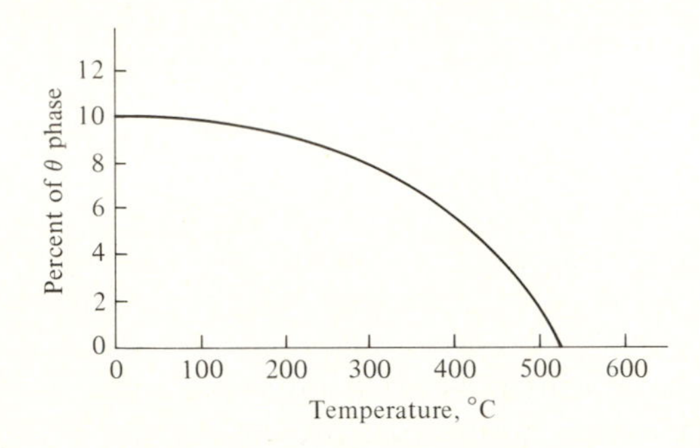

The approximate phase fraction chart is shown above. It is assumed from the phase diagram that no copper is soluble in aluminum at room temperature and that 53% by weight copper is soluble in θ phase at all temperatures. This of course corresponds to 33.3 atomic percent copper to maintain the formula $CuAl_2$. The higher strength in the naturally aged alloy is due to a more desirable uniform distribution of the second phase, whereas the annealed alloy has continuous $CuAl_2$ at the grain boundaries, which is not effective in preventing slip. Furthermore, the aged alloy has coherency, whereas the annealed material does not.

Magnesium Alloys

5.4 General

Magnesium alloys are competitive with those of aluminum because the density of magnesium is two-thirds that of aluminum. Therefore, many aircraft parts are made of magnesium alloys. Also, there has been an interesting competition between magnesium and aluminum alloys for the Volkswagen engine block, since both materials can be used successfully.

We discussed the basic atomic structure of magnesium ($1s^2$, $2s^2$, $2p^6$, $3s^2$) in Chapter 2. Beneath the two valence electrons is a stable ring of eight, making magnesium a very active element. Although fine magnesium powder will burn in air, a melt of liquid magnesium can be poured in air into castings or ingots with a few simple precautions. Magnesium parts will not ignite during heat treatment. However, magnesium will corrode more rapidly than aluminum in many environments, such as sea water, and so outboard motor parts are not made of magnesium.

Another major difference between magnesium and aluminum is that the unit cell of magnesium is HCP. Therefore it has only three slip systems at room temperature, compared with the 12 of aluminum. However, additional systems become operative at higher temperatures. To take advantage of this, magnesium alloys are usually hot-worked rather than cold-worked.

Table 5.3 (page 182) lists alloys. The systems for designating heat treatment (T4, T6, etc.) and working (H24, etc.) are the same as those for aluminum alloys. However, the methods for indicating the composition of the alloys are different. The letters AZ, etc., signify the two most important alloying elements according to the following code:

A = aluminum	K = zirconium	M = manganese
E = rare earths (such as cerium)	Q = silver	S = silicon
H = thorium	Z = zinc	T = tin

The numbers following the letters give the amounts of the elements. The first number gives the percentage of the element shown by the first letter, and the second number the percentage of the second. For example, AZ92 means a 9% Al, 2% Zn alloy.

Magnesium alloys are principally used in aircraft and spacecraft, machinery, tools, and materials-handling equipment. The strength increases with aluminum level by solid-solution strengthening and by the precipitation of a fine lamellar phase, $Mg_{17}Al_{12}$. In both the wrought and cast alloys rare earths are added to minimize flow (plastic deformation) at elevated temperatures (150 to 200°C). The rare earths produce a rigid grain boundary network in the microstructure, such as Mg_9Ce, which resists flow.

Careful control of the network is necessary, since the second phases also limit room-temperature ductility. If they are present in excessive amounts, they may even decrease the tensile strength.

Also, magnesium alloys should play an increasing role as materials because the amount of magnesium in sea water is tremendous, whereas supplies of other important metallic elements, which are usually mined as complex oxides and sulfides, are dwindling.

EXAMPLE 5.3 We have now discussed two families of nonferrous alloys, aluminum- and magnesium-base. What are the essential differences between alloy compositions that might be used for a forging or a casting?

ANSWER In general, a component manufactured by forging or casting is competitive. However, some alloy compositions exhibit low ductility and therefore may only be cast to a final shape. Whereas a forged component requires good ductility, both low- and high-ductility alloys may be cast. In practice, forging stock is produced from cast ingot, bar, or slab.

However, even though the room temperature properties indicate low ductil-

Table 5.3 TYPICAL PROPERTIES OF MAGNESIUM ALLOYS

Alloy Number	*Chemical Analysis, percent*	*Condition*	*Tensile Strength,* psi × 10³*	*Yield Strength,* psi × 10³*	*Percent Elongation*	*BHN*	*Typical Use*
			Wrought Alloys				
AZ31B	3 Al, 1 Zn	F	38	28	9	55	Sheet and
	0.2 Mn	H24	42	32	15	73	plate
AZ80A	8 Al, 0.2 Zn, 0.2 Mn	T5	50	34	6	72	Forgings and extrusions
ZK60A	6 Zn, 0.5 Zr	T5	44	30	16	82	
HK31A	3 Th, 1 Zr	H24	37	29	8	—	Elevated temperatures
			Cast Alloys				
AZ63A	6 Al	As cast	29	14	6	55	General
	3 Zn	Solution heat-treated	40	13	12	50	sand
	0.13 Mn	T6	40	19	5	73	castings
AZ91C	9 Al	As cast	24	14	2	52	High-
	1 Zn	Solution heat-treated	40	12	14	53	tensile
	0.13 Mn	T6	40	19	5	66	castings
QE22A	2 Ag, 2 R.E.†	T6	40	30	4	77	Highest-strength uses
EZ33A	3 R.E., 3 Zn	T5	23	16	3	50	Elevated-temperature uses
EK31A	3 R.E., 1 Zr	T6	31	16	6	55	

*Multiply psi by 6.9×10^{-3} to obtain MN/m^2 (MPa) or by 7.03×10^{-4} to obtain kg/mm^2.
†R.E. = rare earth elements.

ity, it is possible to heat a multiphase alloy to a higher temperature at which a single phase suitable for forging exists. Therefore we must differentiate between hot and cold forming processes.

It is not uncommon to find the same nominal composition available in both cast and wrought shapes. If an alloy is available only as cast products, this suggests its inability to achieve sufficient ductility for forming at all temperatures. For a specific alloy, the appropriate phase diagram and microstructure provide the necessary data for explanation.

Copper and Nickel Alloys

5.5 Copper alloys; general

The copper alloys have a unique combination of characteristics: high thermal and electrical conductivity, high corrosion resistance, generally high ductility and formability, and interesting color for architectural uses. Although the hardness and strength of these alloys do not equal those of the hardest steels, some alloys reach tensile strengths of 150,000 psi (1.035×10^3 MPa).

The atomic structure of copper is $1s^2$, $2s^2$, $2p^6$, $3s^2$, $3p^6$, $3d^{10}$, $4s^1$. Note that the outer electron $4s^1$ does not have a shell of eight beneath it, as do the s electrons of aluminum and magnesium. The energy of this electron is very close to the $3d$ electrons. Therefore the whole group is attracted to the positively charged nucleus. For this reason copper, instead of being an active metal similar to aluminum, is considered a noble, i.e., corrosion-resistant, metal, in the same vertical group in the periodic table as silver and gold. As discussed in Chapter 16, the unique red color of copper is due to selective absorption of the spectrum of white light by interaction with the $3d$ electrons.

At first glance it is easy to be confused by the variety of copper alloys. Over thousands of years the names bronze and brass have been used differently, and other names such as gun metal, admiralty metal, gilding bronze, manganese bronze, and ounce metal have added to the confusion.

We shall take a simple approach based on the microstructures of the alloys involved. As with the light metals, it is possible to divide all the copper alloys into two classes: single-phase and polyphase alloys. We would expect the single-phase alloys to exhibit good ductility because the unit cell is FCC. Table 5.4 (pages 184–185) confirms this. The mechanisms for strengthening the single-phase wrought alloys are the usual solid-solution hardening and combinations of cold work and annealing. In the two-phase alloys, age hardening and other hardening by precipitates or second-phase dispersions are used.

In Table 5.4 we have chosen a few popular alloys to illustrate these points. We begin with ETP (electrolytic tough pitch) copper. This is copper that has been refined electrolytically. The term "tough pitch" refers to the oxygen level of about 0.04% (present as copper oxide), which gives the ingot

Table 5.4 TYPICAL PROPERTIES OF COPPER ALLOYS

Alloy Number	*Chemical Analysis, percent*	*Condition*	*Tensile Strength,* psi × 10^3*	*Yield Strength,* psi × 10^3*	*Percent Elongation*	*Hardness*	*Typical Use*
		Single-phase Wrought Alloys					
C11000	ETP, 99.9 Cu	Annealed	32	10	45	40 R_F	Architectural, electrical
		Cold-worked	50	40	6	85 R_F	
C26800	65 Cu, 35 Zn Yellow brass	Annealed	46	14	65	88 R_F	Plumbing, Grill work
		Cold-worked	74	60	8	80 R_B	
C61400	91 Cu, 7 Al, 2 Fe Aluminum bronze	Cold-worked	82	40	35	90 R_B	Condenser tubing
C71500	70 Cu, 30 Ni Cupronickel	Annealed	44	20	40	37 R_B	Desalinization tubing
		Cold-worked	75	68	12	85 R_B	
		Polyphase Wrought Alloys					
C17200	98 Cu, 2 Be Beryllium copper	Annealed	70	30	42	57 R_B	Springs, tools
		Precipitation-hardened	175	140	7	38 R_C	

Alloy Number	Chemical Analysis, percent	Condition	Tensile Strength,* psi × 10^3	Yield Strength,* psi × 10^3	Percent Elongation	Hardness	Typical Use
			Cast Alloys				
C81100	Cu	As cast	25	9	40	BHN 44	Electrical conductors
C83600	85 Cu, 5 Sn, 5 Zn, 5 Pb	As cast	37	17	30	BHN 60	Valves, bearings
C93700	80 Cu, 10 Sn, 10 Pb	As cast	35	18	20	BHN 60	Bearings, pumps
C96400	70 Cu, 30 Ni	As cast	68	37	28	BHN 140	Marine valves
C82400	98 Cu, 2 Be	Hardened	150	140	1	38 R_C	Dies, tools
C90500	88 Cu, 10 Sn, 2 Ni	As cast	44	22	6	BHN 85	Gears
C95300	89 Cu, 10 Al, 1 Fe	As cast	75	27	25	BHN 140	Gears, bearings
		Heat-treated	85	42	15	BHN 174	

*Multiply psi by 6.9×10^{-3} to obtain MPa or by 7.03×10^{-4} to obtain kg/mm^2.

a unique appearance. This grade of copper is widely used, although it will embrittle if it is heated in an atmosphere containing hydrogen because the hydrogen diffuses through the copper and encounters copper oxide at grain boundaries. Reaction takes place and water vapor is generated. These molecules are large compared to hydrogen. They do not diffuse appreciably, but form voids at the grain boundaries, leading to embrittlement. To avoid this, two other grades of copper are available: phosphorus deoxidized copper (C12200), in which the oxygen is eliminated by a phosphorus addition to the melt, and OFHC (C10200), (oxygen-free high-conductivity) copper, which is melted under special reducing conditions to eliminate oxygen. The mechanical properties of all grades are comparable. The change in yield strength from 10,000 to 40,000 psi (69 to 276 MPa) by cold working is especially important.

5.6 Solid-solution copper alloys

An important solid-solution effect is obtained with a silver content specified as 10 to 25 troy oz/ton. While this is only of the order of 0.01% by weight silver, the softening temperature of the cold-worked copper is raised more than 100°C. This lets the fabricator soft-solder (lead-tin alloy) cold-worked copper without lowering the strength by recovery and recrystallization. The presence of silver does not change the electrical conductivity.

The most widely used solid-solution alloys are those with zinc, which are called *brass*. The commonest range is between 65% Cu, 35% Zn and 70% Cu, 30% Zn. Since copper costs about $1/lb and zinc about $0.40/lb (1980 prices), the higher-zinc brasses are cheaper. The phase boundary of the α field is quite important because the higher-zinc alloys (above 35%) contain the β phase (Fig. 5.6). While this phase is stronger than α, it is more susceptible to a particular type of corrosion called *dezincification,* which is discussed in Chapter 12. The cold working and recrystallization of brass were discussed in Chapter 3.

Considerable strengthening is also accomplished with aluminum and nickel, as shown in the C61400 and C71500 alloys. The C71500 alloy is especially important in applications involving sea water, as in desalinization equipment.

5.7 Polyphase wrought copper alloys

The highest-strength copper alloy is produced by age-hardening a 2% beryllium alloy (C17200). Heating to 800°C gives a single-phase α solid solution (Fig. 5.7, page 188). After being quenched to obtain supersaturated α, the alloy is aged for 3 hr at 315°C to precipitate the γ_2 phase CuBe, which gives a coherent precipitate. This alloy is used widely for springs, nonsparking tools, and parts that require good strength plus high thermal and electrical conductivity. Other precipitation-hardening alloys contain silicon.

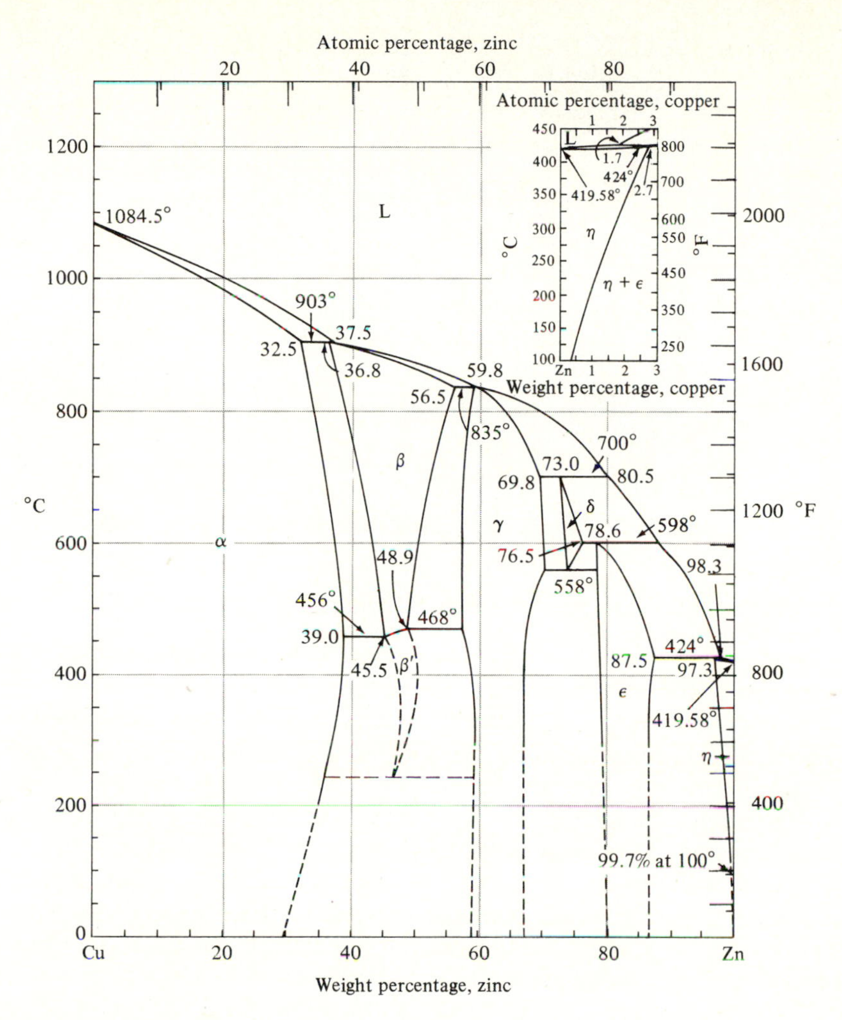

Fig. 5.6 *Copper–zinc phase diagram.* (American Society for Metals Handbook, 8th ed., Vol. 8, *Metallography, Structures and Phase Diagrams*, Metals Park, Ohio, 1973, p. 301)

5.8 Cast copper alloys

The cast alloys offer a wider range of structures because high ductility is not required for working. Of great importance are the high-lead alloys for bearings and the high-tin alloys for gears.

Copper itself is used in castings requiring good electrical and thermal conductivity. It is remarkable that although copper melts at 1084°C, *water-cooled* copper tuyères and lances can be used in the processing of steel, where temperatures reach 1750°C in the atmosphere above the liquid metal. As with the wrought alloys, strong solid solutions are formed with zinc, nickel, and aluminum.

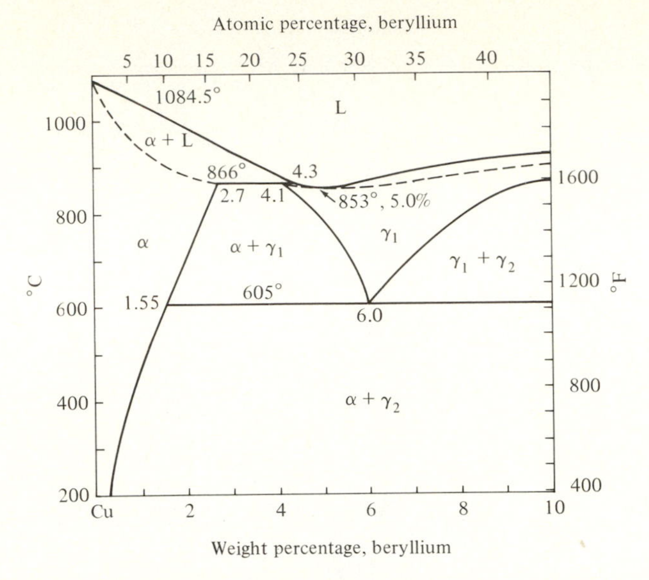

Fig. 5.7 *Copper–beryllium phase diagram.* (American Society for Metals Handbook, 8th ed., Vol. 8, *Metallography, Structures and Phase Diagrams,* Metals Park, Ohio, 1973, p. 271)

The lead alloys are of interest because liquid copper can dissolve lead in unlimited amounts (Fig. 5.8). On cooling, the lead precipitates as metallic lead because it is insoluble in solid copper. Alloys such as 85% Cu, 5% Sn, 5% Zn, 5% Pb are valuable for bearings because of the lubricating effects of the lead droplets (Fig. 5.9*a*, page 190).

The alloys with tin contain a hard intermetallic compound δ ($Cu_{31}Sn_8$). The presence of δ in a ductile α matrix gives an excellent gear bronze because it provides a good mating surface against hardened steel gears. The photomicrograph of Fig. 5.9*b* shows that the δ phrase cracks only after considerable deformation of the surrounding α.

The two-phase alloys include aluminum bronze, in which the aluminum exceeds the solid solubility in α and a hard γ phase is formed on cooling (Fig. 5.4). Under stress the failure occurs through γ regions (Fig. 5.9*c*). The alloy is also used in the heat-treated condition, obtained by heating to the β region and quenching. The β transforms on cooling to a structure called *martensite* (Fig. 5.9*d*), which is not shown on the equilibrium diagram. Chapter 4 discussed the nature of this transformation. Other polyphase alloys include cast beryllium-copper, copper-silicon, and manganese bronze.

5.9 Nickel alloys in general

Nickel is an element somewhat similar to iron in strength, but its alloys have exceptional resistance to corrosion and elevated temperatures, as well as important magnetic properties. This section will discuss only the corrosion-re-

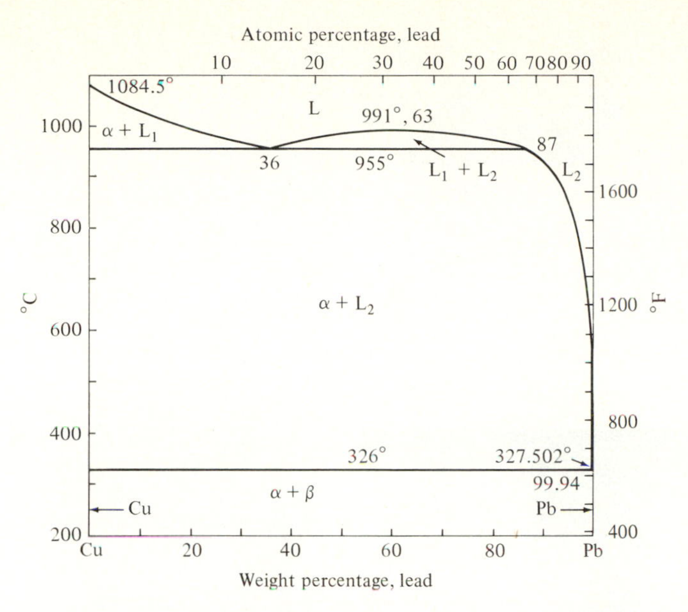

Fig. 5.8 *Copper–lead phase diagram. Copper and lead are essentially insoluble in one another at room temperature.* (American Society for Metals Handbook, 8th ed., Vol. 8, *Metallography, Structures and Phase Diagrams*, Metals Park, Ohio, 1973, p. 296)

sistant alloys; we will reserve discussion of the complex nickel-chromium-iron alloys, together with iron-base heat-resistant alloys, for Chapter 6.

The atomic structure of nickel is related to both copper and iron: $1s^2$, $2s^2$, $2p^6$, $3s^2$, $3p^6$, $3d^8$, $4s^2$. As in copper, there is little difference in energy between the $3d$ and $4s$ electrons, and the metal is relatively noble and corrosion-resistant. The unit cell is FCC, and the lattice parameter a_0 is close to that of copper: 3.52 Å vs. 3.62 Å.

The principal wrought alloys are Monel and Inconel (Table 5.5, page 191). The nickel-copper alloy Monel is an extension of the copper-nickel alloys to the high-nickel side of the phase diagram (Fig. 5.10, page 192).

The rules of solid solubility suggest that copper and nickel should form an extensive series of solid solutions (Chapter 2), and this is the case. Although nickel is stronger than copper, the intermediate nickel-copper alloy (Monel) has a higher yield strength than nickel, another illustration of solid-solution strengthening. This shows that the element added in solid solution will harden and strengthen the solvent metal even though the solute element is soft itself. Both Monel and Inconel may be age-hardened, as shown by the alloys Duranickel 301 and Monel K500.

5.10 Cast nickel alloys

The cast alloys are parallel to the wrought alloys with two exceptions, Monel 505 and Inconel 705 (Table 5.5). These materials contain 4 to 6% silicon,

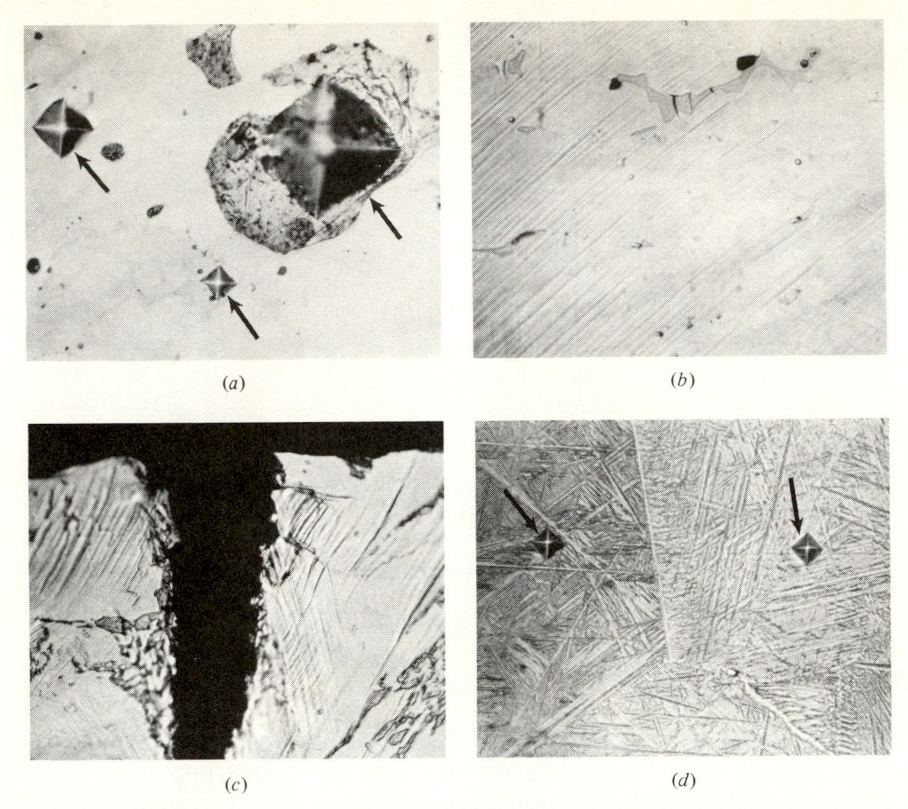

Fig. 5.9 *Typical microstructures of cast copper-base alloys. (a) 85% Cu, 5% Sn, 5% Pb, 5% Zn alloy. Lead phase (dark gray), VHN 16; matrix (white), VHN 98; δ phase (light gray), VHN 293, 500×, chromate etch. (b) 88% Cu, 8% Sn, 4% Zn. Islands of δ phase are in the FCC copper matrix; specimen was stressed after polishing. Ductile matrix shows slip, while brittle δ phase cracked; 500×, chromate etch. (c) 89% Cu, 10% Al, 1% Fe (as cast). Gray γ_1 phase in FCC copper matrix. During stress the ductile matrix showed slip, but failure occurred through the γ_1; 500×, chromate etch. (d) Alloy of (c) heated to 1650°F for 1 hr and water-quenched, forming a martensitic structure. Specimen was not stressed; 500×, chromate etch. (Arrows point to microhardness indentations.)*

which produces a hard intermetallic compound Ni_3Si. This dispersion hardens the cast alloy, and the material is further hardened by aging. Since the elongation is only 3%, these alloys are not available in wrought form.

Titanium, Zinc, and Other Alloys

5.11 Titanium alloys

Titanium alloys have been developed only recently because their great chemical reactivity makes them difficult to melt, cast into ingots, and hot-work.

Table 5.5 TYPICAL PROPERTIES OF NICKEL ALLOYS

Alloy Number	Chemical Analysis, percent*	Condition	Tensile Strength,† psi × 10^3	Yield Strength,† psi × 10^3	Percent Elongation	Hardness	Typical Use
Single-phase Wrought Alloys							
Nickel 200	99.5 Ni	Annealed	65	22	47	BHN 75	Corrosion-resistant parts
		Cold-worked	120	92	8	BHN 230	
Monel 400	66 Ni, 32 Cu	Annealed	72	35	42	BHN 110	Corrosion-resistant parts
		Cold-worked	120	110	8	BHN 241	
Inconel 600	78 Ni, 15 Cr, 7 Fe	Annealed	100	50	35	BHN 170	Corrosion-resistant parts
		Cold-worked	150	125	15	BHN 290	
Polyphase Wrought Alloys							
Dura nickel 301	94 Ni, 4.5 Al, 0.5 Ti	Annealed	105	42	40	90 R_B	Corrosion-resistant parts
		Cold-worked, age-hardened	200	180	8	40 R_C	High-strength parts
Monel K500	65 Ni, 2.8 Al, 0.5 Ti, 30 Cu	Annealed	97	52	35	85 R_B	Corrosion-resistant parts
		Cold-worked, age-hardened	185	155	7	34 R_C	High-strength parts
Single-phase Cast Alloys							
Nickel 210	95 Ni, 0.8 C	As cast	52	25	22	BHN 100	Condensers
Monel 411	64 Ni, 32 Cu, 1.5 Si	As cast	77	38	35	BHN 135	Paper mill equipment
Inconel 610	68 Ni, 15 Cr, 2 Nb, 10 Fe	As cast	82	38	20	BHN 190	Dairy equipment
Polyphase Cast Alloys							
Monel 505	63 Ni, 29 Cu, 4 Si	Aged	127	97	3	BHN 340	Valve seats
Inconel 705	68 Ni, 9 Fe, 6 Si, 15 Cr	Aged	110	95	3	BHN 340	Exhaust manifolds

*These represent the important elements. Other elements may be present.

†Multiply psi by 6.9 × 10^{-3} to obtain MPa or by 7.03 × 10^{-4} to obtain kg/mm^2.

Note: Nickel-base superalloys are discussed with other superalloys in Chap. 6.

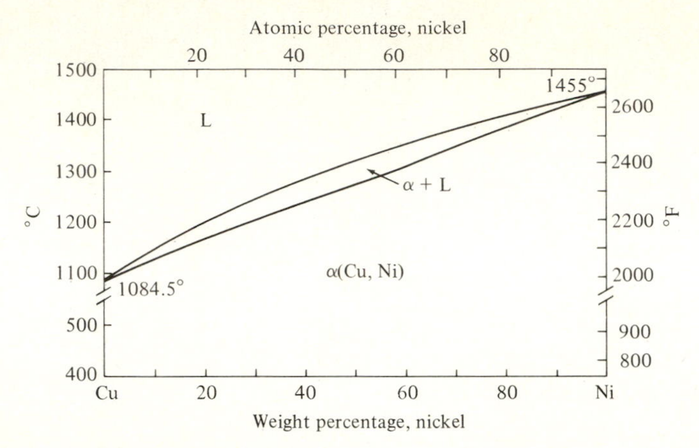

Fig. 5.10 *Copper–nickel phase diagram.* (American Society for Metals Handbook, 8th ed., Vol. 8, *Metallography, Structures and Phase Diagrams,* Metals Park, Ohio, 1973, p. 294)

However, despite their general reactivity, titanium alloys are very corrosion-resistant in certain cases, as discussed in Chapter 12. Also, the relatively low density and high strength of titanium make it a candidate for aircraft applications. In particular, the skin heating at high speed results in aluminum and magnesium alloys losing strength, whereas the titanium alloys can withstand the elevated temperatures.

The atomic structure of titanium is $1s^2$, $2s^2$, $2p^6$, $3s^2$, $3p^6$, $3d^2$, $4s^2$. In contrast to copper and nickel, titanium has only four $(3d + 4s)$ electrons, and these have a stable shell of eight beneath. Therefore these electrons are held rather loosely, and the element is quite active. For example, there is no known crucible material that will not react with liquid titanium, so it is melted in a water-cooled copper crucible. This produces a shell of solid titanium between the liquid and the copper.

Titanium has an HCP structure α at room temperature which transforms to BCC β at 880°C. Certain elements, such as aluminum, stabilize the α phase, giving the "all-α" alloys. Chromium, iron, molybdenum, and vanadium lower the transformation temperature, stabilizing β and leading to α-β or all-β alloys. Table 5.6 gives typical properties.

5.12 Zinc alloys

The principal use of zinc in structures is for die castings. Zinc alloys are ideally suited for this process because of their low melting point and because they do not corrode steel crucibles or dies. As a result, a number of automotive parts, toys, and building hardware are made of zinc alloys (Table 5.7).

Table 5.6 TYPICAL PROPERTIES OF TITANIUM ALLOYS (WROUGHT)

Material	*Chemical Analysis, percent*	*Structure*	*Tensile Strength,* psi $\times 10^3$*	*Yield Strength,* psi $\times 10^3$*	*Percent Elongation*	*BHN*
Ti	99 Ti†	α	38 to 100	22 to 85	17 to 30	115 to 220
Ti, Al, Sn	5 Al, 2.5 Sn‡	α	125	117	18	360
Ti, Al, V	6 Al, 4 V§	α-β	170	155	8	380

*Multiply psi by 6.9×10^{-3} to obtain MPa or by 7.03×10^{-4} to obtain kg/mm^2.
†Variations in properties depending on whether cold-worked or annealed.
‡Annealed sheet.
§Solution heat-treated and aged.

5.13 The less common metals, precious metals

We are entering a period of tremendous growth in the use of the rarer metals. The era began with the discovery that in piston-driven aircraft each part is literally worth its weight in gold. Removing one pound of dead weight makes it possible to carry one more pound of cargo each trip. Over the life of the plane, this extra revenue would no doubt equal the value of a pound of gold. In space travel, the value of a pound of weight is orders of magnitude greater. As a result, parts made of beryllium, rhenium alloys, molybdenum alloys, platinum, gold, silver, and indium are used freely in commerce. A common example is the use of expendable fine wire platinum-platinum, 10% rhodium thermocouples to check each heat of steel. Years ago a thermocouple of this type was a carefully guarded laboratory tool. A detailed discussion of the properties of the less common metals is beyond the scope of this text, but we should mention that the following metals are readily available:

Refractory metals for high-temperature service: molybdenum, tantalum, tungsten, rhenium
Metals for use in nuclear reactors: zirconium, hafnium
Precious metals: gold, silver, platinum, palladium, rhodium, ruthenium, osmium, iridium
Tin and its alloys
Rare earths

Table 5.7 TYPICAL PROPERTIES OF ZINC ALLOYS (DIE-CAST)

Alloy Number	*Chemical Analysis, percent*	*Nominal Tensile Strength,* psi $\times 10^3$*	*Percent Elongation*	*BHN*
AG40A	4 Al, 0.05 Mg	41	10	82
AC41A	4 Al, 1 Cu, 0.05 Mg	48	7	91

*Multiply psi by 6.9×10^{-3} to obtain MPa or by 7.03×10^{-4} to obtain kg/mm^2.

Data such as the analyses and physical properties of these materials are available in the American Society for Metals handbooks and *The Materials Selector* (see References).

EXAMPLE 5.4 The discussion in the chapter suggests that most commercial alloys have compositions close to the extremities of any phase diagram. In other words, compositions such as 50%X, 50%Y are seldom used. Suggest why.

ANSWER Alloys are classified as either single-phase or polyphase. Single-phase alloys contain relatively small amounts of other elements to prevent second phases from forming, except in systems such as the copper-nickel system where there is complete solid solubility (Fig. 5.10). Here we stay near the copper end because of the higher cost of nickel unless we absolutely require a particular property of nickel, such as high-temperature resistance.

In polyphase alloys the center portion of the phase diagram usually shows hard and brittle compounds, which, although wear- and abrasion-resistant, are difficult to maintain in a desirable distribution to maximize strength.

SUMMARY

We devoted our efforts in this chapter to explaining how the properties of a wide range of nonferrous alloys can be understood in terms of their structures.

In the aluminum alloys the wrought materials can be divided into two groups: the single-phase materials, which are strengthened by work hardening, and the polyphase alloys, which are age-hardened and may also be work-hardened. The cast alloys are strengthened by age hardening or by controlling the composition to produce large amounts of a hard dispersed phase such as silicon.

The magnesium alloys are similar in many ways to the aluminum alloys, but the single-phase alloys are not as ductile because of limited slip in the hexagonal structure.

The copper alloys have great ductility in the pure metal and in the single-phase alloys. The polyphase alloys may be age-hardened or strengthened with a martensite reaction.

The nickel and other alloys follow similar reasoning.

DEFINITIONS

Aluminum alloys HX indicates the amount of cold work and annealing for single-phase alloys; see text. TX indicates the combination of age hardening and cold work for two-phase alloys; see text.

Castings Parts made by pouring liquid metal into a mold of the proper dimensions. The mold may be made of sand bonded with clay or resin (sand castings) or metal (die or permanent mold castings).

Copper alloys

Brass An alloy of copper and zinc.

Bronze An alloy of copper and a specified metal, such as tin bronze, aluminum bronze, silicon bronze.

Cupronickel An alloy of copper and 10 to 30% nickel.

Ferrous alloys Alloys that have iron as the base element, such as wrought iron, steel, and cast iron.

Magnesium alloys Described by the same H and T designations used for aluminum alloys, but with a different code for composition.

Nickel alloys

Monel An alloy of nickel and 30% copper.

Inconel An alloy of nickel and 15% chromium.

Nonferrous alloys Alloys that do not have iron as the base element, such as the alloys of aluminum, copper, magnesium, nickel, zinc, and titanium.

Titanium alloys The α-β alloys are a combination of HCP and BCC structures.

Wrought products Parts produced by working a solid metal either hot or cold; for example, by hot and cold forging, rolling, extrusion, spinning, stamping, or wire drawing.

PROBLEMS

5.1 Referring to Table 5.2, explain why the tensile and yield strengths of 3003 in the H18 condition are higher than those for 1060 in the same cold-worked condition. Why is the elongation much lower than in the annealed condition? (Sections 5.1 through 5.4)

5.2 The two most important coherent precipitates in age-hardening aluminum alloys are $CuAl_2$ and Mg_2Si. The solubility of Mg_2Si varies from 1.8% at the eutectic temperature to less than 0.1% at room temperature. Explain why 5050 is listed as a single-phase alloy whereas 6061 can be age-hardened (Table 5.2). (Sections 5.1 through 5.4)

5.3 Calculate the parameter called the "strength-to-weight ratio" or "specific strength" for several of the higher-strength aluminum alloys of Table 5.2, and compare these to high-strength steel with yield strength of 250,000 psi (1.725×10^3 MPa). This value is equal to yield strength divided by density (in pounds per cubic inch). What is the significance of this value? Take 2.7 as the specific gravity of aluminum and 7.8 as that of steel. Density = specific gravity $\times$ 0.0361 lb/in.^{3}. (Sections 5.1 through 5.4)

5.4 Considering only the silicon, calculate the amount of silicon-rich β produced at freezing in aluminum 356.0. Why is this alloy not available in wrought form? (Assume that magnesium has no effect on the aluminum-silicon equilibrium diagram.) (Sections 5.1 through 5.4)

5.5 The density of magnesium is 0.064 lb/in.^{3}. What yield strength would be required to enable a magnesium alloy to compete with the best aluminum alloy on a strength-to-weight basis? (See Prob. 5.3.) (Sections 5.1 through 5.4)

5.6 The engine block of a Chevrolet Vega contains 16% silicon in aluminum. What percentage of area of the cylinder wall contains the β silicon phase? The specific gravity of silicon is 2.33 and that of aluminum is 2.70. (Sections 5.1 through 5.4)

5.7 Calculate the weight of magnesium in a cubic mile of sea water. (There are 1.27 g of magnesium per kilogram of typical sea water; the specific gravity of sea water is 1.01.) (Sections 5.1 through 5.4)

5.8 A student tested two tensile specimens of 1060 aluminum in the annealed and H18 conditions, respectively, and one specimen of 5052 aluminum in the H18 condition, then mixed up the data. Using the values given, place the logical values in the table below. (Sections 5.1 through 5.4)

	Analysis	Condition	Tensile strength, psi	Yield strength, psi	Percent elongation
1060	99.6% min Al	0 (annealed)	________	________	________
1060	99.6% min Al	H18	________	________	________
5052	2.5% Mg, 0.2% Cr, Bal. Al	H18	________	________	________

The tensile test data were:

Tensile strength:	19,000;	10,000;	41,000 psi
Yield strength:	4,000;	18,000;	39,000 psi
Percent elongation:	42;	6;	4

5.9 Copper-nickel alloys are used above 330°C, but copper alloys containing lead are not. Explain this, using phase diagrams. (Sections 5.5 through 5.13)

5.10 Construct fraction charts for the phases you would expect for alloys of copper containing 20% lead and 40% lead slowly cooled. Why is the lead content of commercial castings always below about 30%? (Sections 5.5 through 5.13)

5.11 The range of tensile strength is 125,000 to 160,000 psi (0.863 to 1.105 $\times$ 10^3 MPa) for a 1% beryllium-copper alloy but 190,000 to 215,000 psi (1.31 to 1.48 $\times$ 10^3 MPa) for a 2% beryllium-copper alloy. Explain this quantitatively on the basis of the phases present. (Sections 5.5 through 5.13)

5.12 Why is the yield strength of wrought Inconel greater than that of wrought Monel 400 although the alloy content is lower? (Sections 5.5 through 5.13)

PROBLEMS COVERING OPTIONAL SECTIONS AND PREVIOUS CHAPTERS

5.13 Draw a time-temperature chart for artificial age hardening of an aluminum alloy. Include all steps. (Sections 5.3 and 4.21)

5.14 See Figures 5.3 and 5.4. Indicate whether or not the following alloys are likely candidates for age hardening: (Sections 5.3 and 4.21) (*a*). 90% Al, 10% Mg, (*b*). 97% Mg, 3% Al, (*c*). 80% Al, 20% Mg, (*d*). 90% Cu, 10% Al

5.15 Assume that a large aluminum structure is to be made from an age-hardening alloy. (Sections 5.3 and 4.21)

a. At what stage in the aging process would cold working be done? Why?
b. Why would an aluminum alloy that would exhibit natural aging be selected?
c. Why might welding be difficult?

5.16 The specification for the T6 condition reads "solution heat treated plus artificially aged."

a. Using the phase diagram shown, draw a time-temperature chart specifying the heat treatment you would use to attain this condition in a 6% Cu, 94% Al alloy, assuming that your starting material was slow-cooled from 900°F (482°C) (Sections 5.3 and 4.21)

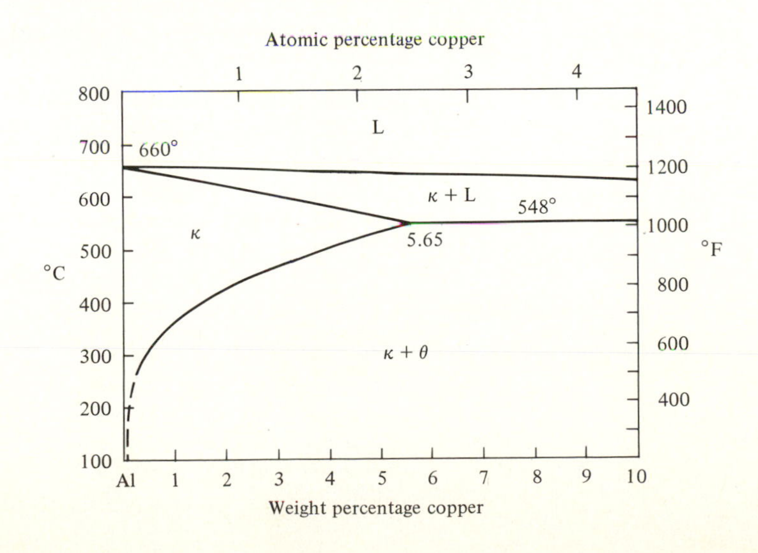

b. Choose the description that best describes the microstructure after treatment: (1) supersaturated κ with separate grains of θ, (2) supersaturated κ, (3) κ with fine precipitate of θ, (4) κ with fine precipitate of θ plus coarse grains of θ.

5.17 At room temperature the slip plane for magnesium is (0001), whereas at elevated temperatures the $(11\bar{2}2)$ and $(10\bar{1}1)$ planes are operative in the unit cell. Sketch these planes in the HCP cell. (Sections 5.4 and 2.8)

5.18 We are going to produce a new solid-solution alloy by adding 5 atomic percent balonium to copper. Indicate how the physical properties might vary (use an X in the appropriate column) (Sections 5.5 and 3.22)

	Increase	Decrease	No Change
Tensile strength	_____	_____	_____
Ductility	_____	_____	_____
Electrical conductivity	_____	_____	_____
Hardness	_____	_____	_____

Possibly important data:

	Pure Copper	Pure Balonium
Tensile strength (psi)	35,000	30,000
Ductility (elongation—%)	50	50
Electrical conductivity (relative to copper—%)	100	90
Hardness	90 BHN	85 BHN

5.19 In copper castings, the mechanisms available for strengthening single-phase alloys are (circle proper answers) (Sections 5.5 and 4.21): Cold rolling; substitutional solid solutions; interstitial solid solution; heat to 1600°F and water quench.

In polyphase copper alloys, the mechanisms for strengthening two-phase alloys are (circle proper answers): Substitutional solid solutions; interstitial solid solutions; martensite formation; age-hardening; controlled isothermal transformation.

5.20

Material	*Chemical Analysis*	*Microstructure*	*Hardening Mechanisms (Use Code)*
(1) Cartridge brass	Cu 70%, Zn 30%	Elongated grains of α showing slip	
(2) Gun metal	Cu 88%, Sn 10%, Zn 2%.	Equiaxed $\alpha + \delta$ phase of CuSn compound (hard)	

Material	*Chemical Analysis*	*Microstructure*	*Hardening Mechanisms (Use Code)*
(3) Aluminum bronze	Cu 90%, Al 10%	Grain boundary precipitate of α plus α + γ_1 eutectoid. γ_1 is CuAl compound (hard)	
(4) Aluminum alloy	4% Cu, balance Al, condition 0 (as annealed)	Particles of $CuAl_2$ + equiaxed α	
(5) Same alloy as in (4)	Condition T9: solution heat-treated, aged, cold-worked	Elongated α showing slip plus very fine (coherent) precipitate of $CuAl_2$ (at 60,000×)	

In the table, list by code letter *all* the hardening mechanisms that have raised the strength and hardness of each material above that of the pure element (copper or aluminum, depending on the case), paying attention to the microstructure given. [*Code:* Mechanisms: A = solid-solution strengthening; B = age hardening; C = cold working; D = dispersion hardening; E = martensite formation.] (Sections 5.3 through 5.8 and 4.17 through 4.21)

5.21 Select from the code below the one hardening mechanism of most importance for the following alloys: (Sections 5.3 through 5.8 and 4.17 through 4.21) 1060 aluminum (99.6% min Al); 2014 aluminum (4.5% Cu, 0.8% Si, 0.8% Mn, 0.5% Mg); C11000 ETP 99.9% Cu; C71500 70% Cu, 30% Zn brass; C17200 (98% Cu 2% Be).
Code: (*a*) work hardening, (*b*) dispersion hardening, (*c*) age hardening, (*d*) solid-solution hardening, (*e*) martensite reaction.

5.22 Shown here is a series of age-hardening curves for one alloy, obtained at different temperatures. Answer the following questions using the letters A, B, C for the three curves. (There may be more than one answer for each question.) (Section 4.21)

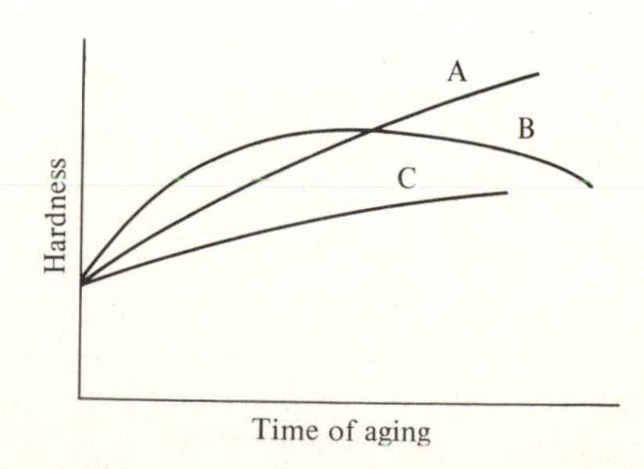

a. Which was aged at the lowest temperature? ______
b. Which shows overaging? ______
c. Which shows solution treatment? ______

5.23 This simple phase diagram shows those alloys that are considered "commercial" as shaded areas. For each group (1, 2, and 3), indicate the principal hardening or strengthening mechanism. (Section 4.21)

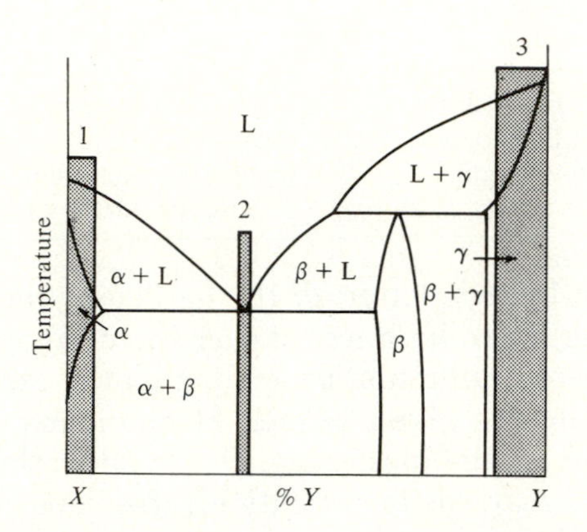

5.24 *a.* Which of the alloys (alloy 1, 2, 3, and/or 4) in the figure below could be strengthened by age hardening? (Assume that β forms a coherent precipitate in α.) (Section 4.21)

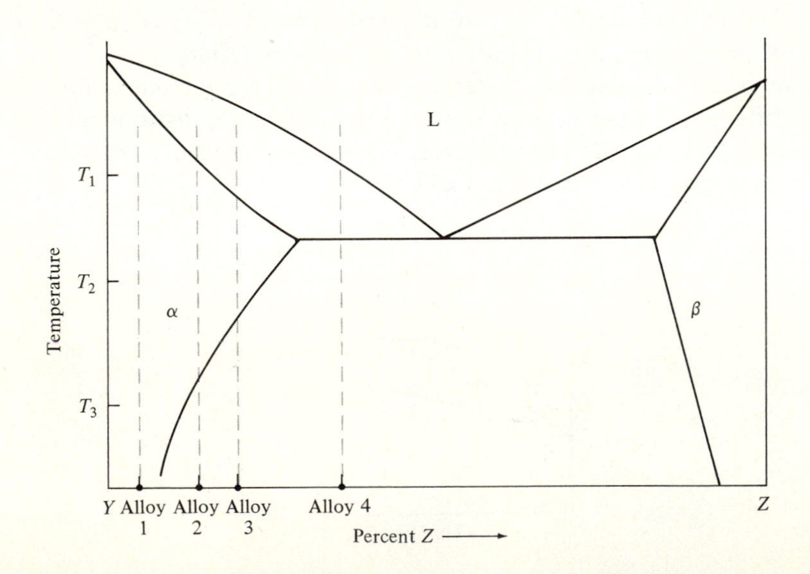

b. In which *one* of the above alloys (list number) could the greatest amount of element Z be dissolved in solid solution?

c. Select a treatment from the following list that would represent the T5 condition, which is defined as "artificially aged only," for alloy 3. (1) Heat to T_1, hold 3 hr, water quench, reheat to T_3, hold 4 hr, air cool; (2) heat to T_3 for 4 hr, air cool; (3) heat to T_2, hold 3 hr, water quench, reheat to T_3, hold 4 hr, air cool.

1μ

6

STEEL, SUPERALLOYS, CAST IRON, DUCTILE IRON, MALLEABLE IRON

IN this chapter we shall take up the second principal group of metallic materials, those based on iron and a few related high-nickel alloys called the "superalloys." Only a few structures are involved, the two crystal forms of iron, BCC and FCC, discussed previously, and a hard compound, iron carbide (Fe_3C)*. By altering the amount, shape, and distribution of the carbide, we can obtain a wide range of properties.*

The electron micrograph shows another, special type of precipitation in a nickel-base superalloy. The matrix is FCC, but the precipitate is $Ni_3(Al,Ti)$*. This gives an exceptional combination of high-temperature strength and oxidation resistance required for gas turbine blades.*

The large photomicrograph shows Ni_3Al (γ') *precipitate in a nickel alloy* (γ) *matrix (approximately 10,000×). The small insert shows dislocations cutting through the* γ' *precipitate (approximately 50,000×).*

6.1 Introduction to iron alloys

Iron and steel are so widely used and inexpensive that at first they seem without glamor compared with the colorful copper alloys and the light alloys that have unusual aircraft and space travel applications. However, some of the most advanced materials—such as special stainless steels for corrosion resistance, alloys for high-temperature service in gas turbines, and ductile iron for castings—are based on iron. The family of iron alloys has the widest range of microstructures and properties, ranging from high-strength steels to the lower-strength but elegantly castable gray cast iron. More than 90% of the tonnage (1980) of metallic materials is based on iron. Among iron-base alloys, the principal difference is in carbon content.

Material	*Carbon, Percent*	*Annual Tonnage, in millions*
Steel (wrought and cast)	0.05 to 2	120
White cast iron	2 to 4.5	1
Malleable iron	2.5 to 3	1
Gray iron	3 to 4	18
Ductile (nodular) iron	3.5 to 4.5	2

To realize the importance of iron alloys, consider the makeup of a car:

Body (doors, sides, frame, fenders): low-carbon steel
Chassis: welded steel
Engine: 95% cast and ductile iron
Gears, drive shaft, axle, torsion bar, springs: alloy steel
Brake disks or drums: cast iron
Wheels: steel
Axle housings, differential housing: malleable iron or ductile iron
Bumpers: steel

We see, therefore, that most parts that bear loads or transmit power are iron alloys. We shall discuss steels first, then the other alloys. After analyzing the high-alloy steels, we shall take up other heat-resistant alloys, including the nickel- and cobalt-base superalloys. Following this, we shall consider cast iron, malleable iron, and ductile iron.

6.2 The iron–iron carbide diagram and its phases

To understand the basic differences among iron alloys and the control of properties, the iron–iron carbide diagram is essential. Figure 6.1 shows the diagram that is commonly used. We note first that the carbon scale only goes up to 6.67% carbon, where we encounter iron carbide, Fe_3C. This should cause us no concern, since we have already used a part of a diagram, the portion of the copper-aluminum diagram extending to $CuAl_2$ (θ), in our discussion of age hardening.

The first step is to examine the individual phases. Beginning with pure iron, we see three solid phases, alpha (α), gamma (γ), and delta (δ), shown at the left side. If we begin with α at 68°F (20°C) and heat slowly, we can expect α to transform to γ at 1670°F (910°C), γ to δ at 2540°F (1393°C), and δ finally to melt at 2802°F (1538°C).

We shall look in vain for β iron; it was lost in antiquity. Naturally, because of its importance the iron–iron carbide diagram was the first to receive attention. Early investigators noted that near 1400°F (760°C) iron lost its ferromagnetism. They considered this to be a phase change and gave the symbol β to iron in the range 1400 to 1670°F (760 to 910°C). It was found, however, that this magnetic effect is not a phase change but is due to a shift in alignment of the atoms. This point will be considered further in Chapter 15.

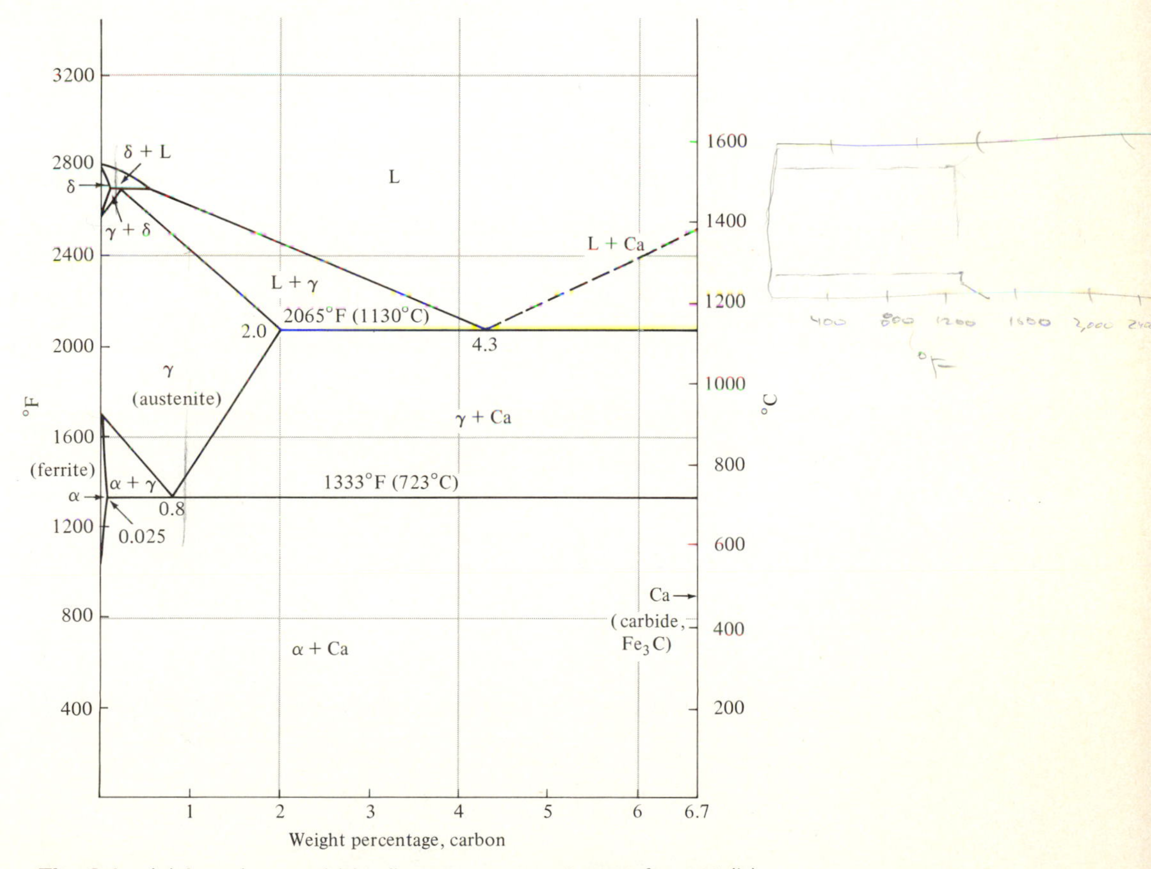

Fig. 6.1 *(a) Iron–iron carbide diagram; see next page for part (b).*

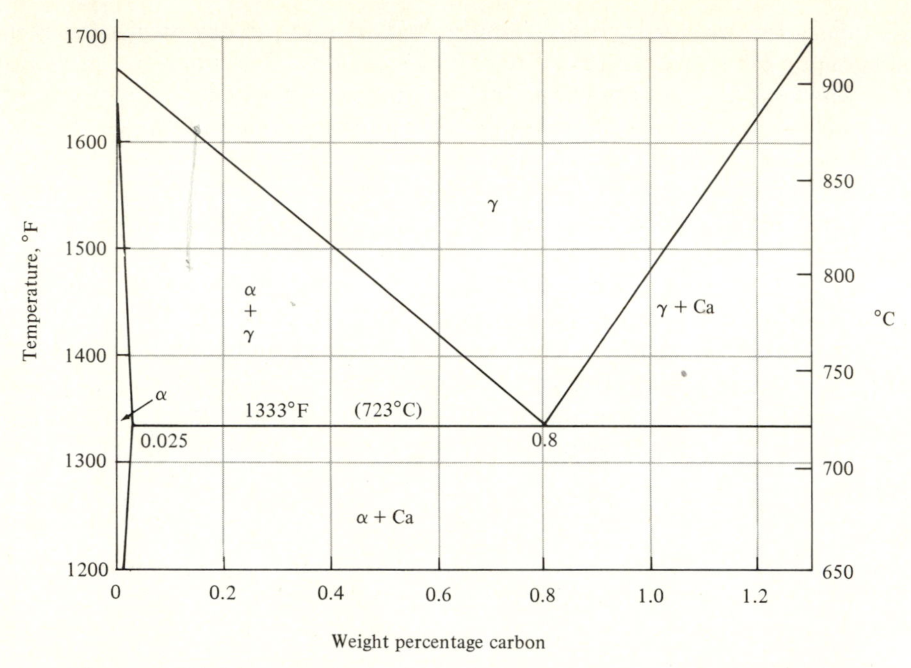

Fig. 6.1 *(b) Enlarged eutectoid portion of the iron–iron carbide diagram*

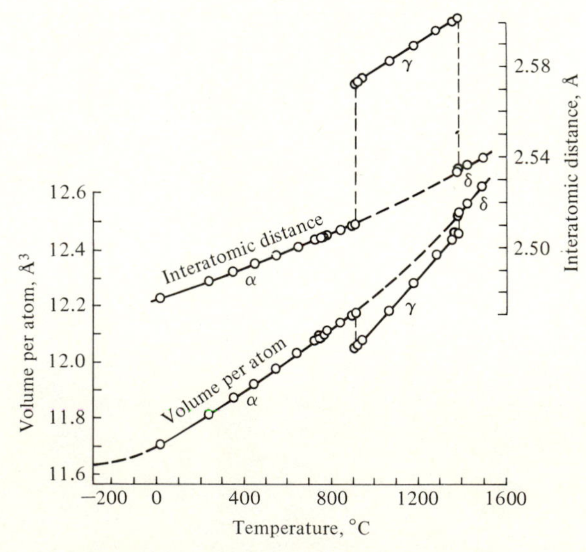

Fig. 6.2 *Lattice parameter of iron as a function of temperature.* (C.S. Barrett and T.B. Massalski, *Structure of Metals,* 3d ed., McGraw-Hill, New York, 1966, Fig. 10.5, p. 232)

This left metallurgists with α, γ, and δ. The picture was further simplified when x-ray diffraction evidence showed α and δ to have the same crystal structure. Both these phases have BCC structures, and the lattice parameter a of δ is the same as that for α if allowance is made for expansion with temperature (Fig. 6.2).

The γ phase is FCC and therefore more densely packed. There is a contraction of about 1% in volume in the $\alpha \rightarrow \gamma$ transformation and an expansion of 0.5% in volume in the $\gamma \rightarrow \delta$ change.

Note that in the phase diagram (Fig. 6.1*a*) the amount of carbon that can be dissolved in γ is 2% maximum (2065°F, 1130°C). This value is many times greater than the maximum solubility in α, 0.025% carbon (1333°F, 723°C). We can see the basis for this if we compare the sizes of the interstitial holes in BCC and FCC unit cells.

EXAMPLE 6.1 Compare the sizes of the largest interstitial holes in the BCC and FCC structures found in iron. How does the size of the carbon atom compare? (The atomic radius of iron is 1.27 Å in FCC and 1.24 Å in BCC.) Calculate the atomic packing factor in both iron structures.

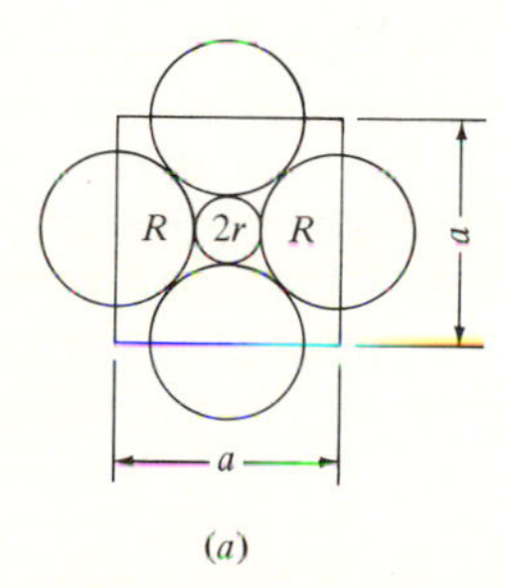

(*a*)

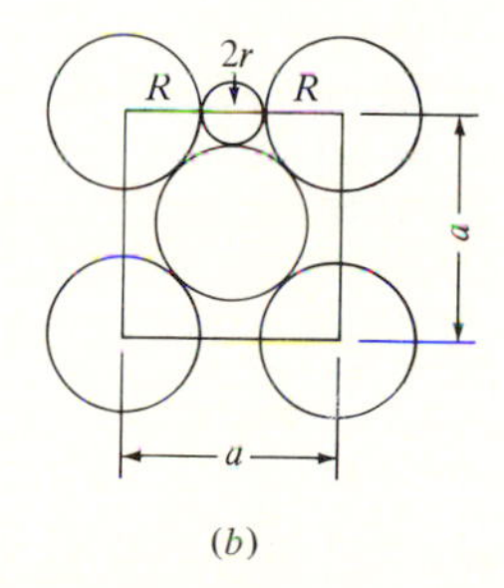

(*b*)

ANSWER A horizontal plane through the center of an FCC structure contains the largest interstitial hole (the same hole can be found in the face of the FCC).

$$a = \frac{4R}{\sqrt{2}} = 2R + 2r \quad \text{or} \quad r = 0.414R = 0.414 \times 1.27 = 0.52\ \text{Å}$$

In the BCC structure the largest interstitial hole occurs at the coordinates $\frac{1}{2}$, 0, $\frac{3}{4}$.

$$(r + R)^2 = (\tfrac{1}{2}a)^2 + (\tfrac{1}{4}a)^2$$

$$r^2 + 2rR + R^2 = \tfrac{5}{16}a^2 = \frac{5}{16}\left(\frac{4R}{\sqrt{3}}\right)^2$$

$$r^2 + 2rR + R^2 = \frac{5R^2}{3}$$

$$r + R = R\sqrt{\tfrac{5}{3}}$$

or

$$r = 0.291R = 0.291 \times 1.24 = 0.36\ \text{Å}$$

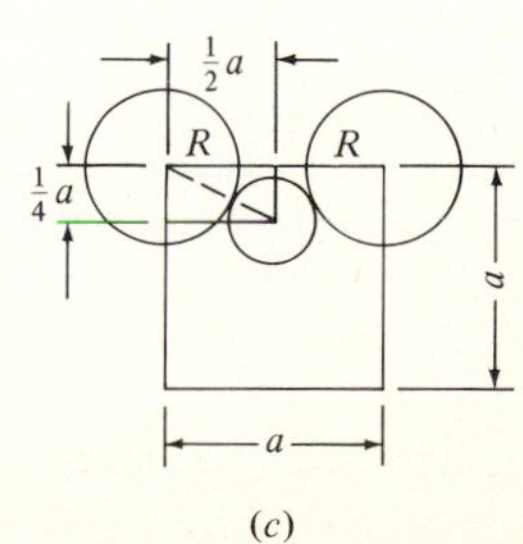

(*c*)

Since the atomic radius of carbon is 0.77 Å, it has a much better chance of interstitial solution in FCC iron. (Note that the radius of the iron atom is slightly different in the BCC and FCC configurations.) The atomic packing factor (APF) is defined as (volume of atoms/unit cell)/volume of unit cell.

$$\text{BCC: APF} = \frac{2 \text{ atoms} \times \frac{4}{3}\pi(1.24 \text{ Å})^3}{\left(\frac{4 \times 1.24 \text{ Å}}{\sqrt{3}}\right)^3} = 0.68$$

$$\text{FCC: APF} = \frac{4 \text{ atoms} \times \frac{4}{3}\pi(1.27 \text{ Å})^3}{\left(\frac{4 \times 1.27 \text{ Å}}{\sqrt{2}}\right)^3} = 0.74$$

Even though the packing of atoms is more efficient in the FCC, the interstitial holes in the FCC are larger than those in the BCC.

In addition to the solid solutions of iron and carbon, the other solid phase in the diagram is iron carbide, Fe_3C, also called *cementite*. This curious name arose from an ancient process in which carbon was diffused into iron by making up a package of layers of iron and carbon compounds, then heating it in a furnace. The term cementite derived from this so-called "cementation" process. It is evident from the diagram that cementite will be encountered at 20°C in any alloy from 0.006 to 6.67% carbon. In sharp contrast to the BCC and FCC iron structures, which are both ductile (over 40% elongation) and soft (BHN 100 to 150), cementite is brittle (0% plastic elongation) and hard [BHN over 700 (VHN 1200)]. We can summarize the phases and their several names as follows.

Temperature Range, Pure Fe	*Phase*	*Crystal Structure*	*a, Å*	*BHN*	*Percent Elongation*
Room temperature to 1670°F (910°C)	α, ferrite	BCC	2.86	~150	~40
1670 to 2540°F (910 to 1393°C)	γ, austenite	FCC	3.60	~150	~40
2540 to 2802°F (1393 to 1538°C)	δ, delta ferrite	BCC	2.89	~150	~40
	Iron carbide (cementite, Fe_3C)	orthorhombic	—	~700	~0

Plain Carbon and Low-alloy Steels: Equilibrium Structures

6.3 Steel

In general, steel contains less than 2% carbon. We shall concentrate our attention on this region. We shall first discuss the steel structures formed under equilibrium conditions, then the nonequilibrium hardening reactions.

6.4 Hypoeutectoid steels

We shall consider first hypoeutectoid steels (up to 0.8% carbon) (Fig. 6.3, page 210), then hypereutectoid steels (0.8 to 2% carbon).† The low-carbon steels are by far the most important of the group, mainly because of their high ductility, both hot and cold, which enables them to be readily fabricated into shapes of excellent toughness and strength. All structural and automobile body steels fall into this group. Most steel castings, such as railroad-car parts, are also in this composition range.

Let us follow the cooling of a typical steel of this type (0.3% carbon) from the liquid state, using the phase diagram as a map and drawing a fraction chart (Fig. 6.4) as discussed in Chapter 4.

The most important changes during cooling occur in the range 1600 to 1300°F (871 to 704°C). Ferrite (α) begins to precipitate at the austenite (γ) grain boundaries at 1550°F (843°C) (Fig. 6.3). This continues to 1334°F (724°C), and the γ changes from 0.3 to 0.8% carbon as a result (tie line changes in the $\alpha + \gamma$ region upon cooling). At 1333°F (723°C) the remaining γ transforms to a mixture of α + carbide by a constant-temperature eutectoid reaction, $\gamma \rightarrow \alpha$ + carbide. The important point is that an interleaved or lamellar mixture of ductile α and hard carbide is formed. This composite is called *pearlite*. It has higher strength but lower ductility than ferrite. Note that the name is given to the unique *lamellar mixture* of the two phases α + carbide.

Let us return to our discussion of the fraction chart (Fig. 6.4). At 1334°F (724°C) we have approximately 63% ferrite and 37% austenite. We shall refer to this ferrite as primary ferrite so as not to confuse it with the ferrite that results from the formation of pearlite. Now at 1333°F (723°C) all the remaining austenite (of 0.8% carbon) transforms to pearlite, and we therefore obtain eutectoid ferrite. If we had started out with 100% austenite at 1333°F (723°C) (this would have required an initial carbon content of 0.8%), we would have obtained $(6.67 - 0.8)/(6.67 - 0.025)$ or 88% ferrite and 12% carbide. However, since we only had 37% austenite, we would obtain $0.88 \times 37 = 32.5\%$ ferrite

---

†Hypoeutectoid means below eutectoid composition, that is, <0.8% carbon. Hypereutectoid means above eutectoid composition, that is, >0.8% carbon.

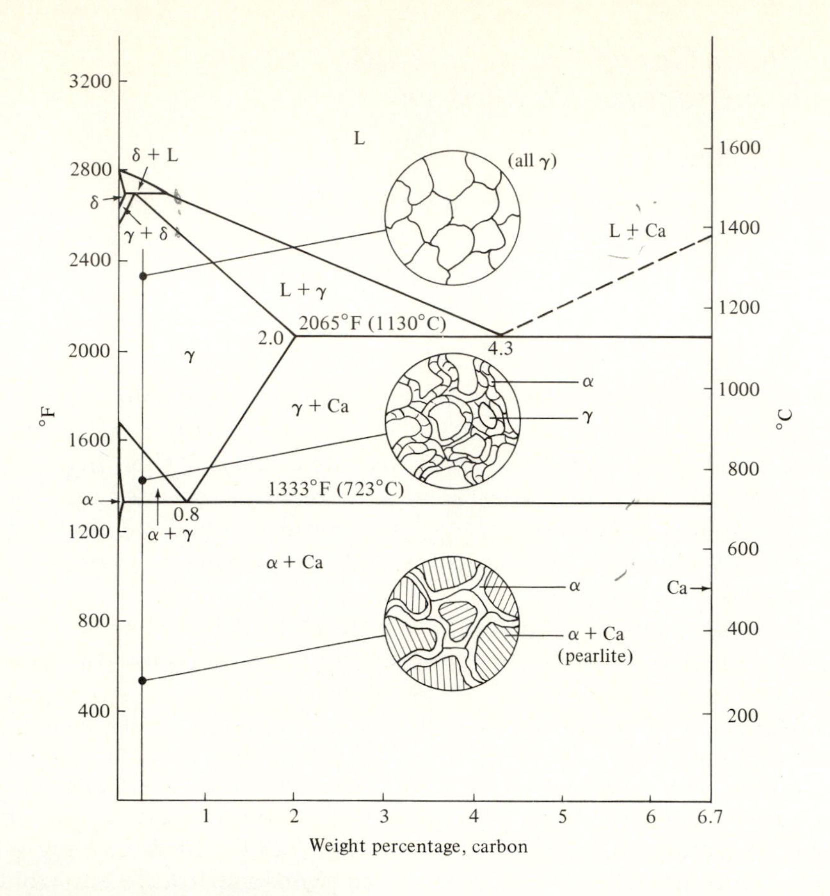

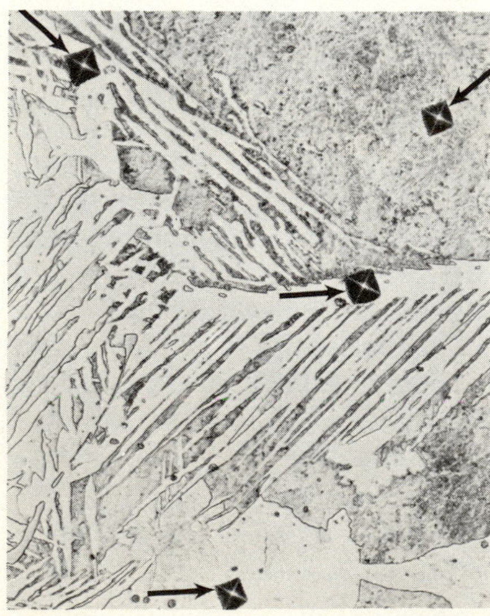

Fig. 6.3 *Changes that occur in microstructure of a hypoeutectoid steel when it is cooled. The photograph is an actual photomicrograph; 500×, 2% nital etch. Ferrite (white), VHN 171; pearlite (gray), VHN 215.*

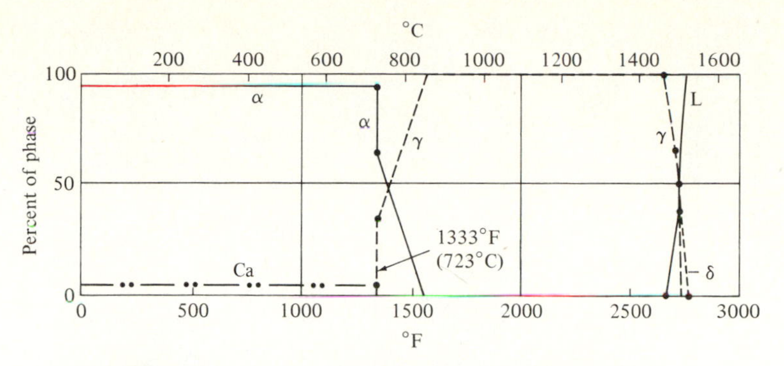

Fig. 6.4 *Phase fraction chart for 0.3% carbon steel*

on slow cooling. The *total* ferrite would therefore be 95.5% (63 + 32.5) at 1332°F (722°C). The fact that Fig. 6.4 shows no change in the amounts of ferrite or carbide when cooled from 1332 to 68°F (722 to 20°C) indicates that a slight decrease in carbon solubility in ferrite (0.025 to <0.01) has a negligible effect on the calculation.

In order to achieve equilibrium, slow cooling must be used. More rapid cooling, such as air cooling, results in a finer pearlite and is called *normalizing*. Slow cooling, such as furnace cooling, through the eutectoid is called *full annealing* and results in a slightly lower strength yet higher ductility than obtained in normalizing. We can get still another variation in grain size and pearlite by reheating or holding for extended periods just below the eutectoid (approximately 1300°F, 704°C). The carbides in the pearlite change from the normal plates to spheres (Fig. 6.5, page 212). This microstructure is called *spheroidite* or *spheroidized pearlite*. It has lower hardness but higher ductility and toughness.

6.5 Hypereutectoid steels

High-carbon steels naturally contain more of the hard carbide phase. They are useful where higher strength, hardness, and wear resistance are needed, as in a knife blade, other cutting tools, and bearings. Let us follow the cooling of a 1% carbon steel on the phase diagram and plot the results as a fraction chart (Fig. 6.6, page 212).

The important temperature range for this material is 1500 to 1300°F (815 to 704°C). It is vital to observe that a brittle phase, iron carbide, precipitates at the austenitic grain boundaries from 1470 to 1333°F (799 to 723°C) (Fig. 6.7), although this primary carbide amounts to only approximately 3.5%. This is in contrast to the ductile primary ferrite that precipitated in the 0.3% carbon steel. The eutectoid reaction is exactly the same in both cases; austenite with 0.8% carbon in solid solution forms the α + carbide in pearlite.

The properties of this high-carbon material in the slow-cooled condition

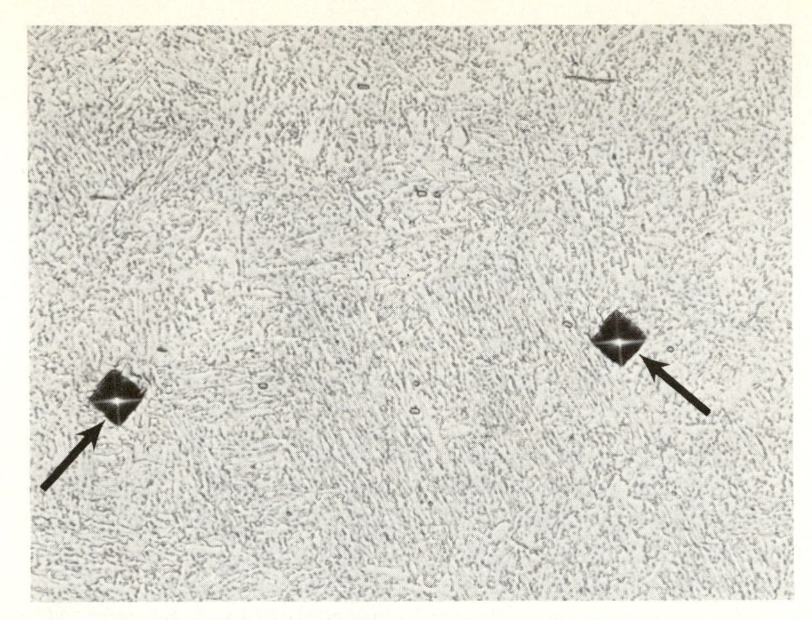

Fig. 6.5 *Spheroidized structure in 0.5% carbon steel. There are very fine spheroids of iron carbide in the α matrix; VHN 170, 500×, 2% nital etch.*

are very poor. Elongation, for instance, is less than 5% because of the brittle carbide network. However, if we heat to 1500°F (815°C) to dissolve the carbide in the austenite, cool rapidly to below 1300°F (704°C) to allow transformation to a fine mixture of α plus carbide, and then reheat to 1300°F (704°C), the coarse-grain boundary carbide does not have time to form on cooling, and we obtain a better carbide shape and distribution with higher ductility. This material is called *spheroidite*. In general, the toughness of these hypereutectoid steels is still lower than that of the hypoeutectoid type because of the greater amount of carbide.

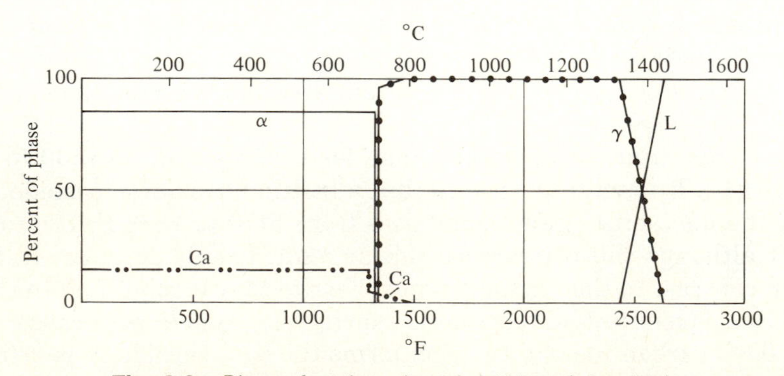

Fig. 6.6 *Phase fraction chart for 1% carbon steel*

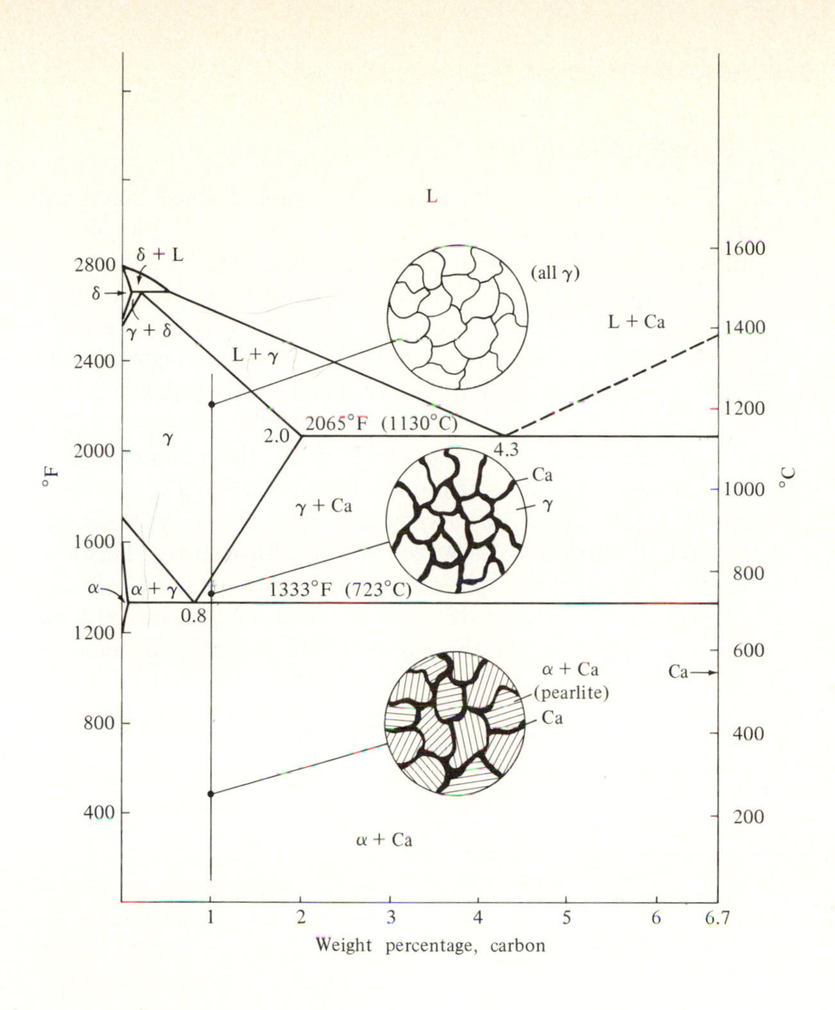

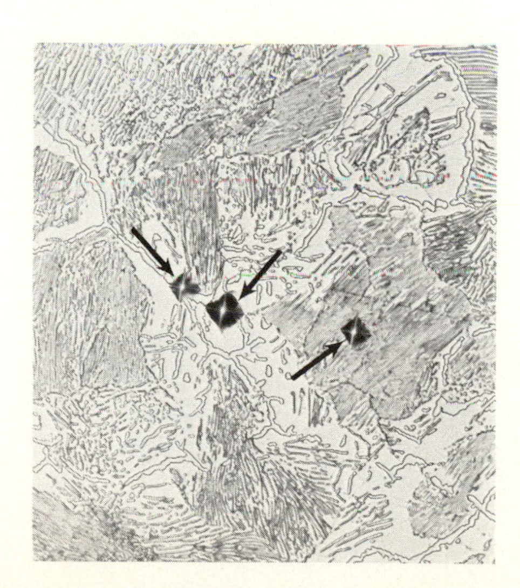

Fig. 6.7 *Changes that occur in microstructure of a hypereutectoid steel when it is cooled. Photograph is an actual photomicrograph; 500×, 2% nital etch. Pearlite (gray), VHN 235. Some ferrite, VHN 160, precipitated next to carbide (white).*

6.6 Specifications

We can now discuss a few specifications and uses of these materials (Table 6.1). Wrought iron and ingot iron both show maximum ductility. Wrought iron is essentially pure iron with slag fibers rolled into the structure. The use of these materials in pipe and architecture is well known, although present-day wrought iron is usually a low-carbon steel. Next we have the so-called plain carbon steels, as distinguished from alloy steels. All contain about 0.5% manganese, however. The code symbol for these steels is 10*xx*, where *xx* designates the percentage of carbon; for example, 1020 steel contains approximately 0.20% carbon. As we increase the carbon, the percentage of pearlite rises and the strength increases from 44,000 psi (299 MPa) at 0.02% carbon to 112,000 psi (774 MPa) at 0.8% carbon. The cold-drawn material shows still higher strength and hardness because of work hardening. However, above 0.5% carbon it is necessary to spheroidize the steel prior to cold drawing to attain sufficient ductility. The yield strength of the cold-worked spheroidized material is higher than that of the corresponding hot-rolled pearlitic structure.

Table 6.1 TYPICAL PROPERTIES OF PLAIN CARBON STEELS

Steel Number	*Carbon Content, percent*	*Condition*	*Tensile Strength,* psi × 10³*	*Yield Strength,* psi × 10³*	*Percent Elongation*	*BHN*	*Typical Use*
Ingot iron†	0.02	Annealed	42	19	48	69	Pipe,
		Hot-rolled	44	23	47	83	architecture
		Cold-drawn	73	69	12	142	
1010	0.10	Hot-rolled	47	26	28	95	Car fenders
		Cold-drawn	53	44	20	105	
1020	0.20	Hot-rolled	55	30	25	111	Structural
		Cold-drawn	61	51	15	121	forms
1040	0.40	Hot-rolled	76	42	18	149	Crankshaft
		Cold-drawn	85	71	12	170	
1060	0.60	Hot-rolled	98	54	12	201	Chisel
		Cold-drawn‡	90	70	10	183	
1080	0.80	Hot-rolled	112	62	10	229	Wear-resis-
		Cold-drawn‡	98	75	10	192	tant parts
1095	0.95	Hot-rolled	120	66	10	248	Cutting
		Cold-drawn‡	99	76	10	197	blades

*Multiply psi by 6.9×10^{-3} to obtain MN/m^2 (MPa) or by 7.03×10^{-4} to obtain kg/mm^2.
†Wrought iron has mechanical properties similar to ingot iron.
‡Spheroidized, then cold-drawn.

EXAMPLE 6.2 Why would automobile fenders be made of a 1010 steel whereas railroad rails might use a 1080 steel?

ANSWER Automobile fenders are relatively thin, and their strength is developed by cold working. Therefore, a low carbon content is required to obtain the necessary ductility for cold forming. Railroad rails, which are rolled sections, must be hot-formed. The high carbon content gives the required wear resistance; however, it limits the cold-working capability and the steel is formed while all austenite.

A higher-carbon steel fender might seem desirable because of the added strength; however, it would have to be hot-formed. The hot-worked steel would show excessive oxide scale formation (Chapter 12) and might not be as strong in the annealed condition as a lower-carbon cold-worked steel. (Compare a cold-drawn 1010 steel and a hot-worked 1020 steel in Table 6.1.)

6.7 Steel—nonequilibrium reactions

If we were limited to the equilibrium structures and the plain carbon steels of the iron–iron carbide diagram, we could not make a great many critical tools and components. Imagine our problems without hardened drills, files, lathe tools, gears, chisels, plows, ball bearings, rolls, razor blades, knives, saws, dies, and hundreds of other parts in which we need hardness and strength! As an example of the hardening operation, let us consider the changes in the properties of SAE 1080 steel that we can accomplish by nonequilibrium cooling. See the table below.

It is not the *act* of quenching but the *effect* of quenching on the structure that produces the change. For example, if we performed the same treatment on a piece of pure iron or on an 18% Cr, 8% Ni stainless steel, there would be no change in hardness.

Before we discuss how this change in hardness takes place, we should recall that all reactions take time because of the requirements for the nucleation of new structures or phases and the diffusion of atoms to allow growth. We shall see in the next section that the rate of transformation of austenite to ferrite and carbide, as in pearlite, is a result of nucleation and growth.

Condition	*Tensile Strength, psi × 10³*	*Yield Strength, psi × 10³*	*Percent Elongation*	*BHN*
Slow-cooled from 1550°F (843°C)	112 (773 MPa)	62 (428 MPa)	10	192
Water-quenched from 1550°F (843°C)	>200 (>1380 MPa)	200 (1380 MPa)	1	680

6.8 Austenite transformation

The key to understanding the variation in hardness is a knowledge of the austenite transformation. Let us consider first the simplest reactions, those that can occur with eutectoid (0.8% carbon) austenite. Later we shall take up the effects of analysis. So far we have discussed only the transformation of austenite under equilibrium conditions to a relatively coarse platelike mixture of carbide and ferrite called pearlite.

What if we avoid the reaction at 1333°F (723°C) by cooling the austenite rapidly from 1500 to 1200 or 800°F (815 to 649 or 427°C) or lower and allowing it to react at these temperatures?

Let us perform an experiment to determine these effects. First we machine a thin specimen of 0.8% carbon steel with two holes that serve as reference points for measurement (Fig. 6.8). Second, we hang the specimen from a hook *A* on a tube of fused silica. The tube itself is rigidly held from platform *B*. Next we place the hook of an inner tube into the lower specimen hole *C*. The specimen is now in light tension, supporting the inner tube, which slides smoothly in the outer tube. Finally we fasten a dial gage onto the top of the inner tube with the point bearing on the support of the outer tube.

This instrument is called a *dilatometer* because it measures change in length or dilation of the specimen. Whether the specimen expands or contracts, the dial shows the movement faithfully. Also, the use of fused silica, a very low-expansion material, reduces any errors that might be caused by expansion of the tubes.

Next let us change the structure of the specimen to *austenite* by heating to 1600°F (871°C). We do this simply by immersing the lower part of the dilatometer with the specimen in a pot of liquid lead at 1600°F (871°C). The fused silica is inert and has excellent thermal shock resistance.

6.9 Pearlite formation

To observe transformation at 1300°F (704°C), we quickly shift another pot of lead at 1300°F (704°C) into the place of the 1600°F (871°C) pot. We record the changes in length of the specimen as time passes and obtain the graph shown in Fig. 6.9*a* on page 218.

By interpreting the graph we find the microstructural changes in the specimen. The first stage (1) is a contraction as the austenite cools. The magnitude of this contraction corresponds exactly with the value expected using the coefficient of expansion (or contraction) of austenite. The specimen is at constant temperature during period (2). Since no change in microstructure is taking place, no length change occurs. At a later time (3) we notice that the length of the specimen is increasing although it is in a bath at constant temperature. This must be due to the change in structure, austenite → pearlite. The length change should not surprise us if we recall the data showing the

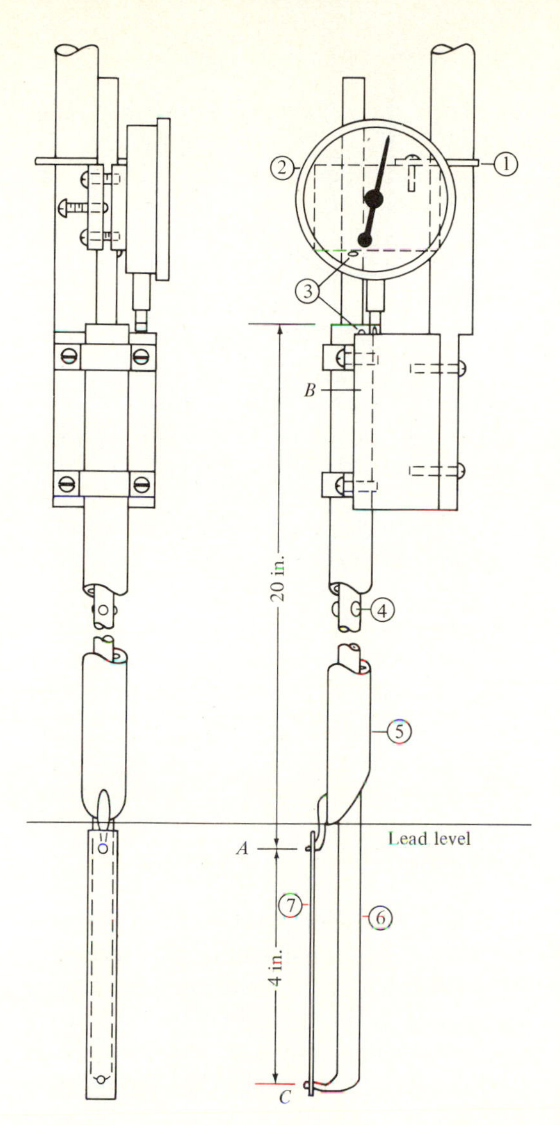

Fig. 6.8 *Dilatometer for measuring transformations. 1 = dial guide, 2 = dial gage, 3 = quartz lugs, 4 = inner-tube guides (quartz, two sets), 5 = outer quartz tube, 6 = inner quartz tube, 7 = specimen (4½ by ½ by 1/32 in.).*

volume change when FCC iron transforms to BCC. The graph tells us, therefore, the time at which transformation begins and ends at this temperature for this steel: 30 and 1000 sec. After the transformation there is no change in dimensions as long as the temperature remains constant.

If we repeat the experiment using a temperature of 1200°F (649°C) for transformation, we obtain different transformation times: 3 and 25 sec (Fig. 6.9*b*).

We now graph the time from beginning to end of transformation as a function of temperature, using a logarithmic scale to accommodate the wide

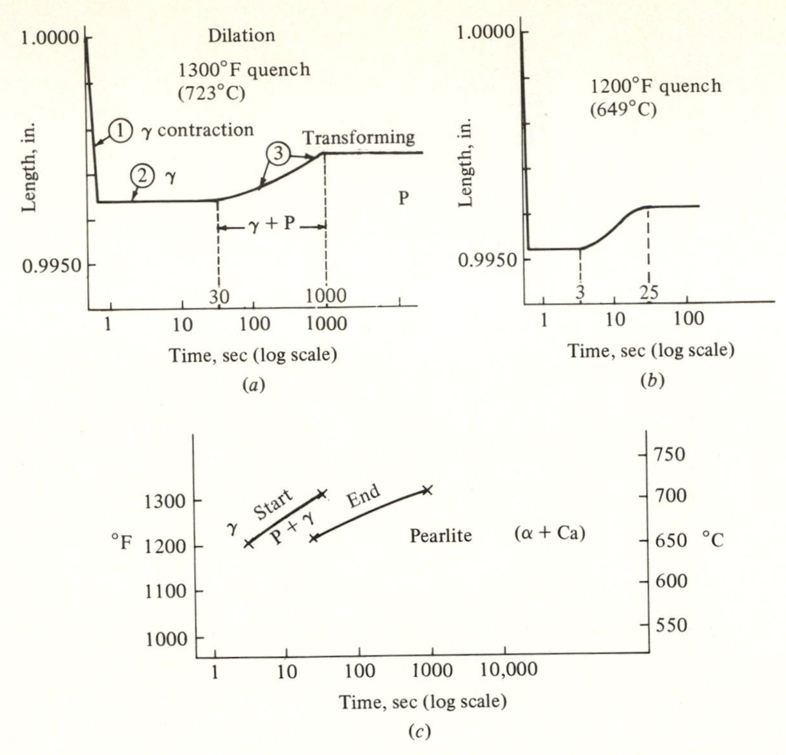

Fig. 6.9 *(a) Change in length when 0.8% carbon steel is quenched to 1300°F (723°C). (b) Change in length when 0.8% carbon steel is quenched to 1200°F (649°C). (c) TTT curve [derived from data from (a) and (b)].*

variation in time (Fig. 6.9*c*). This graph can be called either an isothermal transformation curve or a TTT (time-temperature-transformation) curve.

If we measure the hardness of the steel at room temperature after transformation, we find that the hardness increases as the transformation temperature is lowered. The spacing between the carbide plates in the pearlite is finer, and the hardness and strength are related to the distance between the carbides (Fig. 6.10).

6.10 Bainite formation

Below 1000°F (538°C) the transformed structure changes in appearance from the alternating plates of pearlite to a feathery or acicular structure called *bainite* (Fig. 6.11). The hardness continues to increase because the carbide is becoming increasingly finer, and as a result the distance over which slip can take place in the ferrite is becoming shorter. The transformation start and end times are longer because the diffusion rate is slower. From these data we can obtain more of the isothermal transformation curve (Fig. 6.12, page 220).

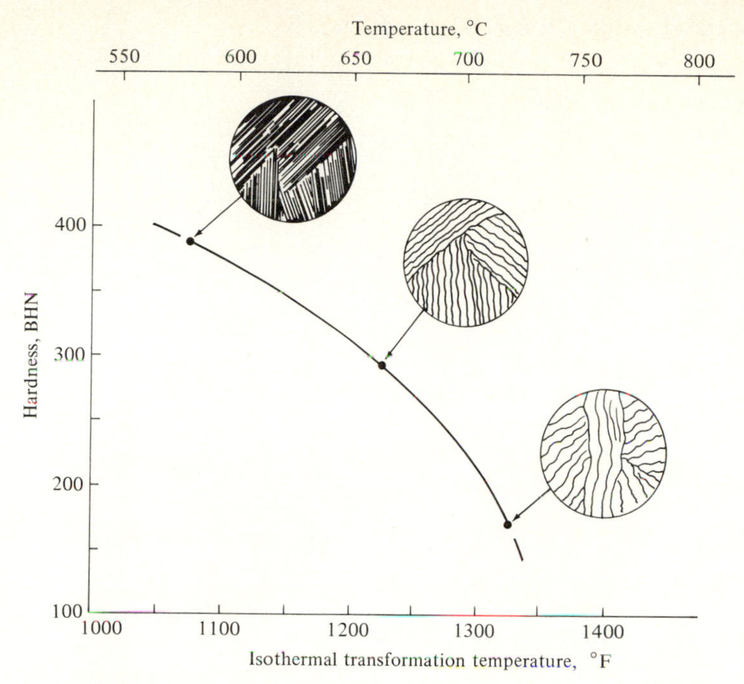

Fig. 6.10 *Relationship between hardness of 100% pearlite (0.80% carbon) and isothermal transformation temperature. Lower temperatures result in finer pearlite.*

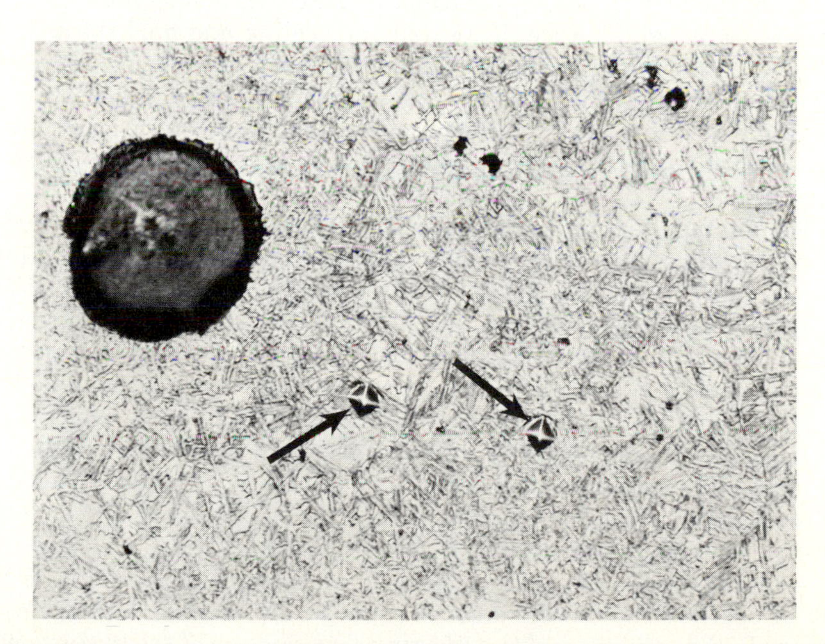

Fig. 6.11 *Bainite produced by isothermal transformation at 600°F (316°C), VHN 400, from 0.7% C austenite; 500×, 2% nital etch. Large spherical particle is graphite (see Section 6.36).*

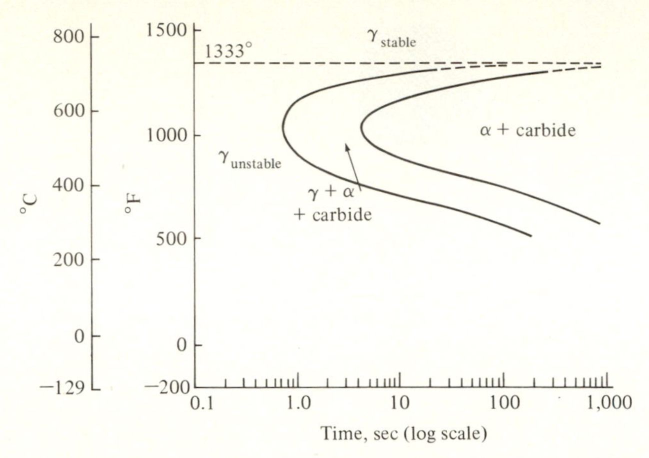

Fig. 6.12 *Construction of isothermal transformation curve (1080 steel)*

6.11 Transformation to martensite

Below 420°F (215°C) we encounter a type of transformation quite different from the isothermal transformations we have just discussed, and the structure is different (Fig. 6.13). As we decrease the temperature below 420°F (215°C), a fraction of the new structure, called *martensite,* forms *instantaneously* with each decrease in temperature. If we halt the cooling and hold the sample at say 300°F (149°C), no further martensite is produced until we resume cooling. The sample is completely transformed when we reach lower temperatures. We call the temperature at which transformation begins the *martensite start temperature,* M_s, and the temperature at which it ends the *martensite finish temperature,* M_f.

The dilatometer curves show this effect clearly (Fig. 6.14). If we quench to just above the M_s, we obtain a normal transformation curve to bainite. If we quench to below the M_s, the decrease in length *on quenching* is less than it would be if we had quenched to 420°F (215°C). Therefore, some expansion resulting from transformation took place *during* cooling. We may now complete our isothermal transformation curve for 1080 steel to include the M_s and M_f, as shown in Fig. 6.15 on page 222.

Martensite is important because it is the hardest structure formed from austenite. X-ray diffraction evidence indicates that the structure is a distorted BCC (tetragonal). The distortion is caused by trapped atoms of carbon, as shown in Fig. 6.16 on page 222.

A mechanism that explains the difference between the nucleation and growth of pearlite and bainite and the rapid transformation to martensite is as follows. In the case of pearlite and bainite, nucleation of the two new phases ferrite and carbide has to occur, and diffusion and long-range movement of atoms take place. In the case of martensite, the structure can be formed by

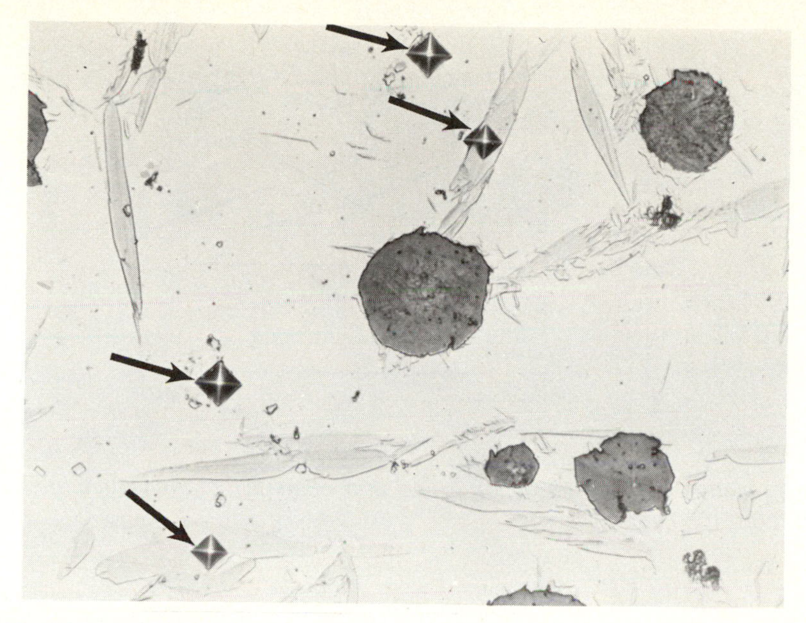

Fig. 6.13 *Martensite produced in high-carbon, high-alloy matrix, VHN 470. The matrix is austenite, VHN 205. The spheres are graphite (see Section 6.36); 500×, 2% nital etch.*

short shearing movements. In Fig. 6.17 we see the similarity between martensite and the parent austenite. The end product can be formed by movements that result in an expansion of the a direction in the tetragonal cell, which is sketched within the FCC lattice of the austenite. Note that the carbon

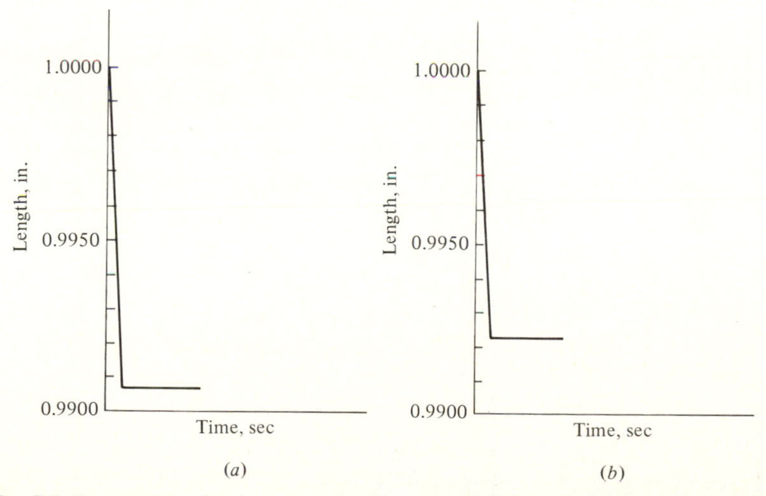

Fig. 6.14 *Dilation curves for lower quenching temperatures (0.8% carbon, 1080 steel). (a) Quenched to 420°F (216°C), M_s. (b) Quenched to 300°F (149°C), γ + martensite.*

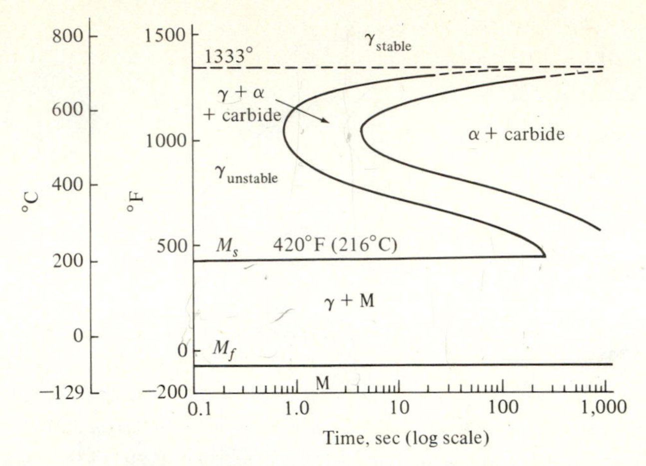

Fig. 6.15 *Isothermal transformation curve and austenite–martensite transformation (0.8% carbon, 1080 steel).*

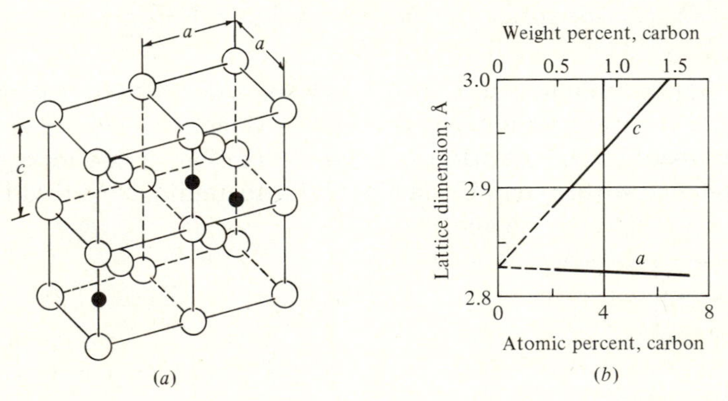

Fig. 6.16 *(a) Location of carbon atoms in martensite. (b) Carbon atoms expand lattice, giving rise to strain and higher hardness.*

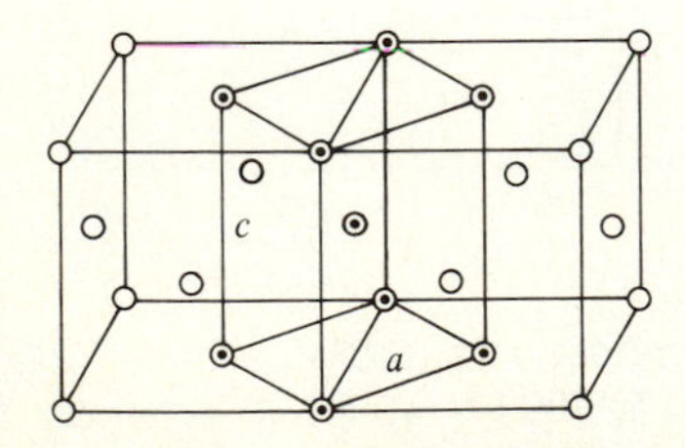

Fig. 6.17 *Relation of unit cell of martensite to parent austenite. Two unit cells of austenite (FCC) and resultant BCT lattice of martensite are shown.*

atoms are trapped along the sides, so that the *c* axis is finally longer than *a*. At the maximum carbon level the *c*/*a* ratio is 1.08, as shown in Fig. 6.16.

6.12 Summary

Let us summarize the austenite transformation of an 0.8% carbon steel shown in Fig. 6.15.

1. Above 1333°C (723°C): Austenite is the stable phase from the equilibrium diagram.
2. 1333 to 1050°F (723 to 566°C): (*a*) Austenite isothermally transforms to α + carbide as pearlite. (*b*) As temperature decreases, the time to transform *decreases* because nucleation is easier. (*c*) More rapid nucleation rate at lower temperatures leads to finer pearlite (BHN 170 to 400).
3. 1050 to 420°F (566 to 215°C): (*a*) Austenite isothermally transforms to α + carbide as bainite. (*b*) As temperature decreases, the time to transform *increases* because although nucleation rates are high, diffusion is slower. (*c*) High nucleation results in finer bainite as temperature decreases (BHN 400 to 580).
4. 420 to −50°F (215 to −46°C), M_s to M_f: Austenite transforms to martensite upon cooling. The hardness of martensite is BHN 680.

Note that once the austenite has transformed to one of the above structures, such as pearlite, it cannot be transformed to another unless the sample is reheated to the equilibrium austenite temperature [above 1333°F (723°C)]. However, a sample can transform to a mixture of transformation products by spending a short interval in the pearlite and bainite transformation ranges, then cooling through the martensite range.

EXAMPLE 6.3 Draw schematic time-temperature diagrams to obtain the following microstructures in 1080 steel. Indicate the microstructures on the diagram at each stage of the treatment. (*a*) Coarse pearlite, (*b*) 50% pearlite and 50% bainite, (*c*) 80% martensite and 20% pearlite

ANSWER

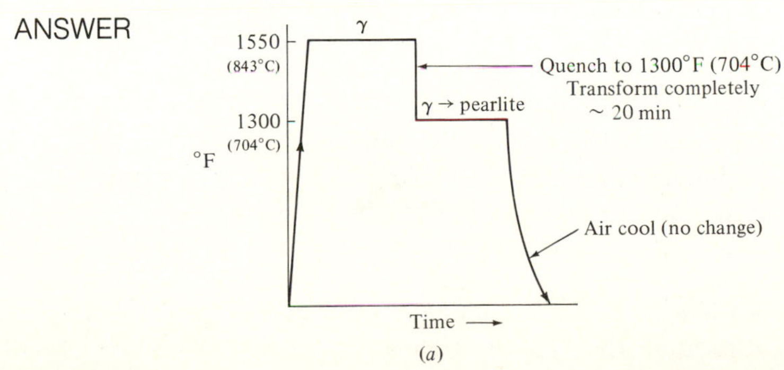

(*a*)

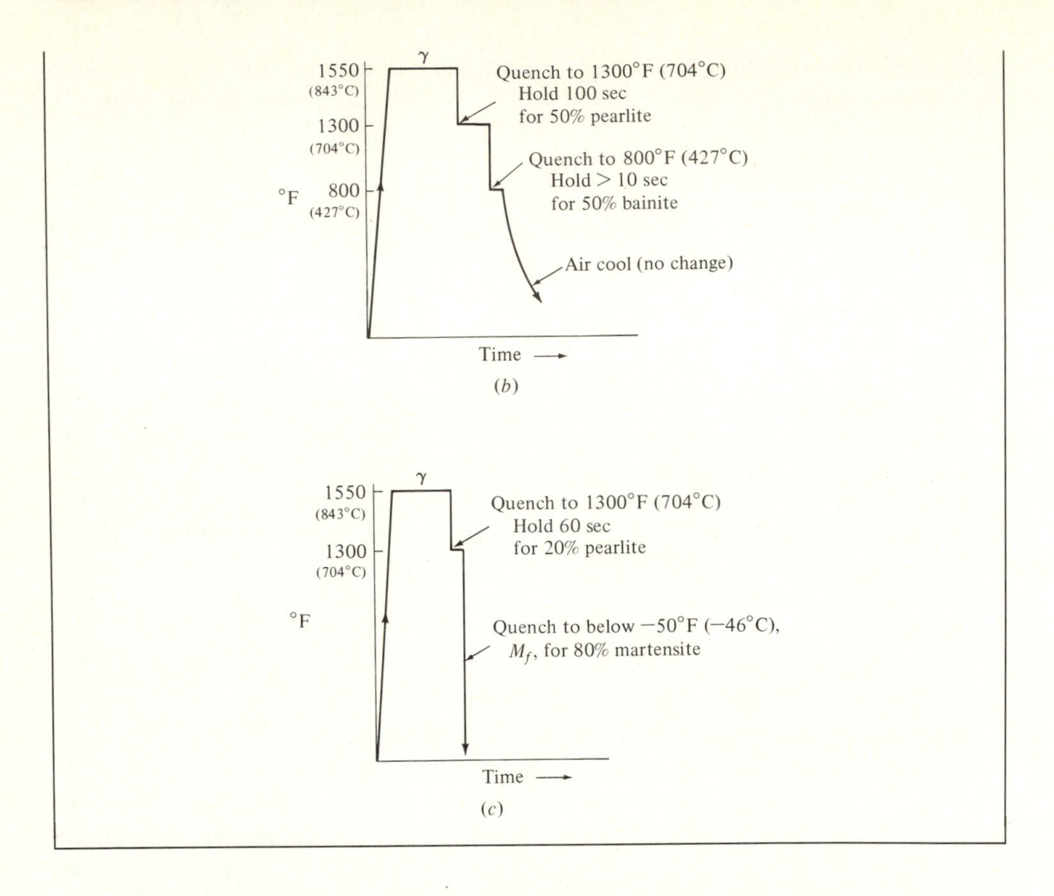

6.13 Uses of the time-temperature-transformation (TTT) curve

Now that we have discussed austenite transformation, let us illustrate some of the uses of the TTT curve.

Suppose that we wish to machine a knife blade from a piece of 1080 steel and then harden it. Suppose further that the hardness of the piece of steel we have on hand is BHN 370 because of its previous history. This is too hard to machine conveniently. We therefore heat the steel to the austenite range, 1450°F (788°C), cool it to 1250°F (677°C) in the furnace, allow it to transform in the furnace for 1 hr, and then air-cool it. The result is *coarse* pearlite with a hardness of BHN 250. We could have obtained a still softer pearlite by using a longer time at 1300°F (704°C), but this is adequate. Also, holding for a long period of time at 1300°F (704°C), will cause the plates of carbide to spheroidize, giving a still softer structure, similar to Fig. 6.5 except at a higher carbon level.

Now we machine the blade and place it in the furnace at 1450°F (788°C) again. (It is usual to hang the part on a wire or to use some support so that thin sections do not sag at the high temperature where the steel is soft.) After austenite is obtained (in less than $\frac{1}{2}$ hr), the blade is quenched in rapidly circulating oil at 68°F (20°C). The quenching must be sufficiently rapid to avoid transformation to either pearlite or bainite so that martensite is produced.

6.14 Tempering

If we test the knife blade, we find that it is at the hardness we wanted, BHN 680 or 64 R_c. However, it is also quite brittle. It will break with slight bending, and the edge will chip with impact. We recall that martensite is a highly stressed supersaturated solid solution of carbon in a distorted ferrite. Therefore, we heat the blade just enough to cause the body-centered tetragonal structure to collapse to the BCC and to allow the carbon atoms to migrate and form some very fine iron carbide crystals. Admittedly the hardness may decrease slightly, but the ductility should improve.

We find the relations shown in Fig. 6.18. The martensite transforms to ferrite plus fine carbide. The higher the tempering temperature, the coarser the carbide, the lower the hardness, and the greater the ductility.† Furthermore, the sample may be quenched from the tempering temperature without changing the effect on hardness.

In many cases we want a balance between hardness and ductility. After a typical claw hammer head is quenched, the striking face is tempered to 60 R_C, but the claws are tempered to 48 R_C because greater toughness is needed in this region. This is accomplished by selective heating with a flame or induction coil.

---

†In some steels certain tempering temperatures are avoided because secondary reactions in the structure cause temper embrittlement.

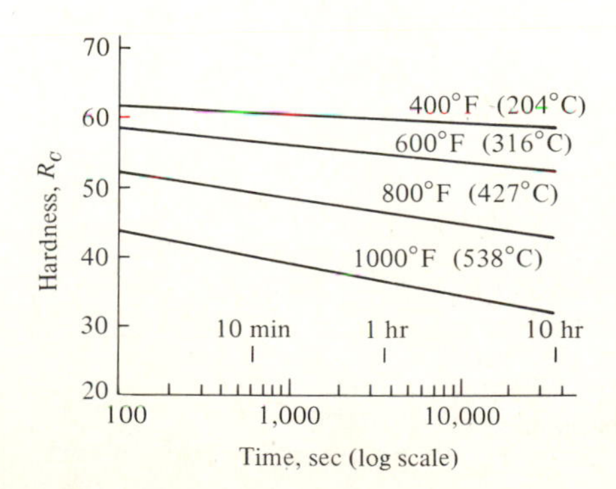

Fig. 6.18 *Relationship between hardness of tempered martensite and tempering temperature and time. Since tempering is a diffusion phenomenon, one might anticipate the dependence on time and temperature (1080 steel).* (L.H. Van Vlack, *Elements of Materials Science*, 2d ed., Addison-Wesley Publishing Company, Reading, Mass., 1964, Fig. 10.29, p. 294)

EXAMPLE 6.4 What are the different microstructures that are mixtures of ferrite and carbide, and how are they obtained?

ANSWER

1. *Pearlite:* normal equilibrium cooling of austenite or isothermal holding of austenite above the nose of a TTT diagram.
2. *Bainite:* isothermal holding of austenite below the nose of a TTT diagram. (Bainite may also be obtained by cooling if two noses are present, with the bainite nose out ahead of the pearlite nose. See Sec. 6.18 and Fig. 6.29.)
3. *Spheroidite:* spheroidization of pearlite by extended holding at approximately 1300°F (704°C) or lower.†
4. *Tempered martensite:* reheating of the distorted body-centered tetragonal martensite to precipitate the carbon as carbide and allow the crystal structure to return to the BCC ferrite plus carbide.

6.15 Marquenching (Martempering)

One of the hazards of quenching is the possibility of distorting and cracking the part. Let us consider these effects in a simple knife blade (Fig. 6.19). To simplify the discussion, let us assume that the thin portion A of the blade will cool first and transform to martensite while the thicker section B remains austenite (above M_s), at say 500°F (260°C). In passing through the martensite transformation, B will expand, causing the knife to bow severely. This effect will be found even in a round bar (Fig. 6.20).

Region A of the bar will cool before B and transform to martensite. Portion B is still austenite, soft and hot, and follows all these changes in length. Later it is B's turn to transform. The expansion cracks the surface layer A because it is brittle.

†Bainite and martensite can also be spheroidized. Whereas it may take 6 to 12 hr to spheroidize pearlite, bainite and martensite may be spheroidized in 1 hr or less.

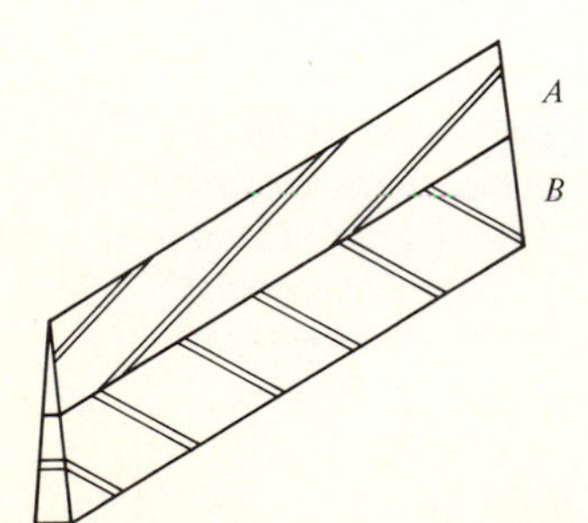

Fig. 6.19 *Schematic representation of cooling and the build-up of stress in a knife blade*

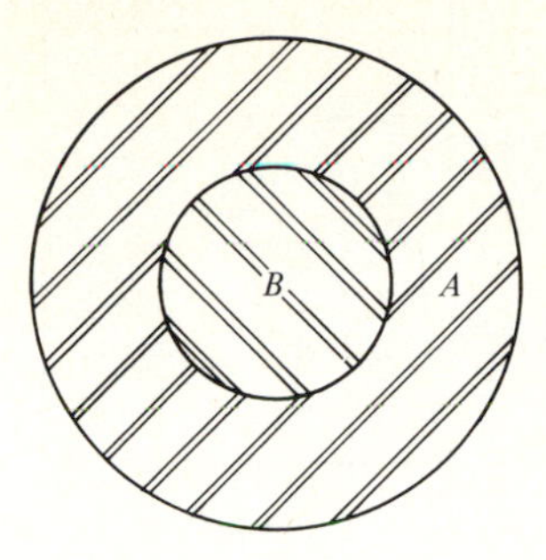

Fig. 6.20 *Schematic representation of cooling and the build-up of stress in a round bar*

The solution to both these problems is to have the austenite-martensite transformation take place in A and B at the same time. This is done by *marquenching* (martempering) as follows:

1. Austenitize (heat at 1450°F, 788°C).
2. Quench in hot oil, liquid metal, or molten salt at just *above* the M_s. Equalize the temperature throughout the part.
3. Air-cool through the M_s-to-M_f range.

The quenching in liquid is necessary to avoid pearlite or bainite formation. The part is removed from the hot bath *before* bainite begins to form and is allowed to transform throughout to martensite. Tempering is still used *after* cooling to below the M_f (Fig. 6.21 on page 228).

6.16 Austempering

This is another process for avoiding distortion and cracking. Here the part is simply transformed to bainite at the desired level of the bainite temperature range. The part is austenitized, quenched in a salt bath above the M_s, allowed to transform, and then air-cooled. Naturally the as-quenched hardness is lower than if martensite were formed. However, since martensite is usually tempered to a lower hardness level, the results are comparable. It is often not practical to use austempering with some alloy steels because of the long time required to form bainite, as discussed later. On the other hand, the advantage of austempering is that the extra tempering step is avoided and in general the distortion is less. We can compare the TTT curves of conventional quenching and tempering, marquenching, and austempering as shown in Fig. 6.21.

6.17 Effects of carbon on transformation of austenite and transformation products

Up to this point we have considered only the simple eutectoid steel 1080. The heat treatment of hypoeutectoid and hypereutectoid steels is also im-

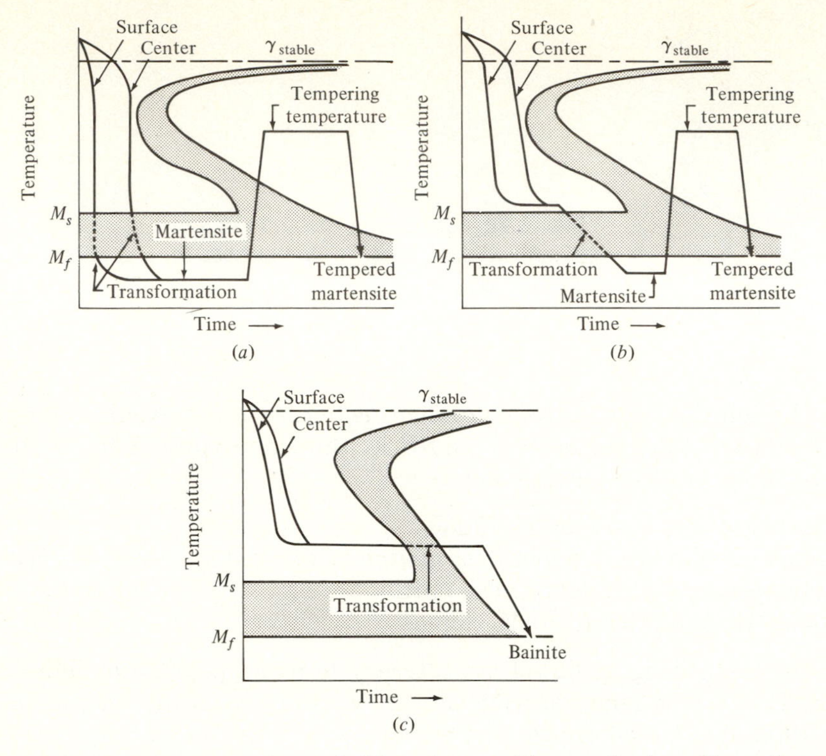

Fig. 6.21 *Schematic comparison of different quenching methods, with cooling curves superimposed on TTT diagrams. (a) Conventional process. (b) Martempering (marquenching). (c) Austempering.* (American Society for Metals Handbook, 8th ed., Vol. 2: *Heat Treating, Cleaning and Finishing*, Metals Park, Ohio, 1964)

portant, and the use of alloys permits the successful hardening of many complex designs that could not be produced otherwise.

To attain maximum hardness in a hypoeutectoid steel, the first step is to austenitize completely. Therefore, the austenitizing temperature for 1040 steel is 150°F (83°C) above that for 1080 steel (Fig. 6.22). The steel may then be hardened by quenching to martensite at rapid cooling rates.

If the steel is cooled slowly from 1500°F (815°C), ferrite will form from 1480 to 1333°F (804 to 723°C), and the austenite will change from 0.4 to 0.8% carbon. Then at 1333°F (723°C) pearlite will form. Now let us see what happens with isothermal transformation. If a sample is quenched from 1500 to 1400°F (815 to 760°C), we encounter start and end times for the formation of the amount of α at equilibrium with austenite at this temperature (Fig. 6.23, page 230). But, until 1333°F (723°C) we have isothermal transformation only to ferrite. If we quench from 1500°F (815°C) to below 1333°F (723°C) to as low as 1000°F (538°C), we find first a period of ferrite formation, then pearlite.

Also, the time for transformation in the pearlite range has been

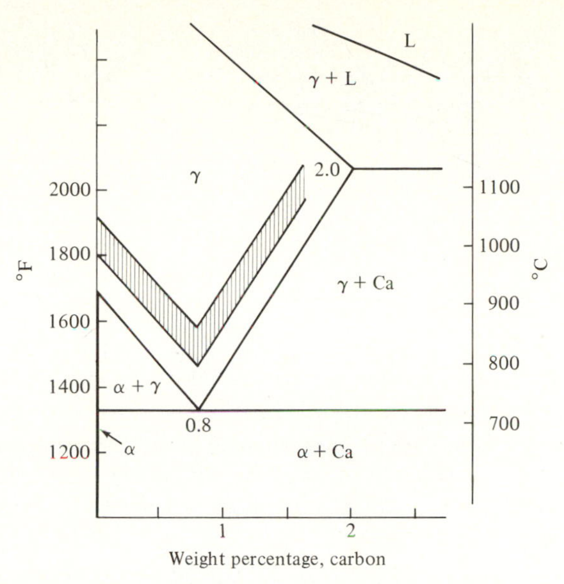

Fig. 6.22 *Austenitizing temperatures for steels containing different amounts of carbon (cross-hatched area)*

shortened compared with that for 1080 steel. The practical consequence is that martensite or bainite can be obtained only in thin sections, which are quenched rapidly.

If the steel is transformed below 1000°F (538°C), separate formation of ferrite does not take place. Therefore the TTT curve shows only the start and end of bainite, just as for 0.8% carbon.

The hardness of the transformation products at any temperature is lower than that of 0.8% carbon steel, because the amount of the hard carbide phase is smaller.

If the steel is quenched rapidly enough to avoid prior transformation to pearlite or bainite, martensite forms from the M_s down to the M_f. Note that as the carbon content decreases, the M_s increases (Fig. 6.24) and the hardness of the martensite decreases (Fig. 6.25 on page 230).

Now let us consider the hypereutectoid steels, which usually fall in the range 0.8 to 1.2% carbon. If we cool a 1.2% carbon steel slowly from the austenite field, hypereutectoid carbide forms first. Then at 1333°F (723°C) the 0.8% carbon austenite transforms to pearlite. Therefore, the TTT curve for a hypereutectoid steel shows curves for the start and end of the carbide formation, then curves for pearlite formation (Fig. 6.26, page 231). The dilation curve for a hypereutectoid steel compared with the curve for a hypoeutectoid steel is interesting, because in the former the precipitation of carbide alone gives a contraction (Fig. 6.27, page 231). This is followed by an expansion when the pearlite transforms.

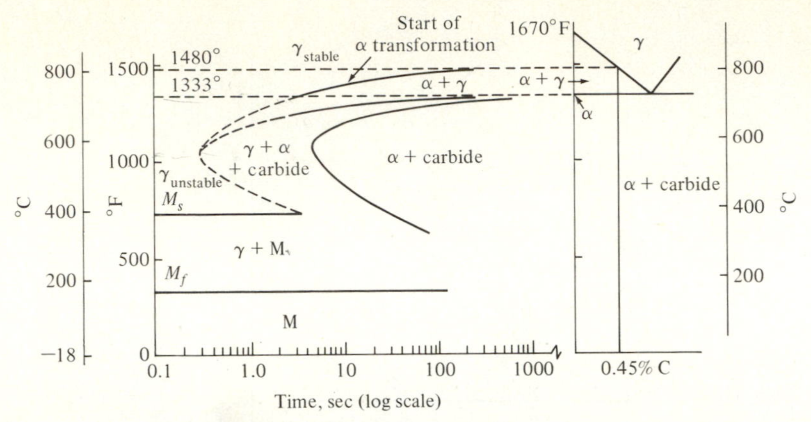

Fig. 6.23 *TTT curve for 0.45% carbon steel. There is an additional region, compared to 1080 steel, above the nose of the curve. A portion of the iron–iron carbide diagram is included to show why primary α occurs.* (L.H. Van Vlack, *Elements of Materials Science*, 2d ed., Addison-Wesley Publishing Company, Reading, Mass., 1964, p. 292)

The differences in the TTT curve for the hypereutectoid steels are a longer time to transformation, a lower M_s than for 0.8% carbon steel, and the addition of the curve for carbide precipitation.

We can summarize the effects of carbon in the TTT curve as follows:

1. Above 1000°F (538°C) the formation of ferrite in hypoeutectoid steel and of carbide in hypereutectoid steel precedes the transformation to pearlite.
2. In the bainite range the product is softer for steel below 0.8% carbon, harder above.
3. In the martensite range the M_s and M_f decrease with increased carbon, as shown in Fig. 6.24.
4. The hardness of the martensite is a function of the carbon content (Fig. 6.25).

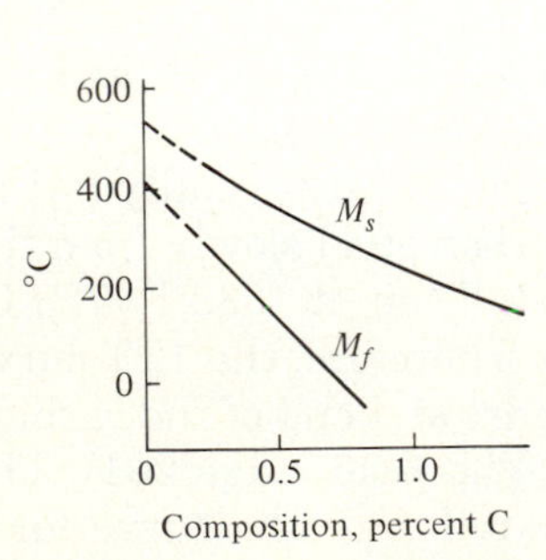

Fig. 6.24 *Effect of carbon content on the M_s and M_f*

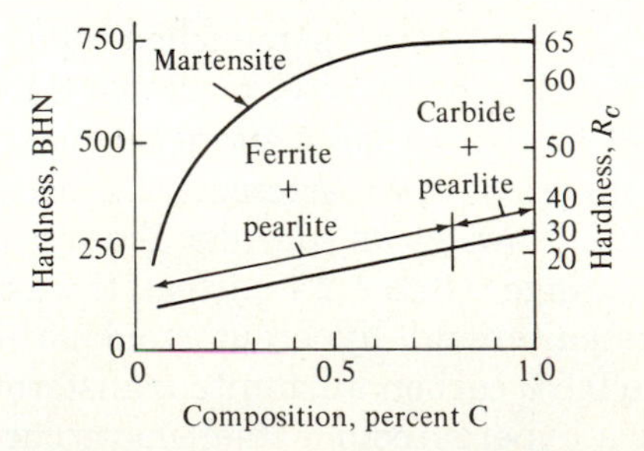

Fig. 6.25 *Hardness of martensite and annealed structures as functions of carbon content.* (L.H. Van Vlack, *Elements of Materials Science*, 2d ed., Addison-Wesley Publishing Company, Reading, Mass., 1964, Fig. 10.21, p. 289)

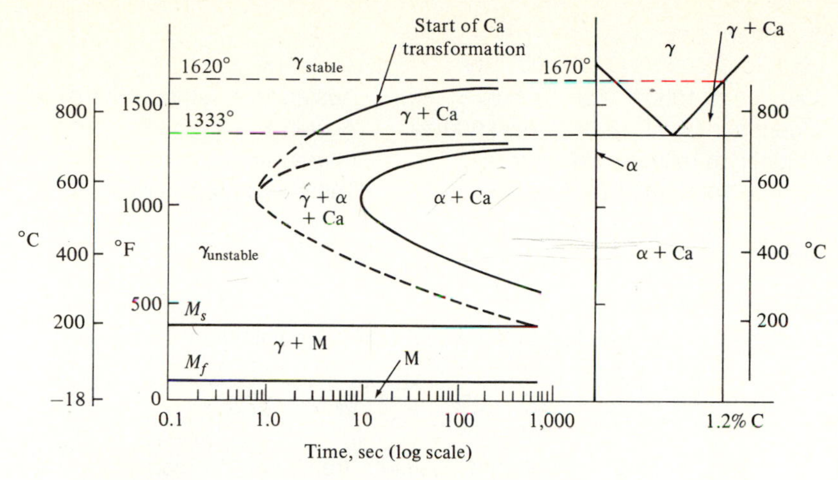

Fig. 6.26 *TTT curve for hypereutectoid steel (1.2% carbon)*

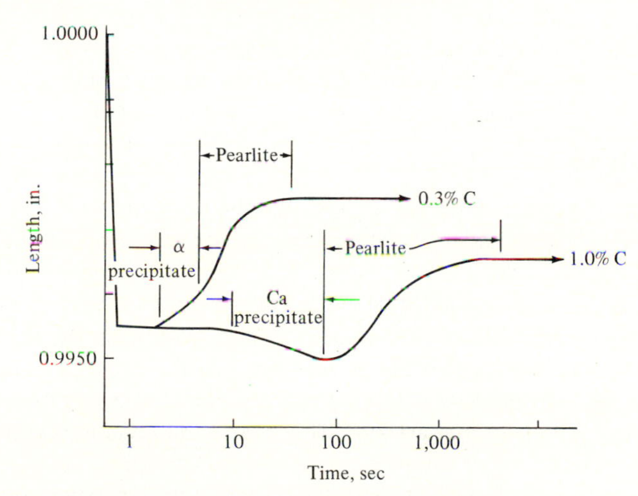

Fig. 6.27 *Dilation curves for isothermal transformation of 0.3% carbon and 1% carbon steels. Transformation temperature is 1300°F (704°C).*

EXAMPLE 6.5 The use of TTT diagrams in conjunction with the equilibrium diagram produces some common errors because of misunderstanding of how the diagrams should be used. Below are several sequential thermal treatments with partially incorrect answers. What would be the correct final microstructure after each step?

1. 1.2% carbon steel	*Microstructure*
a. Heat to 1500°F (815°C), hold 1 hr	Austenite
b. Quench to 0°F (−18°C), hold 1 hr	Martensite
c. Allow to return to room temperature	Martensite + austenite

2. 1.2% carbon steel

a. Heat to 1700°F (927°C), hold 1 hr	Austenite
b. Quench to 1050°F (656°C), hold 5 sec	Austenite + ferrite + carbide
c. Quench to 500°F (260°C), hold 300 sec	Austenite + bainite
d. Quench to room temperature	Martensite
e. Heat to 800°F (427°C), hold 1 hr	Tempered martensite

ANSWER

1. See Figures 6.1*b* and 6.26.

a. Austenite + carbide. In 1 hr we would achieve equilibrium. Therefore

$$\% \text{ Ca} \approx \frac{1.2 - 1.03}{6.7 - 1.03} \times 100 \approx 3\% \quad \text{and} \quad \% \gamma = 97\%$$

b. Martensite + carbide. Only the austenite transforms to martensite; the carbide is unaffected. Note that we must be below the M_f to have complete transformation to martensite.

c. Martensite + carbide. Raising the temperature only a few degrees does not temper the martensite. We cannot obtain austenite again unless we reheat above 1333°F (723°C). *Once the austenite has transformed, the TTT diagram is not used.*

2. See Fig. 6.26.

a. Austenite is correct, since we are well into the γ region.

b. Austenite + fine pearlite. Although the *phases* are α + Ca, the microstructure would be pearlite. It is fine pearlite, since the transformation is near the nose of the curve. Had the transformation temperature been 1100°F (593°C), we would have passed through the γ + Ca region, and after 5 sec we would have had primary carbide + austenite + fine pearlite.

c. Austenite + fine pearlite + acicular bainite. Some of the remaining austenite transforms to bainite, but we are still in a $\gamma + \alpha$ + Ca region after 300 sec. The pearlite remains unchanged.

d. Martensite + fine pearlite + acicular bainite. This presumes room temperature is below the M_f. If not, some austenite may remain.

e. Tempered martensite + fine pearlite + spheroidized bainite. Again nothing will happen to the pearlite. The martensite tempers, and the bainite will begin to spheroidize. Cooling to room temperature will not change the structure.

6.18 Quenching and the need for alloy steels

In previous sections we found that we could produce a hard, strong, wear-resistant structure by heating to form austenite, then transforming the austenite to martensite. We also established that the austenite must be cooled fast enough to avoid transformation to pearlite or bainite.

Figure 6.28 shows the effect of type of quenchant on cooling of 1-in.-diameter bars. Water provides a more rapid quench than salt or oil because the conversion of water to steam at the surface of the bar absorbs a great deal of heat (high latent heat of vaporization). On the other hand, quenches in oil and salt lead to less distortion because of lower thermal gradients.

In many cases in which the section is thick, martensite cannot be obtained even with a water quench. In other parts a water quench cannot be used because of cracking or distortion. The remedy is to increase the time for transformation of austenite to pearlite, so that with a slower cooling rate we still avoid this reaction. We do this by adding alloys such as nickel, molybdenum, chromium, and manganese to the steel.

The TTT curves of Fig. 6.29 demonstrate the basic effects of alloys. Note the great increase in time for pearlite formation compared with plain carbon steels.

These alloy steels give flexibility in heat treatment in two ways. First, an intricate part of thin section can be cooled more slowly from the austenitic field, lessening the susceptibility to cracking. Second, a part of thick section that would cool slowly when quenched can still be produced with a martensitic structure.

Since alloy steels are more expensive than plain carbon steels, some of the effects of alloys acting in combination are worth noting. In general, a triple-alloy steel, such as NiCrMo 8640, will provide a better TTT curve, i.e., longer pearlite formation time, than a single-alloy steel of the same cost. The alloys also affect the M_s. An empirical formula provides an estimate of this point. (Note that all the alloys lower the M_s.)

$$M_s = 930°\text{F} - [540 \times (\%\ \text{C}) + 60 \times (\%\ \text{Mn}) + 40 \times (\%\ \text{Cr}) + 30 \times (\%\ \text{Ni}) + 20 \times (\%\ \text{Mo})]$$

It should be emphasized that the alloys do not change the hardness of the pearlite, bainite, or martensite. They merely increase the transformation time for pearlite and bainite and lower the temperature range for martensite.

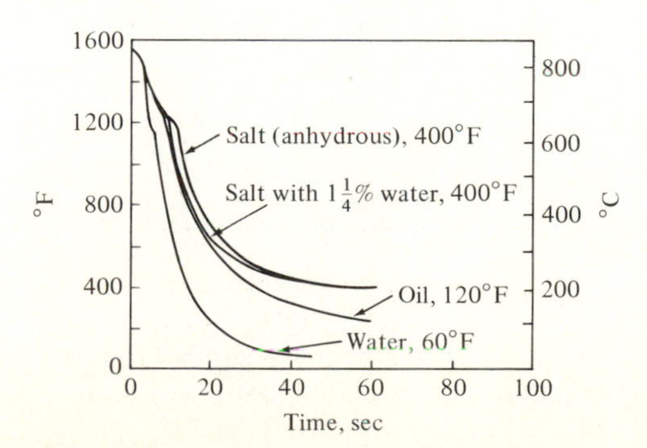

Fig. 6.28 *Effect of various quenchants on cooling rates of 1040 steel bars, 1 in. in diameter by 4 in. long.* (American Society for Metals Handbook, 8th ed., Vol. 2: *Heat Treating, Cleaning and Finishing,* Metals Park, Ohio, 1964)

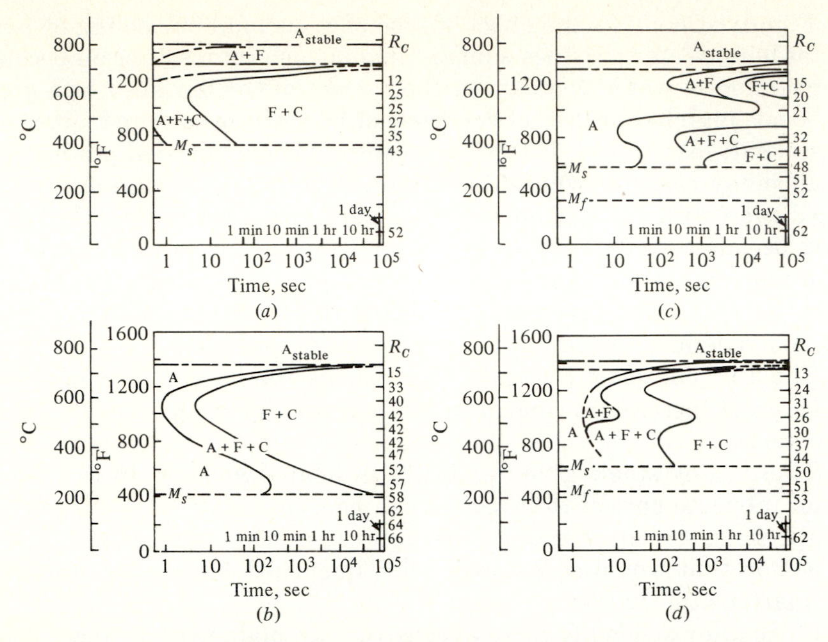

Fig. 6-29 *Effect of alloy on TTT curves. Hardness values to the right of each steel are those obtained after complete isothermal transformation at temperatures indicated. (a) TTT diagram for 1034 steel, 0.34%* C. *(b) TTT diagram for 1090 steel, 0.9%* C. *(c) TTT diagram for 4340 steel, 0.4%* C, *1.7%* Ni, *0.8%* Cr, *0.25%* Mo. *(d) TTT diagram for 5140 steel, 0.4%* C, *0.8%* Cr. *Previous symbols used for* A, F, *and* C *were* γ, α, *and* Ca. (American Society for Metals Handbook, 8th ed., Vol. 2: *Heat Treating, Cleaning and Finishing,* Metals Park, Ohio, 1964, Figs. 4 and 5, p. 38)

In alloy steels the eutectoid temperature and the composition are somewhat different from 1333°F (723°C) and 0.8% carbon, and the correction can be made using the data of Fig. 6.30. The effects at low alloy levels are approximately additive.

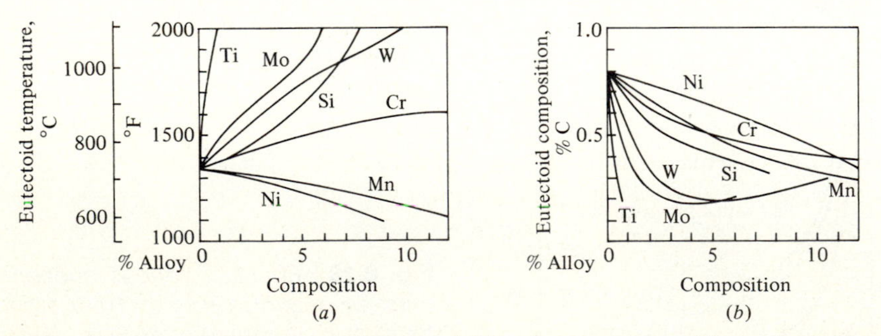

Fig. 6.30 *Effects of additions of alloys on (a) the temperature of the eutectoid reaction, and (b) the carbon content of the eutectoid.* (L.H. Van Vlack, *Elements of Materials Science,* 2d ed., Addison-Wesley Publishing Company, Reading, Mass., 1964)

Finally, Fig. 6.29 shows that two noses may appear in some of the TTT diagrams for alloy steels (4340, for example). The lower nose is called the *bainite nose,* and the upper one is called the *pearlite nose*. In 4340 steel, for example, it is possible to cool at such a rate as to miss the pearlite nose yet encounter the bainite nose and therefore obtain bainite by continuous cooling instead of by the isothermal treatment discussed previously.

EXAMPLE 6.6 Calculate the eutectoid temperature, eutectoid carbon, and M_s for a steel of the following alloy composition: 2% Ni, 1% Cr, 0.5% C.

ANSWER From Fig. 6.30 we find:

Alloying Element	*Change in Eutectoid Temperature*	*Change in Eutectoid Carbon, Percent*
2% Ni	−30°F	−0.05
1% Cr	+50°F	−0.10
	+20°F	−0.15

Approximate eutectoid temperature = 1353°F (1333°F + 20°F) (734°C)

Approximate eutectoid carbon = 0.65% (0.80 − 0.15%)

M_s = 930 − 540 × 0.5 − 40 × 1 − 30 × 2 = 560°F (293°C)

Hardenability; Properties of Plain Carbon and Low-alloy Steels

6.19 General; method for choosing the proper steel for a given part

In the previous sections we established the importance of the time and temperature of austenite transformation on the hardness of the structures produced—pearlite, bainite, and martensite. The basic effect of the carbon and alloy content was shown by TTT curves. With this background we can now come to grips with the important engineering problem of how to select the steel for a given part and heat treatment.

Suppose that we have a simple shaft 2 in. (5.08 cm) in diameter. We wish to obtain a hardness of 50 R_C to at least $\frac{1}{2}$ in. (1.27 cm) beneath the surface. We wish to use an agitated oil quench for hardening, having experienced cracking in tests with a water quench. The question is which steel we should use. This leads us to the concept of hardenability evaluation.

6.20 Hardenability

Many years ago steel producers developed extensive handbooks showing the hardness profiles that could be developed in different steels with different quenching conditions in different section thicknesses. After a little thought we can see that with the variables of composition, section thickness, and quenching medium, literally thousands of curves are needed. In addition, the question of how to check a given heat of steel for its *ability to harden* developed. True, the TTT diagram would be useful, but determining it is time-consuming.

To meet this problem of evaluating the millions of tons of steel used by the automotive industry, Walter Jominy and his associates developed the *hardenability bar*. The idea behind this test is to produce in one bar a wide variety of known cooling rates. Then by measuring the hardness along the bar, we can find the hardnesses obtainable with different cooling rates from the austenitizing temperature.

Figure 6.31 shows the method of making the test. A bar of the steel to be tested is machined to give a cylinder 4 in. (10.16 cm) long and 1 in. (2.54 cm) in diameter with an upper lip. The bar is then austenitized in standard fashion in a furnace and placed in the fixture. The water is quickly turned on and provides a smooth film striking only the end of the bar. The quenched end cools rapidly. The regions away from the end cool at rates proportional to their distances from the quenched portion. After the bar is cool, it is removed from the fixture, a flat is ground along one side, and R_C hardness readings are taken every $\frac{1}{16}$ in. (0.159 cm). Figure 6.32 shows a typical plot. The greatest hardness is at the quenched end, where martensite is formed. The lower hardness farther away is due to softer transformation products.

The most significant point of this plot is not the hardness at a given distance but the hardness at a given cooling rate. (The cooling rates have been measured at the different points and are quite constant for different steels.) A little region in the bar does not know it is in a Jominy end-quench test; it only feels the effect of a given cooling rate. For this reason a position on the Jominy bar gives the hardness that would be obtained at a point in

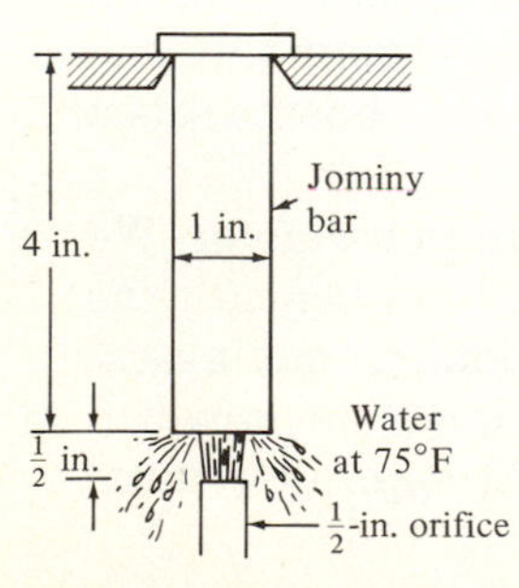

Fig. 6.31 *Jominy end-quench hardenability test.* (A.G. Guy, *Elements of Physical Metallurgy*, 2d ed., Addison-Wesley Publishing Company, Reading, Mass., 1959, Fig. 14-14, p. 484.)

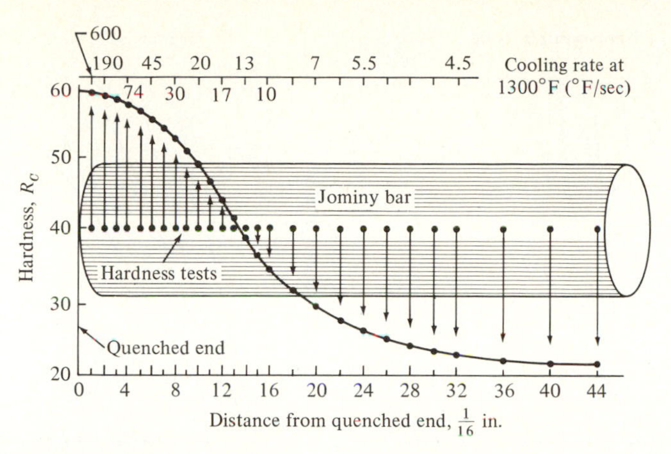

Fig. 6.32 *Typical distribution of hardness in Jominy bars.* (A.G. Guy, *Elements of Physical Metallurgy*, 2d ed., Addison-Wesley Publishing Company, Reading, Mass., 1959, Fig. 14-14, p. 484.)

an oil- or water-quenched bar with the *same cooling rate*. Figure 6.33 (page 238) gives the relationships between given Jominy positions and different locations in actual bars quenched in several media.

With these data and a few Jominy curves for representative steels (Fig. 6.34, page 239), we can now answer our previous problem.

EXAMPLE 6.7 How do we obtain 50 R_C minimum $\frac{1}{2}$ in. beneath the surface of a 2-in.-diameter bar quenched in still oil?

ANSWER From Fig. 6.33, the cooling rate at this position (midradius) is 20°F/sec, and this corresponds to a Jominy position of $\frac{11}{16}$ in. from the quenched end of the bar. We then draw a vertical line at this position on the Jominy curve of Fig. 6.34. We see that of this group only 4340 steel would provide satisfactory hardness, that is, above 50 R_C (4140 would be almost 50 R_C).

EXAMPLE 6.8 We have a complex shape made of 1040 steel. After austenitizing and quenching in agitated oil, the hardness $\frac{1}{8}$ in. below the surface is 30 R_C. This is too soft for our application; 50 R_C at $\frac{1}{8}$ in. below the surface is necessary. What steel would you recommend if the austenitizing treatment and quenching are to remain the same?

ANSWER From the Jominy curves for 1040 steel (Fig. 6.34), 30 R_C appears at $\frac{1}{4}$ in. from the water-quenched end. This means that the material at $\frac{1}{4}$ in. from the water-quenched end of a Jominy test cools at the *same* rate (74°F/sec) as the

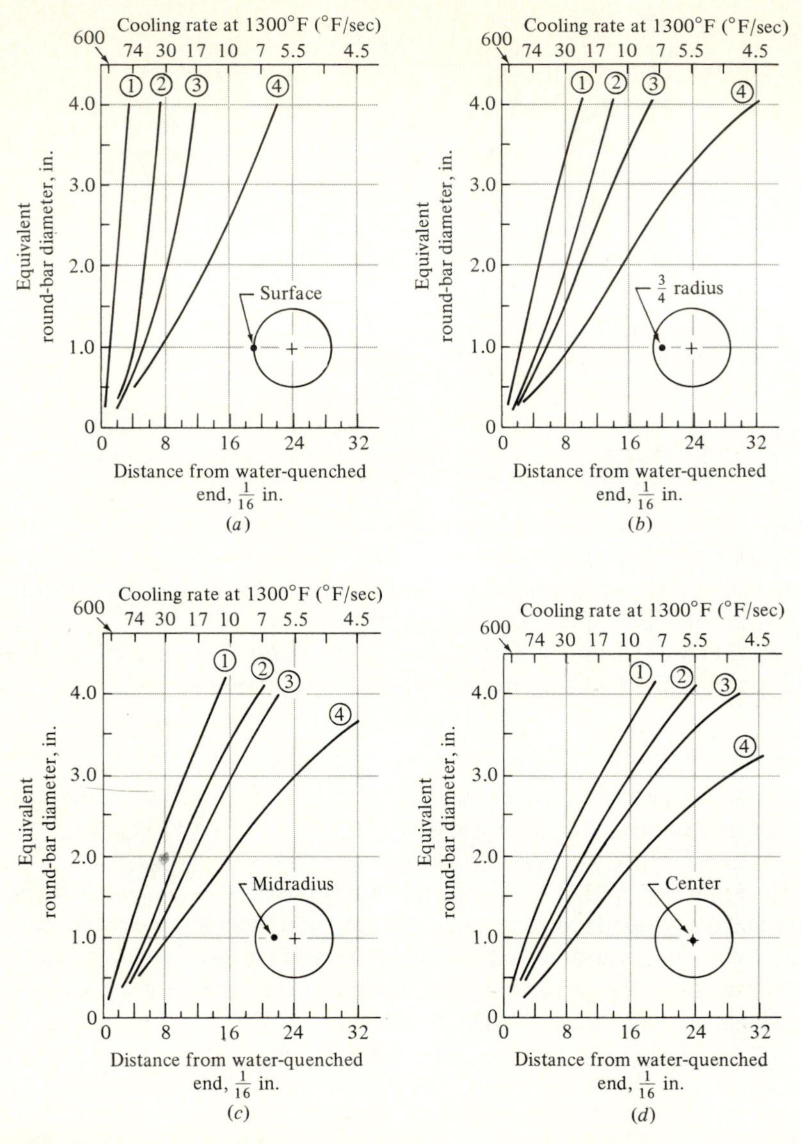

Fig. 6.33 *Relationships between cooling rates in round bars and in Jominy locations. 1 = still water; 2 = mildly agitated oil; 3 = still oil; 4 = mildly agitated molten salt.*

material in our complex shape quenched in oil at *$\frac{1}{8}$ in. below the surface.* Therefore, we look at the hardness obtained for other steels at the same cooling rate ($\frac{1}{4}$ in. station or 74°F/sec) and see that 4340, 4140, and 3140 steels will all give over 50 R_C under these quenching conditions.

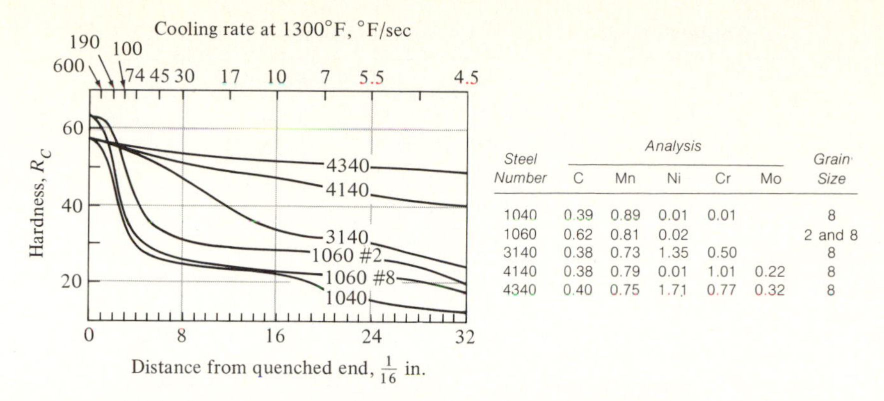

Steel Number	Analysis C	Mn	Ni	Cr	Mo	Grain Size
1040	0.39	0.89	0.01	0.01		8
1060	0.62	0.81	0.02			2 and 8
3140	0.38	0.73	1.35	0.50		8
4140	0.38	0.79	0.01	1.01	0.22	8
4340	0.40	0.75	1.71	0.77	0.32	8

Fig. 6.34 *Hardenability curves for several steels with different compositions and austenite grain sizes (a higher number indicates a finer grain size).* (L.H. Van Vlack, *Elements of Materials Science*, 2d ed., Addison-Wesley Publishing Company, Reading, Mass., 1964, Fig. 11.30, p. 322)

6.21 Analysis and properties of typical low-alloy steels

To simplify the specification of these steels, the Society of Automotive Engineers (SAE) and the American Iron and Steel Institute (AISI) issued a joint specification of SAE-AISI steels (Table 6.2). The key is as follows. The last two digits give the carbon content, as in plain carbon steels; that is, 4340 steel contains 0.40% carbon. In rare cases, such as 52100 ball-bearing steel, five digits are used and the last three represent the carbon, which in this example is 1.00%. The chemical analysis of any given steel will vary somewhat from that in Table 6.2 (p. 240) because of manufacturing variables.

The first two digits indicate alloy content as shown, including the plain carbon steels for completeness. Figure 6.35 shows the combinations of strength, ductility, and hardness obtainable with these steels. Therefore, from a simple hardness test after quenching and tempering it is possible to determine the other properties. In general, the carbon determines the hardness of the martensite in the quenched structure. The alloy content is merely a means of obtaining the desired structure at a given cooling rate.

Cast Steel In the preceding discussions we have generally referred to specifications for wrought steel, that is, for material that is cast and then rolled or forged to shape. However, many important parts are made as steel castings. In this way we can easily obtain many shapes that would be difficult to fabricate. To illustrate this point, let us consider several applications.

Electrical machinery such as generators and motors requires a cast low-carbon steel (0.05 to 0.15% carbon) to obtain optimum magnetic properties.

Table 6.2 SAE-AISI CHEMICAL SPECIFICATIONS

			Nominal Chemical Analysis, percent								
Type	*Name*	*Example*	C	Mn	P *Maximum*	S *Maximum*	Si	Ni	Cr	Mo	*Other*
10*xx*	Plain carbon	1020	0.2	0.4	0.04	0.05					
11*xx*	Free machining	1111	0.1	0.7	0.09 average	0.12 average					
13*xx*	Mn	1330	0.3	1.7	0.04	0.04	0.3				
3*xxx*	NiCr	3140	0.4	0.8	0.04	0.04	0.3	1.0	0.6		
40*xx*	Mo	4042	0.4	0.8	0.04	0.04	0.2			0.25	
41*xx*	CrMo	4140	0.4	0.8	0.04	0.04	0.3		1.0	0.20	
46*xx*	NiMo	4620	0.2	0.6	0.04	0.04	0.3	1.8		0.25	
47*xx*	NiCrMo	4720	0.2	0.6	0.04	0.04	0.3	1.0	0.4	0.20	
48*xx*	NiMo	4820	0.2	0.6	0.04	0.04	0.3	3.5		0.25	
50*xx*	Cr	5015	0.1	0.4	0.04	0.04	0.3		0.4		
52*xx*	Cr	52100	1.0	0.4	0.02	0.02	0.3		1.4		
61*xx*	CrV	6120	0.2	0.8	0.04	0.04	0.3		0.8		0.10 V
81*xx*	NiCrMo	8115	0.15	0.8	0.04	0.04	0.3	0.3	0.4	0.10	
86*xx*	NiCrMo	8650	0.5	0.8	0.04	0.04	0.3	0.5	0.5	0.20	
87*xx*	NiCrMo	8720	0.2	0.8	0.04	0.04	0.3	0.5	0.5	0.25	
88*xx*	NiCrMo	8822	0.2	0.8	0.04	0.04	0.3	0.5	0.5	0.35	
92*xx*	Si	9260	0.6	0.8	0.04	0.04	2.0				
93*xx*	NiCrMo	9310	0.1	0.6	0.02	0.02	0.3	3.0	1.2	0.10	
94*xx*	NiCrMo	94B30*	0.3	0.8	0.04	0.04	0.3	0.4	0.4	0.10	0.0005 B
98*xx*	NiCrMo	9840	0.4	0.8	0.04	0.04	0.3	1.0	0.8	0.25	

*"B" refers to the presence of boron.

Structural parts for the railroad industry, such as side frames, use steel in the range 0.2 to 0.5% carbon to attain a good combination of strength and ductility.

Railroad car wheels and mining equipment components, which require better wear resistance, use eutectoid (0.8% carbon) steel.

Other parts that are to be hardened, such as large gears, use low-alloy steels similar to the SAE-AISI grades already discussed.

6.22 Tempering of alloy and noneutectoid steels

We have already discussed the tempering of 0.8% carbon steel in Sec. 6.14. The other steels follow the same principles.

In general terms, tempering is a reheating operation that leads to precipitation and spheroidization of carbide. There are notable side effects, such as the change in martensite from a stressed tetragonal structure to BCC and often the transformation of retained austenite (untransformed during cooling in high-carbon alloys), but the basic effect on carbide is the most important.

Although we normally refer to the tempering of martensite, the term is also used for the softening of bainitic and even pearlitic structures. If

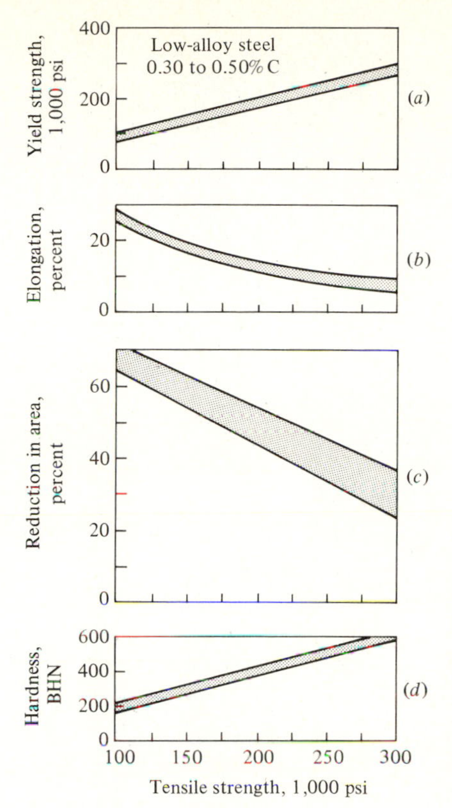

Fig. 6.35 *Properties of tempered martensite (low-alloy steel, 0.3 to 0.5% carbon). (a) Yield strength. (b) Elongation. (c) Reduction in area. (d) Brinell hardness. [To obtain MPa, multiply psi by 6.9×10^{-3}. To obtain kg/mm², multiply by 7.03×10^{-4}.* (American Society for Metals Handbook, 8th ed., Vol. 1: *Properties and Selection of Metals,* Metals Park, Ohio, 1961, Fig. 3, p. 109)

samples of martensite, bainite, and pearlite of the same steel are heated at 1200°F (649°C), for example, all will contain spheroidite (spheroidal carbide plus ferrite) after sufficient time has passed.

Some quantitative relations among alloy content, tempering temperature, and hardness are shown in Fig. 6.36*a*. The alloy steels generally retain their hardness more than the plain carbon steels. The steels shown in Fig. 6.36*a* are all 0.45% carbon, but the effect of other carbon contents can be determined by using Fig. 6.36*b* (page 242).

EXAMPLE 6.9 Find the tempered hardness of a 4330 steel after 1 hr at 600°F (315°C).

ANSWER We read the hardness for 4345 at 50 R_C on Fig. 6.36*a*. Then from Fig. 6.36*b* we find that a subtraction of 6 R_C is needed, giving 44 R_C as the final reading. (*Note:* These corrections should be applied only to steels of similar alloy content.)

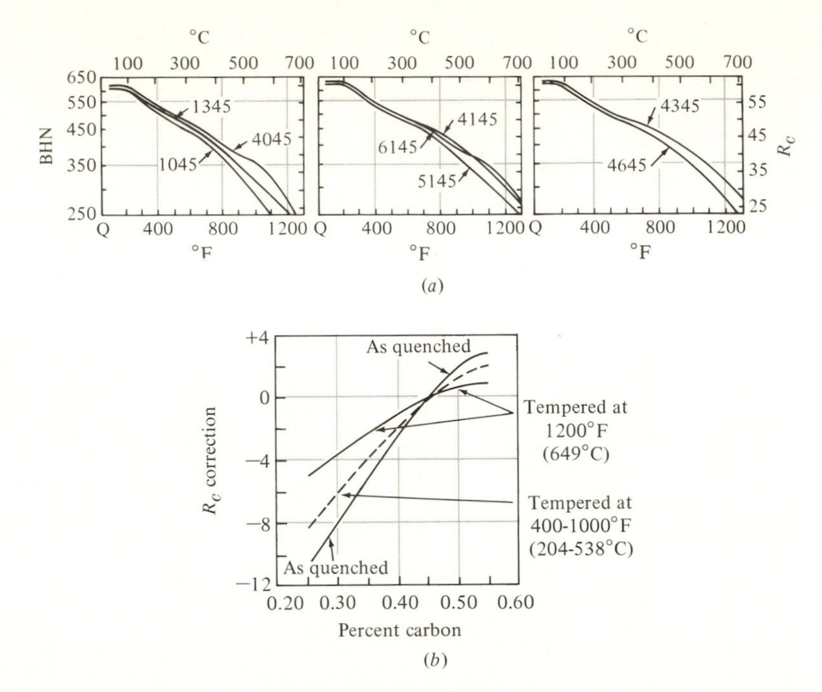

Fig. 6.36 *(a) Tempering characteristics of eight 0.45% carbon alloy steels. Duration of tempering = 1 hr. (b) Effect of carbon content on hardness of quenched and tempered steel, with R_C units to be added or subtracted from the value for 0.45% carbon for different temperatures* (American Society for Metals Handbook, 8th ed., Vol. 1: *Properties and Selection of Metals*, Metals Park, Ohio, 1961, Fig. 2, p. 109)

6.23 High-strength low-alloy steels (HSLA Steels)

With the requirement for higher gasoline mileage in cars, one step that has been taken is to reduce the car weight. Since a good percentage of the weight is in the steel body, intensive effort has been directed toward producing higher-strength, very low-alloy steels that can be used in thinner sections. The conventional "low-alloy" steels that we have just discussed are not as weldable and formable as the usual low-carbon sheet steels with 0.15 to 0.25% carbon. Therefore in the sixties a good deal of effort was directed toward developing low-carbon sheet steels that were "microalloyed" and rolled under controlled conditions to produce high strength and good weldability. Because of the careful control of small amounts of alloy and the processing, these steels are largely proprietary compositions. They are sold on the basis of their mechanical properties rather than their chemical analysis.

Although there are a number of HSLA steels for specific purposes, two grades of a specific type given below indicate the properties that might be anticipated.

Alloy Designation	% C	% Mn	% Si	% Nb	% V	% N
A633-Grade A	0.18	1.00/1.35	0.15/0.30	0.05	—	—
A633-Grade E	0.22	1.15/1.50	0.15/0.50	0.01/0.05	0.05/0.15	0.01/0.03

Alloy Designation	Minimum Tensile Strength† (× 1000 psi)	Minimum Yield Strength† (× 1000 psi)	Minimum Elongation‡ (%)
A633-Grade A	63/83	42	23
A633-Grade E	75/100	55/60	23

High-alloy Steels and Superalloys

6.24 General

Up to this point we have used only relatively small amounts of alloy to change the TTT curve and hardenability. There is another group of steels in which well over 5% alloy is added for various special purposes. These high-alloy steels may be divided into the groups shown in the table you see below.

A number of these steels are called "austenitic," so that one of the effects of the alloy must be to produce a structure that is austenite at room temperature. This can be explained only by a radical alteration of the iron–iron carbide diagram we have studied. The phase relationships in the high-alloy steels must therefore be our first consideration.

Type	Common Chemistry, percent
Stainless steel	
Austenitic	18 Cr, 8 Ni, balance Fe
Ferritic	16 Cr, 0.1 C
Martensitic	17 Cr, 1 C
Precipitation-hardened stainless steel	17 Cr, 7 Ni, 1 Al
Maraging steel	18 Ni, 7 Co
Tool steel; high-speed steel	18 W, 4 Cr, 1 V
Manganese steel, austenitic	12 Mn

†Multiply psi by 6.9×10^{-3} to obtain MN/m^2 (MPa).
‡Elongation in 2 in. The value is 18% in 8 in. (a more common gage length for sheet samples).

6.25 Phase diagrams of the high-alloy steels

The most important effect of alloys is the change in the austenite field. We can classify all alloys on the basis of whether they contract or expand the γ field. Examples are shown in Fig. 6.37. Note that nickel widens the field until, at 32% nickel, γ is encountered at room temperature under *equilibrium* conditions. Conversely, chromium narrows the field. At 13% chromium we have only α as the solid phase at all temperatures. Although the interstitial elements such as carbon and boron do not produce completely open austenitic or ferritic fields, they expand or contract the γ field. Figure 6.38 summarizes the effects of various elements on a periodic basis. It is

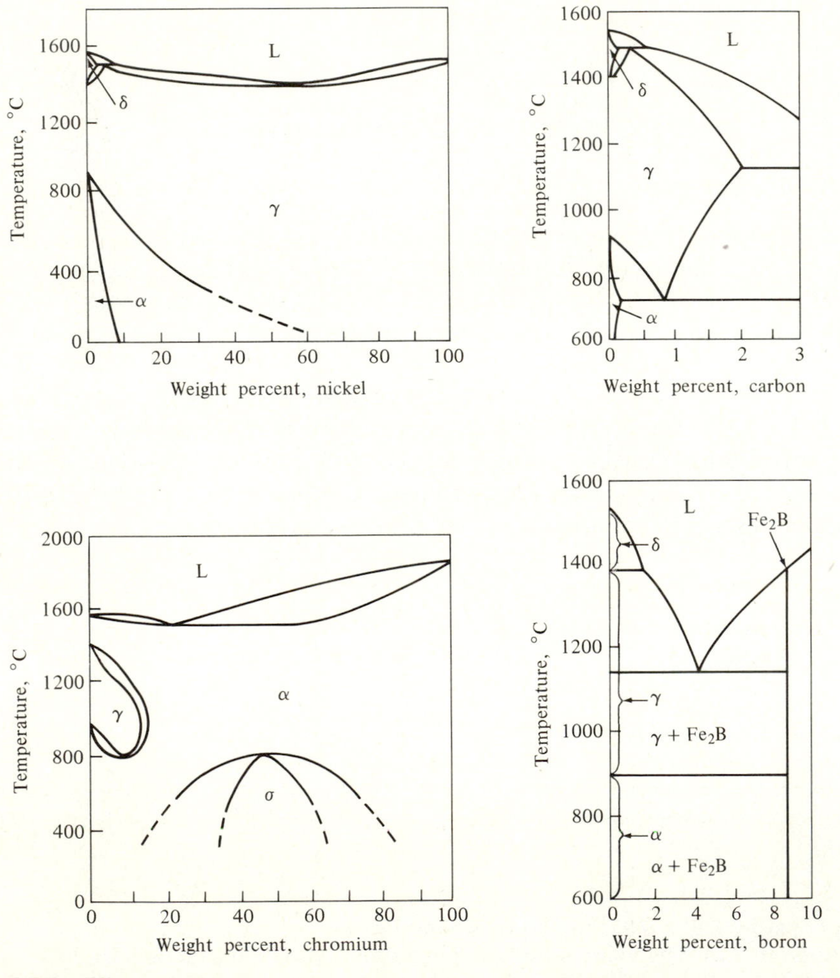

Fig. 6.37 *Effect of alloys on the γ field of iron. Some alloying elements narrow the field, while others widen it.*

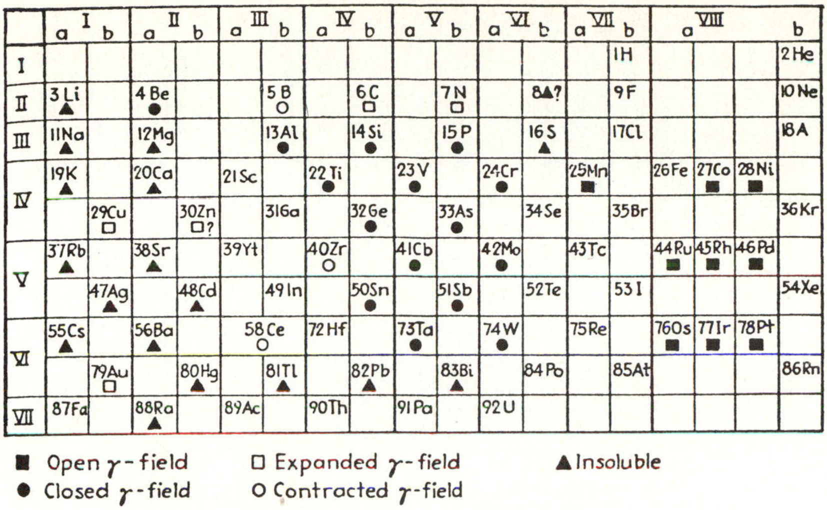

	I a	I b	II a	II b	III a	III b	IV a	IV b	V a	V b	VI a	VI b	VII a	VII b	VIII a			b
I														1H				2He
II	3Li ▲		4Be ●			5B ○		6C □		7N □		8▲?		9F				10Ne
III	11Na ▲		12Mg ▲			13Al ●		14Si ●		15P ●		16S ▲		17Cl				18A
IV	19K ▲		20Ca ▲		21Sc		22Ti ●		23V ●		24Cr ●		25Mn ■		26Fe	27Co ■	28Ni ■	
		29Cu □		30Zn □?		31Ga		32Ge ●		33As ●		34Se		35Br				36Kr
V	37Rb ▲		38Sr ▲		39Yt		40Zr ○		41Cb ●		42Mo ●		43Tc		44Ru ■	45Rh ■	46Pd ■	
		47Ag ▲		48Cd ▲		49In		50Sn ●		51Sb ●		52Te		53I				54Xe
VI	55Cs ▲		56Ba ▲		58Ce ○		72Hf		73Ta ●		74W ●		75Re		76Os ■	77Ir ■	78Pt ■	
		79Au □		80Hg ▲		81Tl ▲		82Pb ▲		83Bi ▲		84Po		85At				86Rn
VII	87Fa		88Ra ▲		89Ac		90Th		91Pa		92U							

■ Open γ-field □ Expanded γ-field ▲ Insoluble
● Closed γ-field ○ Contracted γ-field

Fig. 6.38 *Periodic table of the elements, showing the effects of various elements on the γ field.* (C. Barrett, *Structure of Metals*, 2d ed., McGraw-Hill Book Company, New York, 1952, Fig. 16, p. 251)

interesting that all the group VIII elements plus manganese expand the γ field, whereas the other transition elements contract it.

We shall now discuss the structures encountered in the special steel groups.

6.26 The stainless steels

We pointed out in the introduction that the stainless steels fall into three principal groups—ferritic, martensitic, and austenitic—named according to the predominating structure.

Ferritic Steels We see from the iron-chromium diagram (Fig. 6.37) that the formation of austenite is completely suppressed in a pure iron-chromium alloy above 13% chromium. However, if carbon is present, the chromium will form chromium carbide. In this case 1% of carbon will combine with 17% of chromium. This carbide will precipitate and deprive the iron matrix of a corresponding amount of chromium. For this reason the commercial steel 430 in Table 6.3 (page 246) contains 16% chromium to retain the ferritic structure. The advantage of this material over unalloyed ferrite is its corrosion resistance, as discussed in Chapter 12. It is used in automotive trim and kitchen utensils in the cold-worked condition.

Martensitic Steels When the carbon content of the high-chromium steels is high, as in 440C (Table 6.3), we can heat to the γ field and quench to produce martensite. Also with lower chromium, as in type 410, we can

Table 6.3 TYPICAL PROPERTIES OF STAINLESS STEELS

Number	Chemical Analysis, percent	Condition	Tensile Strength,* psi × 10^3	Yield Strength,* psi × 10^3	Percent Elongation	BHN	Typical Use
			Austenitic Steels				
301	17 Cr, 7 Ni	Annealed	110	40	60	160	Lightweight, high-strength transportation equipment
		Cold-worked	185	140	9	388	
304	19 Cr, 10 Ni	Annealed	85	35	60	149	General chemical equipment
		Cold-worked	110	75	12	240	
347	18 Cr, 11 Ni†	Annealed	90	35	45	160	Welded construction
			Ferritic Steels				
430	16 Cr, <0.1 C	Annealed	80	55	25	140	Automobile trim, kitchen equipment
		Cold-worked	90	80	20	200	
			Martensitic Steels				
410	12 Cr, 0.15 C	Annealed	70	40	30	155	General purpose springs, rules
		Quenched and tempered	140	100	20	300	
440C	17 Cr, 1 C	Annealed	110	65	14	230	Instruments, cutlery, valves
		Quenched and tempered	285	275	2	580	
			Precipitation-hardened Steels				
17-7PH	17 Cr, 7 Ni, 1 Al	Hardened	235	220	6	400	Airframe parts
			Maraged Steels				
Maraging steel	18 Ni, 7 Co‡	Maraged	275	268	11	500	Aircraft components

*Multiply psi by 6.9×10^{-3} to obtain MPa or by 7.03×10^{-4} to obtain kg/mm^2.
†Contains niobium (columbium) to prevent weld embrittlement. Percent Nb = 10 × percent C. Percent C is about 0.08 in usual grades.
‡0.025% C, 0.1% Mn, 0.1% Si, 0.22% Ti, 0.003% B.

produce martensite with only 0.15% carbon. Because of their high alloy content these steels have high hardenability. In some cases it is only necessary to cool them in air to form martensite, easily avoiding pearlite and bainite formation. These steels are excellent for cutlery and dies.

We can summarize the carbon-chromium relationships for the ferritic and martensitic types as follows:

When [% Cr − (17 × % by weight C)] > 13, only ferrite is present. When [% Cr − (17 × % by weight C)] < 13, austenite can be formed and quenched to martensite. Figure 6.39 also shows this.

Austenitic Steels In these steels we find substantial amounts of nickel as well as chromium, as in the familiar 18% Cr, 8% Ni types. The nickel enlarges the austenite field to such an extent that it is stable at room temperature. To illustrate this effect we can use a horizontal section of the iron-nickel-chromium ternary diagram (Fig. 6.40) of the type discussed in Chapter 4. This section represents the phases present at 650°C. (It is simpler to use this temperature to avoid the complications of transformation of the austenite in the low-alloy steels that occur at lower temperature.) We note that the composition 18% Cr, 8% Ni is close to the boundary between the $\alpha + \gamma$ and γ fields. If we want to be sure to have an all-γ structure at room temperature, we use a type 304 composition (Table 6.3), with 10% nickel. However, there is an advantage to type 301 steel, in which some ferrite is encountered. This steel is more responsive to cold working, as shown by the yield strength compared with type 304. Type 347 is used where there will be welding, to avoid the combination of carbon with chromium which leads to poor corrosion resistance. In type 347 the carbon combines with the niobium, instead of with chromium, preventing the corrosion due to chromium depletion of the matrix, as discussed in Chapter 12.

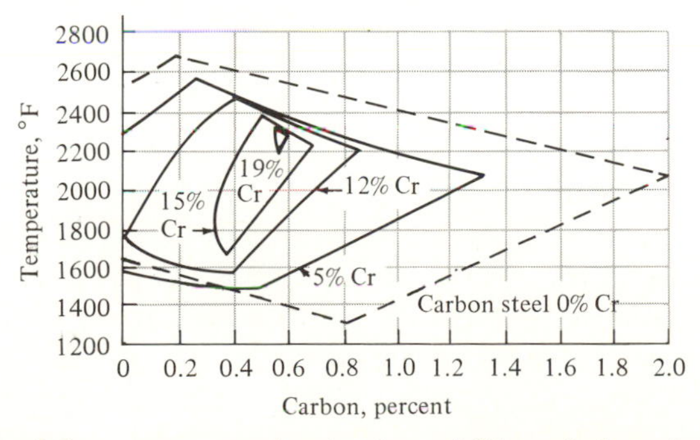

Fig. 6.39 *Closing of the gamma region by the addition of chromium and carbon.* (After Bain, *Alloying Elements in Steel*, 2nd ed., American Society for Metals, Metals Park, Ohio, 1961)

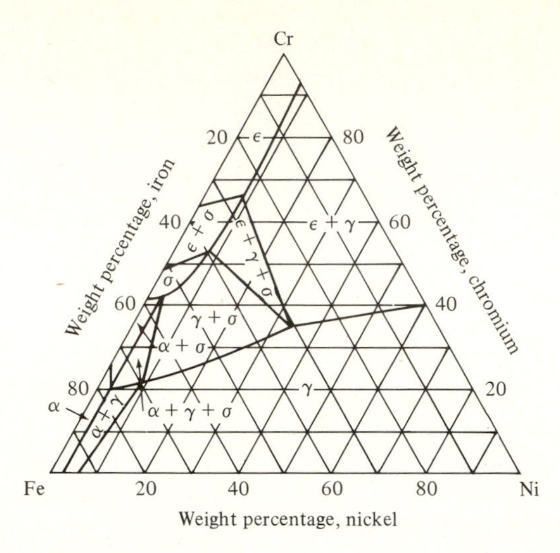

Fig. 6.40 *Iron–nickel–chromium ternary phase diagram at 650°C.* (American Society for Metals Handbook, 8th ed., Vol. 1: *Properties and Selection of Metals,* Metals Park, Ohio, 1961)

6.27 Precipitation hardening and maraging steels

Where maximum strength and some corrosion resistance are required, 17-7PH and maraging 18% nickel steels are used. In the 17-7PH type a solution treatment at 1925°F (1050°C) is followed by aging at 900°F (482°C) to produce a precipitate. In the maraging steels the material is austenitized at 1500°F (815°C) to dissolve precipitated phases, then relatively soft, low-carbon martensite is formed on cooling. The material can be worked or machined in this condition. To strengthen the material, the alloy is heated to 900°F (482°C) for 3 hr, which causes aging. Thus the hardening is due to both martensite and aging precipitates. There is little distortion.

6.28 Tool steels

The tool steels encompass a wide range of compositions. It is possible to produce a hard cutting edge in a high-carbon plain carbon steel. The advantages of the high-alloy steels such as 18% W, 4% Cr, 1% V are twofold. With higher hardenability we can use slower cooling rates during quenching, protecting tools from cracking. Second, the highly alloyed martensite plus excess alloy carbides retain their hardness at the high temperatures generated by fast cutting speeds; i.e., they resist tempering.

The various grades of tool steels are represented in Table 6.4 in order of increasing cost. The plain carbon W1 grade is heat-treated, as is the simple iron-carbon alloy already discussed. The final hardness depends on the carbon content and the tempering temperature. The same hardness can

Table 6.4 TYPICAL CHARACTERISTICS OF TOOL STEELS

Number	*Chemical Analysis, percent**					*Austenitizing Temperature, °F (°C)*	*Quenching Medium*	*Tempering Temperature, °F (°C)*	R_C	*Typical Use*
	C	Cr	Mo	W	V					
W1	0.6 1.4					1550 (843) 1400 (760)	Water	350 (77) 650 (343)	64 50	Tools and dies used below 350°F (177°C)
O1	0.9	0.5		0.5		1500 (815)	Oil	350 (177) 500 (260)	62 57	Tools and dies requiring less distortion than W1
D1	1.0	12.0	1.0			1800 (982)	Air	400 (204) 1000 (538)	61 54	Wear-resistant, low-distortion tools
H11	0.35	5.0	1.5		0.4	1850 (1010)	Air	1000 (538) 1200 (649)	54 38	Hot-working dies
M1	0.8	4.0	8.0	1.5	1.0	2200 (1204)	Air	1000 (538) 1100 (593)	65 60	High-speed tools
T1	0.7	4.0		18.0	1.0	2350 (1288)	Air	1000 (538) 1100 (593)	65 60	High-speed tools

* Balance Fe.

be attained in this steel as in the highly alloyed types. The next grade, O1, can be oil-hardened in simple sections because of the alloy and shows less distortion. The D1 grade, with 12% chromium, resembles the martensitic stainless steels we just discussed at a high-carbon level. The H grade is different from all the others because of the lower carbon level. It is used for hot-working dies and is therefore exposed to thermal shock. A higher-carbon steel would crack under these conditions. To prevent softening of the martensite, a considerable amount of alloy is used: 5% Cr, 1.5% Mo, 0.4% V. The M1 and T1 steels can be considered together as high-speed steels. The T1 type, with 18% tungsten, was developed first, then with the shortage of tungsten in World War II the molybdenum modification was developed.

The most significant feature of the heat treatment of these steels is the need for a very high austenitizing temperature. The phase diagram for the T1 type (Fig. 6.41) explains this point. During the early research on these steels, the hardness obtained after quenching was disappointingly low. This was because with an austenitizing temperature of 900°C (1652°F) the carbon content of the γ is only 0.2%. Therefore a low-carbon martensite is obtained at about 50 R_C. However, when the austenitizing temperature is raised to 1250°C (2270°F), the carbon content of the austenite is 0.6% which gives martensite of 64 R_C on quenching.

The tempering of high-speed steels has received extensive study because of an effect called *secondary* hardening. The hardness rises slightly after tempering at 1000°F (538°C). The as-quenched structure contains some retained austenite, and upon heating to 1000°F (538°C) some alloy carbides precipitate. This effect raises the M_s of the retained austenite (carbides take alloy out of solution), and the structure transforms to martensite

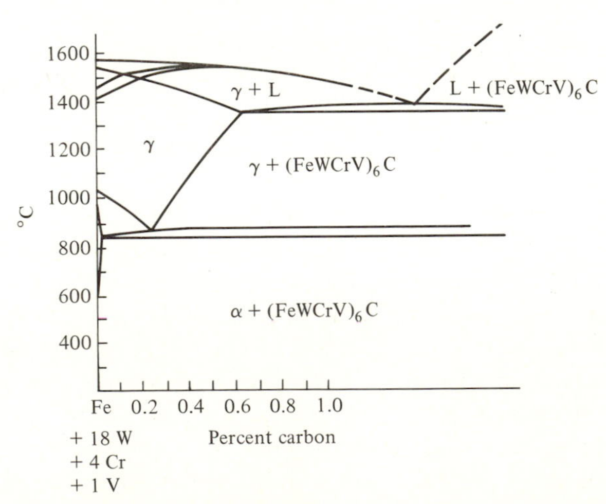

Fig. 6.41 *Phase diagram for T1 type tool steels.* (R.M. Brick and A. Phillips, *Structure and Properties of Alloys*, McGraw-Hill Book Company, New York, 1949)

on cooling from the tempering temperature. Therefore, although the original martensite was softened at 1000°F (538°C), the new martensite raises the overall reading. Retempering often improves the toughness of the structure.

6.29 Hadfield's manganese steel

Manganese, as well as nickel, produces an enlargement of the γ field. We take advantage of this in a very tough, abrasion-resistant austenitic steel containing 12% manganese and 1% carbon. The as-cast structure contains carbide, austenite, and some pearlite. However, after heating to 2000°F (1093°C) and water-quenching, the structure is completely austenitic. A water quench may be used even with complex castings because there are no stresses from transformation of austenite to martensite. Typical properties are: tensile strength: 120,000 psi (828 MPa); yield strength: 50,000 psi (345 MPa); percent elongation: 45; BHN: 160.

The steel is widely used in mining and earth-moving equipment and in railroad-track work.

6.30 Superalloys; general

The name *superalloys* arises from the fact that these materials exhibit far greater strength at high temperatures (1500 to 2000°F) (815 to 1093°C) than conventional alloys. Some of these alloys are even called "exotic" because of the high cost of some of the elements used. The basic reason for the development of these alloys is that the efficiency of a gas turbine increases with the temperature at which the rotor can be operated. However, the problem of strength at elevated temperatures is not merely in the rotor itself but in the vanes, the combustion chamber, and even the compressor section.

There are two principal problems in developing materials for parts operating at high temperatures—oxidation resistance and strength. Oxidation resistance is essentially a problem in gas corrosion. This will be taken up in Chapter 12. At this point we shall consider only the problem of strength.

In testing for strength at high temperatures, simply running a hot tensile test (i.e., testing the conventional 0.505-in.-diameter specimen in a furnace) does not give the required information. To illustrate, suppose that the conventional test gives a yield strength of 50,000 psi (345 MPa). Let us load the specimen to 40,000 psi (276 MPa) and observe it over a period of time at the elevated temperature. First, the sample continues to elongate although it is at constant load. This is called *creep*. Second, the specimen ruptures after a certain time, say 20 hr. These are two new important factors in design: the gradual extension or creep and the lower rupture strength as a function of time of testing. Therefore, if we design a rotor for a gas turbine merely on the basis of yield strength during a tensile test, the blades will either stretch excessively and contact the housing or break after a period of time (see also Section 3.23).

Figure 6.42 shows typical deformation or creep-rupture data for an austenitic stainless steel.

6.31 High-temperature properties of typical superalloys

At first glance the names and analyses of superalloys seem random and complex (Table 6.5). There are, however, certain common features that correlate analysis, structure, and performance.

First, all the important alloys have an FCC structure based on a combination of nickel, iron, and chromium. Therefore, we can view them as a simple extension of the austenitic field of the ternary diagram (Fig. 6.40) to higher-nickel compositions. The chromium, although not added to obtain austenite, is essential to provide oxidation resistance. Thus we have accounted for the iron, nickel, and chromium content. Cobalt is used to replace part of the iron or nickel. It dissolves in solid solution, strengthening the material.

Aluminum and titanium have a special use in providing a very fine precipitate called γ' (read "gamma prime"). The name is derived from the fact that the structure is close to the γ solid solution. γ' has the formula Ni_3Al, and some of the aluminum can be replaced with titanium with only small changes in a_0. The γ' structure has long-range order with aluminum or titanium atoms occupying the corners of the unit cell and nickel atoms in the face centers. The precipitate is in the form of cubes. The magnification of the electron microscope is necessary to resolve it (see the frontispiece for Chapter 6). This precipitate is essential to high strength in these alloys.

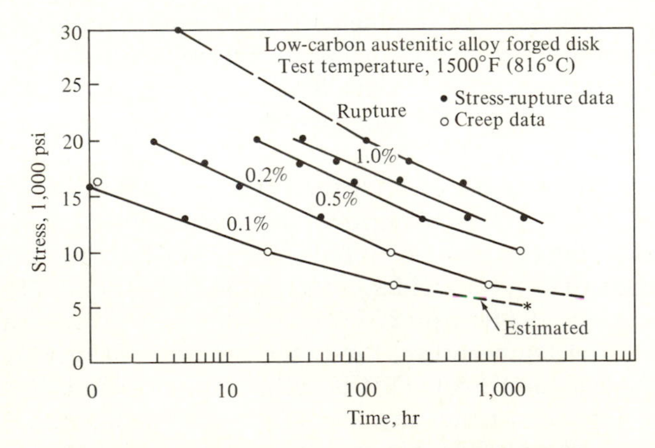

Fig.6.42 *Stress-deformation data for a low-carbon austenitic stainless steel.* (American Society for Metals Handbook, 8th ed., Vol. 1: *Properties and Selection of Metals,* Metals Park, Ohio, 1961)

Table 6.5 SUPERALLOYS: IRON, COBALT, AND NICKEL-BASE ALLOYS

	Chemical Analysis, percent								Stress to Rupture in 100 hr,* psi × 10^3, at Indicated Temperature, °F							Availability	
Code and Name	Fe	Ni	Cr	Co	Al	Ti	Mo	*Other*	*1200*	*1350*	*1400*	*1500*	*1600*	*1700*	*1800*	*Cast*	*Wrought*
HF (18/8 type)	Balance	10	20					0.3 C	30		14		6			×	×
Incoloy 901	Balance	43	13		0.2	3.0	6	0.01 B		50							×
Inconel 718	18	52	19		0.6	0.8	3	5.0 Nb			45					×	×
Waspalloy	2	Balance	20	13	1.3	3.0	4	†				40	25			×	×
Inco 713	2	Balance	13		6.0	0.8	9	2.0 Nb					44			×	
Udimet 500	4	Balance	18	18	3.0	3.0	4	†				45	32	19		×	×
Udimet 700	1	Balance	15	18	4.5	3.5	5	†					42			×	×
In 100	1	Balance	10	15	5.5	5.0	3						56			×	
René 41	5	Balance	19	10	1.5	3.0	10					45	28	17		×	×

*Multiply psi by 6.9 × 10^{-3} to obtain MPa or by 7.03 × 10^{-4} to obtain kg/mm^2.
†Plus small amounts of boron and/or zirconium.

Molybdenum adds to the strength of the matrix and also forms complex carbides that reduce the creep rate. Small amounts of zirconium and boron are also added for their effects on the fine carbides. Usually the carbon content is about 0.1 to 0.2%, but the carbides are carefully controlled.

In general, this field of alloys is changing rapidly because compared with other alloys, the rewards are higher for developing a material with just moderately better performance. As an example, some new alloys have an addition of 0.5% hafnium. As another example, turbine blades are being solidified directionally using an induction field to develop the desired temperature gradient in the mold. Moreover, practically all grades are cast but only a limited number are wrought. The reason is that with increased creep strength the hot working becomes difficult and die life is short. Also, some of the alloys have low ductility; adequate for service but not for rolling. The investment casting process is the most widely used method. It is capable of producing intricate shapes.

White Iron, Gray Iron, Ductile Iron, and Malleable Iron

6.32 General; importance of high-carbon alloys

Gray cast iron is one of the oldest alloys; ductile iron, which is being used more and more, is one of the newest. As an example of the importance of these alloys, 95% of the weight of a typical automotive engine is gray iron and ductile iron. Typical gray iron parts are the block, head, camshaft, piston rings, tappets, and manifolds. The crankshaft and rocker arms are ductile iron. Malleable iron and ductile iron are also used in many other parts of the automobile, such as the differential housings.

The annual production of castings in this group of high-carbon iron alloys is about 22 million tons. This is second only to steel in the field of materials. Nevertheless, the structures and properties of these alloys are not well understood because of a few fundamental differences from steel. To use these alloys properly and to understand their heat treatment, it is important to focus first on the portion of the structure that is different from steel:

White iron: massive carbide
Gray iron: flake graphite
Ductile iron: spheroidal graphite
Malleable iron: clumplike graphite (temper carbon)
+ Steel matrix (ferrite, pearlite, martensite, etc.)

Figure 6.43 shows examples of these microstructures.

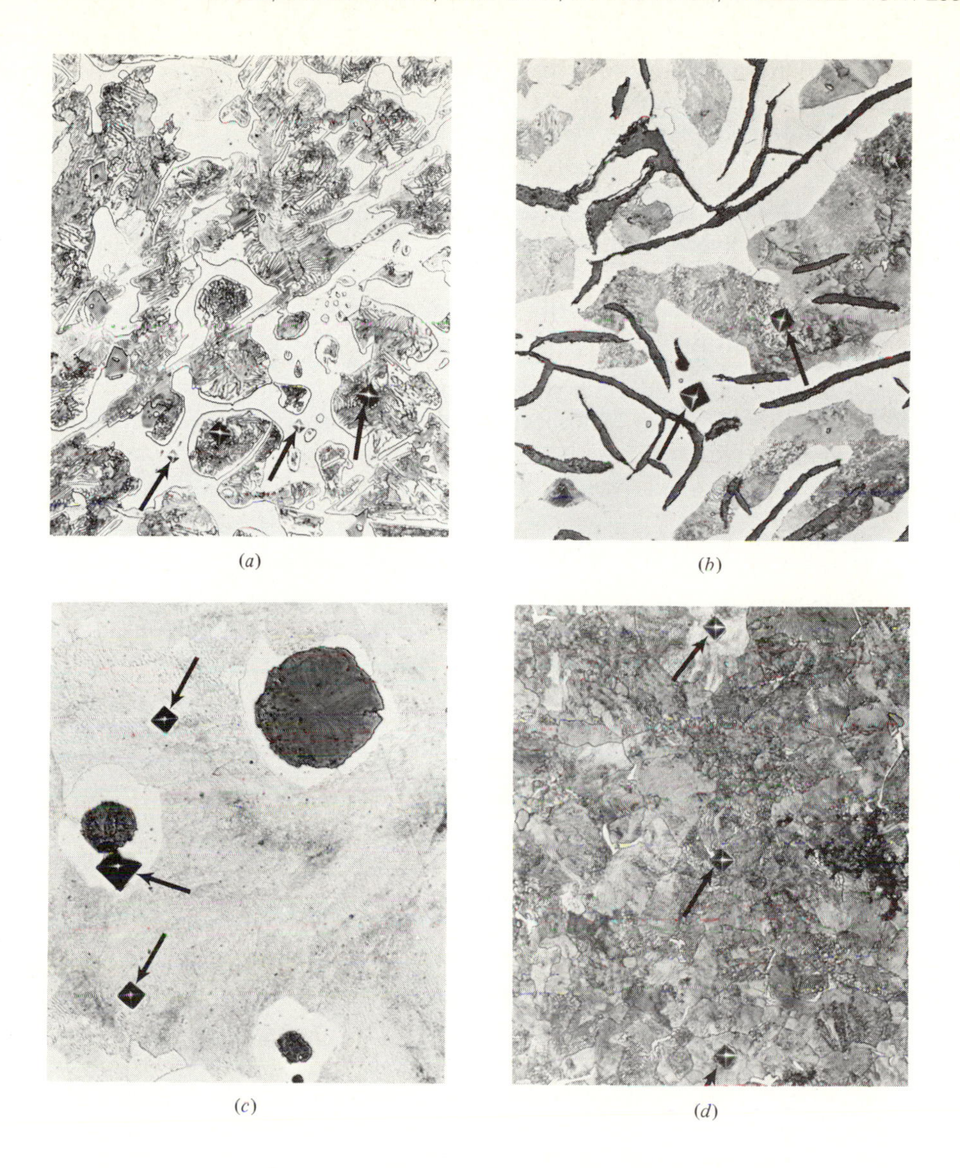

Fig. 6.43 *(a) White cast iron: 3.5% C, 0.5% Si. Massive carbide (white islands), VHN 1300. Pearlite (lamellar), VHN 300 as cast; 500×, 2% nital etch. (b) Gray cast iron: 3.5% C, 2% Si. Flake graphite. Pearlite (lamellar), VHN 285. Ferrite (white), VHN 195 as cast; 500×, 2% nital etch. (c) Ductile cast iron: 3.5% C, 2% Si, 0.05% Mg. Spheroidal graphite. Pearlite (lamellar), VHN 300. Ferrite (white), VHN 190 as cast; 500×, 2% nital etch. (d) Malleable cast iron: 2.5% C, 1% Si. Temper carbon. Pearlite (lamellar), VHN 345. Heat-treated, 1750°F (954°C), 12 hr, air-cooled; 500×, 2% nital etch.*

We are already familiar with the control of austenite transformation to ferrite and other products, so let us study first the factors affecting the massive carbide and graphite structures.

6.33 Relationships between white iron, gray iron, ductile iron, and malleable iron

White Iron We see from the fraction chart of white iron (Fig. 6.44) that carbide precipitates during three important periods (3% carbon iron):

1. 2065°F (1130°C): the eutectic reaction, L $\rightarrow \gamma$ + carbide
2. 2065 to 1333°F (1130 to 723°C): from the eutectic to the eutectoid, $\gamma \rightarrow \gamma\dagger$ + carbide
3. 1333°F (723°C): the eutectoid reaction, $\gamma \rightarrow \alpha$ + carbide (as pearlite)

Further, in reaction 1 large, massive carbides are formed from the liquid. In reaction 2 the carbide crystallizes onto the existing massive carbide, and in reaction 3 pearlite is formed, resulting in the microstructure in Fig. 6.43*a*. The final product, therefore, contains a high percentage of massive carbide and is hard and brittle.

Gray Iron To explain gray cast iron, we need to understand that iron carbide is not basically a stable phase. With abnormally slow cooling (or in the presence of certain alloys such as silicon), graphite (pure carbon)

†The analysis of the austenite changes from 2% carbon to 0.8% carbon with decreasing temperature.

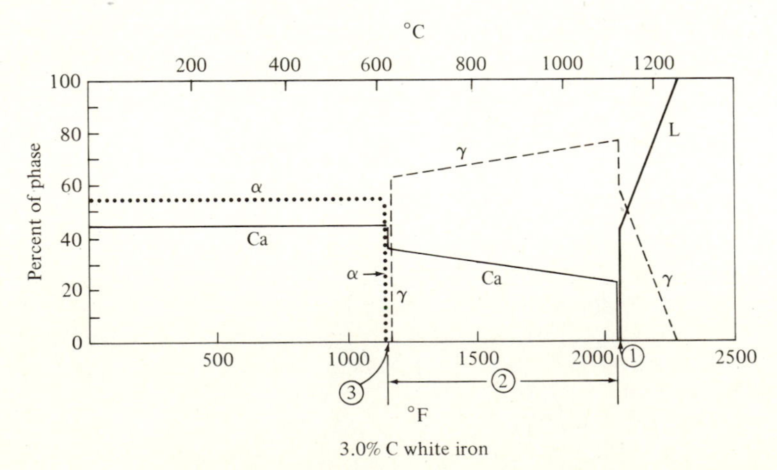

Fig. 6.44 *Fraction chart for white cast iron*

and iron will crystallize (Fig. 6.43). Furthermore, if we heat iron carbide for extended periods, it will decompose: iron carbide → iron + graphite.

In other words, the true equilibrium diagram is the iron-graphite system, which is superimposed on the iron–iron carbide diagram in Fig. 6.45. We do not need to learn a new diagram. For all practical purposes all we need to do is to substitute graphite for carbide in the two-phase fields, as shown in Fig. 6.45, and move the right-hand vertical line to 100% carbon. There are, however, slight differences in eutectic and eutectoid carbon contents and temperatures.

Using this diagram, let us follow the formation of graphite in a ferritic gray cast iron with 3% carbon (i.e., ferrite matrix with flake graphite). To produce graphite in place of carbide, we either cool very, very slowly or

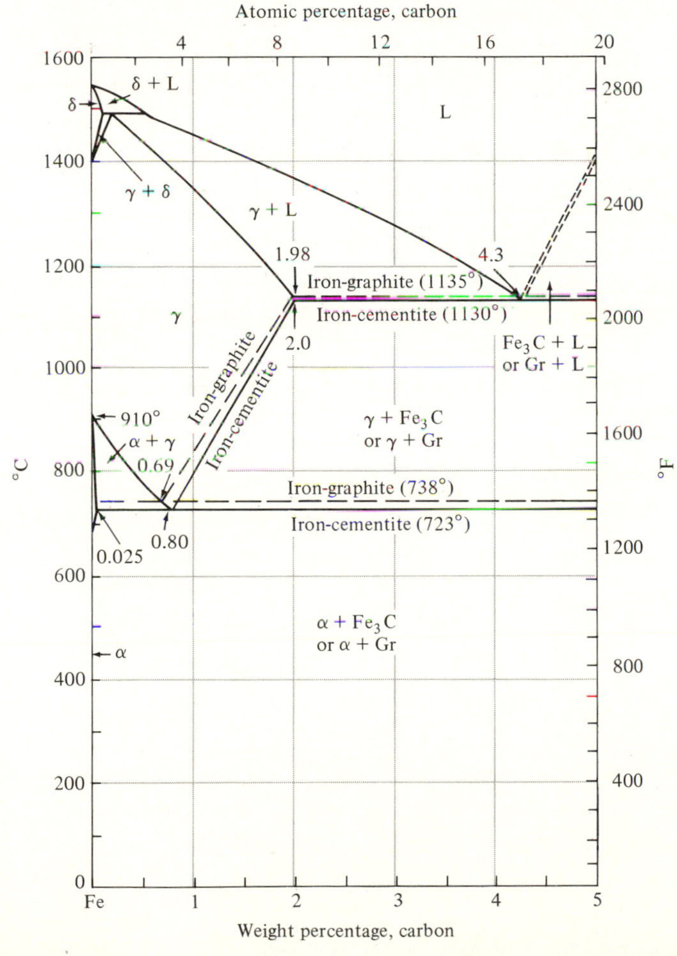

Fig. 6.45 *Iron–iron carbide and iron–graphite phase diagrams*

add silicon, which also promotes graphite formation. We obtain the fraction chart shown in Fig. 6.46.

Now we can contrast the formation of graphite with that of carbide in white cast iron of eutectic composition (4.3% carbon).

Temperature, °F (°C)	*Gray Iron†*	*White Iron*	*Comments*
2065 (1130)	L → γ + graphite	L → γ + carbide	
2065 to 1333 (1130 to 723)	$\gamma \rightarrow \gamma$ + graphite	$\gamma \rightarrow \gamma$ + carbide	Graphite (carbide) precipitates from γ on existing graphite (carbide). Carbon solubility is decreasing.
1333 (723)	$\gamma \rightarrow \alpha$ + graphite	$\gamma \rightarrow \alpha$ + carbide (as pearlite)	The analysis of the austenite is 0.8% carbon.

Therefore, in place of the carbide precipitation in white cast iron, we obtain graphite in the gray cast iron. (We shall modify this position somewhat when we discuss other gray cast iron microstructures. It is possible, for example, by fast-cooling gray iron from 1500°F (815°C) to obtain pearlite or the other austenite transformation products we found in the austenite transformation of steel.)

The final structure in ferritic gray cast iron is therefore flake graphite plus ferrite. The flakes formed at 2065°F (1130°C) grew during cooling to 1333°F (723°C), and at the eutectoid transformation $\gamma \rightarrow \alpha$ + Gr.

Finally, "gray" and "white" not only refer to the microstructure but

†We shall continue to use the iron–iron carbide equilibrium temperatures in this introductory section for simplicity. Later we shall modify these remarks when we discuss the actual commercial alloys containing, for example, silicon.

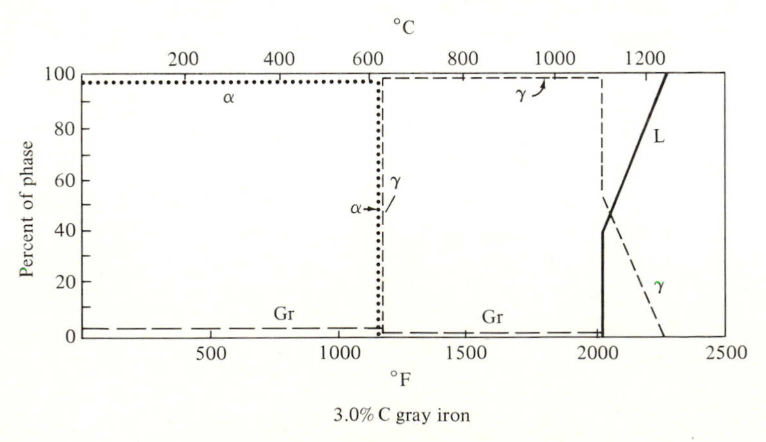

Fig. 6.46 *Fraction chart for gray cast iron*

also describe the appearance of the fracture of the cast irons. Therefore, a fracture test of cast iron will at least tell whether the high-carbon phase is carbide or graphite.

Ductile Iron Although gray iron is excellent for parts such as engine blocks because the graphite flakes give good machinability, a spheroidal graphite shape is better for strength. In 1949 Millis and Gagnebin discovered that dissolving a small (0.05%) amount of magnesium in the liquid iron would cause the formation of spheroidal graphite instead of flake graphite from the melt. Then the graphite that precipitates from 2065 to 1333°F (1130 to 723°C) and at the eutectoid continues to crystallize in the spheroidal form.

The final structure in ferritic ductile iron is therefore *spheroidal* graphite plus ferrite. The phase amounts and analyses are the same as for ferritic gray cast iron, but the different shape of the graphite gives twice the tensile strength and 20 times the ductility.

Malleable Iron The difference between malleable iron and gray or ductile iron is that the *as-cast* structure of malleable iron is the same as that of white cast iron and contains no graphite. In other words, the cooling rate and composition produce white cast iron in the original casting. The cold casting is then heated in a malleablizing furnace at 1750°F (954°C) to change the iron carbide to the equilibrium structure of iron plus graphite. At this point a fair question is: Why not produce the graphite during the original cooling and avoid the heat treatment expense? The answer is that malleable iron was developed many years before ductile iron. Malleable iron was found to be stronger and more ductile than *gray cast* iron because during the heat treatment graphite was formed in clumplike nodules (something like spheroidal graphite); this graphite is called *temper carbon*. Since the discovery of ductile iron, many producers of malleable iron have changed to the new material, although malleable iron is still used for many thin-section castings.

It is interesting to follow the structural changes in the heat treatment cycle for malleable iron. There are two stages:

STAGE 1. Graphitization. Heat at 1750°F (954°C) for 12 hr. The structure of austenite plus carbide changes to austenite plus temper carbon.

STAGE 2. Cool from 1750 to 1400°F (954 to 760°C). Slow-cool at 10°F/hr (5.5°C/hr) to 1200°F (649°C).

Whereas austenite would normally transform to α + carbide, the slow-cooling rate above gives α + graphite, in the eutectoid temperature range. The graphite crystallizes on the existing temper carbon formed at the higher temperatures. The final structure is then ferrite plus temper carbon.

EXAMPLE 6.10 Why is it difficult to produce malleable cast iron in large sections?

ANSWER Malleable cast iron must first be cast white (with carbide as the high-carbon constituent). In large sections the cooling rate is slow enough for graphite to form at the high temperatures during solidification. Since this graphite is in flake form, the desirable properties of the temper carbon structure cannot be obtained. Although it is possible to stabilize the carbide by using a lower percentage of silicon or by adding carbide-stabilizing elements such as chromium, the time required for stage 1 would become prohibitively long and uneconomical.

Now let us consider the variations of properties in these different families as we change the structure of the steel matrix (Table 6.6). Since we have austenite at 1600°F (871°C), we can produce any of the transformation products found in steel: ferrite, pearlite, bainite, martensite, and even retained austenite. We shall now discuss these variations for each type of cast iron.

6.34 White cast iron

The largest tonnage of white cast iron is made with a pearlitic matrix, that is, a structure of pearlite and massive carbide. If alloys are added to suppress the pearlite transformation, martensitic white irons are obtained. Examples are:

	Analysis, percent					
Material	*C*	*Si*	*Cr*	*Ni*	*Mo*	*BHN*
Typical pearlitic white cast iron	3.2	0.5	1.0			450
Martensitic white iron	3.2	0.5	2.5	4.0		600
Martensitic high-chromium iron	2.5	0.5	20.0	1.0	2.0	600

The high alloy content pushes the nose of the TTT diagram far enough to the right so that martensite can even be obtained in as-cast sections. These alloys are used chiefly for abrasion resistance in grinding mills as liners and balls, in cement manufacture and mining equipment, and in rolls for finishing steel.

6.35 Gray cast iron

As in the case of white cast iron it is possible to vary the matrix of gray cast iron. The range of structures in actual use is wider than that of white iron. It includes ferrite, pearlite, bainite, martensite, and even austenite.

Table 6.6 MINIMUM PROPERTIES OF WHITE IRON, GRAY IRON, DUCTILE IRON, MALLEABLE IRON, AND SPECIAL ALLOYS

Name and Number	Chemical Analysis, percent	Condition	Tensile Strength,[a] psi × 10^3	Yield Strength,[a] psi × 10^3	Percent Elongation	BHN	Typical Use
White cast iron, unalloyed	3.5 C, 0.5 Si	As cast	40	40	0	500	Wear-resistant parts
			Gray Iron				
Ferritic class 25	3.5 C, 2.5 Si	As cast	25	20	0.4	150	Pipe, sanitary ware
Pearlitic class 40	3.2 C, 2 Si	As cast	40	35	0.4	220	Machine tools, blocks
Quenched martensitic	3.2 C, 2 Si[b]	Quenched	80	80	0	500	Wearing surfaces
Quenched bainitic	3.2 C, 2 Si[c]	Quenched	70	70	0	300	Camshafts
			Ductile Iron				
Ferritic (60-40-18)	3.5 C, 2.5 Si	Annealed	60	40	18	170	Heavy-duty pipe
Pearlitic (80-55-06)	3.5 C, 2.2 Si	As cast	80	55	6	190	Crankshafts
Quenched (120-90-02)	3.5 C, 2.2 Si	Quenched and tempered	120	90	2	270	High-strength machine parts
			Malleable Iron				
Ferritic (35018)	2.2 C, 1 Si	Annealed	53	35	18	130	Hardware, fittings
Pearlitic (45010)	2.2 C, 1 Si	Annealed	65	45	10	180	Couplings
Quenched (80002)	2.2 C, 1 Si	Quenched and tempered	100	80	2	250	High-strength yokes
			Special Alloy Irons				
Austenitic gray	20 Ni, 2 Cr[d]	As cast	30	30	2	150	Exhaust manifolds
Austenitic ductile	20 Ni[d]	As cast	60	30	20	160	Pump casings
High-silicon gray	15 Si, 1 C	As cast	15	15	0	470	Furnace grates
Martensitic white	4 Ni, 2.5 Cr[e]	As cast	40	40	0	600	Wear-resistant parts, liners
	20 Cr, 2 Mo, 1 Ni[f]	Heat-treated	80	80	0	600	

[a] Multiply psi by 6.9×10^{-3} to obtain MPa or by 7.03×10^{-4} to obtain kg/mm^2.
[b] + 1% Ni, 1% Cr, 0.4% Mo
[c] + 1% Ni, 1% Mo
[d] +3% C, 2% Si
[e] +3.2% C, 0.8% Si
[f] +2.7% C, 0.8% Si

In all cases the structure depends on the alloy content and cooling rate of the austenite.

Pearlite is easily obtained by cooling fast enough through the eutectoid range to avoid the transformation of austenite to α + graphite. An equally important factor is the percentage of silicon and other elements that affect the stability of carbide. For example, a pearlitic gray iron motor block is made with 3.2% carbon and 2% silicon and is air-cooled through the eutectoid after pouring. If the block were shaken out at 1400°F (760°C) and cooled slowly in a furnace at 10°F/hr (5.5°C/hr), the structure would be ferritic. Small amounts of tin or copper result in a pearlite matrix even with cooling in the sand molds.

Bainite can be obtained either by an isothermal heat treatment or by casting an alloy combination that produces a TTT curve with a pronounced bainite nose. For example, to produce a bainite structure in a 1-in.-diameter crankshaft, the following analysis is used: 3.2% C, 2% Si, 0.7% Mn, 1% Ni, 1% Mo. The BHN is 300. The hardness is lower than that of a bainitic steel because of the presence of graphite.

Martensite is usually obtained by heat treating—austenitizing plus oil quenching. A typical analysis for an automotive camshaft is: 3.2% C, 0.7% Mn, 2% Si, 1% Ni, 1% Cr, 0.4% Mo. The BHN is 550. Figure 6.47 shows a section of an induction-hardened shaft. In this case, because of the selective heating, only the surface layers were austenitized before quenching.

To obtain austenite, enough alloy is added to avoid pearlite transformation *and* to lower the M_s below room temperature. This can be done with a 20% Ni, 2% Cr alloy or with a 14% Ni, 6% Cu, 2% Cr alloy. Examples are exhaust manifolds, in which we want a stable heat-resistant structure, and pump parts, for which we want the corrosion resistance of a high-nickel chromium iron.

6.36 Ductile iron

From Table 6.6, the tensile strength of the common grades of gray iron varies from 25,000 to 40,000 psi (173 to 276 MPa) and the elongation is very low. By contrast, the ductile irons have strengths up to 120,000 psi (828 MPa) and high elongation.

The commonest grade is the as-cast pearlitic type 80-55-06. The analysis is controlled to avoid massive carbide, but in addition a pearlitic matrix is formed on cooling. In other words, the silicon is maintained at under 2.5% and the manganese is above 0.4%. Small amounts of copper and tin are also used to promote pearlite.

The ferritic grade is obtained by slow cooling from the casting temperature or by reheating to 1600°F (871°C) and holding, then slow cooling near 1300°F (704°C).

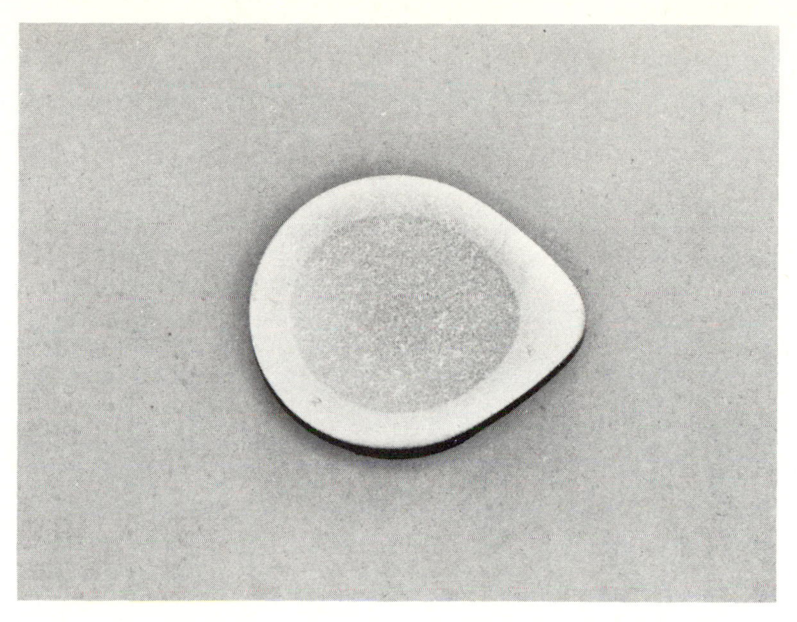

Fig. 6.47 *Cross section of a cam lobe of an automobile camshaft (gray iron). Camshaft was heated on the surface by induction, then quenched. Hardened layer shows as a light etching structure. Actual size; 5% nital etch.*

The quenched and tempered grade is heat-treated in a manner similar to a 0.8% carbon steel. It is austenitized at 1600°F (871°C), quenched, and tempered to the desired hardness.

The austenitic grade has higher strength and toughness than the corresponding gray iron. We could have predicted this, of course, from the differences in graphite shape.

6.37 Malleable iron

The principal grades of malleable iron are ferritic, pearlitic, and quenched. The ferritic grade is obtained by the two-stage graphitization previously described. As in ductile iron, the ferritic grade has the highest ductility (Table 6.6).

The pearlitic grade is made by avoiding the stage 2 graphitization. After stage 1 the casting is air-cooled and the austenite transforms to pearlite. If the pearlite is too hard, it is subsequently tempered (spheroidized).

The quenched grade is also called pearlitic, but this is a misnomer. A martensitic structure is produced by quenching from austenite and then tempering to spheroidite.

EXAMPLE 6.11 Suggest possible iron-base alloys that might be used for the following applications: (1) road grader blade; (2) file cabinet; (3) wood chisel; (4) metal mixing bowl; (5) fire hydrant.

ANSWER

1. *Road grader blade* Because of the wear and abrasion resistance required, a work-hardening material is desirable. A quenched and tempered steel would not have sufficient impact resistance. A Hadfield steel (12% Mn, 1% C) would meet the requirements. An austenitic stainless steel, though highly work-hardenable, would be too expensive.
2. *File cabinet* The need for excellent cold formability and a good surface quality for subsequent painting suggests a low-carbon steel such as a 1010 steel.
3. *Wood chisel* A tool steel is probably not necessary, since the chisel does not heat in service. A quenched and tempered low-alloy steel would be adequate. (Care should be exercised in regrinding a sharp edge, as a high-speed noncooled grinding operation will further temper the steel.)
4. *Metal mixing bowl* Because the bowl will contain food that might be contaminated by corrosion byproducts, a stainless steel is suggested. The most inexpensive stainless steel would be ferritic (e.g., type 430); however, an austenitic grade (type 304) will provide better corrosion resistance.
5. *Fire hydrant* The complex shape suggests a casting. The most inexpensive iron-base alloy would be a gray cast iron such as class 25. Another advantage is easier shearing of the hydrant with less physical harm to a driver if an automobile should collide with it.

SUMMARY

The ferrous or iron-base alloys can be divided into three principal groups:

1. Plain carbon and low-alloy steels
2. High-alloy steels (over 10% alloy)
3. Cast iron, ductile iron, malleable iron

We can predict the equilibrium structures of the first group, the low-alloy steels, from the section of the iron–iron carbide diagram below 2% carbon. Only three important phases are involved: ferrite (α, BCC with up to 0.02% carbon in solid solution), austenite (γ, FCC with up to 2% carbon in solid solution), and iron carbide, Fe_3C (6.67% carbon). The majority of plain carbon steels, such as constructional and automobile body grades, consist simply of ferrite with increasing amounts of carbide as the carbon content is increased. They are specified in the hot-rolled or cold-rolled condition. When the carbon content in unalloyed steel is above 0.3%, it is possible to harden thin sections by heating to produce austenite, followed by rapid quenching in water to form martensite. However, low-alloy steels are used for most heat-treated parts because the hard martensitic structure can be obtained with a less drastic quench and in thicker sections. The TTT curve and the Jominy hardenability bar provide accurate measures of the kinetics of austenite transformation and the relationship between cooling rate and martensite formation. The control of other microstructures, such as pearlite and bainite, employs the same data.

In the second group adding large amounts of alloy alters substantially the iron–iron carbide diagram. One group of elements, such as nickel, gives the open γ (austenite) field, leading to alloys that are austenitic at room temperature, such as austenitic stainless steels. By contrast, the elements that give a closed γ field, such as chromium, tend to produce an α structure at all temperatures, as in the ferritic stainless steels. An intermediate condition occurs in the martensitic stainless steels. Ternary diagrams indicate the quantitative relationships. Tool steels and other specialty grades depend on similar principles.

In the third group, the cast irons and related materials all contain over 2% carbon. Some of this may be present as graphite; to analyze these relationships, we use the iron-graphite diagram. Each of these materials can be considered as a steel matrix (pearlite, bainite, etc.) plus carbide or graphite. In white cast iron no graphite is present, and the product is hard and brittle because of the presence of massive particles of iron carbide. Gray cast iron is soft and machinable because the iron carbide is replaced by graphite flakes, but the ductility is low because of the notch effect of the graphite. Ductile iron exhibits better ductility than gray cast iron because the shape of the graphite is spheroidal. Malleable iron also is ductile because of the temper carbon produced during heat treatment of white cast iron.

DEFINITIONS

Austempering A hardening process involving austenitizing then quenching in a liquid bath above the M_s *and holding* in the bath to transform to bainite, then cooling.

Austenite γ iron, FCC, with a maximum of 2% carbon in solid solution.

Austenitizing Heating to the austenite temperature range (which will vary for the particular steel) to produce austenite.

Bainite A mixture of α and very fine carbides showing a needlelike structure, produced by transformation of austenite between 1000°F (538°C) and above 500°F (260°C).

Carbide In the context of iron-base alloys carbide means iron carbide, Fe_3C. It is also called *cementite.*

Creep Plastic elongation as a function of time in stressed material. It is also related to temperature.

Dilatometer An instrument for measuring the length of a sample during heating and cooling.

Ductile iron An iron-carbon alloy with 3.5 to 4% carbon and graphitizing elements such as silicon. It differs from gray cast iron in that the graphite is in the form of spheres and the ductility is high. The spheroidal graphite is produced by the addition of 0.05% magnesium just before the liquid metal is poured into castings.

Eutectoid composition In the iron-carbon system, this is 0.8% carbon. The eutectoid reaction is γ (austenite) $\rightarrow$ α (ferrite) + iron carbide at 1333°F (723°C). The α + iron carbide mixture is called *pearlite.*

Ferrite α iron, BCC, with a maximum of 0.02% carbon in solid solution.

Gray cast iron An iron-carbon alloy with 2 to 4% carbon and graphitizing elements such as silicon. Most of the carbon is in the form of graphite flakes, so that the material is easily machined but low in ductility.

Hardenability A general term indicating for a given steel the ease of avoiding pearlite or bainite transformation so that martensite can be produced. High hardenability does not mean the same as high hardness. The maximum hardness attainable is a function of the carbon content.

Hardenability curve, Jominy curve A graph showing the hardness attained for a given steel using the Jominy end-quench test.

Hypereutectoid steel Steel with carbon content 0.8 to 2% (plain carbon).

Hypoeutectoid steel Steel with carbon content below 0.8% (plain carbon)

Jominy bar, hardenability bar A 1-in.- (2.54-cm-) diameter by 4-in.- (10.16-cm-) long bar that is austenitized, water-quenched on one end, and then tested for hardness along its length.

M_f The temperature at which austenite-to-martensite transformation finishes.

M_s The temperature at which austenite-to-martensite transformation starts.

Malleable iron An iron-carbon alloy with 2 to 3% carbon. Castings of white cast iron are produced and then heat-treated at 1750°F (954°C). The massive iron carbide is converted to nodules of graphite plus iron called *temper carbon.* The properties are similar to those of ductile iron.

Maraging steel A highly alloyed steel that is hardened both by martensite and by age hardening.

Marquenching (*Martempering*) A hardening process involving austenitizing then quenching in a liquid bath at above the M_s, followed by air cooling through the martensite transformation range.

Martensite A nonequilibrium phase consisting of body-centered tetragonal iron with carbon in supersaturated solid solution.

Nose of TTT curve The temperature range in which isothermal transformation is fastest.

Pearlite A mixture of α and carbide phases in parallel plates, produced by the transformation of austenite between 1333°F (723°C) and 1000°F (538°C).

Spheroidite α plus spheroidized carbide, produced by heating pearlite, bainite, or martensite at elevated temperatures.

Stainless steel A steel that is highly alloyed with chromium and often nickel to improve corrosion resistance.

Steel An iron-carbon alloy with 0.02 to 2% carbon. The most common range is 0.05 to 1.1% carbon. Steel can be hot- or cold-worked.

Stress rupture The value of stress needed to cause rupture in a specified time at a given temperature.

Superalloys Alloys used for high-temperature service. FCC alloys with nickel, cobalt, iron, and chromium are usually strengthened with an $Ni_3(Al,Ti)$ precipitate.

Tempered martensite Martensite that has been heated to produce BCC iron and a fine dispersion of iron carbide.

TTT curve A time-temperature-transformation curve that indicates the time required for austenite of a given composition to transform to α plus carbide at different temperatures (isothermally).

White cast iron An iron-carbon alloy with 2 to 4% carbon. The most common range is 3 to 3.5% carbon. Because of the presence of large amounts of brittle iron carbide, the material is not hot-worked or machined.

Wrought iron Iron with very low carbon content (0.02%) that is easily worked into intricate forgings while hot.

PROBLEMS

6.1 What would be the maximum solubility in percent by weight if all the positions like $\frac{1}{2}, \frac{1}{2}, \frac{1}{2}$ in γ iron were occupied by carbon? Why is this not attained? (If 2% by weight carbon is soluble in the FCC iron, what is the number of unit cells of iron per carbon atom?) [*Hint:* See Fig. 2.17.] (Sections 6.1 through 6.6)

6.2 Draw fraction charts showing the amount of each phase present as a function of temperature for the following alloys: (Sections 6.1 through 6.6)

a. SAE-AISI 1010 steel (0.1% carbon)—neglect peritectic
b. SAE-AISI 1080 steel (0.8% carbon)
c. 2.5% carbon white cast iron

6.3 If a sample of 3.3% carbon white cast iron were heat-treated so that only ferrite and graphite were present, what would be the percent by *volume* of graphite in the sample? (The specific gravity of iron is 7.87 and that of graphite is 2.25.) (Sections 6.1 through 6.6)

6.4 A railroad track (1080 steel) has welded joints between each rail, laid end to end. The track is laid in winter (32°F or 0°C). What magnitude of stress could be expected in the middle of summer in the sunlight (132°F or 55.6°C)? The coefficient of linear expansion is 6.0×10^{-6} in./in./°F (10.8 m/m/°C). (Sections 6.1 through 6.6)

6.5 From the accompanying phase fraction chart for a plain carbon steel, (Sections 6.1 through 6.6)

a. Is the steel hypo- or hypereutectoid?
b. Determine the carbon content.
c. Determine the amount of pearlite.

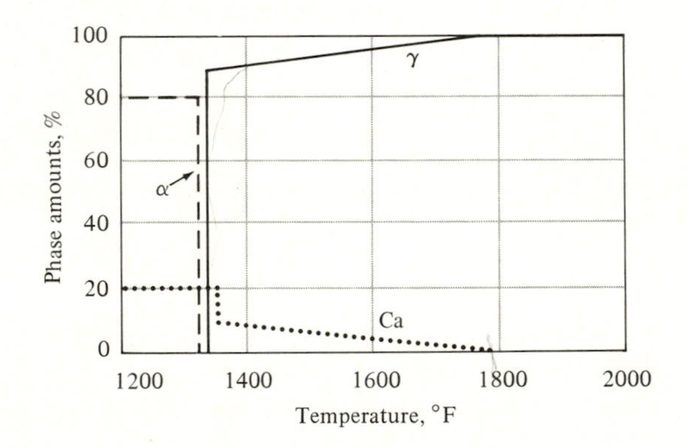

6.6 Using the TTT curve for 1080 steel, give the microstructure present in a thin [0.010-in. (0.0254-cm)] strip after each step of the following treatments. The original material is hot-rolled. (See Fig. 6.15.) (Sections 6.7 through 6.16)

a. 1600°F (871°C), 1 hr; quench in lead at 1300°F (704°C), hold 20 min; quench in lead at 900°F (482°C), hold 1 sec; water-quench.
b. 1300°F (704°C), 6 hr; water-quench.
c. 1600°F (871°C), 1 hr; quench in salt at 800°F (427°C), hold ½ hr; heat rapidly to 1100°F (593°C), hold 1 hr; water-quench.
d. 1600°F (871°C), 1 hr; quench in salt at 500°F (260°C), hold 1 min; air-cool. What is the name of this treatment?
e. 1600°F (871°C), 1 hr; quench in salt at 600°F (315°C), hold 1 hr; air-cool. What is the name of this treatment?

6.7 A high tonnage of plain carbon steel is sold in the "hot-rolled" condition, that is, rolled at above 1600°F (871°C) and air-cooled. (Sections 6.7 through 6.16)

a. Hot-rolled bars of 1020 and of 1080 steel would have the same phases present at 70°F (21°C). Name the phases and give their chemical analyses.
b. What are the percentages of phases present in both steels? What is the percentage of pearlite present in both steels?
c. Both grades of steel are available in a higher-strength grade because of a difference in processing not requiring heat treatment or alloy additions. What are these grades called?

6.8 The age-hardening process for aluminum alloys and the hardening of a 0.5% C steel by martensite formation have similar heat treatment steps, but there are some differences in the structural changes taking place within the alloy in each step. For each step listed, select the code number of the phrase from the group below which accurately describes the structural change. (Sections 6.7 through 6.16 and 4.21)

a. In both cases the *first* step is: Heat to elevated temperature and quench. During quenching, what takes place in the aluminum alloy? In the steel?
b. In both cases the *second* step is to heat to a temperature considerably lower than the first temperature. What structural change takes place in the aluminum alloy? In the steel?
c. Indicate the code numbers below those structures that depend on diffusion for their rate of formation.

Code

1. Bainite forms as the principal constituent.
2. No precipitation occurs.
3. Precipitation of a new phase occurs with decreasing temperature.
4. Pearlite is formed.
5. Precipitation of a new phase from the phase formed at high temperature takes place as a function of time.
6. Alteration of the new phase that was formed during quenching takes place.

6.9 From the following group, choose heat-treating cycles to produce the following microstructures in a 0.010-in.-diameter (0.0254-cm) bar of 1080 steel: (1) 100% pearlite; (2) 100% bainite; (3) 50% pearlite, 50% bainite; (4) 50% martensite, 50% bainite. (Sections 6.7 through 6.16)

Code letters

A 1100°F (593°C), 1000 sec, air-cool.
B 1600°F (871°C), 500 sec, quench in lead bath at 700°F (371°C), hold 1000 sec, air-cool.

C 1100°F (593°C), 10 sec, quench in lead at 700°F (371°C), hold 1000 sec, air-cool.

D 1600°F (871°C), 500 sec, quench in lead bath at 1250°F (677°C), hold 100 sec, quench in lead bath at 800°F (427°C), hold 1000 sec, water-quench.

E 1600°F (871°C), 500 sec, quench in water to 0°F (−18°C), heat to 800°F (427°C), hold 5 sec, water-quench.

F 1600°F (871°C), 400 sec, quench in lead at 600°F (312°C), hold 300 sec, water-quench.

G 1600°F (871°C), 500 sec, quench in lead at 1100°F (593°C), hold 1000 sec, air-cool.

PROBLEMS COVERING OPTIONAL SECTIONS

6.10 Draw schematic time-temperature diagrams to obtain the following structures in an SAE 1045 steel, giving the structure at each step. (*Percentages are approximate.*) (See Example 6.3) (Sections 6.17 through 6.18)

a. 45% ferrite, 55% pearlite
b. 20% ferrite, 80% martensite
c. 100% martensite
d. 100% spheroidite
e. 90% bainite, 10% ferrite
f. Calculate the approximate M_s temperature for the martensite in part *b*.

6.11 Draw schematic time-temperature diagrams to obtain the following structures in a 1.2% carbon steel. Give the structure after each step. (Sections 6.17 through 6.18)

a. Spheroidized primary carbide + pearlite
b. Grain boundary carbide + pearlite
c. Spheroidized carbide + tempered martensite

6.12 Indicate whether the following steels might be hypoeutectoid or hypereutectoid. [*Hint:* The alloy content changes both the percent carbon and the temperature associated with the eutectoid.] (Sections 6.17 through 6.18)

a. 0.6% C, 1% Mo, 1% Cr
b. 0.6% C, 1% Ni, 1% Cr

6.13 The figure on the opposite page gives the TTT curve for an alloyed 0.30% carbon steel. The hardness data are for fully transformed structures. (Sections 6.17 through 6.18)

a. A foundry finds that castings made of this steel are hard and unmachinable (400 BHN) in the as-cast condition. Name two microstructures that could be responsible.
b. The same foundry hears that a competitor is annealing its castings

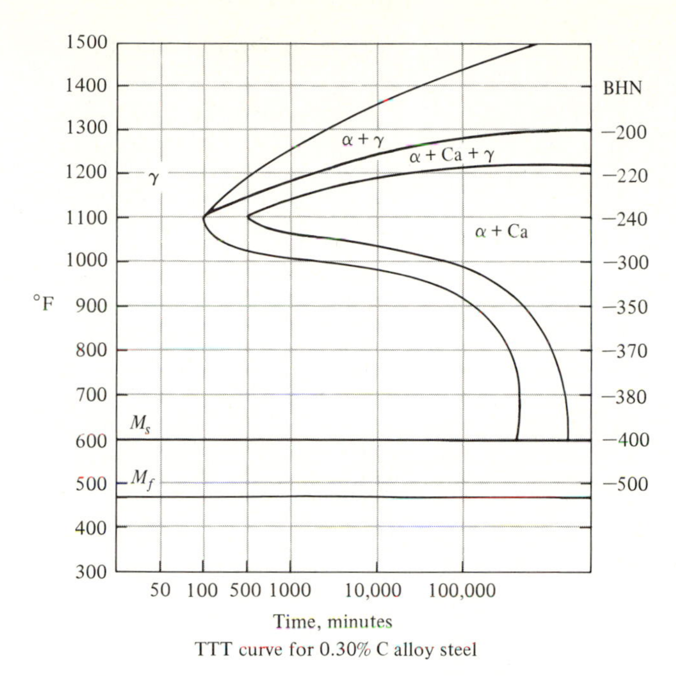

TTT curve for 0.30% C alloy steel

by a short cycle called an *isothermal anneal.* This involves heating the castings, followed by isothermal transformation to a structure of 250 BHN max. Draw a time-temperature chart giving this result, *labeling temperatures* and isothermal transformation *time* accurately.

6.14 It was assumed in Example 6.6 that the effect of alloying elements on the eutectoid temperature and eutectoid carbon content is additive. Check the validity of this assumption for the relatively high alloy content of the T1 tool steel in Figure 6.41. (Assume that V acts like Mo and that the W curves may be linearly extrapolated.) Suggest reasons for any discrepancies. (Sections 6.17 through 6.18)

6.15 An isothermal transformation curve for a 4130 steel is shown on page 272. Describe the microstructure (using only the given words) after each of the following heat treatments. [*Words:* austenite, ferrite, graphite, liquid, pearlite, bainite, martensite, tempered martensite, spheroidite] (Sections 6.17 through 6.18)

a. A thin section of hot-rolled 4130 steel is heated to 1400°F, held 1 hr, and water-quenched to room temperature.

b. A thin section of hot-rolled 4130 is heated to 1550°F, held 1 hr, quenched in a molten salt pot at 800°F, held 1000 seconds, and slowly cooled in air.

c. A thin section of hot-rolled 4130 is heated to 1550°F, held 1 hr, quenched in a salt pot at 1200°F, held 1000 seconds, and water-quenched.

d. Which material (*a*, *b*, or *c*) will be the softest?

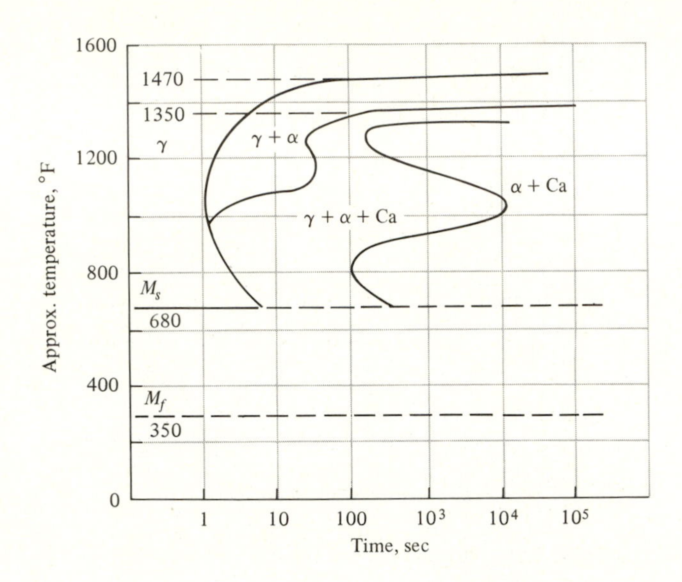

6.16 Although a single specimen of steel will give a single line as a hardenability curve, results of the tests of commercial steel will fall within a hardenability band such as that shown for the 86*xx* series steels in Fig. 6.48 on page 274 and the table on page 275. (Sections 6.19 through 6.23)

a. Why is the hardness band at the $\frac{2}{16}$-in. station higher in 8650H than in 8630H? What is the microstructure in each case?

b. What will be the microstructure of 8620H at the $\frac{18}{16}$ position compared with 8640H? Are there several structures that might be encountered at the $\frac{18}{16}$ position in 8640H?

c. What is the principal element responsible for the bandwidth at the $\frac{2}{16}$ position in all cases?

6.17 A 1-in.-diameter shaft is made from a sample of 8630H with hardenability that falls in the middle of the band (Fig. 6.48). The shaft is austenitized at 1600°F (871°C) and oil-quenched. A hardness survey shows 40 R_C $\frac{1}{8}$ in. (0.317 cm) beneath the surface. A hardness of 35 R_C minimum is desired at this position, but minimum carbon (that is, the minimum curve from those steels in Fig. 6.48) is recommended. Which 86*xx* steel should be used? (Sections 6.19 through 6.23)

6.18 A midradius hardness of 50 R_C minimum, 62 R_C maximum is required in a 2-in. (5.08-cm) bar. Which of these steels would meet the specification and how should it be quenched (Fig. 6.48)? (Sections 6.19 through 6.23)

6.19 The following readings are obtained from bars of an 8627 steel. Is the hardenability within specification? (Sections 6.19 through 6.23)

	R_C, midradius	R_C, center
A 2-in. (5.08-cm) bar quenched in still water	45	39
A 2-in. (5.08-cm) bar quenched in agitated oil	36	35

6.20 Which of the eight steels shown in Fig. 6.36 would make the best drill bit? Explain. (Sections 6.19 through 6.23)

6.21 Differentiate between hardness and hardenability. (Sections 6.19 through 6.23)

6.22 The center hardness of 6 bars of the same steel is indicated below. From these data, plot the hardenability curve for the steel. (Show points and be accurate.) (Sections 6.19 through 6.23)

1-in.-diameter (2.54-cm) still water quench	58 R_C
1-in.-diameter (2.54-cm) agitated oil quench	57 R_C
2-in.-diameter (5.08-cm) still water quench	54 R_C
2-in.-diameter (5.08-cm) agitated oil quench	44 R_C
4-in.-diameter (10.16-cm) still water quench	34 R_C
4-in-diameter (10.16-cm) agitated oil quench	30 R_C

6.23 A $3\frac{1}{2}$-in. (8.89-cm)-diameter automatic transmission gear is sketched below. Specifications call for oil quenching to give as-quenched hardness values of (1) at least R_C 50 near the surface (point A); (2) no greater than R_C 45 at point B. A trial gear was made from AISI 1060 steel, austenitized at 1550°F, and quenched in oil. Hardnesses were recorded as: point A, R_C 42; point B, R_C 34 (Sections 6.19 through 6.23). (Problem continues on page 276.)

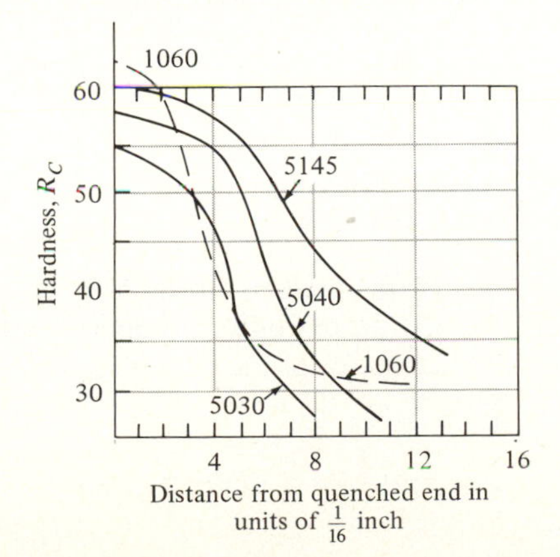

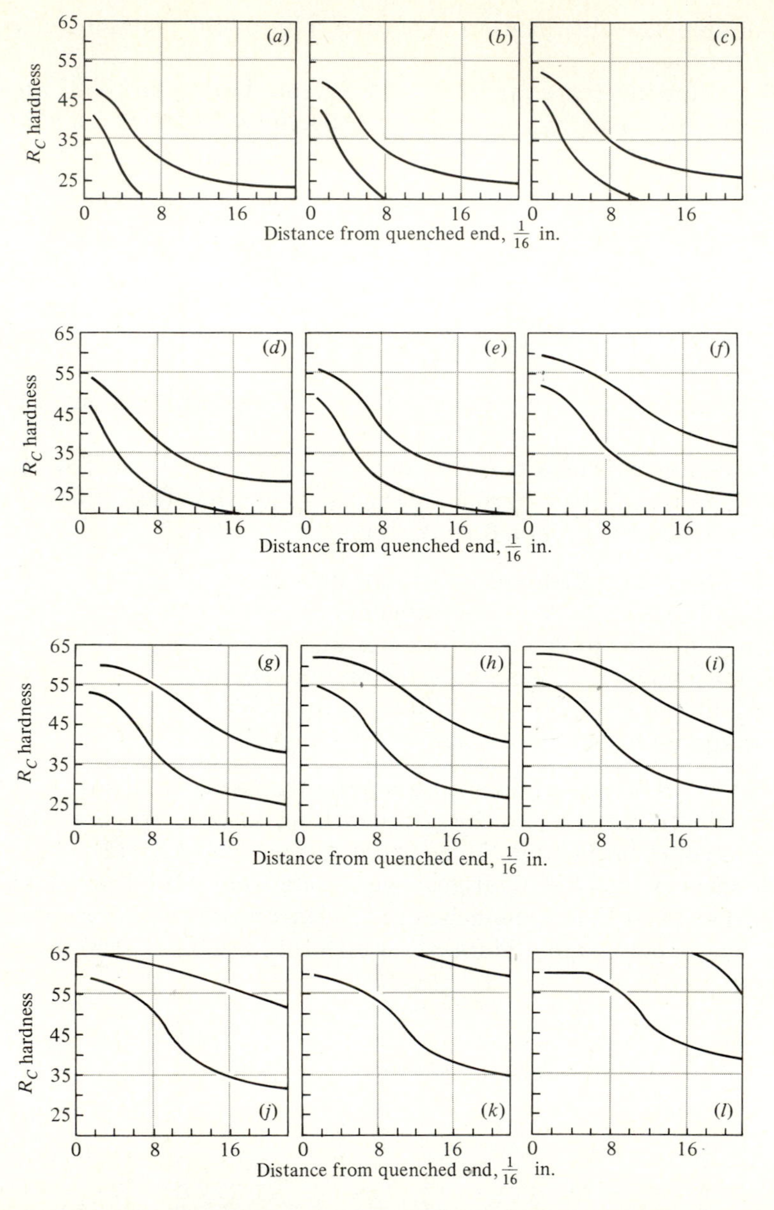

Fig. 6.48 *Hardenability of 8600 series steels: (a) 8620H, (b) 8622H, (c) 8625H, (d) 8627H, (e) 8630H, (f) 8637H, (g) 8640H, (h) 8642H, (i) 8645H, (j) 8650H, (k) 8655H, (l) 8660H.* (American Society for Metals Handbook, 8th ed., Vol. 1: *Properties and Selection of Metals*, Metals Park, Ohio, 1961)

Table for Fig. 6.48

SAE-AISI Steel	Chemical Analysis, percent						Normalizing Tempera-ture, °F	Austenitizing Tempera-ture, °F
	C	Mn	Si	Ni	Cr	Mo		
8620H	0.17 to 0.23	0.60 to 0.95	0.20 to 0.35	0.35 to 0.75	0.35 to 0.65	0.15 to 0.25	1700	1700
8622H	0.19 to 0.25	0.60 to 0.95	0.20 to 0.35	0.35 to 0.75	0.35 to 0.65	0.15 to 0.25	1700	1700
8625H	0.22 to 0.28	0.60 to 0.95	0.20 to 0.35	0.35 to 0.75	0.35 to 0.65	0.15 to 0.25	1650	1600
8627H	0.24 to 0.30	0.60 to 0.95	0.20 to 0.35	0.35 to 0.75	0.35 to 0.65	0.15 to 0.25	1650	1600
8630H	0.27 to 0.33	0.60 to 0.95	0.20 to 0.35	0.35 to 0.75	0.35 to 0.65	0.15 to 0.25	1650	1600
8637H	0.34 to 0.41	0.70 to 1.05	0.20 to 0.35	0.35 to 0.75	0.35 to 0.65	0.15 to 0.25	1600	1550
8640H	0.37 to 0.44	0.70 to 1.05	0.20 to 0.35	0.35 to 0.75	0.35 to 0.65	0.15 to 0.25	1600	1550
8642H	0.39 to 0.46	0.70 to 1.05	0.20 to 0.35	0.35 to 0.75	0.35 to 0.65	0.15 to 0.25	1600	1550
8645H	0.42 to 0.49	0.70 to 1.05	0.20 to 0.35	0.35 to 0.75	0.35 to 0.65	0.15 to 0.25	1600	1550
8650H	0.47 to 0.54	0.70 to 1.05	0.20 to 0.35	0.35 to 0.75	0.35 to 0.65	0.15 to 0.25	1600	1550
8655H	0.50 to 0.60	0.70 to 1.05	0.20 to 0.35	0.35 to 0.75	0.35 to 0.65	0.15 to 0.25	1600	1550
8660H	0.55 to 0.65	0.70 to 1.05	0.20 to 0.35	0.35 to 0.75	0.35 to 0.65	0.15 to 0.25	1600	1550

a. AISI steels 5145, 5040, 5030, and 1060 are under consideration for this part. Which steel or steels would be acceptable?
b. What would be the hardness at point *A* using the selected steel?
c. What would be the hardness at point *B* using the selected steel?

6.24 In a certain chisel the desired hardness is *minimum* R_C 50 at $\frac{1}{8}$ in. (0.32 cm) from the point (in the center of the chisel), *maximum* R_C 30 at 2 in. (5.08 cm) from the point. An investigator tries to meet these specifications with steel *A* with an oil quench, but the hardness is R_C 45 at $\frac{1}{8}$ in. (0.32 cm) from the chisel point and R_C 35 at 2 in. (5.08 cm) from the point. To meet the specification, would you use steel *B*, *C*, or *D*? State the hardness you would obtain at the points specified. The Jominy end-quench curves for these steels are shown. Which steel has the highest carbon content? (Sections 6.19 through 6.23)

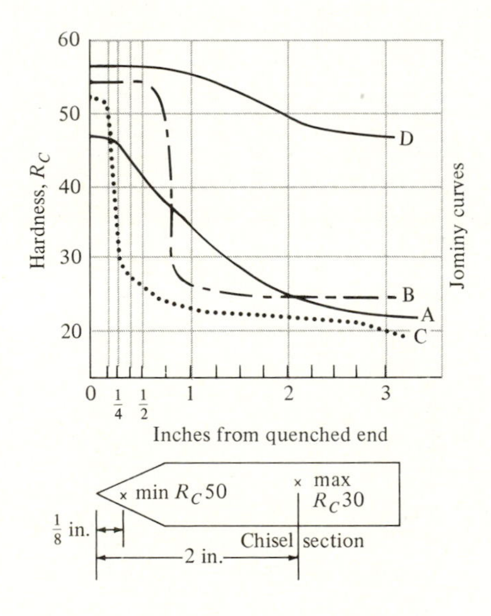

6.25 An unidentified bar of steel is known to be from one of four lots. The hardenability curve for each of the four lots is given. The unknown bar has the hardness traverse shown (page 277). (Sections 6.19 through 6.23)

a. Which of the four steels is the unknown bar? (Indicate your reasoning.)
b. Estimate the carbon content of these four steels.
c. Sketch on the first figure a possible hardenability curve for steel *A* when it has a coarser γ grain size (no change in composition).
d. Draw the hardness traverse of the unknown bar if it were reheated and quenched in water.

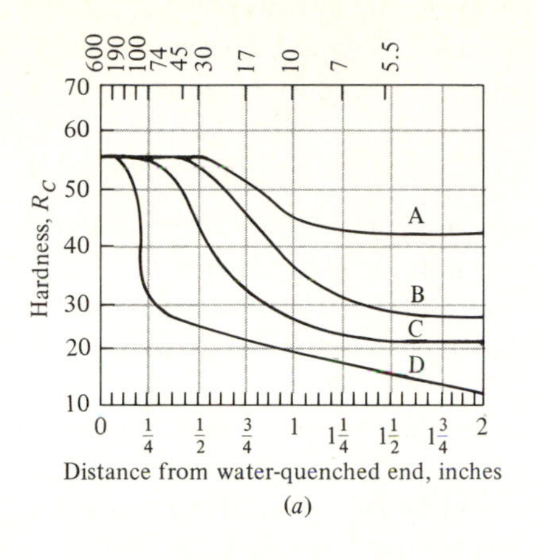

(*a*)

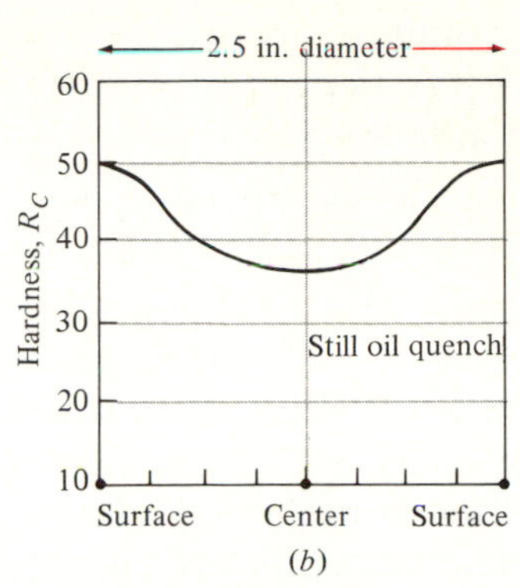

(*b*)

6.26 Given a stainless steel with 15% chromium, what is the maximum carbon content it can contain and still remain ferritic? (Sections 6.24 through 6.31)

6.27 Hadfield's manganese steel (12% Mn, 1% C) when quenched will be all austenite, yet when slowly cooled may contain extensive amounts of martensite. Explain what might take place. [*Hint:* The alloy content of the austenite determines the M_s.] (Sections 6.24 through 6.31)

6.28 Which steel(s) of Table 6.3 would you specify for the following applications? Explain the deficiencies of the others. (Sections 6.24 through 6.31)

a. A kitchen knife of intricate blade shape for preparing grapefruit
b. A blade for carving that would hold its edge best
c. Automobile trim
d. Corrosion-resistant aircraft parts with the best strength-to-weight ratio
e. A cast alloy for exhaust valves contains 15% Ni, 15% Cr, 1% C. From the point of view of structure, why is this more resistant to deformation than one of the austenitic steels listed?

6.29 Calculate the alloy cost per pound of the tool steels given in Table 6.4 using $4/lb for chromium, $9/lb for molybdenum, $12/lb for tungsten, and $7/lb for vanadium (1980 prices). (Sections 6.24 through 6.31)

6.30 Why are the austenitizing temperatures for the M1 and T1 types much higher than for other grades of low-alloy steel? (Sections 6.24 through 6.31)

6.31 Why are higher tempering temperatures used for the M1 and T1 types than for other low-alloy steels? (Sections 6.24 through 6.31)

6.32 From the creep data in Fig. 6.42, calculate the following (Sections 6.24 through 6.31).

a. Minimum creep rate (percent strain per hour) at a stress of 15,000 psi

(103 MPa). (It may be necessary to replot the data. Use only the straight-line portion of the curve.)

b. Stress to rupture in 1000 hr.

c. Total creep in 1000 hr.

6.33 Which of the following data would you use for the applications given below for operation at elevated temperatures (over 1000°F, 538°C)? (1) Stress rupture; (2) Second-stage creep (minimum creep rate per hour); (3) Total elongation vs. time; (4) Hot tensile test; (5) Room temperature tensile and elongation values.
In some cases failure is determined by excessive deformation, in others by time to rupture. Example: In a superheater tube considerable distortion is allowable, but rupture cannot be tolerated. By contrast, in a valve stem the part would not function properly after deformation. (Sections 6.24 through 6.31) Applications:

a. Bolts *b.* Steam valves *c.* Oil still tubes *d.* Turbine rotors
e. Turbine casings, steam turbine blading

6.34 As mentioned in the text, nickel combines with aluminum and titanium to form the compound Ni_3X, where X is aluminum or titanium or both. From the analysis in Table 6.5 for Udimet 700, calculate the percent by weight of this compound present, assuming all aluminum and titanium are in the compound. Now convert to percent by volume. At what content of aluminum weight percent or titanium weight percent would the structure be 100% γ'? (Sections 6.24 through 6.31)

6.35 Here are three compositions of stainless steel and their potential applications. Indicate for each whether we would expect ferritic, austenitic, or martensitic stainless steel (Sections 6.24 through 6.31).

a. 16% chromium; 0.5% carbon (cutter for a compost grinder)
b. 16% chromium; 10% nickel; 0.03% carbon (vessel to hold corrosive fluids)
c. 16% chromium; 0.1% carbon (decorative trim)

6.36 From the vocabulary, name all the mechanisms you could use to harden each of the steels. Use only the given vocabulary. (Sections 6.24 through 6.31)

1. Hadfield manganese steel; 12% Mn, 1% C, equiaxed grains
2. Austenitic stainless steel; 18% Cr, 8% Ni, 0.02% C
3. Ferritic stainless steel; 14% Cr, 0.05% C
4. Martensitic stainless steel; 14% Cr, 0.7% C, 200 BHN
5. 18% W, 4% Cr, 1% V tool steel

Which steel(s) would be more sensitive to cracking by the heat treatment 1900°F, 2 hr, water-quench? [*Vocabulary:* A, work hardening; B, quenching from austenitizing temperature and temper; C, solution heat-treating and age hardening]

6.37 In "stainless steel" of the Fe-Ni-Cr system, the predominant phases are α-BCC, γ-FCC, and a brittle intermetallic compound σ. (Sections 6.24 through 6.31)

a. Using the ternary diagram of Fig. 6.40, indicate which of the following analyses would be difficult to form or bend at 650°C.

% Alloy	% Cr	% Ni	(*balance* Fe)
1.	20	20	
2.	28	2	
3.	15	35	
4.	40	20	

b. An engineer wants a steel with a structure of 100% austenite at 650°C. Which (name all) of the compositions given in (*a*) would be satisfactory?

6.38 Superalloys as found in Table 6.5 often contain many elements and may exhibit age hardening. Suggest why these alloys might not overage in service. (Sections 6.24 through 6.31)

6.39 Draw fraction charts for a 3% carbon iron (showing percentage of phases as a function of temperature) that is cooled under different conditions (Sections 6.32 through 6.36)

Case A. Cooled at an intermediate rate to produce pearlite plus flake graphite.

Case B. White cast iron reheated to 1750°F (954°C), held 12 hr, and air-cooled. Assume that no graphite formed on cooling.

Case C. White cast iron reheated to 1750°F (954°C), held 12 hr, slowly cooled to 1300°F (704°C), held 12 hr, and air-cooled

Case D. Case C reheated to equilibrium at 1600°F (871°C), and water-quenched. (Omit chart, give final percentage phases.)

6.40 The effect of appreciable silicon (over 0.5%) on the iron-iron carbide diagram is important. Figure 6.49 shows the phase diagram for 2% silicon. (Sections 6.32 through 6.36)

a. According to the phase rule, why can α + γ + Ca (three phases) exist over a range of temperature?

b. The eutectic composition is no longer 4.3% carbon, but follows the formula: eutectic = 4.3 − (0.3) (% Si). Verify this for the diagram shown in Fig. 6.49 at the top of page 280.

c. What is the carbon content of a pearlitic region in a 2% silicon cast iron?

d. How does the austenitizing temperature prior to quenching compare with that of a 0.8% carbon steel?

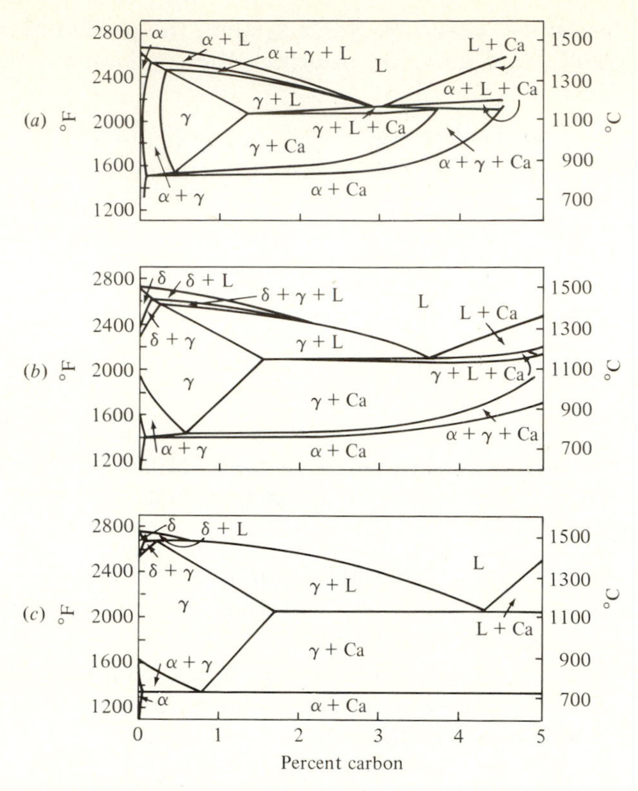

Fig. 6.49 *Iron–iron carbide diagram at different levels of silicon. (a) 4% silicon. (b) 2% silicon. (c) 0% silicon.* (After Greiner, et al.)

6.41 Which material(s) could be used to fill the following specifications, and what analyses and heat treatments should be used? (See Table 6.6.) (Sections 6.32 through 6.36)

Case A. Automobile crankshaft: maximum section 2 in. (5.08 cm), BHN 220, endurance limit in fatigue 20,000 psi (138 MPa). This is discussed later and can be taken as 0.25 × tensile strength.

Case B. Engine head: sections $\frac{1}{4}$ to 1 in. (0.635 to 2.54 cm) in diameter, BHN 200, good machinability, tensile strength 30,000 psi (207 MPa).

Case C. Tappet: 1 in. (2.54 cm) in diameter by 2 in. (5.08 cm) long. One end to be BHN 400, free from graphite to depth of $\frac{1}{4}$ in. (0.635 cm). Balance of casting BHN 250 maximum, machinable.

Case D. Differential housing: sections $\frac{1}{4}$ to 1 in. (0.635 to 2.54 cm) in diameter, tensile strength 50,000 psi (345 MPa), good toughness to resist road conditions, good machinability.

Case E. Camshaft: good machinability. Cam lobes 50 R_C (477 BHN), structure: tempered martensite, flake graphite, 5% massive carbide.

Case F. Disk brake: good machinability, structure: flake graphite plus pearlite, BHN 220.

Case G. Cylinder block: flake graphite plus pearlite. (A plain gray iron 3.2% C, 0.5% Mn, 2% Si contains 15% ferrite, balance pearlite, and is not satisfactory because of the effect of wear on ferrite.)

Case H. Grinding mill roll: BHN 550 minimum on surface, no graphite.

6.42 Determine the phases present, the phase analyses, and the amounts of phases at 70°F (21°C) for the following materials. (Sections 6.32 through 6.36)

a. SAE 1080 steel, annealed (1600°F or 871°C) 5 hr, slow-cooled

b. Martensitic gray iron (3% C), quenched from 1500°F (816°C) (fully martensitic)

c. White cast iron (3% C) (pearlite + massive carbide)

d. Ductile cast iron [ferrite + nodular (spheroidal) graphite], 3.5% C, 2.0% Si

6.43 Using the following vocabulary, describe the microstructures that you would expect to find in these specimens. [*Vocabulary:* austenite, martensite, ferrite, bainite, massive carbide, spheroidal graphite, pearlite, tempered martensite] (Sections 6.32 through 6.36)

a. *Analysis:* 0.8% C, 0.5% Mn, 0.2% Si, balance iron.
A 1 in. × 1 in. × $\frac{1}{4}$ in. (2.54 × 2.54 × 0.635 cm) specimen is heated to 1550°F (843°C), held $\frac{1}{2}$ hr, and water-quenched, giving BHN 600. It is then heated to 300°F (149°C), held $\frac{1}{2}$ hr, and cooled (BHN 550).

b. *Analysis:* 3.5% C, 2% Si, 0.05% Mg, balance iron. The liquid is cast into a 1-in. (2.54 cm)-diameter, 12-in. (30.48 cm)-long bar in a sand mold.

c. *Analysis:* 0.1% C, 0.5% Mn, 0.02% P, 0.02% S, balance iron. A 1-in. (2.54 cm) cube is heated to 1750°F (954°C) for 1 hr and cooled at 10°F/hr (5.6°C/hr) to 1000°F (538°C), then water-quenched.

d. *Analysis:* 3% C, 4% Ni, 3% Cr, balance iron. The alloy is cast in a 1 in. × 14 in. × 32 in. (2.54 × 35.56 × 81.28 cm) piece in a mold made of graphite. BHN 650 as cast.

e. Which of the preceding alloys (*a*, *b*, *c*, or *d*) would you use for: (1) a hand file; (2) a liner for a ball mill grinding iron ore; (3) a car fender; (4) a car crankshaft; (5) a gasoline can?

6.44 Seventy pounds (31.8 kg) of an iron-carbon alloy containing a 3.0% carbon is equilibrated at 2070°F (1132°C). (Sections 6.32 through 6.36) *a.* What phase(s) is (are) present? What is the percent of carbon in each phase? What is the amount of each phase? Show all calculations.

b. The alloy is held at 2000°F (1093°C) until all the Fe_3C has graphitized. How many pounds (kg) of graphite will the 70-pound (31.8 kg) casting then contain? Show calculations.

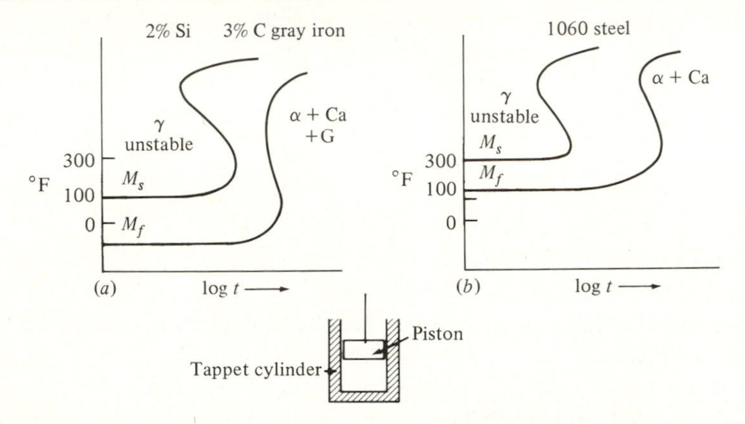

6.45 A hydraulic tappet consists of a hollow cylinder and a piston (see sketch). The tappet cylinder is made of gray cast iron with the TTT curve shown, and the piston of steel with the TTT curve shown. Both are austenitized and quenched to 70°F (21°C). The quench is rapid enough to avoid pearlite and bainite. What is the microstructure in each case? (*a*) Gray cast iron; (*b*) steel (assume 0.6% carbon).

The castings were precision ground after heat treatment, so that a sliding fit was obtained at the plant in Chicago. Then they were shipped to Detroit in January during a severe cold spell. On arrival the fit of the pistons was found to be too loose in the bore of the cylinder. Explain in less than 30 words how a structural change could have caused this. (Sections 6.32 through 6.36)

6.46 Assume that we have a cast iron of chemistry (3.0% C, 2.0% Si) that is all liquid and can be treated in several ways. Indicate the final matrix and high-carbon phase expected for each treatment. [*Matrix:* ferrite or pearlite; *high-carbon phase:* carbide, flake graphite, spheroidal graphite, or temper carbon.] (Sections 6.32 through 6.36)

Treatment	*Matrix*	*High-carbon phase*
1. Cast in a large section, slowly cooled in the mold to 1300°F (704°C), then air-cooled	_____	_________
2. Cast in a thin section and air-cooled from 1500°F (816°C)	_____	_________
3. Cast in a large section after treatment with magnesium and air-cooled from 1500°F (816°C)	_____	_________

6.47 Describe the microstructures of the following alloys completely, using as many of the numbers of the given phrases as needed. The same number may be used in several answers. (In all cases the balance of the composition is iron.) [(1) martensite and austenite, (2) massive carbide, (3) spheroidite, (4) bainite, (5) pearlite, (6) spheroidal graphite, (7) nodular graphite (clusters), (8) flake graphite.] (Sections 6.32 through 6.36)

a. A 3% carbon iron is poured in small (0.1-in. (0.254 cm)-diameter) droplets into ice water.
b. A 1-in. (2.54-cm) diameter bar of an alloy containing 3% C, 2% Si, and 0.7% Mn is cast in sand, cooled to 1500°F (816°C) in sand, then cooled in air.
c. A 1-in. (2.54-cm) diameter bar of 2.5% C alloy is cast in a metal mold. After cooling to 70°F (21°C), it is reheated to 1800°F (982°C), held for 18 hr, and air-cooled.
d. A 3.5% C, 2% Si, 0.05% Mg alloy is cast in a 1-in. (2.54-cm) diameter bar in sand.

Using the microstructures found in Problem 6.47 *a*, *b*, *c*, and *d*, select the best structure that is practical in the following applications.

e. Pressure vessel with 4-in. (10.16-cm) walls weighing 3 tons (2724 kg)
f. Engine block in an automobile
g. Clutch pedal in an automobile
h. Metal shot for shot blasting

6.48 The figure gives several schematic heating and cooling curves that would be useful in obtaining the given microstructures. Indicate *by letter* the correct heat-treatment cycle for each microstructure. Do not worry about actual times and temperatures, as the curves are relative and more than one microstructure may be obtainable from the same type of cycle. [*Note:* A vertical line means quench.] (Covers all sections in Chapter 6.)

Microstructures: (1) bainite in plain carbon steel, (2) martensite, (3) tempered martensite, (4) age hardening, (5) normalizing a plain carbon steel, (6) ferritic malleable cast iron, (7) recrystallization of cold-worked steel.

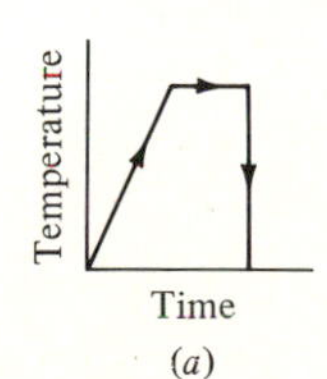

(*a*)

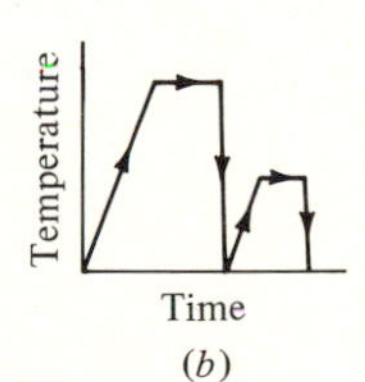

(*b*)

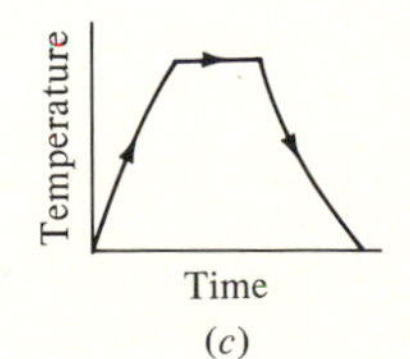

(*c*)

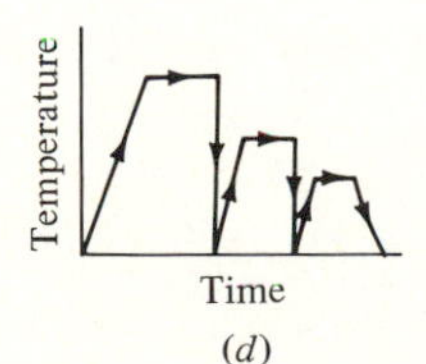

(*d*)

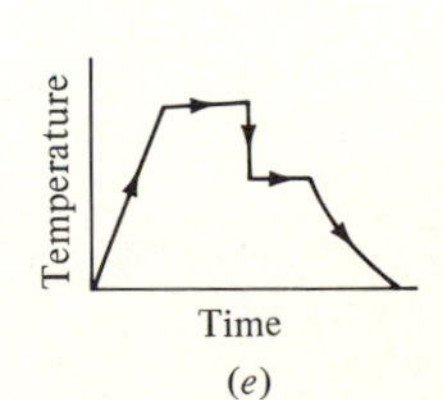

(*e*)

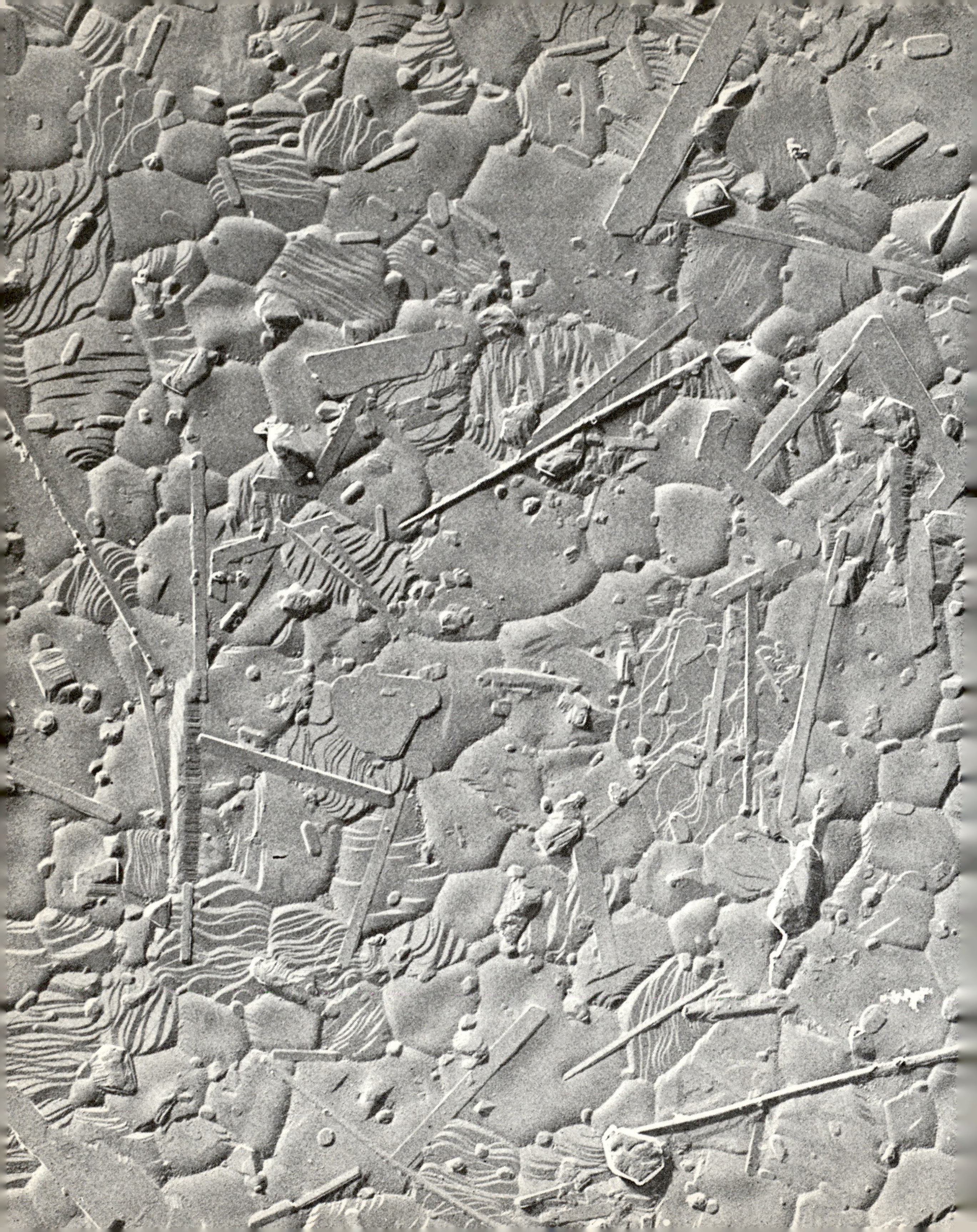

CERAMIC STRUCTURES AND THEIR PROPERTIES

THE microstructure of a newly invented ceramic called "Pyroceram" is shown in the photomicrograph. This ceramic is typical of the important advances taking place in the very old field of ceramics.

There is still some argument as to whether glass should be considered a ceramic, but the new field of glass ceramics of which Pyroceram is a typical product provides an answer.

In this photomicrograph, taken with an electron microscope at 25,000×, the long bladelike crystals of TiO_2 *(rutile) have assisted in the nucleation of the spodumene from a glassy matrix (see text). (Corning Glass Laboratory, Dr. J. H. Munier.)*

In this chapter we shall first study glass and glass processing, then simple crystalline ceramic materials. In the following chapter we shall study ceramic products (other than glass).

7.1 Ceramics and related materials

The field of ceramics and related materials covers the widest range of objects. Some are very beautiful, such as statues and fine china; others very useful, such as the ceramic magnets in TV sets; others very common, such as wash-bowls and sewer pipes. What can a Ming vase have in common with a water closet that we should discuss them in the same chapter? What does the word "crystal" really mean when applied to glassware? What is the difference between fine porcelain and ordinary china?

The approach that makes it possible to cover the important ceramic materials in only two chapters is similar to that we used to study the metals—all ceramics have similarly bonded structures. The ceramics are a little more complex than the metallic structures, which is why we took up the metals first. We shall find that we can take advantage of many facts we learned about metals to shorten the theoretical portion of our work so that we can advance rapidly to a study of the actual materials and their properties.

We can divide our study of ceramics into three principal parts:

1. Glass (noncrystalline ceramics) and glass processing
2. Synthetic and natural simple crystalline materials
3. Ceramic products (other than glass) Chapter 8

Before we discuss the first two topics, we should point out that electrical, magnetic, and optical properties of ceramics are discussed in Chapters 14, 15, and 16. Also, concrete has been classified as a composite material and is discussed in detail in Chapter 11.

7.2 Glass products: general

The glassy state has certain properties quite different from those of metals and alloys. These are very useful in processing. Recall that, in the case of a pure metal, when the liquid cools to the freezing point a crystalline solid precipitates. With a glassy material, however, as the liquid is cooled it becomes more and more viscous, turns to a soft plastic solid, and finally becomes hard and brittle. There is no sharp melting or freezing point. One way of showing this is to plot the specific volume (1/density) as a function of temperature. There is a point where the curve changes slope, Fig. 7.1. This is the temperature at which the material becomes more like a solid than a liquid. It is called the *glass transition temperature*.

To express these relations in more detail, let us review a typical curve of glass viscosity vs. temperature (Fig. 7.2). Four important levels are defined as the glass cools:

1. The working point, viscosity = 10^4 poises (10^3 N · s/m^2). In this temperature range the glass is readily drawn or pressed.

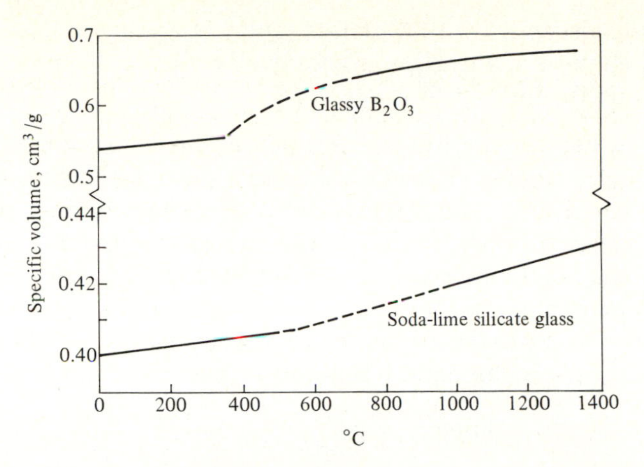

Fig. 7.1 *Relationship of temperature of glass to specific volume of glass. The point at which the curve changes slope is called the* glass transition temperature. (After R.H. Doremus, *Glass Science,* John Wiley & Sons, Inc., New York, 1973, Fig.1, p. 115)

2. Softening point, viscosity = 10^8 poises (10^7 N · s/m^2). At this temperature the glass still deforms under its own weight.
3. Anneal point, viscosity = 10^{13} poises (10^{12} N · s/m^2). Above this point the glass still creeps and elastic strain can be converted rapidly to plastic, thereby relieving stress.

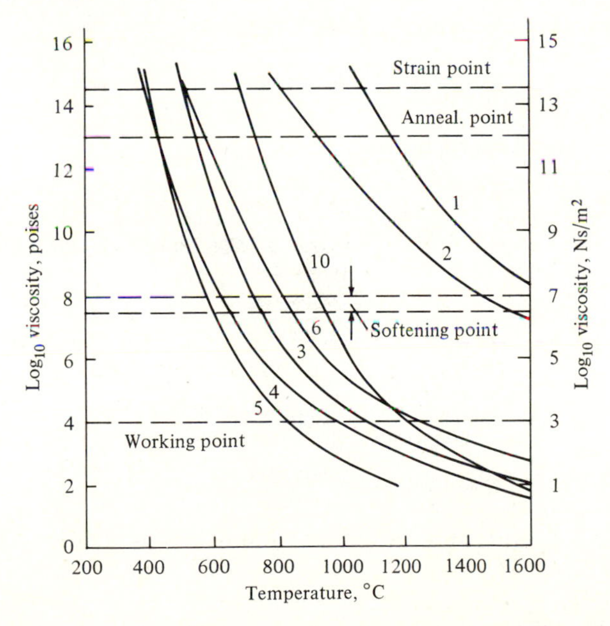

Fig. 7.2 *Change in viscosity with temperature for several varieties of glass. Numbers refer to the chemical compositions of glasses given in Table 7.1 on page 292.* (O. H. Wyatt and D. Dew-Hughes, *Metals, Ceramics, and Polymers,* Cambridge University Press, 1974, p. 259)

4. Strain point, viscosity = $10^{14.5}$ poises ($10^{13.5}$ N · s/m^2). Below this temperature the behavior of the glass is essentially elastic. No permanent deformation can occur without fracture.

Although Fig. 7.2 indicates exact points for working a glass (working point), we can use higher or lower viscosities for specific processes. For example, rapid processing, such as blowing a glass shape, may require lower viscosity (or higher fluidity, which is the inverse of viscosity). Conversely, slower processing, such as pressing, will tolerate higher viscosity and therefore lower temperatures.

The exact viscosity requirement is determined by experience. It is related to temperature by an equation not unlike the diffusion relationship:

$$\text{Fluidity} = \frac{1}{\text{Viscosity}} = Ae^{-B/T}$$

where A and B are constants and T is temperature in kelvin.

EXAMPLE 7.1 A particular glass is known to have a viscosity of $10^{1.7}$ N · s/m^2 at 1000°C (1273 K) and 10^3 N · s/m^2 at 835°C (1108 K). A particular forming operation requires a maximum viscosity of 2×10^2 N · s/m^2 in order to prevent failure during forming. What is the minimum temperature of forming?

ANSWER We may find the temperature by first deriving a general relationship between viscosity and temperature:

$$\frac{1}{10^{1.7}} = Ae^{-B/1273} \quad \text{and} \quad \frac{1}{10^3} = Ae^{-B/1108}$$

If we divide the first equation by the second,

$$10^{1.3} = e^{-B(1/1273 - 1/1108)} \quad \text{or} \quad \ln 10^{1.3} = -B(1/1273 - 1/1108)$$

from which $B = 2.56 \times 10^4$. Substituting this back into the first equation gives

$$\frac{1}{10^{1.7}} = Ae^{-2.56 \times 10^4/1273} \quad \text{and} \quad A = 1.08 \times 10^7$$

Therefore

$$\frac{1}{\text{Viscosity}} = 1.08 \times 10^7\, e^{-2.56 \times 10^4/T}$$

Substituting to find our unknown temperature:

$$\frac{1}{2 \times 10^2} = 1.08 \times 10^7\, e^{-2.56 \times 10^4/T} \quad \text{and} \quad T = 1191 \text{ K} \quad \text{or} \quad 918^\circ\text{C}$$

The reason for glassy behavior is related to the structure. If we fuse pure silica (SiO_2), it forms a glass on cooling called *vitreous silica*. This is very useful in chemical glassware because it resists thermal shock. The basic unit of the structure is the silica tetrahedron, Fig. 7.3. This is composed of a silicon nucleus (valence = 4+) surrounded by four equidistant oxygen atoms. Since each oxygen atom has a valence of 2−, the charge is shared with adjacent SiO_4^{4-} tetrahedra (Fig. 7.4, page 290). This produces a network in space of chains of silica tetrahedra. At high temperatures these chains slide easily past each other because of the thermal vibrations. However, as the melt cools, the structure becomes rigid. Note that silica, SiO_2, is found in nature in the nonglassy crystalline state also shown in Fig. 7.4. For example, the commonest form of silica is the mineral quartz, found in sandstone and silica sand.

Silica is the most important constituent of glass, but other oxides are added to lower the melting point to simplify processing or to change the physical characteristics, such as the index of refraction for optical or decorative glass. There are three major constituent groups.

1. Other glass formers, such as boron oxide, B_2O_3. The valence of the metal ion is usually 3 or greater, and the ion is small.
2. Modifiers, or low-valence elements such as sodium and potassium. They tend to break up the continuity of the chains, but can only be added in limited amounts (Fig. 7.5, page 290). This leads to lower melting temperatures and simplifies the processing.
3. The intermediate oxides do not form glasses by themselves but join the silica chain to maintain a glass (Fig. 7.6, page 291). An example is lead oxide. It is added in large amounts (up to 60%) and produces an ornamental glass of great brilliance. Lead oxide can both become a part of the chain

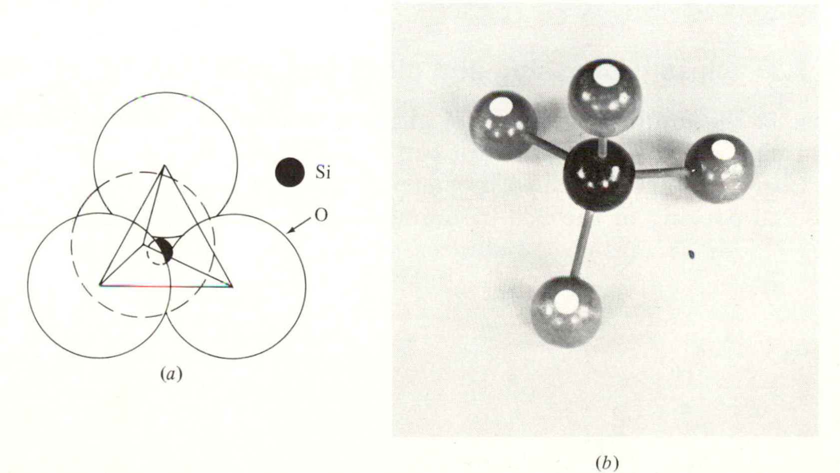

Fig. 7.3 *Silica tetrahedron. (a) Sketch. (b) Photograph of a model of a silica tetrahedron. Atoms with light-colored dots indicate unsatisfied oxygens to which other ions or molecules may attach.*

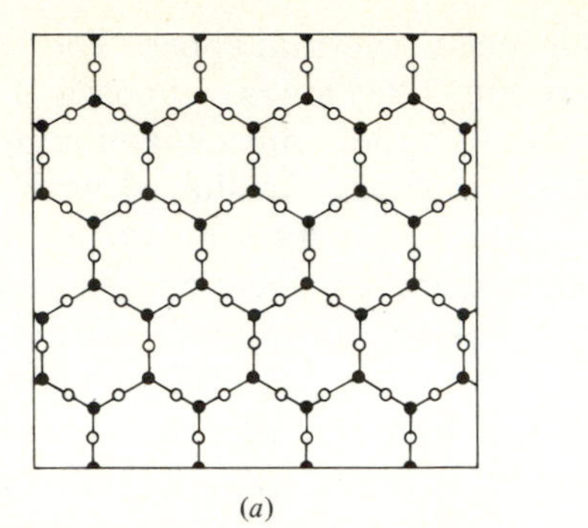

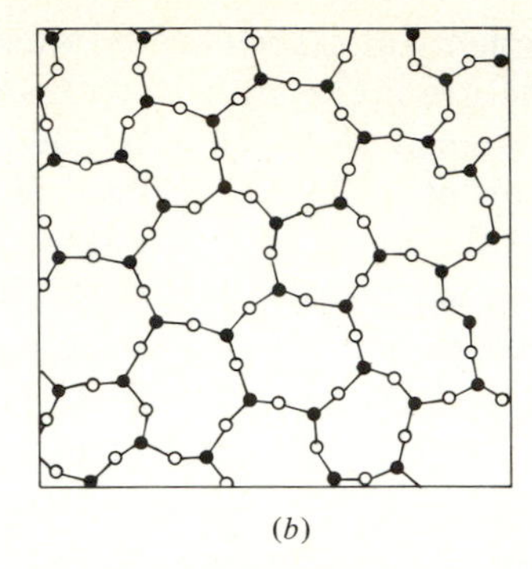

Fig. 7.4 *Two-dimensional sketch of a silica network in (a) crystalline, and (b) low-order glass form.* (After Zachariasen and Warren)

and modify the structure at internal positions. This explains why we can add large quantities and still maintain a glassy structure, as shown in Fig. 7.6.

Figure 7.2 and Table 7.1 (page 292) show compositions of different types of glass and their viscosity-temperature relationships. This figure demonstrates the important influence of the additions. If we compare 96% silica glass (Vycor) (2) with fused silica (1), we see that the softening point of the latter is about 200°C lower. Passing on to window glass, we see that the softening point is about 750°C, compared with 1400°C for the Vycor. As a result, window glass can easily be rolled into large plates, whereas Vycor is available only in smaller shapes and is far more costly. With this background, we may take up glass-processing methods.

7.3 Glass processing and glass products

Because of the unique properties of glass, it can be cast, rolled, drawn, and pressed like a metal. In addition, it can be blown.

Casting is accomplished by pouring the liquid into a mold. A famous case is the pouring of the 200-in.-diameter telescope disk by Corning Glass Works. Under normal circumstances, if the disk were to reach the elastic

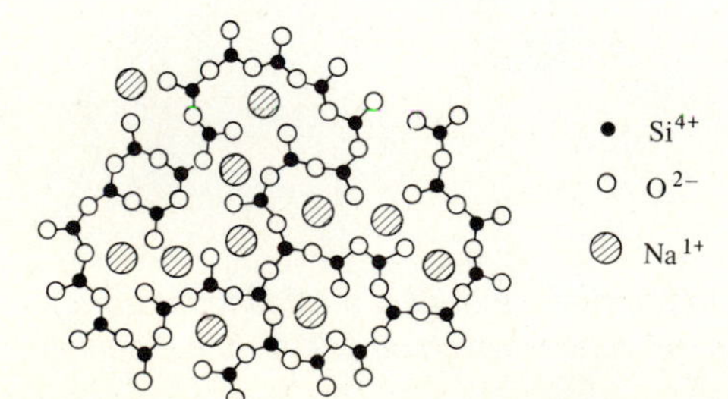

Fig. 7.5 *Effect of a modifier (Na^+) in breaking up the continuity of silica glass*

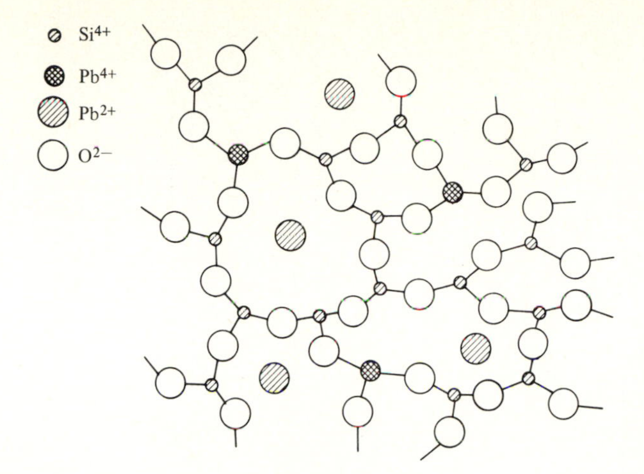

Fig. 7.6 *Lead oxide, considered to be an intermediate oxide, occurs both in the network (as Pb^{4+}) and as a modifier in internal positions (as Pb^{2+}).*

range with temperature gradients present, the material would crack as these equalized later on. Also, as material was removed in grinding and polishing, the surface would distort. Therefore, the disk was cooled over a very long period of time to avoid temperature differences in the mass.

Rolling is widely used to produce window glass and plate glass. The raw materials are melted at one end of a large furnace called a *tank furnace,* and the liquid flows to the other end over a period of time to allow bubbles to float out. The temperature at the end where the rolls are located is controlled, so that the glass is of the right viscosity to be rolled into a sheet. The sheet then passes through a long annealing furnace called a *lehr,* where residual stresses are removed. The rolled material is usable for ordinary window glass in its original form, but plate glass needs extensive grinding and polishing.

A novel method for forming plate glass has become quite important (Fig. 7.7, page 293). The glass flows from the melting furnace onto a float bath of liquid tin, which is covered by a refractory roof. A controlled, heated atmosphere is maintained to prevent oxidation. In the float chamber both surfaces of the glass become mirror-smooth. The sheet then passes into the annealing furnace, which has smooth rollers to avoid harming the finish. This has replaced the costly grinding and polishing operations of the old plate-glass roller method.

Centrifugal casting is used to make the funnels at the back sides of television tubes. A gob of glass is dropped into a metal mold. The mold is then rotated so that the glass rises by centrifugal force. The upper edge is trimmed, then the glass faceplate is sealed to the funnel using a special low-melting-point solder glass.

Drawing of glass tubing is similar to rolling. Glass of the proper viscosity flows directly from the melting furnace around a ceramic tube or mandrel pulled by asbestos-covered rollers. Air blowing through the mandrel

Table 7.1 COMPOSITION OF SOME GLASSES

	Glass	SiO_2	Na_2O	K_2O	CaO	MgO	BaO	PbO	B_2O_3	Al_2O_3	*Remarks*
1	(Fused) silica	99.5+									Difficult to melt and fabricate but usable to 1000°C. Very low expansion and high thermal shock resistance.
2	96% silica (Vycor)	96.3	<0.2	<0.2					2.9	0.4	Fabricate from relatively soft borosilicate glass; heat to separate SiO_2 and B_2O_3 phases; acid leach B_2O_3 phase; heat to consolidate pores.
3	Soda-lime:plate glass	71–73	12–14		10–12	1–4				0.5 –1.5	Easily fabricated. Widely used in slightly varying grades, for windows, containers and electric bulbs.
4	Lead silicate: electrical	63	7.6	6	0.3	0.2		21	0.2	0.6	Readily melted and fabricated with good electrical roperties. High lead absorbe X-rays* high refractive used in achromatic lens. Decorative crystal glass.
5	high lead	35		7.2				58			
6	Borosilicate: low expansion (Pyrex)	80.5	3.8	0.4					12.9	2.2	Low expansion, good thermal shock resistance and chemical stability. Widely used in chemical industry.
7	low electrical loss	70.0		0.5				1.2	28.0	1.1	Low dielectric loss
8	Aluminoborosilicate: standard (apparatus)	74.7	6.4	0.5	0.9		2.2		9.6	5.6	Increased alumina, lower boric oxide improves chemical durability.
9	low alkali (E-glass)	54.5	0.5		22				8.5	14.5	Widely used for fibers in glass resin composites
10	Aluminosilicate	57	1.0		5.5	12			4	20.5	High temperature strength, low expansion
11	Glass-ceramic	SiO_2 40–70; Al_2O_3 10–35; MgO 10–30; TiO_2 7–15									Crystalline ceramic made by devitrifying glass. Easy fabrication (as glass), good properties. Various glasses and catalysts.

O.H. Wyatt and D. Dew-Hughes, *Metals, Ceramics, and Polymers*, Cambridge University Press, 1974, p. 261

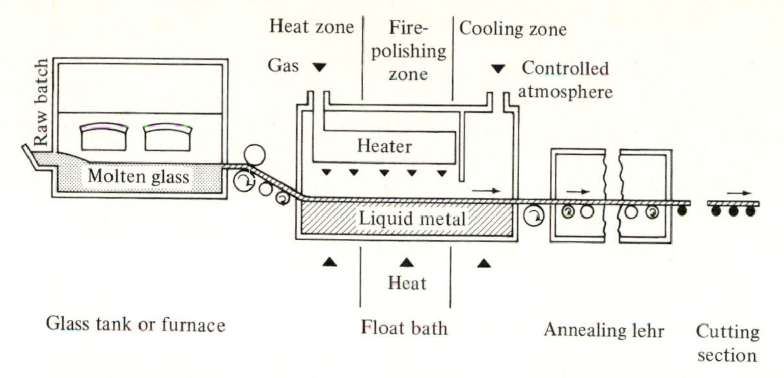

Fig. 7.7 *Method of making "float" glass (Pilkington process)*

keeps the tube from collapsing after it passes the mandrel. Annealing is necessary, as for plate.

Pressing is accomplished by metering a gob of glass into a metal mold, compressing it, and removing it for annealing.

The *press-and-blow* method shown in Fig. 7.8 is widely used to make containers. A gob is fed into a mold and pressed. Then the bottom half of the mold is removed and a mold of the final shape is substituted. The blow operation gives the desired contour. The partly formed glass is called a *parison*.

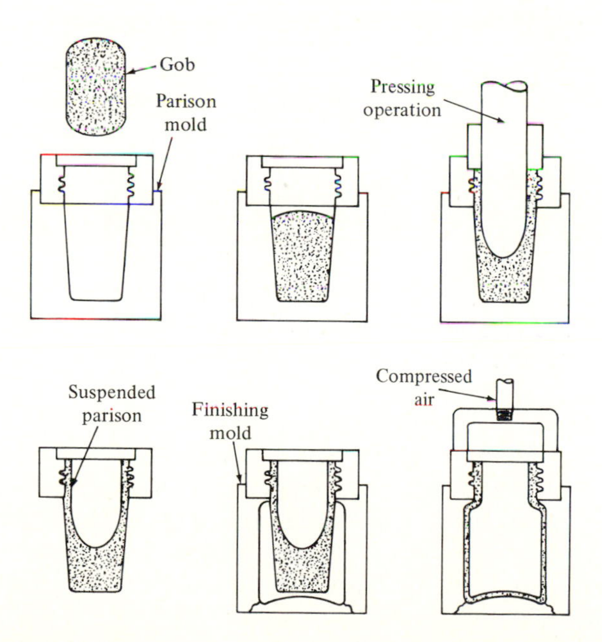

Fig. 7.8 *Press-and-blow method for forming a wide-mouth container.* (W.D. Kingery, *Introduction to Ceramics*, John Wiley & Sons, Inc., New York, 1960, Fig. 3.25, p. 67)

The fiber for fiberglass is now a very important product. One method of making this involves remelting glass marbles that flow through a heated platinum plate with orifices, giving filaments. Traction is provided by rotating the winding tube at surface speeds up to 12,000 ft/min (61 m/sec). Sizing material to separate and lubricate the fibers is applied as they are wound. The use of fiberglass with plastic is discussed in Chapters 10 and 11.

The specifications for glass products vary widely, depending on the end use. For window and plate glass the chief requirements are flatness, transparency, and freedom from bubbles and harmful stresses that may cause not only breakage but also distortion. For containers accuracy of volume is usually important. In chemical ware it is necessary to maintain compositions that will not corrode. When thermal shock is a consideration, the coefficient of expansion is important and silica or high-silica glass is specified. In optical glasses the index of refraction is most important (Chapter 16), whereas in the electrical industry the dielectric constant is of high importance (Chapter 14).

Finally, the current interest in recycling has been readily applied to glass products. Because of the fabrication methods, it is advantageous to begin with a "prealloyed" glass, that is, one with a controlled chemistry and viscosity. Rather than starting from raw silica and adding other materials to reduce the softening point, manufacturers need only remelt recycled glass and reform it to the necessary shape. This can result in a considerable saving.

7.4 Crystalline materials in ceramics: general

We began the study of metals with a review of the metallic elements and the formation of the metallic bond. By contrast, we shall see that most ceramics, instead of a single metal or metallic phases, usually contain both metallic and nonmetallic elements with ionic or covalent bonds. Therefore, we need to consider not only the structure of the metallic atom but also that of the nonmetallic atom, often oxygen, as well as the balance of charges given by the valences.

As with the metals, the unit cell is a key point in describing ceramic structures. The cubic and hexagonal cells are still the most important. In addition, however, the difference in radii between the metallic and nonmetallic ions plays an important part in the arrangement of ions in the unit cell.

In the metals the regular arrangement of metallic atoms into densely packed planes led to the occurrence of slip under stress, giving metals their characteristic ductility. In ceramics, by contrast, both the arrangement of the atoms and the type of bonding are different. In general, we encounter brittle fracture rather than slip. However, the regular arrangement of the atoms determines the path of the fracture. The *cleavage,* the plane of fracture, is closely related to the makeup of the planes of atoms.

The phase diagram is of great importance in understanding the formation and control of the microstructure of polyphase ceramics, just as it is for polyphase metallic materials. Also, nonequilibrium structures are even more prevalent in ceramics because the more complex crystal structures are more difficult to nucleate and to grow from the melt.

With this introduction, let us first consider the special features of bonding in ceramic structures and some of the physical characteristics of commercial ceramics.

7.5 Bonding forces

In the metallic bond, the metal atoms give up their outer (valence) electrons and take positions as positive ions in a space lattice. The electrons provide a bonding force as an electron gas. In contrast, let us now review ionic and covalent bonding in ceramics.

In the *ionic bond* of sodium chloride the single electron of the outer ring of the sodium is attracted to the outer shell of the chlorine, which contains seven electrons. There is a strong driving force for this transfer because this added electron gives chlorine a stable outer shell of eight and leaves a stable inner ring for the sodium. We now have to determine how an assortment of equal quantities of Na^+ and Cl^- ions comes to equilibrium. By the same x-ray diffraction technique used for metals we find the arrangement shown in Fig. 7.9. As we would expect, the negative ions are clustered around the positive ions and, conversely, the positive around the negative. We can also say that this arrangement is obtained because ions of the same charge repel each other. The bonding force is related to how electropositive, i.e., ready to give up electrons, the sodium is and how electronegative, i.e., ready to accept electrons, the chlorine is. In general, ionic bonds are stronger than metallic bonds, as Chapter 2 pointed out. Finally, note that there are no NaCl "molecules." Each Na^+ is surrounded by and attracted equally to six equidistant Cl^- ions.†

The *covalent bond* is also of great importance in ceramics. In solids a diamond crystal is a perfect example of the covalent bond. Each carbon atom has four electrons in its outer shell and needs eight for completion. It shares

†More distant ions also contribute to the bonding.

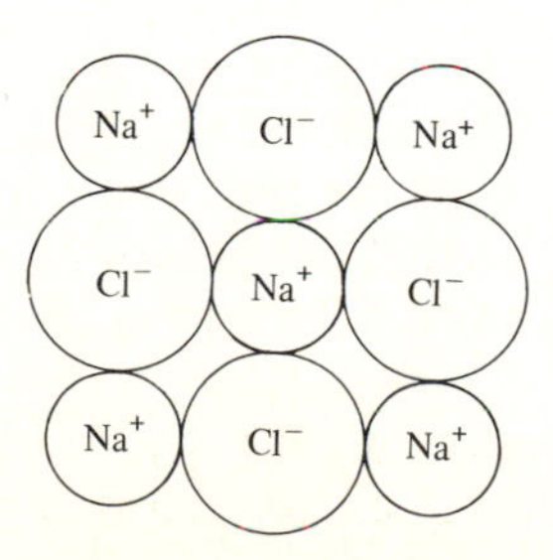

Fig. 7.9 *Two-dimensional sketch of the sodium chloride structure. Note that the center Na^+ is situated near six Cl^- (only four of which are shown, the others being above and below).*

one of its electrons with each of four equally spaced neighbors (Fig. 7.10). Thus the covalent bond is produced by electron sharing, the electrons remaining equidistant between the atoms. This bond is of the highest strength. Since there are very few free electrons or even mobile ions, the best electrical insulators are made of ceramics of this type.

In fact, most ceramics have bonding that is partly ionic and partly covalent. Since electrons are in motion, we can say that in a substance such as silica sand, SiO_2, the bonding is divided between ionic and covalent depending on the position of the bonding electron relative to the ions. In addition to the ionic and covalent bonds, *van der Waals forces* are active in bonding, but they are smaller in magnitude. As an example, the ionic and covalent bonds are strongest in the plane of a layer of clay, and weaker van der Waals forces hold adjacent layers together. Therefore, the cleavage is perfectly parallel to the layers.

7.6 Unit cells

In the metals we considered only cubic, hexagonal, and tetragonal unit cells. In ceramic structures we encounter additional types. For completeness we show all the crystal systems again (Fig. 7.11). However, the cubic, hexagonal, and tetragonal systems are the most important in the ceramics, just as they are in the metals.

In ceramic materials certain groups of crystal structures are known by key names such as the "sodium chloride structure" and the "calcium fluoride structure." The chemicals of the key structures are not very important ceramic materials in themselves, but they represent large structural groups. For example, the very important FeO, MgO, and CaO crystal structures are the same as the sodium chloride structure. Another way of looking at this is to point out that there are some frequently occurring chemical formulas in ceramics: AX, AX_2, A_2X, ABX_3, A_2X_3, and AB_2X_4. In each case A and B are metals and X is a nonmetal. If we know the few characteristic structures of AX compounds in general, the entire field is much simpler. Therefore, we use the term "sodium chloride type structure" to describe a whole group of AX structures. Let us examine these key structures.

```
C : C : C : C
••  ••  ••  ••
C : C : C : C
••  ••  ••  ••
C : C : C : C
••  ••  ••  ••
C : C : C : C
```

Fig. 7.10 *Two-dimensional sketch of covalent bonding in diamond. Compare with Fig. 7.18.*

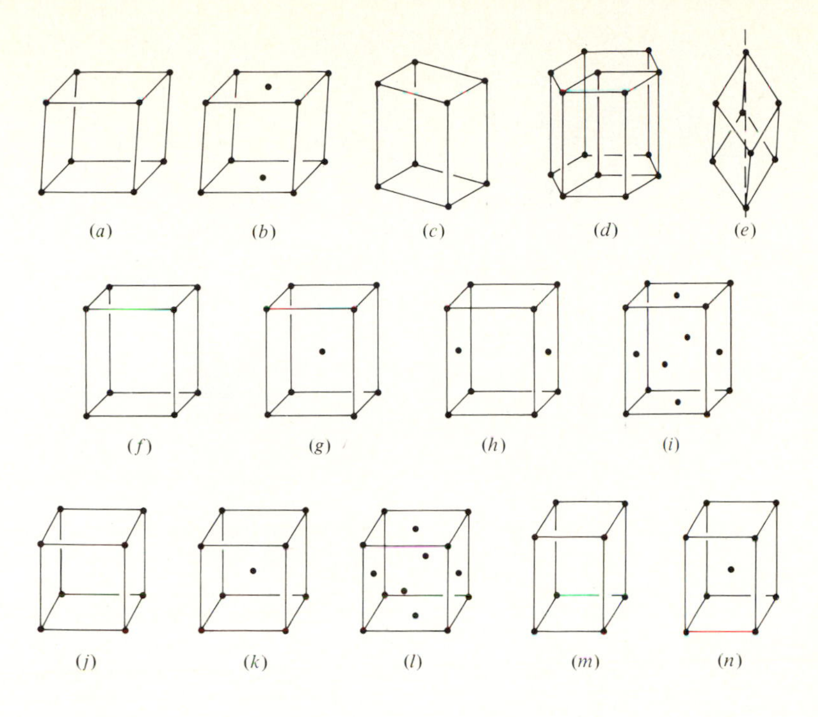

System	*Axes*	*Axial Angles*
Cubic	$a_1 = a_2 = a_3$	All angles = 90°
Tetragonal	$a_1 = a_2 \neq c$	All angles = 90°
Orthorhombic	$a \neq b \neq c$	All angles = 90°
Monoclinic	$a \neq b \neq c$	Two angles = 90°; one angle ≠ 90°
Triclinic	$a \neq b \neq c$	All angles different; none equals 90°
Hexagonal	$a_1 = a_2 = a_3 \neq c$	Angles = 90° and 120°
Rhombohedral	$a_1 = a_2 = a_3$	All angles equal, but not 90°

Fig. 7.11 *The 14 crystal lattices and their geometric relationships. The lattices continue in three dimensions. (a) Simple monoclinic. (b) End-centered monoclinic. (c) Triclinic. (d) Hexagonal. (e) Rhombohedral. (f) Simple orthorhombic. (g) Body-centered orthorhombic. (h) End-centered orthorhombic. (i) Face-centered orthorhombic. (j) Simple cubic. (k) Body-centered cubic. (l) Face-centered cubic. (m) Simple tetragonal. (n) Body-centered tetragonal.*

The Sodium Chloride Structure This cell is made up of equal numbers of sodium and chlorine ions. It is described as an FCC grouping of Cl^- ions with Na^+ ions touching each one of a group of six Cl^- ions, as shown in Fig. 7.12 on page 298.

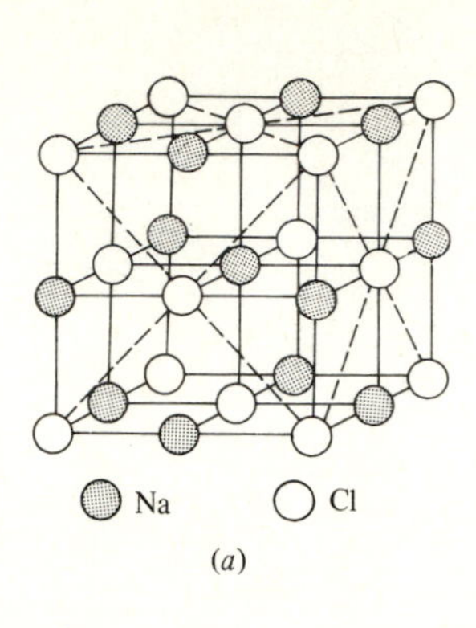

(a)

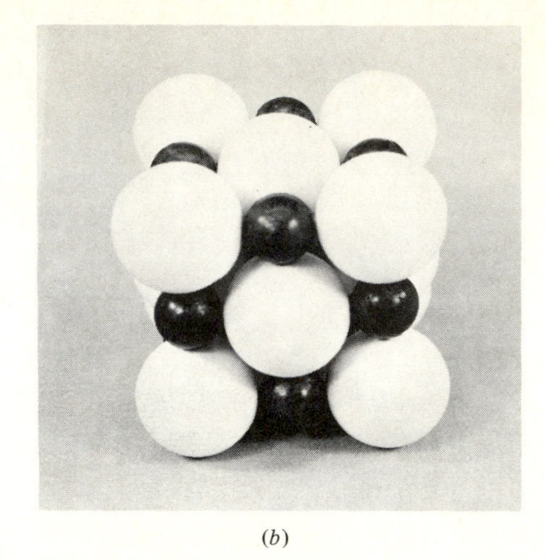
(b)

Fig. 7.12 *Sodium chloride structure. (a) The ion configuration. (b) Photograph of a hard-sphere model. Note that ions do not touch along what might be considered a facial diagonal. (Larger spheres are Cl^-.)*

EXAMPLE 7.2 Calculate the density of sodium chloride from the crystal structure (Fig. 7.12) and the atomic weights of Na^+ and Cl^-.

ANSWER We know that

$$\text{Density} = \frac{\text{mass}}{\text{volume}} = \frac{\text{mass of a unit cell}}{\text{volume of a unit cell}}$$

In Fig. 7.12 the larger ions (Cl^-) seem to form an FCC-like arrangement. From Chapter 3 we know that the FCC structure has four atoms per unit cell. Therefore, sodium chloride has four Cl^- ions, and to maintain the 1 : 1 stoichiometry it must have four Na^+ ions.

$$\frac{\text{Mass}}{\text{Unit cell}} = \frac{4\ Na^+ \text{ ions} \times 23 \text{ g/at. wt.} + 4\ Cl^- \text{ ions} \times 35.45 \text{ g/at. wt.}}{6.02 \times 10^{23} \text{ atoms/at. wt.}}$$
$$= 3.88 \times 10^{-22} \text{ g}$$

Even though the structure in Fig. 7.12 resembles an FCC structure, the ions touch along an edge, as in a simple cubic, *not* along a face diagonal. (For ionic radii see Table 7.2.)

$$\text{Volume/unit cell} = (2 \times r_{Cl^-} + 2 \times r_{Na^+})^3$$
$$= (2 \times 1.81 \times 10^{-8} \text{ cm} + 2 \times 0.98 \times 10^{-8} \text{ cm})^3$$
$$= 1.737 \times 10^{-22} \text{ cm}^3$$

$$\text{Density} = \frac{3.88 \times 10^{-22} \text{ g}}{1.737 \times 10^{-22} \text{ cm}^3} = 2.24 \text{ g/cm}^3$$

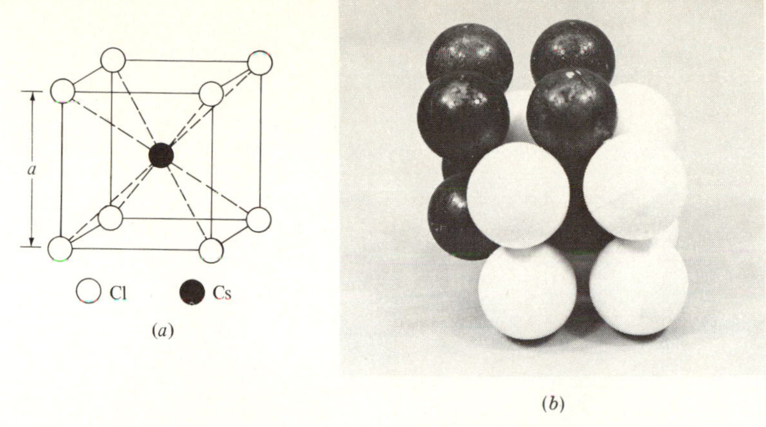

(b)

Fig. 7.13 *Cesium chloride structure. (a) The ion configuration. (b) Photograph of the hard-sphere model. (Larger ions are Cl^-.) The structure looks like two interpenetrating simple cubics. The ions touch along the body diagonal.*

The Cesium Chloride Structure Cesium has the same valence as sodium, but it forms a different structure with chlorine (Fig. 7.13). In this case each cesium ion touches *eight* chloride ions. This is called a *coordination number* (CN) of 8 compared with a CN of 6 in sodium chloride. The basic reason for the difference is that the cesium ion is larger and can touch more Cl^- ions at the same time. By contrast, a sodium ion would "rattle around" (Fig. 7.14*a*, page 300) inside a cube made up of eight Cl^- ions. It can touch only six at the same time, as shown in Fig. 7.12.

Coordination Number (CN) Let us interrupt our discussion of unit cells to point out the relations of CN to the ionic radii in the structure. Table 7.2 (pages 301–303) gives ionic radii. You should note that in general the negative (nonmetallic) ions have the larger radii. This is because the positive (metallic) ions give up outer electrons, and the ionic radius therefore becomes smaller than the atomic radius. In contrast, negative ions add electrons to the outer shell. This reduces the force of the nucleus on the individual electrons, and so the ionic radius is greater than the atomic radius. See, for example, Fe^{2+} vs. Fe^0 and Cl^- vs. Cl^0.

To achieve a stable grouping of larger ions around a smaller ion, i.e., a given CN, the smaller ion must touch all the larger ones. The greater the ratio of the radius r of the smaller ion to the radius R of the larger (i.e., the closer the ratio is to 1), the greater the number of ions that can be grouped around the smaller ion. In the unit cells found in nature CN can equal 2, 3, 4, 6, 8, and 12. Figure 7.14*b* illustrates the minimum radius ratio, r/R, for each of these groupings.

(a)

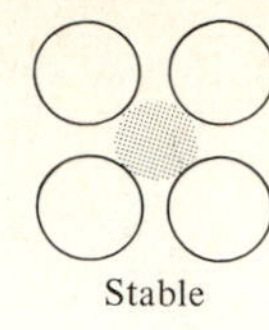

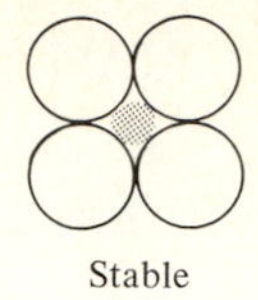

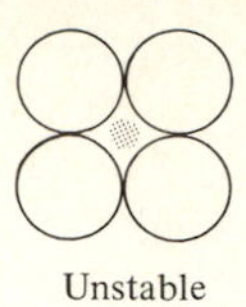

Ratio of Cation Radius to Anion Radius	*Disposition of Ions about Central Ion*	*CN*
1	Same as FCC structure, Chap. 2	12 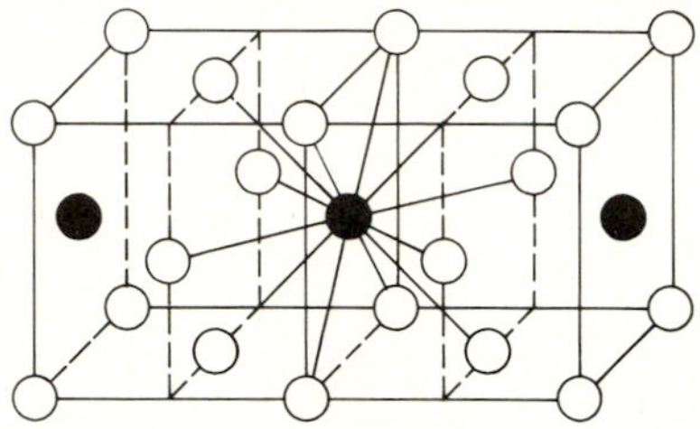
1 to 0.732	Corners of cube	8 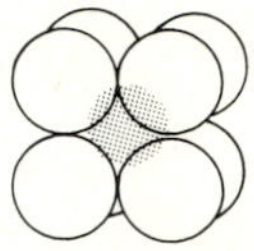
0.732 to 0.414	Corners of octahedron	6 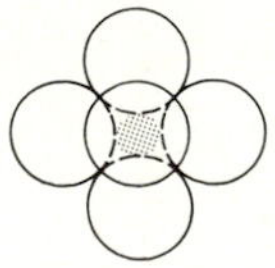
0.414 to 0.225	Corners of tetrahedron	4 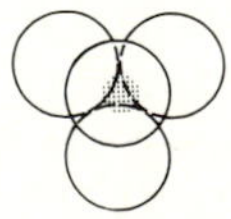
0.225 to 0.155	Corners of triangle	3 
0.155 to 0	Single pair	2 

(b)

Fig. 7.14 *(a) Stable and unstable ionic coordination configurations. Instability arises when the ion (cation in this instance) "rattles around" between the anions. (b) Radius ratios for various atom arrangements in ionic bonding. The ratios are cation/anion* (r^+/r^-). (C.R. Barrett, W.D. Nix, and A.S. Tetelman, *Principles of Engineering Materials,* Prentice-Hall, Inc., Englewood Cliffs, N.J. 1973)

Table 7.2 ATOMIC AND IONIC RADII OF THE ELEMENTS

		As Element				*As Ion*	
Atomic Number	*Symbol*	*Type of Structure**	*Coordination Number*	*Interatomic Distances, Å†*	*Atomic Radii, Å*	*State of Ionization*	*Goldschmidt Ionic Radii, Å*
1	H	HCP	6, 6	—	0.46	H^-	1.54
2	He	—	—	—	—	—	—
3	Li	BCC	8	3.03	1.52	Li^+	0.78
4	Be	HCP	6, 6	2.22; 2.28	1.14	Be^{2+}	0.54
5	B	—	—	—	0.97	B^{3+}	0.2
6	C	Dia. Hex.	4 3	1.54 1.42	0.77 —	C^{4+}	<0.2
7	N	Cubic	—	—	0.71	N^{5+}	0.1 to 0.2
8	O	Orthorh.	—	—	0.60	O^{2-}	1.32
9	F	—	—	—	—	F^-	1.33
10	Ne	FCC	12	3.20	1.60	—	—
11	Na	BCC	8	3.71	1.86	Na^+	0.98
12	Mg	HCP	6, 6	3.19; 3.20	1.60	Mg^{2+}	0.78
13	Al	FCC	12	2.86	1.43	Al^{3+}	0.57
14	Si	Dia.	4	2.35	1.17	Si^{4-} Si^{4+}	1.98 0.39
15	P	Orthorh.	3	2.18	1.09	P^{5+}	0.3 to 0.4
16	S	FC Orthorh.	—	2.12	1.06	S^{2-} S^{6+}	1.74 0.34
17	Cl	Orthorh.	—	2.14	1.07	Cl^-	1.81
18	Ar	FCC	12	3.84	1.92	—	—
19	K	BCC	8	4.62	2.31	K^+	1.33
20	Ca	FCC HCP	12 6, 6	3.93 3.98; 3.99	1.97 2.00	Ca^{2+}	1.06
21	Sc	FCC HCP	12 6, 6	3.20 3.23; 3.30	1.60 1.64	Sc^{2+}	0.83
22	Ti	HCP	6, 6	2.91; 2.95	1.47	Ti^{2+} Ti^{3+} Ti^{4+}	0.76 0.69 0.64
23	V	BCC	8	2.63	1.32	V^{3+} V^{4+} V^{5+}	0.65 0.61 ~0.4
24	Cr	BCC (α) HCP (β)	8 6, 6	2.49 2.71; 2.72	1.25 1.36	Cr^{3+} Cr^{6+}	0.64 0.3 to 0.4
25	Mn	Cubic (α) Cubic (β) FCT (γ)	— — 8, 4	2.24 to 2.96 2.36 to 2.68 2.58; 2.67	1.12 1.18 ~1.37	Mn^{2+} Mn^{3+} Mn^{4+}	0.91 0.70 0.52
26	Fe	BCC (α) FCC (γ)	8 12	2.48 2.52	1.24 1.26	Fe^{2+} Fe^{3+}	0.87 0.67
27	Co	HCP (α) FCC (β)	6, 6 12	2.49, 2.51 2.51	1.25 1.26	Co^{2+} Co^{3+}	0.82 0.65
28	Ni	HCP (α) FCC (β)	6, 6 12	2.49; 2.49 2.49	1.25 1.25	Ni^{2+}	0.78
29	Cu	FCC	12	2.55	1.28	Cu^+	0.96
30	Zn	HCP	6, 6	2.66; 2.91	1.33	Zn^{2+}	0.83
31	Ga	Orthorh.	—	2.43 to 2.79	1.35	Ga^{3+}	0.62
32	Ge	Dia.	4	2.44	1.22	Ge^{4+}	0.44

(Continued)

Table 7.2 ATOMIC AND IONIC RADII OF THE ELEMENTS (*Continued*)

		As Element				*As Ion*	
Atomic Number	*Symbol*	*Type of Structure**	*Coordination Number*	*Interatomic Distances, Å†*	*Atomic Radii, Å*	*State of Ionization*	*Goldschmidt Ionic Radii, Å*
33	As	Rhomb.	3, 3	2.51; 3.15	1.25	As^{3+} As^{5+}	0.69 ~0.4
34	Se	Hex.	2, 4	2.32; 3.46	1.16	Se^{2-} Se^{6+}	1.91 0.3 to 0.4
35	Br	Orthorh.	—	2.38	1.19	Br^{-}	1.96
36	Kr	FCC	12	3.94	1.97	—	—
37	Rb	BCC	8	4.87	2.51	Rb^{+}	1.49
38	Sr	FCC	12	4.30	2.15	Sr^{2+}	1.27
39	Y	HCP	6, 6	3.59; 3.66	1.81	Y^{3+}	1.06
40	Zr	HCP BCC	6, 6 8	3.16; 3.22 3.12	1.58 1.61	Zr^{4+}	0.87
41	Nb	BCC	8	2.85	1.43	Nb^{4+} Nb^{5+}	0.74 0.69
42	Mo	BCC	8	2.72	1.36	Mo^{4+} Mo^{6+}	0.68 0.65
43	Tc	—	—	—	—	—	—
44	Ru	HCP	6, 6	2.64; 2.70	1.34	Ru^{4+}	0.65
45	Rh	FCC	12	2.68	1.34	Rh^{3+} Rh^{4+}	0.68 0.65
46	Pd	FCC	12	2.75	1.37	Pd^{2+}	0.50
47	Ag	FCC	12	2.88	1.44	Ag^{+}	1.13
48	Cd	HCP	6, 6	2.97; 3.29	1.50	Cd^{2+}	1.03
49	In	FCT	4, 8	3.24; 3.37	1.57	In^{3+}	0.92
50	Sn	Dia. Tetra.	4 4, 2	2.80 3.02; 3.18	1.58 —	Sn^{4-} Sn^{4+}	2.15 0.74
51	Sb	Rhomb.	3, 3	2.90; 3.36	1.61	Sb^{3+}	0.90
52	Te	Hex.	2, 4	2.86; 3.46	1.43	Te^{2-} Te^{4+}	2.11 0.89
53	I	Orthorh.	—	2.70	1.36	I^{-} I^{5+}	2.20 0.94
54	Xe	FCC	12	4.36	2.18	—	—
55	Cs	BCC	8	5.24	2.65	Cs^{+}	1.65
56	Ba	BCC	8	4.34	2.17	Ba^{2+}	1.43
57	La	HCP FCC	6, 6 12	3.72; 3.75 3.75	1.87 1.87	La^{3+}	1.22
58	Ce	HCP FCC	6, 6 12	3.63; 3.65 3.63	1.82 1.82	Ce^{3+} Ce^{4+}	1.18 1.02
59	Pr	Hex. FCC	6, 6 12	3.63; 3.66 3.64	1.83 1.82	Pr^{3+} Pr^{4+}	1.16 1.00
60	Nd	Hex.	6, 6	3.62; 3.65	1.82	Nd^{3+}	1.15
61	Pm	Hex.	—	—	—	Pm^{3+}	1.06
62	Sm	Rhomb.	—	—	1.81	Sm^{3+}	1.13
63	Eu	BCC	8	3.96	2.04	Eu^{3+}	1.13
64	Gd	HCP	6, 6	3.55; 3.62	1.80	Gd^{3+}	1.11
65	Tb	HCP	6, 6	3.51; 3.59	1.77	Tb^{3+} Tb^{4+}	1.09 0.89

Table 7.2 ATOMIC AND IONIC RADII OF THE ELEMENTS (*Continued*)

		As Element				*As Ion*	
Atomic Number	*Symbol*	*Type of Structure**	*Coordination Number*	*Interatomic Distances, Å†*	*Atomic Radii, Å*	*State of Ionization*	*Goldschmidt Ionic Radii, Å*
66	Dy	HCP	6, 6	3.50; 3.58	1.77	Dy^{3+}	1.07
67	Ho	HCP	6, 6	3.48; 3.56	1.76	Ho^{3+}	1.05
68	Er	HCP	6, 6	3.46; 3.53	1.75	Er^{3+}	1.04
69	Tm	HCP	6, 6	3.45; 3.52	1.74	Tm^{3+}	1.04
70	Yb	FCC	12	3.87	1.93	Yb^{3+}	1.00
71	Lu	HCP	6, 6	3.44; 3.51	1.73	Lu^{3+}	0.99
72	Hf	HCP	6, 6	3.13; 3.20	1.59	Hf^{4+}	0.84
73	Ta	BCC	8	2.85	1.47	Ta^{5+}	0.68
74	W	BCC (α) Cubic (β)	8 12; 2, 4	2.74 2.82; 2.52, 2.82	1.37 1.41	W^{4+} W^{6+}	0.68 0.65
75	Re	HCP	6, 6	2.73; 2.76	1.38	Re^{4+}	0.72
76	Os	HCP	6, 6	2.67; 2.73	1.35	Os^{4+}	0.67
77	Ir	FCC	12	2.71	1.35	Ir^{4+}	0.66
78	Pt	FCC	12	2.77	1.38	Pt^{2+} Pt^{4+}	0.52 0.55
79	Au	FCC	12	2.88	1.44	Au^{+}	1.37
80	Hg	Rhomb.	6	3.00	1.50	Hg^{2+}	1.12
81	Tl	HCP BCC	6, 6 8	3.40; 3.45 3.36	1.71 1.73	Tl^{+} Tl^{3+}	1.49 1.06
82	Pb	FCC	12	3.49	1.75	Pb^{4-} Pb^{2+} Pb^{4+}	2.15 1.32 0.84
83	Bi	Rhomb.	3, 3	3.11; 3.47	1.82	Bi^{3+}	1.20
84	Po	Monocl.	—	2.81	1.40	Po^{6+}	0.67
85	At	—	—	—	—	At^{7+}	0.62
86	Rn	—	—	—	—	—	—
87	Fr	—	—	—	—	Fr^{+}	1.80
88	Ra	—	—	—	—	Ra^{+}	1.52
89	Ac	—	—	—	—	Ac^{3+}	1.18
90	Th	FCC	12	3.60	1.80	Th^{4+}	1.10
91	Pa	—	—	—	—	—	—
92	U	Orthorh.	—	2.76	1.38	U^{4+}	1.05
93	Np	—	—	—	—	—	—
94	Pu	—	—	—	—	—	—
95	Am	—	—	—	—	—	—
96	Cm	—	—	—	—	—	—

*Abbreviations: HCP = hexagonal close-packed, BCC = body-centered cubic, Dia. = diamond, Hex. = hexagonal, Orthorh. = orthorhombic, FCC = face-centered cubic, FCT = face-centered tetragonal, Rhomb. = rhombohedral, Tetra. = tetragonal, Monocl. = monoclinic.

†Some elements have two values because calculations in different directions in the unit cell give different numbers.

EXAMPLE 7.3 Calculate the minimum r/R for CN = 3.

ANSWER Note that in Fig. 7.14*b* the smaller ion has just reached a radius at which it is touching all three larger ions. We can calculate r/R by geometry as follows. From the figure we see that

$$AD = r + R \qquad \tfrac{1}{2}AC = R$$

Since ABC is an equilateral triangle (each angle = 60°) and the smaller ion is at the center, α must equal 30°.

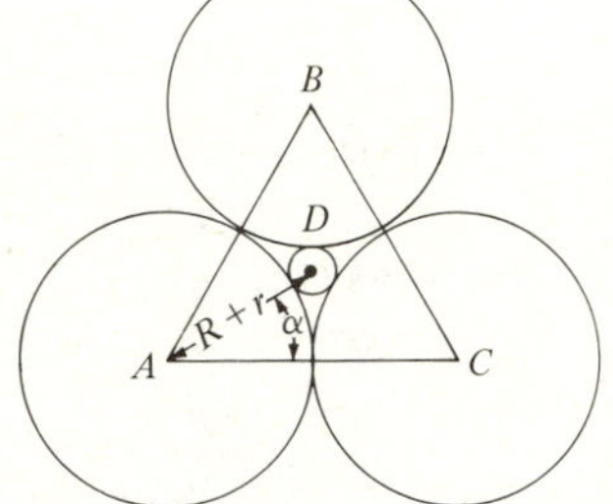

Then

$$\cos 30° = \frac{\tfrac{1}{2}AC}{AD}$$

$$0.866 = \frac{R}{r + R}$$

$$r/R = 0.155$$

We can solve the other cases in a similar way, beginning by finding $r + R$ as a function of R from the geometry of the figure.

In considering the effect of the radius ratio in this way on the shape of the unit cell, we find that in boron nitride we have a simple triangle of larger N^{3-} ions with a B^{3+} ion in the center. If we formed a tetrahedron with four touching N^{3-} ions, the B^{3+} ion would rattle around. Therefore the configuration would be unstable ($R_{N^{3-}} = 1.17$ Å, $r_{B^{3+}} = 0.2$ Å).

As we progress to the radius ratio of silicon to oxygen, we find that the important SiO_4^{4-} complex ion is made up of a silicon atom at the center of four oxygen atoms. In another electronically balanced case we have ZnS. (In general practice the words "ion" and "atom" are used interchangeably, just as they are in discussions of metal structures. However, we must use the *ionic* radius in calculations for ceramic structures.)

Now let us look specifically at the ionic radius ratios for sodium chloride and cesium chloride, which have CN's of 6 and 8, respectively. For sodium chloride we have 0.98/1.81 = 0.54, which is in the range 0.732 to 0.414 (CN = 6). For cesium chloride we have 1.65/1.81 = 0.91, which is in the range 1 to 0.732 (CN = 8). The sodium chloride coordination is described as octahedral. Although the Cl^- ions form an octahedral figure around the Na^+, there are only six ions, not eight, and CN = 6.

Let us sum up the effect of CN on the unit cell structure. For a given CN the radius ratio will usually be large enough so that the smaller atom touches all the larger atoms indicated by the CN. For example, for CN = 8, r/R should be *greater* than 0.732. However, we may have a structure in which the CN is lower than that permitted by the radius ratio. In this case the smaller atom is indeed touching all the larger atoms, but additional larger atoms could be packed around the smaller atom. For example, a carbon atom in diamond (Fig. 7.18, page 307) has only four nearest neighbors because of the arrangement providing four shared electrons.

Interstitial Sites While we have before us the models of tetrahedral and octahedral groups of ions around a central ion for CN = 4 and CN = 6, respectively, let us discuss the terms *octahedral site* and *tetrahedral site*. These are voids in the unit cell in which we can place different atoms. We recall that inside the FCC structure of iron we could place a carbon atom at $\frac{1}{2}, \frac{1}{2}, \frac{1}{2}$. This is called an octahedral site because the six nearest atoms, the iron atoms at the centers of the faces, form an octahedron around the void (Fig. 7.15, page 306). The size of the site (or void) is expressed as the diameter of the sphere that would just touch the iron atoms (0.41 times the radius of the iron atom). There are also tetrahedral sites in the same structure (Fig. 7.16, page 306). An atom at $\frac{1}{4}, \frac{1}{4}, \frac{1}{4}$ is equidistant from four iron atoms, making up a tetrahedron, and $r = 0.225R$. [r = small atom, R = large (iron) atom.]

In a larger region of the space lattice, including adjacent unit cells, we find that there are octahedral sites at the center of each cube edge. On the average there are four octahedral sites per FCC unit cell, or one site for each FCC atom, and eight tetrahedral sites per cell, or two sites per atom. Thus close-packed structures have *twice* as many tetrahedral as octahedral sites.

HCP structures also have octahedral and tetrahedral sites in the same ratios of sites to atoms as FCC structures. One of the most important uses of this understanding of sites is in the ferrimagnetic ceramics. We shall see in Chapter 15 that magnetic field strength is directly related to the way in which atoms are placed in these sites. Now let us continue our discussion of the important unit cells.

The Calcium Fluoride Structure The easiest way to visualize this structure is to consider the Ca^{2+} ions as making up an FCC-like structure with a simple cube of eight F^- ions inside it (Fig. 7.17, page 307). The F^- ions occupy $\frac{1}{4}, \frac{1}{4}, \frac{1}{4}$ type positions in the "FCC structure," the tetrahedral sites just discussed. Of course, because of their large size the F^- ions force apart the Ca^{2+} ions making up the corners of the tetrahedral site. Note that Ca^{2+} ions do not touch along a "facial diagonal." Since twice as many F^- ions as Ca^{2+} ions are required (CaF_2), the F^- ion positions cannot be octahedral. This structure is also characteristic of AX_2 compounds in which the cations occupy the F^- and the anions the Ca^{2+} positions.

The Diamond Structure If we omit four of the eight F^- ions that form the cube inside the CaF_2 structure, we form a tetrahedron (Fig. 7.18, page 307). Here the ratio of the atoms inside the cell to those forming the FCC "box" is 1. The inner atoms are still at $\frac{1}{4}, \frac{1}{4}, \frac{1}{4}$ locations relative to the corners. This structure is found in the important semiconductor materials silicon and germanium as well as in diamond, tin, and some AX compounds.

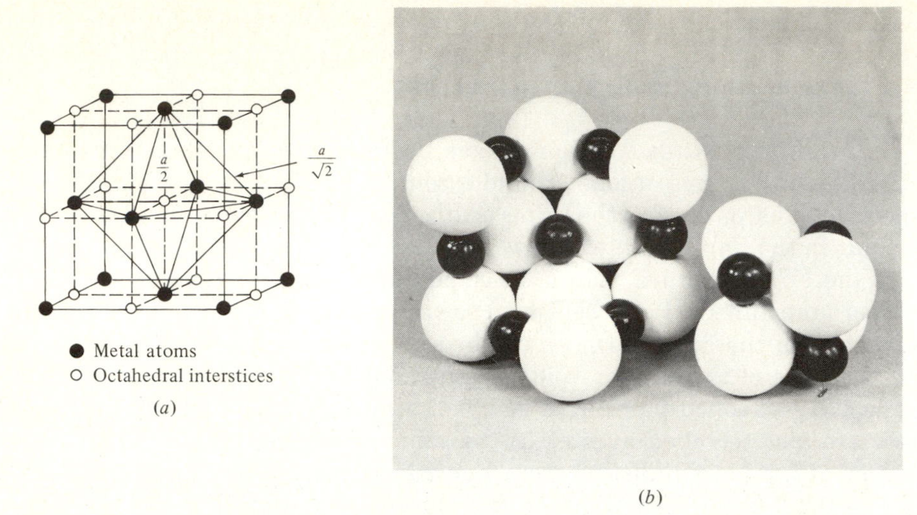

Fig. 7.15 *Octahedral sites (CN = 6). (a) Sites in an FCC structure. (b) Photograph of a hard-sphere model of an FCC structure with octahedral sites filled. Corner atoms have been removed to show site in center of structure.*

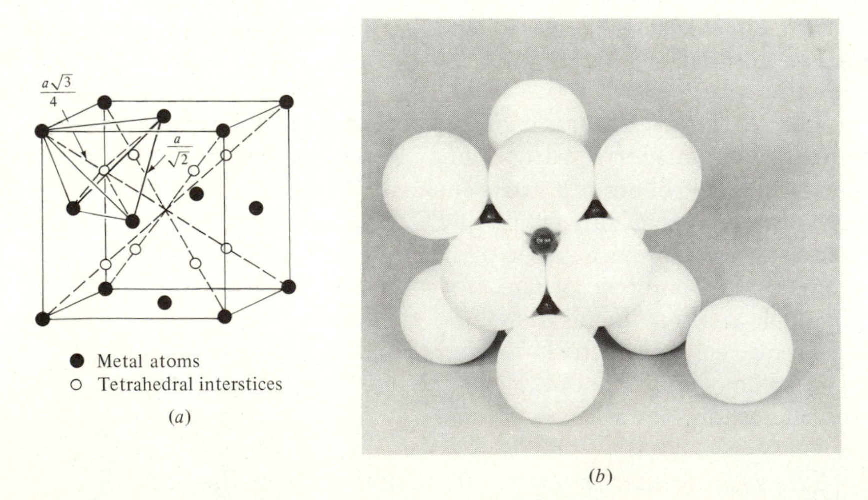

Fig. 7.16 *Tetrahedral sites (CN = 4). (a) Sites in an FCC structure. (b) Photograph of a hard-sphere model of an FCC structure, showing only four tetrahedral sites. Corner atom has been removed. Note small size of these sites compared to size of octahedral sites (Fig. 7.15).*

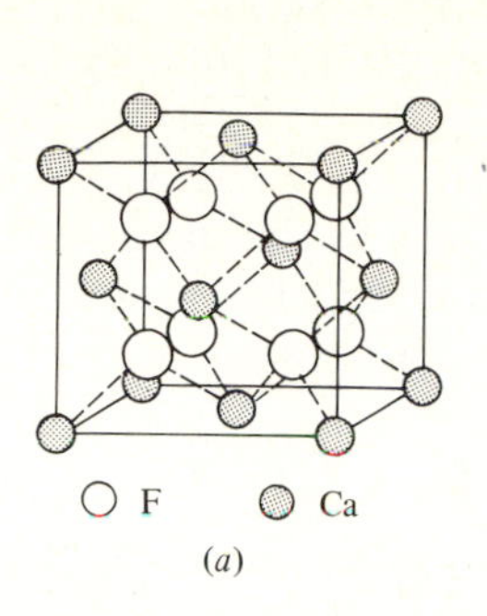

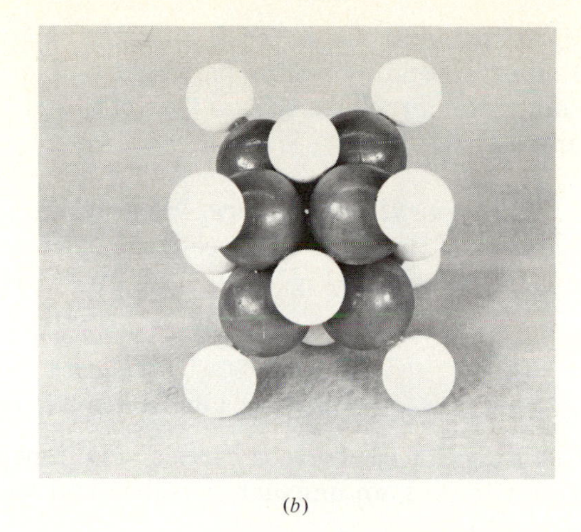

(b)

Fig. 7.17 *Calcium fluoride structure. (a) The ion configuration. (b) Photograph of a hard-sphere model of CaF_2. Note how F^- ions (larger ions) at tetrahedral sites have expanded the structure to create large distances between Ca^{2+} ions (along a facial diagonal).*

The Corundum Structure The mineral corundum is Al_2O_3. Depending on small amounts of impurities, it may have a red color as in the ruby or a blue color as in the sapphire. Common, clear corundum (alumina) is an important abrasive and refractory. The structure can be built up by starting with an HCP-like structure of O^{2-} ions. The number of octahedral sites in this lattice (Fig. 7.19) is equal to the number of oxygen atoms. However, since the aluminum ion is trivalent, we can have only two Al^{3+} to three O^{2-}. As a result, only two-thirds of the sites are filled.

The Spinel Structure The basic formula for spinel is $A^{2+}B_2{}^{3+}O_4$, where A and B are metal ions ($MgAl_2O_4$, for example). The oxygen ions form

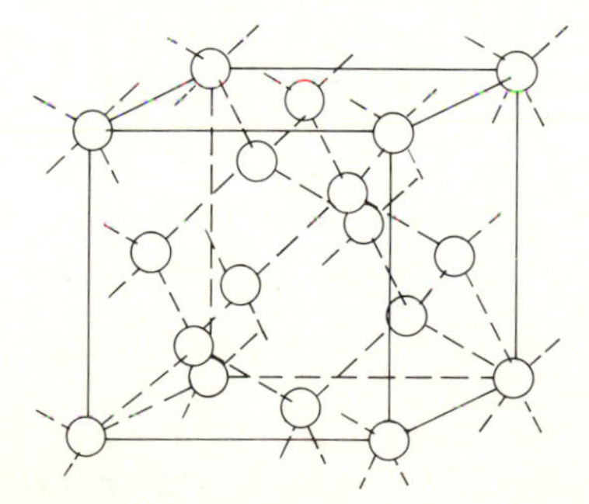

Fig. 7.18 *Structure of diamond, shown as a three-dimensional array of carbon atoms*

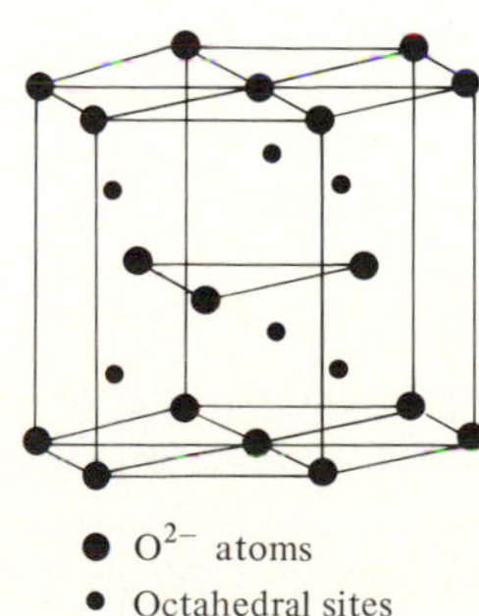

Fig. 7.19 *Structure of corundum*

an FCC lattice, and the A and B ions can be found in tetrahedral and octahedral sites, depending on the particular spinel. This is of great importance in the magnetic ferrites, a special type of spinel that we will take up in detail in Chapter 15.

The Perovskite Structure In perovskite, $CaTiO_3$, the oxygen ions are at the centers of the faces of the cell, the calcium ions at the corners, and the titanium in the center (Fig. 7.20). This gives the proper formula:

$$6O^{2-} \times \tfrac{1}{2}\ \text{ion/face} = 3$$
$$8Ca^{2+} \times \tfrac{1}{8}\ \text{ion/corner} = 1$$
$$1Ti^{4+} \times 1\ \text{ion/center of body} = 1$$

This type of structure is particularly important in the similar structure of barium titanate, an unusual dielectric and ferroelectric material discussed in Chapter 14.

The Silica and Silicate Structures These are among the most complicated structures. We shall go into the details in the next section when we discuss commercial materials. However, we should realize here that all these structures can be simplified greatly if we note in each case the common building block, the SiO_4^{4-} tetrahedron (Fig. 7.3). In other words, the silicon atoms are always bonded to four oxygen atoms. The complexity arises from the different attachments made to the other four unsatisfied bonds extending from the oxygen atoms outward from the tetrahedron.

In the simplest case, SiO_2, the bonds are all satisfied by adjacent silicon atoms, forming an SiO_2 network (Fig. 7.21). (It must be appreciated that the

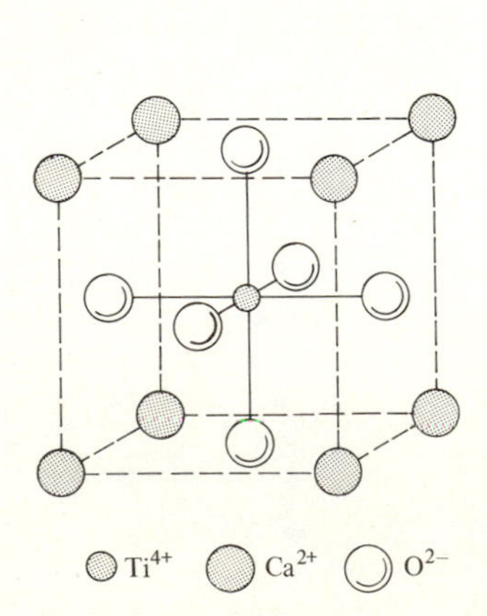

Fig. 7.20 *Structure of perovskite (also of barium titanate)*

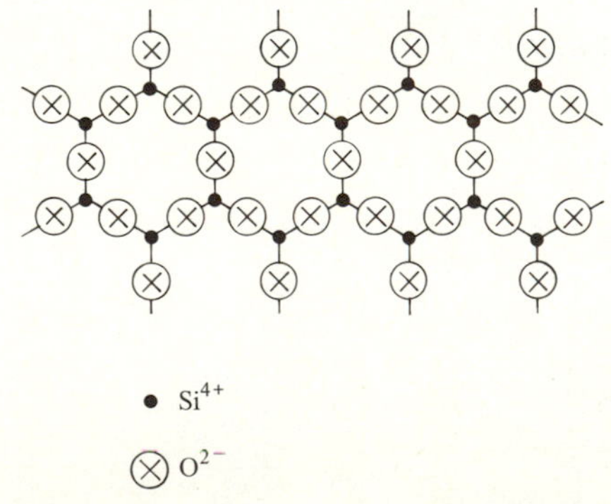

Fig. 7.21 *A silica network. The fourth bond of the silicon is not shown. It is normal to the plane of the paper and connects to the fourth oxygen (no foreign atoms).*

structure is really three-dimensional.) At the other extreme all four bonds can be satisfied by foreign atoms such as magnesium (Fig. 7.22). This is called an *island structure* because the SiO_4^{4-} tetrahedra, as in Fig. 7.3, are isolated from each other by foreign atoms (Mg^{2+}). The resultant formula is Mg_2SiO_4. (One-half of each Mg^{2+} is shared by each O^{2-} in the SiO_4^{4-} tetrahedron.)

7.7 Solid solutions

Up to this point we have discussed only pure compounds. However, we also encounter solid solutions in ceramics as in metal systems, although additional principles affect their formation. For example, an important series of minerals occurs between pure magnesite, MgO, and iron oxide, FeO. Both have the sodium chloride structure, and we find a continuous series of solid solutions similar to the copper-nickel system in metals. However, other points such as similar valence are equally important in obtaining a continuous series of solid solutions.

EXAMPLE 7.4 Olivine sands are usually given the formula $(Mg,Fe)_2SiO_4$, which suggests complete solid solution of magnesium and iron. Justify the high solid solubility and indicate whether lithium ions might also substitute.

ANSWER Mg^{2+} (0.78 Å) and Fe^{2+} (0.87 Å) have the same charge and are very close in size, which are the criteria for a high degree of solid solubility. Although Li^+ (0.78 Å) is also the same size, the charge is not balanced. Some solid solution is still possible, however, by the following techniques:

1. $Li^+ + Fe^{3+}$ (0.67 Å) substitute for two Mg^{2+}.
2. Two Li^+ substitute for one $Mg^{2+} + \square$. (cation vacancy)

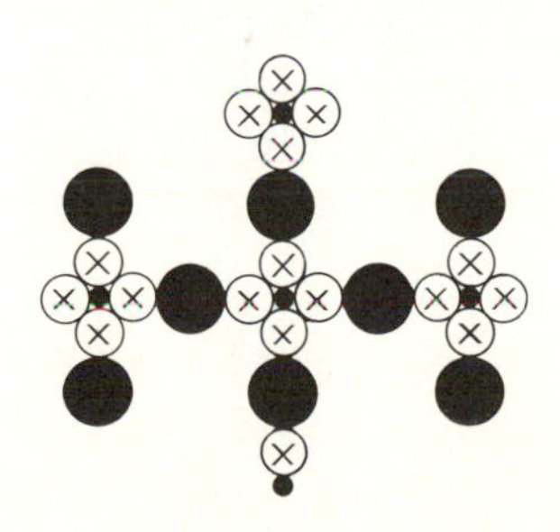

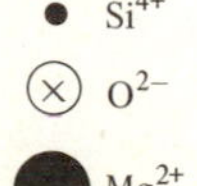

Fig. 7.22 *Satisfying unshared oxygen ions in the SiO_4^{4-} tetrahedron (island structure) with Mg^{2+} ions. Each O^{2-} is bonded to an Si^{4+} and an Mg^{2+}.*

Types of Transformation As in the metals, in the ceramics we encounter transformations of the nucleation and growth type and of the diffusionless type. When the transformation of a structure involves the breaking of bonds, it is called *reconstructive*. If only small shifts in position are involved, it is called *displacive* and is similar to the martensite type of transformation in metals. Examples of displacive transformation are $\alpha \rightarrow \beta$ quartz and cubic $\rightarrow$ tetragonal $BaTiO_3$. When a large volume change accompanies displacive transformation, as in quartz, severe cracking can take place, since there is no ductility.

Also, as in metals, the kinetics of reactions are very important. The transformation to pearlite, for example, takes time, and this permits the formation of martensite. TTT diagrams exist in ceramics also. They have been used to determine the crystallization time of glasses. As we might expect, adding "alloying elements" shifts the nose of the diagram, as it does in metals. In general, most reactions are more sluggish in ceramics because several ions must move in combination to maintain electroneutrality, a problem not encountered in metals. These unusual features of ceramics will be discussed further in subsequent sections.

7.8 Defect structures, lattice vacancies

The point defects or vacancies in the lattices that we found in metals also occur in ceramic crystals. Vacancies are especially important when a foreign ion of different valence is dissolved in the standard structure, and some compensation is needed to obtain a charge balance over the crystal as a whole. A typical case is iron oxide crystals in which both Fe^{2+} and Fe^{3+} are present. The normal FeO structures is like that of NaCl, but if we make up the crystal with some Fe^{3+} ions present, one Fe^{2+} is absent to balance two Fe^{3+} ions (Fig. 7.23).

The symbol □ signifies an ion vacancy where an Fe^{2+} was omitted to balance the charge surplus caused by two Fe^{3+}. This is called a *cation* vacancy. If, on the other hand, we substitute a cation of lower than normal valence, say K^+, then *anion* vacancies are needed to balance the charge. (In this case some O^{2-} would be left out.) Defect structures are used in voltage rectifiers, as discussed in Chapter 14.

O^{2-} Fe^{2+} O^{2-} Fe^{2+} O^{2-} Fe^{2+} O^{2-} Fe^{2+}

Fe^{2+} O^{2-} Fe^{2+} O^{2-} Fe^{2+} O^{2-} Fe^{2+} O^{2-}

O^{2-} Fe^{3+} O^{2-} Fe^{2+} O^{2-} □ O^{2-} Fe^{2+}

Fe^{2+} O^{2-} □ O^{2-} Fe^{3+} O^{2-} Fe^{3+} O^{2-}

O^{2-} Fe^{3+} O^{2-} Fe^{2+} O^{2-} Fe^{2+} O^{2-} Fe^{2+}

Fe^{2+} O^{2-} Fe^{2+} O^{2-} Fe^{2+} O^{2-} Fe^{2+} O^{2-}

Fig. 7.23 *The defect structure of* Fe_1O. *The symbol □ denotes an ion vacancy.*

EXAMPLE 7.5 In Fig. 7.23 (wüstite) what fraction of cation (iron) sites will be vacant if there are 10 Fe^{3+} ions to every 100 Fe^{2+} ions?

ANSWER In a problem of this sort we must first decide whether the anion or cation lattice will be perfect. In this case it is the anion (oxygen) lattice.

$$100\ Fe^{2+} + 10\ Fe^{3+} = 110 \text{ cation sites filled}$$

For every two Fe^{3+} there must be one Fe^{2+} vacancy for the charge to balance. Therefore there are five Fe^{2+} vacancies, since we have 10 Fe^{3+} ions.

The anion sites are all filled, however, with oxygen, and there are equal numbers of anion and cation sites in FeO.

$$100\ Fe^{2+} + 10\ Fe^{3+} + 5\,\square = 115 \text{ cation sites total}$$

or

$$\text{Anion sites} = 115 \text{ (all filled)}$$

$$\text{Percent vacant cation sites} = \tfrac{5}{115} \times 100 = 4.35$$

We cannot select a few common properties as a basis for comparison of ceramics in the same way we did for metals. For example, the tensile test, a common denominator for comparing metals, is rarely performed on ceramics. In many cases the insulating or optical properties are more important than the mechanical properties. Therefore, we shall discuss some of the properties that are important in ceramics, then take up the individual materials.

7.9 Some properties of ceramics

A number of ceramics are used as refractories. This is a broad term meaning in general materials with good resistance to heat. However, the temperature of operation of the equipment and the degree of thermal shock are important. A good refractory for a kitchen oven might contain considerable asbestos, but this combination would melt at steel-making temperatures (1600°C). For this reason the melting point or solidus temperature of ceramics is important.

Another important factor in using refractories to resist high temperatures is the thermal coefficient of expansion α, also called the *linear coefficient of expansion*. If we heat a bar of length l from T_1 to T_2, the length will increase by Δl. We define α as

$$\alpha = \frac{\Delta l/l}{T_2 - T_1} \quad \text{or} \quad \alpha = \frac{\Delta l}{l\Delta T}$$

The units are therefore $(°F)^{-1}$, $(°C)^{-1}$, or $(K)^{-1}$, and the numerical values of °C and K will be the same. If the coefficient is high, the surface layers will expand greatly relative to the unheated interior. Since ceramics are brittle, these heated layers will flake off. This is called *spalling*.

EXAMPLE 7.6 Calculate the elastic strain and then the stress developed when the surface layer of a fire clay brick is heated from 100 to 600°F (38 to 315°C).

ANSWER Assume that $\alpha = 2.5 \times 10^{-6}$ (°F)$^{-1}$ [4.5×10^{-6}(°C)$^{-1}$] and the modulus of elasticity $E = 15 \times 10^6$ psi (1.035×10^5 MPa). These levels are reasonably constant in this temperature range. If the layer is unrestrained, the expansion ϵ is

$$\epsilon = 2.5 \times 10^{-6}(°F)^{-1} \times 500°F \; [4.5 \times 10^{-6}(°C)^{-1} \times 277°C] = 12.5 \times 10^{-4}$$

If the layer is completely restrained by surrounding material and there is no heat transfer between layers, then

$$\sigma = E\epsilon = 18{,}750 \text{ psi} \qquad (130 \text{ MPa}) \qquad (\text{a maximum value})$$

In reality the heated layer tends to expand, developing a shearing stress between the backing layers, and a thin layer spalls off the surface.

It should be added that in a noncubic crystal the expansion coefficient is different in different directions (Table 7.3).

In some materials the coefficient of expansion for polycrystalline material is not the mean of these values, because the difference in expansion in different directions in various grains leads to cracking. This is encountered in magnesium and aluminum titanates, for example. There is a general relationship of coefficients in different families of ceramics:

Corundum type (Al_2O_3, Cr_2O_3, Fe_2O_3): 8 to 12 $\times 10^{-6}$(°C)$^{-1}$
Zircon type ($ZrSiO_4$): 4 $\times 10^{-6}$(°C)$^{-1}$
MgF_2 type (TiO_2): 9 to 11 $\times 10^{-6}$(°C)$^{-1}$

The coefficient for glasses is not predictable. It varies from 0.5×10^{-6} (°C)$^{-1}$ for fused silica to 7.6×10^{-6}(°C)$^{-1}$ for fused GeO_2. A major breakthrough in control of the coefficient was attained in one variety of Pyroceram. The basic formula is related to the mineral β spodumene, $Li_2O \cdot Al_2O_3 \cdot 4SiO_2$,

Table 7.3 COEFFICIENT OF EXPANSION IN SOME CERAMICS

	Coefficient of Expansion, (°C)$^{-1}$ $\times 10^{-6}$	
Crystal	*Normal to c Axis*	*Parallel to c Axis*
Alumina, Al_2O_3	8.3	9.0
Mullite, $3Al_2O_3 \cdot 2SiO_2$	4.5	5.7
$ZrSiO_4$	3.7	6.2
$CaCO_3$	−6.0	25.0
SiO_2	14.0	9.0
Graphite	1.0	27.0

which has a negative coefficient in the c direction, $-16.9 \times 10^{-6}(°C)^{-1}$ and a positive coefficient in the a direction, $8.11 \times 10^{-6}(°C)^{-1}$. If we control the amount of this material in a glassy matrix, we can obtain an overall coefficient close to zero.

Another factor in determining thermal shock resistance and heat loss through refractories is thermal conductivity k. It is essentially this characteristic that determines the heat transmitted between the end walls of a unit volume per second. The equation for heat flux is

$$J = k\frac{\Delta T}{\Delta X}$$

(See Chapter 4 and Fig. 4.18.)

If the conductivity is high, the temperature difference between the surface and the underlying layers will be small. Therefore, this will reduce the tendency to spall. The excellent resistance to thermal shock of graphite is due to its high thermal conductivity, whereas that of fused silica is due to the low coefficient of expansion.

Hardness and strength are important in some cases. We should review the measurement of these properties in ceramics. Two of the methods for testing hardness in metals, the Brinell and Rockwell tests, are not used for ceramics because the material would fracture. However, the Vickers test, discussed in Chapter 3, is used. A small, carefully polished diamond pyramid indenter is used with very light loads (below 100 g). Below 500 the Vickers scale is about the same as the Brinell scale, but above 500 the corresponding Brinell number is lower (Fig. 7.24, page 314). Before the development of the microhardness test, the Mohs scratch hardness method was used. Ten minerals ranging from diamond (10) to talc (1) were selected to cover the known spectrum of hardness (Fig. 7.24). The values were chosen on the basis that a specimen of a higher number would scratch one of a lower number; for example, diamond (10) scratched sapphire (alumina, 9). Mineralogists and some ceramists still use this scale.

The tensile strength of a ceramic is usually difficult to determine because of the sensitivity of the material to small cracks, which are almost always present in specimens of appreciable size. Griffith demonstrated long ago that fine glass fibers had tensile strengths many times that of the bulk material because it was possible to produce the fibers relatively free from defects. In brittle materials such as most ceramics, a crack propagates easily under stress because no energy is dissipated in plastic deformation ahead of the notch (as in a metal). There is even a large difference between the strength of specimens with rough surfaces that have been abraded to produce notches and specimens with smooth surfaces. We say, therefore, that the *fracture toughness* of the usual ceramic is low. As we shall see, in some cases, such as Pyroceram, the controlled growth of a very fine crystal network has improved the situation. (See Chapter 13 for a discussion of fracture toughness.)

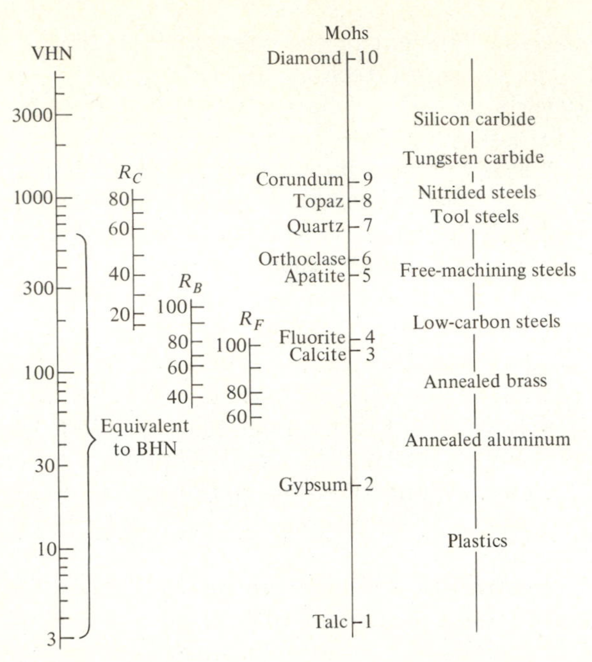

Fig. 7.24 *Interrelationship of different hardness scales. (Readings on carbides were obtained with special indenters.)*

An alternative to the tensile test for brittle, hard materials is a bending test. This test uses a small beam about 1 by 0.4 by 0.4 in. (2.54 by 1.02 by 1.02 cm), loaded at the center and supported beneath near the ends. The breaking stress at the outermost fibers can be calculated from the simple beam formula, which is based on the absence of plastic deformation of the material in the test (Fig. 7.25). This is called the *transverse rupture strength*. It is roughly comparable to the tensile strength. Also, although the tensile strength may not be high, the compressive strength can be excellent. As we shall see in the next chapter, high compressive and low tensile strengths lead to materials such as tempered glass and prestressed concrete. Ancient civilizations did not have the advantage of many of our modern materials; however, they did know how to use ceramics. Since ceramic structures would easily fail in tension, they developed the arch, which was architecturally pleasing and from an engineering standpoint was loaded in compression.

Now let us consider a few of the more common ceramics.

7.10 Simple ceramic materials

Silica and silicates are by far the most important ceramic materials. Therefore, just as we gave iron a good deal of attention as a base for many metallic materials, we should investigate silica carefully. Under equilibrium condi-

Fig. 7.25 *Bend test for specimens of ceramics and powder metals. Centers of the two bottom supports are 1 in. (2.54 cm) apart. Load is applied to top member, which places the bottom of the beam in tension. The tensile stress is given approximately by the formula* $S = 3PL/(2bh^2)$, *where* S = *transverse rupture strength, psi* (kg/mm^2); P = *load required to fracture, lb (kg);* L = *length of span, in. (mm);* b = *width of specimen, in. (mm); and* h = *thickness of specimen, in. (mm).*

tions and atmospheric pressure we encounter three allotropic forms as we cool liquid silica: cristobalite at 3110 to 2678°F (1710 to 1470°C), tridymite at 2678 to 1598°F (1470 to 870°C), and quartz below 1598°F (870°C). These phase changes are extremely sluggish, because the atomic rearrangements are far more complex than in the $\gamma \rightarrow \alpha$ transformation in iron. We have already mentioned that if we cool liquid silica at a moderately fast rate, we obtain none of the equilibrium structures but glass instead. It is also possible to retain a high-temperature phase such as cristobalite at room temperature or, conversely, to melt quartz without transforming to the intermediate phases.

We need to distinguish clearly between these sluggish reactions involving extensive rearrangement and simple *displacive* transformations, which take place rapidly within the three crystal structures at relatively fixed temperatures (Fig. 7.26, page 316). The $\beta \rightarrow \alpha$ quartz transformation is rapid and will shatter a roof of silica brick if uncontrolled. The unit cells of the crystalline forms are fairly complex, and we don't need to cover them in detail. As an example, Fig. 7.27 (page 316) shows structures of β cristobalite and α quartz. Note first that in cristobalite the silicon atoms are in the diamond

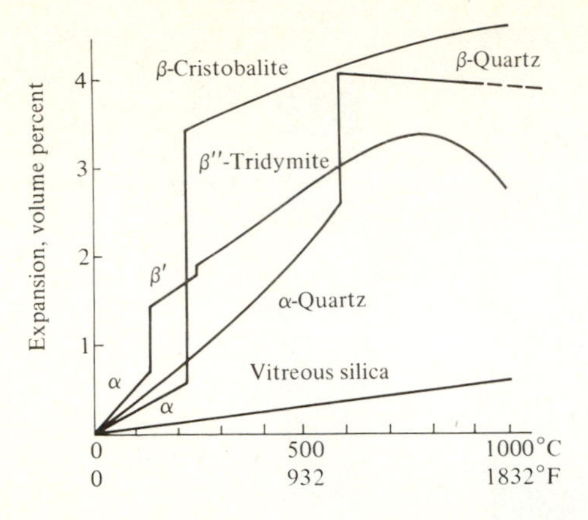

Fig. 7.26 *Changes in volume with temperature in silica structures*

structure (Fig. 7.18), and then that there are four oxygen atoms around each silicon atom in the characteristic tetrahedron.

Silica This material, SiO_2, is used in refractory brick in furnaces where high temperatures are encountered, as in steel making (3000°F, 1649°C). The brick is made from silica. If a small amount (2%) of $Ca(OH)_2$ is added, it forms a liquid with a lower melting point. This cements the grains during firing and also catalyzes the transformation of some of the quartz to a mixture of cristobalite and tridymite. The result is better expansion characteristics than if the brick were composed of all quartz. However, even this mixture has to be heated carefully to avoid cracking from thermal shock.

Quartz is also used for oscillating crystals of known frequency, such as those used in some radio equipment (see Chapter 14). Although a large quantity of quartz is mined, the Bell Telephone Laboratories have developed methods for growing excellent crystals.

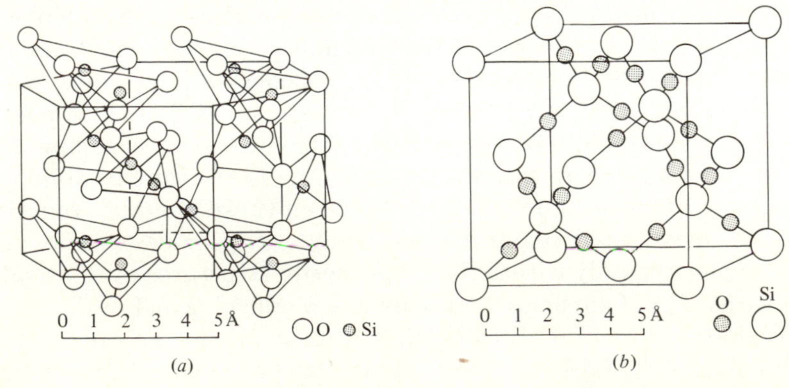

Fig. 7.27 *(a) Structure of α quartz (a form of SiO_2). (b) Structure of β cristobalite (a form of SiO_2).* [Part (a) after W.L. Bragg and Gibbs. Part (b) after Wyckoff]

Alumina Al_2O_3 (Fig. 7.19) is widely used as a raw material in ceramic mixtures as well as in the pure state. Single crystals of pure alumina or of ruby and sapphire, which are colored by minute quantities of chromium, iron, and titanium, are encountered in nature. However, the material for optical uses, as in the ruby laser (Chapter 16), is produced by melting alumina powder in an oxyacetylene flame and building up a single crystal from the droplets.

Because of its high melting point, alumina is widely used as a refractory. Grains of relatively high-purity alumina are pressed into a shape and sintered at high temperatures to produce brick, tubes, and crucibles. Small amounts of flux are used. In more recent applications alumina tubes have been used for sodium vapor illuminating lamps because of their resistance to attack and their optical transmission of 90%. Also, alumina has been used in cutting tools for machining metal because of its high hardness.

Magnesia MgO is also an important raw material and refractory. The unit cell is the sodium chloride structure (Fig. 7.12). In many cases the refractories are prepared by heating the mineral dolomite, a carbonate of magnesium and calcium, $(Mg,Ca)CO_3$, to eliminate CO_2. This class of brick is known as *dolomite refractory*. The advantage is that in making steel it is often necessary to use a slag which is high in CaO on the surface of the metal. If the walls of the furnace or ladle are made of SiO_2, they are fluxed by the CaO to form a low-melting-point glass and therefore are gradually dissolved. However, there is little reaction with dolomite refractories, since no low-melting-point materials are formed.

Silicates Silica reacts with Al_2O_3, MgO, CaO, and FeO, as well as with Na_2O and K_2O, to form many useful compounds including the glasses, which will be discussed last. To illustrate the range of properties and how they depend on the structure, let us consider the familiar minerals asbestos, mica, and clay.

ASBESTOS (COMPLEX HYDROUS MAGNESIUM SILICATES) The original building block for silica is the tetrahedron SiO_4^{4-} (Fig. 7.3). Now consider adding several of these islands to make a single or double chain (Figs. 7.28 and 7.29, pages 318 and 319). Unsatisfied oxygen atoms lie only at the edges of chains and can be bonded with foreign atoms. In asbestos, Mg^{2+} ions can join these chains, and these chains can also be terminated by OH^- ions (from water). It is natural, therefore, for the structure to be fibrous. When stress is applied, the chains do not fracture, although failure between chains can occur.

If the material is being considered for high-temperature vacuum systems, the fact that asbestos is hydrous and hence can be dehydrated at elevated temperatures should be remembered.

MICA This material is a complex silicate in which the bonding produces a sheet structure, as for example in muscovite, $[KAl_3Si_3O_{10}(OH)_2]_2$. The double chain (Fig. 7.29) is extended into a plane (Fig. 7.30, page 320). Thus there is

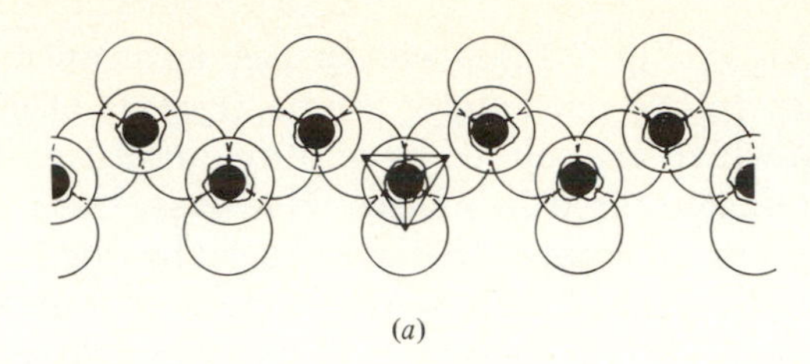

(*a*)

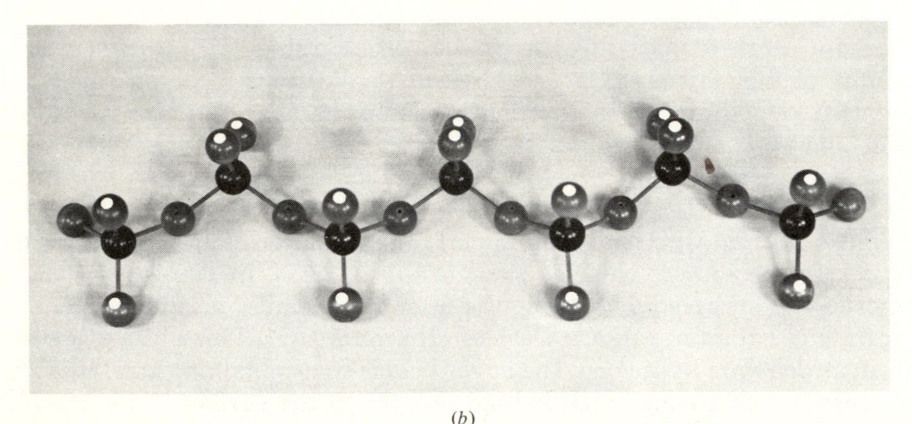

(*b*)

Fig. 7.28 *A single-chain silicate structure. (a) Sketch. (b) Photograph of silicate model. Oxygen atoms with white dots have bonds that are still unsatisfied.* [Part (a) from L.H. Van Vlack, *Elements of Materials Science*, 2d ed., Addison-Wesley Publishing Company, Reading, Mass., 1964, Fig. 8-11, p. 212]

strong silicon-oxygen bonding in two dimensions rather than in a line. The extra bonds for the oxygen atoms above the sheet plane are again satisfied by foreign ions. In this case we have platelike cleavage, enabling the mica to be split into very thin sheets.

CLAY This material has a sheet structure like that of mica. Kaolinite, $[Al_2Si_2O_5(OH)_4]_2$ (Fig. 7.31, page 320), for instance, has alternating layers of Si^{4+} ions and Al^{3+} ions, but the $SiO_4^{\ 4-}$ tetrahedra are still distinct. The complex unit cell leads to the sheetlike cleavage and greasy feel of clay. This structure also accounts for the excellent moldability of this mineral when water is added. Water molecules, being polar, can attach themselves to the clay layers by van der Waals forces, providing plasticity. Too much water allows for water-to-water bonds and thus a loss in plasticity or a soupy consistency.

Mullite Important ceramics are encountered at various percentages of silica and alumina, as the phase diagram in Fig. 7.32 (page 321) shows. When a fire clay (kaolin) is heated, water is given off and other complex changes take place, so that above 1595°C the mineral mullite and a liquid are present. This limits the use of alumina-silica refractories with less than 72% Al_2O_3 to temperatures below about 1550°C. However, above 72% Al_2O_3 we have mullite

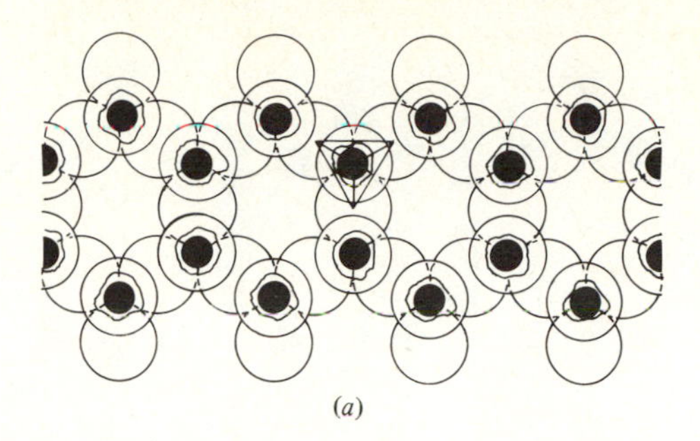

(a)

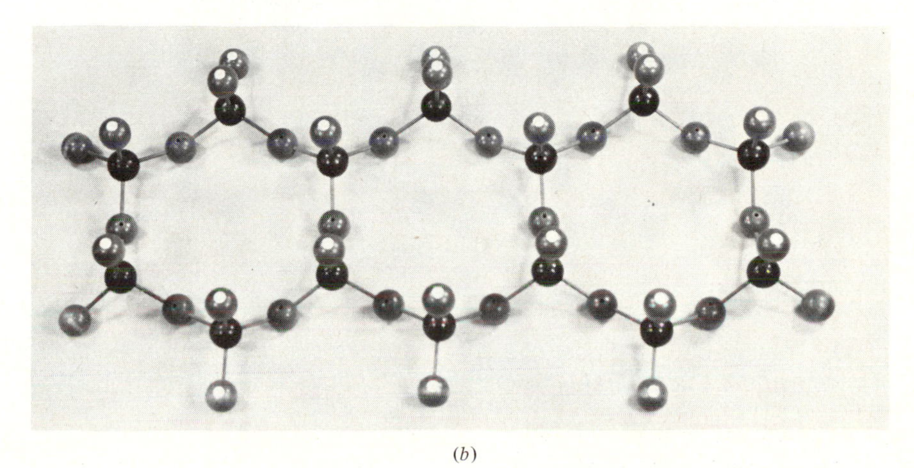

(b)

Fig. 7.29 *A double chain of* SiO_4 *tetrahedra. (a) Sketch. (b) Photograph of silicate model. Again the unsatisfied oxygen atoms are shown with white dots.* [Part (a) redrawn from L.H. Van Vlack, *Elements of Materials Science*, 2d ed., Addison-Wesley Publishing Company, Reading, Mass., 1964, Fig. 8-12, p. 212]

($3Al_2O_3 \cdot 2SiO_2$) or mullite plus corundum (Al_2O_3) with a solidus at 1840°C. Therefore, the high-alumina refractories are widely used for steel making (1600°C). It should be emphasized that a relatively slight change in Al_2O_3, from 70 to 80%, for example, changes the solidus by 255°C in a very important temperature range. For this reason mullite is an important refractory.

Spinels There are two broad uses of spinels, $A^{2+}B_2^{3+}O_4$ (where A and B are metal ions): in refractories and in the electrical industry. In refractories the phase $MgAl_2O_4$ is found in the MgO-Al_2O_3 diagram (Fig. 7.33, page 321), at an equal molar ratio of MgO and Al_2O_3, just as we find mullite in the Al_2O_3-SiO_2 system at a 3:2 molar ratio. The material has a greater resistance to thermal shock than MgO. It also has good resistance to most slags.

The electrical industry is interested in the spinel structure because of its magnetic properties. Magnetite, $FeFe_2O_4$, or lodestone is a naturally occurring spinel. In contrast to the refractories, the electrical industry is inter-

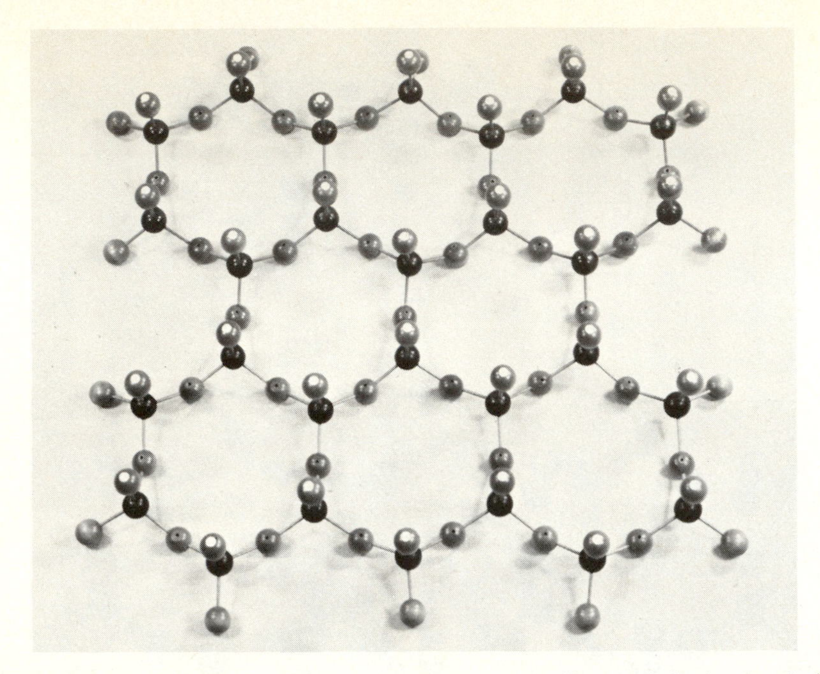

Fig. 7.30 *Photograph of a model of a sheet-structure silicate. Only the fourth (upper) oxygen atoms have unsatisfied bonds (light-colored dots).*

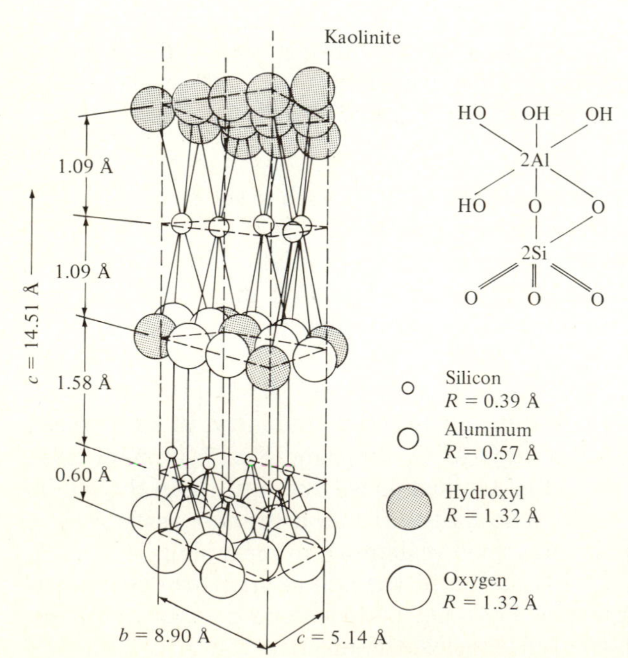

Fig. 7.31 *Sheet structure of kaolinite clay. (Note the planes of potential easy cleavage.)* (After W.E. Hauth)

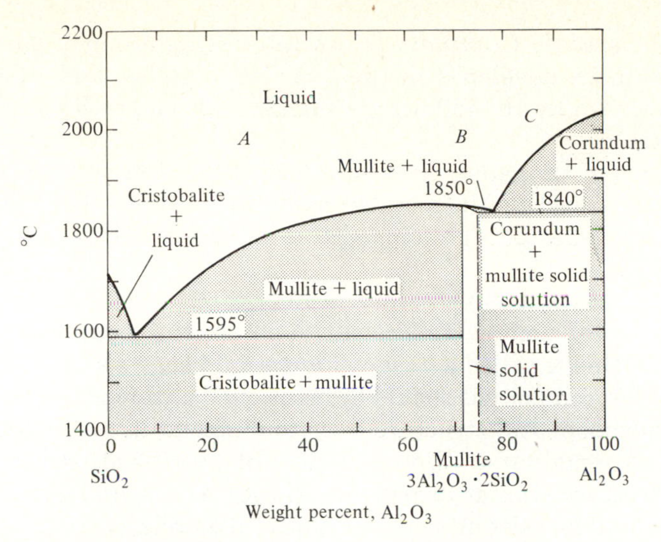

Fig. 7.32 $SiO_2-Al_2O_3$ *phase diagram. A = fire clay composition, B = mullite refractory, C = high-alumina refractory.*

ested only in those spinels with transition elements that lead to the formation of magnetic materials such as $CoFe_2O_4$, discussed in Chapter 15.

Barium Titanate This material, $BaTiO_3$, has the perovskite structure discussed earlier (Fig. 7.20) with barium instead of calcium. Its electrical uses depend on the fact that the Ti^{4+} ion does not fit at the center of the unit cell formed by barium and oxygen ions. This results in an off-center charge for the cell. We shall see how this leads to use in transducers in Chapter 14.

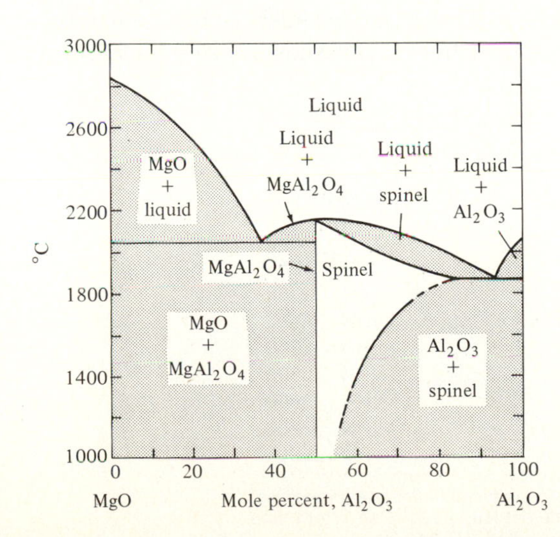

Fig. 7.33 *Phase diagram for* $MgO-Al_2O_3$. *The spinel-type structure is important as a refractory and in ferrimagnetic materials.*

Other Ceramic Compounds An interesting group of hard refractory materials has been developed in the carbides, nitrides, and silicides and in graphite, as shown by the following examples, although the total tonnage is not very large.

Carbides of tungsten, titanium, tantalum, and chromium are made commercially by reacting the metal or oxide with carbon. The others are usually produced by special high-temperature processes. Table 7.4 gives typical hardness values.

In many cases it is necessary to heat a mixture of the carbides, etc., with a softer metal such as 10% cobalt to obtain the toughness necessary in a cutting tool. This is called a *cermet*. Figure 7.34 shows a typical microstructure. A few other compounds should be mentioned at this point. Boron carbide, B_4C, has a rhombohedral structure close to diamond in microhardness, VHN 3700, and is extremely inert. Boron nitride, BN, on the other hand, has a soft hexagonal structure similar to graphite. However, it is an electrical insulator and more oxidation-resistant. Like graphite, it is relatively easy to machine because of its structure. Other carbides and nitrides have generated interest because their high melting points and high hardnesses have suggested possible uses ranging from nose cones of space vehicles to refractories and cutting tools.

Carbon This can be produced in many forms important in ceramics. Particles of carbon can be bonded with pitch, and the mixture can be heated to decompose the pitch to "amorphous carbon." When this mixture is heated to a higher temperature, as by having heavy electric currents passed through it, graphite crystallizes in the mass. The layered structure has strong covalent bonds within the sheets (Fig. 7.35) but semimetallic bonds between the sheets. This arrangement leads to cleavage between the sheets, giving excellent properties as a lubricant as well as semimetallic properties such as high electrical and thermal conductivity parallel to the sheets. Between these sheets there

Table 7.4 MICROHARDNESSES OF SEVERAL CERAMIC MATERIALS

Material	*Microhardness**
Hardened tool steel	VHN 600
Quartz	VHN 1250
WC	VHN 2400
Al_2O_3	VHN 2800
TiC	VHN 3200
SiC	VHN 3500
TiB_2	VHN 3400
B_4C	VHN 3700
Diamond	VHN ~8000

*These values are somewhat different from those given in Fig. 7.24 because of the different load used. Also, micro- and macrohardness values are different (Chap. 3).

Fig. 7.34 *Cobalt-bonded tungsten carbide material for making tools. Gray crystals are tungsten carbide; matrix is cobalt. To make this structure, tungsten carbide crystals were placed in a ball mill with cobalt powder. After milling, 1% paraffin was added. Then the mixture was pressed in a die and sintered for 0.5 hr at 2650°F (1452°C). The cobalt melted and the entire mass was sintered; 1500×, alkaline ferrocyanide etch.*

are barriers to electron motion. In the diamond form the perfect covalent bonding throughout the mass leads to the highest hardness known. However, the cleavage on (111) planes is perfect, so that a crystal may be split with a sharp edge. Nonetheless, diamond tools and wheels impregnated with diamond are invaluable in shaping other hard ceramics.

Glass Ceramics (Pyroceram) We have discussed the pronounced effects of small cracks in lowering the strength of glass. A crack is particularly disastrous in glass because there are no "crack stoppers" such as grain boundaries and the mass is homogeneous. Recently much stronger materials have been developed by producing a glass shape first, then heat-treating to develop

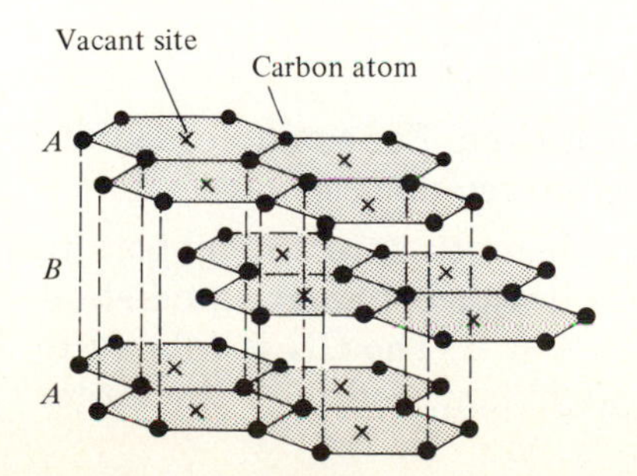

Fig. 7.35 *The layer structure of graphite (carbon)*

a very fine crystalline phase to strengthen the glass. The material can be made very resistant to thermal shock as well if the precipitate has a very low coefficient of expansion. This is a new field called *glass ceramics*. The Pyroceram materials are outstanding examples of products that have been developed. To obtain the fine crystalline precipitate, nucleants such as titanium dioxide may be used. The frontispiece for this chapter is a photomicrograph showing nucleation of a Pyroceram by rutile (TiO_2).

SUMMARY

The bonding in ceramics differs from that in metals in two important aspects. Ionic and covalent bonds, instead of metallic bonds, are predominant. The role of the coordination number is important because most ceramics are made up of different size atoms.

The density of a ceramic crystal can be calculated from its unit cell dimensions, as can the density of a metal.

The unit cells of many ceramic materials can be described by reference to a few key structures: sodium chloride, cesium chloride, calcium fluoride, alumina, perovskite, spinel, and silicate groups.

The glasses are another important family of ceramics. Athough their structure is a well-bonded three-dimensional network in space, the side-by-side unit cell grouping of a crystal is absent. The small units such as SiO_4^{4-} tetrahedra are present and may be bonded to each other or to ions of other metals at the corners.

Solid solutions are encountered in crystals and in glasses. Lattice vacancies are important and can be related to the presence of ions of different charges.

Thermal properties are often important in conjunction with mechanical properties because of the possibility of spalling and the use of ceramics for insulation.

Solid-state transformations occur. Two types are distinguished: the displacive, which requires only minor shifts in ion positions, and the nucleation and growth of a new phase, such as quartz to cristobalite.

The structure and properties of simple ceramic materials are described.

DEFINITIONS

Cleavage The fracture path when a material is broken. For example, sodium chloride cleaves into cubes, whereas glass has no regular cleavage.

Coordination number, CN The number of equidistant nearest neighbors to a given atom in a unit cell. The sodium ion, for instance, has six equidistant chloride ions in sodium chloride; therefore, CN = 6. The CN is related to the radius ratio: ionic radius of the smaller atom/ionic radius of the larger atom.

Glass A noncrystalline solid that softens on heating rather than showing a sharp melting point. The structure is composed of small units such as SiO_4^{4-} tetrahedra, linked to each other or by other ions in a mostly random three-dimensional network.

Glass characteristics

Working point The temperature at which glass is easily formed, 10^4 poise.

Softening point The temperature at which glass sags appreciably of its own weight.

Annealing point The temperature at which locked-up stresses can be relieved.

Strain point The temperature at which glass becomes rigid.

Transition temperature The temperature at which, on cooling, the rate of contraction changes to a lower value.

Octahedral void or site The space within a group of *six* atoms which, when connected, form an octahedron. For example, the hole in the center of an FCC unit cell is an octahedral void. The *tetrahedral void* is the center of four atoms making a tetrahedron.

Refractory A material that resists exposure to high temperatures, as in a furnace lining or a gas turbine blade.

Spalling The cracking and breaking away of surface layers resulting from exposure to heat or cold, sometimes combined with a chemical action.

Viscosity coefficient The force required to slide one layer of a liquid past another under specified conditions. This will affect the flow rate of liquid glass in a mold or tube because the layer next to the tube is stationary and the inner layers must flow past it. The unit is the poise (1 poise = 0.1 Pa-sec = 0.1 N · s/m²).

PROBLEMS

7.1 Determine the annealing temperature for the glass in Example 7.1. (Sections 7.1 through 7.3)

7.2 We said in Sec. 7.2 that the viscosity–temperature relationship for glass was similar to the diffusion coefficient–temperature relationship.

$$D = D_o e^{-Q/RT} \qquad \frac{1}{\text{Viscosity}} = Ae^{-B/T}$$

For glass 6 in Fig. 7.2 (low-expansion borosilicate), develop the general viscosity-temperature equation by using the work point and the annealing point. Check the equation by using the upper softening point (10^7 N · s/m²) to determine the softening temperature. (Sections 7.1 through 7.3)

7.3 From the discussion of a nonfixed working point for glass as in Fig. 7.2, suggest why the annealing point may also not be a fixed point. (Sections 7.1 through 7.3)

7.4

Type	SiO_2	$Na_2O + CaO$	Al_2O_3	B_2O_3	PbO
A	99.9	—	—	—	—
B	96.	—	—	4	—
C	73.6	25.4	1.0	—	—
D	35.0	7.0	—	—	58

Using the code numbers of the compositions above (may be used twice), which glass would have: (Sections 7.1 through 7.3)

1. the lowest cost in a finished part (not necessarily the lowest raw material cost)
2. the highest melting point
3. the highest index of refraction
4. the best thermal shock resistance
5. which compositions have glass formers (list all)?
6. which compositions have modifiers (list all)?
7. which compositions have intermediates (list all)?

7.5 Calculate the density of CsCl (Fig. 7.13). Use the ionic radii given in Table 7.2. (Sections 7.4 through 7.6)

7.6 Calculate the density of CaF_2 (Fig. 7.17). [*Note:* There are twice as many F^- ions per unit cell as Ca^{2+} ions.] Would you expect your value to be higher or lower than the handbook value? Why? (Sections 7.4 through 7.6)

7.7 Check the limiting value for CN's of 6 and 8. (Refer to Fig. 7.14.) (Sections 7.4 through 7.6)

7.8 Check the CN for CsCl (Fig. 7.13). What is the atomic packing factor (i.e., what percentage of space is occupied by the ions)? What other ions might substitute for Cs^+ and still give the same CN? (Sections 7.4 through 7.6)

7.9 What type of structure, that is, NaCl, etc., would you expect for the following materials: KCl, CsBr, $NiAl_2O_4$, Fe_3O_4, UO_2, $CaZrO_3$? Indicate which structure given in the text each material would resemble. (For example, KCl has either the NaCl or CsCl structure. Since the radius ratio fits the CN = 6 group, KCl has the NaCl structure.) (Sections 7.4 through 7.6)

7.10 Calculate the hole size, i.e., sphere diameter, that would fit inside the barium titanate ($BaTiO_3$) unit cell if the barium and oxygen ions touched across the face diagonal. [*Hint:* Calculate a_0 and note that the diameter of the hole equals $a_0 - 2r_{O^{2-}}$.] Compare the hole size with the ionic diameter Ti^{4+}. How can the titanium position adjust to the misfit? (See Chapter 14 on electrical properties for further explanation and the utility of this phenomenon.) (Sections 7.4 through 7.6)

7.11 Titanium nitride (TiN) has the same crystal structure as NaCl. The side of the unit cell of TiN is 4.235 Å. (0.4235 nm). What is its density? (Sections 7.4 through 7.6)

7.12 Study the characteristics of strontium oxide (SrO). Show all calculations. (Sections 7.4 through 7.6)

a. What is the coordination number? *b.* What is the lattice parameter? *c.* Calculate the density.

7.13 CaO might be expected to form a CsCl structure. (Sections 7.4 through 7.6)

a. Show by calculation the coordination number. *b.* Indicate two other divalent ions that might substitute for Ca^{2+} and still maintain the same coordination number. *c.* Calculate the density of CaO.

7.14 From the following data calculate the ionic radius of fluorine: (Sections 7.4 through 7.6)

Density CaF_2 = 3.18 g/cm^3 At. wt. Ca = 40.1
r (ionic) Ca = 1.06 Å At. wt. F = 19.0
$N_A = 6.02 \times 10^{23}$ atoms/at. wt

7.15 From the following data calculate the ionic radius of barium in cubic $BaTiO_3$, density 5.97 g/cm^3. (Sections 7.4 through 7.6)

	Atomic Weight	*Ionic Radius (Å)*
Ba	137.3	*X*
O	16	1.32
Ti	47.9	0.64

$N_A = 6.02 \times 10^{23}$ atoms/at. wt.

Assume Ba^{2+} at cube corners; Ti^{4+} at $\frac{1}{2}, \frac{1}{2}, \frac{1}{2}$; oxygen at face centers.

7.16

	Ionic Radius (Å)	*Atomic Radius (Å)*	*Atomic Number*	*Atomic Weight*
U^{4+}	1.05	1.38	92	238
O^{2-}	1.32	0.60	8	16

UO_2 crystallizes in the CaF_2 type structure. Calculate the density in grams per cubic centimeter. What is the coordination number of the U ions? What is the coordination number of the O ions? (Sections 7.4 through 7.6)

7.17 Mg_2Si has an "anti" CaF_2 structure. It is called "anti" because the metallic ions (Mg) are at the positions of the nonmetallic ions (F) in CaF_2 and the silicon ions are at the calcium positions. Calculate the density (g/cm^3) of Mg_2Si from the following data.

	Ionic Radius, Å	*Atomic Weight*
Mg	0.78	24.32
Si	0.39	28.09

What is the CN of a silicon ion at 0,0,0, for example? Illustrate by a simple sketch. Which ions touch the silicon? (Sections 7.4 through 7.6)

7.18 The ceramic compound TiC is even harder than martensite. Therefore it is very valuable as a cutting tool in industry. It is cubic. The sketch (page 328) shows the atom arrangements on the (100), (010), and (001) planes

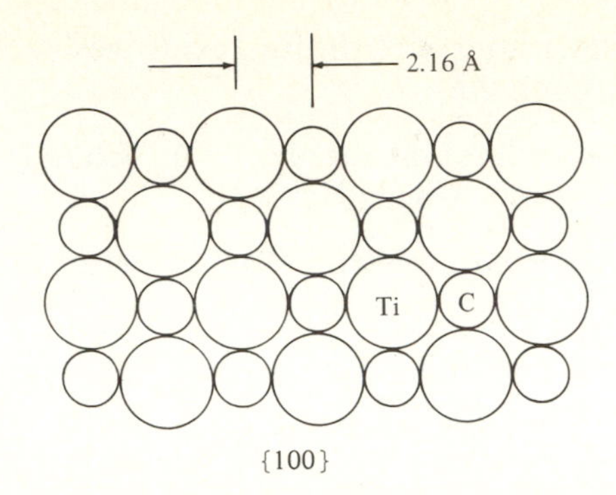

(facial planes). The center-to-center distance between the closest titanium and carbon atoms is 2.16 Å. Determine the following.

a. Is the arrangement of the titanium atoms in this structure simple cubic, BCC, or FCC?
b. The coordination number for titanium in this unit cell.
c. The unit cell volume in $Å^3$.
d. The closest center-to-center approach of two titanium atoms in angstroms. Each titanium atom is related to how many other titanium atoms at this same distance?
e. Is the packing similar to CsCl, NaCl, or CaF_2?
f. Why is the material so hard? (Sections 7.4 through 7.6)

7.19 Sketch the structure of lead-zirconium-titanate (PZT). (Sections 7.7 through 7.10)

7.20 Make a sketch and show whether Fe_2SiO_4 (fayalite) and $MgSiO_3$ (enstatite) are island, chain, or sheet type silicate structures. (Sections 7.7 through 7.10)

7.21 Zirconia (ZrO_2) is often stabilized with calcium to provide an important refractory. The basic cell is ZrO_2 with 1 Ca^{2+} ion present for every 10 Zr^{4+} ions. Will the vacant sites be anion or cation? What percentage of the total number of all sites will be vacant? (Sections 7.7 through 7.10)

7.22 Draw a graph showing the softening temperature (temperature at which liquid begins to appear) as a function of composition in refractories made of different amounts of SiO_2 and Al_2O_3. Indicate the significance of these values. (Sections 7.7 through 7.10)

7.23 Kaolinite clay, $Al_2(Si_2O_5)(OH)_4$, is heated to drive off the hydrogen as water. (Sections 7.7 through 7.10)

a. What is the percentage loss in weight?
b. What would be the liquidus and solidus temperatures of the resultant mixture of Al_2O_3 and SiO_2 (Fig. 7.32)?

7.24 Referring to Fig. 7.26, draw a graph showing the thermal expansion of a silica brick made up of approximately equal quantities of quartz, tridymite, cristobalite, and vitreous silica. Would this be better than an all-quartz brick? Why? (Sections 7.7 through 7.10)

7.25 A spalling resistance index for refractories is given by the formula

$$\text{Spalling resistance} = \frac{kS}{\alpha E C_p \rho}$$

where k = thermal conductivity, S = tensile strength of the material, α = coefficient of thermal expansion, E = Young's modulus (modulus of elasticity), C_p = specific heat, and ρ = density.

Explain how each term contributes to a reduction or increase in the tendency to spall. (Sections 7.7 through 7.10)

7.26 The following general rules apply to cleavage:

1. With layer lattices such as those in mica and MoS_2, the cleavage is parallel to the layers.
2. Cleavage planes are the most widely spaced.
3. Cleavage does not cut through radicals or ionic complexes.
4. Cleavage occurs so as to expose planes of anions if this does not violate rule 3.
5. In AX crystals, cleavage occurs on the (100) planes.

Explain the following observations using rules 1 to 5. (Sections 7.7 through 7.10)

a. The cleavage of NaCl is cubic.
b. The cleavage of CaF_2 is octahedral.
c. The cleavage of graphite is parallel to the basal plane.
d. The cleavage of zinc is on the (0001) planes.

7.27 Many minerals and glasses with complex formulas such as $LiAlSiO_4$ are called "stuffed derivatives" of SiO_2. First, Li^+ and Al^{3+} substitute for one Si^{4+}. Second, the Al^{3+} ion takes the place of the Si^{4+} in the unit cell and the Li^+ ion is "stuffed" into a hole. In the allotropic form of silica called tridymite, important stuffed derivatives are $KNa_3Al_4Si_4O_{16}$ (nepheline) and $KAlSiO_4$. Which ions replace silicon in the tridymite lattice and which are stuffed?

One grade of Pyroceram has the formula of β spodumene, $Li_2O \cdot Al_2O_3 \cdot 4SiO_2$, plus quartz. Which are the stuffed ions? (Sections 7.7 through 7.10)

7.28 Both kaolinite clay $[Al_2(OH)_4\ Si_2O_5]_2$ and talc $Mg_3(OH)_2(Si_2O_5)_2$ are sheet-type structures, yet talc is softer and does not have the same "plastic" properties when water is added. Explain. (Sections 7.7 through 7.10)

7.29 Schematically show how bonding occurs in the single-chain silicate structure $MnSiO_3$. (Sections 7.7 through 7.10)

7.30 Forsterite is a mineral compound of MgO and SiO_2 that can be treated as $2MgO \cdot SiO_2$ or Mg_2SiO_4. Sketch this "island structure." (Sections 7.7 through 7.10)

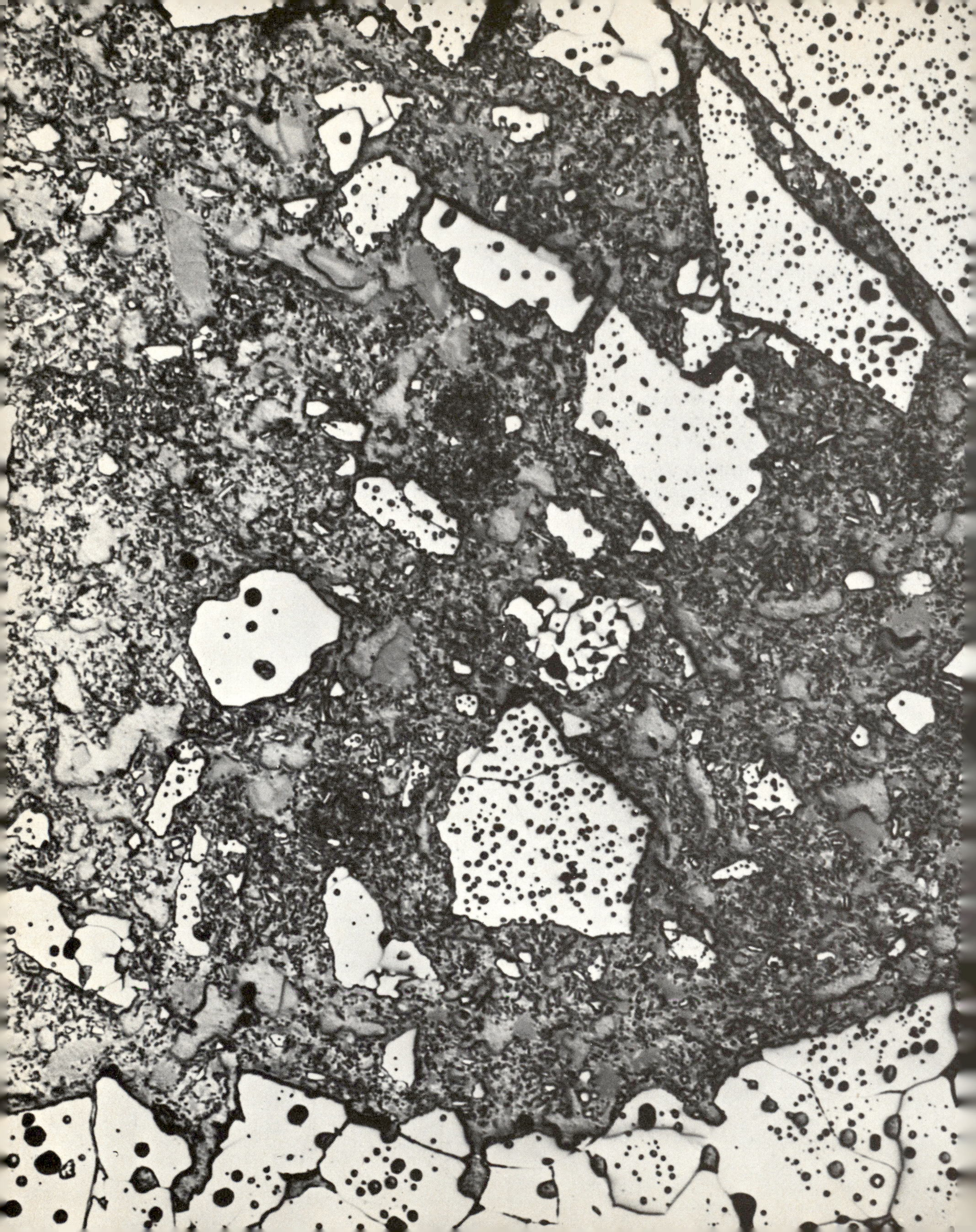

8

PROCESSING, SPECIFICATIONS, AND APPLICATIONS OF CERAMICS

THIS illustration shows a typical 90% alumina refractory. The brick consists of coarse synthetic alumina (white) bonded by a glassy aluminosilicate matrix. Magnification 100×.

In this chapter we shall discuss common processing methods for different types of ceramic products and then the structures and uses of the different products.

8.1 General

Ceramics have a tremendous range of applications, from common building brick to delicate porcelain to special optical glass. Discussing the processing and applications of these products in one chapter may seem difficult, but since we are now familiar with most of the important structures, we need only see how these structures or mixtures of them are assembled into usable shapes.

We need to study the processing of ceramics in more detail than that of metal products because, in general, ceramics are not machinable or forgeable from bar or plate stock into a finished part. In most cases we have to take the raw materials, press them into a shape such as a brick or plate or magnet, then develop the desired structure in the shape by heating. If we reflect on this a little, we will see that the reason is that most ceramics are hard, precluding machining, and brittle, eliminating cold working. (There are a few exceptions, such as the machining of pressed ceramic before firing in the "leather hard" condition and soft materials such as graphite and boron nitride. Also, the formability of glass while it is hot is extremely useful.)

Having noted these exceptions, we must conclude that in most cases to use a ceramic successfully, we must produce both the finished part and the desired structure in a closely organized sequence. One more example: Beer can manufacturers buy steel or aluminum sheet from another company and form cans at their leisure, whereas bottle makers must start with raw materials, melt them, and then form the bottles to exact dimensions in one sequence.

This chapter will discuss first the manufacture and production of materials that are not glass—brick, refractories, art ware, cookware, dinnerware, tile, chemical ware, plumbing fixtures, ceramic bonded abrasives, magnetic ferrites, and ferroelectrics. There is a group of standard processes for the nonglasses, such as pressing and sintering.

Let us list the different processing methods and see how the various products are involved (Table 8.1). Note from this list that the same processes are used for a variety of products. Let us therefore discuss the processes listed, not in great mechanical detail, but to obtain a feel for how these can affect the final structure and shape.

8.2 Molding followed by firing

In this group a shape is formed by different methods and then fired to give it strength.

Slip casting is an interesting and rather unusual method. A suspension of clay in water is poured into a mold (Fig. 8.1). The mold is usually made of plaster of Paris *with controlled porosity,* so that some of the water of the suspension enters the mold wall. As the water content of the suspension decreases, a soft solid forms. The remaining fluid is poured out,

Table 8.1 PROCESSES USED FOR DIFFERENT CERAMIC PRODUCTS

	Molding plus Firing Processes						
Product	*Slip Casting*	*Wet Plastic Forming*	*Dry Powder Pressing*	*Hot Pressing*	*Viscous Fluid*	*Chemical Bonding*	*Single Crystal*
Cemented products		×				×	
Brick		×	×		×		
Refractory and insulation		×	×	×	×	×	
Whiteware	×	×					
Vitreous enamelware		×					
Abrasive wheels			×				
Molds for metal castings	×	×				×	
Special (magnets, laser crystals, etc.)			×	×			×

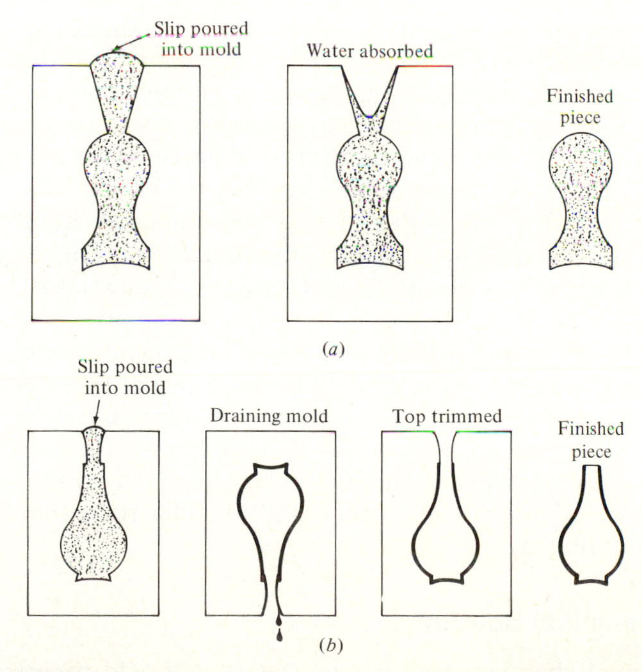

Fig. 8.1 *Slip casting of ceramic shapes.* (W.D. Kingery, *Introduction to Ceramics,* John Wiley & Sons, Inc., New York, 1960, Fig. 3.17, p. 52)

then the hollow form is removed from the mold. The bond at this point is clay-water. Next the part is fired. Although this method is usually used for low production rates, it is important because it is now being employed with fine dispersions of refractory *metal* powders on curved surfaces. A similar technique is used to form a refractory shell on the outside of wax patterns. In this case the pattern is melted away and the mold is fired, leaving a precise cavity in the refractory for subsequent casting of metal.

Wet plastic forming is done by several methods. In one case a wet or damp refractory is rammed into a mold and then forced out to give the required shape. In another case extrusion is used for simple shapes such as brick or pipe (Fig. 8.2). The plastic mass is forced through a die to give a long shape, which is then sliced to the desired lengths. On the other hand, when circular shapes such as plates are to be formed, a mass of wet clay is placed on a rotating potter's wheel and shaped by a tool. This old process has now been highly mechanized.

Dry powder pressing is accomplished by loading a die with a charge of powder and pressing it. The powder usually contains some lubricant such as stearic acid or wax. A variety of shapes can be made by thoughtful design of multipart dies (Fig. 8.3).

After any of the preceding processes, the "green" part is placed in a kiln and fired. During firing the water and volatile binders are driven off, low-melting-point fluxes melt and bond the refractory, and sintering of the refractory grains takes place at 700 to 2000°C. We discussed sintering briefly under powder metallurgy in Chapter 5. Remember that this is a diffusion process. It is especially important in ceramics because the high melting point of the material often makes it impossible to actually melt the material to bond the grains together. Sintering provides the solution, since the diffusion of material results in true bonding between grains.

Hot pressing involves both pressing and sintering simultaneously. The advantages are greater density and finer grain size. The problem is to obtain adequate die life at elevated temperatures. Protective atmospheres are often used.

Figure 8.4 (page 336) shows cold powder pressing and sintering. Here it was necessary to produce a crucible with an internal thermocouple protection tube that could withstand superheated magnesium under an inert gas pressure to prevent its boiling. An MgO-MgF_2 mixture was selected. The shrinkage was very severe and had to be accounted for in the original design. The use of a graphite external die allowed for induction heating to accomplish the sintering.

8.3 Chemical bonding

The advantage of chemical bonding is that it is a cold process and can produce precise dimensions. The most common and important bond is the

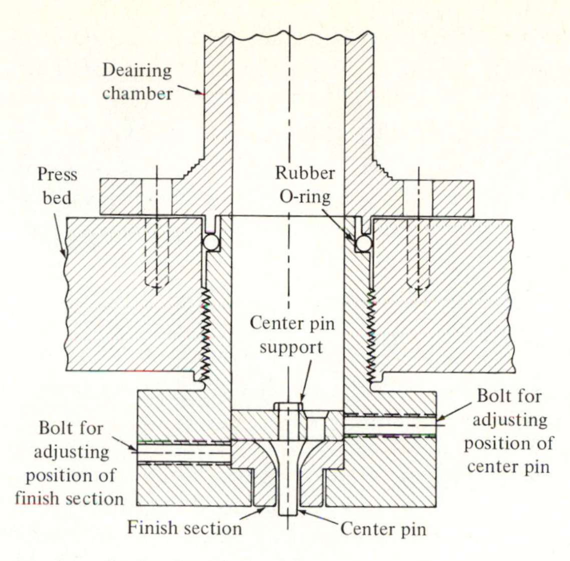

Fig. 8.2 *Piston-extrusion die for forming tubing.* (W.D. Kingery, *Introduction to Ceramics,* John Wiley & Sons, Inc., New York, 1960, Fig. 3.14, p. 48)

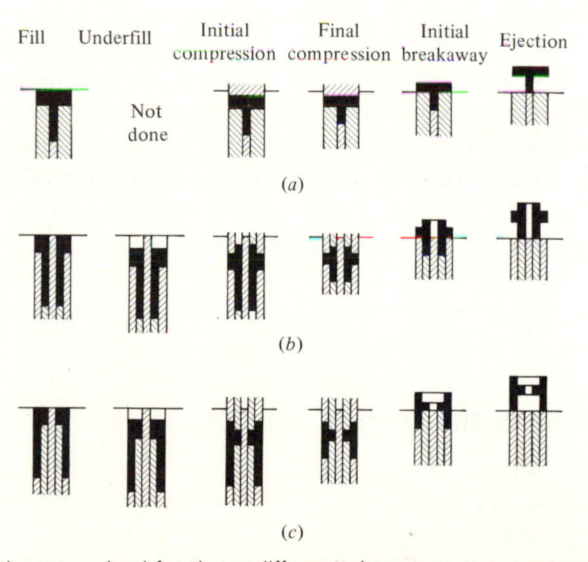

Fig. 8.3 *Die motions required for three different shapes.* (W.D. Kingery, *Introduction to Ceramics,* John Wiley & Sons, Inc., New York, 1960, Fig. 3.9, p. 42)

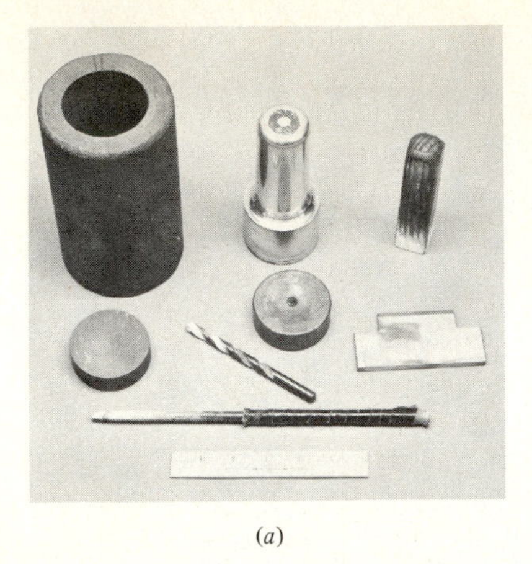

(*a*)

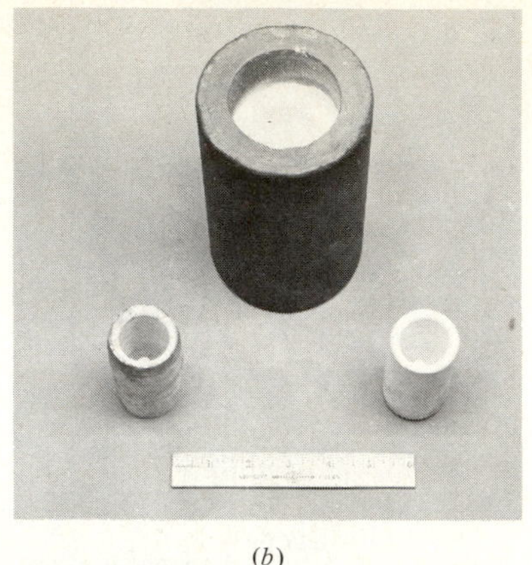

(*b*)

Fig. 8.4 *Objects used in making ceramic shapes* (MgO) *by pressing and sintering. (a) Components of crucible die and fabricating tools. (b) A powdered compact in the graphite mold-susceptor (center). At left, a sintered crucible after removal from the mold. At right, a crucible after resintering in an oxidizing atmosphere. Note the shrinkage of the crucible from its "green" state in the mold-susceptor to the sintered shape.* (Courtesy of Guichelaar, Flinn, and Trojan, University of Michigan)

setting of Portland cement, but other processes involve plaster of Paris, ethyl silicate, or phosphate cements. Section 8.5 describes the structures produced in detail.

8.4 Single crystals

To obtain desired optical, electrical, or magnetic properties, it is often essential to have a single crystal instead of a polycrystalline material. In all commercial cases the solid is crystallized from a melt under controlled conditions. It is most important to nucleate the crystal at a colder point and to avoid the formation of other nuclei. In one case the melt is contained in a tube that passes slowly from the furnace. In another the crystal is started from a nucleus on a rod touching the bath, then literally pulled from the bath. In powder methods for corundum crystals the powder is fused and crystallized on an existing nucleus.

8.5 Cements and cement products

In many ceramic shapes the crystalline phases are held together by the glass phase, so that we can consider glass a special high-temperature cement. In another large class of cements, however, a mixture may be formed into a shape at low temperature. Because of interaction with water, a hydraulic bond develops. This is the field of common cements and plasters.

In general, when water is added, the existing minerals in these cements either decompose or combine with water, and a new phase grows throughout the mass. Examples are the growth of gypsum crystals in plaster of Paris and the precipitation of a silicate structure from Portland cement. It is usually quite important to control the amount of water to prevent an excess that would not be part of the structure and would therefore weaken it.

Portland Cement and High-alumina Cement

There are several grades of these cements, as shown in the ternary diagram in Fig. 8.5. A typical Portland cement will contain 19 to 25% SiO_2, 5 to 9% Al_2O_3, 60 to 64% CaO, and 2 to 4% FeO, whereas a high-alumina cement will be mostly CaO and Al_2O_3. Both types are prepared by grinding clays and limestone of the proper mixture, firing in a kiln, and regrinding.

Several minerals are present in Portland cement: Tricalcium silicate (C_3S), 3 CaO · SiO_2, dicalcium silicate (C_2S), and tricalcium aluminate (C_3A) (see Fig. 8.5). Tetracalcium aluminum ferrite, 4 CaO · Al_2O_3 · FeO, is also present.

The reaction is solution, recrystallization, and precipitation of a silicate structure.† To retard setting, 2% gypsum ($CaSO_4 \cdot 2H_2O$) is added. This reacts with the tricalcium aluminate and reduces shrinkage. The heat of hydration (heat of reaction in the adsorption of water) in setting can be large and can damage massive structures. Low-heat cements are made by reducing the amount of tricalcium aluminate.

High-alumina cement has approximately 35 to 42% CaO, 38 to 48% Al_2O_3, 3 to 11% SiO_2, and 2 to 15% FeO. Setting is produced by the formation of hydrated alumina crystals from the tricalcium aluminate. The

†This is a sheet structure of silicate groups with calcium and oxygen ions in the interstices. Water molecules separate the sheets.

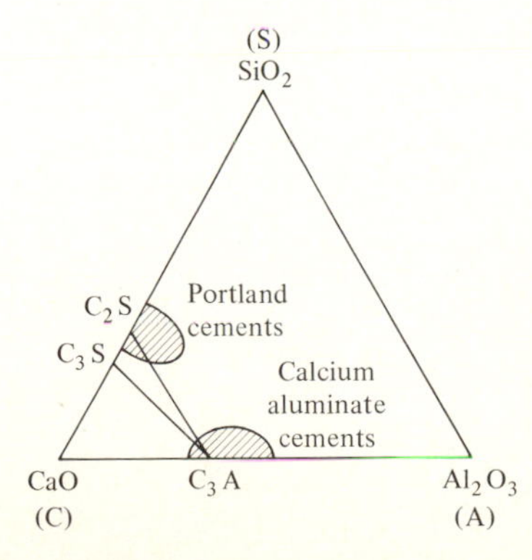

Fig. 8.5 *Cement compositions. [Ternary diagrams are explained in Chapter 4 (see in particular Fig. 4.13).]*

cement is quick-setting and attains in 24 hr the same strength that Portland cement attains in 30 days.

Mortar used in bricklaying combines the effect of Portland cement with a lime reaction. When lime (CaO) is mixed with water, it forms hydrated lime, $Ca(OH)_2$. This reacts with air to form $CaCO_3$, and also reacts with sand to form a calcium silicate. The formation of these new crystals causes bonding. A typical mortar is one part Portland cement, two parts lime hydrate, and eight parts sand by volume.

One of the most important uses of Portland cement is as a binder in the manufacture of concrete. This is treated as a composite material in Chapter 11.

Silicate cements are made from sodium silicate and fine quartz. The quartz is bonded by the formation of a silica gel from the breakdown of the sodium silicate. The process is used in the manufacture of molds and cores in the foundry. The reaction is accelerated by passing carbon dioxide gas through the sand to form sodium carbonate and silica rapidly.

Gypsum cement is used widely in wall construction. The mineral gypsum is heated to change the double hydrate to the hemihydrate: $CaSO_4 \cdot 2H_2O \rightarrow CaSO_4 \cdot \frac{1}{2}H_2O + \frac{3}{2}H_2O$ (gas). The hydrogen and oxygen atoms are an integral part of the crystal structure. When the hemihydrate is mixed with water, the gypsum phase reforms and hardens the shape.

Other cements involving the same type of reaction include magnesium oxychloride and phosphate minerals.

8.6 General packing in solids

Although we have mentioned solid packing before, in discussing powdered metals and earlier in this chapter, we have deferred the analysis of it until now, since it has important applications in the manufacture of ceramic products. Basically we have to precisely define several terms that are illustrated in Fig. 8.6, in which we have some limestone rock in a box.

1. *Porosity*
   Open pores: Spaces into which water can penetrate
   Closed pores: Spaces into which water cannot penetrate
   True porosity: Volume of (open + closed pores)/total volume of box
   Apparent porosity: Volume of open pores/total volume of box
2. *Volume*
   Bulk volume: True volume of rock + (open + closed pores)
   Apparent volume: True volume + closed pores only
3. *Density*
   Bulk density: Mass/bulk volume
   True density: Mass/true volume
   Apparent density: Mass/apparent volume

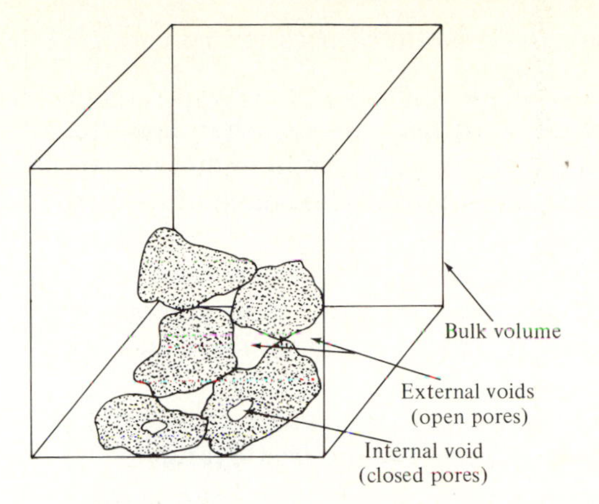

Fig. 8.6 *Relationship between bulk volume, external voids, and internal voids*

or Apparent density: Mass/(true volume + closed-pore volume)

4. *Packing factor*
   True packing factor: True volume/bulk volume or bulk density/true density
   Apparent packing factor: Apparent volume/bulk volume or bulk density/apparent density
5. *Archimedes' Principle*
   Weight in fluid = dry weight − buoyant force, where buoyant force = weight of fluid displaced.

The normal engineering problem is to find methods of optimizing or controlling the packing factor or, more importantly, the porosity, which is 1 minus the packing factor.

EXAMPLE 8.1 In the packaging industry, container shape is important. What is the apparent porosity if basketballs are shipped inflated so that they just fit into the box?

ANSWER With a radius r and a cubic box of side length a,

$$2r = a$$

$$\text{Packing factor} = \frac{\frac{4}{3}\pi r^3}{a^3} = \frac{\frac{4}{3}\pi r^3}{8r^3}$$

$$= \pi/6 = 0.524$$

$$\text{Porosity} = 47.6\%$$

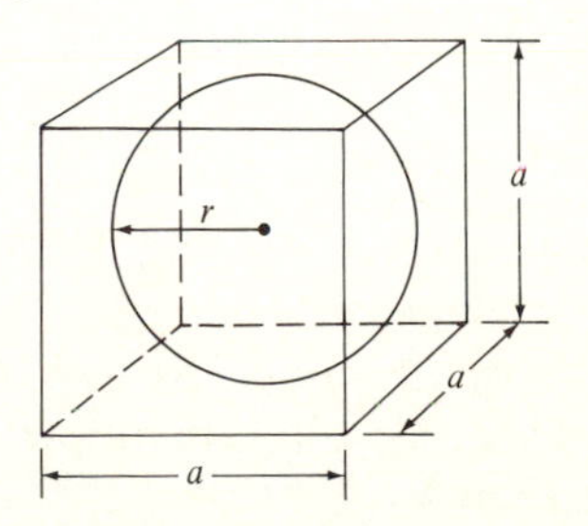

The answer would be the same for any spherical shape placed in a cubic container into which it would just fit.

The next question is: "How can the amount of porosity be changed?" There are essentially three different ways.

1. Shape changes. Bricks can be packed with less porosity than spheres. A greater weight of pencils of hexagonal cross section than of dowels can be placed in a box. Even the honeybee has optimized the shape of its honeycomb.
2. Compression and impregnation. The exclusion of air pockets, as happens when one stuffs a sleeping bag into a tote bag or compresses garbage with a garbage compactor, increases packing. The use of wood fibers and resins to make plywood is an example of impregnation.
3. Mixed sizes. Small spheres can be used to fill the spaces between large spheres. The manufacture of concrete is probably the best example.

EXAMPLE 8.2 We are to produce a small, flat grinding stone by compacting 100 g of fine alumina powder (true specific gravity 3.80) with 10 g of a low-melting powdered glass (true specific gravity 2.20). The green compact measures $10.0 \times 4.75 \times 1.0$ cm. After sintering slightly above the softening point of the glass and cooling, the grinding stone is suspended in water and is found to weigh 72.5 g.

a. What is the percentage true porosity in the green state?
b. What are the apparent density and the closed-pore volume after sintering?
c. If the volume shrinkage in sintering is 10%, what is the open-pore volume?
d. What is the true porosity after sintering?

ANSWER

a. The total weight is 100 g + 10 g = 110 g. The true volume of each component is

Al_2O_3 $$\frac{100 \text{ g}}{3.80 \text{ g/cm}^3} = 26.32 \text{ cm}^3$$

Glass $$\frac{10 \text{ g}}{2.20 \text{ g/cm}^3} = 4.55 \text{ cm}^3$$

$$\text{Total true volume} = 30.87 \text{ cm}^3$$

$$\text{Bulk volume} = 10.0 \times 4.75 \times 1.00 \text{ cm} = 47.5 \text{ cm}^3$$

$$\text{True porosity} = 1 - \text{true packing factor} = 1 - \frac{\text{true volume}}{\text{bulk volume}}$$

$$= 1 - \frac{30.87 \text{ cm}^3}{47.5 \text{ cm}^3} = 0.35 \text{ or } 35\%$$

b. $$\begin{aligned} \text{Weight in fluid} &= \text{dry weight} - \text{buoyant force} \\ &= \text{dry weight} - \text{weight of fluid displaced} \\ &= \text{dry weight} - (\text{density} \times \text{volume})_{\text{fluid displaced}} \end{aligned}$$

where volume of fluid displaced = apparent volume of sintered compact.

$72.5\text{ g} = 110\text{ g} - (1\text{ g/cm}^3 \times \text{apparent volume})$ or Apparent volume $= 37.5\text{ cm}^3$

This is the apparent volume, not the true volume, because the water fills only the open pores. Therefore,

$$\text{Apparent density} = \frac{\text{weight}}{\text{apparent volume}} = \frac{110\text{ g}}{37.5\text{ cm}^3} = 2.93\text{ g/cm}^3$$

and

$$\begin{aligned}\text{Closed-pore volume} &= \text{apparent volume} - \text{true volume}\\ &= 37.5\text{ cm}^3 - 30.87\text{ cm}^3 = 6.63\text{ cm}^3\end{aligned}$$

c. By definition, 10% volume shrinkage means:

$$-0.10 = \frac{(\text{bulk volume})_{\text{final}} - (\text{bulk volume})_{\text{initial}}}{(\text{bulk volume})_{\text{initial}}} = \frac{(\text{bulk volume})_{\text{final}} - 47.5}{47.5}$$

or

$$(\text{bulk volume})_{\text{final}} = 42.75\text{ cm}^3$$

Since Bulk volume = true volume + closed-pore volume + open-pore volume or
Bulk volume = apparent volume + open-pore volume

$$\text{Open-pore volume} = 42.75\text{ cm}^3 - 37.5\text{ cm}^3 = 5.25\text{ cm}^3$$

d. $$\begin{aligned}\text{True porosity after sintering} &= 1 - \frac{\text{true volume}}{\text{bulk volume}} = 1 - \frac{30.87\text{ cm}^3}{42.75\text{ cm}^3}\\ &= 0.28 \text{ or } 28\%\end{aligned}$$

The relatively high porosity in the sintered condition is advantageous to the grinding operation even if grinding is done by hand (see Sec. 8.10).

Ceramic Products

8.7 Brick and tile

We can now discuss the special features of ceramic products, beginning with brick and tile. For construction brick, a low-cost, easily fused clay is used as the base. This contains high-silica, high-alkali, high-FeO, gritty materials found in natural deposits. The products are formed by either dry or wet pressing. They are fired at relatively low temperatures.

The specifications are simple, involving only compressive strength ranging from 2,000 to 8,500 psi (13.8 to 58.6 MPa) (depending on grade) and dimensional tolerances.

Building tile (unglazed), clay pipe, and drain tile are processed similarly.

8.8 Refractory and insulating materials

Furnace and ladle linings use either brick or monolithic (rammed in place) linings. In dealing with liquid metals and slag it is essential to distinguish between the acidic, neutral, and basic refractories. In general, the refractories composed of MgO and CaO are called "basic" because the water solutions are basic. The refractories based on SiO_2 give very weak acid solutions. The really important effect is that acid slags will attack basic refractories, and the converse is also true.

The most important characteristics of these bricks are resistance to slag, resistance to temperature effects, and insulating ability. Figure 8.7 shows the relation of thermal conductivity to temperature for various materials.

Acidic bricks are less expensive, but in many furnaces slags high in CaO and MgO are used to refine the metal (remove phosphorus and sulfur). These slags react with SiO_2 to form low-melting-point materials that erode

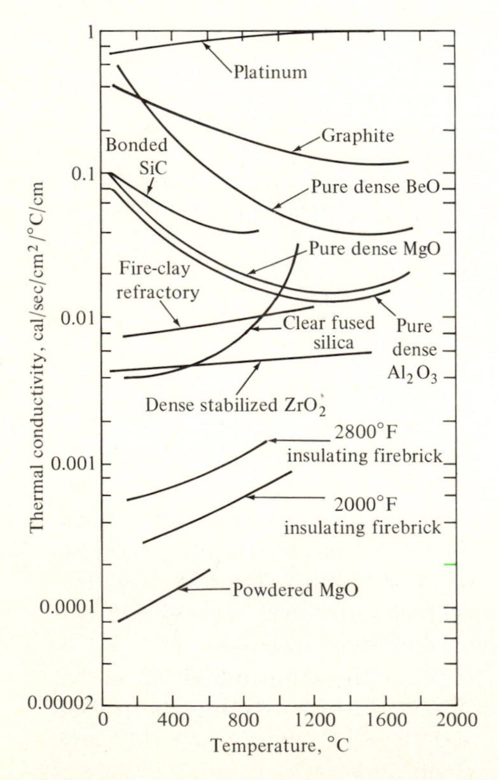

Fig. 8.7 *Thermal conductivity (logarithmic scale) of ceramic materials over a wide range of temperatures.* (W.D. Kingery, *Introduction to Ceramics*, John Wiley & Sons, Inc., New York, 1960, Fig. 14.37, p. 507)

the bricks. Therefore, when basic slags are used, the refractories must be basic. In intermediate cases alumina and chromite brick are used. The furnace or ladle linings can be formed from brick or by ramming a stiff refractory mud in place and firing it with a gas flame or coke fire.

Insulating brick contains a great deal of pore space. It is therefore not as resistant to slag as the inner lining of a vessel. It is usually used behind the working lining. Insulating brick for lower temperatures is made of asbestos and plaster mixtures.

8.9 Earthenware, stoneware, china, ovenware, and porcelain

This group includes a variety of products, from simple earthenware shapes to fine china and electrical ceramics. We shall review first the raw materials used to make each product, then the reactions that occur during processing.

Earthenware is made of clay (kaolin, for example). However, in some cases silica (SiO_2) and feldspar, such as $K(AlSi_3)O_8$, are present. The important feature is that it is fired at a low temperature compared with the other products in this group. This gives a relatively porous, earthy fracture. For example, the cup shown in Fig. 8.8*a* (page 344) is made today by Indians in Chile simply by mining clay from the nearby hillside and using a wood-fired kiln. Although the cup is quite serviceable, it is not high in strength because there has been very little fusion or sintering. A good deal of widely used clay "soil pipe" for drains has similar characteristics.

Finer grades called *semivitreous earthenware* are made using clay-silica-feldspar mixtures. Because of the three constituents, these are called *triaxial*. The firing temperature is higher, resulting in the formation of some glass, lower porosity, and higher strength. All grades may be unglazed or coated with a separate glassy-surface-forming material that gives a glazed surface.

Stoneware differs from earthenware in that the firing temperature is higher, giving a porosity of less than 5% compared with 5 to 20% for earthenware (Fig. 8.8*b*). The composition is usually controlled more carefully than that of earthenware, and the unglazed product has the matte finish of fine stone. In some variations, such as the well-known Wedgwood jasper stoneware, barium compounds are added. This is an excellent material for ovenware or chemical tanks and coils. It is essentially unattacked by most acids but is corroded by alkalis.

China is obtained by firing the triaxial mixture mentioned above or other mixtures to a high temperature to obtain a translucent object. This is why an expert will hold a plate toward the light to see how clearly the shadow of a hand comes through. China is translucent because a large share of the mixture of quartz, clay, and feldspar crystals has been converted to a clear glass. The expression "soft paste" porcelain is sometimes

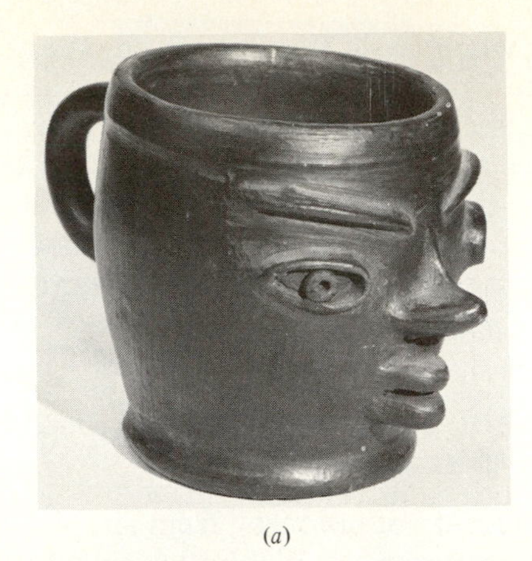

(*a*)

(*b*)

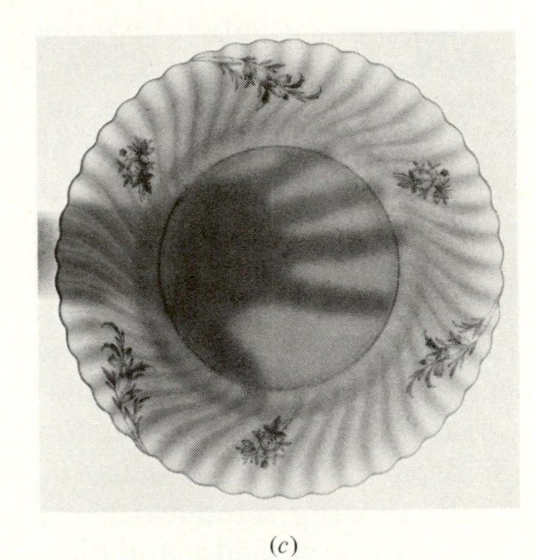

(*c*)

Fig. 8.8 *(a) Earthenware cup, made in Chile. Potters obtain the clay from nearby hills and shape it, then bake it in a charcoal-fired kiln. Beneath the glaze, the structure is relatively porous. (b) Stoneware mug, made in Annapolis, Md. The clay–silica mix is fired in an electrically heated kiln at 2100°F (1149°C). The product, which is partly fused, is stronger than earthenware. (c) English bone china (Minton), made in England. The mixture of flint, clay, and bone ash is fired at a temperature high enough to produce a glassy phase; this gives it a translucent appearance.*

used. The firing temperature is lower than for "hard" porcelain because a small amount of CaO is present as a flux. English bone china (Fig. 8.8*c*) is different in composition, containing about 45% bone ash (from cattle bones) and 25% clay, with the balance feldspar and quartz. The calcium phosphate from the bones gives a lower-melting-point material, and the phosphate group substitutes for part of the silica as a glass former. Another well-known china, "Beleek," is made by adding glass to the original mixture as a flux, giving the final product high translucency.

The specifications for *ovenware* and flameware are interesting. It has been found that ceramics with a coefficient of expansion of about $4 \times 10^{-6}\ (°C)^{-1}$ will withstand heating in an oven and cooling in air, but parts in direct contact with a flame, such as a frying pan, need a coefficient below $2 \times 10^{-6}(°C)^{-1}$. As a result, some triaxial compositions with coefficients of about $4 \times 10^{-6}(°C)^{-1}$ are used in the oven. To obtain the lower coefficients, it is necessary to add Li_2O as contained in the mineral spodumene or cordierite.

Porcelain is fired at the highest temperatures of the group and is closely related to china. In general, the avoidance of fluxes and the higher temperatures give a very hard, dense product.

Since the clay-silica-feldspar system is of such great importance to all these products, the ternary phase diagram should be reviewed briefly. Figure 8.9 is essentially a contour map showing the decrease in liquidus temperature as we go from hard porcelain, which is high in silica and alumina, to dental porcelain, which is high in alkali (K_2O). The low melting point of the *dental porcelain* makes it possible to fuse most of the tooth and convert it to a glassy translucent material similar to a real tooth.

Ceramics for Electrical Use Electrical ceramic products may be divided into large insulators, such as those used for power lines, and small electronic components, such as capacitors and magnets.

Large insulators (Fig. 8.10) are produced from a triaxial porcelain such as 60% kaolin, 20% feldspar, and 20% silica, as shown on the ternary

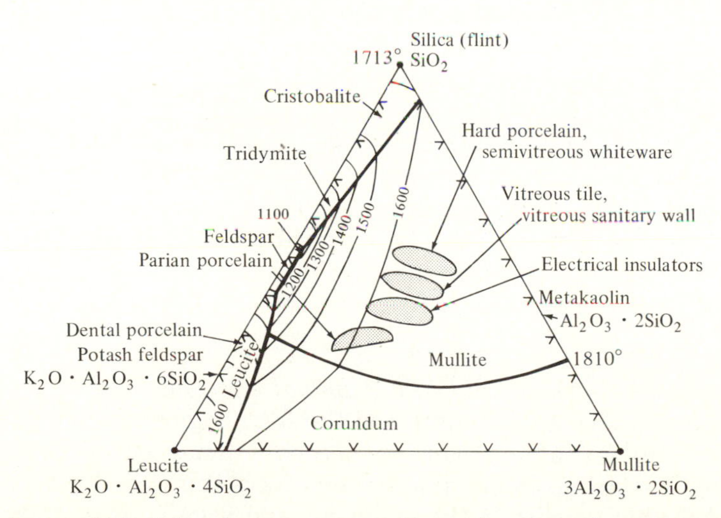

Fig. 8.9 *Areas of triaxial whiteware compositions shown on the silica-leucite-mullite phase equilibrium diagram.* (W.D. Kingery, *Introduction to Ceramics*, John Wiley & Sons, Inc., New York, 1960, Fig. 13.10, p. 419)

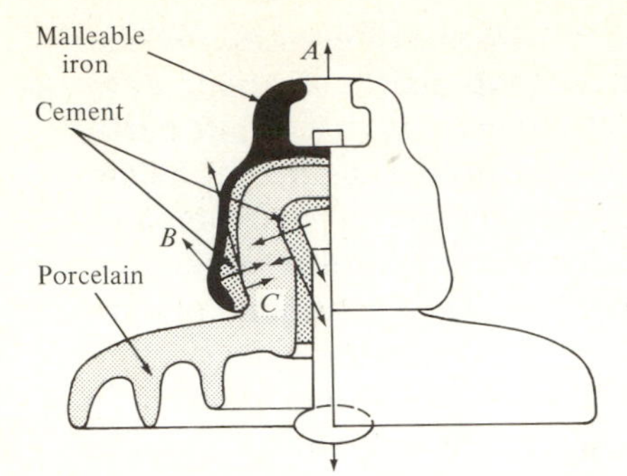

Fig. 8.10 *Cross section of a disk insulator. The mechanical strength of porcelain, like that of most ceramic materials, is greater in compression than in tension. Therefore, disk insulators, which are in tension during service, must be designed to ensure that the main stresses on the porcelain component are of a compressive nature.* (Courtesy of Doulton Insulators, Ltd.)

diagram in Fig. 8.9, using either specialized slip casting or plastic forming. Special glazes are used for two reasons. First, the coefficient of expansion of the glaze is lower than that of the porcelain. This puts the glaze in compression to strengthen the surface. Second, a semiconducting glaze is used in some cases to equalize the charge. Also, a glaze lowers the surface porosity where water might collect and cause an electrical short.

EXAMPLE 8.3 The tensile strength of porcelain is far less than the compression strength, and also less reliable. Porcelain is needed, however, to insulate high-voltage wires from the towers. The design shown in Fig. 8.10 uses the ductility of malleable iron and the compressive strength of porcelain. Without this design the porcelain would have to be in tension. Analyze the direction of stresses (tension or compression) at points *A*, *B*, and *C*. The cement bonds firmly to the insulator and may be considered part of it. Why is it used? Why does the ceramic have a large outside diameter and the deep sheds or "petticoats" (ripples on bottom)?

ANSWER Because of the design of the part, only the metal is in tension. The ceramic is in compression, as shown in the figure. The use of the cement permits the metal part to be inserted into the insulator first. The large outside diameter gives an electrical discharge a long path to travel on the surface and therefore prevents the discharge. The deep sheds are to avoid build-up of a continuous sheet of water or ice that could cause a short circuit.

In many ways the production of *special electronic ceramics,* the ferroelectrics such as barium titanate and the ferrimagnetics such as cobalt-iron spinel, is similar to the production of conventional ceramics employing dry pressing and firing. In other ways the processing is different, because of the need for high purity in the materials and usually for the avoidance of a glassy phase in sintering. Also, dimensional tolerances are extremely limited in many parts, to the order of 0.005 in. (0.0127 cm).

The most common material, $BaTiO_3$, is produced by reacting $BaCO_3$ and TiO_2, eliminating CO_2. A small amount of SiO_2 or Al_2O_3 (2%), which might be tolerated in a normal refractory, lowers the dielectric constant appreciably. In the production of capacitors for miniaturized circuits, wafers are pressed or slip-cast only 0.008 in. (0.0203 cm) thick. After firing at 1300°C, a conducting circuit consisting of a powdered silver-glass mixture is printed on the surface. Resistors are added by printing on a carbon mix. The whole part is then fired at low temperature, leads are soldered on, and the part is encased in water-resistant plastic or wax. A complete circuit may be constructed in a package smaller than a postage stamp.

In producing magnetic ferrites with a spinel structure, the pure single oxides such as MnO, CoO, and Fe_2O_3 are mixed in a blender and then fired in a tunnel kiln at 650 to 1048°C, depending on the composition for solid-state reaction. After the desired compound is obtained, the product is reground, pressed, and sintered under carefully controlled conditions (± 1°C) to control grain size, which determines magnetic characteristics.

Chapters 14 and 15 will consider these materials further.

8.10 Abrasives

In abrasive wheels or papers the idea is to grip the hard particles firmly enough so that they do not leave the wheel until they have become rounded. This is accomplished by two methods: using a softer matrix that is worn away, and providing porosity to weaken the support. The porosity is also important in conveying coolant to prevent "burning" of the part. In the case of steel, for example, heavy grinding heats the surface to the austenitic range. The resulting structure depends on the cooling rate and the composition. Many a carefully heat-treated part has been ruined by lack of coolant during finish grinding, because a brittle fresh martensite layer has been formed.

About 85% of the abrasive wheels manufactured are synthetic alumina and 15% are silicon carbide. The carbide is harder but more fragile, so that there is a great deal of competition, depending on the application. The bond may be either a ceramic glass, resin, or rubber. The softer bonds wear faster, but the wheels can be operated at higher speeds and produced in thinner sections, such as cutoff wheels.

8.11 Molds for metal castings

Although not formally recognized as a branch of ceramics, the preparation of ceramic molds for 20 million tons of metal castings deserves attention because of both the volume of ceramics used and the intricacy of the problems of metal-mold reactions. The problems are really at the interface between ceramics and metallurgy. The objective of the ceramist is to produce an expendable mold that will provide adequate surface finish and dimensional accuracy. From the metallurgical point of view, the action of the metal on the mold should be as neutral as possible. For example, the result of the reaction of metal with a mold containing excess water is a solution of hydrogen in the metal and gas holes in the casting. The combination of aluminum dissolved in liquid steel with SiO_2 in the mold wall can affect the casting surface, as shown in Fig. 8.11. A wide variety of ceramic materials and molding methods have been developed, including clay-bonded green sand (i.e., undried), dry sand, oil-bonded sand, sodium silicate-bonded sand, as well as mixtures using alumina, zirconia, and olivine, $(Mg,Fe)_2SiO_4$. The different "mixes" provide variation in casting surface finish and dimensional tolerances.

8.12 Residual stresses and contraction

The development and control of residual stresses are more important in ceramics than in metals because in ceramics there is no ductility to allow plastic deformation. However, residual stresses, if properly controlled, can improve performance, as pointed out in previous sections.

In glasses we can control stress by both mechanical and chemical methods. In tempered glass we take advantage of the fact that the compressive strength of glass is many times that in tension. Therefore, in a sheet of glass that is subjected to bending, we would like to have residual compression in the surface layers. We accomplish this by quenching (normally by an air jet)

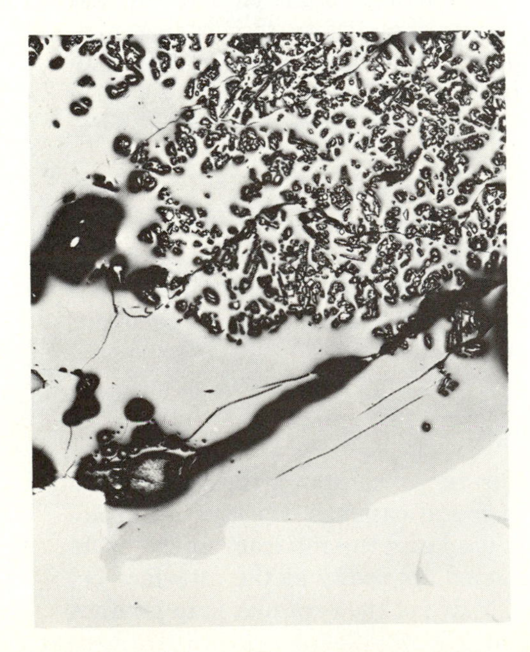

Fig. 8.11 *Photomicrograph of a cross section of an inclusion attached to a steel casting. Aluminum dissolved in the liquid steel reacted with SiO_2 from silica sand mold, producing Al_2O_3 (corundum) crystals as a well as a glassy iron silicate material; 100×, unetched.* (After G.A. Colligan)

the surfaces while the glass is in the plastic state. The surfaces are at a lower temperature as a result of the quench. Immediately after the quench there is no residual stress because the core is plastic. However, when it cools, the core will attempt to contract a greater amount than the surface because it falls through a greater temperature interval. When the glass reaches room temperature, there is tension in the core and compression in the surface (Fig. 8.12).

The chemical method is to expose the surface of a glass containing sodium ions to a solution of (larger) potassium ions. Chemical exchange takes place, and the "wedging in" of the larger ions causes surface compression.

Present government safety regulations require tempered glass in many applications. Windows other than windshields in automobiles are tempered glass. When such a window fails, it breaks up into many small, almost dustlike pieces rather than large splinters that could become projectiles in an accident. On the other hand, windshields are laminated with a polymer to contain any glass splinters, since the use of highly tempered glass might result in a total loss of transparency during an accident.

In reinforced concrete, prestressed beams have been developed to take advantage of the high compressive strength of the concrete. Before the concrete is poured, steel rods are positioned in the mold in the side of the beam under tension. The steel rods are stressed in tension by jacks, then the concrete is poured. After the concrete is set, the jacks are removed. The steel contracts, producing residual compression in the concrete. When the beam is loaded, the first portion of the load merely reduces the stress in the concrete on the tension side of the beam to zero. Therefore, the total load can be greater before the capacity of the beam is exceeded. (See also Chapter 11.)

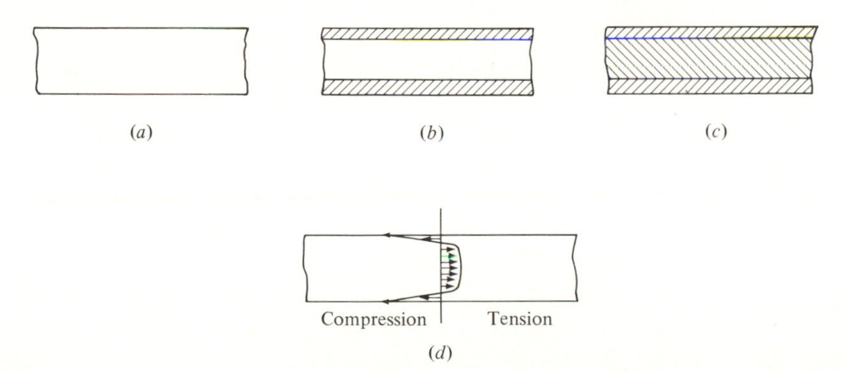

Fig. 8.12 *Changes in dimension of glass as it is tempered. (a) Hot glass; no stresses. (b) Surface cooled quickly; surface contracts, center adjusts, only minor stresses. (c) Center cools, center contracts, surface is compressed, center in tension. (d) Surface compression of tempered glass. These compressive stresses must be overcome before the surface can be broken in tension.* (L.H. Van Vlack, *Elements of Materials Science*, 2d ed., Addison-Wesley Publishing Company, Reading, Mass., 1964, Fig. 8-36, p. 228)

Stresses due to thermal expansion and spalling are related. When the surface layers of a refractory are heated, they expand and tend to spall off the surface. Two techniques have been developed to combat this problem.

In one method, glazes that have a lower coefficient of expansion than the base material are applied and fired. Since the base material tends to contract more than the glaze, the glaze is in compression and heating merely lowers the stress level. The other method is the use of a glass with a very low or negative coefficient of expansion, such as Pyroceram.

Therefore, even though ceramics inherently have low tensile strengths, they can still be used in tensile applications by imposing residual surface compressive stresses.

EXAMPLE 8.4 The following components are often made of ceramics. Indicate the characteristics you would consider desirable for each application and suggest a material that would meet these requirements. (Several answers may be appropriate for each application; however, an engineering solution requires some optimization. There are normally competitive materials for a given component.) (1) "Oven-proof" casserole; (2) coffee cup; (3) cap for a tooth; (4) fibrous insulating material exposed to 2000°F (1093°C).

ANSWER

1. *"Oven-proof" casserole.* Good thermal shock resistance, high thermal conductivity, and resistance to thermal spalling are desirable characteristics. From Table 7.1, fused silica or 96% silica would be excellent; however, processing costs are high. Therefore, consider borosilicate glass. Pyroceram is competitive and has the advantage of decorative finishes.
2. *Coffee cup.* At first it would seem that the materials discussed in (1) would be the correct selections. However, the temperature is not as high and high thermal conductivity may be undesirable if the coffee is to be kept hot. An earthenware cup with some porosity that provides lower thermal conductivity may be a better choice, although it has low strength.
3. *Cap for a tooth.* Strength, ability to be cemented to a tooth, and coloration are most important. Figure 8.9 gives dental porcelain compositions.
4. *Fibrous insulating material.* The requirement is for a fibrous material packing in a container for maximum insulation accomplished by dead air space. Asbestos, though fibrous, is toxic. It is also a hydrate, so that it may lose weight by dehydration and become more compacted. We must determine the availability of ceramic fibers with decomposition or softening temperatures above 2000°F (1093°C). Pure silica, alumina, or any of their intermediate compositions would make an excellent choice (see phase diagram, Fig. 7.32).

SUMMARY

To produce a part of a ceramic material, we must press or form the part from powder or paste, then fire the part to sinter the grains together and eliminate moisture. In many cases, such as brick, pipe, and whiteware, it is uneconomical to attempt to change dimensions after firing, and so careful processing is needed.

In the production of shapes from concrete no firing is required. The strength is due to a hydraulic bond; that is, crystallization of the cement develops a new rigid structure. In the calculation of amounts of material for production of masses of concrete, the concepts of packing and pore volume are important.

Glass processing takes advantage of the plastic nature of the material between the softening point and the working point temperatures.

Residual stresses may be eliminated by annealing to allow conversion of elastic strain to plastic strain, or they may be controlled at desired values by prestressing steel bars in concrete or by controlling cooling rates.

DEFINITIONS

Bulk density The mass of a loosely packed container of material per unit volume.

Chemical bonding A process in which a ceramic shape is bonded by the formation of a new structure between the grains (usually a hydraulic bond).

Closed pores The internal cavities in a material that are not penetrated when the material is immersed in a liquid.

Dry powder pressing A process in which ceramic powder plus lubricant and binder are pressed into a die.

Hot pressing A method similar to dry pressing, except that the heated die causes sintering of the part.

Open pores The spaces in a container of rock or other material that are filled when a liquid is added.

Portland and high-alumina cement Fine powders produced by sintering mixtures of clays and limestone followed by grinding. A strong sheet structure is formed when water is added.

Residual stress A part at rest that contains elastic strain. If the elastic strain is known, we can calculate the residual stress from the modulus of elasticity.

Slip casting A shape-forming method in which a suspension of a solid, such as clay in water, is poured into a porous mold. The water diffuses from the layers next to the mold surface, leaving a solid shape. The liquid is poured out of the interior of the mold.

True density The mass per volume of material without any voids.

True packing factor The true volume (no voids)/bulk volume or bulk density/true density.

Wet plastic forming A method in which a wet plastic refractory mix is shaped by extrusion or other forming.

PROBLEMS

8.1 An insulating brick measures 9 by 4 by 2.5 in. (22.86 by 10.16 by 6.35 cm). When placed in a container of water, it is found to displace 2.4 lb (1.09 kg) of water. What would be the apparent porosity? (Sections 8.1 and 8.6)

8.2 A ceramic tube is made by compaction of a powder. In the "green" state it is 5 cm OD by 3 cm ID by 5 cm high and is allowed to absorb water and become saturated. It absorbs 35 g of water. If the final porosity after sintering is 2%, what would be the final dimensions if the shrinkage is uniform in all directions? (Sections 8.1 and 8.6)

8.3 An insulating brick (true specific gravity = 2.58) weighs 3.90 lb (1.77 kg) dry, 4.77 lb (2.17 kg) when the open pores are saturated with kerosene, and 2.59 lb (1.18 kg) when suspended in kerosene (specific gravity = 0.82). What are (*a*) The true volume, apparent pore volume, and bulk volume; (*b*) The closed porosity? (Sections 8.1 and 8.6)

8.4 A 15.0-cm length of sintered tube of the material in Prob. 8.18 weighs 410 g. The OD is 6.0 cm, and the ID is 5.0 cm. When open pores are saturated with water, the weight is 425 g.

a. What is the bulk volume in cm^3?
b. What is the open-pore volume in cm^3?
c. What is the closed-pore volume in cm^3, given that the true specific gravity is 3.70? (Sections 8.1 and 8.6)

8.5 Carbonated beverage bottles have been known to explode, particularly if the gas pressure increases as the result, for example, of leaving a full bottle in the sun. As a manufacturer, how might you minimize this possibility? (Section 8.12)

8.6 If a piece of heated glass is subjected to a cold air blast, it will sometimes crack several days afterward. Explain. (Section 8.12)

8.7 You are going to put a "ceramic enamel" on a cast-iron bathtub. How could you do it and what precautions are necessary? (Section 8.12)

8.8 A champagne glass is filled with ice-cold liquid to within $\frac{1}{2}$ in. of the top. After about 10 sec, a circumference crack occurs. How does the stress that causes the crack develop? Be specific about direction and sign.

The crack develops only in glasses that are not cooled slowly from the plastic state. Discuss the way in which residual stress that would lead to the failure could develop. (Section 8.12)

8.9 A student has a choice of two glazes to use with a given underbody. The coefficients of expansion are: glaze 1: 4×10^{-6}/°C; glaze 2: 6×10^{-6}/°C. The underbody has a coefficient of 5×10^{-6}/°C. What type of stress (tensile or compressive) is present in each glaze? Which glaze is more likely to crack, and why? (Section 8.12)

8.10 A ceramic tile used on the walls of a shower stall is generally glazed on only one side. (Section 8.12)

a. Should the glaze have a higher or lower thermal coefficient of expansion than the underbody? Give reasons.

b. The tile contains both open and closed porosity. What are the advantages of each type of porosity?

PROBLEMS COVERING OPTIONAL SECTIONS AND PREVIOUS CHAPTERS

8.11 A brick is made from sand (SiO_2) and 10% sodium metasilicate ($Na_2SiO_3 \cdot 9H_2O$). Sodium metasilicate is also known to lose $6H_2O$ at 100°C. The brick weighs 3 lb at room temperature. How much would it weigh after heating at a temperature slightly above 100°C? (Sections 8.7 through 8.11)

8.12 A buyer comparing some small pure Al_2O_3 crucibles notes that there is a significant difference in physical properties among manufacturers. Suggest why this might occur. Also, why might one manufacturer have size tolerances of $\pm 2\%$, whereas another's might be twice that high? (Sections 8.2 through 8.3)

8.13 What is the basic reason for the good thermal shock resistance of borosilicate and fused silica glasses? (Sections 8.2 and 7.2, 7.9, 7.10)

8.14 An important family of refractories has varying percentages of silica and alumina. Draw a graph showing the maximum usable temperature of these refractories as a function of percent of Al_2O_3 (assuming that the maximum usable temperature is the solidus). (Sections 8.8 and 7.10)

8.15 *a.* For a 40% SiO_2, 60% Al_2O_3 brick, calculate the percentage of liquid present at 1591°C under equilibrium conditions.

b. In an 80% Al_2O_3, 20% SiO_2 brick, calculate the percentage of liquid at 1800°C.

c. If you were to specify the Al_2O_3 content for a brick to be used to resist liquid metal at 1650°C, which of these compositions would you use (bearing in mind that Al_2O_3 is more expensive than SiO_2)? (1) 40% Al_2O_3; (2) 60% Al_2O_3; (3) 70% Al_2O_3; (4) 80% Al_2O_3; (5) 99% Al_2O_3. (The balance of the composition is silica.) (Sections 8.8 and 7.10)

8.16 An investigator observes a sample of refractory brick under the microscope and determines that 70% of the area is mullite and 30% corundum. Assuming equilibrium conditions, what is the lowest temperature at which the brick will liquefy completely? Assume that the density of corundum is 4.0 g/cm^3 and that of mullite 3.6 g/cm^3. (Sections 8.8 and 7.10)

8.17 Given two ceramic materials and considering only the following characteristics, which would have the greater tendency to fail by spalling? Be quantitative; that is, estimate how many times greater, using the spalling index. Show calculations. (Sections 8.12 and 7.9; Problem 7.25)

	Material A	*Material B*
a_0 (Å)	5.06	4.09
Tensile strength, psi	100,000	50,000
Modulus of elasticity, psi	10×10^6	20×10^6
Coefficient of thermal expansion, °F^{-1}	8×10^{-6}	4×10^{-6}
Density, g/cm^3	3	5

8.18 You want to produce a ceramic tube by compacting and sintering MgO with the addition of 10% by weight CaF_2. (Sections 8.6, 7.6, and 7.10)

a. What is the structure of MgO? Support your answer.

b. The objective in sintering is to have the CaF_2 break down according to the reaction

$$CaF_2 + \tfrac{1}{2}O_2 \rightarrow F_2\ (\text{gas}) + CaO$$

(1) With the initial addition of 10 wt % CaF_2, what is the weight percentage of CaO in the MgO-CaO mixture?

(2) From the accompanying CaO-MgO phase diagram, find the liquidus and solidus temperatures. Are these changed significantly from those of pure MgO?

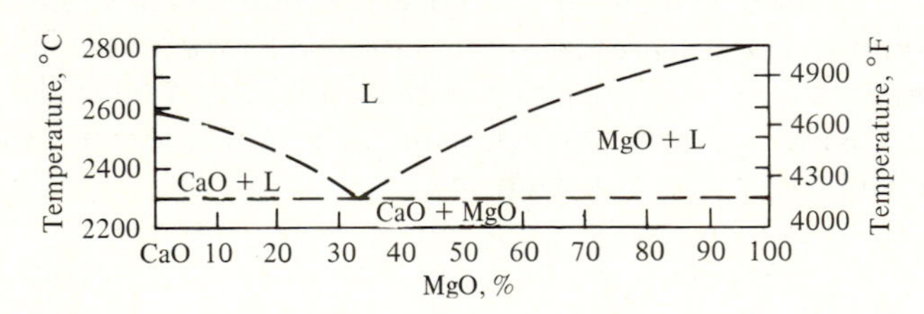

(3) Will the Ca^{2+} ions replace the Mg^{2+} ions in the MgO structure? Support your answer.

8.19 What process and materials would you use to make the following?

a. A Coke bottle

b. Window glass for a bank window

c. A thermocouple protection tube exposed to liquid steel at 3000°F

d. A ceramic hot plate 0.5 by 6 by 6 in. (1.27 by 15.24 by 15.24 cm)

e. A crucible for melting 100 lb (45.4 kg) of steel at 3000°F (1649°C) with basic slags

f. A resin-bonded grinding wheel
g. A vitreous- (glass)-bonded grinding wheel
h. A three-floor parking structure for 600 cars
i. A cut-glass decanter
j. A mold for pouring a 50-ton (45,400-kg) steel roll for a steel mill
k. Insulating brick for an oven to operate at 1000°F (540°C)
l. A high-voltage power line insulator with a glazed surface
(Summary of Chapters 7 and 8)

8.20 Some common ceramic materials are as follows. Draw the structure of each material in two dimensions and select the term(s) that describes an outstanding characteristic(s) of the material. (Summary of Chapters 7 and 8)

Materials	*Descriptions*
a. SiO_2 (quartz)	1. High thermal shock resistance
b. SiO_2 (fused silica)	2. High index of refraction
c. $MgSiO_3$ (enstatite)	3. Hexagonal structure
d. Lead-alkali glass (54% SiO_2, 11% CaO, 35% PbO)	4. Fibrous cleavage
	5. Basal cleavage
	6. Low melting (below 1000°F or 538°C)
	7. Gives off H_2O when heated

Example: Graphite, (ring of six C atoms: C—C above, C at left and right, C—C below), 3 and 5.

8.21 The following list shows several ceramic materials and their possible characteristics. Match the correct ceramic(s) with the characteristic. There may be more than one correct answer and all should be included. (Summary of Chapters 7 and 8)

a. Quartz
b. SiO_2 containing Na_2O
c. Vitreous silica
d. MgO
e. Al_2O_3
f. Graphite

Characteristic

1. Lowest melting point
2. Highest thermal conductivity
3. Lowest coefficient of expansion
4. Crystalline
5. Transparent
6. Withstand 3000°F (1650°C) air

8.22 As noted in the text, barium titanate is formed from $BaCO_3$ and TiO_2. What is the percentage loss in weight in this process? (Section 7.6)

9

PLASTICS (HIGH POLYMERS)—STRUCTURES, POLYMERIZATION TYPES, PROCESSING METHODS

THE intertwined strings of beads in the illustration portray the salient difference between the high polymers or plastics we shall discuss in this chapter and the predominantly crystalline materials we have studied thus far.

Although there are important occurrences of crystallinity in the polymers, the basic building block is the molecule, and the nature and interaction of the molecules have the dominant influence on properties. We shall encounter interesting analogies between the structures of glass and high polymers.

9.1 Introduction

The high polymers are fascinating not only because of their increasing engineering importance, but also because their structures can be altered and tailor-made to provide a wide spectrum of color, transparency, and formability into different shapes.

We should mention for completeness that the study of high polymers includes two major fields: the biological large molecules that are the basis of life and food, and the engineering polymers such as the synthetic materials—plastics, fibers, and elastomers—and the natural materials—rubber, wool, and cellulose. We shall be concerned here with only the engineering materials, principally the plastics and elastomers (rubbers).

Since the polymers are different in some ways from metals and ceramics, let us take a general look at these materials before going into detail.

The backbone of every organic material is the chain of carbon atoms, so let us first review carbon. There are four electrons in the outer shell (Fig. 9.1*a*), giving a valence of four. Each electron can form a covalent bond to another carbon atom or to a foreign atom (Fig. 9.1*b*). The elements encountered most frequently and their valences are: valence 1: H, F, Cl, Br, I; valence 2: O, S; valence 3: N; valence 4: C, Si.

At the heart of the polymer structure is the fact that two carbon atoms can have one, two, or three common bonds (i.e., share one, two, or three pairs of electrons) as well as bonding with other atoms. This is often shown in two dimensions (Fig. 9.2), but remember that we have a three-dimensional structure similar to silicon in the center of the SiO_4^{4-} tetrahedron (Fig. 9.3), which we studied in ceramics. Just as we needed the three-dimensional picture to understand the sheet and fiber structures of the ceramics, we need a three-dimensional model to understand the properties of the plastics. However, the simpler two-dimensional view shows many important features.

What are the results of bonding carbon atoms and forming chains of increasing length? If we start with ethane, C_2H_6, with two carbon atoms per molecule, we have a gas. As we increase the number of carbon atoms in the chain to several hundred, we pass through liquids of increasing boiling points to greases and waxy solids. Finally, when we reach over 1,000 carbon atoms, we obtain materials with the characteristics of plastics, a combination of strength, flexibility, and toughness. These phenomena are due to the basic point that as the length of the molecules increases, the total binding force between molecules rises. There are both van der Waals forces between molecules and mechanical entanglements between chains.

In view of these properties of the large molecules, let us continue our study of the carbon backbone of the polymers. A straight line is usually used to represent the structure in two dimensions:

$$-\overset{|}{\underset{|}{C}}-\overset{|}{\underset{|}{C}}-\overset{|}{\underset{|}{C}}-\overset{|}{\underset{|}{C}}-$$

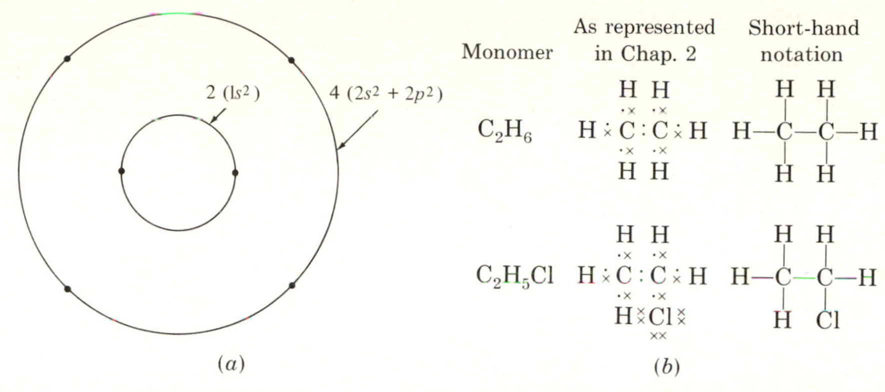

Fig. 9.1 *(a) The electrons of carbon. Note that four valence electrons are available for covalent bonds. (b) Carbon bonds (sharing of electron pairs) in the simple monomers C_2H_6 and C_2H_5Cl. Note that carbon may bond to another carbon atom, to an electropositive element such as hydrogen, or to an electronegative element such as chlorine.*

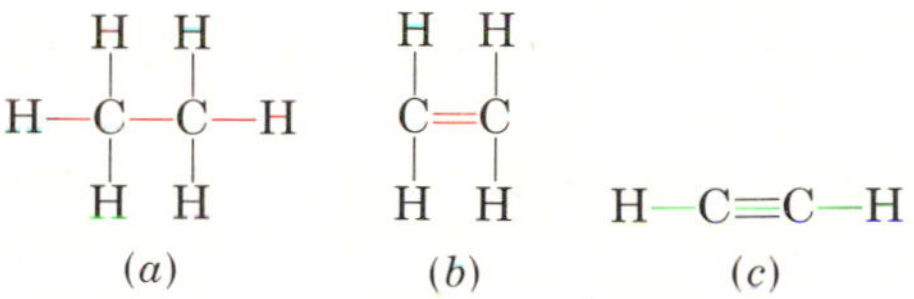

(a) H—C—C—H (each C also bonded to two H) (b) H₂C=CH₂ (c) H—C≡C—H

Fig. 9.2 *Single, double, and triple carbon bonds: (a) ethane, (b) ethene (also called ethylene), (c) acetylene*

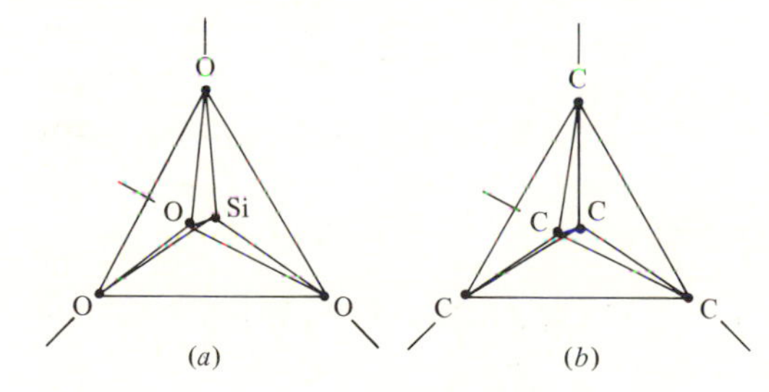

Fig. 9.3 *Comparison of tetrahedra in (a) silicate and (b) carbon structures*

This is not the actual situation. Each bond is at 109° to the next. Thus the carbon backbone extends through space like a twisted string of Tinkertoys with sockets that have cylindrical holes at 109° to each other (Fig. 9.4, page 360). When stress is applied, these intertwined molecules stretch to provide elongation that can be thousands of times greater than it would be in a typical crystal of a metal or ceramic.

A vital point in making the long carbon chains is that we can take two molecules of a material such as ethylene, in which there are two carbon-to-carbon bonds, open up one of the bonds in each, and join the molecules (Fig. 9.5, page 360). This is the source of the word *polymer* and the key to polymer chemistry. The two original units are called *monomers*. Therefore, the large

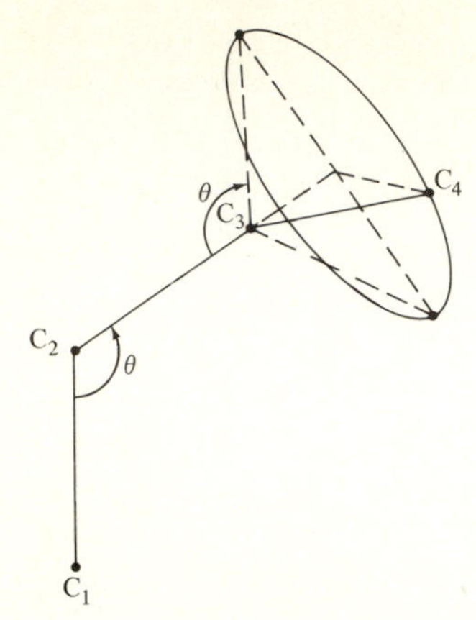

Fig. 9.4 *Formation of the carbon backbone. Atoms C_1, C_2, and C_3 define a plane; atom C_4 may lie at any one of several preferred positions on the circle shown.*

molecule is a polymer. After the two mers are joined, there are still two free bonds for joining to other mers. The process goes on, linking mers together until it is stopped by the addition of another chemical called a *terminator,* which satisfies the bonds at the ends of the molecules.

Thus in this simple case we form a number of large molecules of a linear polymer, polyethylene. We have shown the molecule as a straight line and it is called a *linear polymer,* but we should emphasize that actually the molecules are not straight but can be thought of as a mass of worms randomly thrown into a pail. There is considerably more intertwining than there would be with worms, because if we scaled up the typical polymer molecule to a diameter of 0.25 in. (0.635 cm), it would be 20 ft (6.1 m) long!

We can continue our analogy a little further. We know that segments of the worms continually coil and uncoil, and we note a similar motion with the polymer molecules. If we pulled slowly on the mass of worms, we would

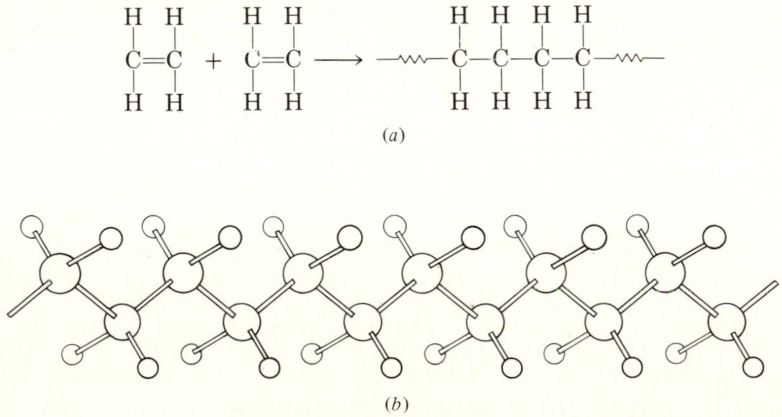

$$\begin{array}{ccc} \mathrm{H}\ \ \mathrm{H} & & \mathrm{H}\ \ \mathrm{H} \\ \mathrm{C{=}C} & + & \mathrm{C{=}C} \\ \mathrm{H}\ \ \mathrm{H} & & \mathrm{H}\ \ \mathrm{H} \end{array} \longrightarrow \begin{array}{c} \mathrm{H}\ \ \mathrm{H}\ \ \mathrm{H}\ \ \mathrm{H} \\ \sim\mathrm{C{-}C{-}C{-}C}\sim \\ \mathrm{H}\ \ \mathrm{H}\ \ \mathrm{H}\ \ \mathrm{H} \end{array}$$

(*a*)

(*b*)

Fig. 9.5 *(a) Polymerization of ethylene by the opening of double bonds. (b) The geometry of a polyethylene chain.*

find a higher percentage with the long axis parallel to the direction of tension. The same effect occurs with nylon molecules, and after the alignment the material is stronger but less ductile.

The effects of temperature on a linear polymer are striking. They are related to the behavior of a typical glass, discussed in Chapter 7. We can visualize a soda-lime glass made up of molecules of chains of SiO_4^{4-} tetrahedra terminated by sodium atoms. When the glass is cooled, the thermal agitation of the molecules decreases and the material becomes viscous. Finally, the chains become locked in place and the glass is brittle. The same phenomenon occurs in polymers. The temperature at which chain movement decreases to a low value is quite aptly called the *glass transition temperature,* designated T_g. In addition to the plastic and brittle ranges, some polymers have another temperature range of behavior because of the ability of the molecules to uncoil elastically; this is called the *rubbery* range. The whole basis for plastic, rubbery, or brittle behavior, therefore, is related to the amount of molecular movement in a given material.

So far we have introduced only one group of plastics, called the *thermoplastics* because they become plastic when heated. There is another group (Fig. 9.6) in which a single large network, instead of many molecules, is formed during polymerization. Since this is usually accomplished by heating the basic materials together, this group is called the *thermosetting plastics.* An example is a billiard ball made up of a thermosetting plastic; the molecular weight would be about 10^{28}! Since there is essentially one giant molecule, there is no movement between molecules once the mass has set. Furthermore, the material will not become plastic on heating, since there is no motion

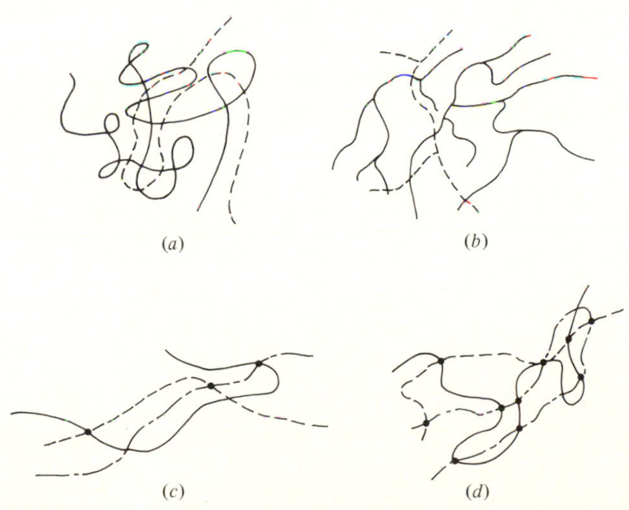

Fig. 9.6 *Different types of carbon backbone structures: (a) linear, (b) branched, (c) loose network, (d) tight network*

between molecules as there is in the thermoplastics. We shall see that for the network structure to form, the mers must have more than two places for bonding; otherwise we can obtain only a linear polymer. Both structures are useful. The thermoplastics are more easily formed and deformed elastically and can be remelted, whereas the thermosetting polymers are more rigid and generally have higher strength.

One final point: We have spoken of the carbon atom and the grouping of carbon and other atoms into many jointed chains (and even rings) giving the *molecules*. Now it may come as a surprise when we say that these chains, often containing 10,000 atoms, may be folded back and forth to form *crystals*. This is only possible in some cases, but where it occurs, the properties of the crystalline polymer are quite different from those of the disorganized structure of the amorphous material. The crystallization is usually only partially complete, but it is still possible to obtain x-ray diffraction measurements of the repetitive atom spacings.

Now that we have discussed the structures of polymers in general, let us plan our discussion in detail.

First we shall consider how the molecular structures are built up. In the metals the nature, amount, size, shape, distribution, and orientation of phases determined the properties of a given material. In the polymers we shall find that in a related way the properties are functions of the mers of which the polymer is made up, of the arrangement of different mers in the polymer, of the size and shape of the molecules, and of their distribution (alignment, crystallinity).

Next we shall discuss the processing of the polymers, which makes possible rapid low-cost production of components with excellent surfaces and closely controlled dimensions.

Finally, in Chapter 10 we shall summarize the mechanical properties of polymers and take up some typical applications of polymeric materials.

Polymerization Methods

9.2 Review of the building blocks—the mers

Before we study methods of polymerization, we should take a look at the mers, the building blocks we are going to use. A shelf of the most frequently used chemicals would look like Table 9.1 on pages 364–365.

The nomenclature is based on the common saturated hydrocarbon in the majority of cases. With one carbon atom we have the methyl group, with two carbon atoms the ethyl group, and so forth. The bonds are all considered covalent, so that there is no question of sign, only of the quantity of shared electrons. Again these valences are 1: H, F, Cl, Br, I; 2: S, O; 3: N; 4: C, Si.

The most important group in the table, the *olefins,* is basic to the polymer industry. These differ from the saturated hydrocarbons by the double bond. This

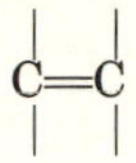

$$\overset{|\ \ |}{\underset{|\ \ |}{C{=}C}} \quad \text{bond can open to become} \quad \times\!-\!\overset{|\ \ |}{\underset{|\ \ |}{C{-}C}}\!-\!\times$$

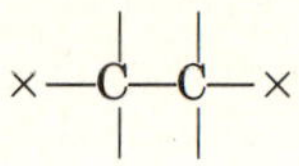

and other mers can join at the points marked × to form the polymer. We shall take up numerous examples shortly. Note the structural differences among the principal categories: that is, "ene" signifies an unsaturated hydrocarbon, "ol" an alcohol (with the OH group), "aldehyde" a —HC=O group, "acid" an H—O—C=O group, and "ester" a combination of an alcohol and an acid to yield the ester plus water. For example, ethyl alcohol plus acetic acid (C_2H_5OH + CH_3COOH) gives ethylacetate and water. The amide, amine, benzene, phenol, vinyl, and styrene groups are also of great importance to the generation of specific polymer families.

9.3 Bond strengths

Now let us consider the elements of bonding these structures into plastics. First we should review the table of bond strengths, Table 9.2, page 366.

EXAMPLE 9.1 Suppose that we are to make polyethylene from ethylene. How much energy would be interchanged? Would the reaction be spontaneous?

ANSWER The reaction can be written in the following way:

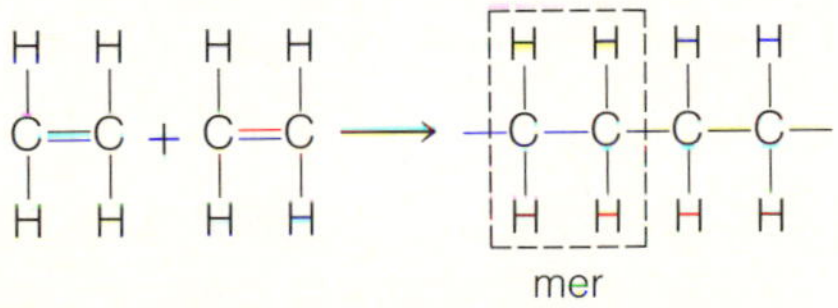

$$\begin{matrix} H & H \\ | & | \\ C{=}&C \\ | & | \\ H & H \end{matrix} + \begin{matrix} H & H \\ | & | \\ C{=}&C \\ | & | \\ H & H \end{matrix} \longrightarrow \underbrace{\begin{matrix} & H & H \\ & | & | \\ + & C{-}&C + \\ & | & | \\ & H & H \end{matrix}}_{\text{mer}} \begin{matrix} H & H \\ | & | \\ C{-}&C{-} \\ | & | \\ H & H \end{matrix}$$

or $$2\ C{=}C \text{ bonds} \longrightarrow 4\ C{-}C \text{ bonds}$$

[If more than two monomers (C=C bonds) are used, twice as many C—C bonds will always be formed.] Therefore,

$$2(162 \text{ kcal/mole}) \longrightarrow 4(88 \text{ kcal/mole})$$

or $$324 \text{ kcal/mole} \longrightarrow 352 \text{ kcal/mole}$$

Since we obtain 28 kcal/mole more for the final products than for the reactants, we get more energy out of the bonds formed than we put in to break the bonds. Therefore, the reaction is spontaneous.

We shall use the same reasoning later in more complex reactions where we want to decide which of two possible processes will prevail.

In addition to the covalent bonds within a molecule, there are van der

Table 9.1 SOME COMMON CHEMICALS (MONOMERS) USED IN PLASTICS

Family	*Characteristic Bond in Family*	*Compounds*			
Saturated hydrocarbons	 —C—	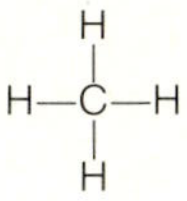 H—C—H Methane	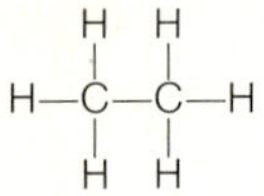 H—C—C—H Ethane	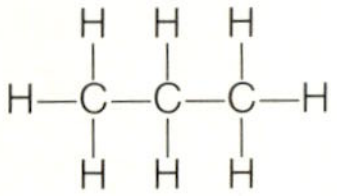 H—C—C—C—H Propane	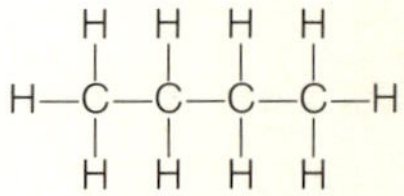 H—C—C—C—C—H Butane
Unsaturated hydrocarbons (olefins)	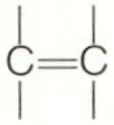 C=C	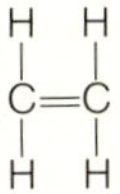 C=C Ethylene	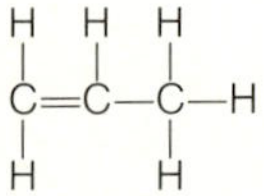 C=C—C—H Propylene	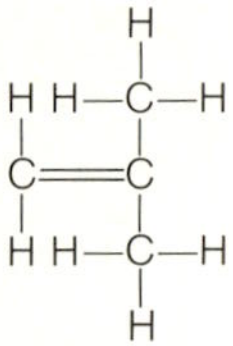 C=C Butylene	
		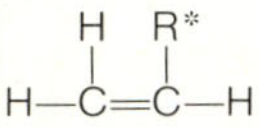 H—C=C—H (H, R*) Vinyl	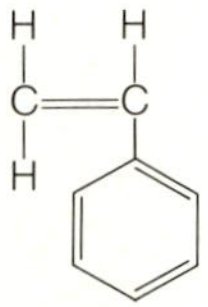 C=C (with benzene ring) Styrene	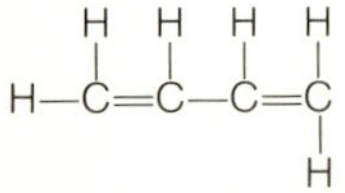 H—C=C—C=C Butadiene	
Alcohols	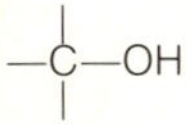 —C—OH	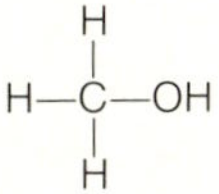 H—C—OH Methyl alcohol	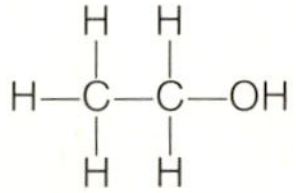 H—C—C—OH Ethyl alcohol	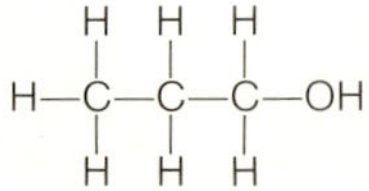 H—C—C—C—OH Propyl alcohol	
Aldehydes and ketones	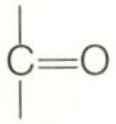 C=O	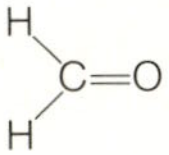 H₂C=O Formaldehyde	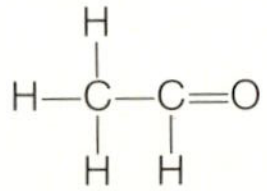 H—C—C=O Acetaldehyde	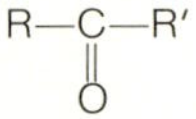 R—C—R′ (C=O) Ketone	

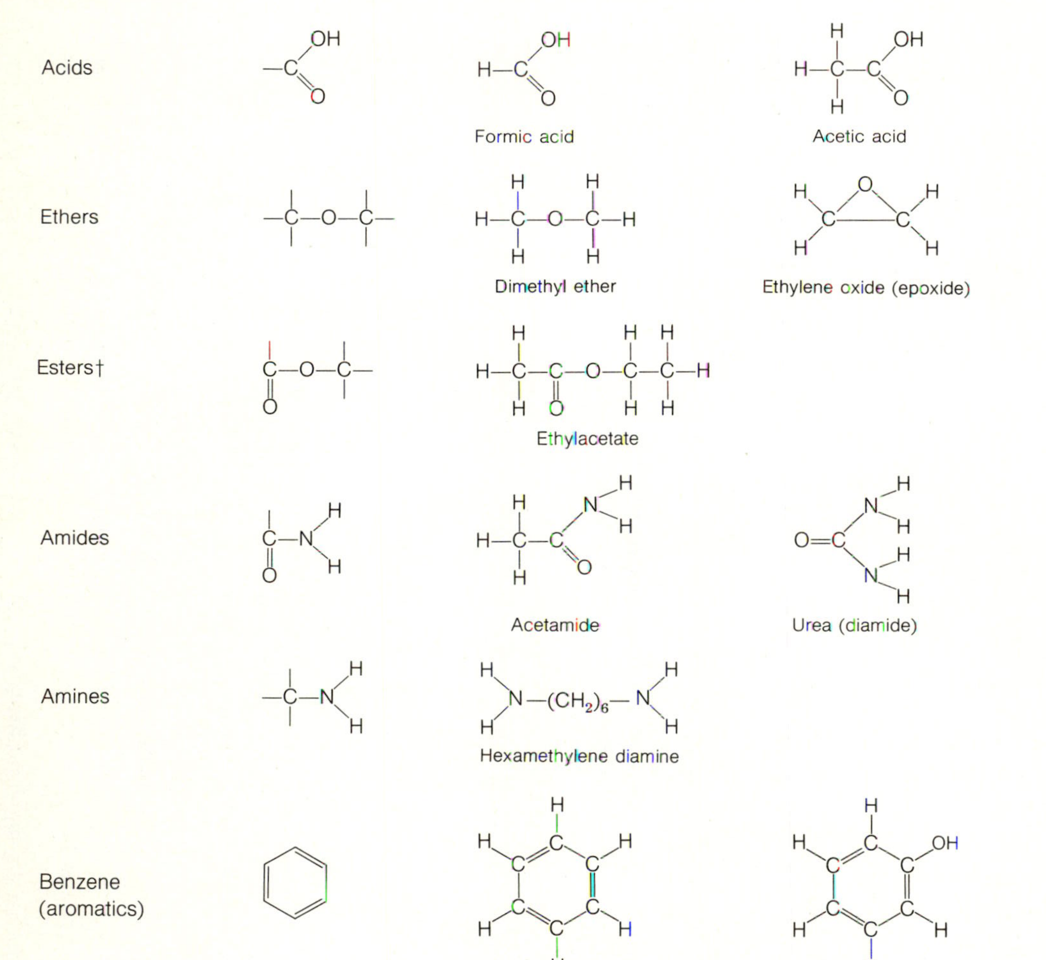

*R stands for Cl, ⬡ OH, etc.

†From acid and alcohol, with H_2O as byproduct

Table 9.2 BONDING ENERGIES IN ORGANIC MOLECULES

Bond	*Bond Energy,* kcal/g-mole (approx.)*	*Bond Length, Å*	*Bond*	*Bond Energy,* kcal/g-mole (approx.)*	*Bond Length, Å*
C—C	88	1.5	O—H	119	1.0
C=C	162	1.3	O—O	52	1.5
C≡C	213	1.2	O—Si	90	1.8
C—H	104	1.1			
C—N	73	1.5	N—H	103	1.0
C—O	86	1.4	N—O	60	1.2
C=O	128	1.2			
C—F	108	1.4			
C—Cl	81	1.8	H—H	104	0.74

*The values vary somewhat with the type of neighboring bonds; e.g., the C—H is 104 kcal/mole in methane, H—C(H)(H)—H, but only 98 kcal/mole in ethane, H—C(H)(H)—CH_3, and 90 kcal/mole in trichloromethane, H—C(Cl)(Cl)—Cl. To obtain J/g-mole, multiply kcal/g-mole by 4.186×10^3.

Source: L.H. Van Vlack, *Materials Science for Engineers,* Addison-Wesley, Reading, Mass., 1970.

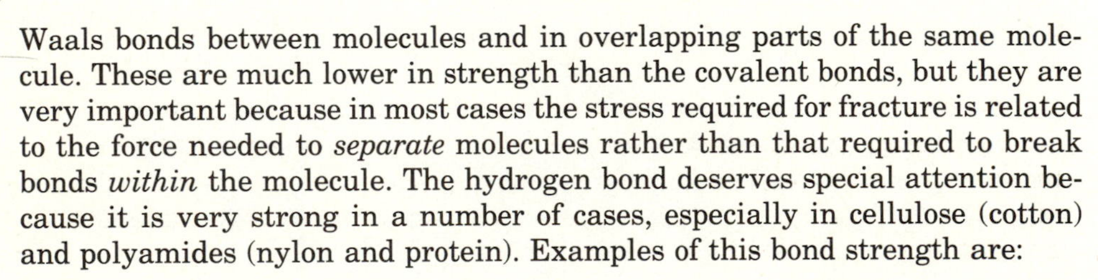

Waals bonds between molecules and in overlapping parts of the same molecule. These are much lower in strength than the covalent bonds, but they are very important because in most cases the stress required for fracture is related to the force needed to *separate* molecules rather than that required to break bonds *within* the molecule. The hydrogen bond deserves special attention because it is very strong in a number of cases, especially in cellulose (cotton) and polyamides (nylon and protein). Examples of this bond strength are:

Bond	*—H . . . x—Bond Length, Å*	*Energy, kcal/g-mole†*
O—H . . . O—	2.7	3 to 6
N—H . . . O—	2.9	4
N—H . . . N—	3.1	3 to 5

This bond does not have the strength of a covalent bond because no electrons are shared. It arises from the fact that at the H side of an OH radical, for example, there is a positive polarity because the electron from the hydrogen is attracted closely to the oxygen. Similarly, an oxygen that has accumulated an electron from another source is of negative polarity.

9.4 Bonding positions on a mer, functionality

The mer must have a sufficient number of bonds that can be opened up for attachment to other mers of the same or different formula. This number is

†To obtain J/g-mole, multiply kcal/g-mole by 4.186×10^3.

called the *functionality*.† The ethylene mer, therefore, is bifunctional, because the C═C bond can react with two neighboring monomers. For polymerization the mers must be at least bifunctional. Not only is a double carbon bond bifunctional, but we can split the N—H bond in an amine, the O—H bond in an alcohol, and the C—OH bond in an acid to form another bond. There are two or more of these bonds in amines and dialcohols. For example, in ethylene glycol (a dialcohol of ethylene), $H—O—CH_2—CH_2—O—H$, we find bifunctionality for polymerization by splitting both OH groups.‡ Therefore, we may form polymers without a carbon double bond in the mer. When the monomers are trifunctional or greater, we can form network or thermosetting polymers.

The *degree of polymerization* (DP) is the molecular weight of the polymer divided by the molecular weight of the mer. It tells the number of mers in the molecule, or its average length. It is important because, as we have already seen, larger molecules result in higher bond strengths and therefore higher melting or softening points. There is always a variation in the DP in a given batch of polymers because not all chains start growing at the same time. The time of growth of individual chains is also variable. This variation can be measured and expressed statistically.

9.5 Polymerization mechanisms

Now let us take up the important polymerization mechanisms, since the type of structure is intimately related to the mechanism involved.

Addition Polymerization We said earlier that in ethylene, for example, we break the C═C bond to form the polymer. We need to explain how we do this and also how the growth of a polymer molecule stops. The sequence is as follows.

An *initiator* is added to the polymer. The initiator, such as a radical R with either a free electron or an ionized group, attracts one of the electrons of the carbon double bond (Fig. 9.7, page 368). The other electron of the broken bond is unsatisfied. It attracts an electron from another mer; and so the molecule grows. Finally the chain is stopped when two growing segments meet or when a growing segment meets another R, the *terminator*. The rate of addition polymerization can be very rapid; for example, the polymerization of isobutylene using H_2O_2 can take place in a few seconds, yielding polymers made up of thousands of mers. For this reason the process is also called *chain-reaction polymerization*. A variety of chemicals, including organic substances,

---

†This discussion is so greatly simplified that it is not possible to predict many cases of functionality from these simple premises. See a text on organic chemistry for a fuller explanation.

‡This happens provided that other bifunctional molecules *having groups that can react with dialcohols in this manner* are present.

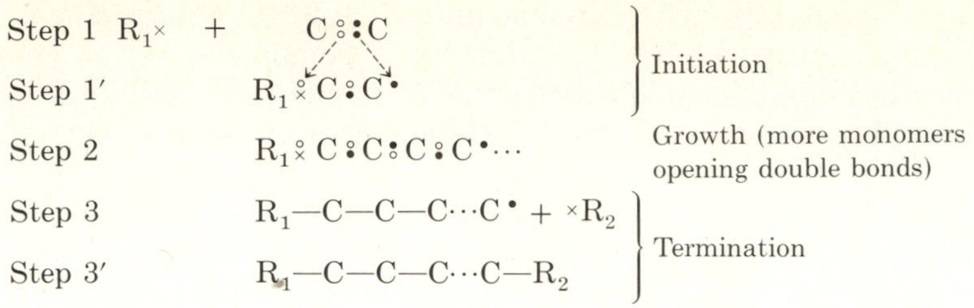

Step 1	$R_1\times$ + C::C	Initiation
Step 1′	R_1 ⁘ C : C•	
Step 2	R_1 ⁘ C : C : C : C•···	Growth (more monomers opening double bonds)
Step 3	R_1—C—C—C···C• + $\times R_2$	Termination
Step 3′	R_1—C—C—C···C—R_2	

Fig. 9.7 *Atom bonding in addition polymerization. For simplicity, the hydrogen or other atoms bonded to the sides of the carbon chain are not shown.* R_1 *and* R_2 *are the initiator and terminator groups, respectively.*

can be used as initiators and retarders to modify the reaction rate. Addition polymerization can also take place between two different mers, such as ethylene and vinylchloride. This is called *copolymerization.*

> EXAMPLE 9.2 Sketch how the polymerization of styrene takes place and indicate whether it is thermoplastic or thermosetting.
>
> ANSWER 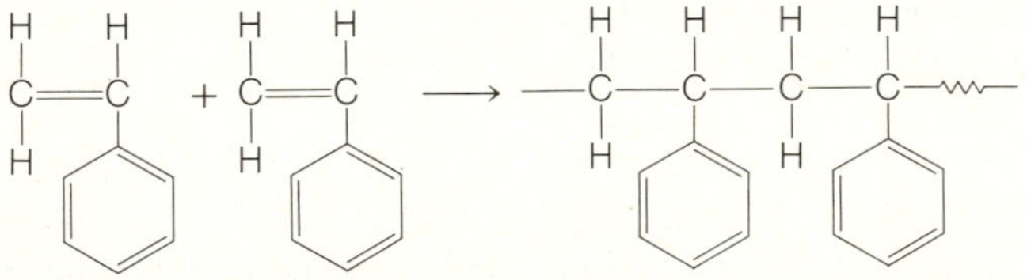
>
> The polymer is thermoplastic because it is linear and will not polymerize between the benzene rings to form a network.

Condensation Polymerization This process is also called *step-reaction polymerization* because two or more different molecules have to get together in each step if the molecule is to grow. For this reason and because of the difference in kinetics, this process is much slower than the zipperlike addition polymerization. Furthermore, there is usually a byproduct that must condense; hence the name. This reaction can produce *either* a chainlike molecule, 6/6 nylon, *or* a network structure, phenol-formaldehyde (Fig. 9.8). In general, linear polymers formed by condensation have a lower degree of polymerization. The chains are generally stiffer because of the backbone elements bonded between the carbon atoms, and the attractions between chains are greater because of hydrogen bonding.

One of the mers must be trifunctional or greater for the network structure to be formed. This can involve adjacent ring groups, as shown by a new

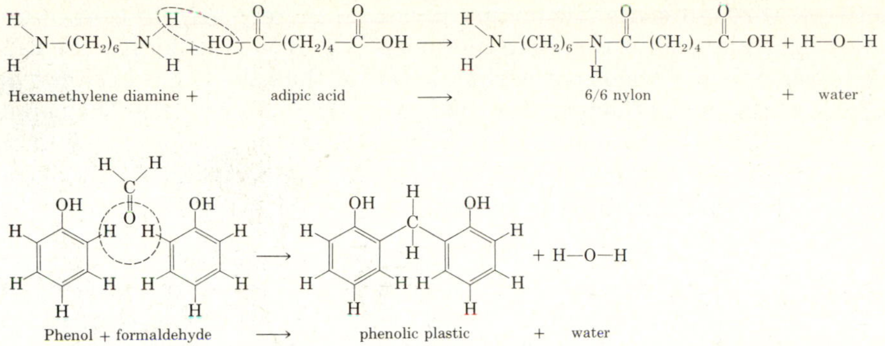

Hexamethylene diamine + adipic acid ⟶ 6/6 nylon + water

Phenol + formaldehyde ⟶ phenolic plastic + water

Fig. 9.8 *Formation of nylon and phenol-formaldehyde polymers. In the latter case, further polymerization gives a network.* [Note: *The condensation reaction for phenol-formaldehyde is more complex than shown, and involves several intermediate steps.*]

development in heat-resistant materials, the formation of a ladder structure (Fig. 9.9). This material, derived from polyacrylonitrile, can be held in an open flame without change. These properties are due to the rigidity of the C—N ring backbone.

9.6 Polymer structures

We have discussed the formation of linear and network structures in relation to polymerization processes. We should emphasize that in either addition or condensation polymerization, if the molecules are only bifunctional, we form a linear or chainlike polymer. This is the thermoplastic polymer. On the other hand, if one of the molecules is trifunctional or greater, we can form a three-dimensional network by either addition or condensation polymerization. This is a thermosetting polymer.

As an example of the first case, we form long, chainlike polyethylene molecules. When these are heated, the van der Waals forces between them weaken and the material melts (thermoplastic). As an example of the second case, phenol-formaldehyde can continue to bond at other locations and form

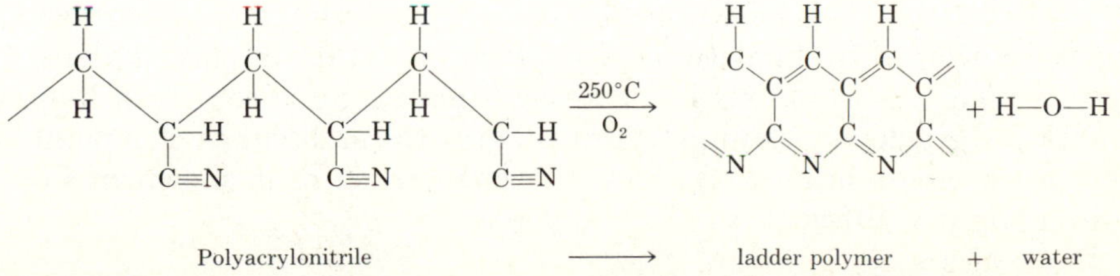

Polyacrylonitrile ⟶ (250°C, O_2) ladder polymer + water

Fig. 9.9 *Formation of a ladder polymer from polyacrylonitrile, which still has unsaturated positions. The original polyacrylonitrile was formed by an addition reaction (no byproduct).*

a network through space. Heat (and pressure) merely help to continue the bonding. The material is therefore thermosetting. If the temperature is raised further, the primary bonds of the chains, rather than the bonds between molecules, are broken and the process called *degradation* takes place. The polymer is no longer useful. The material used in an electrical wall socket is a good example of a thermosetting plastic. It may char, but it certainly will not melt in service (phenol-formaldehyde type).

EXAMPLE 9.3 The reaction between urea and formaldehyde is somewhat like the phenol-formaldehyde reaction (Fig. 9.8) but more complex. It takes place in two steps: first the formation of a methylol compound and then the condensation step.

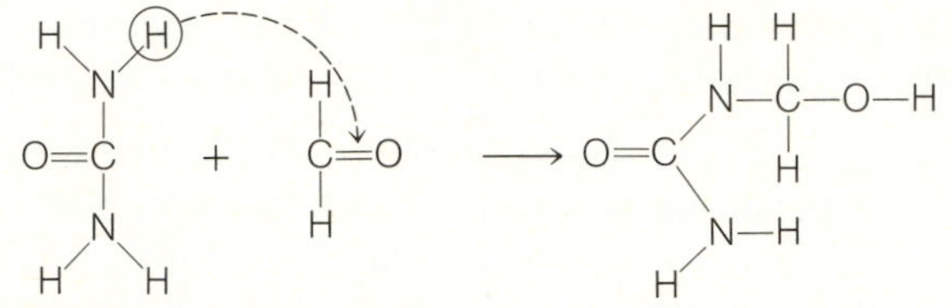

Urea + formaldehyde ⟶ methylol compound

How does the condensation step take place, and is the product thermoplastic or thermosetting?

ANSWER

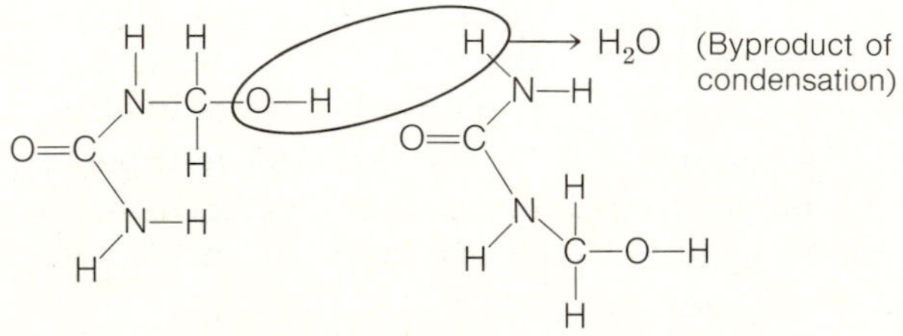

⟶ H_2O (Byproduct of condensation)

(The reaction is repeated at other N—H bonds.)

A complete network is formed, which makes the polymer thermosetting. The same network would inhibit crystallinity (see Sec. 9.7).

Branching This modification can take place in a linear, thermoplastic polymer as well as in the network type. It is important in polyethylene. One way in which this can occur is for an addition agent to remove a hydrogen atom from the side of a chain, whereupon growth can occur at this point. The difference between branching and a network structure is shown in Fig. 9.6 and in a slightly different way in Fig. 9.10.

Crosslinking Another modification of polymerization is crosslinking, which produces a network structure. It is especially important in obtaining the desired hardness and toughness in rubber (Fig. 9.11). The degree of cross-

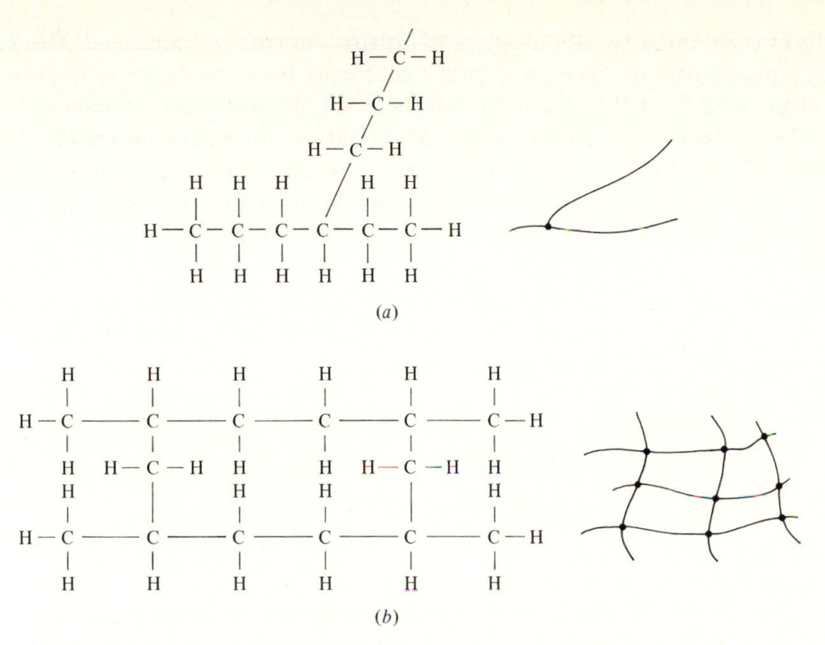

Fig. 9.10 *(a) Branching structure. (b) Network structure.*

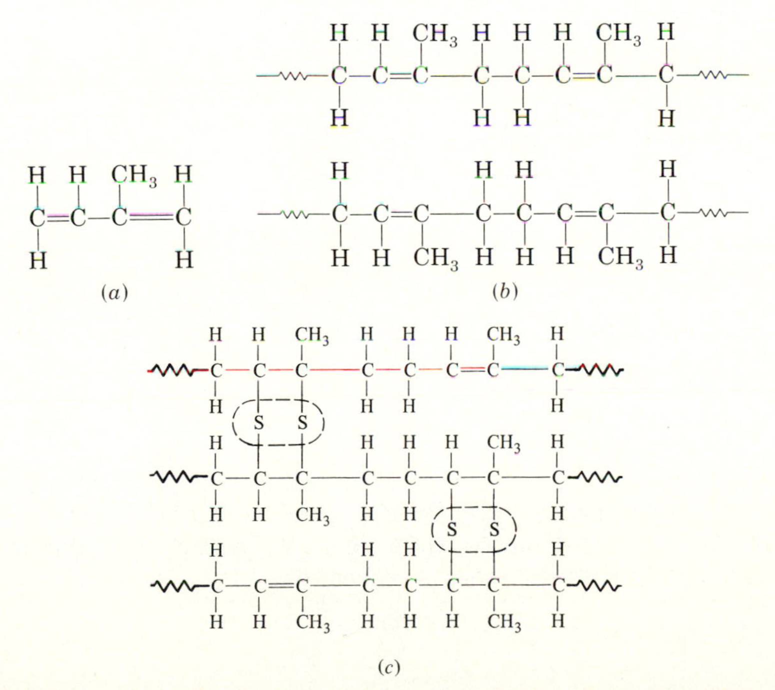

Fig. 9.11 *Cross-linking of rubber by sulfur (vulcanization): (a) isoprene monomer, (b) isoprene polymer, (c) three polymer coil-like chains cross-linked with sulfur (outlined with dashed lines)*

linking is controlled by the amount of sulfur. As this is increased, the rubber changes progressively from a gummy material to a tough, elastic substance as in tires, and finally to hard rubber as in combs and battery cases.

The reason rubber can crosslink is that an unsaturated carbon bond is available. Not all of these are lost in the primary polymerization, as shown in Fig. 9.11 for polyisoprene. The available positions for crosslinking are seldom completely utilized, as the rubber would be too hard.

Ring Scission This is related to addition polymerization and also to crosslinking in that two molecules are joined by a third. However, in this case a ring structure, such as an epoxide group,

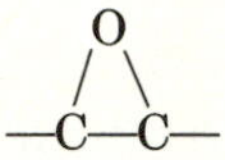

```
  O
 / \
—C—C—
```

is broken by combination with the linking reagent.

EXAMPLE 9.4 As many people know, epoxy cement comes in two tubes. In one tube is the epoxy itself, and in the other is a material that will crosslink the epoxy molecules. Show how polymerization could take place by the breaking of one of the rings (Table 9.4, page 387). Also explain how the reaction may lead to a tough, crosslinked network.

ANSWER

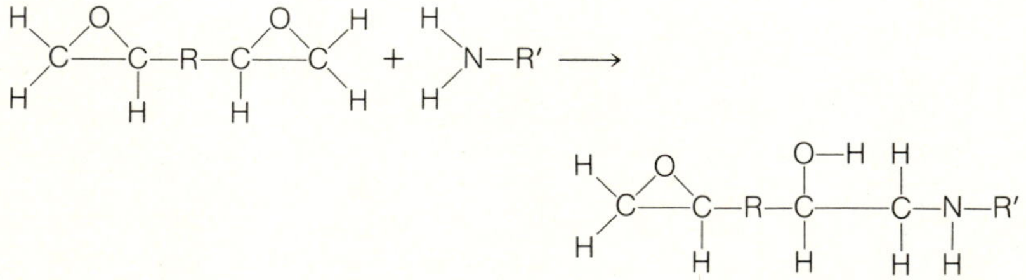

```
H    O          O    H      H
 \  / \        / \  /        \
  C————C—R—C————C     +       N—R' ——→
 /     |    |     \          /
H      H    H      H        H

                    H    O          O—H  H
                     \  / \         |    |
                      C————C—R—C————————C—N—R'
                     /     |    |        |  |
                    H      H    H        H  H
```

If we looked only at the

```
  O
 / \
—C—C—
```

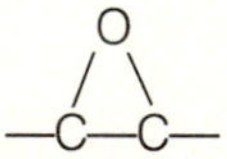

ring (there are two per molecule), we could not explain crosslinking or the formation of a network owing to bifunctionality. However, crosslinking may take place between molecules at the R and R′ groups, which are usually multifunctional. Another possibility is to use a monomer that has more than two

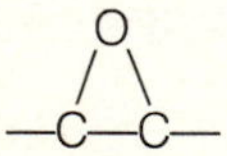

```
  O
 / \
—C—C—
```

rings per molecule. Several epoxy grades actually do this.

Location of Atom Groups It is important to control the symmetry or distribution of a given group or element on the sides of the chain. There are three possibilities, and we shall use polypropylene as an example (Fig. 9.12). In polypropylene the *atactic* (random side group) material is a waxlike product of little use, softening at 74°C. However, both the *isotactic* (same side of chain) and *syndiotactic* (opposite side of chain) types are useful, tough plastics, partly crystalline and melting near 175°C. Bulky side groups may also provide *stereo interference* that limits the symmetry configuration.

Other Important Configurations The most important other configurations are *trans* and *cis* structures. These can be illustrated by the different configurations of the same chemical formula, gutta-percha and natural rubber (Fig. 9.13, page 374).

Note that in the *trans* structure the CH_3 and H are on opposite sides of the double bond, giving a relatively straight carbon chain. In rubber the chain curves (there cannot be later rotation of the *double* bond), leading to a helical molecule that is springlike compared with gutta-percha, which is brittle.

9.7 Crystallinity

Up to this point we have considered only two types of polymer structures: (1) the random arrangement of linear molecules, and (2) the network structure consisting of one large molecule with an amorphous (noncrystalline) structure. A third type of structure, the crystalline, is important in the linear polymers because it leads to stiffer, stronger materials and usually makes an

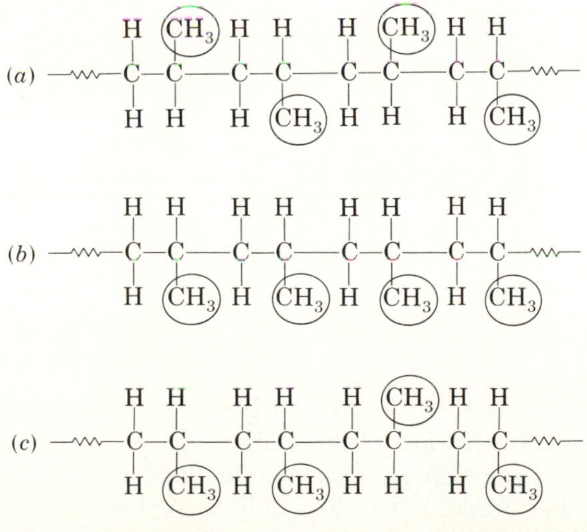

Fig. 9.12 *(a) Syndiotactic (opposite side), (b) isotactic (same side), and (c) atactic (random sides) structures for polypropylene*

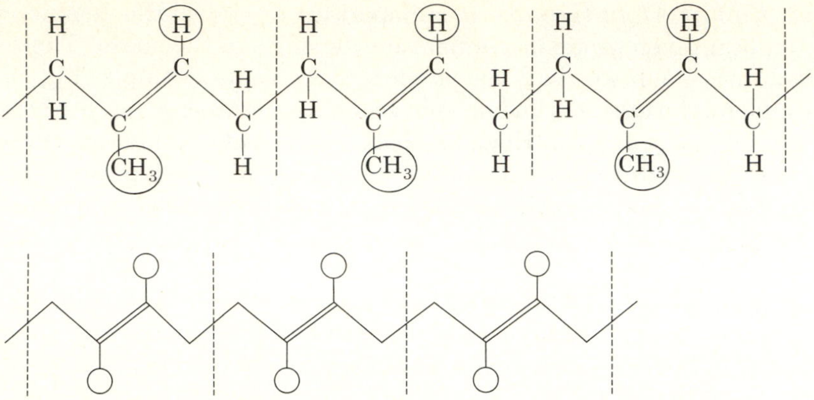

(*a*) trans-polyisoprene
[Possible rotations about indicated C—C bonds (----------) retain the linear structure of the molecule.]

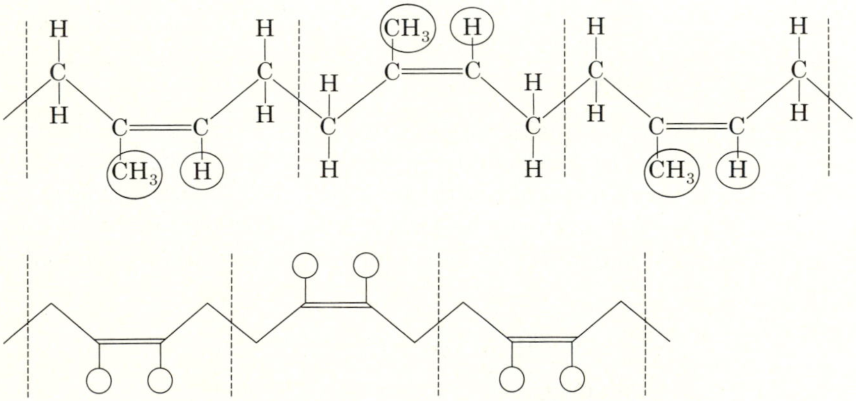

(*b*) cis-polyisoprene
[Possible rotations about indicated C—C bonds (----------) allow the molecule to coil upon itself.]

Fig. 9.13 *Trans and cis configurations. (a) Trans: gutta-percha. (b) Cis: natural rubber.* [Note: *The structure cannot pivot around a C═C bond.*]

amorphous material translucent or opaque because of the light scattering at grain boundaries.

The earliest model of the crystalline structure is called the *fringed micelle* (Fig. 9.14*a*). In this model the molecules were considered to lie side by side in some crystalline regions (the micelles) and to be at random in others. This model is rather crude, but it does help to explain the strengthening that occurs when a fiber such as nylon is drawn: The molecules in the amorphous regions become stretched out and aligned in the direction of the applied stress, raising the yield strength. Recent research using electron microscopy has given more detailed knowledge of the crystalline regions. Researchers have grown single crystals of a number of polymers and have found that the chain structures of the molecules fold back and forth in regular fashion to build up the crystal (Fig. 9.14*b*). They have also found that they can increase the size either by annealing at elevated temperatures or by increasing the temperature of crystallization, as in the crystals of metals. For example, a polyeth-

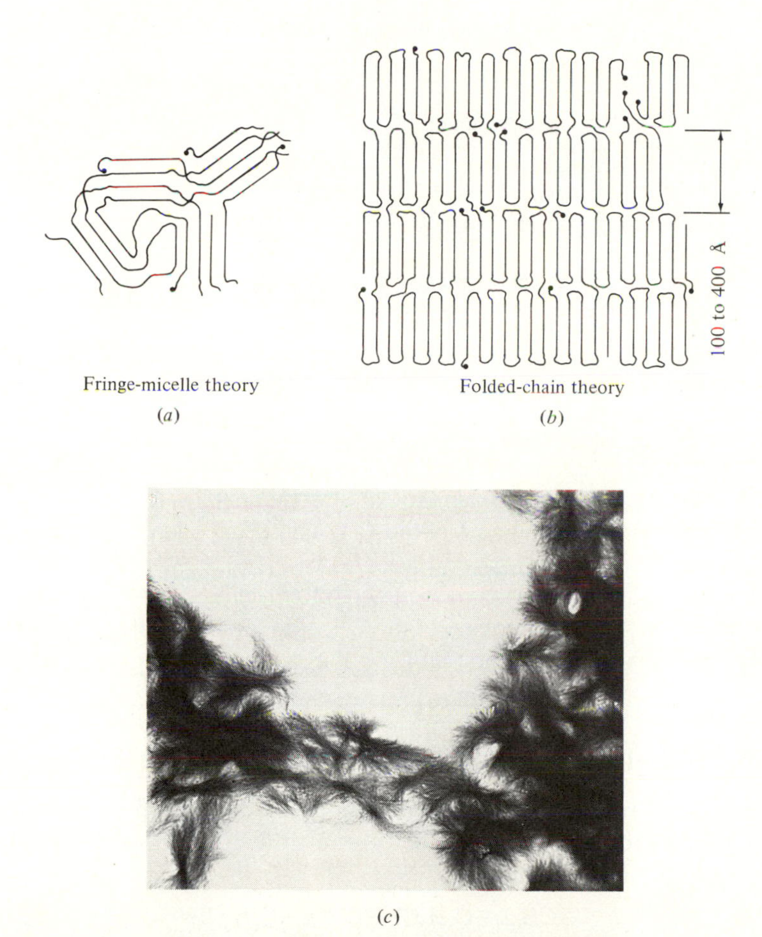

Fig. 9.14 *(a) Fringed-micelle model of crystallinity. (b) Crystallinity created by the folding back of chains. The growth mechanism involves the folding over of the planar zigzag chain on itself at intervals of about every 100 chain atoms. A single crystal may contain many individual molecules. (c) Spherulites in polycarbonate polymer; transmission electron micrograph, 66,000×.* [Parts (a) and (b) from L.E. Nielsen, *Mechanical Properties of Polymers*, Litton Educational Publishing, Inc., New York, 1962. Reprinted by permission of Van Nostrand Reinhold Company. Part (c) courtesy Jim de Rudder, University of Michigan.]

ylene crystal at 100°C might have a typical thickness of 100×10^{-8} cm. Heating at 130°C for several hours will increase it to 400×10^{-8} cm. Figure 9.14*c* shows formation of spherulitic crystals.

The tendency to crystallize is very important and is closely related to the structure and polarity of the molecule. Regular molecules without bulky side groups or branches show strong tendencies to crystallize. It should be emphasized that partial rather than complete crystallization is obtained at best.

EXAMPLE 9.5 Explain the relative percent crystallinity in the following polymers for each numbered example.

Polymer	*Percent Crystallinity*
1. Linear polyethylene	90
Branched polyethylene	40
2. Isotactic polypropylene	90
Atactic polypropylene	0
3. Random copolymers of linear polyethylene and isotactic polypropylene	0
4. *Trans*-1,4-polybutadiene	80
Cis-1,4-polybutadiene	80
Random *cis* and *trans* forms	0

ANSWER

Case 1: The branches interfere with regular arrangement.

Case 2: The random arrangement of the CH_3 side groups leads to noncrystalline material.

Case 3: Introducing a nonregular assortment of ethylene and polypropylene mers leads to a random spacing of the CH_3 side groups of the polypropylene.

Case 4: The situation is similar to part (2) in the random case.

9.8 Copolymers, blending, plasticizers

It is possible to synthesize useful polymers by joining different mers. For example, if we alternate mers of ethylene and vinylchloride, we form the ethylene-vinylchloride copolymer:

$$\text{—}(CH_2)_2\text{—}C_2H_3Cl\text{—}(CH_2)_2\text{—}C_2H_3Cl\text{—}$$

This is an alternating copolymer. Other types of copolymers are random, block (meaning an insert of a number of mers of one kind), and graft (in which the copolymer is added as a branch) (Fig. 9.15). Crystallization in copolymers is only partial at best. Blending or alloying is the blending of two or more distinct polymer molecules such as polyethylene and nylon to form a new product, in this case with unusual permeability.

Blending with a low-molecular-weight (approximately 300) material is

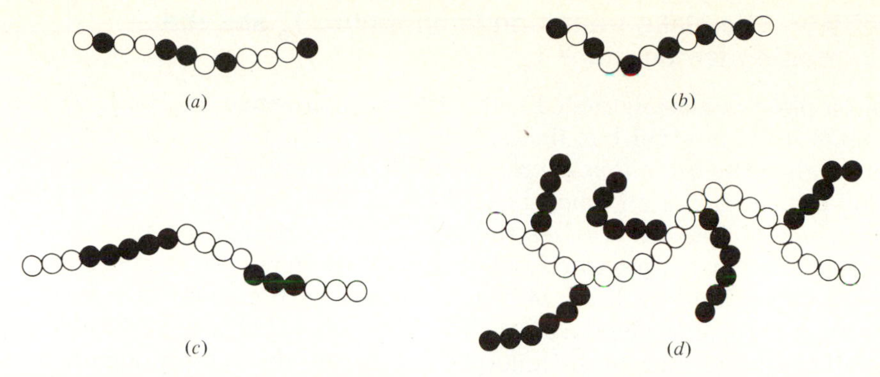

Fig. 9.15 *Different types of copolymers: (a) random, (b) alternate, (c) block, (d) graft*

called using a *plasticizer* because this is a common way to soften a polymer or make it flexible. Just as adding water to clay will plasticize it, adding a small polar molecule can affect the van der Waals forces between the polymer chains. If we add too much plasticizer, the bond is chiefly between the small molecules and we have a liquid. An example is a common paint in which excess plasticizer allows brushing. Then the plasticizer evaporates and the paint dries. This is usually accompanied by some polymerization and crosslinking with oxygen. The residual plasticizer makes the film tough and flexible. With time the combination of further oxidation and loss of plasticizer can result in brittleness and flaking of the paint.

EXAMPLE 9.6 Draw a possible structure of ABS (acrylonitrile-butadiene-styrene) if it is described as "graft of styrene and acrylonitrile on a butadiene backbone" (Table 9.4, page 387). Indicate whether the material is thermosetting or crystalline and whether it could form a network.

ANSWER

```
                          More acrylonitrile
                          ↑
                      H—C—C≡N
                          |
     H        H   H—C—H   H       H   H   H   H
     |        |           |       |   |   |   |   |
 —∿∿—C————————C———————————C———————C———C———C═C—C—∿∿—
     |        |           |       |   |           |
     H    H—C—H           H       H   H           H
              |
  ⟨benzene⟩—C—H                       Butadiene
              ↓
          More styrene
```

The material is thermosetting or thermoplastic, depending on whether it crosslinks to form a network. Since there can still be unsaturated carbon positions in the butadiene (as outlined in the second mer), crosslinking is possible. The material is too irregular to be crystalline (copolymers show little crystallinity).

9.9 The glass transition temperature T_g and the melting temperature T_m

When a piece of glass is cooled below the melting range, it is still plastic for a temperature interval but finally reaches a temperature where it is rigid. This is called the *transition temperature.* We encounter the same phenomenon during the cooling of an amorphous thermoplastic material. Curiously, however, when the *plastic* reaches the transition temperature, it is called the "*glass* transition temperature" T_g. Figure 9.16 shows these characteristics. The specific volume (volume of 1 g) is plotted with change in temperature. Note that there is a sharp contraction at the melting point if the material crystallizes. In the case of the amorphous material the volume decreases more rapidly for a while below T_m, but below T_g the amount of contraction is small.

9.10 Elastomers

The rubbers are essentially amorphous polymers with a glass transition temperature below the service temperature. Their structure is characterized by highly flexible kinked segments that allow freedom of motion. Most commercial rubbers are based on dienes, but many other polymers have rubber-like characteristics. These include the acrylate, fluorocarbon, polysulfide, polyurethane, and silicone rubbers. Despite the looseness of the structure of these materials, all become rigid at some interval below the individual glass transition temperature. This is serious in low-temperature applications. A simple experiment is to immerse a sample of rubber in liquid air and then remove it; it will shatter with a hammer blow.

We have already discussed the *cis* nature of the carbon chain and the use of sulfur as a crosslinking agent. Many other linkages are also used. In fact, the crosslinking of rubber with oxygen and the resultant loss in elasticity is a common aging process in most rubber products.

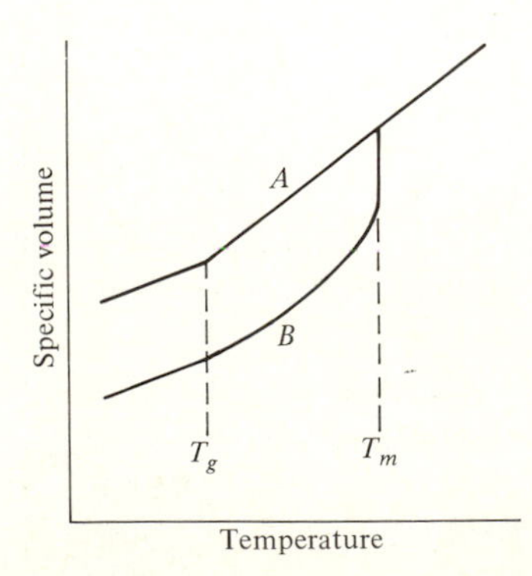

Fig. 9.16 *The glass transition temperature for a polymer. Specific volume vs. temperature for A, an amorphous polymer, and B, a partly crystalline polymer. (T_g = glass transition temperature; T_m = melting temperature.)*

9.11 Fillers

A high percentage of plastics are made with fillers. Phenolic and amino resins are almost always compounded or filled with substances like wood flour, short-fiber cellulose, powdered mica, and asbestos. These materials improve strength and dimensional stability and also decrease the cost of the polymers. The use of glass fibers in polyester resins is of great importance. Finally, carbon black is an important filler and strengthener of rubber. In a common tire mix the actual amount of polymer may be only 60%, with the remainder essentially made up of fillers. Clearly, a discussion of the properties of polymers should carefully distinguish whether plasticizers or fillers have been used (Chapter 10).

Processing of Polymers

9.12 Processing of plastics

The processing of plastics has some of the features of the processing of both metals and ceramics. It is possible to buy sheets, rods, and fibers of many polymers, whereas, in contrast, hard, brittle ceramic shapes must usually be bought in their finished form. On the other hand, in many cases in the fabrication of a plastic part, a chemical mixture is compounded in the same equipment that forms it. For this reason we shall take up the fabrication processes in this chapter while the polymerization methods are still before us.

Fabricators of plastics use practically all the methods we have discussed for metals and ceramics plus a few of their own. One of the key advantages of the plastics industry is the ability to produce accurate components with excellent surface at low cost and high speed. Part of this ability is due to the lower temperatures at which plastics are liquid compared with metals and ceramics. (However, as we shall see in the next chapter, this same property limits their use at elevated temperatures.)

We shall discuss here only the processes for making typical components, leaving specialized processes for fibers, adhesives, etc., for the applications section of the next chapter.

The process we use depends largely on whether we have a thermoplastic or thermosetting material. Table 9.3 (page 380) lists some groups of plastics. We see that injection and extrusion molding are predominant for the thermoplastics, and compression and transfer molding are most important for the thermosetting resins.

Figure 9.17 (page 381) shows compression molding. A slight excess of the material in a preformed blank is placed in the die; the mold is closed. Under heat and pressure the material becomes plastic and the excess is squeezed out as "flash." If this is used for a thermoplastic material, the

Table 9.3 SUMMARY OF PROCESSING METHODS FOR VARIOUS POLYMERS

Polymer	*Compression* Pressure,* psi × 10^3	*Compression* °C	*Injection Molding* Pressure,* psi × 10^3	*Injection Molding* °C	*Extrusion,* °C	*Transfer Molding* Pressure,* psi × 10^3	*Transfer Molding* °C
Thermoplastics							
Polyethylene			5 to 22	135 to 143	80 to 94		
Polypropylene			10 to 22	204 to 288	193 to 221		
Polystyrene			10 to 24	162 to 243	90 to 107		
Polyvinylchloride	0.5 to 2	140 to 175	7 to 15	160 to 175	162 to 204		
Tetrafluorethylene†							
ABS‡		162 to 190	6 to 30	218 to 260			
Polyamides			10 to 20	271 to 343	277 to 299		
Acrylics			10 to 20	160 to 260	177 to 232		
Acetals		200 to 245	15 to 25	193 to 215	188 to 204		
Cellulosics	0.5 to 5		8 to 32	215 to 254	215 to 232		
Polycarbonates			15 to 20	274 to 330	247 to 304		
Thermosetting Polymers							
Phenolics	1.5 to 5	143 to 193	4 to 8	160 to 171		2 to 10	135 to 171
Urea-melamine	2 to 5	149 to 171				6 to 20	149 to 165
Polyesters						1 to 5	121 to 177
Epoxies						0.1 to 2	143 to 177
Silicones§	1 to 3	177				0.5 to 10	177

*Multiply psi by 6.9×10^{-3} to obtain MPa or by 7.03×10^{-4} to obtain kg/mm^2

†Requires preforming of particles and fusion at 370°C.

‡Acrylonitrile-butadiene-styrene.

§Glass-reinforced.

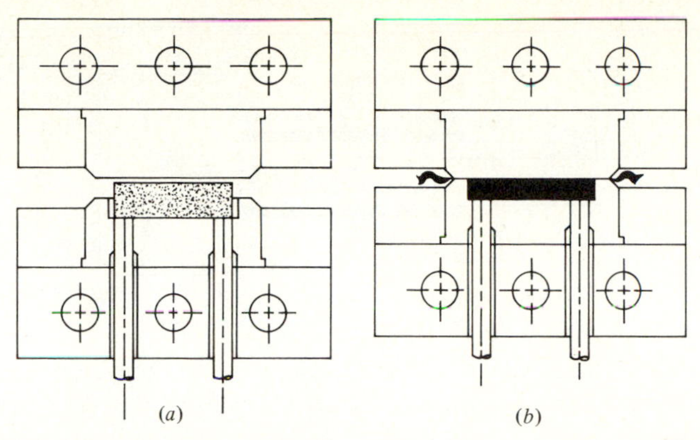

Fig. 9.17 *Compression mold. (a) Open, with preform in place. (b) Closed, with flash forced out between mold halves.* (*Modern Plastics Encyclopedia*, McGraw-Hill, Inc., New York, 1966)

mold must be cooled. Therefore, most thermoplastics are formed by injection instead. However, for a thermosetting material the part may be ejected hot.

In transfer molding (Fig. 9.18), used for thermosetting compounds only, a partly polymerized material is heated to a high enough temperature to flow but not to crosslink in a premolding chamber. The material is then squeezed into the mold itself in the same equipment, where crosslinking takes place at a higher temperature and pressure.

Injection molding (Fig. 9.19, page 382) is done by feeding polymer powder into a barrel, melting it, and then injecting it into a mold. The process resembles the die casting of metals.

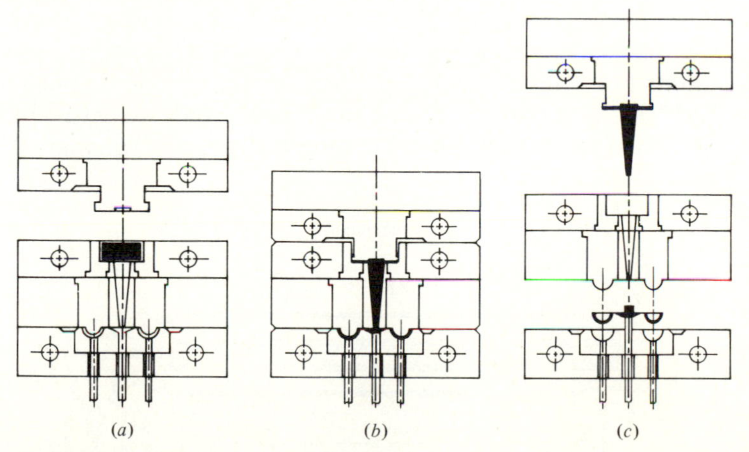

Fig. 9.18 *Typical transfer molding operation. Material is fed into the pot of the transfer mold and then forced under pressure when hot through an orifice and into a closed mold. After the polymer has formed, the part is lifted by ejector pins. The sprue remains with the cull in the pot.* (*Modern Plastics Encyclopedia*, McGraw-Hill, Inc., New York, 1966)

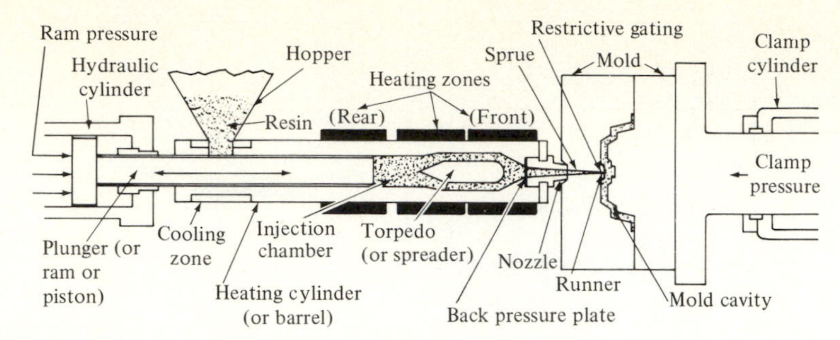

Fig. 9.19 *Schematic cross section of typical plunger (or ram or piston) injection molding machine.* (*Petrothene Polyolefins—A Processing Guide,* 3d ed., U.S. Industrial Chemicals Co., New York, 1965)

Extrusion of tubing, rod, and other shapes of constant cross section is similar to the same process for metals (Fig. 9.20). However, it is necessary to cool the part to near T_g to gain dimensional stability. This is done by running the part into cooling water or by an air blast. In the case of rubber extrusion the rubber is carefully compounded so that it can be formed to the desired shape before it is finally cured to stiffen it.

Plastic bottles are made by a process similar to that used for glass. First a parison is formed; then it is blown into a mold of the final desired shape, as shown in Chapter 8.

SUMMARY

The backbone of the structure of high polymers consists of chains of carbon atoms. In thermoplastic materials the chains have long linear structures, and as a result the materials soften and flow upon heating. In thermosetting materials the chains are connected in a network structure and do not melt.

The backbone structure is made by joining small structural elements called *mers*. When there are only two bonding positions in the mer, the linear

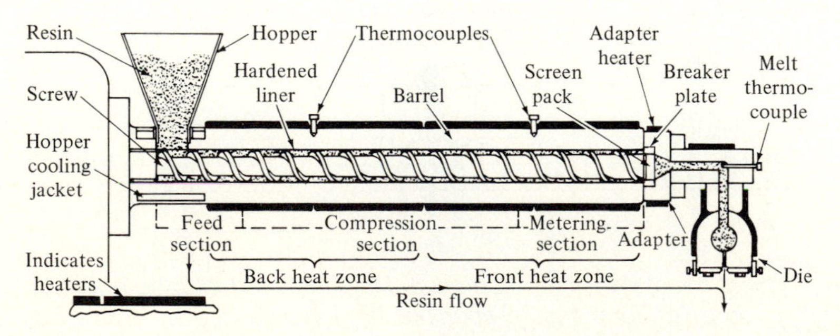

Fig. 9.20 *Schematic cross section of a typical extruder.* (*Petrothene Polyolefins—A Processing Guide,* 3d ed., U.S. Industrial Chemicals Co., New York, 1965)

thermoplastics are formed; whereas if at least one variety of mer in the mix is more than bifunctional, a network will form.

The polymer can crystallize if the mers are linear and possess no side groups on the chain or small, regularly oriented side groups (isotactic or syndiotactic).

Amorphous polymers exhibit behavior similar to that of glasses, passing from a viscous to a rigid structure at the glass transition temperature T_g. They also have a rubbery range above T_g. When a linear polymer is made of several different mers, it is called a *copolymer* and generally will not crystallize.

Most processing is accomplished by compression, transfer, injection, extrusion, and blow molding.

DEFINITIONS

Addition polymerization Bonding between similar and/or different monomers by linking at functional positions without the formation of a condensation product. Also called *chain-reaction polymerization*. When monomers are different, this is called *copolymerization*.

Atactic structure A structure in which the side groups such as CH_3 of a molecule are arranged nonpreferentially on the sides of the chain.

Bifunctional Having two bonding positions. ("Tri" signifies three positions and "tetra" signifies four.)

Blending An intimate mechanical mixture of several polymers.

Blowing A process in which a gob of plastic is formed and then blown into a mold of the desired shape.

Bond strength The energy required to break a particular bond.

Branched polymer A linear polymer with forked branches.

Cis structure A curved carbon backbone produced, for example, by positioning the H and CH_3 groups on the same side of the chain.

Compression molding A process in which a metered amount of plastic is placed in a heated die and compressed to the desired dimensions. It is an important process for thermosetting resins.

Condensation polymerization Bonding between mers that is effected by a reaction in which the bonding takes place, with the emission of a byproduct such as H_2O, NH_3 gas, etc. For example, phenol + formaldehyde gives phenol-formaldehyde resin + water.

Crosslinking Connections made between polymer molecules by a crosslinking agent. For example, butadiene rubber can be crosslinked with sulfur (vulcanization) or oxygen (oxidation).

Crystallinity The alignment of a molecule or molecules in a regular array for which a unit cell can be described. Crystallinity is generally not complete.

Degree of polymerization The molecular weight of a polymer divided by the molecular weight of the mer.

Extrusion The forcing of liquid or plastic resin through a die to obtain the required shape, such as a rod.

Filler Foreign material used to strengthen or modify properties of polymers.

Functionality The number of positions on a monomer at which bonding to another monomer can take place.

Glass transition temperature, T_g The temperature at which a high polymer becomes rigid. T_g is similar to the transition temperature for glasses.

Initiator A material that when added to a monomer acts to initiate polymerization.

Injection molding The injection of liquid resin into a die, followed by solidification and ejection.

Isotatic structure A structure in which the side groups are all arranged on one side of the chain.

Linear polymer A polymer in which the mers are joined in a line rather than a network.

Mer A unit consisting of relatively few atoms, joined to other units to form a polymer.

Monomer The same unit as a mer standing alone, i.e., not part of a polymer.

Network structure The structure formed when one or more of the monomers is tri- or polyfunctional.

Plasticizer A chemical of lower molecular weight added to a polymer to soften or liquefy it.

Polymer A molecule made up of repeating structural groups or mers. For example, polyethylene is made up of $—CH_2—CH_2—$ groups.

Ring scission The break up of a ring structure to provide bonds for polymerization.

Solvent A liquid that dissolves the polymer, for example, in a paint.

Syndiotactic structure A structure in which the side groups are regularly arranged on alternate sides of the chain.

Terminator A material that reacts with the end of a growing polymer chain to terminate growth.

Thermoplastic A high polymer that flows and melts when heated. Scrap may be recovered by remelting and reusing.

Thermosetting plastic A high polymer that sets into a rigid network. Such a polymer does not melt when heated but chars and decomposes, a process called *degradation*. These polymers are not reusable.

Trans structure A relatively straight carbon backbone produced, for example, by positioning the H and CH_3 on alternate sides of the chain.

Transfer molding A process in which material is fed into the pot of a transfer mold, heated, then forced into an adjoining cavity of the same mold under pressure.

PROBLEMS

Let us devote our efforts to synthesizing the important polymers not previously treated as examples. Many of the materials and chemicals needed are included in Tables 9.1 (page 364) and 9.4 (page 387). Most are relatively simple, but a few are complex.

9.1 Another amino resin can be formed from melamine and formaldehyde. The structure of melamine is:

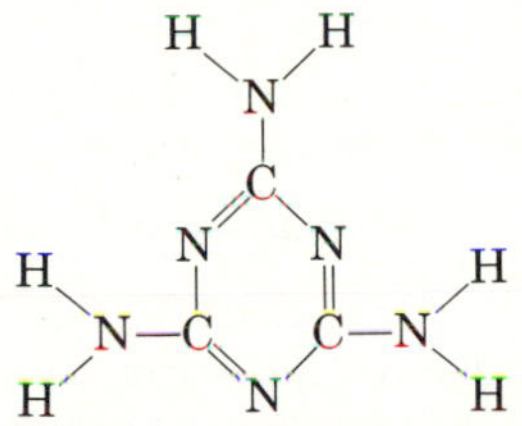

```
            H     H
             \   /
               N
               |
               C
             //  \
 H          N     N          H
  \         |     ||        /
   N — C           C — N
  /         \\   /          \
 H             N             H
```

Show the polymerization reaction of melamine-formaldehyde and indicate the functionality of melamine if all hydrogens are replaceable. (Sections 9.1 through 9.5)

9.2 Table 9.4, pages 387–389, shows the cellulose mer. Despite the complex structure, the regularity leads to crystallinity. Cellulose esters were the most important group of plastic materials before 1950. Remembering that an ester is an alcohol (OH group) plus an acid, sketch the formation of the ester cellulose acetate. (Sections 9.1 through 9.5)

9.3 The most important nylons are 6/6, 6, and 6/10. The numbers refer to the number of carbon atoms in the mers that are joined by a condensation reaction. Let us take 6/6 nylon, which is formed from adipic acid and hexamethylene diamine (Fig. 9.8). Actually the condensation takes place by combining the H of the NH_2 group with an OH to form water. Calculate the energy of the bonds broken and of the new bonds formed. Compare the alternative possibility of breaking off an NH_2 group and the H of the OH group to form NH_3 gas as a condensation product. Is the nylon thermoplastic or thermosetting? (Sections 9.1 through 9.5)

9.4 Polytetrafluorethylene (TFE, Teflon) is an important plastic. What is the structure of the mer? Is the polymer thermoplastic or thermosetting? Suggest how it is attached to frying pans. (Sections 9.1 through 9.5)

9.5 The basis of the *silicone* structure is silane, SiH_4, which is analogous to methane, CH_4. Silicon hydrides are named according to the number of silicon atoms in the chain; thus Si_3H_8 is trisilane. When chlorine replaces hydrogen, we have chlorosilane. It is possible to produce a backbone structure analogous to the acetal structure consisting of alternating silicon and oxygen atoms. These are the silicones. They are prepared by first reacting chlorosilanes with water to form the trihydroxy silane in Table 9.4. Show how these hydroxyl compounds can undergo condensation polymerization, giving a silicone. (Sections 9.1 through 9.5)

9.6 What would be the degree of polymerization if the average molecular weight of a styrene polymer was 73,000 g/mol. wt.? (Sections 9.1 through 9.5)

9.7 Given the average molecular weight of polyvinyl chloride as 27,500 g/mol. wt., what is the average degree of polymerization? (Sections 9.1 through 9.5)

9.8 *a.* Sketch the polymerization process in the formation of polyacrylonitrile. (Sections 9.1 through 9.5).
b. The average molecular weight of the polymer is 24,350 g/mol. wt. What is the average degree of polymerization?
c. What would be the advantage if the degree of polymerization were doubled?

9.9 From the calculation of bond energies, what is the possibility of the following chemical reaction? $CH_4 + H_2O \rightarrow CH_3OH + H_2$ (Sections 9.1 through 9.5)

9.10 In the polymerization of urea formaldehyde (Example 9.3), how many pounds of water are evolved per pound of polymer formed? (Sections 9.1 through 9.5)

9.11 A researcher has suggested the following condensation polymerization between ethylene glycol and acetone: (Sections 9.1 through 9.5)

H—O—C(H)(H)—C(H)(H)—O—H + H—C(H)(H)—C(=O)—C(H)(H)—H → H—O—H

Ethylene glycol   Acetone

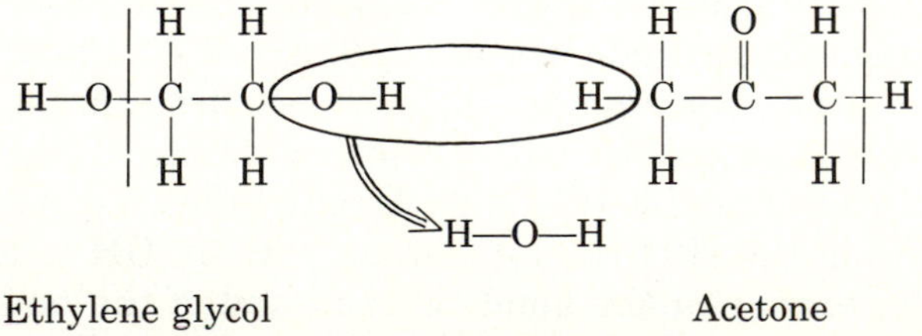

a. Show by calculating the energy involved whether the proposed reaction may be expected to take place.
b. Calculate how many pounds of water are formed per pound of mer (outlined above).

Table 9.4 SUMMARY OF IMPORTANT POLYMERS

Group I. Thermoplastics

Polymer	*Percentage of Market*	*Monomer(s) Used*
Polyethylene	28	$H_2C{=}CH_2$
Polyvinylchloride	17	$H_2C{=}CHCl$
Polystyrene	16	$H_2C{=}CH(C_6H_5)$ (benzene ring is benzene, C_6H_6)
Polypropylene	5	$H_2C{=}CH(CH_3)$
ABS	2	$H_2C{=}CH(C{\equiv}N)$ Acrylonitrile (graft) $H_2C{=}CH{-}CH{=}CH_2$ Butadiene (chain) $H_2C{=}CH(C_6H_5)$ Styrene (graft)
Acrylics (example: polymethyl methacrylate, Lucite)	2	$H_2C{=}C(CH_3){-}C({=}O){-}O{-}CH_3$
Cellulosics	1	Cellulose ring: $-O-$CH(H)$-$ ... ring of C–O with CH_2OH, H, OH, OH substituents (mer of cellulose)

Table 9.4 SUMMARY OF IMPORTANT POLYMERS (*Continued*)

Polymer	Percentage of Market	Monomer(s) Used
Group I. Thermoplastics (Continued)		
Acetals	<1	$\sim\sim CH_2-O \sim\sim$ (mer)
Nylons	<1	$H_2N-(CH_2)_6-NH_2$ $HO-C(=O)-(CH_2)_4-C(=O)-OH$
Polycarbonates	<1	$\sim\sim O-C(=O)-O-C_6H_4-C(CH_3)_2-C_6H_4 \sim\sim$ (mer)
Fluoroplastics (example: polytetrafluorethylene)	<1	$F_2C=CF_2$
Polyester, thermoplastic type [Example: polyethylene-terephthalate (dacron)]	3	$HO-CH_2-CH_2-O-H$ $HO-C(=O)-C_6H_4-C(=O)-OH$
Group II. Thermosetting Polymers		
Phenolics (Example: phenol formaldehyde)	6	C_6H_5OH $O=CH_2$ C_6H_5OH
Amino resins (Example: urea formaldehyde)	4	$H_2N-C(=O)-NH_2$ $H_2C=O$ $H_2N-C(=O)-NH_2$
Polyesters, thermoset type	1	$H_2C(OH)-HC(OH)-H_2C(OH)$ $HO-C(=O)-(CH_2)_x-C(=O)-OH$
Epoxies	1	$H_2C(-O-)CH-R-CH(-O-)CH_2$; H_2N-R' (R and R′ are complex polyfunctional molecules)

Table 9.4 SUMMARY OF IMPORTANT POLYMERS (*Continued*)

Group II. Thermosetting Polymers (Continued)			
Polymer	*Percentage of Market*	*Monomer(s) Used*	
Polyurethane, also thermoplastic	1	OCN—R—NCO + HO—R′—OH (diisocyanate)	(R and R′ are complex polyfunctional molecules)
Silicones	1	CH_3–Si(Cl)(Cl)–Cl: Cl—Si—Cl with CH_3 above and Cl below	Trichlorosilane
		H—O—Si—OH with CH_3 above and OH below	Trihydroxy silane

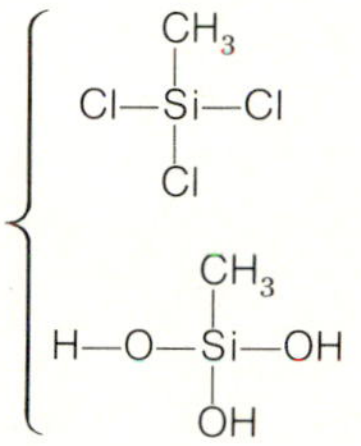

9.12 Figure 9.8 shows the polymerization of phenol-formaldehyde. Show by calculation whether the following is also possible. (Sections 9.1–9.5)

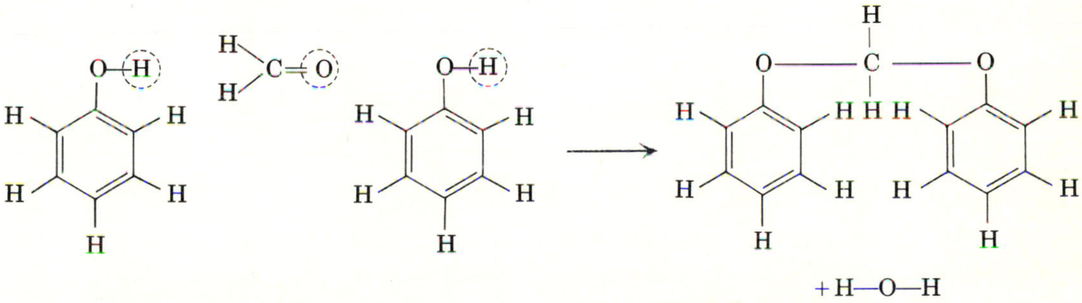

+ H—O—H

9.13 Two possibilities are shown for a condensation polymerization between melamine and formaldehyde. Show by calculation which reaction has the greater chance of occurrence (calculate the bond energies associated with the two reactions). (Sections 9.1 through 9.5)

a.

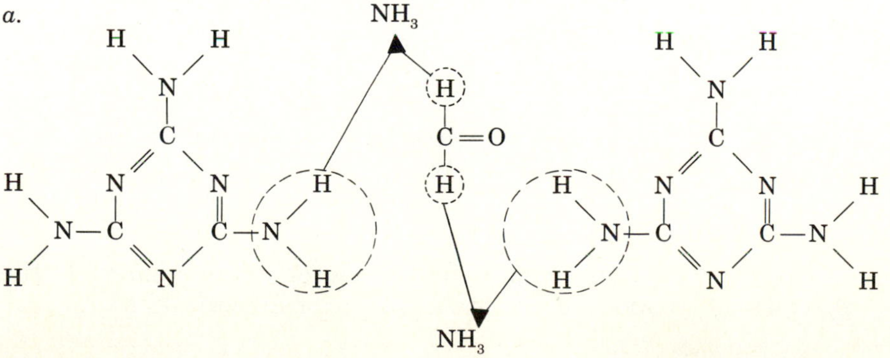

NH_3

NH_3

to form [structural formula]

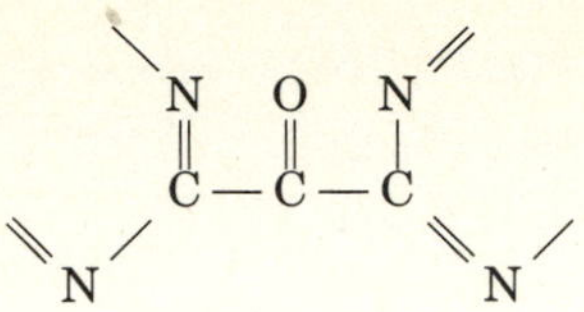

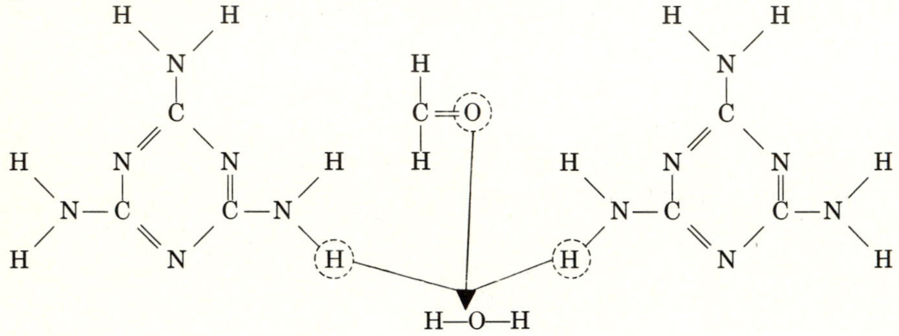

b. [structural formulas: two melamine molecules and formaldehyde, H—O—H eliminated]

to form [structural formula]

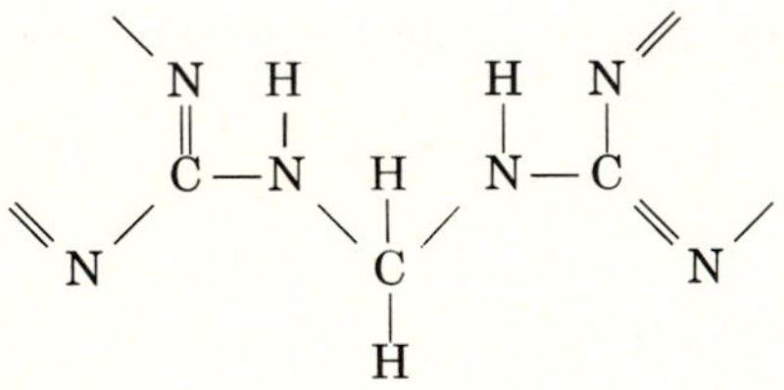

c. Would the final polymer be thermosetting or thermoplastic?

9.14 It has been suggested that the following polymerization will take place:

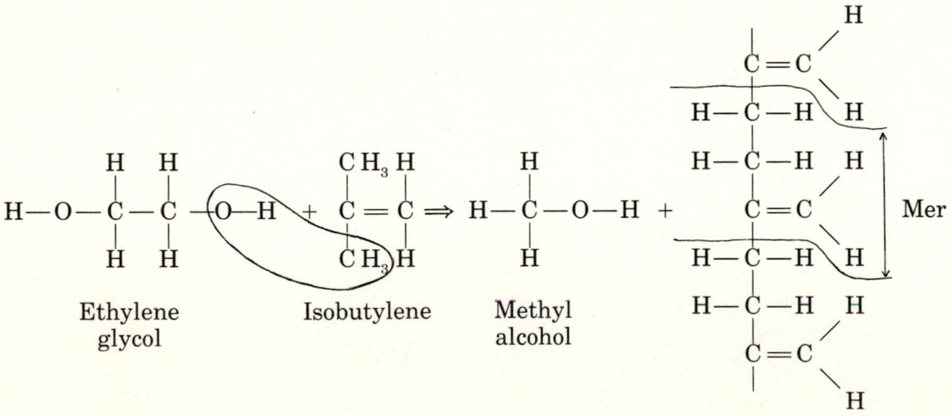

$$\text{H—O—}\underset{\text{H}}{\overset{\text{H}}{\text{C}}}\text{—}\underset{\text{H}}{\overset{\text{H}}{\text{C}}}\text{—O—H} + \underset{\text{CH}_3}{\overset{\text{CH}_3}{\text{C}}}=\underset{\text{H}}{\overset{\text{H}}{\text{C}}} \Rightarrow \text{H—}\underset{\text{H}}{\overset{\text{H}}{\text{C}}}\text{—O—H} + [\text{polymer chain}]$$

Ethylene glycol | Isobutylene | Methyl alcohol | Mer

a. Is the polymer a thermoplastic, thermoset, or elastomer? Is the polymerization addition, condensation, or copolymerization?

b. Calculate the energy exchange in the polymerization.
c. Calculate the number of grams of methyl alcohol produced per gram of ethylene glycol plus isobutylene.
d. What is the average degree of polymerization if the average molecular weight of the polymer is 34,500 g/mol. wt.? (Sections 9.1 through 9.5)

9.15 Polyethylene is sold in two forms, low-density (20% of the market) and high-density (8% of the market). The low-density form shows a large amount of branching, whereas the high-density type is relatively free of branching and is ordered. (Sections 9.6 through 9.7)

a. Sketch both types.
b. Which type will show greater crystallization?
c. Is it thermoplastic or thermosetting?

9.16 Consider polyvinylchloride (PVC). (Sections 9.6 through 9.7)

a. Sketch the polymer and the copolymer with ethylene.
b. Could either crystallize? (*Hint:* Polyvinylchloride is syndiotactic.)
c. Is it thermoplastic or thermosetting?

9.17 The polymer made from phenol and formaldehyde is an example of the phenolics. Would it be: (*a*) Crystalline? (*b*) Thermoplastic or thermosetting? Sketch the structure. (Sections 9.6 through 9.7)

9.18 Sketch the polypropylene molecule with (*a*) Atactic structure, (*b*) Isotactic structure, (*c*) Syndiotactic structure. Which would be amorphous and which could crystallize? (Sections 9.6 through 9.7)

9.19 The term *ester* is applied to the product resulting from the reaction of an alcohol and an acid. If we let R and R′ stand for the remainder of the molecule, such as C_2H_5, etc., the reaction is

$$R \cdot OH + R'COOH \longrightarrow R' \cdot \underset{\underset{\displaystyle O}{\|}}{C} - O - R + H_2O$$

Vinylacetate is considered an ester because it has the vinyl group from vinyl alcohol† and the acetate group from acetic acid. The polymer is polyvinylacetate. Sketch its structure. Would this be thermoplastic or thermosetting? Is it crystalline?

For us to use the term polyester *resin,* the oxygen must be part of the carbon backbone. Therefore, the polyester resins are more complex. A diacid (two COOH groups) and a dialcohol (two OH groups) are reacted to give a linear polymer. However, if there are also double bonds, the polymer can be crosslinked, using the unsaturated bonds to give a network structure as shown in Fig. 9.11. Also, a trialcohol or polyalcohol may be used with a diacid to give a network structure. Thus there is a

†Vinyl alcohol does not exist as a monomer, but polyvinyl alcohol is a real material.

wide variety of materials in this polyester family. (Sections 9.6 through 9.7)

9.20 The most common acrylic is polymethylmethacrylate, sold as Lucite or Plexiglas. Table 9.4 gives the methylmethacrylate structure. Sketch the structure of the polymer. Is it linear or network? Is it crystalline? Thermoplastic? (Sections 9.6 through 9.7)

9.21 After over 50 million dollars in research a method was found for making acetal resin by polymerization of formaldehyde,

O
‖
H—C—H

Show how this is formed from formaldehyde. Is the material thermoplastic or thermosetting? Is it crystalline? Compare the structure with polyethylene (see Table 9.4). (Sections 9.6 through 9.7)

9.22 Table 9.4 shows the structure of a polycarbonate. The molecule is regular in structure and can be crystallized by evaporation or long heating. Would you expect it to be thermoplastic? (Sections 9.6 through 9.7)

9.23 If all the isoprene in Fig. 9.11 is crosslinked with sulfur, what percent by weight sulfur would have been added? (weight of sulfur/weight of sulfur + isoprene)? (Sections 9.6 through 9.7)

9.24 By calculation of energy values, show that polymerization of epoxy, as shown in Example 9.4, will indeed take place. (Sections 9.6 through 9.7)

9.25 Using only the following chemicals, illustrate with *structural*† *formulas* the following reactions. Be sure that all bonds satisfy valences. Show the unsatisfied bonds of a continuing polymer thus: —∿—C═C═C—∿— (Sections 9.6 through 9.7)

Chemicals:

CH_2O, C_6H_5OH, C_2H_4, C_2H_3Cl, CH_3NH_2, H_2O_2,

H O O H
\ / \ / \ /
C — C—R—C — C
/ | | \
H H H H

1. Formation of copolymer by addition polymerization
2. Formation of thermosetting resin by condensation polymerization
3. Polymerization by ring scission
4. Formation of a *cis* structure
5. Formation of a syndiotactic copolymer

†Example:

H H H H
\ / \ /
C—C H C—C
// \\ \ // \\
H—C C—H + O → H—C C—O—H + H—H
\ / / \ /
C═C H C═C
/ \ | |
H H H H

9.26 Rubber will oxidize by crosslinking at unsaturated (—C=C—, with a single bond drawn up from each C) positions. If the O=O bond is 118 kcal/mol, calculate the energy exchange for oxidation of rubber. How is the oxidation tendency decreased in, for example, a rubber tire? (Sections 9.6 through 9.7)

9.27 Illustrate the following phenomena by drawing the molecules, using only the materials listed below (the same material may be used more than once). (Sections 9.6 through 9.7) *Phenomena:* Atactic structure, branching, copolymer, polymer crystal (specify polymer used in drawing). *Materials:* Polyethylene, phenol-formaldehyde, melamine, polystyrene.

9.28 Fiberglass is used in boat hulls and automobile bodies. Everyone is interested in this application of glass plus plastic. Although a number of basic steps are involved, we have built up the background to understand them. (Sections 9.8 through 9.11)

First the glass must be chosen. There are two types of glass: low-alkali and high-alkali (15%) borosilicate. Which gives better weathering characteristics and is more expensive (see the discussion of ceramics)?

Next the glass has to be prepared for later blending with the resin. It is pretreated with a silane. Vinyl trichlorosilane, $Cl_3SiCH{=}CH_2$, is hydrolyzed in the presence of glass fiber. The OH groups of the $(HO)_3SiCH{=}CH_2$ react with the OH groups on the glass surface, leading to condensation of water and an —O— bond to the glass. The vinyl group is on the other side of the —O— bond, ready to bond with the polyester. Sketch how these reactions give the structure in Fig. 9.21 (page 394).

Now let us consider a typical polyester formula, which appears complex at first: 0.2 mole phthalic anhydride, 0.2 mole maleic anhydride, 0.2 mole propylene glycol, 0.2 mole ethylene glycol, 0.2 mole styrene, trace hydroquinone (0.02%).

This mixture is compounded carefully, giving a ready-to-use liquid. It really has two parts, the polyester and the monomer. We recall that an ester is made of two parts, an acid and an alcohol. In this case two acids (dehydrated to give the anhydrides) are present. One is an unsaturated acid, so that crosslinking into a network structure can be obtained later, and the other is saturated to avoid an excess of crosslinking, which would lead to brittleness. Figure 9.21 (page 394) shows formation of the polyester, expressing anhydrides as acids. The hydroquinone stabilizes the mixture until it is to be used. When the glass mat is ready, an initiator (peroxide) and accelerator are added to the polyester-styrene mixture. The mixture polymerizes, first gelling and then hardening. Sketch how the styrene can: (*a*) polymerize, (*b*) bond to the polyester, (*c*) bond to the glass-silicone face.

9.29 In Fig. 9.21, calculate the pounds of water released per pound of polyester formed [refer to part (*b*); maleic acid + ethylene glycol + phthalic acid provide one mer]. (Sections 9.8 through 9.11)

9.30 In polymers, what structural differences exist above and below the glass transition temperature? (Sections 9.8 through 9.11)

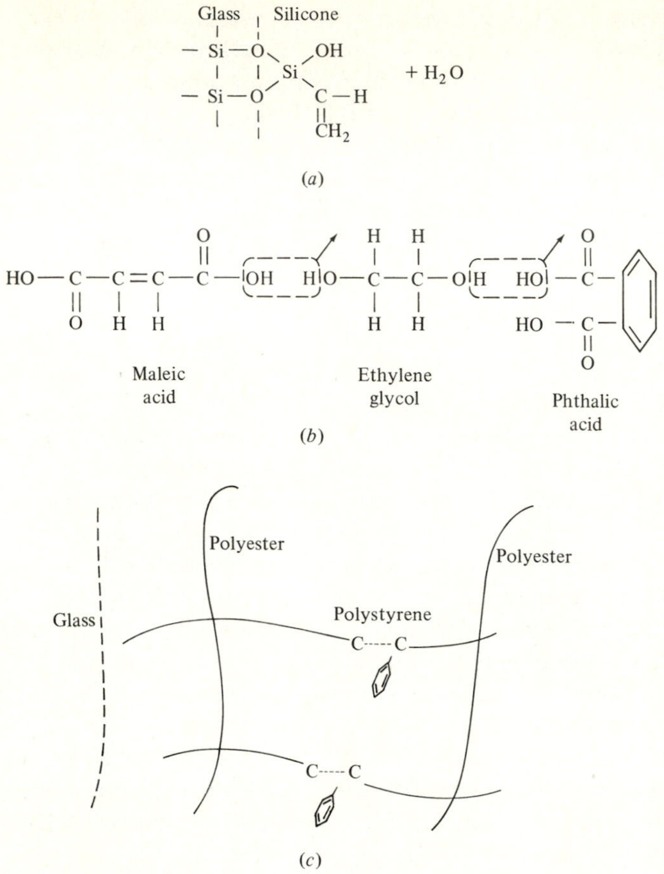

Glass Silicone

$+ H_2O$

(a)

Maleic acid

Ethylene glycol

Phthalic acid

(b)

Polyester

Polyester

Glass

Polystyrene

(c)

Fig. 9.21 *Formation of bonds in polyester-styrene-fiberglass: (a) glass treated with silicone, (b) formation of polyester, (c) final structure*

9.31 The data below give a group of chemicals that are to be used alone or in combination to illustrate the formation of the following components. Draw the structures of the components and indicate how they combine to form the following:

a. A transparent thermoplastic copolymer
b. A translucent thermoplastic linear polymer
c. A highly elastic, crosslinked elastomer
d. An opaque network polymer (thermoset)
e. A thermosetting *polyester* made by condensation polymerization (Sections 9.8 through 9.11)

1. Formaldehyde, CH_2O
2. Sulfur

3. Carbon black
4. Ethylene, C_2H_6
5. Vinylchloride, C_2H_5Cl
6. Ethyl alcohol, C_2H_5OH
7. Styrene, $C_2H_5\ (C_6H_5)$
8. Phenol, C_6H_5OH
9. Tetrafluoroethylene, C_2F_4

10. A trialcohol,

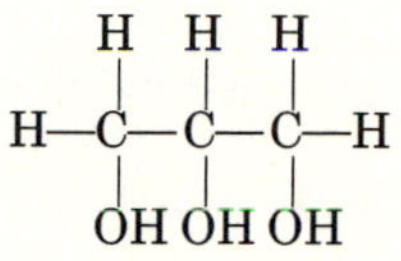

$$\begin{array}{ccccccccc} & & H & & H & & H & & \\ & & | & & | & & | & & \\ H & - & C & - & C & - & C & - & H \\ & & | & & | & & | & & \\ & & OH & & OH & & OH & & \end{array}$$

11. A diacid,

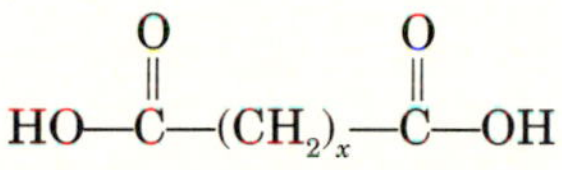

$$HO-\overset{\overset{\displaystyle O}{\|}}{C}-(CH_2)_x-\overset{\overset{\displaystyle O}{\|}}{C}-OH$$

12.

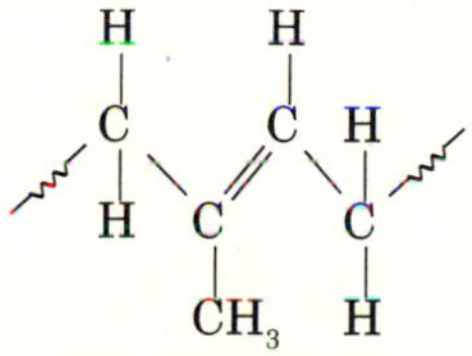

$$\begin{array}{ccccccc} & H & & H & & & \\ & | & & | & & & \\ & C & & C & H & & \\ \sim / & | & \diagdown & /\!/ & \diagdown & | & / \sim \\ & H & C & & & C & \\ & & | & & & | & \\ & & CH_3 & & & H & \end{array}$$

13.

$$\begin{array}{ccccccc} & H & & & & H & \\ & | & & & & | & \\ & C & & & & C & \\ \sim / & | & \diagdown & & / & | & \diagdown \sim \\ & H & C & = & C & H & \\ & & | & & | & & \\ & & CH_3 & & H & & \end{array}$$

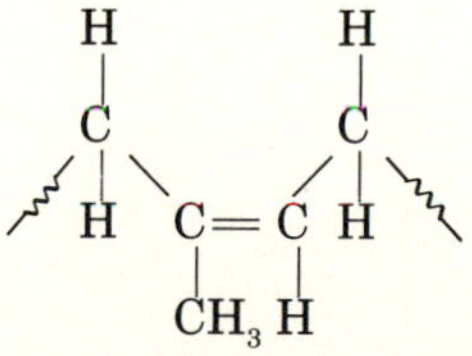

9.32 From the choices given by the sketches, select *one* that illustrates each of the following characteristics: (You may use the same symbol several times.) (Sections 9.8 through 9.11)

1. Linear polymer
2. Condensation polymerization
3. Atactic structure
4. Isotactic structure
5. Syndiotactic structure
6. Copolymer
7. Trifunctional mer essential for thermosetting polymer
8. Mer for thermoplastic polymer
9. Polymer that could have crystalline structure
10. Polymer that could not have appreciable crystallinity
11. Polymer with high strength because of crosslinking
12. Elastomer
13. Crosslinking agent
14. Mer for polymer strengthened by chain stiffening

a.

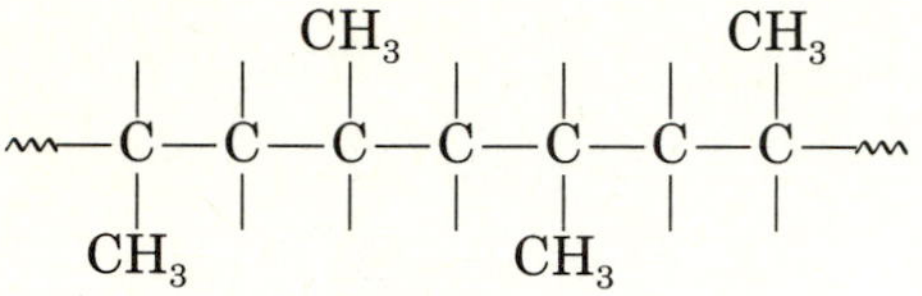

```
              CH3                 CH3
    |    |    |    |    |    |    |
 ~~-C -- C -- C -- C -- C -- C -- C-~~
    |    |    |    |    |    |    |
    CH3                 CH3
```

b.

```
    |    |    |    |    |    |    |
 ~~-C -- C -- C -- C -- C -- C -- C-~~
    |    |    |    |    |    |    |
    CH3       CH3       CH3       CH3
```

c.

```
                        CH3
    |    |    |    |    |    |    |
 ~~-C -- C -- C -- C -- C -- C -- C-~~
    |    |    |    |    |    |    |
    CH3       CH3                 CH3
```

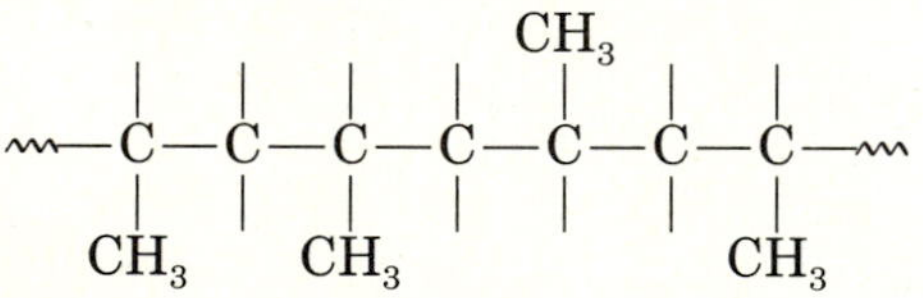

d. $n(CH_2O) + n(C_6H_5OH)$; *e.* Sulfur; *f.* Phenol

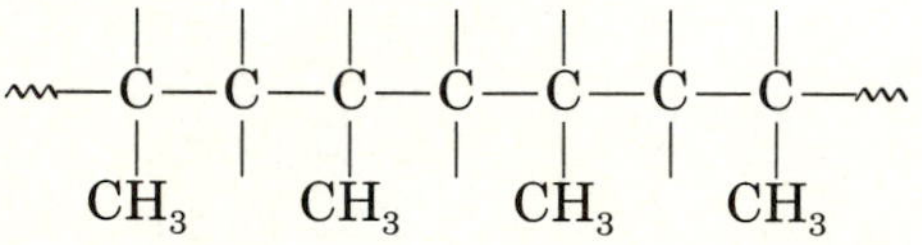

g.

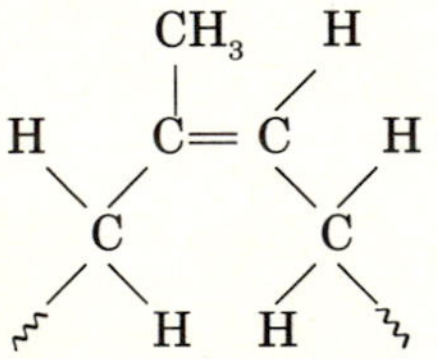

```
          CH3   H
           |   /
    H      C = C     H
     \    /     \   /
       C          C
     /   \      /   \
   ~~     H    H     ~~
```

h.

```
          CH3   H
           |    |
    H      C    C
     \    /  \\/ | \
       C       C H  ~~
     /   \     |
   ~~     H    H
```

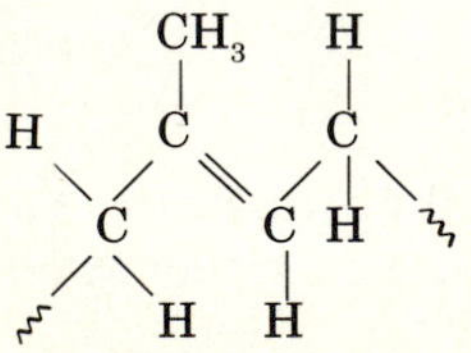

i.

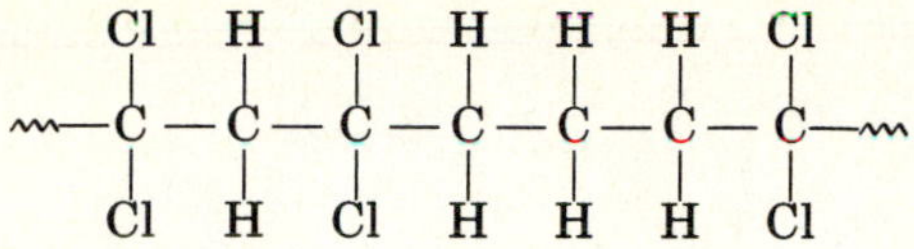

```
     Cl   H   Cl   H   H   H   Cl
     |    |   |    |   |   |   |
  ~~-C -- C - C -- C - C - C - C-~~
     |    |   |    |   |   |   |
     Cl   H   Cl   H   H   H   Cl
```

j. Vinyl chloride; *k.*

```
H  CH3 O       H
|  |   ||      |
C==C - C - O - C - H
|              |
H              H
```

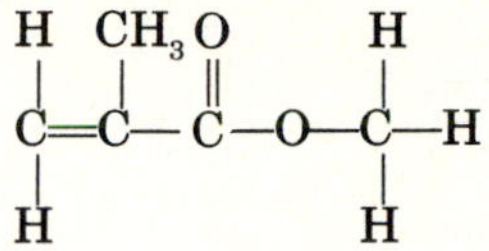

9.33 One or more of the listed terms relates to each of the following sketches of polymer structures. For each structure, indicate the appropriate term(s) that describe the polymer. (Sections 9.8 through 9.11)

a. atactic
b. bifunctional
c. branching
d. condensation
e. copolymer
f. crosslinking
g. isotactic
h. monomer
i. polyfunctional
j. silicone
k. thermoplastic
l. thermoset
m. vinyl
n. vulcanization

1.

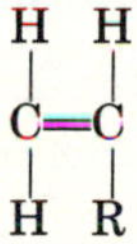

```
H  H
|  |
C==C
|  |
H  R
```

2.

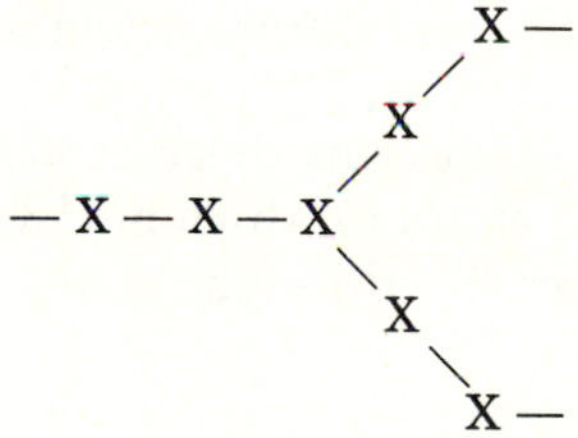

```
                    X —
                   /
                 X
                /
— X — X — X
              \
               X
                \
                  X —
```

3.

```
      CH3              CH3
      |                |
H)O — Si —(OH  H)O — Si — O(H
      |                |
      O                O
      H                H
```

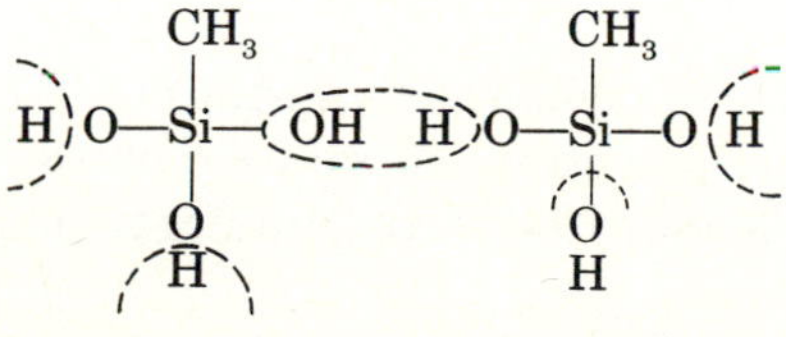

4.

```
—#—#—#—#—#—
      |
      ⊗
      |
  —#—#—#—#—#—
```

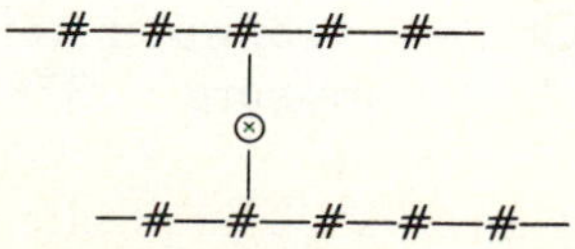

(Continued on page 398)

5. —N(H)(H) ··· HO—C ----

6. —X—Y—X—X—X—Y—X—Y—Y—

7. 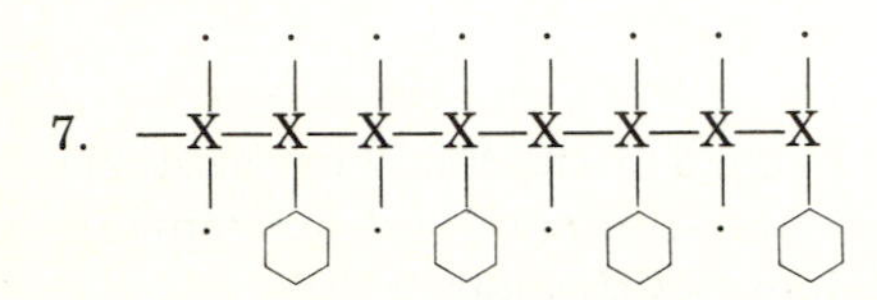

9.34 *a.* Show by a sketch how butadiene rubber [$CH_2(CH)_2CH_2$] can polymerize and crosslink with oxygen.
b. Show by *calculation* whether the crosslinking with oxygen would be a spontaneous reaction. (O═O bond is 118 kcal/mol)
c. What would be the increase in weight if oxygen crosslinking were to occur at all available positions? (Use units of pounds increase per pound of butadiene.)
d. Explain how oxygen crosslinking is minimized. (Sections 9.8 through 9.11)

9.35 ABS is described as a backbone or chain of butadiene to which has been grafted acrylonitrile and styrene. (The product is a copolymer.)

a. From the above description, sketch a possible configuration, using A, B, and S to represent the individual mers.
b. Would you expect the polymer to be thermoplastic or thermosetting?
c. Why would you not expect to have a styrene chain to which butadiene and acrylonitrile had been grafted? (Sections 9.8 through 9.11)

PROBLEMS COVERING OPTIONAL SECTIONS

9.36 Which processing method would you use to make the following? (*a*) Saran Wrap sheet, (*b*) polypropylene rope, (*c*) polyethylene squeeze bottle, (*d*) dish of melamine, (*e*) nylon fishing leader (0.007 in. diameter, clear) (Section 9.12)

9.37 How many gallons of water are produced from the complete combustion of 10 lb of methane (CH_4)? (Specific gravity of CH_4 = 0.415; water weighs 8.33 lb/gal; 1 lb mole of any gas at 60°F and 1 atm pressure = 378 ft^3. (Requires chemistry)

9.38 The local dump is planning to burn polymeric refuse. The public has expressed concern about airborne chlorine as a result of the combustion process. If we assume complete combustion of hydrogen and carbon to H_2O and CO_2, respectively, how much chlorine would it be possible to obtain per pound of vinyl chloride burned? If we assume that HCl is formed rather than chlorine, how much of the acid would be formed per pound of vinyl chloride? (Requires chemistry)

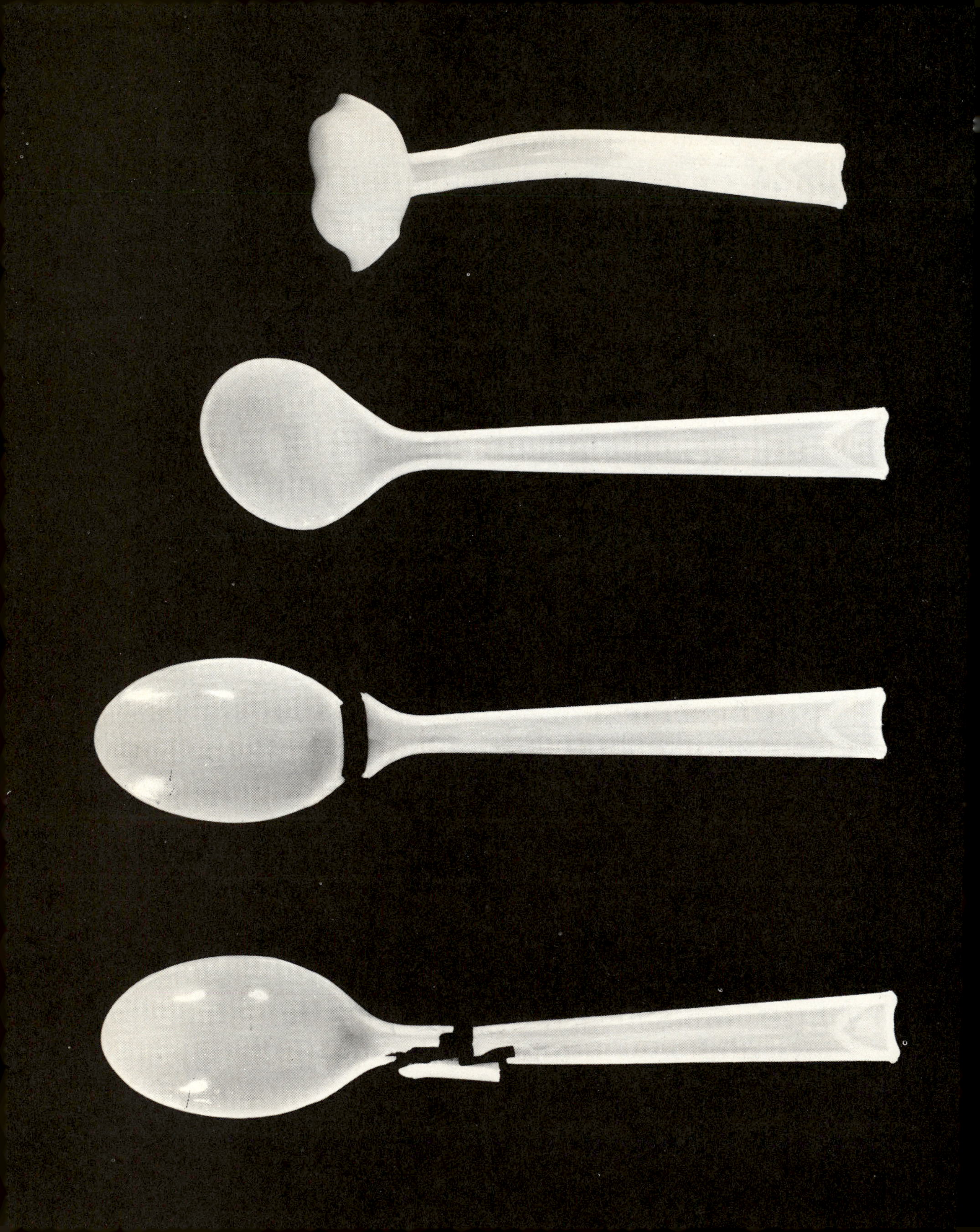

10

PROPERTIES AND APPLICATIONS OF PLASTICS

THE spoons in the illustration were bent after immersion in water at different temperatures to show the very important effect of temperature on the mechanical properties of a thermoplastic material. The spoon at the top was immersed in boiling water; it curled up readily when lightly stressed. The next spoon, immersed at 150°F (66°C), bent less readily. The third spoon, immersed at 120°F (49°C), fractured, and the bottom spoon, immersed at 70°F (21°C), splintered.

In this chapter we shall compare the mechanical and thermal properties of high polymers and take up the reasons for transparent and translucent structures.

10.1 General

In this chapter we shall correlate the mechanical properties of the plastics, such as strength and elongation, with the structures we have discussed. We shall find that there are predictable relationships, just as there are in metals and ceramics. We shall discuss corrosion resistance and the electrical properties later, when we deal with these special features of all materials (Chapters 12, 14, and 15).

We have arranged the structures of the representative thermoplastic and thermosetting materials (Tables 10.1 and 10.2, pages 404–406) in the order in which they were summarized in Table 9.4, adding certain important properties. Before we take up these specific properties, let us consider in a general way what we would expect from the different structures.

Just as we were able to anticipate and interpret the properties of the metals and ceramics in terms of the nature, amount, size, shape, distribution, and orientation of the phases, it is important to follow the changes in the molecules and molecular arrangements of the plastics to understand their properties.

In the thermoplastics we begin with the simple polyethylene structure, a carbon backbone with hydrogen atoms. The variations in properties are due to the degree of polymerization (i.e., the size of the molecule) and to grafting, a shape effect.

As we advance through polypropylene, polystyrene, polyvinylchloride, and Teflon, we see that we are changing the nature of the molecule by replacing hydrogen atoms at the sides of the carbon backbone. This strengthens and even changes the chemical reactivity, as for example in Teflon.

Next we graft on additional molecules and add copolymers in the chain, as in acrylonitrile-butadiene-styrene (ABS). Then we stiffen the chain, as in nylons, acrylics, acetals, cellulosics, and polycarbonates. When polar groups are added, as in nylon, hydrogen bonding gives further strengthening. Finally, using crosslinking, as in polyesters and rubber, we can approach the rigidity of the thermosetting group.

In the thermosetting group the dominating effect is the network structure, which gives essentially one giant molecule. The properties can, of course, be modified with fillers or plasticizers.

10.2 Effects of temperature and time

First let us consider the change in a typical property such as the modulus of elasticity E with falling temperature for an amorphous thermoplastic material (Fig. 10.1, page 407). From right to left in the figure, as we lower the temperature from the liquid state, at first the material is *plastic,* flow is easy, and the molecular chains slip by each other. Next we encounter the *rubbery* range, a plateau where the extension is very great but elastic. The chains uncoil but go back to their original positions after the stress is removed. At lower temperatures we pass T_g, and now the thermal agitation of the chains is less. When we are well into the glassy range, only small elastic movements are possible, and fracture takes place with very little elongation although the modulus is higher. The region between the glassy portion and the rubbery plateau is usually referred to as the *leathery* region.

Now let us consider the effects of speed of testing or stressing, again using the elastic modulus as a measure of bond strength. When strain takes place, molecules must move, and such movement takes time. Therefore, Fig. 10.1 applies to one strain rate. It is possible to modify the results by changing the strain rate. Also, the effects noted at higher temperatures in Fig. 10.1 would be obtained at lower temperatures with lower strain rates. Conversely, at a given temperature a short-time test results in a higher modulus and brittle or glassy behavior for a given polymer. On the other hand, the same specimen would be rubbery or even flow if the material were given sufficient time to react to the applied stress. Silly Putty is an excellent example of this varied behavior. This familiar material bounces like a ball (short-time stress) but flows like a liquid if left for a long time on a table top. In an extreme case, with very rapid stress application (a hammer blow), it will shatter.

The term that describes the time-strain-temperature interdependence in polymers is *viscoelasticity*. The relationship is literally a combination of elastic and viscous flow characteristics. This is not the first time we have encountered the phenomenon; recall our discussion of the viscosity of glass in Chapters 7 and 8.

In an ideal solid material, application of a load within the elastic region gives the result shown in Fig. 10.2*a* (page 407). When stress is applied at $t = 0$, a corresponding elastic strain occurs (point 1). The value of the strain depends on the stress and the modulus of elasticity (strain = stress/E). The elastic strain remains constant with time (point 2), and upon unloading all the elastic strain is recovered (point 3). An example of this behavior would be steel at room temperature even if it has been under load for several years.

Now let us consider a polymer that may undergo some viscoelastic flow at room temperature (Fig. 10.2*b*). Upon loading at $t = 0$, the strain increases to point 1. The material reacts in an elastic fashion because *immediate* unloading would return the sample to zero strain. However, over time the strain increases to point 2 even though the load is held constant. Upon removal of the load, we find that there has been some permanent or plastic deformation (point 3). We would never have observed this had we not held the sample under load for an extended period. We may now make the following observations:

1. The modulus of elasticity, stress divided by strain, depends on the strain value, which may change with time.
2. The plastic strain component varies with temperature, since the increase

Table 10.1 PROPERTIES OF THERMOPLASTIC RESINS

Name and Structure[a]	Use, percent	Price,[b] dollars/lb	Tensile Strength,[c] psi	Percent Elongation	Rockwell Hardness, R	Impact,[d] izod, ft-lb	Modulus,[c] psi × 10^3	Specific Gravity	Coefficient of Expansion, $(°F)^{-1} \times 10^{-6}$ $[(°C)^{-1} \times 10^{-6}]$	Heat Distortion,[e] °F (°C)	Burning Rate,[f] in./min	Typical Applications
Polyethylene												Clear
High density	8	0.40 to 0.42	4,000	15 to 100	40	1 to 12	120	0.95	70 (120)	120 (49)	1	sheet,
Low density	20	0.36 to 0.38	2,000	90 to 800	10	16	25	0.92	100 (180)		1	Bottles
—w—C—C—w—												
Polypropylene —w—C—C(—C)—w—	5	0.40 to 0.42	5,000	10 to 700	90	1 to 11	200	0.91	50 (90)	150 (66)	1	Sheet, pipe, coverings
Polystyrene —w—C—C(—C₆ ring)—w—	16	0.46 to 0.48	7,000	1 to 2	75	0.3	450	1.05	38 (68.5)	180 (82)	1	Containers, foams
Polyvinylchloride (rigid) —w—C—C(—Cl)—w—	17	0.50 to 0.52	6,000	2 to 30	110	1	400	1.40	30 (54)	150 (66)	<1	Floors, fabrics
Polytetrafluoroethylene (Teflon) —w—CF₂—CF₂—w—	<1	3.25	2,500	100 to 350	70	4	60	2.13	55 (99)	270 (132)	0	Chemical ware, seals, bearings, gaskets
ABS, acrylonitrile-butadiene-styrene copolymer	2	0.60	4,000 to 7,000	20 to 80	95	1 to 10	300	1.06	50 (90)	210 (99)	1	Luggage, telephones

Polyamides (6/6 nylon) —w—N(—H)—C(=O)—w—	<1	1.00	11,800	60	118	1	410	1.10	55 (90)	220 (104)	Low	Fabric, rope, gears, machine parts
Acrylics (Lucite) —w—C(—CH₃)—C(—w—)—C(=O)—O—CH₃	2	0.60	8,000	5	130	0.5	420	1.19	40 (72)	200 (93)	1	Windows
Acetals —w—C—O—C—O—w—	<1	0.80	10,000	50	120	2	520	1.41	44 (79)	255 (124)	1	Hardware, gears
Cellulosics —w—O—C(ring: C—O, C—C)C—w—	1	0.60 to 0.80	2,000 to 8,000	5 to 40	50 to 115	2 to 8	500 to 4,000	1.25	75 (135)	115 to 190 (46 to 88)	1.4	Fibers, films, coatings, explosives
Polycarbonates —w—O—C(=O)—O—R—w—	<1	1.22	9,000	110	118	14	350	1.2	25 (45)	275 (135)	<1	Machine parts, propellers
Polyesters —w—C(=O)—O—w—	3	0.52	8,000	300	117	1	340	1.3	33 (60)	130 (56)	Low	Magnetic tape, fibers, films

[a] A continuing bond is indicated by —w— . R may be any complex molecule. In general, hydrogen atoms are not shown.
[b] 1980 prices.
[c] Multiply by 6.9×10^{-3} to obtain MN/m^2 (MPa) or by 7.03×10^{-4} to obtain kg/mm^2.
[d] Multiply by 0.138 to obtain kg-m.
[e] Loaded at 264 psi 1.82 MN/m^2 (MPa).
[f] Multiply by 2.54 to obtain cm/min.

Table 10.2 PROPERTIES OF THERMOSETTING RESINS

Name and Structure[a]	Use, percent	Price,[b] dollars/lb	Tensile Strength,[c] psi	Percent Elongation	Rockwell Hardness, R	Impact,[d] izod, ft-lb	Modulus,[c] psi × 10³	Specific Gravity	Coefficient of Expansion, (°F)⁻¹ × 10⁻⁶ [(°C)⁻¹ × 10⁻⁶]	Heat Distortion,[e] °F (°C)	Burning Rate,[f] in./min	Typical Applications
Phenolics (phenol-formaldehyde) (structure: two benzene rings linked by CH₂, with continuing bonds —ww—)	6	0.40	7,500	0	125	0.3	1,000	1.4	45 (81)	300 (149)	<1	Electrical equipment
Urea-melamine —ww—N—C(=O)—N—ww—	4	0.57	7,000	0	115	0.3	1,500	1.5	20 (36)	265 (129)	0	Dishes, laminates
Polyesters —ww—C(=O)—O—ww—	<1	0.40	4,000	0	100	0.4	1,000	1.1	42 (75.5)	350 (177)	1.4	Fiberglass composite, coatings
Epoxies —C—O—C— (three-membered ring)	1	0.70	10,000	0	90	0.8	1,000	1.1	40 (72)	350 (177)	1	Adhesives, fiberglass composite, coatings
Urethanes —ww—O—C(=O)—NH—ww—	In change		5,000					1.2	32 (57.5)	190 (88)	<1	Sheet, tubing, foam, elastomers, fibers
Silicones —ww—Si—O—ww—	<1	1.50	[1,000] 3,500	[400] 0	89	0.3	1,200	[1.25] 1.75	[280 (500)] 20 (36)	360 to 900 (177 to 482)	<1	Gaskets, adhesives, elastomers

Properties are for glass-filled silicone. Elastomer properties are shown in brackets.

[a] A continuing bond is indicated by —ww— . R may be any complex molecule. In general, hydrogen atoms are not shown.
[b] 1980 prices.
[c] Multiply by 6.9×10^{-3} to obtain MN/m² (MPa) or by 7.03×10^{-4} to obtain kg/mm².
[d] Multiply by 0.138 to obtain kg-m.
[e] Loaded at 264 psi 1.82 MN/m² (MPa).
[f] Multiply by 2.54 to obtain cm/min.

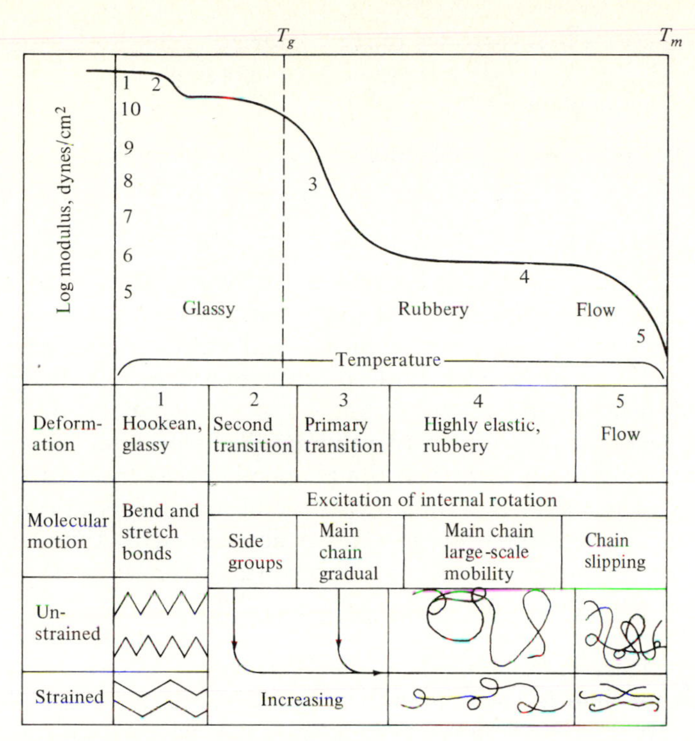

Fig. 10.1 *Molecular motions of amorphous polymer chains under stress at various temperatures.* (*Modern Plastics Encyclopedia*, McGraw-Hill, Inc., New York, 1968)

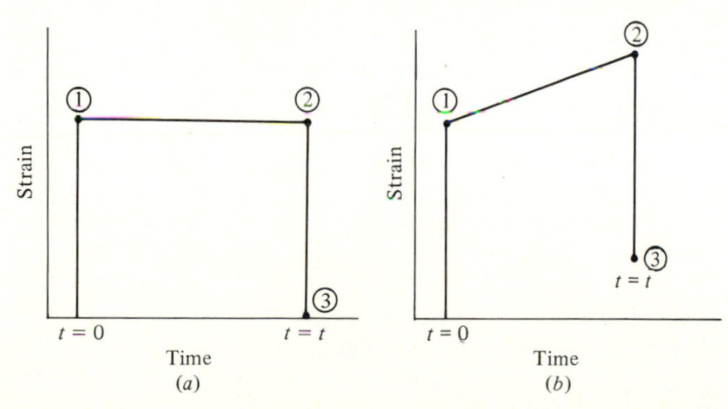

Fig. 10.2 *When a material is loaded elastically, it may return to its original zero strain condition, as in (a), or it may exhibit viscoelastic flow with a permanent residual strain, as shown in (b).*

in plastic strain is a response of the material to the load. Atoms or molecules must move, and their motion is temperature-dependent in addition to being load-dependent.

Therefore the strain rate–temperature interdependency applied to the polymer characteristics in Fig. 10.1 is explained by the viscoelastic response of these materials.

Figure 10.2*b* has been somewhat idealized for our discussion. In actual practice we would find varying degrees of curvature between points 1 and 2. Also, upon unloading, we may find exponential decay to point 3. It is possible that all the strain may be recovered, but only after a long period of time. These observations have been treated mathematically by using analogs of springs and dashpots in various arrangements. The springs are used to represent the elastic strain component, whereas dashpots represent viscous flow.

Similarly we may approach the phenomenon as the ability of the material to adjust to its stress. That is, if we had maintained a constant strain on our material, the stress would show a decrease with time. A number of polymers show this stress decay to be exponential according to the relationship

$$\sigma = \sigma_0 e^{-t/\lambda}$$

where σ_0 = stress at time 0, σ = stress at time t, and λ = relaxation time. The relaxation time is therefore the time necessary to reduce the stress to $1/e$ of its original value. The relationship can be derived from a single spring and dashpot in series.

In our example of Fig. 10.2*a*, the relaxation time approaches infinity and we have a perfectly elastic material. However, in polymers the relaxation time may be finite. The reduction of stress or correspondingly the increase in strain then becomes important to the selection of materials.

EXAMPLE 10.1 The relaxation time for a particular polymer is known to be 1 year at 20°C. How long would it take to reduce the original stress on a component made of the same polymer by 25%? How might we determine the stress relaxation at 100°C if the relaxation time is 8 months at 50°C?

ANSWER

$$\sigma = \sigma_0 e^{-t/\lambda} \qquad 0.75\sigma_0 = \sigma_0 e^{-t/1}$$

$$\ln 0.75 = -t \qquad \text{or } t = 0.29 \text{ yr (approximately } 3\tfrac{1}{2} \text{ months)}$$

The dependence of relaxation on temperature would be expected to follow the same relationship as that for diffusion and viscosity, or

$$\frac{1}{\lambda} = Ae^{-B/T} \qquad \frac{1}{12} = Ae^{-B/293} \qquad \frac{1}{8} = Ae^{-B/323}$$

Solving simultaneously, A = 6.58 and B = 1280. Therefore

$$\frac{1}{\lambda} = 6.58e^{-1280/T} \text{ for } \lambda, \text{ in months} \quad \text{and} \quad \frac{1}{\lambda_{373\ K}} = 6.58e^{-1280/373} = 0.213$$

or

$$\lambda_{373\ K} = 4.7 \text{ months} \quad \text{and} \quad 0.75\sigma_0 = \sigma_0 e^{-t/4.7}$$

$$t = 1.35 \text{ months}$$

Relaxation requires a much shorter time period at the higher temperatures.

We shall attempt to use parameters such as modulus of elasticity, tensile strength, and hardness consistent with our discussions of metals and ceramics. However, remember that, because of the viscoelastic characteristics of polymers, other measures of mechanical properties that reflect the viscoelastic behavior are sometimes used.

To proceed further with our discussion of general properties, consider the variation with crosslinking, as shown in Fig. 10.3. With an increasingly crosslinked network there is a disappearance of the flow and rubbery ranges. In the case of natural rubber there is a great elasticity with only a small percentage of crosslinking, but when 70 to 80% of the available crosslinks are used, we have the hard, rather brittle material used in some combs and battery cases.

Figure 10.4 (page 410) shows schematically the effect of crystallinity for a thermoplastic material. Again the polymer becomes stronger but more brittle. Toughness can be increased by having partial crystallinity at a sacrifice in strength.

The toughness of the material can be related to the area under the stress–strain curve, which will be discussed further in Chapter 13. In effect, for some polymers, a higher modulus of elasticity results in a lower overall strain at fracture and hence a lower area under the stress–strain curve, or lower toughness.

There is quite a variation in the degree to which the modulus and strength fall off with temperature for a given strain rate, as shown in Fig.

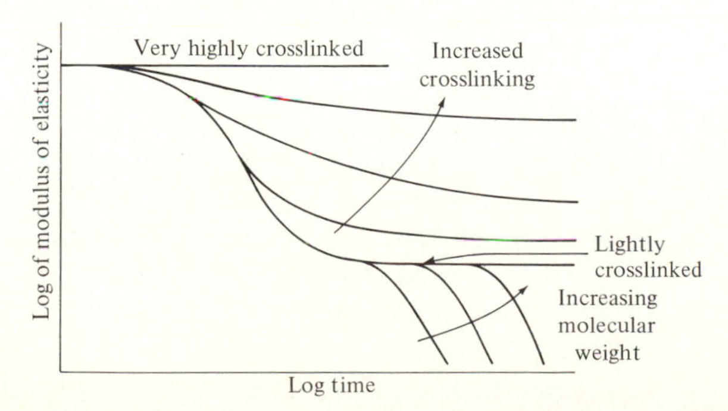

Fig. 10.3 *Effect of crosslinking on the modulus of elasticity*

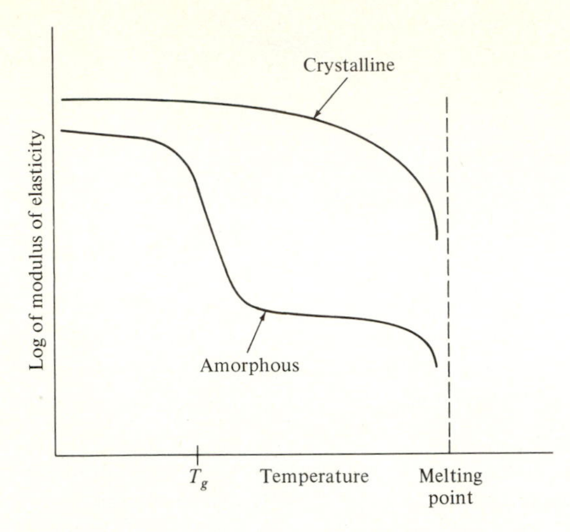

Fig. 10.4 *Effect of crystallinity on the modulus of elasticity. (See text, pp. 409, 412.)*

10.5. The chains with the simplest carbon skeleton, such as polyethylene, fall off severely, whereas complex structures, such as polycarbonate and ABS, hold up quite well with increasing temperature.

10.3 Review of mechanical properties of polymers

With these qualifications in mind, let us take an overall view of the properties. In each case we shall discuss the thermoplastic group (Table 10.1) first, then the thermosetting group (Table 10.2).

Tensile Strength In the thermoplastics the highest tensile strengths are obtained in groups with stiffened carbon chains, such as the nylons, acrylics, acetals, and polycarbonates. The general level of 9,000 to 10,000 psi (62 to 69 MPa) for this group is well above that of the polyolefins at 2,000 to 7,000 psi (13.8 to 48.3 MPa). Also, the heat-distortion temperature is higher for the chain-stiffened polymers. [The heat-distortion temperature is the temperature at which considerable deflection takes place under a light stress (264 psi, 1.82 MPa).]

The thermosetting resins show higher levels of tensile strength than the simple olefins, as would be expected from the network structure. The strengths of the epoxies and urethanes equal those of the best thermoplastics. These materials also have much higher heat-distortion temperatures. This is because the network structure is so strong that these materials do not melt. They finally decompose at high temperatures.

Percent Elongation In the polyolefins there is a sharp difference between the ductile polyethylene and polypropylene and the brittle polystyrene.

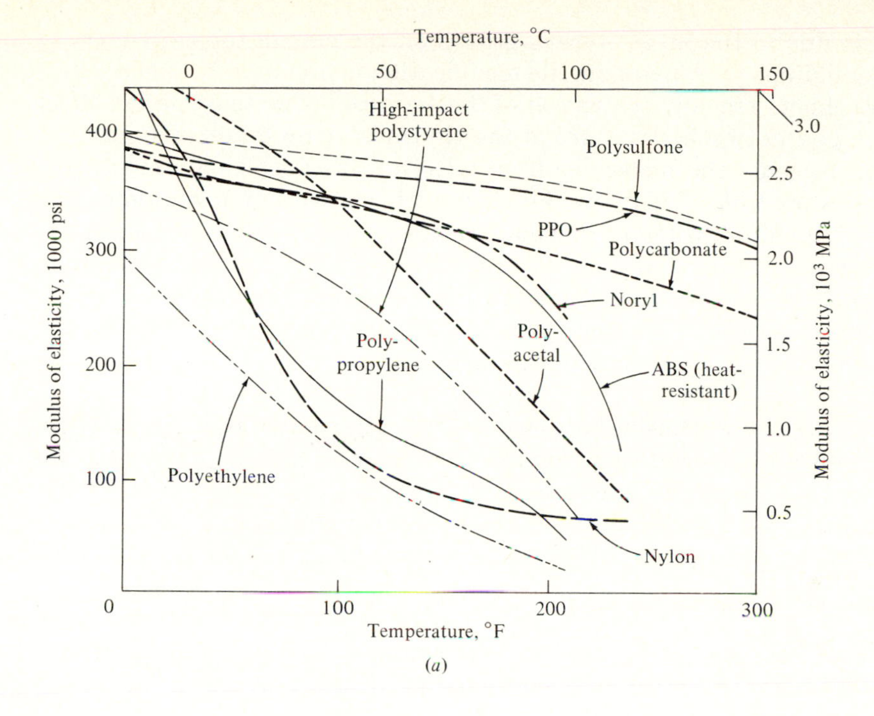

(*a*)

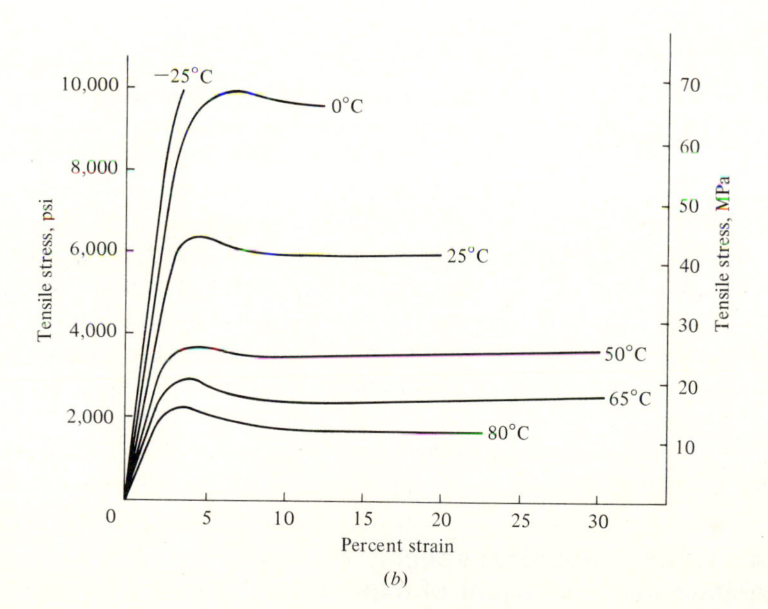

(*b*)

Fig. 10.5 *(a) Effects of temperature on properties of various resins. (b) Effect of temperature on the stress–strain curve of cellulose acetate.* [Part (a) from *Modern Plastics Encyclopedia*, McGraw-Hill, Inc., New York, 1968. Part (b) after T.S. Carswell and H.K. Nason]

This is due to the large styrene groups on the side of the carbon chain that prevent slippage. A more ductile modification is produced by blending styrene with rubber. The low elongation of the acrylics is due to a similar effect.

The network structure of the thermosetting materials, on the other hand, prevents the molecules from uncoiling because they are anchored by the network links. Therefore, all show little elongation at all temperatures.

In general, when thermoplastics are tested at or below their glass transition temperature, low elongation is obtained (Fig. 10.5*b*). Furthermore the introduction of high rates of strain will also tend to produce lower elongation. The basic feature is whether uncoiling or slip can occur during the time interval of the test at a particular temperature.

Hardness Hardness is customarily determined by a Rockwell instrument but on a special scale for plastics. As with metals, higher hardness is roughly indicative of higher strength. However, hardness is not an accurate indicator of wear resistance. For example, nylon and acetal have outstanding wear resistance, but do not have a correspondingly higher Rockwell reading.

Impact Impact tests were discussed in detail in Chapter 3. It is only necessary to state here that an impact test is a high-strain-rate test and provides a good illustration of how a material that is ductile at the strain rate in a tensile test may become brittle at higher rates. Note, for example, that ABS shows 20 to 80% elongation but low impact strength.

Modulus of Elasticity In general, in metals and ceramics the modulus of elasticity is fairly constant regardless of time of testing and over a reasonably wide temperature range near room temperature. However, as noted earlier, in plastics the modulus changes drastically with time and temperature. Also there is a great difference in moduli among the plastics. Even in one material, such as polyethylene, the modulus can vary. For example the high-density crystalline form of polyethylene has a modulus of 120×10^3 psi (828 MPa), whereas the branched, less crystalline form shows a value of 25×10^3 psi (172 MPa). Note that these values are hundreds of times smaller than those of the metals and ceramics. Therefore, care must be taken if a plastic is to be substituted for a metal in a closely mating part subject to deflection. The moduli of the rigid thermosetting materials are considerably higher than those of the thermoplastics.

10.4 Other properties: specific gravity, transparency, coefficient of expansion

The low specific gravity of the plastics compared with other classes of materials is important. This leads to favorable strength-to-weight and stiffness ratios. Some examples are shown in the table on page 413.

Specific gravity is a function of the weight per volume of the individual molecules and the way they pack. Hydrocarbons are made of light atoms. Therefore, the density is generally low. The density of simple hydrocarbon polymers—polyethylene, polypropylene, polystyrene—is 0.9 to 1.1 g/cm^3.

Material	*Tensile Strength,† psi*	*Strength/Weight,‡ psi/(lb/in.^{3})*
Cold-drawn SAE 1010 steel	53,000	156
ABS	6,500	170
6/6 nylon	11,800	290
Polyethylene (high density)	4,400	127

Substituting relatively heavy atoms such as chlorine or fluorine for hydrogen gives polyvinylchloride with a density of 1.2 to 1.55 g/cm^3 and Teflon with 2.1 to 2.2 g/cm^3. The acetals are denser because of the C—O—C chain packing.

The crystalline form of a plastic is always denser than the amorphous form because of more efficient packing. The difference in density is important in determining transparency, since the index of refraction is proportional to density. If the densities of the amorphous and crystalline materials are close, there is little dispersion as the light passes through a mixture of the two forms and the material is transparent. On the other hand, if there is a substantial difference between the densities, the material is opaque.

As an example, let us observe the series polyethylene, polypropylene, polypentene. In polyethylene the difference is great (specific gravity is 0.85 for the amorphous form and 1.01 for the crystalline). Therefore, where substantial crystallization takes place, the material is opaque. With polypropylene the difference is smaller (0.85 amorphous vs. 0.94 crystalline), and parts are at least translucent. With polypentene (—C_5H_{10}—) the densities are similar, and the moldings are transparent.

However, a crystalline material may be transparent if it is cooled rapidly so that the crystallites are very small, shorter than the wavelength of the light to be transmitted.

Also, although amorphous polymers are transparent, fillers such as asbestos and carbon black are often used. These give an opaque product.

The most important point concerning the coefficient of expansion is that in plastics it is from 2 to 17 times as great as it is in a typical metal such as iron. This means that if a plastic is substituted for a metal in a part with close tolerance, due allowance should be made. Furthermore, if a composite metal-plastic part is molded of a brittle plastic, separation and cracking may take place. This is often noted on automobile steering wheels that have been through severe temperature changes.

---

†Multiply by 6.9×10^{-3} to obtain MN/m^2 (MPa) or by 7.03×10^{-4} to obtain kg/mm^2.
‡Yield strength of metals or tensile strength of plastics (in psi) divided by density (in lb/in.^{3}).

EXAMPLE 10.2 Show by calculation that there is sufficient energy in ultraviolet radiation (wavelength 2000 Å) to break the hydrogen-oxygen bond in hydrocarbons.

ANSWER From Table 9.2,

$$\text{O—H} = 119 \text{ kcal/g-mole} \quad \text{and} \quad \frac{119 \text{ kcal/g-mole}}{6.02 \times 10^{23} \text{ bonds/g-mole}}$$

$$= 1.98 \times 10^{-19} \text{ cal/bond}$$

The energy associated with the radiation is governed by the relationship:

$$E = \frac{hc}{\lambda}$$

where h = Planck's constant = 6.62×10^{-27} erg·sec, c = speed of light = 3×10^{10} cm/sec, and λ = wavelength.

$$E \frac{\left(6.62 \times 10^{-27} \text{ erg·sec} \times 10^{-7} \text{ J/erg} \times \frac{1 \text{ cal}}{4.186 \text{ J}}\right)(3 \times 10^{10} \text{ cm/sec})}{2000 \text{ Å} \times 10^{-8} \text{ cm/Å}}$$

$$= 2.37 \times 10^{-19} \text{ cal}$$

Since the energy is greater than the bond strength, there is sufficient energy available. The significance is that polymers may undergo degradation upon exposure to strong sunlight.

10.5 Elastomers (rubbers)

In the preceding chapter we saw that rubberlike polymers are amorphous polymers with low glass transition temperatures, so that the molecules can uncoil under stress. This behavior is illustrated in Fig. 10.6, which shows how an amorphous material with a high T_g (polystyrene) has a high modulus of elasticity compared with natural rubber.

EXAMPLE 10.3 Why is it usual to find a thermoplastic with a *higher* modulus of elasticity than an elastomer (rubber)?

ANSWER The modulus of elasticity is defined as the load divided by the strain within the "elastic region." In rubbers a small load provides a high elastic strain and therefore a low modulus. Confusion usually arises when we emphasize the word "elasticity" rather than "modulus" in the definition. Highly crosslinked elastomers show an increase in the modulus of elasticity with a decrease in elasticity (the maximum amount of elastic strain).

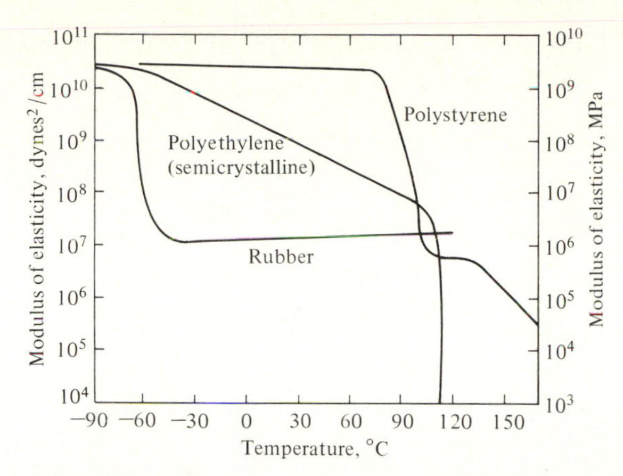

Fig. 10.6 *Comparison of the effect of temperature on the modulus of elasticity of rubber vs. the modulus of elasticity of thermoplastics.* (*Modern Plastics Encyclopedia*, McGraw-Hill, Inc., New York, 1968)

Table 10.3 (page 416) lists some commercial rubbers. Natural rubber, polyisoprene, and butadiene-styrene copolymer are widely used, but have poor resistance to oil and gasoline and a limited temperature range. The more complex types such as nitrile have extended temperature ranges and better resistance to oil and gasoline.

10.6 Special plastic products

Before we consider specific fields, we should discuss a few important special plastic products. These include coatings such as paint and adhesives, fibers, films, and foams.

Coatings There are two broad classifications of plastic coatings, depending on whether the plastic is used alone or dissolved in a solvent that subsequently volatilizes. In fluid bed coating, an unheated bed of powdered plastic is kept in motion with air, i.e., fluidized, then previously heated metal parts are dipped into the bed. A layer of plastic adheres to the part and spreads over the surface. We can use the same process to apply a thermosetting plastic if we adjust the temperature of the part to produce the thermosetting reaction.

A good example of the other type of coating, i.e., coating containing a solvent, is typical automobile touch-up paint. This material contains an acrylic resin in a solvent plus a very volatile solvent, such as CCl_2F_2, which acts as a propellant. In selecting a plastic we want T_g to be somewhat above room temperature so that the film will not be too soft, but not so high that the film will not be able to adjust to the movement of the metal underneath as the temperature changes.

Table 10.3 PROPERTIES OF DIFFERENT RUBBERS (ELASTOMERS)

Common Name	*Chemical Name*	*Tensile Strength,* psi*	*Percent Elongation*	*Resistance to Oil, Gas*	*Useful Temperature Range, °F (°C)*
Natural rubber	Cis-polyisoprene	3,000	800	Poor	−60 to 180 (−51 to 82)
GR-S or Buna S	Butadiene styrene copolymer	250	3,000	Poor	−60 to 180 (−51 to 82)
Isoprene	Polyisoprene	3,000	400	Poor	−60 to 180 (−51 to 82)
Nitrile or Buna N	Butadiene acrylonitrile copolymer	700	400	Excellent	−60 to 300 (−51 to 149)
Neoprene (GR-M)	Polychloroprene	3,500	800	Good	−40 to 200 (−40 to 93)
Silicone	Polysiloxane	700	300	Poor	−178 to 600 (−117 to 315)
Urethane	Diisocyanate polyester	5,000	600	Excellent	−65 to 240 (−54 to 115)

*Multiply psi by 6.9×10^{-3} to obtain MN/m^2 (MPa) or by 7.03×10^{-4} to obtain kg/mm^2.

Adhesives These can be classified as permanently liquid, solidifying by physical processes, and solidifying by chemical processes. In Scotch tape the bond is a permanent liquid that sticks to the cellophane backing and, with pressure, to the paper or other surface. For this type of adhesive to work, the liquid must wet the surface and form strong bonds.

Examples of the second type are sealing wax or tar. They are applied hot and then solidify. The modern electric glue gun utilizes a thermoplastic that hardens by this technique. As another example, airplane cement or rubber cement hardens by evaporation of a solvent. The cement for polyvinylchloride (PVC) pipe is in reality a solvent.

A good example of the third type is a thermosetting material such as epoxy. The liquid resin and the amine hardener are mixed and applied before setting, resulting from chemical action, can occur.

Possibly the most important adhesives are hybrids that solidify by both physical and chemical means.

Fibers These materials are produced from acrylics, rayon (derived from cellulose), cellulose acetate, spandex (polyurethane), Teflon, polyester, polyacrylonitrile, nylon, the olefins (polyethylene, polypropylene), cotton (α cellulose), and wool (a protein structure). In one method of making synthetic fibers a polymer melt is extruded through a die and put in filament form by cooling. After solidification the yarn is drawn at a temperature between T_g and T_m to orient the crystals in a favorable direction. To produce fabric with a memory, i.e., wash and wear, cotton is treated with melamine, which tends to make the fibers regain their original position after washing or creasing because the deformation was elastic rather than permanent.

Films Films are made by extrusion through a die [as wide as 10 ft (3.05 m)]. The still-plastic mass is then led through a series of polished rolls that orient the molecules by drawing.

Foams These products are important for insulation and padding. Closed-cell foams are more desirable for life jackets, whereas open-cell foams are better for furniture. The polymer from which a foam is made can be a thermosetting plastic or a thermoplastic.

Foamed polyurethane, rubber, and vinylchloride are very important in the furniture market. To make the foam, polyurethane is compounded to form CO_2 bubbles, then the mixture is poured into a moving slab mold. Rigid foam is made by increased crosslinking. It can be injected (foamed into place).

10.7 Industrial applications of polymers

Since the field of plastics is growing rapidly, it is interesting to review the applications in general and then to study one specific area, automotive applications, in detail. The approximate quantities used by various industries in 1978 are given in Table 10.4 on page 418.

Appliances In addition to serving as shapes for doors, cabinets, housings, and baskets, plastics are used for a good many parts where high corrosion resistance is needed, such as impellers, gears, and nozzles. Many good uses involve parts with limited service demands and small size.

Construction More than 1.5 million tons of plastics are used in paints, coatings, and adhesives and another 1.5 million in structural items such as glazing, panels, pipe, and flooring.

Electronics Polyethylene and vinyl coatings for cable and special Teflon-impregnated glass cloth for circuitry are important uses of plastics.

Table 10.4 INDUSTRIAL USE OF POLYMERS, 1978

Industry	*Polymers Used in 1000 Tons ($kg \times 10^9$)*
Packaging	3565 (3.237)
Construction	3160 (2.869)
Furniture and housewares	1030 (0.935)
Automobile	790 (0.717)
Electrical and electronics	780 (0.708)
Toys	285 (0.259)

Furniture and Housewares In furniture, polyurethane foam and polyvinylchloride coverings are important. In housewares, polypropylene and polyethylene are commonly used for containers.

Packaging In 1978 approximately 3.5 million tons (3.18×10^9 kg) of plastics were used in packaging, with more than half going into sheet less than 10 mils (0.0254 cm) thick. Polyethylene, polyvinylchloride, and polypropylene were the most frequently used. Specialties such as polyester were used in "boil-in-bag" applications. Bottles use polyethylene, polyvinylchloride, and acrylic copolymers.

The Automobile Industry This industry deserves close scrutiny because of its special engineering demands. To show the importance of the better-strength materials, let us compare the consumption by the automobile industry in 1978 with that of industry in general in the same year (Table 10.5).

Most parts in automobiles have service requirements from -40 to 180°F (-40 to 82°C) and higher. This eliminates many plastics that would either embrittle or become too soft.

Table 10.5 AUTOMOBILE INDUSTRY'S USE OF POLYMERS COMPARED WITH OVERALL CONSUMPTION, 1978

Material	*Automobile Industry, in 1000 Tons ($kg \times 10^9$)*	*Overall Market in 1000 Tons ($kg \times 10^9$)*	*Cost, Dollars/lb*
ABS	70 (0.064)	508 (0.461)	0.48–0.64
Acrylic	20 (0.018)	249 (0.226)	0.61
Nylon	24 (0.022)	123 (0.112)	1.16–2.33
Phenolic	23 (0.021)	660 (0.599)	0.40
Polypropylene	150 (0.136)	1341 (1.218)	0.30–0.36
Polyurethane	170 (0.154)	840 (0.763)	0.40–0.65
Polyvinylchloride	130 (0.118)	2617 (2.376)	0.30–0.43
Reinforced Polyester	160 (0.145)	280 (0.254)	1.10
Other	40 (0.036)	8806 (7.996)	—

EXAMPLE 10.4 Table 10.6 (pages 420–425) sums up the uses of polymeric materials in cars. Of course, in many applications plastics are in competition with other materials. Assuming the competitive material to be metal, either steel or aluminum, cite the advantages and disadvantages of the competing parts for door handles, carburetor components, and impact-bumper parts.

ANSWER

Application	*Material*	*Advantages*	*Disadvantages*
Door handles	Vinyl	Formability	Poor low-temperature impact
	Chrome-plated steel	Impact and wear resistance	Cost
Carburetor components	Acetals	Formability, toughness	Poor wear resistance
	Aluminum die casting	Strength and wear resistance	Cost
Impact-bumper parts	Polycarbonates	Toughness, corrosion resistance	Cost
	Chrome-plated steel	Strength, ease of repair	Poor corrosion resistance

SUMMARY

In discussing the properties of polymers, it is helpful to separate the materials into thermoplastics and thermosetting plastics.

The thermoplastics vary in tensile strength from the polyethylenes at 2,000 to 5,000 psi (13.8 to 34.5 MPa) to polyamides (6/6 nylon) at 11,800 psi (81.5 MPa) as chain stiffening and hydrogen bonding are increased. The elongation of this group is generally good (above 5%) except in the case of polystyrene. Materials with amorphous structures, such as Lucite and polystyrene, are transparent, whereas crystalline materials are translucent to opaque. On cooling from the melt most thermoplastics pass through five stages: liquid, viscous, rubbery, leathery, glassy.

The thermosetting plastics are stronger than the polyolefins and retain their rigidity to higher temperatures than the thermoplastics. They finally degrade rather than melt when heated and do not exhibit plastic or rubbery behavior. The impact strength is lower and the modulus higher than for the thermoplastics.

Elastomers are polymers with molecular structures that uncoil readily, exhibiting very high values of elastic elongation.

Adhesives are divided into three groups, based on whether they operate by permanent liquid films, application of a liquid that solidifies, or polymerization in place.

Table 10.6 SUMMARY OF APPLICATIONS OF PLASTICS IN TYPICAL AUTOMOBILES

Generic Name	*Typical Trade Names*	*Typical Uses*	*Advantages*	*Disadvantages*	*Average Weight per Automobile, U.S. Production, 1970, lb (kg)*
			Thermoplastics		
Vinyl	Geon Elvax Vygen Vinoflex	Wire insulation, hoses, heel pads, seat trim, door panel trim, armrest skins, headlining convertible tops, landau tops, exterior and interior moldings, crash-pad skins, door-locking knobs, coat hooks, horn pads, bumper fillers, seat-belt-retractor housings, steering wheels,* exterior door handles,* vinyl-covered seats and interior trim, vinyl-coated metal	Stable colors, weather-resistant, flexibility	Stiff at low temperatures, fogging (unless specially compounded)	18 (8.16)
Polyolefins Polyethylene	Marlex Alathon Hi-fax Petrothene	Splash shields, spring interliners, ducts, floor-pan plugs, water shields, protective seat covers, fuel tanks,* windshield washer bottles	Low cost, low friction, low-temperature impact	Not paintable, tendency to warp, tendency for stress cracking, low stiffness	17 (7.71) (polyethylene and polypropylene combined)
Polypropylene	Tenite Marlex Pro-fax	Fender liners, accelerator pedals, interior trim, door panels, steering wheels, fan shrouds, fresh-air down spouts, cowl-trim panels, seat backs, wheelhouse and spare-tire covers, fuse blocks, electric connectors, battery cases,* cowl screens, lamp housings	Low cost, colorability, low friction, heat-resistant	Not paintable, poor low-temperature impact (Homopolymer only)	

ABS	Cycolac Kralastic Lustran Abson	Instrument bezels, post covers, armrest bases, consoles, knobs, radiator grilles, air-conditioner outlets, instrument-panel trim plates, seat backs, push buttons, heater push-button housings, lamp housings,* interior trim (plated), deck-lid spoilers	Paintable, adequate impact, electroplatable, vacuum-metalizable heat-resistant, compounding latitude	Opaque, not weatherable	11 (5)
Polyvinyl Butyral	Saflex Butacite	Interlayer for safety glazing			2 (0.908)
Acrylic	Plexiglas Lucite Perspex	Lamp lenses, medallions, instrument facings, side markers	Clarity, weather-resistant	Poor heat resistance; impact not quite adequate, except for some special compounds	3.5 (1.59)
Acrylic fiber bundles	Crofon	Interior signals, instrument lights*			
Cellulosic Acetate Butyrate Propionate	Tenite Forticel	Knobs, trim strips, steering wheels, push buttons	Good impact, clarity, colorability	Poor heat resistance; odor	2.25 (1.02)
Nylon	Zytel Ultramid	Bushings, slides, gears, fasteners, timing chain sprocket, fuel filter housing, cable liners, fender extensions*	Low friction; heat-resistant; fuel-, oil-, and chemical-resistant; good toughness	High water absorption, brittle when dry	2.25 (1.02)
Acetal	Delrin Celcon	Same as nylon. Also housings, shear pins for energy-absorbing steering column, gears, carburetor components*	Low friction, heat-resistant (but lower than nylon), moisture-resistant, rigid, high pull-out strength of threaded studs, impact below nylon	Problems with paintability	2 (0.908)
Polycarbonate	Lexan Merlon	Interior lamp bases (indicator), lenses, exterior impact-bumper parts,* housings	Clarity, very high impact, heat-resistant	High cost, tendency to solvent crazing, poor weatherability	1 (0.454)

Table 10.6 SUMMARY OF APPLICATIONS OF PLASTICS IN TYPICAL AUTOMOBILES (*Continued*)

Generic Name	*Typical Trade Names*	*Typical Uses*	*Advantages*	*Disadvantages*	*Average Weight per Automobile, U.S. Production, 1970, lb (kg)*
			Thermoplastics (Continued)		
Glass fiber-filled thermoplastic: Polystyrene, ABS, Polyolefin, Nylon, Acetal		Air-conditioner heater push-button housing cover (nylon), instrument panels (styrene*), glove box door (styrene*), transmission cover for under-pan (propylene) housings,* switch parts	Increased strength, stiffness, heat-resistant, dimensional stability, low thermal expansion and contraction	High cost	1 (0.454)
			Thermosetting Plastics		
Phenolic	Bakelite, Durez, Genal	Terminal blocks, coil towers, fuse retainers, insulators, carburetor spacers, switch components, brake components, distributor caps*	Temperature-resistant, electrical insulator, good strength and impact (in some special filled compounds)	Colorability, tracking	3 (1.36)
Phenolic laminate	Norplex, Formica	Terminal blocks, electrical insulators, spacers, bushings, transmission thrust washers, printed circuit boards	Temperature-resistant, electrical insulator, high strength, high impact	High cost, must be machined	0.5 (0.227)

Alkyd plastic	Glaskyd Duraplex	Brush holders (starters, wipers, heater blowers), slip-ring insulators, distributor caps	Arc- and tracking-resistant, heat-resistant, fast molding cycles	High cost, relatively short shelf life, limited colorability	2 (0.908)
Reinforced polyesters					
Mineral-filled premix (Bulk molding compound: polyester plus mineral fillers plus catalyst)		Exterior body components, fender extensions, fender skirts, bumper closures, hood scoops, door scoops, instrument panels,* molded semistructural components, rear-window frames, tail-light assembly*	Freedom from warping, good stud retention, excellent molded surface, dimensional accuracy, paintable without sanding, minimum property variation with temperature, flexural strength—14,000 psi	Material cost, limited production capacity per mold, mold flow limitations, subject to creep	5 (2.27)
Glass mineral-filled (sheet molding system: polyesters plus fillers plus glass fibers plus catalysts)		Air vanes, truck body panels,* upper grille panels (front-end sheet metal replacement)*	Freedom from warping, good stud retention, excellent molded surfaces, minimum property variation with temperature, dimensional accuracy, paintable without sanding, high physical properties, flexural strength—20,000 psi	Material cost, limited production capacity per mold, mold flow limitations, subject to creep	1 (0.454)
Foamed-in-place Trim Components					
Flexible ABS film (vacuum-formed skins)		Instrument-panel trim pads, door trim panels, trim bolsters	Capability to produce complex trim shapes, wide selection of skin materials, can be foamed to a defined impact performance	Skin shrinkage, poor edge retention, loss of grain for deep draws	Skins: 1.5 (0.681) Foam: 1.5 (0.681)

Table 10.6 SUMMARY OF APPLICATIONS OF PLASTICS IN TYPICAL AUTOMOBILES (*Continued*)

Generic Name	*Typical Trade Names*	*Typical Uses*	*Advantages*	*Disadvantages*	*Average Weight per Automobile, U.S. Production, 1970, lb (kg)*
		Foamed-in-place Trim Components (Continued)			
Plastisol and Drysol (slush-molded skin)		Instrument-panel trim pads, horn pads, headrests,* armrests, console covers	Nonshrink skins, good grain, good color match, semi–self-locking skins	Uneven skins, delicate tooling	Skins: 2.25 (1.02)
Rotocast skin		Horn pads, headrests, armrest pads, trim pads*	Nonshrink skins, good grain, good color match, self-locking skins	Uneven skins, delicate tooling	Foam: 3 (1.36)
Vinyl skin (Injection-molded. May also be assembled.)		Horn pads, armrests, trim pads*	Uniform skin thickness, more flexible skin, good color match and pattern reproduction, rapid production rate	Requires parting lines, not economical for low-volume parts	0.3 (0.136)
Urethanes					
Self-skinning foam		Horn pads, armrests,* small trim pads, steering wheels, bumpers, side scoops	Capable of producing grained and embossed surfaces; paintable; soft feel; freedom of styling in shape, fit, and body match	Not economical for high-volume parts, cannot mold undercuts, solid-metal engraved tools required	1 (0.454)
Other		Seats	Complex shapes are feasible, economical		12 (5.45)

Miscellaneous			
Polyester film	Mylar†	Dial faces, trim components	Total: 3 (1.36)
Polyvinylidene chloride	Saran	Fuel filters	
Epoxy potting compounds	Epon, Araldite, Bakelite, Epi-Rez	Voltage regulators	
Solid urethane	Texin, Vibrathane, Estane	Bushings, bearings	
Cellular polyolefins	Ethafoam	Gaskets, visors	
Polytetrafluoroethylene	Teflon	Seals, bushings, cable liners	
Polyphenylene oxide	Noryl		
Ethylene-vinylacetate copolymer	Ultrathene		
Polysulfone	Astrel		

* Plastics in competition with other materials.
† Vacuum-metalized externally.

Fibers are produced by extrusion, filament formation, and drawing at controlled temperatures.

Key industrial applications of the polymers are discussed for the appliance, construction, electronics, furniture, packaging, and automotive industries.

DEFINITIONS

Adhesive A molecular structure that bonds two other materials by either physical or chemical means.

Chain stiffening The inclusion of foreign atoms in or along the carbon chain, resulting in increased bonding.

Elastomer A high polymer with a coiled structure, resulting in rubberlike characteristics.

Fiber A polymer in a finely drawn form.

Foam A polymer containing gas bubbles that add to the insulating value and also "sponginess," as in a car seat.

Heat-distortion temperature The temperature at which a polymer softens severely.

Impact The testing of specimens with high strain rate.

Leathery range The temperature range in which chains are more rigid and fracture takes place at less than 10% elongation.

Rubbery range The temperature range in which very great elastic elongation is obtained. The chains uncoil but return to their original positions.

PROBLEMS

10.1 Compare Tables 9.4, 10.1, and 10.2. The strengthening mechanisms include chain stiffening, through the addition of other atoms and side groups, crosslinking, network structuring, and crystallization, to mention a few of the more important ones. Make a table and compare each material in the figures listed (Tables 9.4, 10.1, and 10.2) with low-density polyethylene. Indicate which of the strengthening mechanisms are most important in each use. (Sections 10.1 through 10.3)

10.2 *a.* Why does nylon creep more in high humidity?

b. How does paint "dry"?

c. Does butadiene rubber have a higher or lower modulus of elasticity than copper? Explain. (Sections 10.1 through 10.3)

10.3 Establish a set of correctly labeled axes and show the following relationships schematically: (Sections 10.1 through 10.3)

a. Variation in tensile strength with molecular weight of a linear polymer.

b. Variation in elastic modulus with temperature of a linear polymer at a constant strain rate.

c. Variation in elastic modulus with strain rate of a linear polymer at constant temperature.

10.4 Using the data from Example 10.1, determine the time necessary to reduce the stress by 50% and 75% of the original value at 50°C. Suppose that we used the polymer as bonding to hold together a wooden crate. What would be the significance of your results? (Sections 10.1 through 10.3)

10.5 What would be the significance of a material exhibiting viscoelastic characteristics, such as that in Fig. 10.2*b,* if we were to use it as a fishing line? (Sections 10.1 through 10.3)

10.6 Suppose that we established a "relaxation modulus" as

$$E_t = E_0 e^{-t/\lambda}$$

where E_0 is the modulus at $t = 0$, E_t is the modulus at $t = t$, and λ = relaxation time. How might the relaxation modulus vary with temperature? What effect would it have on the data in Fig. 10.5*a*? (Sections 10.1 through 10.3)

10.7 Rubber products sometimes undergo degradation over extended periods of time. Explain why the following phenomena occur in rubber. (Sections 10.4 through 10.5)

a. A rubber band fails after having been wrapped very tightly around an object for several months.

b. An automobile heater hose bursts.

c. A rubber washer in a water faucet will no longer seal.

10.8 Polyethylene as used in containers is normally translucent. If a piece of this material is pulled in tension, it transmits less light (becomes chalky white). Explain. (Sections 10.4 through 10.5)

10.9 Explain the behavior of polyethylene in the tensile specimen shown in Fig. 10.7. Why doesn't failure take place at the reduced section in (*b*) instead of the necking down continuing? (Sections 10.4 through 10.5)

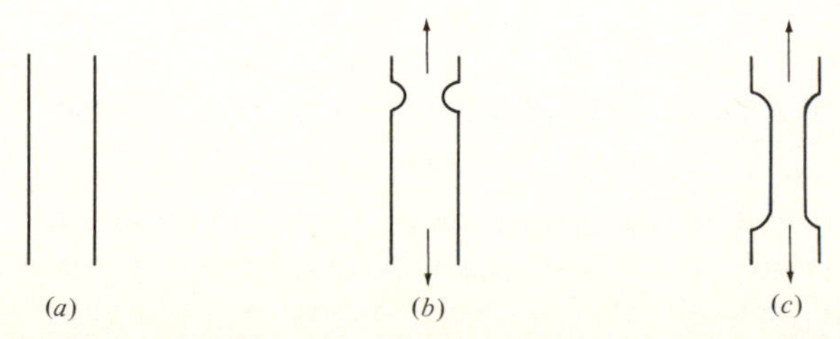

Fig. 10.7 *Behavior of polyethylene: (a) unstressed, (b) necking down begins, (c) necking down continues, with more stress*

10.10 The properties of five high polymers at 25°C are given here. Identify the polymers, using names from the following list.

a.	*Tensile Strength, psi (MPa)*	*Percent Elongation*	*Impact Strength, Izod ft lb (newton-meter)*	*Modulus of Elasticity, psi × 10^3 (MPa × 10^3)*	*Burning, in./min*
(1)	9000 (62.1)	110	14 (19.04)	350 (2.415)	<1
(2)	7500 (51.8)	0	0.3 (0.41)	1000 (6.90)	<1
(3)	4000 (27.6)	70	3 (4.08)	120 (0.828)	1
(4)	10,000 (69.0)	0	0.8 (1.09)	1000 (6.90)	1
(5)	2500 (17.3)	200	4 (5.44)	60 (0.414)	0

Names: epoxy, tetrafluoroethylene, polyethylene, phenol formaldehyde, polycarbonate

b. Name all the above resins that would not be transparent. Why? (Sections 10.4 through 10.5)

10.11 An advertisement attached to an article made of the polycarbonate thermoplastic polymer LEXAN reads: "A better name than metal, LEXAN resin is as tough as metal but the similarity ends there. Won't rust or corrode. Won't conduct heat or electricity. In short, LEXAN is the toughest name you can drop. Or pound. Or whack. Or buy."

a. Indicate whether the properties of LEXAN would be greater than, less than, or about the same as the following engineering properties of cold rolled 1020 steel: (1) modulus of elasticity, (2) tensile strength, (3) yield strength, (4) creep strength at 800°F (427°C), (5) dielectric constant (see Chapter 14).

b. In which of the following applications would you substitute LEXAN because of its better engineering properties for the application described: (1) a ductile iron crankshaft in a car; (2) a steel tackle box for salt-water fishing; (3) a carburized and hardened steel gear; (4) a steel ball in a car wheel bearing; (5) an aluminum frying pan; (6) a fiberglass football helmet (Sections 10.4 through 10.5)

10.12 Consider the data from Example 10.2 and Table 9.2. Are there bonds available that would be resistant to ultraviolet radiation? (Show by calculation.) (Sections 10.4 through 10.5)

PROBLEMS COVERING OPTIONAL SECTIONS AND PREVIOUS CHAPTERS

10.13 Listed below are several automotive applications involving competitive materials, as discussed in Example 10.4. Comparing the polymers with other materials indicated, cite advantages and disadvantages of each. (*a*) Steering wheels: PVC vs. polypropylene, (*b*) Fuel tanks: steel stamping vs. polyethylene, (*c*) Battery cases: polypropylene vs. hard rubber, (*d*) Lamp housings: polypropylene vs. sheet steel, (*e*) Fender extensions: nylon vs. sheet steel. (Sections 10.6 through 10.7)

10.14 Most adhesives cannot be used in all types of environments. Explain. What type of glue could be used where it would be exposed to weathering? (Sections 10.6 through 10.7)

10.15 Table 9.4 lists a number of monomer structures and how they polymerize. For each of the following applications, select one of the polymers as the most likely candidate. (*a*) Frying pan handle, (*b*) children's toy truck, (*c*) container resistant to organic solvents, (*d*) window for a storm door, (*e*) outdoor trash container. (Sections 10.6 through 10.7)

10.16 *a*. From the following polymers (as sketched), select the one that best fits each of the applications listed.

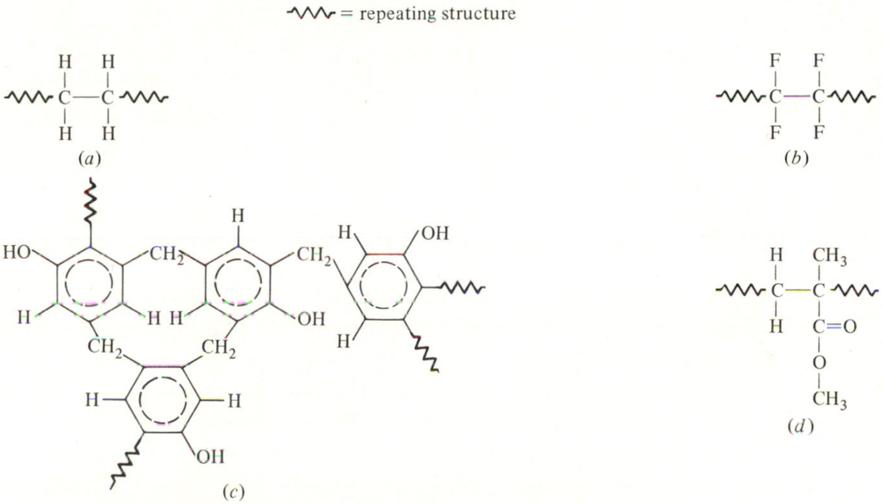

= repeating structure

(*a*)

(*b*)

(*c*)

(*d*)

(1) Rigid electrical outlet; (2) aircraft window; (3) nonflammable chemical tubing; (4) squeeze bottle.

b. List all the above polymers that are thermoplastic.

c. List all the above that formed a byproduct during polymerization. (Sections 10.6 through 10.7)

10.17 Several monomers are given.

a. Sketch the polymer structure of each.

b. From the following suggestions, indicate a typical use for each polymer: rope, electrical outlets, chemical valves, squeeze bottles, baking dishes, microscope lenses, ball bearing. (Sections 10.6 through 10.7)

Monomer(s)

1. C_2H_4 2. C_2F_4 3. $\left.\begin{matrix} C_6H_5OH \\ CH_2O \end{matrix}\right\}$

(continued on page 430)

```
        O  H  H  H  H  O
        ‖  |  |  |  |  ‖
    HO—C—C—C—C—C—C—OH
           |  |  |  |
           H  H  H  H
```

4.

```
  H     H  H  H  H  H  H     H
   \    |  |  |  |  |  |    /
    N—C—C—C—C—C—C—N
   /    |  |  |  |  |  |    \
  H     H  H  H  H  H  H     H
```

10.18 A certain type of rubber has a true specific gravity of 1.40. This type of rubber is used in the manufacture of a foamed rubber that weighs 0.015 lb/in^3 (415.6 kg/m^3) when dry, and 0.025 lb/in^3 (692.6 kg/m^3) when saturated with water. *a.* What is the apparent porosity of the foamed rubber? *b.* What is the true porosity? (Sections 10.5 and 8.6)

10.19 A particular cellular polymer is being considered for use as a sponge. The dry sponge weighs 30 lb/ft^3 (481 kg/m^3). The specific gravity is 1.34. How many pounds (kg) of water could a sponge 6 in. $\times$ 6 in. $\times$ 3 in. (0.152 $\times$ 0.152 $\times$ 0.076 m) hold if all the pores were open and available to be filled with water? (Sections 10.6 and 8.6)

10.20 Using only the following monomers, indicate which monomer or corresponding polymer would best fit each description.

(a) Ethylene, (*b*) vinylchloride, (*c*) propylene, (*d*) styrene, (*e*) butadiene, (*f*) urea formaldehyde. (Section 10.6 and Chapter 9)

1. One that would form the strongest polymer
2. Example of one that might form various stereo arrangements
3. A combination that might form a copolymer
4. One that might crosslink with another atom
5. One that might crystallize
6. One that would be most difficult to glue
7. One that might be easy to recover and reuse as scrap
8. One that would have the lowest glass transition temperature
9. One that could not be plasticized
10. One that might be classified as an elastomer

10.21 Anyone who has ever owned a car will recognize the following situations. Suggest what might be occurring in the structure of the material. (Sections 10.6 and 9.8)

a. The insides of the windows develop a haze that resembles, but is not, a smoke haze.

b. Vinyl seats crack after a period of time. Those areas exposed to sunlight are especially susceptible.

10.22 Five polymers and five ceramics are listed below. Choose *one* of each (ceramic *and* polymer) to meet the indicated material requirement.

(There may be more than one correct answer, but only one is requested. The first requirement is given as an example.) (Chapters 7, 8, 9, 10)

Polymers	*Ceramics*
1. polyethylene	6. soda-lime glass
2. polyvinylchloride	7. vitreous silica
3. polymethyl-methacrylate	8. alumina (Al_2O_3)
4. phenol formaldehyde	9. Pyroceram
5. chloroprene rubber	10. Portland cement

	Requirement	*Polymer*	*Ceramic*
	Processing uses more than one component	4	10 (also 9)
a.	Is most transparent		
b.	Is most crystalline		
c.	Is most temperature-resistant		
d.	Is easiest to recycle		
e.	Is the hardest		

10.23 A grinding wheel is to be a composite of Al_2O_3 particles in a polymer matrix. (Chapters 9, 10, and Section 8.6)

a. Why is Al_2O_3 used rather than SiO_2?

b. Is the polymer likely to be thermoplastic or thermosetting?

c. The wheel is intentionally made to contain porosity. Why is some porosity desirable?

d. The specific gravity of Al_2O_3 is 3.80. The grinding wheel is 35.0 cm in diameter and 4.0 cm wide (assume that it has no hole in the center), and 8,000 g of Al_2O_3 is present in the wheel. Calculate the apparent packing factor for the Al_2O_3. (Assume that no polymer has been added as yet, and that there is no closed-pore porosity.)

e. The specific gravity of the polymer is 1.65, and the wheel contains a final apparent 20% porosity. Calculate the amount of polymer in grams added to the Al_2O_3 if all the porosity is within the polymer.

10.24 Some polymers are potentially toxic if they are burned and the fumes are allowed to become airborne. How much HCl could be obtained per pound of combusted PVC? (Section 10.8 and knowledge of chemistry)

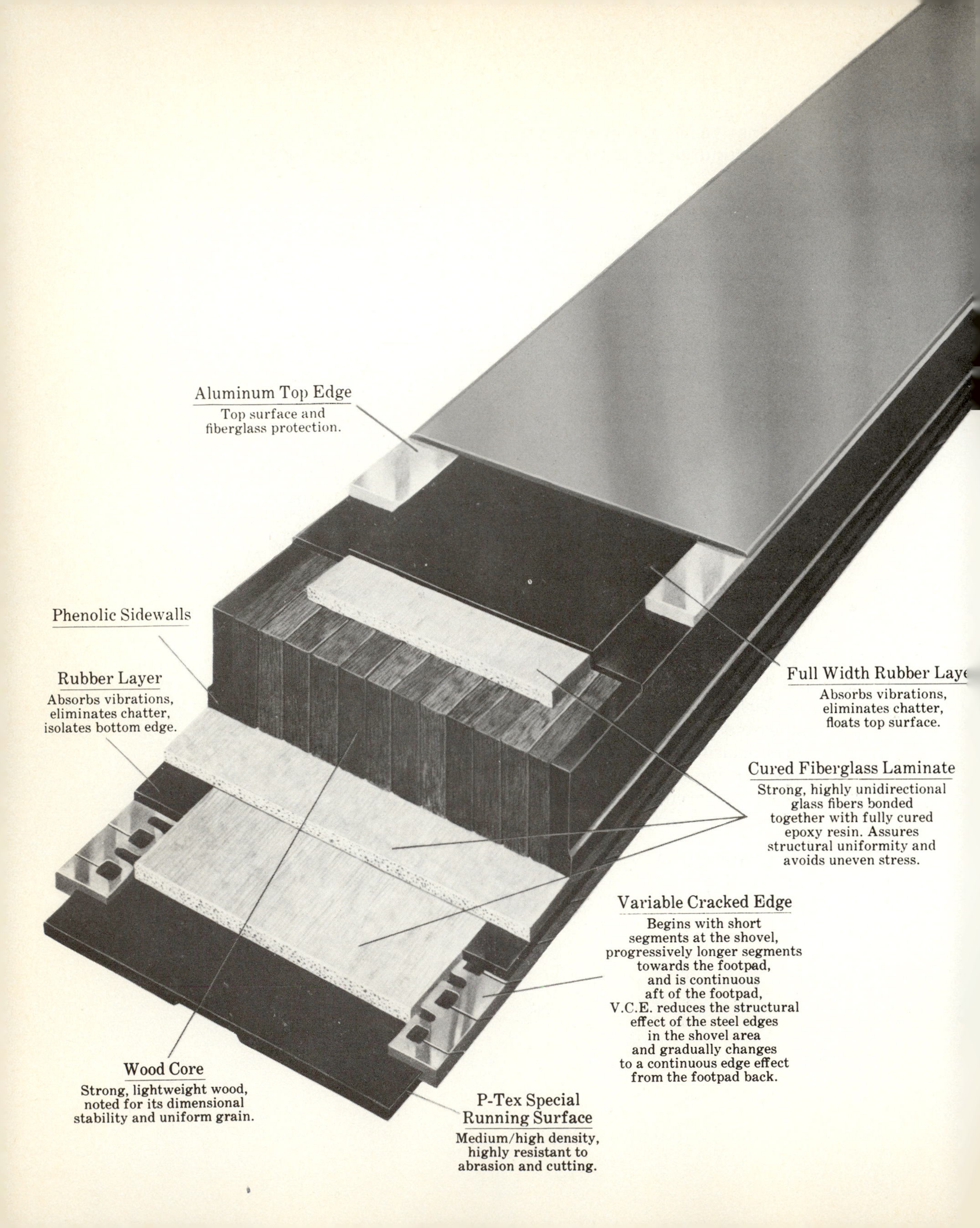
Aluminum Top Edge
Top surface and fiberglass protection.
Phenolic Sidewalls
Rubber Layer
Absorbs vibrations, eliminates chatter, isolates bottom edge.
Full Width Rubber Laye
Absorbs vibrations, eliminates chatter, floats top surface.
Cured Fiberglass Laminate
Strong, highly unidirectional glass fibers bonded together with fully cured epoxy resin. Assures structural uniformity and avoids uneven stress.
Variable Cracked Edge
Begins with short segments at the shovel, progressively longer segments towards the footpad, and is continuous aft of the footpad, V.C.E. reduces the structural effect of the steel edges in the shovel area and gradually changes to a continuous edge effect from the footpad back.
Wood Core
Strong, lightweight wood, noted for its dimensional stability and uniform grain.
P-Tex Special Running Surface
Medium/high density, highly resistant to abrasion and cutting.

11

COMPOSITE MATERIALS, INCLUDING CONCRETE AND WOOD

THE modern ski shown in the illustration exhibits an interesting combination of materials.

The wood core is made of African mahogany because this material has low density, high shear strength, excellent flexibility, and good machinability, which make it a good material for precision parts. No plastic has characteristics that are the same—as wood's are—at all temperatures to which the ski is subjected. The wood core, used in conjunction with fiberglass laminates, provides desirable vibrational characteristics.

The top edges are 7075-T6 aluminum, for the best strength-to-weight ratio. The edges are a combination of 1065 (R_C 38–40) and 1095 (R_C 48–50) steel, for toughness and resistance to wear. The top surface is made of phenolic, which is chip- and scratch-resistant. The fiberglass laminates are 70% E glass fibers and 30% epoxy resin.

In this chapter we shall discuss this combination.

11.1 General

A *composite* can be broadly defined as a combination of two or more materials, each of which has its own distinctive properties. In a way, we studied composite materials on a micro scale when we investigated multiphase structures in the metals, ceramics, and polymers. However, when we speak of a composite, we generally mean two or more different materials assembled in a mechanical way, whether by people, such as bonding fiberglass with plastic, or by nature, such as assembling cellulose fibers and lignin in wood. Using mechanical assembly accompanied by chemical bonding, we can attain properties superior to those we could produce with a single melt or starting material. We can do this mechanical assembly on a micro scale, however, as when we mix tungsten carbide powder with cobalt powder, press the product into a tool shape, and sinter. The tungsten carbide retains its identity. We could not obtain the same structure by making a melt of tungsten, carbon, and cobalt of the same overall composition.

In this chapter we shall discuss two broad classes of composites, synthetic and natural. We shall not try to cover all possibilities, but rather we shall show how the properties depend on the structure of the assembly.

11.2 Synthetic composites

In many cases we can obtain an outstanding combination of properties by using two or more synthesized materials together. Perhaps the most common example is a fiberglass fender for a car. The polymer used alone would have insufficient strength and excessive deflection, and the glass alone would obviously be too brittle.

However, fiberglass is only one type of synthetic composite. Before proceeding with a more complete discussion of fiberglass, we shall briefly consider the general types, which can be divided into three categories.

1. *Dispersion strengthened* The matrix is the major load-bearing constituent.
2. *Particle reinforced* The load is shared by the matrix and the particles.
3. *Fiber reinforced* The fiber is the primary load-bearing component.

11.3 Composites strengthened by dispersion and particle reinforcement

Consider a metal matrix with a fine distribution of secondary particles. Since deformation in the matrix is accompanied by slip and dislocation movement, the degree of strengthening achieved is proportional to the ability of the particles to impede the dislocation movement. It follows that a finer dispersion of particles results in greater strengthening. The objective is to have the particles small enough and spaced closely enough so that dislocations cannot easily move between them.

In *dispersion strengthening* it can be shown that with particle diameters less than 0.1 micron (1 micron = 10^{-6} meter) and volume concentrations of 1 to 15%, dislocation movement can be effectively impeded. The strengthened matrix becomes the main load-bearing constituent.

If we have a greater percentage of dispersion and larger particles, we obtain *particle reinforcement,* where the load is shared by the matrix and particles. In general, in a particle-reinforced composite, particle diameters are greater than 1 micron and volume concentrations are greater than 25%.

An example of a *dispersion-strengthened* alloy is SAP: sintered aluminum powder. If we make a composite of fine Al_2O_3 particles in an aluminum matrix (by compacting and sintering the powders), we can significantly increase the high-temperature properties of aluminum alloys since the composite does not overage.

Another example, increased thermal resistance of iron-Al_2O_3 composites, is shown in Fig. 11.1.

Possibly the best examples of *particle-reinforced* composites are the cemented carbides discussed in Chapter 7. Here the stress is supported by both the matrix and the particles, and the volume fraction of the particles is large (see Fig. 7.34). Similarly, most commercial ceramics, ranging from bricks to grinding wheels, and many filled polymers are particle-reinforced composites. Figure 11.2 (p. 436) shows effects of rubber particle fillers on stress–strain curves of normally brittle polystyrene. The high strain rate is intended to approximate the strain imposed by impact loading in service.

11.4 Fiber reinforcement

Of all the composite materials, fiberglass is certainly the best known. As stated previously, the fiber becomes the major load-bearing constituent. There are two important classes (Fig. 11.3 on page 436).

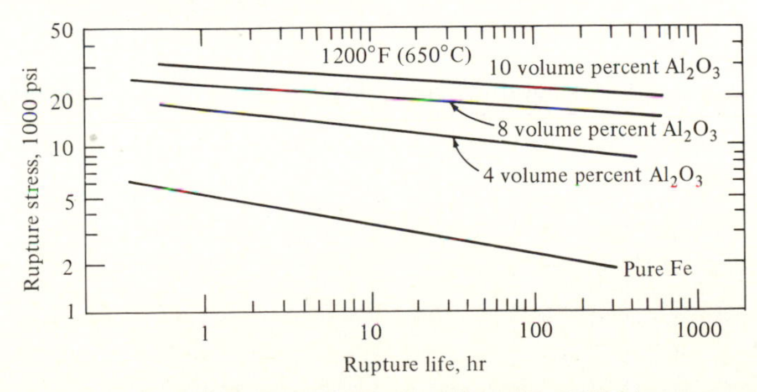

Fig. 11.1 *The increase in elevated temperature (1200°F, 650°C) stress–rupture properties obtained by dispersion strengthening iron with Al_2O_3 particles (to obtain MPa, multiply psi by 6.9 × 10^{-3}).* (From L. J. Broutman and E. H. Krock, *Modern Composite Materials*, Addison-Wesley, Reading, Mass., 1967, p. 487)

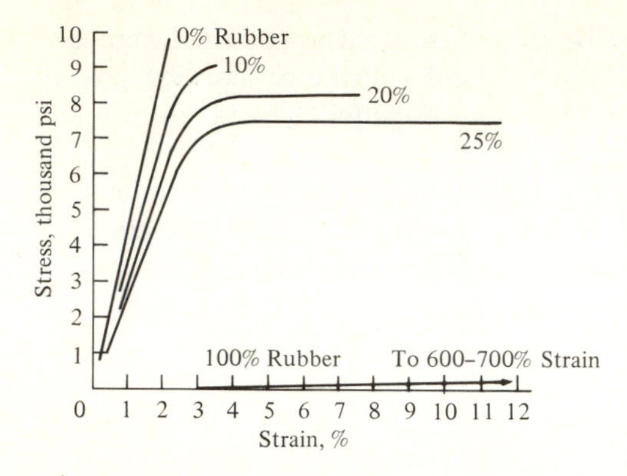

Fig. 11.2 *The influence of rubber particles on the strength of polystyrene. The stress–strain curve was obtained at high rates of strain (133 in/in. or m/m per min). (To obtain MPa, multiply stress by 6.9 × 10^{-3}.)* (From S. L. Rosen, *Fundamental Principles of Polymeric Materials for Practicing Engineers*, Barnes and Noble, New York, 1971, p. 254)

1. Composites in which the fibers of the strengthener are continuous
2. Composites in which the fibers of the strengthener are discontinuous and may be merely chopped-up pieces

We shall analyze only the factors that make up the overall strength and modulus of elasticity of the composite in the first case, which gives the strongest combination.

Composites with Continuous Fibers in a Matrix For example, take the simple case of a bar 0.5 in.^{2} (3.23 cm^2) in cross section in which continuous glass fibers are in a matrix of polyester, a typical fiberglass. What effect will increasing the amount of glass have on the strength and modulus of elasticity of the composite? Assume the following values:

Material	*Tensile Strength,† psi × 10^3*	*Yield Strength,† psi × 10^3*	*Modulus,† psi*	*Percent Plastic Elongation*
E fiberglass	250	250	10×10^6	0
Polyester	5	5	4×10^5	~0

†Multiply by 6.9 × 10^{-3} to obtain MN/m^2 (MPa) or by 7.03 × 10^{-4} to obtain kg/mm^2.

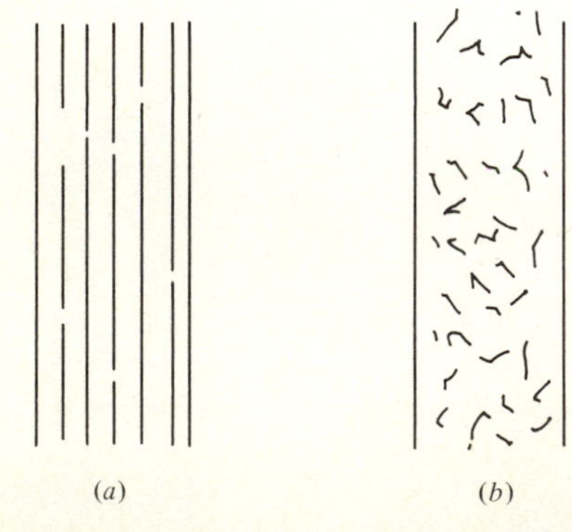

Fig. 11.3 *Variations in distribution of fiber in a composite material such as fiberglass. (a) Continuous or practically continuous fibers. (b) Chopped-up, discontinuous fibers.*

The glass has a stiff stress–strain curve showing little deformation, whereas the polyester shows somewhat more (Fig. 11.4).

We shall consider the glass as the fiber and the polyester as the matrix surrounding the fiber. To enable us to understand the loads in the fiber and matrix, let us conduct the following analysis.

First, we apply a load P that will be carried by both the fiber and the matrix. We assume a good bond between fiber and matrix so that both stretch the same amount and the load direction is parallel to the fiber alignment direction. This loading is called *isostrain,* for reasons that follow.

We can write

$$P_c = P_f + P_m \tag{1}$$

where P_c = total load on composite, P_f = load carried by fiber, and P_m = load carried by matrix.

Also, we know that the stress in the fiber σ_f is the load divided by the cross-sectional area of the fiber:

$$\sigma_f = \frac{P_f}{A_f} \qquad \text{or} \qquad P_f = \sigma_f A_f \tag{2}$$

Writing similar equations for the composite and the matrix, and substituting in Eq. (1), we have

$$\sigma_c A_c = \sigma_f A_f + \sigma_m A_m \tag{3}$$

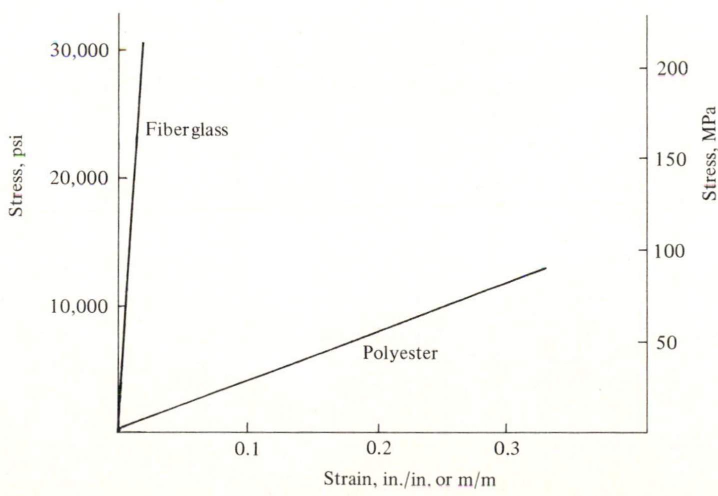

Fig. 11.4 *Stress–strain curves for fiberglass and polyester. At 0.003 strain, the stress in the fiberglass is 30,000 psi (207 MPa). At 0.003 strain, the stress in the polyester is 1200 psi (8.3 MPa).*

Dividing by A_c, the total cross-sectional area of the composite,

$$\sigma_c = \sigma_f \frac{A_f}{A_c} + \sigma_m \frac{A_m}{A_c} \tag{4}$$

A_f/A_c is the fraction of the total area occupied by the fiber. In this case it is also the volume fraction of the fiber, since the lengths of the fiber, composite, and matrix are all the same. If we let C_f denote the area (volume) fraction of fiber and C_m the area (volume) fraction of matrix, then $C_f + C_m = 1$, and rewriting Eq. (4), we obtain

$$\sigma_c = \sigma_f C_f + \sigma_m C_m \tag{5}$$

Now let us look at the strain produced by the load. If we call the overall strain ϵ_c, the strain in the fiber ϵ_f, and that in the matrix ϵ_m, then

$$\epsilon_c = \epsilon_f = \epsilon_m$$

because we assumed a good bond, and all are strained equally under load P_c. Hence the loading condition is called *isostrain.*

Let us divide the terms of Eq. (5) by ϵ_c, ϵ_f, and ϵ_m, separately:

$$\frac{\sigma_c}{\epsilon_c} = \frac{\sigma_f}{\epsilon_f} C_f + \frac{\sigma_m}{\epsilon_m} C_m \tag{6}$$

We can now substitute the overall modulus of the composite E_c for σ_c/ϵ_c, the modulus of the fiber E_f for σ_f/ϵ_f, and the modulus of the matrix E_m for σ_m/ϵ_m:

$$E_c = E_f C_f + E_m C_m \tag{7}$$

This allows us to determine the deflection or strain.

EXAMPLE 11.1 Suppose that we have a 0.5-in^2 (3.23-cm^2) bar that is 60% glass and 40% polyester by volume. What are the stresses and strains in the individual components? Assume loading parallel to the fibers or isostrain loading.

ANSWER

$$E_c = (10 \times 10^6 \text{ psi})(0.6) + (4 \times 10^5 \text{ psi})(0.4)$$
$$= 6.16 \times 10^6 \text{ psi } (0.425 \times 10^5 \text{ MPa})$$

The strain will be σ_c/E_c or, since the bar is 0.5 in^2 (3.225 cm^2),

$$\epsilon = \frac{P_c}{0.5E_c} = \frac{2P_c}{E_c}$$

Next let us find the factors affecting the loads carried by the glass and the plastic.

The load carried in the fiber will be the stress in the fiber times its area, and similarly in the matrix. We can write

$$\frac{\text{Load carried in fiber}}{\text{Load carried in matrix}} = \frac{\sigma_f A_f}{\sigma_m A_m} \tag{8}$$

or, using the relation $\sigma = E\epsilon$,

$$\frac{P_f}{P_m} = \frac{E_f \epsilon_f A_f}{E_m \epsilon_m A_m} \tag{9}$$

But $\epsilon_f = \epsilon_m$ and $A_f/A_m = C_f/C_m$, so

$$\frac{P_f}{P_m} = \frac{E_f}{E_m}\frac{C_f}{C_m} \tag{10}$$

From this expression we can calculate the loads carried by the fiber and the matrix and predict the overall strength of the composite.

Let us take our 60% glass–40% polymer mixture and assume a total load of 10,000 lb (4,540 kg). From Eq. (10), the ratio of the loads will be

$$\frac{P_f}{P_m} = \frac{10 \times 10^6}{4 \times 10^5}\frac{0.6}{0.4} = 37.5$$

The load on the polymer will be

$$P_m = P_o - P_f = 10{,}000 \text{ lb} - 37.5\, P_m = 260 \text{ lb (118 kg)}$$

$$\sigma_m = \frac{P_m}{A_m} = \frac{260 \text{ lb}}{(0.5 \text{ in.}^2)(0.4)} = 1300 \text{ psi (9.0 MPa)} \qquad P_f = 9740 \text{ lb (4,422 kg)}$$

$$\sigma_f = \frac{P_f}{A_f} = \frac{9740}{(0.5 \text{ in.}^2)(0.6)} = 32{,}500 \text{ psi (224 MPa)}$$

Checking that $\epsilon_m = \epsilon_f$, we have

$$\epsilon_m = \frac{\sigma_m}{E_m} = \frac{1300}{4 \times 10^5} = 3.25 \times 10^{-3} \qquad \epsilon_f = \frac{\sigma_f}{E_f} = \frac{32{,}500}{10^7} = 3.25 \times 10^{-3}$$

We have therefore accomplished our objective for Case 1, composites with continuous fibers. Note that the modulus of the composite is merely the weighted sum of the moduli of the fiber and matrix, and that the load carried in the fiber is affected by the *ratio* of the moduli:

$$E_c = E_f C_f + E_m C_m \qquad \text{from Eq. (7)}$$

$$\frac{\text{Load in fiber}}{\text{Load in matrix}} = \frac{E_f}{E_m}\frac{C_f}{C_m} \qquad \text{from Eq. (10)}$$

We can graphically show the relationship in Eq. (10). For modulus ratios less than 100, Figure 11.5 gives the load ratios for various fiber

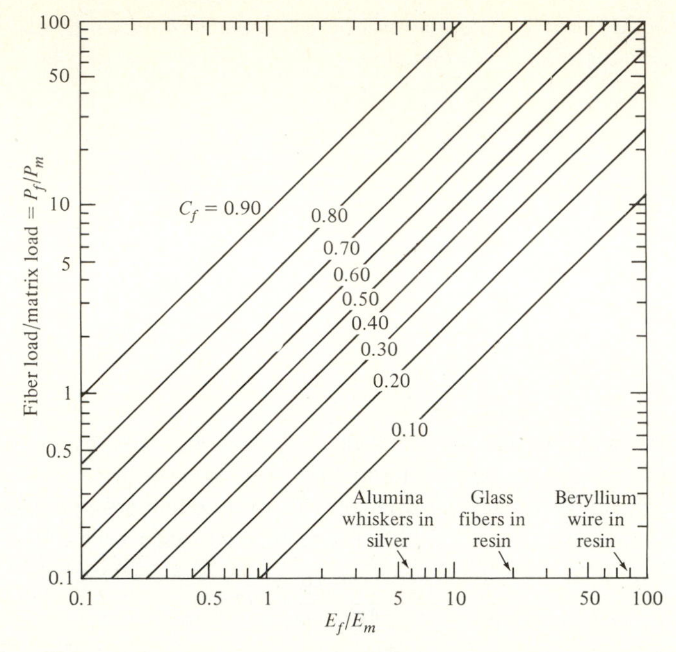

Fig. 11.5 *The interdependency of elastic moduli, load-bearing ratios, and volume fractions in fiber-reinforced composites that are to undergo isostrain loading.* (From L. J. Broutman and E. H. Krock, *Modern Composite Materials*, Addison-Wesley, Reading, Mass., 1967, p. 18)

volume fractions. The advantage of greater differences in the moduli between the matrix and the fiber (such as in Example 11.1) is that lower volume fractions of fiber are necessary to make the fiber the main load-bearing component.

Loading Perpendicular to Fiber Assume in Fig. 11.3 that the composite thickness is equivalent to the fiber diameter and loading is perpendicular to the fiber axis. The load is supported by the series resistance of the fiber and the matrix. Therefore the stress in the matrix, fiber, and composite are equal. We have what is termed an *isostress* condition.

$$\sigma_c = \sigma_m = \sigma_f \qquad \text{or} \qquad \sigma = E_c \epsilon_c \tag{11}$$

and

$$\epsilon_c = C_m \epsilon_m + C_f \epsilon_f \tag{12}$$

Therefore

$$\epsilon_c = \frac{\sigma}{E_m} C_m + \frac{\sigma}{E_f} C_f \tag{13}$$

Substituting Eq. (13) into Eq. (11), we obtain

$$\frac{1}{E_c} = \frac{C_m}{E_m} + \frac{C_f}{E_f} \quad \text{or} \quad E_c = \frac{E_m E_f}{C_f E_m + C_m E_f} \tag{14}$$

EXAMPLE 11.2 Calculate the modulus of the composite in Example 11.1, given that loading is perpendicular to the fibers.

ANSWER

$$E_c = \frac{E_m E_f}{C_f E_m + C_m E_f} = \frac{(4 \times 10^5)(10 \times 10^6)}{(0.6)(4 \times 10^5) + (0.4)(10 \times 10^6)}$$

$$= 9.4 \times 10^5 \text{ psi } (6.5 \times 10^3 \text{ MPa})$$

Comparison of the two loading methods, parallel vs. perpendicular to the fibers, shows that in the above examples parallel loading provides a composite modulus that is approximately 6.5 times higher. Figure 11.6 shows this same effect schematically. The advantage of isostrain loading, or loading along the fiber length, is apparent.

The isostrain and isostress conditions provide the limits of the design. Most experimental results fall between these limits.

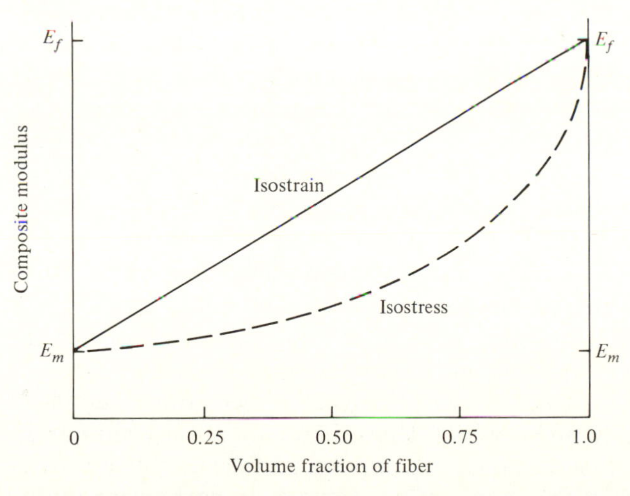

Fig. 11.6 *Schematic representation of isostrain vs. isostress loading of a composite material to which fiber has been added. For isostrain loading, less fiber is required to obtain the same modulus of elasticity of the composite.*

Discontinuous Fibers In composites with discontinuous fibers, the critical factors are the length-to-diameter ratio of the fiber, the shear strength of the bond between the fiber and the matrix, and the amount of fiber. All these variables affect the strength of the composite. However, a good deal of fiberglass is made with chopped-up pieces of glass because of the ease with which it is formed into complex shapes even when the length-to-diameter ratio is unfavorable.

Adherence of Fiber to Matrix In the preceding section we assumed that the strength of the glass-to-polymer bond was greater than the strength of the plastic. This is not always the case, and considerable research is in progress on this point.

Figure 11.7*a* and *b* shows photomicrographs of a polycarbonate–glass fiber composite showing poor bonding in contrast with a well-bonded composite. Table 11.1 (page 444) gives the properties.

In the material molded at 190°C the tensile strength is lower in the material containing glass fibers, whereas after treatment at higher temperatures, which improves the bond, the strength and elongation are higher in this material.

Further examination indicates that treatment at higher temperatures results in the formation of tiny crystalline regions in the resin that are accompanied by improved bonding. The material with the higher-temperature treatment also wets the fibers. It may also provide a layer of intermediate modulus, improving stress transfer without fracture.

Recently there has been great interest in the use of carbon fibers to reinforce a polymer matrix. Figure 11.8 (page 444) compares carbon fibers to long and short glass fibers in a nylon matrix. The flexural properties refer to the characteristics of the material in bending. Since many applications, such as automobile components, use thin sections, the flexural characteristics are of great importance, and carbon fiber composites exhibit the highest properties. However, the costs of carbon fiber, its availability for large-volume production items, and the speed of processing polymer-fiber composites are limitations that must be overcome before these composites are widely accepted.

Finally we should comment about other materials often called composites that we have treated in earlier sections or that we shall consider later. In most cases they are not true composites, as they do not fit our previous definitions, but are rather *duplex materials*.

Surface coatings may have either a mechanical, adhesive-type bond or a diffusion bond in which there is a continuous change in structure. Examples of adhesive bonds are painted surfaces, rubber coatings, and platings. In contrast, there are diffusion bonds in carburized and nitrided surfaces, bimetal castings, hot galvanized pipe, and brazed and welded assemblies.

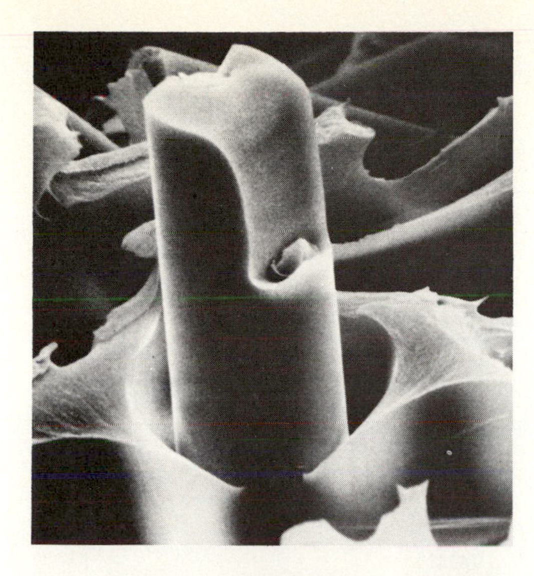

Fig. 11.7 *(a) Tensile fracture (as seen with scanning electron microscope) of an epoxy glass–polycarbonate composite molded at 190°C (374°F). Note the void around the fibers; 9100×.*

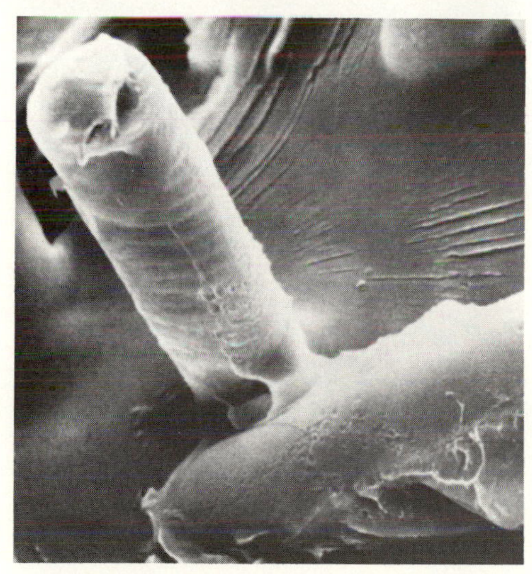

Fig. 11.7 *(b) Same as in Fig. 11.7(a), but molded at 275°C (527°F) followed by annealing at 245°C (473°C) for 3 hr; 8600×.* (Photographs courtesy J.L. Kardos, F.S. Cheng, and T.L. Tolbert, Washington University/Monsanto Association, St. Louis, Missouri)

Laminated construction may also involve either mechanical or diffusion bonds. In plywood the weakness and warpage of wood in one direction are overcome by gluing pieces in different orientations. In Alclad sheet an age-hardened aluminum, a 4.5% copper alloy core with poor corrosion resistance, is covered with a layer of pure aluminum. The heating during and after bonding must be controlled so that the copper from the core does not diffuse to the surface, lowering the corrosion resistance of the sheet.

Now let us consider a combination of natural and synthetic materials, exemplified by concrete.

Table 11.1 PHYSICAL PROPERTIES OF TYPE E GLASS–POLYCARBONATE COMPOSITE

Material	*Tensile Strength,* psi × 10³*	*Percent Elongation*	*Modulus of Elasticity,* psi × 10⁵*
	Molded at 190°C		
Resin alone	9.0	5.1	3.25
Resin and glass	7.6	1.3	8.19
	Molded at 275°C plus 3 hr at 245°C		
Resin alone	8.9	4.9	3.11
Resin and glass	11.2	2.3	9.28

*Multiply psi by 6.9×10^{-3} to obtain MN/m² (MPa) or by 7.03×10^{-4} to obtain kg/mm²

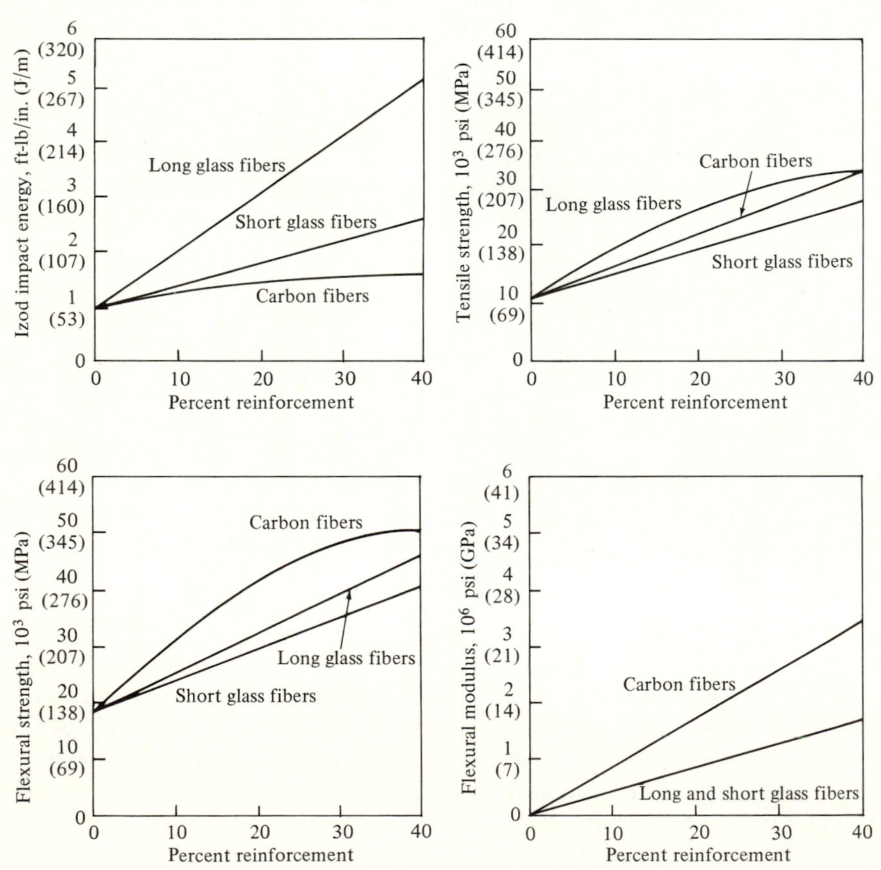

Fig. 11.8 *The effect of glass and graphite fibers on the mechanical properties of nylon 6/6.* (From *Metal Progress*, American Society for Metals, Metals Park, Ohio, Vol. 116, No. 6, Nov. 1979, p. 39)

11.5 Concrete: general

Concrete is a major construction material. The annual tonnage of concrete is greater than that of all metals combined. In addition, concrete offers the designer an exceptional flexibility in design and form because the material can be poured in place at room temperature, and even under water. This flexibility has its hazards, however. A lot of steel girders can be tested before they are installed, but the hydraulic bond in concrete depends on the skill of the contractor.

There are other sharp differences between the metals and concrete that are important, such as the low tensile strength, high compressive strength, and lack of ductility of concrete. In view of these differences, it is essential to study carefully not only the role of the constituents of concrete, but also the ambient conditions during curing, and during service.

Concrete is a mixture of a paste (Portland cement, water, and sometimes entrained air) with an aggregate. Variations of our present-day concrete practice were used by the early Egyptians and improved by the Greeks and Romans. However, Portland cement as we know it was not developed until the mid-1800s. In recent years there has been considerable research on concrete, leading to the development of many modern construction techniques.

Although concrete is widely used, it has a few limitations that must be recognized if we are to guarantee maximum service performance. The principal limitations are low tensile strength, thermal movements, shrinkage, creep under load, and permeability.

However, before we see how we can minimize these disadvantages, we must consider the contributions of the individual components in the concrete.

11.6 Components of concrete

As mentioned above, concrete is a mixture of a paste and an aggregate. We shall now consider the properties of each of the ingredients found in a concrete mixture: (1) cement, (2) water, (3) air, (4) aggregate, and (5) special additions.

Portland Cement In Chapter 8 we described the minerals present in Portland cement:

C_3S Tricalcium silicate $3CaO \cdot SiO_2$
C_2S Dicalcium silicate $2CaO \cdot SiO_2$
C_3A Tricalcium aluminate $3CaO \cdot Al_2O_3$
C_4AF Tetracalcium aluminoferrite $4CaO \cdot Al_2O_3 \cdot Fe_2O_3$

The American Society for Testing Materials (ASTM) recognizes five main grades of Portland cement. These are used for different purposes, as shown in Table 11.2, page 446.

Table 11.2 TYPICAL PORTLAND CEMENT COMPOSITIONS†

ASTM Grade	Use	Typical Composition			
		C_3S	C_2S	C_3A	C_4AF
I	General purpose or normal	50	24	11	8
II	Some sulfate protection;‡ less heat generation than Type I	42	33	5	13
III	High early strength	60	13	9	8
IV	Low heat generation	26	50	5	12
V	High sulfate‡ resisting	40	40	4	9

†Adapted from data of Portland Cement Association, Skokie, IL.
‡Certain soils contain sulfates that react with concrete, causing deterioration.

The compositions in Table 11.2 do not add up to 100% because other impurities are present, notably MgO and compounds of sulfur.

Cement hardens by a hydration reaction with the formation of a gel and crystals. It is vital to understand that water is an important part of the final structure, and cement should not be allowed to dry during setting. This will stop the setting reaction! The compounds form at different rates, which is the reason for the several compositions. The characteristics of the four major components are as follows.

C_3S Becomes jellylike in a few hours with considerable generation of heat. It is important to the early strength, developed in periods of less than 14 days.

C_2S A slow hydration reaction with low heat generation. It is responsible for the long-time strength development or durability.

C_2A Hydrates rapidly with high heat generation. It is responsible for initial stiffening, but offers the least contribution to long-time strength.

C_4AF Appears to have little effect on cement performance. It provides little strength and is added to decrease the processing temperature of the cement.

The fineness as well as the chemistry of the cement is important to the rate of hydration. Finer cement tends to react more quickly (surface areas are larger as particles become finer); therefore Type III, high early strength, tends to be finer to further decrease the setting time.

Table 11.3 indicates the relative concrete strengths that can be developed using the different types of Portland cement.

Under the same conditions of concrete mix, temperature, and moisture, all grades of Portland cement should achieve the same level of compressive strength after three months.

Mixing Water Although any drinkable water can be used for making concrete, water unsuitable for drinking may also be used. Table 11.4 gives

Table 11.3 COMPRESSIVE STRENGTH OF CONCRETE FOR DIFFERENT PORTLAND CEMENTS*

ASTM Grade	*Fraction of Compressive Strength Developed at Indicated Time, Based on Unity for Type I*			
	1 day	*7 days*	*28 days*	*3 months*
I	1.0	1.0	1.0	1.0
II	0.75	0.85	0.90	1.0
III	1.90	1.20	1.10	1.0
IV	0.55	0.55	0.75	1.0
V	0.65	0.75	0.85	1.0

*Adapted from data of Portland Cement Association, Skokie, IL.

levels of water constituents that when exceeded may affect both set time and strength. When there is some doubt about water quality, tests should be conducted on concrete mixes.

Air Entrainment The primary purpose of air entrainment is to improve the workability of concrete and increase its resistance to freeze–thaw cycles. Although a certain amount of air is entrapped in all concrete, air entrainment is the deliberate addition of air bubbles by mixing with an agent such as a resin. The amount of resin is usually low (less than 0.1% by weight of the cement). The bubbles are small in size, normally 0.001 to 0.003 in. (0.0025 to 0.0075 cm), and may range from 3 to 9% by volume of the concrete mix. The important point is that the bubbles are not interconnected and are well distributed. See Fig. 11.9 on page 448.

Aggregates The aggregate is very important to the serviceability of the concrete, since it normally makes up 60 to 75% of the total volume, as

Table 11.4 TOLERABLE LIMITS OF COMPONENTS IN CONCRETE MIXING WATER*

Permissible Concentration (ppm)	*Components*
Less than 50	Sanitary sewage
50–500	Bicarbonates of Ca and Mg; salts of Mn, Sn, Zn, Cu, Pb; sugar
500–1000	Carbonates and bicarbonates of K and P
1000–5000	Silt; industrial waste
5000–20,000	NaCl; Na_2SO_4; acid water
20,000–40,000	$MgSO_4$; $MgCl_2$; iron salts; sea water, as salt content

Others: Oil is tolerable in small amounts; algae reduce strength; alkaline water less than 0.5% by weight of cement; $CaCl_2$ up to 2% by weight of cement can be added to accelerate hardening and strength gain.

*Adapted from data of Portland Cement Association, Skokie, IL.

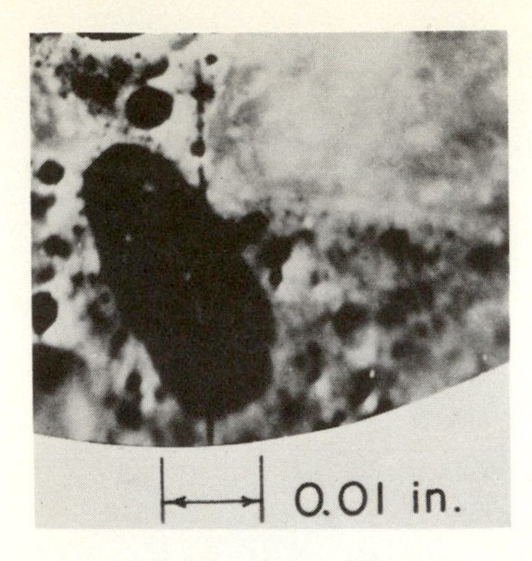

Fig. 11.9 *Enlarged and polished cross section of air-entrained concrete.* (From *Design and Control of Concrete Mixtures*, 11th ed., 1968, Portland Cement Assoc., Skokie, Ill.)

shown in Fig. 11.10. The aggregate is generally characterized as fine (less than $\frac{1}{4}$ in., 0.635 cm, usually sand) and coarse (greater than $\frac{1}{4}$ in., 0.635 cm, usually gravel). For optimum strength and workability the correct aggregate proportions must be used and the characteristics of the aggregate must be known and controlled. The aggregate variables are size, shape, porosity, specific gravity, moisture absorption, resistance to freeze–thaw, strength, abrasion resistance, and chemical stability. We have already discussed a number of these variables in relation to packing of solids. The terminology can be found in Sec. 8.6.

As might be expected, the amount of cement and water required to coat the aggregate depends on the size distribution of the aggregate. Therefore a *fineness modulus* has been developed to characterize the aggregate. The cumulative percentage obtained on a fixed sequence of sieves is determined. The added cumulative percentage divided by 100 gives the fineness modulus. Table 11.5 gives the sequence of sieves and their openings.

It is common to apply the fineness modulus to only the fine aggregate and use the following guidelines for the *maximum size* of the coarse aggregate:

1. One-fifth the dimension of nonreinforced members
2. Three-fourths the clear dimension between reinforcing bars or nets
3. One-third the depth of nonreinforced slabs on the ground

Table 11.5 STANDARD SIEVES FOR CLASSIFYING AGGREGATE

Fine Aggregate (Sand)			*Coarse Aggregate (Gravel)*		
Sieve No.	*Opening: in.*	*(mm)*	*Sieve No.*	*Opening: in.*	*(mm)*
4	0.187	(4.75)	6	6	(152.4)
8	0.093	(2.36)	3	3	(76.2)
16	0.046	(1.18)	$1\frac{1}{2}$	$1\frac{1}{2}$	(38.1)
30	0.024	(0.6)	$\frac{3}{4}$	$\frac{3}{4}$	(19)
50	0.012	(0.3)	$\frac{3}{8}$	$\frac{3}{8}$	(9.5)
100	0.006	(0.15)	4	$\frac{3}{16}$	(4.75)

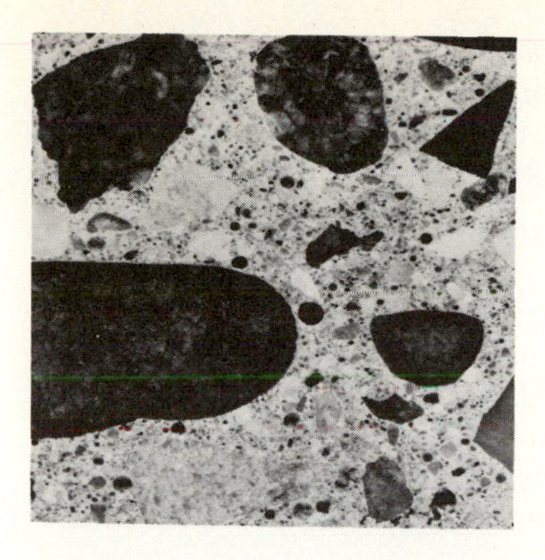

Fig. 11.10 *Polished macrosection of normal concrete, showing aggregate and distribution of paste.* (From *Design and Control of Concrete Mixtures,* 11th ed., 1968, Portland Cement Assoc., Skokie, Ill.)

EXAMPLE 11.3 A 500-g sample of fine aggregate has weights retained on each sieve as given here. Determine the fineness modulus.

Sieve No.	*Grams Retained*
4	11.3
8	74.5
16	100.0
30	80.3
50	120.1
100	98.7
Pan	14.3
	499.2

ANSWER The fines passing through the no. 100 sieve may be ignored. The initial 500-g sample usually does not yield 500 g in the final analysis because of weighing errors and inevitable losses.

Sieve No.	*Grams Retained*	*Percent Retained*	*Cumulative Percent Retained*
4	11.3	2.2	2.2
8	74.5	14.9	17.1
16	100.0	20.0	37.1
30	80.3	16.1	53.2
50	120.1	24.1	77.3
100	98.7	19.8	97.1
Pan	14.3	2.9	Total 284.0

$$\text{Fineness modulus} = \frac{284}{100} = 2.84$$

This sand is not well graded, since the percent retained shows a double peak.

Other Additions (Admixtures) In addition to Portland cement, aggregate, water, and entrained air, other agents may be added to concrete. A description of several admixture agents and their purpose follows. The list is not complete; however, it gives examples of common additions.

1. *Accelerators* decrease the set time necessary at low temperatures. Calcium chloride is the most common.
2. *Retarders* increase the set time necessary in very hot weather. They are similar to water-reducing agents, described below.
3. *Water reducers* (plasticizers) provide good workability at lower water-to-cement ratios. Lignosulfonate (a byproduct of wood pulp) is an example.
4. *Pozzolans* react with lime [$Ca(OH)_2$] released during setting. They have no cementitious value and retard the set time. Pulverized ash from burned coal is a common pozzolan.
5. *Super-plasticizers* increase the workability or flowability of the concrete mix. They allow lower water-to-cement ratios, giving higher strength while still maintaining workability. Several organic sulfonated condensates are used.

11.7 Properties of concrete

We are now ready to consider the mechanical and physical properties of concrete. We want to maximize the strength at minimum cost. Furthermore, the strength of concrete is normally higher at lower water-to-cement ratios, yet we must arrive at a reasonable compromise to attain satisfactory workability.

As with all ceramic materials, concrete has a higher compressive strength than tensile strength. In service, therefore, the primary loading is in compression. We can increase the resistance to tensile loads by reinforcing the concrete with steel or by prestressing, to be discussed in a later section.

We shall first treat the typical compressive strength of concrete, followed by workability, moisture effects, temperature effects, shrinkage, creep properties, and abrasion resistance.

Compressive Strength The hydration reaction of Portland cement paste is time dependent. Figure 11.11 shows typical time–compressive strength curves for air- and non-air-entrainment concrete mixtures. Low water-to-cement ratios increase the compressive strength significantly. Compression test coupons are normally 6 in. in diameter $\times$ 12 in. high (15.24 $\times$ 30.48 cm). However, cores or trepanned specimens can also be cut out of existing structures.

Since air entrainment appears to offer lower compressive strength at a given water-to-cement ratio, its utilization may seem questionable. However, air entrainment makes concrete more durable, especially under freeze-thaw conditions, because the voids are discontinuous (Fig. 11.9). Furthermore,

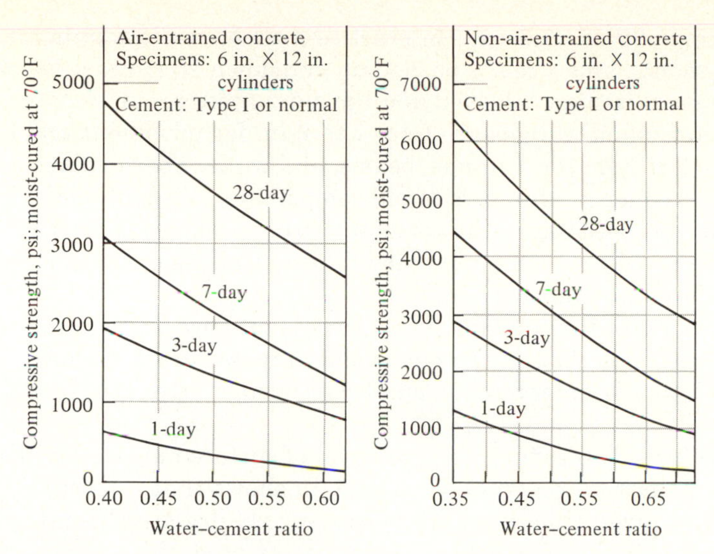

Fig. 11.11 *Typical compressive strengths for air-entrained and non-air-entrained concretes as related to curing time.* (From *Design and Control of Concrete Mixtures*, 11th ed., 1968, Portland Cement Association, Skokie, Ill.)

air entrainment allows good workability at lower water-to-cement ratios. Hence we may be able to obtain equivalent compressive strengths and equivalent workability as determined by the slump test, to be discussed next, in two concretes using two different water-to-cement ratios.

Workability A slump test, consisting of a truncated cone of 8 in. (20.3 cm) lower diameter, 4 in. (10.2 cm) upper diameter, 12 in. (30.5 cm) high, is used as an index of concrete workability in the field. Figure 11.12 shows a

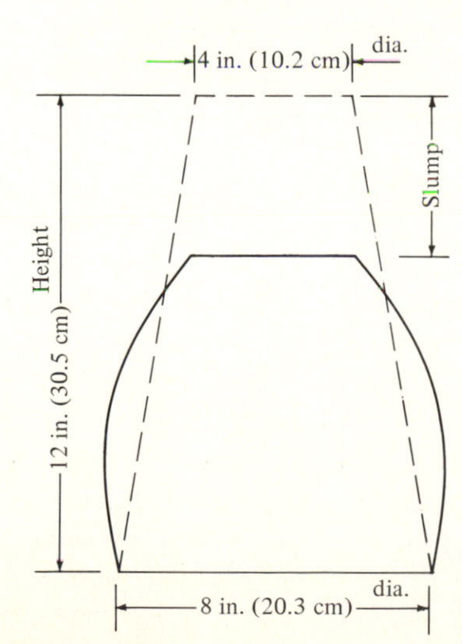

Fig. 11.12 *The slump test for concrete. (Slump is idealized; it is usually not this symmetric.)*

typical slump of 3 in. (7.6 cm). In practice a sheet steel mold is filled in one-third increments, and a rod is pushed up and down 25 times after each increment. The mold is then removed and the slump measured.

Typical slump ranges are from 1 to 2 in. for pavement and heavy construction to 3 to 5 in. for columns, beams, and walls. These values assume the use of high-frequency vibrators to aid compaction. Without such aids, higher slumps may be necessary. Unfortunately, the slump test does not measure the *work* required for compaction, and different mixes of the same slump can require different amounts of work.

The workability must not be so great as to cause segregation or bleeding of the concrete. *Bleeding* is the movement of water to the surface. This leads to higher water-to-cement ratios and lower surface strengths and durability.

Moisture The reduction or removal of surface moisture will slow down or stop the hydration reaction. Figure 11.13 shows that interruption of moist curing after a given time interval by exposure to dry air ultimately stops the curing. Interestingly, if moist-air curing is restored, the strength will again increase. For example, after 28 days a dry-air-cured concrete achieves only 55% of the compressive strength of a moist-cured concrete, whereas if the sample is moist-cured for the first 3 days and then dry-air-cured the compressive strength is 80%. If after 28 days we expose both the dry-air and 3-day samples to moist air, the strength again increases.

Temperature As with many chemical reactions, the hydration reaction releases heat, and the rate of hydration is higher at higher temperatures.

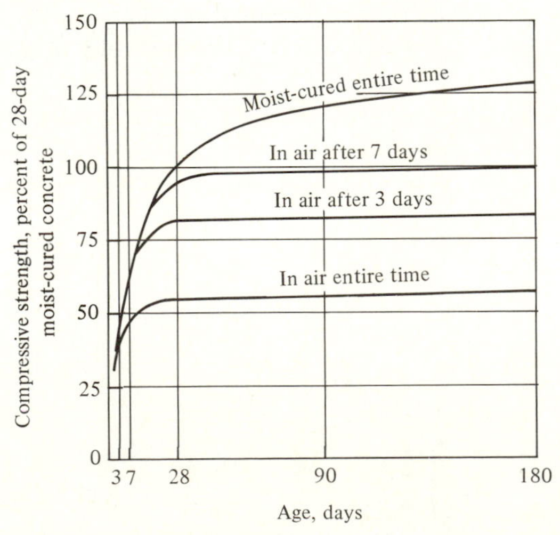

Fig. 11.13 *Absence of water in the air lowers the strength of concrete. Maximum strength depends on length of time moisture is present.* (From *Design and Control of Concrete Mixtures,* 11th ed., 1968, Portland Cement Association, Skokie, Ill.)

Therefore the correct type of cement paste, water-to-cement ratio, and treatment to achieve optimum strength differ depending on the ambient temperature. We shall treat this further when we discuss special concretes for high- and low-temperature setting.

Temperature plays a very important role in the control of shrinkage. When severe, shrinkage can result in the concrete cracking before it has reached an optimum strength.

Shrinkage Shrinkage can occur in two stages. In the first stage, it occurs while the concrete is still in a plastic state. This stage is water, temperature, and time dependent. There is water loss to the forms and evaporation plus water of hydration, with a net effect of a decrease in volume. Meanwhile the temperature rise often masks the shrinkage due to the coefficient of expansion. An extreme case can result in *plastic cracking*. To minimize this, we would like to have the rate of water loss be equivalent to the natural bleeding of water to the surface.

Figure 11.14 shows the drying shrinkage that may be anticipated for concrete stored in air. It is evident that larger water-to-cement ratios result in higher shrinkage. This is another reason for minimizing the ratio.

The second stage of shrinkage occurs after initial hardening of the paste. It is due to further hydration and cooling of the mass. This may cause little trouble, except that a concrete mass may not harden uniformly because of nonuniformity of moisture in the surroundings, such as above and below ground. If this happens, complex stresses may be set up and the concrete mass may crack even after a year or more of curing.

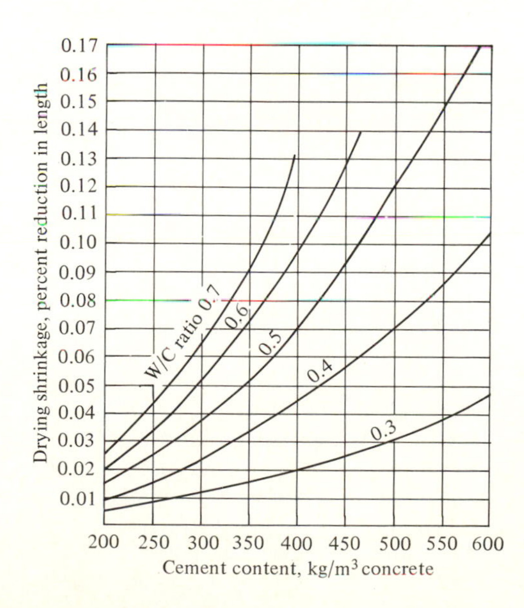

Fig. 11.14 *Average values for shrinkage of concrete stored in air, on the basis of the cement content and the water-to-cement ratio.* (From L.J. Hurdock and K.H. Brook, *Concrete Materials and Practice*, John Wiley & Sons, Inc., New York, 1979, p. 17)

Creep Creep was discussed in Chapter 3 and stress relaxation and viscous flow in Chapters 10 and 7, respectively. Creep is not always detrimental, as it may relieve the stresses imposed by drying shrinkage. In general, the creep rate is lower for the following conditions.

1. At higher concrete strengths; therefore at lower water-to-cement ratios and for longer curing times
2. At lower total volume of cement paste
3. With larger aggregate

Figure 11.15, the creep data for concrete cylinders loaded to 600 psi (3.8 MPa) after a 7-day cure, shows several of these variables.

Abrasion Resistance and Durability Abrasion becomes very important in roads, concrete floors, and dam spillways. As might be expected, a stronger concrete has better wear resistance, as shown in Fig. 11.16. However, other agents have an effect on the durability of concrete. For example, some soils are high in sulfate. For these soils, a Portland cement higher in tricalcium aluminate (Type V) is recommended. Sea water, salt on roads, and alternate freezing and thawing all shorten the life of concrete. Even atmospheric pollution takes its toll in destruction of the hydrated bond. This has led to international concern over the potential loss of concrete buildings and statues left by early civilizations. The problem is most acute in urban areas.

11.8 Special concretes

We shall now briefly identify some of the concretes used for special applications.

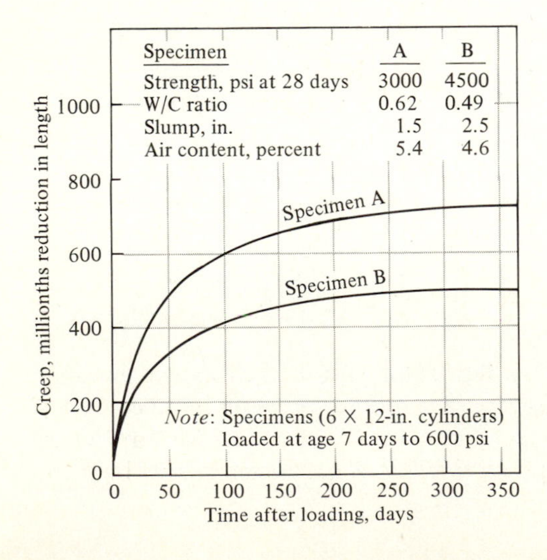

Fig. 11.15 *When stressed at 600 psi (4.14 MPa) after a 7-day age, concrete with higher compressive strength shows a lower creep rate.* (From *Design and Control of Concrete Mixtures*, 11th ed., 1968, Portland Cement Association, Skokie, Ill.)

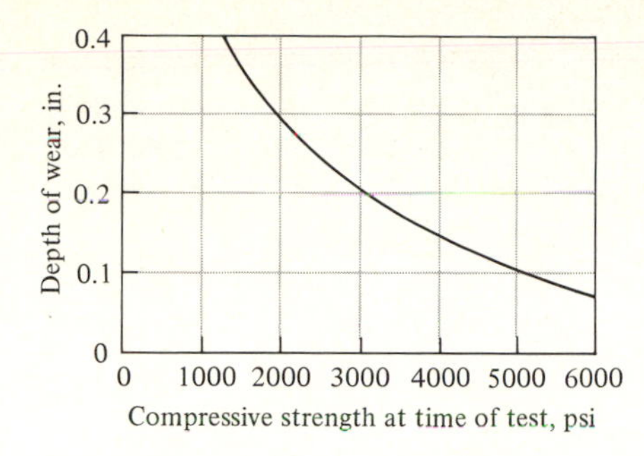

Fig. 11.16 *Concretes with higher compressive strength show a greater resistance to wear.* (From *Design and Control of Concrete Mixtures*, 11th ed., 1968, Portland Cement Association, Skokie, Ill.)

Air-entrained Concrete We have already discussed air entrainment. However, it is important to differentiate between entrainment and entrapment. All concretes contain entrapped air that cannot be removed because of the viscosity of the paste. In general, finer aggregate results in more entrapped air. Entrained air is in the form of very fine bubbles retained by the addition of an organic chemical (a neutralized vinsol resin). The air content recommended by the Portland Cement Association for severe exposure conditions is as follows.

Maximum Size Coarse Aggregate, in. (cm)	*Air Content, Percent by Volume*
$1\frac{1}{2}$ (3.81); 2 (5.08); $2\frac{1}{2}$ (6.35)	5 ± 1
$\frac{3}{4}$ (1.91); 1 (2.54)	6 ± 1
$\frac{3}{8}$ (0.95); $\frac{1}{2}$ (1.27)	$7\frac{1}{2} \pm 1$

Lightweight Concrete These concretes may be classified in two distinct categories.

1. *Lightweight structural concrete.* 28-day compressive strength in excess of 2500 psi (17.25 MPa) and an air dry density of less than 115 lb/ft^3 (1843 kg/m^3). The light weight is obtained by using heat-expanded lightweight aggregates such as shale, clay, slate, blast furnace slag, or fly ash.
2. *Lightweight insulating concrete.* 28-day compressive strength of 100 to 1000 psi (0.69 to 6.9 MPa) and an air-dry density of 15 to 90 lb/ft^3 (240 to 1443 kg/m^3). Various porous aggregates and/or incorporation of a cement paste into a cellular matrix with air voids give the low densities. The advantages in terms of our present and future concerns with energy conservation are shown by the thermal conductivity, Fig. 11.17 on page 456.

Heavyweight Concrete We can obtain protection from x rays, γ rays, and neutrons by using high-density aggregates to give a density of concrete up to 400 lb/ft^3 (6412 kg/m^3). In some cases metal punchings or shot are used for a portion of the aggregate, although oxides of iron, titanium, and barium are more common.

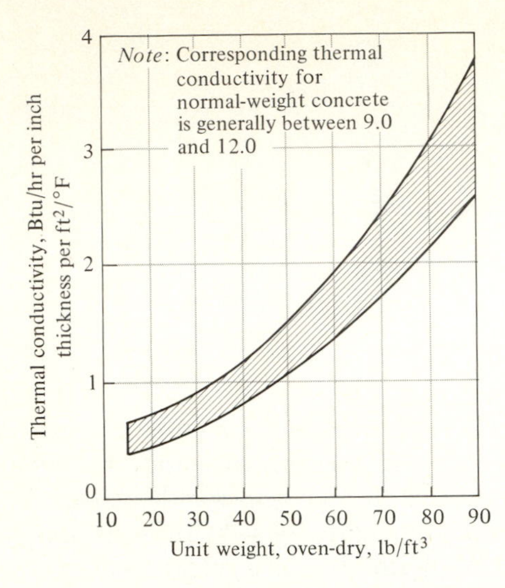

Fig. 11.17 *The thermal conductivity of lightweight (insulating) concrete is lower for low unit weights (lb/ft³).* (From *Design and Control of Concrete Mixtures*, 11th ed., 1968, Portland Cement Association, Skokie, Ill.)

Hot-weather Concrete Some parts of the country have very hot, dry weather year around and special procedures are necessary. More water is required to compensate for evaporation, and the concrete has a greater tendency to crack. Therefore a low-heat-generation cement such as II or IV is preferred (Table 11.2).

Cold-weather Concrete At low temperatures the cure rate is reduced and high early strength cements can be used, as shown in Fig. 11.18. A higher-temperature concrete mix can be poured and will provide initial setting even if the ambient temperature is below freezing. The important point is to develop strength quickly, using the internal heat of the initial mix plus the heat of hydration. The concrete can often be protected by heated plastic bubble

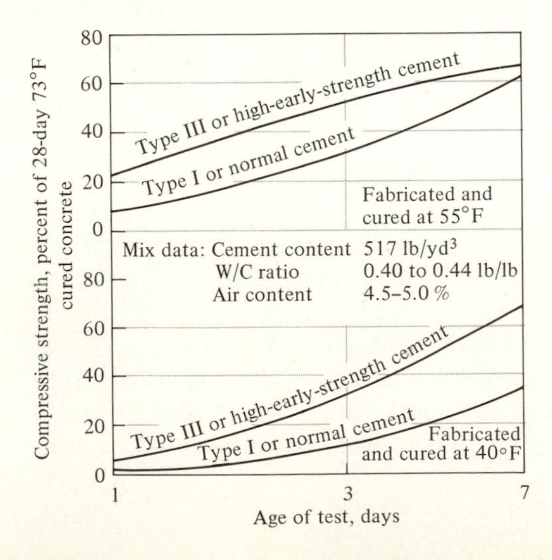

Fig. 11.18 *When cured at low temperatures, type III cement gives higher early compressive strengths.* (From *Design and Control of Concrete Mixtures*, 11th ed., 1968, Portland Cement Association, Skokie, Ill.)

covers or by use of steam-accelerated curing. Calcium chloride (up to 2% maximum) accelerates curing.

11.9 Reinforced and prestressed concrete

The tensile strength of concrete is approximately one-tenth of its compressive value; therefore, design is primarily in compression. Older civilizations recognized this, and this led to the development of the arch. However, there are many applications in which tensile stresses are developed. Examples are bending of beams and loading of tall columns, where potential buckling gives rise to tensile stresses. Therefore it is common practice to apply steel reinforcement at the portion of the beam in tension, as shown in Fig. 11.19.

Steel is used rather than other metals for the obvious reason of its higher yield strength. Not so obvious is the fact that steel has a coefficient of expansion approximately the same as that of concrete. Also concrete is alkaline, which causes minimal corrosion to steel.

Reinforced concrete may crack as a result of buildup of tensile stresses during curing and shrinkage. In most cases, if there is good bonding between the cement and the reinforcement, the width of the cracks can be minimized. One major cause of cracking is lack of concrete cover over the reinforcement. Under mild weather and protected conditions, a covering of $\frac{1}{2}$ to $\frac{3}{4}$ in. (1.27 to 1.91 cm) may be adequate. However, under severe conditions and exposure to de-icing salt, 2 to $2\frac{1}{2}$ in. (5.08 to 6.34 cm) is required.

In road building, with a steel wire mesh used for reinforcement, a single layer of reinforcement is placed at the centerline. Because of frost upheaval, movement of the roadbed, and weight of traffic, tension may develop at both the upper and lower surfaces, and a single layer of reinforcement placed at the centerline is justified. If two layers are used (above and below the centerline), the concrete must be thicker to satisfy the cover requirement.

Although air entrainment produces discontinuous pores and hence less penetration by road salt, there is still considerable difficulty with road breakup. Using calcium chloride as an accelerator for cold-weather curing only adds to the steel corrosion.

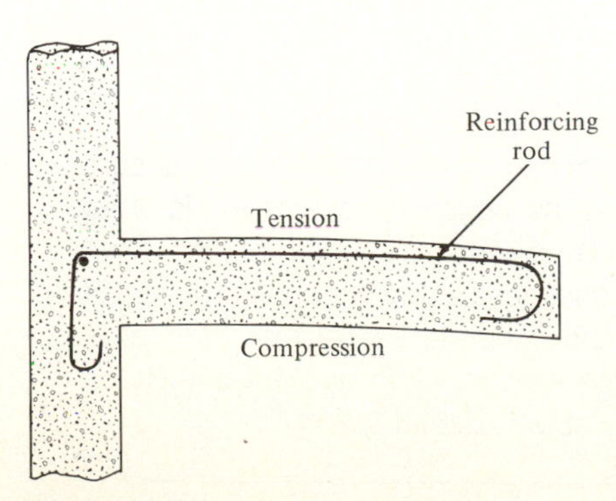

Fig. 11.19 *A cantilevered concrete beam, showing a reinforcing rod on the side to be in tension (deflection has been accentuated)*

The use of polymers to patch and lengthen the life of concrete roads, bridges, and parking structures has gained some popularity. For example, parking garage decks have been patched successfully by cleaning the concrete down to the reinforcing rods and applying two-part epoxy primers and sand-filled epoxy resins.

The corrosion of reinforcing rods in road bridge decks has led to considerable research into ways of coating and protecting the surfaces. One method requires drying the concrete at 500 to 600°F (260 to 315°C) with heaters, pressure impregnation with a monomer such as methylmethacrylate, and finally polymerization by hot water or steam. Depths of impregnation of up to 4 in. (10.16 cm) have been obtained. However, the cost is high, $7.00/ft^2 (1979 prices).

The poor tensile strength and higher compressive strength of concrete suggest the use of residual compression that must be overcome before the component can fail in tension. This result, found in *prestressed concrete,* is accomplished in two ways.

Pretensioning is placing steel rods or wire in tension, pouring the concrete, and removing the tension after curing of the concrete to pull the concrete into compression. The technique requires a good bond between the steel and concrete and end-tapered or anchored configurations for the steel to prevent stress relief.

Post-tensioning requires no bond between the concrete and the steel. In its simplest application, steel wire is placed within a tube and concrete is cast around it. After curing, the steel is placed in tension and anchored at its ends to the concrete. When the tension is removed, the anchored steel places the concrete in compression.

The amount of residual compression is limited by the compressive strength of the concrete and the attainable elastic strain in the steel. Ideally the amount of tensile elastic strain in the steel is equivalent to the compressive elastic strain in the concrete. However, geometric placement, surface adhesion, and ambient conditions inhibit this ideal condition.

EXAMPLE 11.4 The residual stress in a steel rod in prestressed concrete is found to decrease with time. What characteristics of the concrete suggest that this is normal?

ANSWER The principal reason is continued curing of the concrete and accompanying shrinkage, which relieves the strain. An associated phenomenon is creep of the concrete. Both of these may occur over a considerable length of time, since total curing may take years depending on the thickness of the concrete. The similar coefficients of expansion between the steel and concrete suggest no thermal effects. However, thermal and mechanical fatigue may result in poorer bonding between the two components and hence some loss of residual stress.

11.10 Proportioning of concrete mixtures

For small amounts of concrete, the simplest—but not the most precise—method is to use empirical volume ratios. The following guideline ratios for cement/fine aggregate/coarse aggregate often give satisfactory results.

Structure	*Ratios*
Reinforced concrete	1/2/4
Large concrete mass	1/3/6
Pavement and sidewalks	1/2/3

A more realistic approach is to design for a specific strength requirement in the following way.

1. Establish the conditions of exposure and geometry. Table 11.6 provides an initial guideline to maximum water-to-cement ratios.
2. Determine the minimum strength requirement after a specific cure time. Figure 11.20 provides average compressive strengths for two types of Portland cement after a 7-day and a 28-day cure.
3. Optimize the workability (slump requirements) and water-to-cement ratios and establish a trial batch knowing the fineness modulus and the maximum coarse aggregate size. Table 11.7 (pages 460–461) gives typical trial batches.
4. On the basis of measured slump, workability, and air content, modify the mix to obtain the required characteristics.

The technique can best be shown by working out an example. However, the calculations will be carried out using units common to the concrete industry (lb, yd^3, sacks of cement, etc.). We shall give the conversion to metric units after the calculation.

The following conversion factors and properties are necessary to the calculations (page 462).

Table 11.6 MAXIMUM PERMISSIBLE WATER/CEMENT RATIOS†

Structure	*Severe Application (Air-entrained Only)*			*Mild Application (Seldom Below Freezing)*		
	In Air	*In Fresh Water*	*In Sea Water*	*In Air*	*In Fresh Water*	*In Sea Water*
Thin sections	0.49	0.44	0.40	0.53	0.49	0.40
Bridge decks	0.44	0.44	0.40	0.49	0.49	0.44
Moderate sections	0.53	0.49	0.44	‡	0.53	0.44
Heavy sections	0.58	0.49	0.44	‡	0.53	0.44
Concrete slabs in ground contact	0.53	—	—	0.53	—	—
Pavements	0.49	—	—	0.53	—	—

†Adapted from *Recommended Practice for Selecting Properties for Concrete* (ACI613-54)
‡Ratio selection based on workability, but should not be less than 470 lb cement/cubic yard.

Table 11.7 SUGGESTED TRIAL MIXES FOR AIR-ENTRAINED CONCRETE OF MEDIUM CONSISTENCY (3-to 4-in. slump*)

					With Fine Sand Fineness Modulus = 2.50			With Coarse Sand Fineness Modulus = 2.90		
Water–Cement Ratio, lb per lb	Maximum Size of Aggregate, in.	Air Content, Percent	Water, lb per yd³ of Concrete	Cement, lb per yd³ of Concrete	Fine Aggregate, Percent of Total Aggregate	Fine Aggregate, lb per yd³ of Concrete	Coarse Aggregate, lb per yd³ of Concrete	Fine Aggregate, Percent of Total Aggregate	Fine Aggregate, lb per yd³ of Concrete	Coarse Aggregate, lb per yd³ of Concrete
0.40	$\frac{3}{8}$	7.5	340	850	50	1250	1260	54	1360	1150
	$\frac{1}{2}$	7.5	325	815	41	1060	1520	46	1180	1400
	$\frac{3}{4}$	6	300	750	35	970	1800	39	1090	1680
	1	6	285	715	32	900	1940	36	1010	1830
	$1\frac{1}{2}$	5	265	665	29	870	2110	33	990	1990
0.50	$\frac{3}{8}$	7.5	340	680	53	1400	1260	57	1510	1150
	$\frac{1}{2}$	7.5	325	650	44	1200	1520	49	1320	1400
	$\frac{3}{4}$	6	300	600	38	1100	1800	42	1220	1680
	1	6	285	570	34	1020	1940	38	1130	1830
	$1\frac{1}{2}$	5	265	530	32	980	2110	36	1100	1990
0.60	$\frac{3}{8}$	7.5	340	565	54	1490	1260	58	1600	1150
	$\frac{1}{2}$	7.5	325	540	46	1290	1520	50	1410	1400
	$\frac{3}{4}$	6	300	500	40	1180	1800	44	1300	1680
	1	6	285	475	36	1100	1940	40	1210	1830
	$1\frac{1}{2}$	5	265	440	33	1060	2110	37	1180	1990
0.70	$\frac{3}{8}$	7.5	340	485	55	1560	1260	59	1670	1150
	$\frac{1}{2}$	7.5	325	465	47	1360	1520	51	1480	1400
	$\frac{3}{4}$	6	300	430	41	1240	1800	45	1360	1680
	1	6	285	405	37	1160	1940	41	1270	1830
	$1\frac{1}{2}$	5	265	380	34	1110	2110	38	1230	1990

0.40	3/8	3	385	965	50	1240	1260	54	1350	1150
	1/2	2.5	365	915	42	1100	1520	47	1220	1400
	3/4	2	340	850	35	960	1800	39	1080	1680
	1	1.5	325	815	32	910	1940	36	1020	1830
	1 1/2	1	300	750	29	880	2110	33	1000	1990
0.50	3/8	3	385	770	53	1400	1260	57	1510	1150
	1/2	2.5	365	730	45	1250	1520	49	1370	1400
	3/4	2	340	680	38	1100	1800	42	1220	1680
	1	1.5	325	650	35	1050	1940	39	1160	1830
	1 1/2	1	300	600	32	1010	2110	36	1130	1990
0.60	3/8	3	385	640	55	1510	1260	58	1620	1150
	1/2	2.5	365	610	47	1350	1520	51	1470	1400
	3/4	2	340	565	40	1200	1800	44	1320	1680
	1	1.5	325	540	37	1140	1940	41	1250	1830
	1 1/2	1	300	500	34	1090	2110	38	1210	1990
0.70	3/8	3	385	550	56	1590	1260	60	1700	1150
	1/2	2.5	365	520	48	1430	1520	53	1550	1400
	3/4	2	340	485	41	1270	1800	45	1390	1680
	1	1.5	325	465	38	1210	1940	42	1320	1830
	1 1/2	1	300	430	35	1150	2110	39	1270	1990

*Increase or decrease water per cubic yard by 3 percent for each increase or decrease of 1 in. in slump, then calculate quantities by absolute volume method. For manufactured fine aggregate, increase percentage of fine aggregate by 3 and water by 15 lb per cubic yard of concrete. For less workable concrete, as in pavements, decrease percentage of fine aggregate by 3 and water by 8 lb per cubic yard of concrete.

From *Design and Control of Concrete Mixtures*, Portland Cement Association, Skokie, IL, 11th ed.

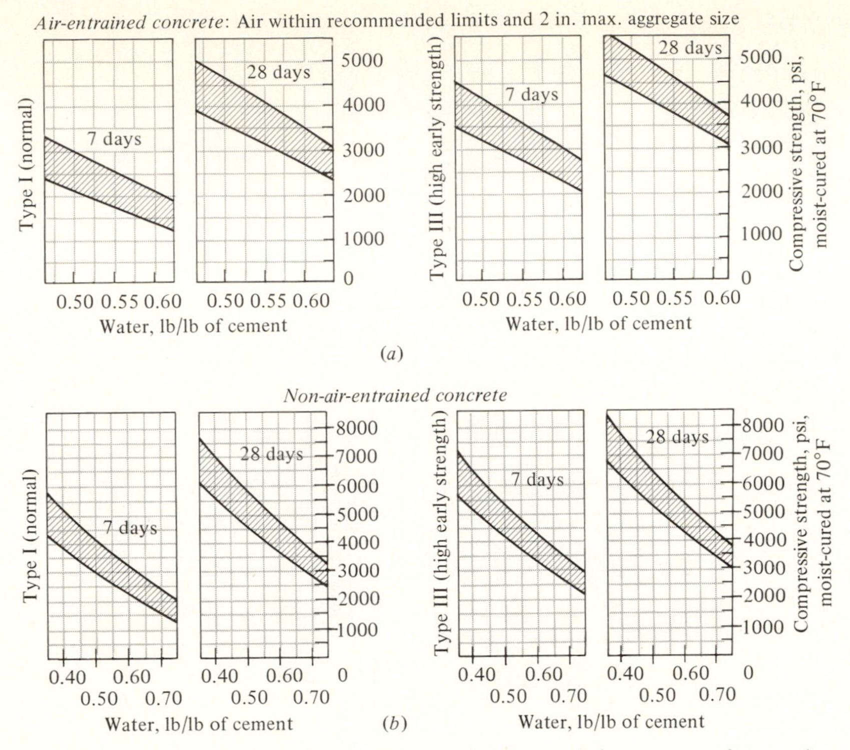

Fig. 11.20 *Development of average compressive strength in concrete for varying water/cement ratios, cure time, and type of cement.* (From *Design and Control of Concrete Mixtures*, 11th ed., 1968, Portland Cement Association, Skokie, Ill.)

1. Water: density 62.4 lb/ft^3 = 1685 lb/yd^3; weight 8.33 lb/gal
2. Portland cement: apparent density 3.15 × 62.4 lb/ft^3 = 196.6 lb/ft^3; 1 sack = 94 lb
3. Aggregate: specific gravity range 2.4 to 2.9; average true specific gravity of sand, 2.65; average true specific gravity of gravel, 2.60
4. Air: volume in cubic feet/cubic yard concrete ~0.27 times air content in percent.

EXAMPLE 11.5 We wish to make a concrete slab in contact with the ground subjected to mild application where air entrainment is not necessary. A slump of 3 to 4 in. is recommended, with a coarse aggregate size of $\frac{3}{4}$ in. and a fineness modulus for the sand of 2.50. The coarse aggregate contains 2% moisture and the fine aggregate 5% moisture. The proposed 28-day cure strength using non-air-entrained type I Portland cement is 4000 psi.

ANSWER

1. From Table 11.6, the recommended *maximum* water-to-cement ratio is 0.53.
2. A 4000-psi concrete must be given a factor of safety, normally 15%.

Therefore $4000 \times 0.15 + 4000 = 4600$ psi concrete required. From Fig. 11.20, a water-to-cement ratio of 0.56 would place us in the middle of the band. However, since 0.53 is the maximum recommended ratio, we must use this number in our calculations and accept the consequence that the strength will be greater than requested.

3. Table 11.7 gives us trial batches. Interpolation is necessary for differences in slump, fineness modulus, and water-to-cement ratio. The trial batch would be as follows.

Water		340 lb/yd^3
Cement	340/0.53 =	640 lb/yd^3
Fine aggregate		1130 lb/yd^3
Coarse aggregate		1800 lb/yd^3
Total		3910 lb/yd^3 (2323 kg/m^3)

The water must not merely be added as liquid, since the aggregate contains water in its open pores.

$$\text{Water in sand} = 1130 \times 0.05 = 55 \text{ lb}$$
$$\text{Water in gravel} = 1800 \times 0.02 = 35 \text{ lb}$$
$$\text{Total} = 90 \text{ lb } (40.9 \text{ kg})$$
$$\text{Water to be added as liquid} = 340 - 90 = 250 \text{ lb/yd}^3 \ (149 \text{ kg/m}^3)$$

Had we neglected this, the true water-to-cement ratio would have been $(340 + 90)/640 = 0.67$, which would have resulted in lower strength.

We may also calculate the volume contribution of each component as follows (assuming average densities for the aggregates).

$$\text{Water} \quad \frac{340 \text{ lb}}{1685 \text{ lb/yd}^3} = 0.20 \text{ yd}^3$$

$$\text{Cement} \quad \frac{640 \text{ lb}}{196.6 \text{ lb/ft}^3 \times 27 \text{ ft}^3/\text{yd}^3} = 0.12 \text{ yd}^3$$

$$\text{Fine aggregate} \quad \frac{(1130 - 55) \text{ lb}}{2.65 \times 1685 \text{ lb/yd}^3} = 0.24 \text{ yd}^3$$

$$\text{Coarse aggregate} \quad \frac{(1800 - 35) \text{ lb}}{2.60 \times 1685 \text{ lb/yd}^3} = 0.40 \text{ yd}^3$$

$$\text{Total} = 0.96 \text{ yd}^3 \ (1.26 \text{ m}^3)$$

Since we should have 1.00 yd^3, the remaining 0.04 yd^3 is made up of 0.02 yd^3 anticipated entrapped air (Table 11.7) and 0.02 yd^3 of closed porosity in the aggregate.

4. Let us assume that our trial mixture gave a slump that was too great. A decrease in slump of 1 in. requires 3% less water (see footnote to Table 11.7). It is important to maintain the same water-to-cement ratio; therefore the dry solids are increased.

$$\text{Initial dry solids} = 640 + (1130 - 55) + (1800 - 35) = 3480 \text{ lb/yd}^3$$
$$\text{New water content} = 340 - 0.03 \times 340 = 330 \text{ lb/yd}^3$$

As a first approximation, the ratio of water decrease should be inversely proportional to the dry solids increase.

or
$$\text{Dry solids} = 3480 \times 340/330 = 3585 \text{ lb/yd}^3$$
$$\text{Portland cement} = 330/0.53 = 625 \text{ lb/yd}^3$$

$$\text{Fine aggregate} = \left(\frac{1130}{1130 + 1800}\right) \times 3585 = 1385 \text{ lb/yd}^3$$

$$\text{Coarse aggregate} = \left(\frac{1800}{1130 + 1800}\right) \times 3585 = 2200 \text{ lb/yd}^3$$

$$\text{Liquid water addition} = 330 - (1385)(0.05) - (2200)(0.02) = 215 \text{ lb/yd}^3$$

11.11 Asphalt

Since some of our previous discussion of concrete concerned its use for road building, it is appropriate that we also consider asphalt. Asphalt is a bitumen. Although it occurs naturally, it is most often obtained as a byproduct of petroleum refining. In road building, approximately 6% asphalt is used to bond together an aggregate. Furthermore, it is a thermoplastic material. Therefore it is applied hot and made to flow into place with the aid of heavy rollers.

Although it does not require a cure time as concrete does, these same thermoplastic characteristics cause difficulty with roads. For example, in high summer temperatures asphalt roads undergo viscous flow or creep. In winter they become brittle (below the glass transition temperature), and uneven stress can lead to fracture, causing the familiar pothole.

The aggregate sizing in asphalt roads is as important as with concrete. Various attempts have been made to use fillers that enhance the properties. The addition of ground glass, particularly at intersections, increases traction and gives better reflectivity at night. Addition of rubber particles gives better resiliency, especially at low temperatures (the application is not unlike that shown in Fig. 11.2 for additions of rubber to polystyrene).

This concludes our discussion of concrete and asphalt or combinations of synthetic and natural materials to produce a composite. We shall now continue with the last section on wood, a completely natural composite material.

11.12 Wood—general

Wood is a beautiful, varied, complex, widely distributed construction material. The annual tonnage of wood used in the United States is greater than the combined tonnage of steel and concrete—more than 300,000,000 tons. More than 60 native woods are in common use, and 30 are imported. Many more composites, such as plywood, particle board, and paper, are of great importance.

We have postponed a discussion of wood to this point because it is the most complex of the composite polymeric materials and because its properties are difficult to describe. We cannot use the simple approach we used for the metals and ceramics, starting with unit cells as modules, stacking these neatly together to form grains, then showing how the properties depend on slip and rupture of these assemblies. We shall see instead that wood is a composite honeycomb structure made up of different biological cells, and that the cell walls are made up of complex arrays of cellulose fibers. These are reinforced with a matrix of polymers such as lignin and other organic compounds plus varying amounts of inorganic crystals, hard enough in certain cases to thwart the teeth of the toredo worm.

To enable the engineer to select wood for design in a variety of circumstances, from siding for a house to massive beams for an industrial building, we shall present our discussion in three parts.

1. *Macrostructure.* In contrast to other materials, the macrostructure is the most important feature of wood. Many properties vary by a factor of 20 depending on the location in the log and the direction of testing.
2. *Microstructure.* An examination of the microstructure gives details essential for understanding certain features of the macrostructure and the properties.
3. *Testing and correlation of properties* with structure. With the background of 1 and 2, the great directionality in properties can be understood and allowed for in design.

The section concludes with a discussion of some typical uses of wood and wood products.

11.13 Wood macrostructure

The major point we must keep in mind in analyzing the structure of wood, in contrast to a billet of steel or an extruded blank of plastic, is that the complex honeycomb structure of a tree was not produced in order to make homogeneous, isotropic lumber, but in response to a basic force to grow and survive. Therefore, if we are to use wood in construction, we must not become impatient with its anisotropy, but instead strive to understand how this variation in properties with direction developed in response to growth conditions.

Let us begin by examining the cross section of a typical tree, as shown in Fig. 11.21. The most important feature lies at the intersection between the bark and the wood. This is a very thin layer called the *cambium,* which is the source of both wood and bark cells. Each year the cambium grows new wood cells on its inside surface and new bark cells on its outside surface. Immediately after its formation by cell division, the wood cell begins to enlarge in both diameter and length. During this period the cell has a very thin, pliable primary wall. Once the cell attains full size, the wall is thickened by addition of a secondary wall.

The new cells add to the circular† band of *sapwood,* which is generally light in color, in contrast to the *heartwood,* which is darker. The sapwood is made up of living cells that carry fluids plus some older dead cells.

The *heartwood* is composed entirely of dead cells. Its darker color is due to greater deposits of tarlike materials and minerals. It is denser, stronger, and more decay-resistant than the sapwood.

Yearly growth rings are usually present. They result from differences in the thickness of the cell walls formed in different seasons. In some tropical regions where there is little change in weather, the growth rings are poorly defined. These cells will be seen clearly in the microstructure.

Wood rays are horizontal radial canals connecting the various layers from the center to the bark for storage and transfer of food.

Pith is the original soft tissue around which the first wood growth takes place. Since the tree grows vertically as well as horizontally, the pith core is found throughout the length.

†The growth is not only in a radial direction but also vertically. As a result, the tree trunk has a slender conical shape; it is not a perfect cylinder.

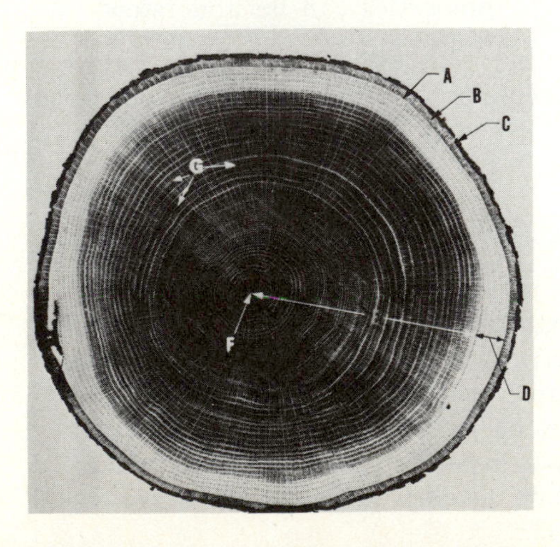

Fig. 11.21 *Cross section of a typical tree. A = cambium; B = inner bark; C = outer bark; D = sapwood; E = heartwood; F = pith; G = wood rays.* [U.S. Dept. of Agriculture Handbook No. 72 (revised Aug. 1974), p. 2-2]

Hardwoods and Softwoods These designations are about as important to wood specifications as the divisions into ferrous and nonferrous alloys is to the metals industry. "Hardwoods" are *usually* harder and stronger than the "softwoods," but there are some exceptions. Softwoods such as Douglas fir and long-leaf pine are harder than the hardwoods bass and aspen. Woods are divided into classes in two ways. First, the softwoods, especially in the United States, are conifers, "evergreens" such as pine and spruce with needlelike leaves and exposed seeds. By contrast, the hardwoods lose their leaves, have true flowers, and have seeds that are covered in a fruit such as a nut. In the next section, dealing with microstructure, we shall see that there is a structural difference.

Before we leave the topic of macrostructure, we should orient ourselves regarding directions in the tree, so that the correlation with microstructure will be clear. The three axes used are shown in Fig. 11.22. L designates the longitudinal axis in a log, which is the vertical axis of the tree, R designates radial direction, and T designates the tangential direction (normal to the radial direction).

11.14 Wood microstructure

By viewing the microstructure of wood, we can understand the makeup of many of the features we saw in the macrostructure, such as growth rings and the role of the cells. In addition, we can understand the differences between softwood and hardwood and the variation in properties in different directions. Just as with other materials, the scanning electron microscope is useful at intermediate magnifications (around 300×) for analyzing the overall features and at high magnification (about 1000×) for viewing cell structure.

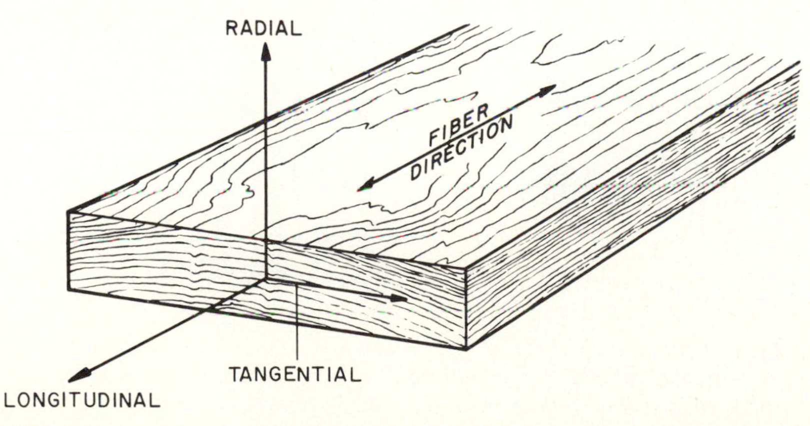

Fig. 11.22 *Axes used to specify directions in wood.* [U.S. Dept. of Agriculture Handbook No. 72 (revised August 1974) p. 4-2]

We shall begin with observations of softwood structure. Then we shall cover the differences we encounter in the hardwoods, and finally the cell itself.

Microstructure of Softwood The most important and striking feature of softwood structure (Fig. 11.23*a* and *b*) is that it is composed of long, tubelike cells running in the longitudinal direction with some other cells normal to this direction (radial). We shall first explore the characteristics of this giant honeycomb, then examine the different types of cells in detail.

A new layer of longitudinal cells (called *tracheids*) is grown each year from the cambium. The early-season (spring) cells (5) are larger and thinner-walled than the late wood cells (6). A typical cell is 4 mm long and 0.04 mm wide. When a log is sawn in cross section, this difference is seen as a growth ring. The radial rays we pointed out in the macrostructure are made up of cells (*parenchyma*) oriented in a radial direction (4). These store food. Sometimes the ray also contains a resin duct, as shown in (8) and (10). Resin ducts may also be vertical (9). As would be expected, resin ducts are only found in some softwoods. Pitlike features are visible in the cell walls (4 and 12). These are actually orifices that the cells can close like valves to control the flow of fluids through the cells. In softwoods, sap flows through the longitudinal cells (tracheids).

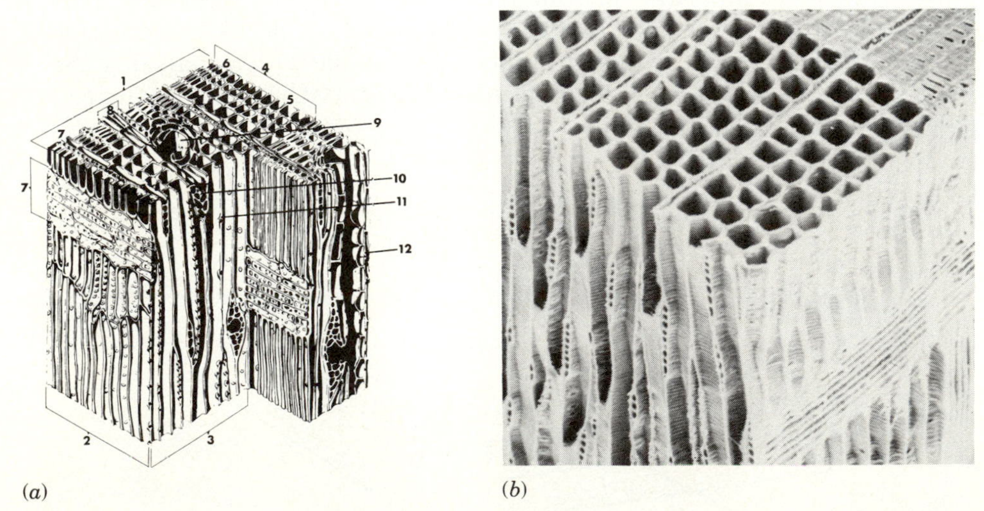

Fig. 11.23 *(a) Soft-wood structure (schematic). 1 = cross-sectional face; 2 = radial face; 3 = tangential face; 4 = annual ring; 5 = early wood; 6 = late wood; 7 = wood ray; 8 = fusiform ray; 9 = vertical resin duct; 10 = horizontal resin duct; 11 = bordered pit; 12 = simple pit.* [*Classroom Demonstrations of Wood Properties*, U.S. Dept. of Agriculture, Forest Products Laboratory, PA900, p. 2] *(b) Douglas fir; scanning electron micrograph, 300×.* (From slide of U.S. Forest Products Laboratory, Madison, Wis.)

Microstructure of Hardwoods The microstructure of hardwood (Fig. 11.24) is similar in many ways to that of softwood. Again, the majority of the cells have their long dimensions parallel to the longitudinal axis of the tree, and early wood and late wood are present, giving growth rings. These are called *fibers*. The wood rays are also similar, and the cells are called *parenchyma*. Resin ducts are not present.

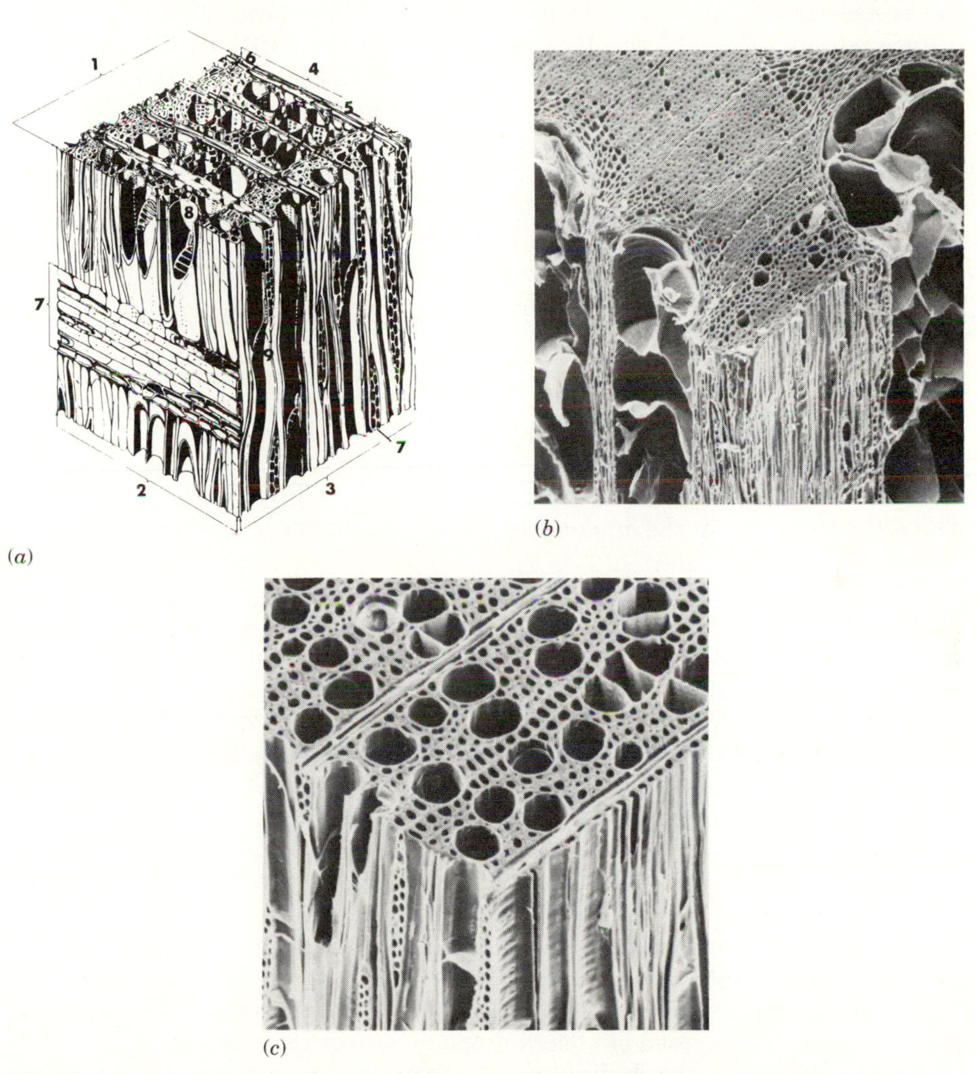

Fig. 11.24 *(a) Hardwood structure (schematic). For meaning of numbers 1 through 7, see legend for Fig 11.23. Here 8 = a vessel; 9 = perforation plate.* [*Classroom Demonstrations of Wood Properties*, U.S. Dept. of Agriculture, Forest Products Laboratory, PA 900, p. 4] *(b) Diffuse porous structure in hardwood (yellow poplar). Scanning electron micrograph, 300×.* (From slide of U.S. Forest Products Laboratory, Madison, Wis.) *(c) Ring porous structure in hardwood (white oak). Scanning electron micrograph, 300×.* (From slide of U.S. Forest Products Laboratory, Madison, Wis.)

The most important difference is the addition of large longitudinal tubes called *vessels* (8), which transport fluids. At the ends of these vessels there is a perforation plate. Variations in these structures are used to identify hardwoods. Average vessel length is 0.5 mm and width 0.07 mm.

The other important feature of the vessels is their size in a given year's growth. In the illustration shown (yellow poplar), the diameter of the vessels is the same in both early wood and late wood. This is called *diffuse porous* (Fig. 11.24*b*). By contrast, in white oak, the wide vessels appear in rings. This is called *ring porous* (Fig. 11.24*c*). These vessels can be seen with a hand magnifying glass, and the bands contribute an attractive feature to the wood. There is an important variation in the amounts of different types of cells in hardwoods. For example, large amounts of vessels lower the apparent specific gravity:

Species	*Percent Vessels*	*Percent Fiber (long cells)*	*Percent Parenchyma (food cells)*	*Specific Gravity*
Basswood	56	36	8	0.32
Hickory	6	67	22	0.64

Cell Structure Finally, let us go to higher magnification to observe the details of cell structure. Despite the many variations in macrostructure and microstructure, the *basic* biological cells of wood have similar features. The differences are in the *amounts* of the constituents used to make up the structure.

We can compare the cell structure roughly to building up different layers of fiberglass and resin in constructing the hull of a boat. There are two principal divisions, the primary wall and the secondary wall, as shown in Fig. 11.25. Let us consider first the framework of cellulose fibers. (In Chapter 9 we illustrated the structure of cellulose as a strong linear polymer.) In the primary wall the cellulose fibers are in a loose, irregular, flexible network. This is built up first, followed by the three layers of the secondary wall: S_1, a crisscross network; S_2, a parallel, spiral-type network; and finally S_3, another network. S_2 makes up 70 to 90% of the cell wall. The closer the fibers approach the longitudinal direction, the stronger the cell is in this direction.

Whereas the strength in the longitudinal direction is strongly influenced by the strength of the cellulose, the transverse strength is related to the *lignin* and *hemicellulose* deposits, which form a matrix between the fibers. Lignin is composed of phenol-propane network structures similar to the thermosetting network discussed earlier. Hemicellulose is made up of shorter branched cellulose molecules with a degree of polymerization of 150 to 200 compared to 5000 to 10,000 for the cellulose of the network.

In addition to these three constituents, there are two other important components. There is a group of oil-like hydrocarbons, called *extractives*

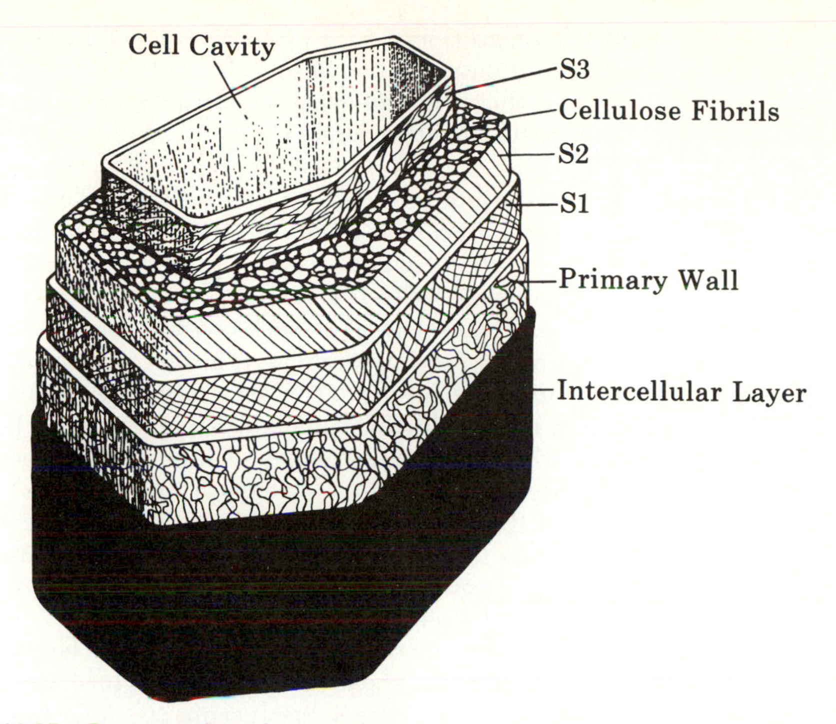

Fig. 11.25 *Cross section of a woody cell.* (*Classroom Demonstrations of Wood Properties*, U.S. Dept. of Agriculture, Forest Products Laboratory, PA900, p. 6)

because they can be dissolved from the wood chemically, in contrast to the polymeric structures. These can act as powerful deterrents to bacterial action and weathering, as in the case of cedar and cypress. The other component is minerals such as silica, which can add hardness and resistance to borers by the formation of a hard mineral network.

11.15 Properties of wood

In describing the mechanical properties of wood, we must include some additional data not required for the metals. For example, we would not expect pieces of 0.2% carbon steel from different sources to vary appreciably in specific gravity or to shrink when exposed to dry atmospheres. However, two completely dry pieces of wood of the same species may differ greatly in specific gravity if growth conditions produced different amounts of early wood cells (specific gravity 0.2) and late wood cells (specific gravity 0.80). The dimensions of a given piece also vary. There is very little dimensional change in the longitudinal direction because of the continuous cellulose chain network. Radial shrinkage or expansion accounts for about 40% of the volume change, and tangential expansion about 60%. Actual shrinkage in passing from the green to oven-dry condition varies from 2 to 8% in the radial direction and from 4 to 11% in the tangential direction.

For these reasons, wood should be tested at a stabilized known moisture content, and the specific gravity should be determined.

Orthotropic Properties From the previous discussions of wood structure, we expect to find different properties along the three axes: L, longitudinal, R, radial, and T, tangential. This anisotropy is given a special name, *orthotropic,* among wood technologists to indicate that there is aniso*tropy* along the three *ortho*gonal axes (axes normal to each other).

As an example, we can test specimens cut in different directions, and from the straight-line portion of the curve we can determine the modulus of elasticity E. The variation in values with direction is striking. Example: Douglas fir: $E_L = 13{,}400$, $E_T = 670$, $E_R = 911$, where L, T, and R are directions and all values are in megapascals.

The high value of E in the longitudinal direction is due to the continuous cellulose network, whereas the lower values in the other directions are caused by the voids and lack of continuity. The higher value in the radial direction is produced by the rodlike cell structures of the rays.

A simple model will help us to understand these variations better. We may consider the tree as made up of a stack of soda straws glued together. (The diameter-to-length ratio of a soda straw represents the long tubelike shapes of the tracheids and the vessel cells.)

To describe mechanical properties completely, we would need to make measurements in all three orthotropic directions. Because of practical limitations and the fluctuation in any property, the measurements are generally made "across the grain," i.e., normal to the longitudinal direction, and also parallel to the grain.

Table 11.8 gives typical data for some common softwoods and hardwoods. The first line of each set refers to the green condition and the second with 12% water. All specimens are selected clear grain. We shall deal with the role of defects later.

The most common test is in bending because this is the typical use of structural lumber and because it is not possible to cut a "typical" tensile specimen, owing to the inhomogeneous nature of wood. The modulus of rupture is not really a modulus, but reflects the maximum load-carrying capacity of a beam. The compression and shear properties are important in the design of load-bearing areas. Hardness is tested by measuring the load required to press a steel ball to a specified depth; therefore the higher the load, the greater the hardness.

We can draw several general conclusions from the data of Table 11.8. The typical hardwoods listed are indeed harder and stronger than the softwoods. There are great differences within each group, however; the modulus of rupture of Douglas fir is 40 to 50% greater than that of white pine or cedar.

Table 11.8 TYPICAL MECHANICAL PROPERTIES OF WOODS GROWN IN THE UNITED STATES

Species	Specific Gravity	*Static Bending* Modulus of Rupture, lb/in²*	*Static Bending* Modulus of Elasticity, 10^6 lb/in²*	Compression Parallel to Grain; Maximum Crushing Strength, lb/in²*	Compression Perpendicular to Grain; Fiber Stress at Prop. Limit, lb/in²*	Shear Parallel to Grain; Maximum Shearing Strength, lb/in²*	Side Hardness Load Perpendicular to Grain, lb_f[(2)]
Softwoods							
Eastern	0.34 Gr[(1)]	4,900	0.99	2440	220	680	290
White pine	0.35 KD	8,600	1.24	4800	440	900	380
Douglas fir	0.45 Gr	7,700	1.56	3780	380	900	500
(Coast)	0.48 KD	12,400	1.95	7240	800	1130	710
Western	0.31 Gr	5,200	0.94	2770	240	770	260
Red cedar	0.32 KD	7,500	1.11	4560	460	990	350
Redwood	0.34 Gr	5,900	0.96	3110	270	890	350
(Young growth)	0.35 KD	7,900	1.10	5220	520	1110	420
Hardwoods							
White ash	0.55 Gr	9,600	1.44	3990	670	1380	960
	0.60 KD	15,400	1.74	7410	1160	1950	1320
Yellow	0.55 Gr	8,300	1.50	3380	430	1110	780
birch	0.62 KD	16,600	2.01	8180	970	1880	1260
Black	0.47 Gr	8,000	1.31	3540	360	1130	660
cherry	0.50 KD	12,300	1.49	7110	690	1700	950
Hickory	0.60 Gr	9,800	1.37	3990	780	1480	1310
(pecan)	0.66 KD	13,700	1.73	7850	1720	2080	1820
Maple	0.44 Gr	7,400	1.10	3240	450	1110	620
(big leaf)	0.48 KD	10,700	1.45	5950	750	1730	850
Oak	0.66 Gr	8,300	1.25	3560	670	1250	1060
(white)	0.68 KD	15,200	1.78	7440	1070	2000	1360

(1) Gr = green state; KD = kiln-dried to 12% moisture
(2) Load (lb_f) to make standard diameter impression; multiply by 4.44 to obtain newtons
*Multiply by 6.90×10^{-3} to obtain MPa.

This explains why Douglas fir is used for structural timber. However, cedar and redwood are preferred for siding because of the "extractives," which give better weathering characteristics and resistance to decay. Eastern white pine is extensively used for objects when an easy-to-work material is needed.

There are large differences among the hardwoods as well. Hickory is the hardest of the woods listed, correlating with its use for baseball bats. Oak, birch, and ash show both high strength and hardness, supporting their use for furniture and flooring. Cherry and maple are lower in these properties than the other hardwoods. However, they are also much used in furniture because of their beauty and workability.

Finally we should consider briefly the effect of method of sawing the log on the properties of the lumber. Figure 11.26 shows two common methods, plain sawing and quarter sawing. In plain sawing all the boards are cut par-

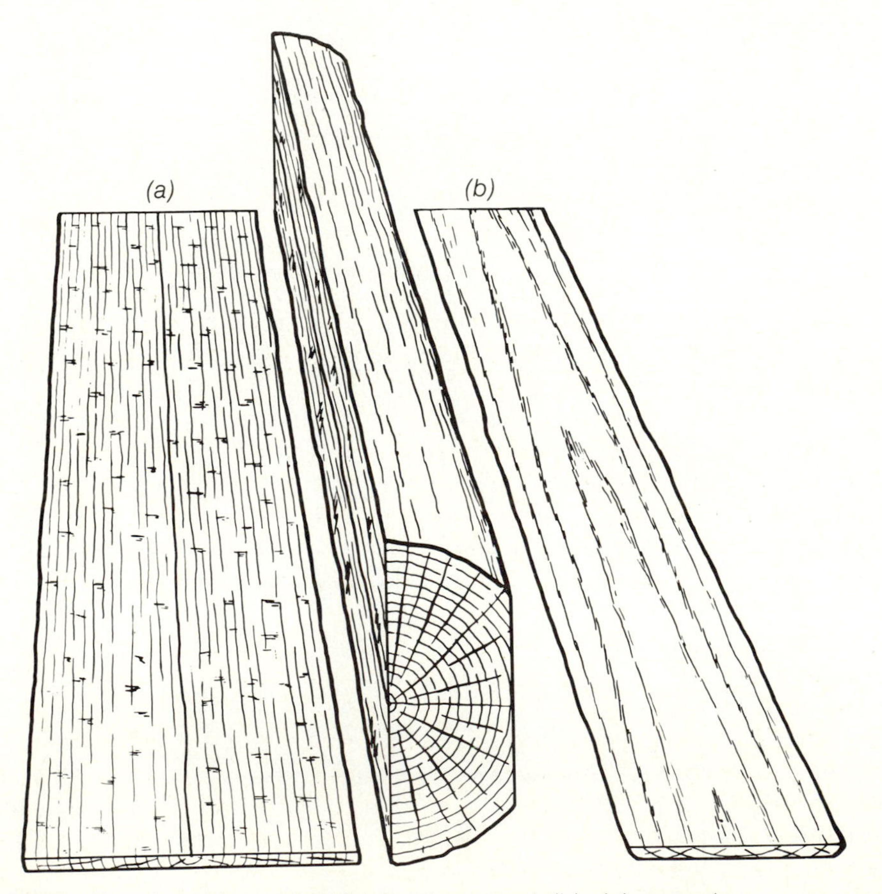

Fig. 11.26 *Boards cut from a log. (a) Quarter-sawed, (b) plain-sawed.* [U.S. Dept. of Agriculture Handbook No. 72 (revised August 1974), p. 3-1]

allel to the same plane, whereas in quarter sawing the log is rotated 90° four times to produce the grain shown.

Plain-sawed lumber is lower in cost, but quarter-sawed lumber shrinks and swells less in width and wears more evenly. On the other hand, plain-sawed lumber shrinks and swells less in thickness and shows figure patterns due to annual rings more conspicuously.

11.16 Role of defects in lumber

It is highly important to realize that the data covered were obtained on clear-grained specimens. In large pieces of lumber there may be a number of defects present, such as knots, checks, and so forth.

If we compare the properties of no. 2 lumber with those of clear-grained wood of the same species, we obtain the following data:

	Modulus of Rupture, MPa	*Modulus of Elasticity, GPa*	*Compressive Strength, MPa*
Wood (clear)	97.5	13.0	53.5
No. 2 lumber	19.4	11.0	31.9

The property showing the greatest decrease as a result of defects is modulus of rupture, one of the commonest design values. The modulus of elasticity is not greatly affected by imperfections, and the compressive strength is severely reduced.

Restructuring to Reduce Effect of Defects The simplest method of restructuring is to glue sections of wood together to produce *laminated beams*. The position of knots can be controlled, and much larger lengths and cross sections can be produced. Most of the adhesives used are thermal setting, using temperatures ranging from room temperature to those requiring hot pressing.

Plywood is produced by machining sheets from logs and gluing the plies with the grain at 90° to each other. In this way the material is nearly isotropic and swelling is equalized.

Small pieces and shavings that were previously discarded can be used for *particle board and fiberboard*. Particle board can be glued with thermosetting resin to place the larger chips in the surface layers for better appearance and wear. In the case of fiberboard, the chips are converted to a pulp. In the wet process, the pulp is squeezed and dried. It derives its strength from mechanical interlocking of the fibers and bonding by lignin. In dry bonding, a resin is used as a glue.

SUMMARY

The composite materials have been categorized as completely synthetic (such as fiberglass), completely natural (such as wood), or combinations of synthetic and natural materials (such as concrete and asphalt).

Particle-reinforced and dispersion-strengthened materials are classified according to the size of the second phase. Dispersion strengthening finds many analogies in the metallurgical principles discussed in earlier chapters. Possibly the best example of particle reinforcement is sintered carbides, in which volume fractions of particles are large.

In fiber reinforcement, a lower volume fraction of fiber is required if loading is parallel to the direction of the fibers rather than perpendicular to the direction of the fiber. In all cases, however, the volume fractions of fibers, particles, or the dispersed phase must be controlled to obtain the desired properties.

Similarly, the amounts of aggregate and cement paste determine the properties of concrete. The objective is to obtain a concrete mass at minimum cost, and therefore the aggregate makes up a large volume fraction. Concretes with higher compressive strengths are obtained at lower water-to-cement ratios. However, the limitation may well be the concrete workability as measured by a slump test.

Since concrete, as a ceramic composite, has low tensile strength, the use of steel reinforcing rods in areas of high tension significantly increases the serviceability. Similarly, the concrete may be placed in residual compression at positions of tensile loading through the use of prestressing with steel reinforcement.

Asphalt uses a bitumen binder for the aggregate rather than the Portland cement found in concrete. The thermoplastic bitumen sets more rapidly than concrete, with its somewhat slow hydration reaction which is highly moisture and temperature sensitive. However, asphalt also has the limitations of most thermoplastics, such as excessive creep at high temperatures and brittleness at low temperatures.

Wood can be considered to be a composite of cellulose fibers bonded by a matrix of polymers, primarily lignin. The macrostructure is made up of sapwood, which carries nutrients, and heartwood, which is composed of dead cells.

The two primary groups of woods are hardwoods and softwoods. The softwoods have needles and exposed seeds, whereas hardwoods lose their leaves, have flowers, and bear fruit or nuts. The microstructures of these two types of wood are similar except that softwood has resin ducts and hardwood has large longitudinal fluid transport vessels.

The mechanical properties of wood depend on the type, the amount of moisture, and the directional orientation of the specimen. The most common test is in bending, with the hardwoods showing somewhat higher values. The presence of defects such as knots materially decreases the strength of wood.

DEFINITIONS

Aggregate Natural filler in concrete, normally sand and gravel

Air entrainment Addition of resin to concrete to give discontinuous small voids

Asphalt Aggregate in a bitumen (thermoplastic) matrix

Cambium Interface layer between the bark and sapwood of a tree

Cement paste Portland cement, water, and entrained air

Concrete A mixture of cement paste and aggregate

Dispersion strengthening Even distribution in a matrix of particles that are small (less than 0.1 micron) and 0.01 to 0.15 volume fraction. The matrix is the main load-bearing component.

Fiber reinforcement The fiber is the main load-bearing component. Volume fraction requirements depend on the modulus difference between the matrix and the fiber.

Fineness modulus Sieve analysis to characterize an aggregate

Hardwood Trees that lose their leaves, bear fruit, and have true flowers

Heartwood Dead wood cells near the center of a tree

Heavyweight concrete Concrete made using heavyweight (high specific gravity) aggregates

Isostrain Loading of a fiber composite parallel to the fiber alignment

Isostress Loading of a fiber composite perpendicular to the fiber alignment

Lightweight concrete Concrete made using lightweight (expanded) aggregates

Lignin Phenol-propane polymer

Orthotropic Anisotropy or directional properties

Parenchyma Radial wood cells

Particle reinforced The load is shared by the matrix and particles. Particles are greater than 1 micron in size, and volume concentrations are greater than 20%.

Portland cement Mixtures of silicates and aluminates that harden by a hydration reaction

Post-tensioning Pulling steel not adhering to concrete in tension and anchoring before removing the load

Prestressed concrete Concrete into which residual compression has been introduced by pretensioning or post-tensioning steel rods

Pretensioning Pouring fresh concrete around steel already in tension

Reinforced concrete Concrete with steel added to tension areas

Sapwood Live wood cells near the surface of a tree that carry nutrients

Slump test One method of measuring workability; the amount of collapse of a standard-size truncated cone of fresh concrete

Softwood Trees that have needles and exposed seeds

Tracheids Longitudinal wood cells in softwoods

Vessels Longitudinal wood cells in hardwoods

Wood Natural composite of cellulose fibers in a polymer matrix (primarily lignin)

Wood rays Horizontal radial rays connecting the tree center to the bark

Workability Work required to place fresh concrete

PROBLEMS

11.1 In dispersion-strengthened composites the following general relationship is found to hold:

$$D_p = (2d^2/3V_p)(1 - V_p)$$

where d = particle diameter in microns, V_p = volume concentration of particles (volume fraction), and D_p = interparticle separation in microns. The normal limits are d = 0.01 to 0.1 micron and V_p = 0.01 to 0.15. Within these constraints, what is the range of interparticle separation necessary to inhibit dislocation movement through the matrix? (Sections 11.1 through 11.3)

11.2 From the data in Fig. 11.1, determine the rupture stress to give a 100-hour rupture life for the different Al_2O_3 levels. Plot the stress versus volume percent Al_2O_3, and from this project the advantage of using quantities of Al_2O_3 larger than 10 volume percent. (Sections 11.1 through 11.3)

11.3 From the data in Fig. 11.2, determine the modulus of elasticity and plot it versus percent rubber. On the same plot, using another scale, plot elongation versus percent rubber, assuming that the end point of the stress–strain curve gives the elongation. From these data cite the advantages or disadvantages of adding rubber particles to polystyrene. (Sections 11.1 through 11.3)

11.4 Why is straw sometimes used to cover fresh concrete, especially when the weather is cold? (Sections 11.5 through 11.9 and 11.11)

11.5 A 600-g sample of fine aggregate has the sieve analysis shown. Determine the fineness modulus. Also plot the percent retained versus the sieve opening size and determine the nature of the size distribution. (Sections 11.5 through 11.9 and 11.11)

Sieve No.	4	8	16	30	50	100	Pan
Grams Retained	13.8	78.4	134.1	156.2	107.7	85.6	23.8

11.6 In steel-reinforced concrete:

a. Why are maximum aggregate sizes specified for a steel net or mesh like that used for a road?
b. Why is the amount of concrete cover important?
c. Why is aluminum reinforcement not used?
(Sections 11.5 through 11.9 and 11.11)

11.7 In prestressed concrete, some ductility (elongation) in the steel is desirable. From the general knowledge of a stress–strain diagram, what would be the advantage of some plastic deformation in the steel? Why would a steel rod never be stressed beyond the tensile strength? (Sections 11.5 through 11.9 and 11.11)

11.8 Explain why concrete is generally not recycled, whereas asphalt can be reused. (Sections 11.5 through 11.9 and 11.11)

11.9 Why does plywood normally have an uneven number of wood plies? (Sections 11.12, 11.15, 11.16)

11.10 Referring to Table 11.8, on properties of wood, show that hardwoods are not always harder and stronger than softwoods. What are the structural differences between hard and softwoods? (Sections 11.12, 11.15, 11.16)

11.11 Which properties of wood, as given in Table 11.8, make it unsuitable for design of structural members? (Sections 11.12, 11.15, 11.16)

11.12 Explain the differences in percentage expansion due to absorption of moisture you would expect in the *L*, *T*, and *R* directions in a piece of oak, compared with a well-cemented piece of plywood. (Sections 11.12, 11.15, 11.16)

PROBLEMS COVERING OPTIONAL SECTIONS AND PREVIOUS CHAPTERS

11.13 It has been suggested that we produce a fiber composite of alumina fibers in an epoxy matrix or in a copper matrix and load it in an isostrain condition. It is stated that with the epoxy matrix less fiber is necessary to have the load carried by the fiber than with the copper matrix. Is this statement correct or incorrect? (Section 11.4)

11.14 From Prob. 11.13 we have the following material properties. Determine the modulus of the composite when loaded in both the isostrain and isostress condition for each matrix material. Which material combination gives the least difference between the two methods of loading? (Section 11.4)

	Modulus of Elasticity
Al_2O_3 ($C_f = 0.4$)	50×10^6 psi (3.45×10^5 MPa)
Epoxy ($C_m = 0.6$)	8×10^5 psi (5.52×10^3 MPa)
Copper ($C_m = 0.6$)	20×10^6 psi (1.38×10^5 MPa)

11.15 A high-strength composite material is made by bonding silica glass fibers (all aligned in the same direction) with an epoxy. The composite contains 45% (by weight) glass. The data are as follows. (Section 11.4)

Epoxy: Density = 1.3 g/cm^3 Young's modulus = 300 × 10^3 psi (0.207 × 10^4 MPa)
Glass: Density = 2.6 g/cm^3 Young's modulus = 10 × 10^6 psi (6.9 × 10^4 MPa)

a. Calculate Young's modulus for the composite measured parallel to the fibers.
b. In order to make a more rigid glass–epoxy composite, we could increase the volume fraction of glass. If each fiber of glass were 0.0005 in. (0.00127 cm) in diameter, and if each fiber were coated with an epoxy film 0.0001 in. (0.000254 cm) thick before bonding with more epoxy, what would be the maximum volume fraction of glass that could be achieved in a unidirectionally aligned composite? Assume that a cross section shows that the fibers are in an HCP structure.
c. What would Young's modulus be in part *b*?

11.16 Figure 11.14 gives data on drying shrinkage for concrete. For the final mix used in Example 11.5, water-to-cement ratio = 0.53 and Portland cement = 625 lb/yd^3 (371 kg/m^3). (Section 11.10)

a. Determine the drying shrinkage.
b. If the concrete was restrained at the edge after pouring, determine the stress set up by the shrinkage. The modulus of elasticity may be taken as 6 × 10^6 psi (4.14 × 10^4 MPa).
c. If the tensile strength is one-tenth of the compressive strength, would the concrete crack from the stress in part *b*?
d. Suggest why this tensile stress would not be as high as calculated.

11.17 Using the data from Example 11.5, assume that the slump of the initial mix was not great enough and we wanted to increase it by 1 in. (2.54 cm). Calculate the new mix required. (Section 11.10)

11.18 Determine the proportions for a concrete that has to meet the following specifications for a bridge deck. (Section 11.10)

Mild service: 7-day compressive strength of 4000 psi using type I Portland cement; non-air-entrained
Slump: 3 to 4 in.
Coarse aggregate: 1 in. maximum; 1.5% contained water
Fine aggregate: Fineness modulus 2.75; 3.5% contained water

11.19 Assume that the bridge deck in Prob. 11.18 is to be made during cooler weather and will be subjected to severe service. Slump, fine aggregate, and coarse aggregate remain the same. The change will be to severe-service, 7-day compressive strength of 3500 psi using type III Portland cement, air-entrained. Calculate the proportions of the concrete and the volume of all components, including air, for a 1 yd^3 mix. (Section 11.10)

11.20 In refinishing a large fine wood surface, such as a table top, why is it recommended that the same number of varnish coats be placed on both the top and the bottom? (Sections 11.13 through 11.14)

11.21 Which woods possess unique characteristics that render them more resistant to weathering and decay? Where in the structure are the chemicals that prevent weathering? (Sections 11.13 through 11.14)

11.22 Recalling that asphalt uses a polymer binder, explain why shipping hot asphalt paving in an uncovered truck makes laying more difficult. Assume that the asphalt does not cool and that we observe that the difficulty does not arise when the asphalt is covered or kept under a nitrogen blanket. (Section 11.11 and Chapter 9)

11.23 Given a 1 ft^3 piece of oak that weighs 32 lb dry and 45 lb when saturated with water. (*a*) What is the bulk density? (*b*) What is the volume of the open pores? (*c*) What is the apparent density? (*d*) The oak is to be impregnated with a thermosetting polymer whose specific gravity is 1.40. As a means of quality control, the 1-ft^3 block of impregnated oak is to be weighed. Calculate the new weight, given that all the open porosity is to be filled with polymer. (Sections 11.15 and 8.6, and Chapter 9)

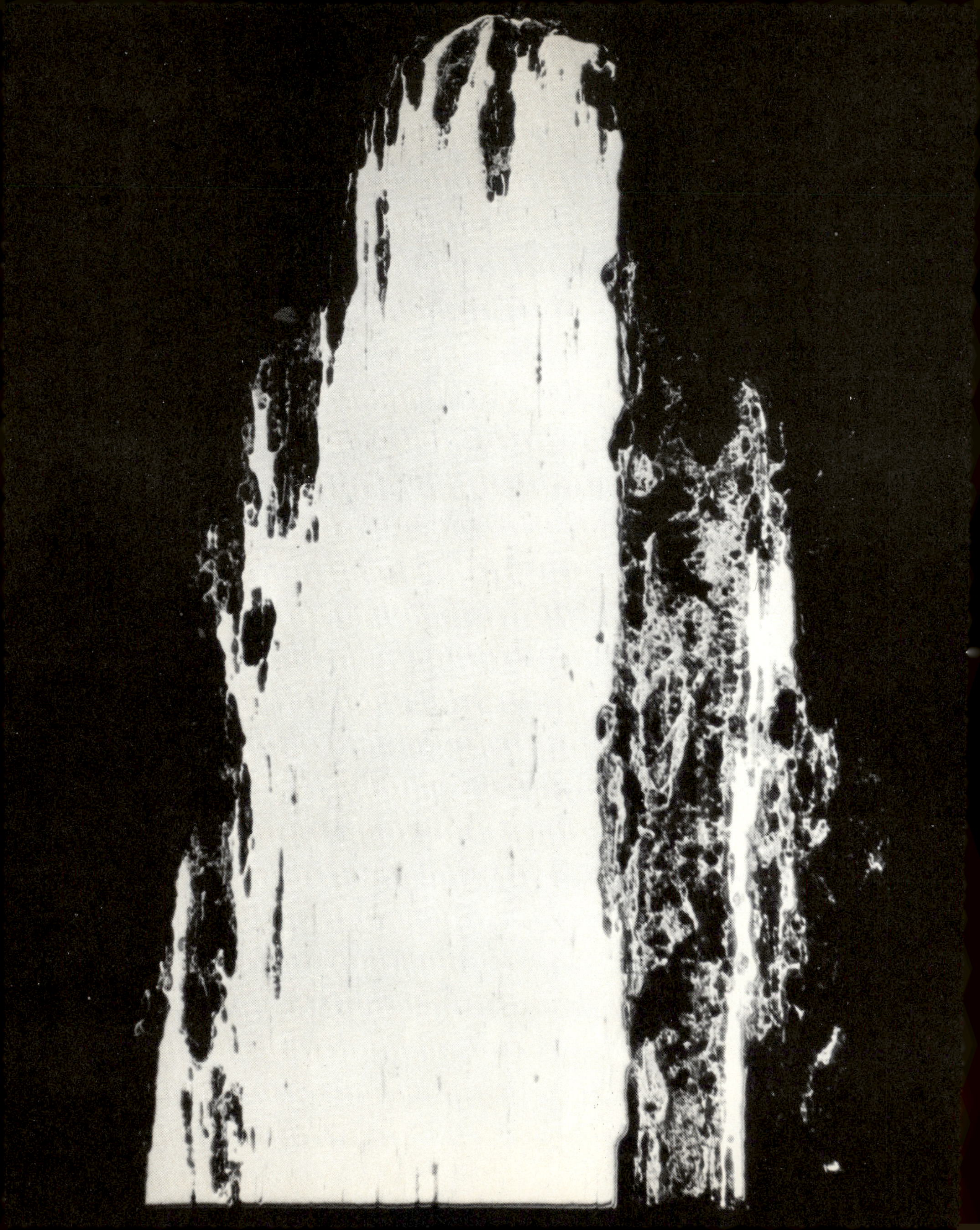

12

CORROSION OF MATERIALS

THIS photomicrograph at 8× shows the longitudinal section of a nail collected from a wreck more than 100 years old on the shore of Cape Hatteras, North Carolina. The material is wrought iron, as shown by the large gray fibers of slag in a ferrite matrix. This is an example of the most important cause of corrosion: the iron (in this case next to the slag fibers) dissolves to form iron ions. These migrate to react with OH^- *ions which are liberated in the oxygen-rich regions to form rust.*

In this chapter we take up the basic reactions occurring in corrosion and their relation to the behavior of materials in corrosive environments.

12.1 Introduction

Corrosion is not a favorite subject of engineers. Many a proud designer or project engineer has developed a new component or process with outstanding performance only to have it fail prematurely because of corrosion. Furthermore, despite active research by corrosion engineers, a visit to the local scrap yard shows that large percentages of cars and domestic appliances still fail because of corrosion. As a result, the annual cost of corrosion and corrosion protection in the United States is of the order of $8 billion, far more than the annual profit of the nation's larger corporations.

On the positive side, there is real satisfaction in designing a component that can resist punishing service conditions under which other parts fail. A major area in the development of the gas turbine is a corrosion (oxidation) application. The key to the problem is understanding the nature of corrosion and how different structures react to it. We shall see that the understanding of the nature of the different phases of materials developed in earlier chapters is of vital assistance in surmounting the corrosion problem.

In most cases we cannot completely prevent corrosion—we cannot afford platinum bodies for automobiles—but we can try to avoid obsolescence of the component due to corrosion. For example, for a long time the average automobile was scrapped because of body corrosion. Now its combination of coatings has doubled its corrosion life.

Finally, metals, ceramics, and plastics are all attacked by different atmospheres and solutions. However, we shall soon learn that the corrosion of metallic materials is electrochemical in nature, whereas the nonmetals usually involve a simple solution. Therefore, we shall first consider the corrosion of metallic materials.

12.2 Corrosion of metals

In discussing the corrosion of metals we shall cover four areas:

1. *Chemical principles:* the driving force for corrosion, anode and cathode reactions in a corrosion cell, the solution tendency of metals and alloys, effects of concentration, inhibitors, passivity
2. *Corrosion phenomena* based on chemical principles: galvanic action, selective leaching, hydrogen embrittlement, oxygen corrosion cells, pit and crevice corrosion; and combined mechanical-corrosive effects: stress corrosion, corrosion fatigue, corrosion erosion, cavitation
3. *Corrosion environments:* the reaction of metals to different atmospheres, fresh and salt water, and chemicals
4. *Corrosion in gas* at elevated temperatures, scaling, and growth

Chemical Principles

12.3 Does the metal react?

In analyzing corrosion, the first question is: Does the metal react with its environment? If so, what is the nature of the corrosion product? A copper wire placed in distilled water simply does not react. Therefore there is no problem. An aluminum wire reacts, but the corrosion product, aluminum oxide, is so adherent that there is no further reaction. An iron wire reacts more slowly than the aluminum wire at first, but reaction continues because the product, rust, is nonprotective. Therefore, in each case, we must study the reaction involved and the types of products generated.

12.4 Anode and cathode reactions (half-cell reactions)

Whether the corrosion is spectacularly fast, such as zinc dissolving in hydrochloric acid, or quiet, such as rust insidiously forming on the back side of an automobile rocker panel, the basic types of reactions are the same.

1. There is an *anode* reaction, at which point metal goes into solution as an ion; i.e., it corrodes. For example, in the reactions just mentioned,

$$Zn \longrightarrow Zn^{2+} + 2e \qquad\qquad Fe \longrightarrow Fe^{2+} + 2e$$

where e = electron, or, in general,

$$M \longrightarrow M^{n+} + ne$$

where M = metal.

2. The electrons flow through the metal part until they reach a point where they can be used up (*cathode* reaction). Again we use the examples given above.

In the case of zinc in acid, the electrons combine with hydrogen ions at the surface. Atomic hydrogen is formed. Most of this combines to form molecular hydrogen, which bubbles off, but some dissolves in the metal. This is important in cases of hydrogen attack, discussed later:

$$2H^{+} + 2e \longrightarrow 2H \longrightarrow H_2 \text{ (gas)}$$

In the case of iron the solution is neutral, and we have a reaction involving oxygen and water using up the electrons from the anode to form hydroxyl ions:

$$\underset{\text{(Dissolved)}}{O_2} + 2H_2O + 4e \longrightarrow 4OH^{-}$$

Iron does not corrode in pure water in the absence of dissolved oxygen.

For corrosion to progress, it is essential to have both anode and cathode reactions; otherwise a charge builds up, stopping corrosion. The anode reaction is generally the simple case of metal going into solution, but a variety of cathode reactions are encountered, depending on the conditions (Table 12.1). Note that in all the cathode reactions electrons are *absorbed.*

Several points should be emphasized here: At the anode electrons are left in the metal as the metal ions leave. The metal in this region is negative. At the cathode electrons are absorbed by ions. This region is positive because positively charged ions accumulate to receive electrons.

12.5 Cell potentials

Now that we have defined corrosion cell action, we can go on to consider the driving force of a cell. Why, for example, will zinc corrode in nonoxidizing dilute acid such as HCl whereas copper will not? Each metal has a different driving force for solution that can be measured as a voltage (Table 12.2). On the one hand, elements such as zinc have a strong negative voltage with respect to hydrogen, indicating that the metal will go into solution and the hydrogen will come out. On the other hand, with the noble metals such as gold, silver, and copper there is no driving potential to replace hydrogen. However, with an oxidizing acid, copper dissolves because of the potential difference between the reaction $O_2 + 4H^+ + 4e \rightarrow 2H_2O$ and copper.

It is useful to review the method of determining these potentials because it gives us another insight into the nature of corrosion.

12.6 Hydrogen half-cell

To establish standard conditions, we place an electrode of the metal to be tested in a 1-molar solution of its ions (Fig. 12.1). A semipermeable membrane divides the cell. For the other half of the cell a platinum electrode is placed in a 1-molar solution of hydrogen ions (produced by an acid). A stream of

Table 12.1 POSSIBLE CATHODE REACTIONS IN DIFFERENT GALVANIC CELLS

Cathode Reaction	*Example*
$2H^+ + 2e \longrightarrow 2H^0$	Acid solutions; see text.
$O_2 + 2H_2O + 4e \longrightarrow 4OH^-$	Neutral and alkaline solutions
$O_2 + 4H^+ + 4e \longrightarrow 2H_2O$	Using both O_2 and H^+ in acid solutions
$M^{3+} + e \longrightarrow M^{2+}$	This is encountered when ferric ions are reduced to ferrous.
$M^{2+} + 2e \longrightarrow M^0$	When iron is placed in a copper-salt solution, the electrons from solution of the iron reduce copper ions to metallic copper.

Table 12.2 STANDARD OXIDATION-REDUCTION POTENTIALS FOR CORROSION REACTIONS*

	Corrosion Reaction	*Potential, Volts vs. Normal Hydrogen Electrode†*
↑ More cathodic	$Au \longrightarrow Au^{3+} + 3e$	+1.498
	$2H_2O \longrightarrow O_2 + 4H^+ + 4e$	+1.229
	$Pt \longrightarrow Pt^{2+} + 2e$	+1.200
	$Pd \longrightarrow Pd^{2+} + 2e$	+0.987
	$Ag \longrightarrow Ag^+ + e$	+0.799
	$2Hg \longrightarrow Hg_2^{2+} + 2e$	+0.788
	$Fe^{2+} \longrightarrow Fe^{3+} + e$	+0.771
	$4(OH)^- \longrightarrow O_2 + 2H_2O + 4e$	+0.401
	$Cu \longrightarrow Cu^{2+} + 2e$	+0.337
	$Sn^{2+} \longrightarrow Sn^{4+} + 2e$	+0.150
	$H_2 \longrightarrow 2H^+ + 2e$	0.000
More anodic ↓	$Pb \longrightarrow Pb^{2+} + 2e$	−0.126
	$Sn \longrightarrow Sn^{2+} + 2e$	−0.136
	$Ni \longrightarrow Ni^{2+} + 2e$	−0.250
	$Co \longrightarrow Co^{2+} + 2e$	−0.277
	$Cd \longrightarrow Cd^{2+} + 2e$	−0.403
	$Fe \longrightarrow Fe^{2+} + 2e$	−0.440
	$Cr \longrightarrow Cr^{3+} + 3e$	−0.744
	$Zn \longrightarrow Zn^{2+} + 2e$	−0.763
	$Al \longrightarrow Al^{3+} + 3e$	−1.662
	$Mg \longrightarrow Mg^{2+} + 2e$	−2.363
	$Na \longrightarrow Na^+ + e$	−2.714
	$K \longrightarrow K^+ + e$	−2.925

*Measured at 25°C. Reactions are written as anode half-cells. Arrows are reversed for cathode half-cells.

†In some chemistry texts the signs of the values in this table are reversed; for example, the half-cell potential of zinc would be +0.763 volt. This depends on whether we consider charge motion in the external circuit or within the cell. In any case it is agreed that zinc goes into solution with respect to the hydrogen half-cell, and we shall adopt the convention that corrosion is at the more negative electrode of the cell.

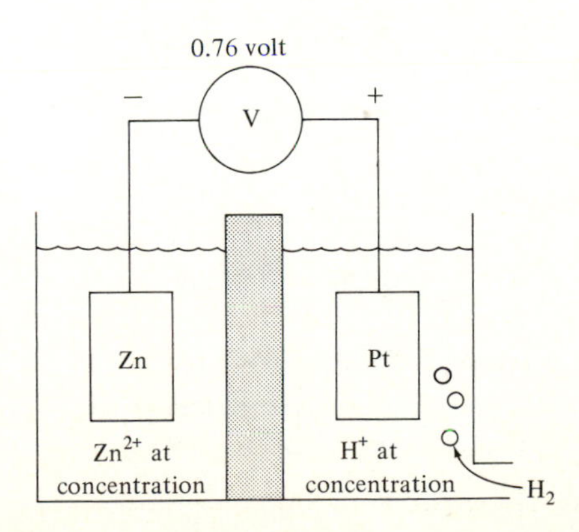

Fig. 12.1 *Diagram of a half-cell for measurement of potential*

hydrogen gas is bubbled around the platinum electrode. Therefore the platinum merely serves to adsorb the gas and does not take part in the reaction.

If the metal is more reactive than hydrogen, we have the following half-cell reactions when the electrodes are connected:

$$\mathrm{M} - ne \longrightarrow \mathrm{M}^{n+} \qquad n\mathrm{H}^{+} + ne \longrightarrow n\mathrm{H}^{0}$$

and the negative potentials in Table 12.2 are developed. On the other hand, if the metal is nobler (less reactive) than hydrogen, we have

$$n\mathrm{H}^{0} - ne \longrightarrow n\mathrm{H}^{+} \qquad \mathrm{M}^{n+} + ne \longrightarrow \mathrm{M}^{0}$$

and the positive potentials in Table 12.2 are developed.

With these examples in mind we should visualize the situation if the metal is not surrounded by a 1-molar concentration of its ions. If the concentration of the ions is less than 1 molar, the driving force to dissolve will be greater because there are fewer ions available for the reverse reaction to $\mathrm{M}^{+} + \mathrm{e} \rightarrow \mathrm{M}^{0}$. The change in electrode voltage for each tenfold change in concentration, to 0.1 molar for example, is 0.0591 volt. This is for the case in which there is a transfer of one electron per ion going into solution. In the case of divalent ions, the voltage change is 0.0295 volt, and for trivalent ions 0.0197 volt. If the concentration is above 1 molar, the driving voltage is correspondingly less.

These effects are derived from the well-known Nernst equation, which can be simplified to suit our case at 25°C:

$$E = E_0 + \frac{0.0592}{n} \log C_{\text{ion}}$$

where E = new electromotive force (emf), E_0 = standard emf of Table 12.2, n = number of electrons transferred as $\mathrm{M}_0 - ne \longrightarrow \mathrm{M}^{n+}$, and C_{ion} = molar concentration of ions.

EXAMPLE 12.1 What would be the new half-cell voltages for zinc and copper if the ion concentrations were changed to 0.1 molar?

ANSWER

$$\text{Zinc: } E = -0.763 + \frac{0.0592}{2} \log 0.1 = -0.763 - 0.0296 = -0.793 \text{ volt}$$

$$\text{Copper: } E = +0.337 + \frac{0.0592}{2} \log 0.1 = +0.337 - 0.0296 = +0.307 \text{ volt}$$

The result for zinc shows a greater tendency to corrode. For copper we see a greater tendency to go into solution, but the potential of a copper-zinc cell would be less because the copper is the electrode at which ions plate out.

12.7 Cell potentials in different solutions

Even more serious corrections must be made to the driving voltage of a given corrosion couple because the metals are rarely in solutions of their own salts. For example, Table 12.3 gives the order of half-cell potentials in salt water for a group of commercial alloys. Note that aluminum and zinc have changed places compared with the standard potential sequence.

We may now review the anode, the cathode, and their electrode potentials, as they can be difficult to identify.

Consider a piece of aluminum and a piece of iron placed in salt water, but *not connected* (Fig. 12.2*a*, p. 490). Both metals tend to corrode, owing to local anodes and cathodes within each material. These arise from inhomo-

Table 12.3 HALF-CELL POTENTIALS OF VARIOUS ALLOYS IN SALT WATER*

Metal or Alloy† (↑ More anodic)	*Potential,‡ Volts, 0.1 N Calomel Scale*
Magnesium	−1.73
Zinc	−1.10
7072, Alclad 3003, Alclad 6061, Alclad 7075	−0.96
520-T4	−0.92
5056, 7079-T6, 5456, 5083, 514.0, 518.0	−0.87
5154, 5254, 5454	−0.86
5052, 5652, 5086, 1099	−0.85
3004, 1185, 1060, 1260, 5050	−0.84
1100, 3003, 6053, 6061-T6, 6062-T6, 6063, 6363, Alclad 2014, Alclad 2024	−0.83
413, cadmium	−0.82
7075-T6, 356.0-T6, 360.0	−0.81
2024-T81, 6061-T4, 6062-T4	−0.80
355.0-T6	−0.79
2014-T6, 850.0-T5	−0.78
308.0	−0.77
380.0F, 319.0F	−0.75
296.0-T6	−0.72
2014-T4, 2017-T4, 2024-T3 and T4	−0.68 to −0.70§
Mild steel	−0.58
Lead	−0.55
Tin	−0.49
Copper	−0.20
Bismuth	−0.18
Stainless steel (series 300, type 430)	−0.09
Silver	−0.08
Nickel	−0.07
Chromium	−0.4 to +0.18

*53 g NaCl + 3 g H_2O_2 per liter, 25°C
†The potential of all tempers is the same unless temper is designated.
‡Data from Alcoa Research Laboratories
§The potential varies with quenching rate.

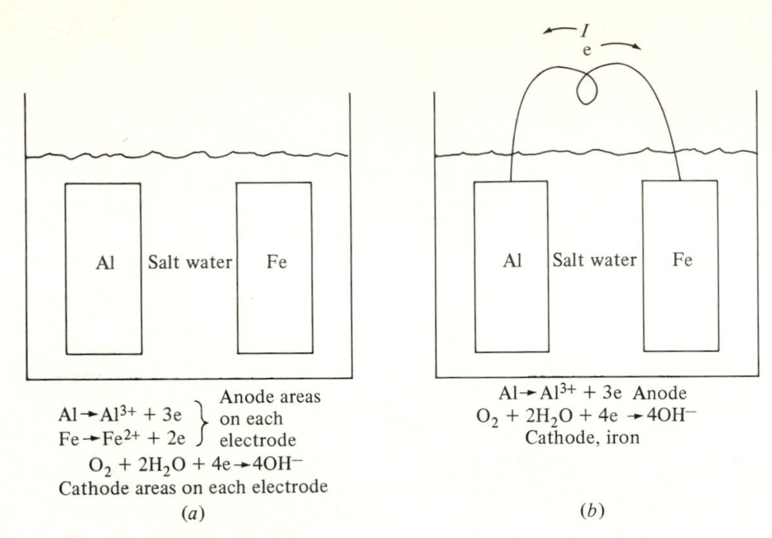

Fig. 12.2 *(a) Electrode reactions that take place at aluminum and iron electrodes that are not connected. (b) When electrodes are connected, the iron does not corrode and the aluminum becomes the anode.*

geneities such as impurities and grain boundaries, to be discussed in Sec. 12.13. Upon *connection* of the metals, the aluminum becomes anodic to the iron (Fig. 12.2*b*). This can be determined from Table 12.3 by comparing 1100 aluminum and mild steel. We would find accelerated corrosion of the aluminum and little or no corrosion of the iron.

The aluminum is therefore the anode, and delivers electrons to the external circuit and hence to the iron. By convention, current is positive and must flow in an opposite direction. However, if the aluminum is to continue to deliver electrons to the cathode, the electrons must be "used up." This is accomplished at the iron-electrolyte surface by the cathode reaction in Table 12.1: $O_2 + 2H_2O + 4e \rightarrow 4OH^-$.

To prevent difficulty with polarity and hence identification of the anode and cathode, we must differentiate between internal and external circuits. For example, the cathode accepts electrons from the *external* circuit, and hence appears positive. However, it accepts positive ions from the *internal* circuit (as in acid solutions with H^+ ion migration) and therefore appears negative. This concept becomes important in batteries. They are labeled plus and minus in terms of how they will be used in an external sense. Therefore the anode is the negative electrode and the cathode is positive, the same convention we have adopted for corrosion cells.

With these concepts in mind, we may now look at the corrosion rate.

12.8 Corrosion rates

We have introduced the concept of solution potential, and we find with further study that equilibrium and this emf are closely associated. That is, the emf indicates the *tendency to corrode* as well as the anode and cathode, but it does not directly govern the corrosion *rate*. For example, a battery is merely a case of controlled corrosion. If we were to ask for a 1.5-volt battery in a store, we could choose from dozens of sizes, all of which would provide 1.5 volts; the life in ampere-hours of the batteries would be decidedly different, however. Thus the corrosion rates would be different because of the different battery sizes and different materials in the batteries. In effect, the corrosion rate depends on Faraday's law:

$$\text{Weight of metal dissolving (g)} = kIt$$

where $k = \dfrac{\text{at. wt. of metal}}{\text{no. of electrons transferred} \times 96{,}500 \text{ A-sec}}$

I = current (A) and t = time (sec)

In corrosion we are usually interested in the loss in weight from a given area per unit time. Therefore, dividing both sides of the above equation by the area gives us the corrosion rate as equal to a constant times the current density [current (A)/area (cm^2)]. In other words, the rate of weight loss from a given area is proportional to the current density.

When we determine the current density for an aluminum-iron corrosion cell, we find that the initial current is relatively high (4 mA/cm^2), and that the current gradually falls off until it reaches 1 mA/cm^2 after 6 min (Fig. 12.3, page 492). Also, the emf changes during this period. Such a decrease in corrosion rate with time is caused by *polarization* and is very important to corrosion control.

An example is the copper-zinc cell shown in Fig. 12.4*a*, page 492. The zinc acts as an anode, and its emf tends to become more cathodic (positive) as the current or current density increases, as shown in Fig. 12.4*b*. Meanwhile, the copper is cathodic and tends to become more anodic, as shown by the emf–current density relationship. The two curves intersect, giving us a corrosion emf of less than 1.10 volts and a corresponding current density that determines the corrosion rate.

Therefore, if we want to lower the corrosion rate, we must find a way to decrease the current density of the cell. This involves the concept of polarization. Polarization is really composed of two effects: *activation polarization* and *concentration polarization*. The latter is easier to describe first, since it follows directly from our discussion of the effect of concentration on cell potential.

Let us again consider the copper-zinc corrosion cell in operation. As zinc goes into solution, the electrons are delivered to the copper. In order for

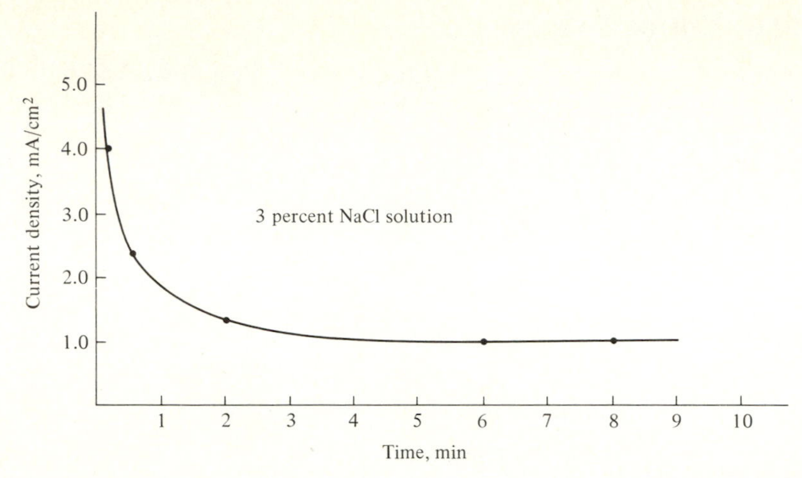

Fig. 12.3 *Decrease in corrosion current in an aluminum-iron galvanic cell (25°C, 77°F)*

the electrons to be absorbed, positive ions are needed at the copper cathode. In Fig. 12.4 these are copper ions. After current has passed for a short time, there is a scarcity of positive ions at the cathode because of the time needed for diffusion of ions within the solution. This is shown schematically in Fig. 12.5. The greater the requirement for positive ions, as with very high initial corrosion rates, the sooner an ion-depleted layer appears at the cathode. The net effect, of course, is to decrease the corrosion rate.

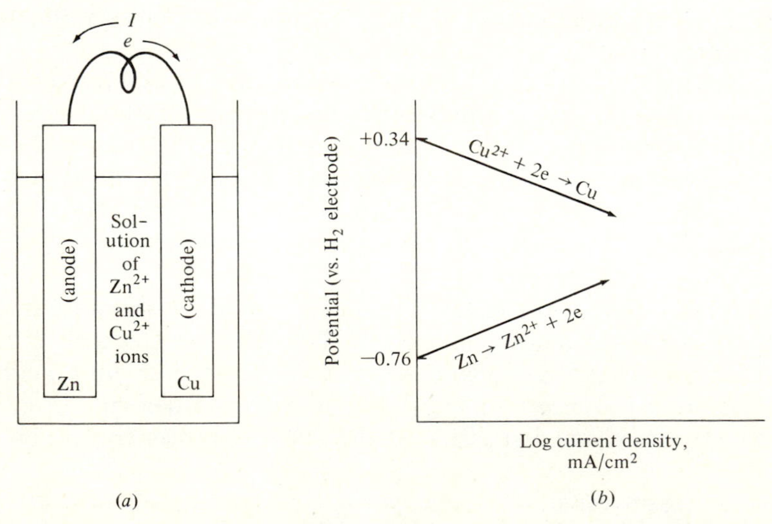

Fig. 12.4 *(a) Polarization in a copper-zinc galvanic corrosion cell (25°C, 77°F). (b) Potential vs. log of current density.*

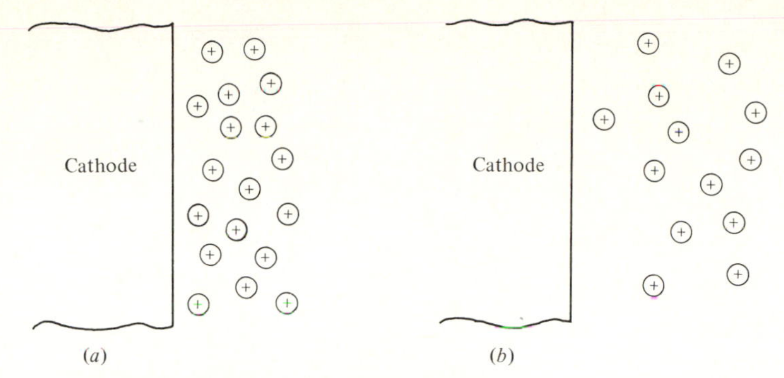

Fig. 12.5 *Concentration polarization (ion depletion) due to a high rate of reaction at the cathode. (a) Time = 0. (b) Time > 0.*

Furthermore, as we saw in Example 12.1, the cathode voltage becomes more negative at lower ion concentrations. Therefore depletion of ions at the cathode changes the cathode voltage, as shown in Fig. 12.6, and lowers the corrosion rate.

A high concentration of Zn^{2+} ions at the anode interface might also inhibit more zinc from going into solution and hence might cause concentration polarization at the anode. However, this concentration polarization is usually negligible compared with that at the cathode, since a good deal of metal surface is exposed and can ionize.

Finally, it must be emphasized that concentration polarization is related to diffusion. Therefore conditions that change diffusion rates also affect concentration polarization. As an example, stirring the liquid reduces the concentration gradient of positive ions. We can expect similar effects leading to lower concentration polarization and increased corrosion rates with an increase in temperature and an increase in the concentration of the ions in solution.

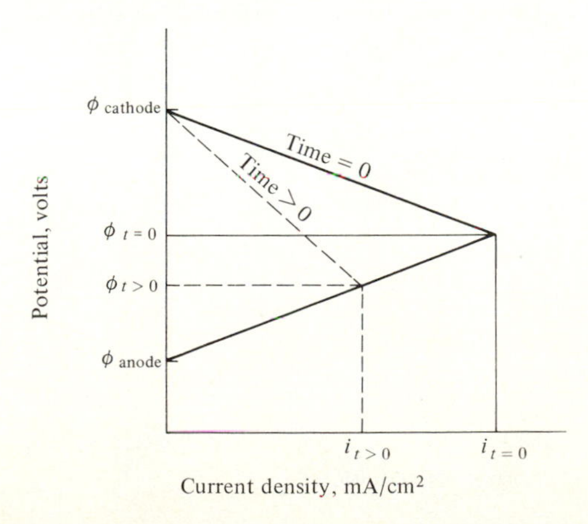

Fig. 12.6 *Effect of concentration polarization at the cathode in lowering rate of corrosion (decrease in current density in a cathode-controlled polarization).*

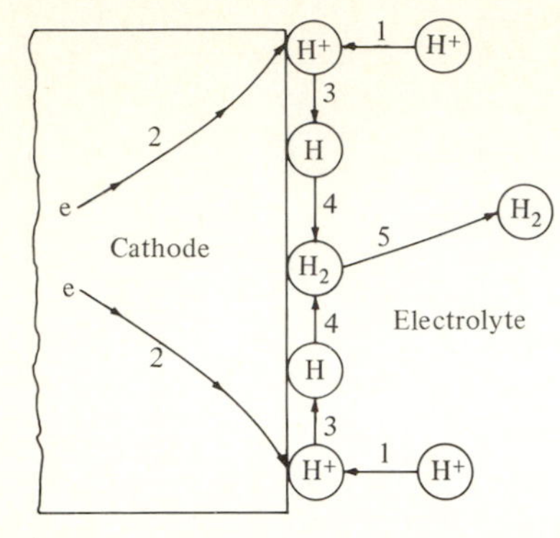

Fig. 12.7 *Steps in formation of hydrogen gas at the cathode. (Any of the steps can be rate-limiting due to activation polarization.) Step 1: migration of hydrogen ion to interface. Step 2: motion of electrons. Step 3: formation of atomic hydrogen. Step 4: formation of H_2 gas. Step 5: detachment of gas bubble from surface of cathode.*

Now let us consider activation polarization, the common variety of polarization when the reaction rate is low. A good illustration of this effect is the evolution of hydrogen at the cathode, as shown in Fig. 12.7. The hydrogen is first formed as atomic hydrogen, then two atoms must form molecular hydrogen. Next the molecules must form a bubble, and finally the bubble rises. A slowdown in any step can retard the entire sequence. The term "activation polarization" is used to give the idea of the "activation energy" needed to overcome the resistance of the slowest step. There is also activation polarization at the anode. This is the barrier the metal atom or ion encounters in leaving the metal specimen and entering the solution. This value is larger for the transition metals iron, cobalt, nickel, and chromium than for silver, copper, and zinc.

Let us summarize these effects. The concentration polarization is more of a chemical barrier, whereas the activation polarization is more of a physical or electrical type barrier. Both reduce corrosion rate.

The polarization effects at anode and cathode are rarely equal. When the polarization results in a greater decrease in cathode potential, the reaction is said to be *cathodically controlled,* as shown in Fig. 12.6. The term *anodic control* is used similarly. We should again emphasize that current density is our major concern, as it dictates the corrosion rate.

EXAMPLE 12.2 Write the anode and cathode reactions (half-cell reactions) for the following conditions:

a. Copper and zinc in contact and immersed in sea water
b. As in *a* with the addition of HCl
c. As in *a* with the addition of copper ions
d. Copper immersed in fresh water
e. Iron immersed in fresh water
f. Cadmium-plated steel scratched and immersed in sea water

ANSWER

a. Anode: $Zn \rightarrow Zn^{2+} + 2e$
Cathode (copper): $O_2 + 2H_2O + 4e \rightarrow 4OH^-$ (see Table 12.1)

b. Anode: $Zn \rightarrow Zn^{2+} + 2e$
Cathode (copper): $O_2 + 4H^+ + 4e \rightarrow 2H_2O$
(Since H^+ ions can help use up electrons generated by corrosion of zinc, the corrosion rate is higher.)

c. Anode: $Zn \rightarrow Zn^{2+} + 2e$
Cathode: $Cu^{2+} + 2e \rightarrow Cu$

d. Little or no corrosion because copper is noble. Compare the emf for a solution of copper with the equation for $O_2 + 2H_2O + 4e \rightarrow 4OH^-$

e. Anode: $Fe \rightarrow Fe^{2+} + 2e$
Cathode: $O_2 + 2H_2O + 4e \rightarrow 4OH^-$

f. Anode: $Cd \rightarrow Cd^{2+} + 2e$
Cathode: $O_2 + 2H_2O + 4e \rightarrow 4OH^-$
[*Note:* Cadmium is cathodic to iron in the standard oxidation-reduction potentials (Table 12.2) but anodic to iron in salt water.]

12.9 Inhibitors

We have given this detailed account of polarization because one of the most important methods for controlling corrosion is the use of inhibitors. The action of these chemicals is related to polarization. The corrosion rate can be reduced drastically by inhibiting the reaction at *either* the anode or the cathode. In many cases the role of the inhibitor is to form an impervious, insulating film of a compound on either the cathode or the anode. A common case is the use of chromate salts in automobile radiators. The iron ions liberated at the anode surface combine with the chromate to form an insoluble coating. Another case is the use of gelatin, which is adsorbed and limits ion reaction with the electrode.

12.10 Passivity

A logical extension of inhibition is the development of a passive film on the surface of a metal. A classic example is the corrosion of iron in nitric acid. If a piece of iron is placed first in concentrated nitric acid and then in dilute nitric acid, no appreciable corrosion occurs. A thin adherent *passive* film of iron oxide is produced in the concentrated acid. If the sample is then scratched, the iron corrodes rapidly in the dilute acid because of the rupture of the film. When the sample is not immersed first in concentrated acid, both the initial and continuing corrosion are rapid in the dilute acid.

The importance of the experiment is that *self-repairing passive films* are formed when more than 10% chromium is present in the iron. In general, iron, chromium, nickel, titanium, and aluminum alloys can form passive layers.

We can also explain passivating with the aid of a polarization curve obtained by measuring cell current for the material as a function of voltage. A material that passivates exhibits the anode polarization curve shown in Fig. 12.8. As the anode is driven to higher current densities (generally by exposure to an oxidizing acid), it reaches a maximum value called $i_{critical}$. On reaching this value, the anode forms a protective layer, and its corrosion decreases to a value called $i_{passive}$. It is important to realize that if the initial corrosion rate is not great enough, $i_{critical}$ will not be exceeded, and this can cause some peculiar effects. For example, stainless steel will corrode more readily in a weak oxidizing acid ($i < i_{critical}$) than in a concentrated oxidizing acid (i reaches $i_{passive}$).

The polarization of the cathode can determine whether the anode reaches the passive condition. In Fig. 12.8, for instance, two possible *cathode* polarization curves have been superimposed on the anode polarization curve. In the case of cathode 1, the intersection with the anode is in the passive region. Cathode 2 polarizes at a higher rate (such as through concentration polarization), and it intersects the anode polarization curve within the active region, giving a higher corrosion rate. Therefore care must be exercised with

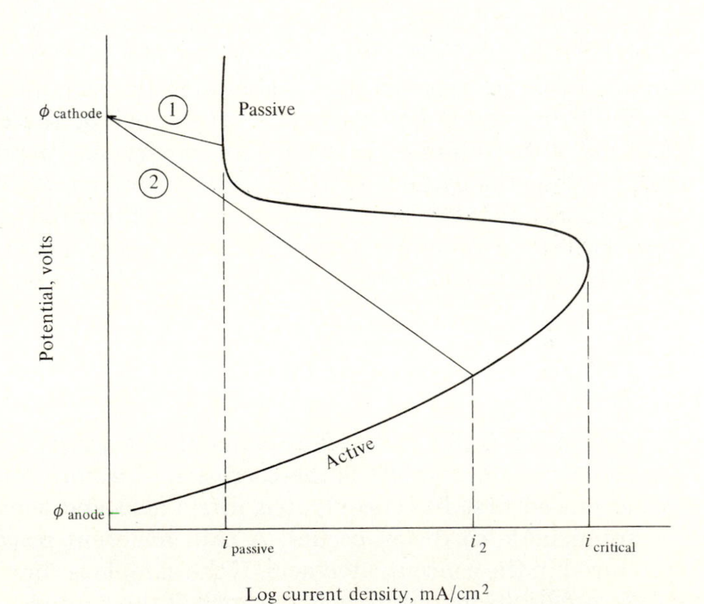

Fig. 12.8 *Polarization curve of a passive metal. High cathode polarization results in high corrosion rate* i_2*; low cathode polarization maintains a passive anode and lower corrosion rate* ($i_{passive}$).

coupling passive metals to highly polarizable cathodes because passivation may be destroyed and a high corrosion rate obtained.

Corrosion Phenomena

12.11 Types of corrosion

The basic principles we have just discussed can explain practically any case of corrosion. However, there are a number of corrosion situations or phenomena that are encountered often and that are given special names and deserve careful analysis. Also these common cases are good illustrations of how to apply the basic principles. The phenomena to be covered are:

1. Corrosion units
2. Galvanic corrosion, macroscopic and microscopic cases
3. Selective leaching (dezincification, etc.)
4. Hydrogen damage
5. Oxygen-concentration cells, water-line attack
6. Pit and crevice corrosion
7. Combined mechanical-corrosive effects
   a. Stress corrosion
   b. Corrosion fatigue
   c. Liquid velocity effects: corrosion erosion, cavitation

12.12 Corrosion units

In the standard corrosion tests, which we discuss later in detail, samples of a metal are measured and weighed before and after immersion for a given period of time. The following common units are used:

mdd: milligrams lost per square decimeter per day
ipy: inches corroded per year
mpy: mils corroded per year (1 mil = 0.001 in.)

The last two terms are preferable because they allow us to visualize easily the long-range effect. These values apply to uniform corrosion. They may be used in design only if nonuniform corrosion, such as pit corrosion, is not present.

12.13 Galvanic corrosion

In a general way all corrosion depends on galvanic action, but the term *galvanic corrosion* means specifically a type of corrosion that occurs because two materials of different solution potential are in contact. We shall also include galvanic effects on the microscopic scale after we discuss the macroscopic case.

Macroscopic Cases of Galvanic Corrosion Perhaps the best illustration of this phenomenon was given by an inspired sculptor who produced a figure for a central plaza in New York City. To express himself he used a bronze body, an aluminum crown, an iron sword, and a stainless steel base. The result was a fine collection of galvanic cells, and the original form did not last long.

Many less spectacular cases of galvanic action exist. Often the more reactive member of the couple is deliberately used as a *sacrificial anode*. In the common case of galvanized steel or wire, the steel is coated with zinc either by being dipped into molten zinc or by electroplating. The effects in Table 12.4 were obtained in an experiment in which zinc and steel samples of equal size were tested separately for the same time period in the attached or coupled condition.

Note that, when uncoupled, both zinc and steel corrode in the solutions, but when coupled, the zinc protects the steel, corroding at an accelerated rate as a sacrificial anode.

Another common couple is found in the "tin" can, which is composed of steel covered with a thin layer of tin. If a cut section is allowed to corrode, the steel rather than the tin usually corrodes. This confirms a prediction that might be made from Table 12.2, since iron has a greater solution potential. The question might then be asked: Why aren't cans made of galvanized steel? The answer is that although the zinc would protect the steel, the ions would be present in the food at an undesirable level. In the manufacture of tin cans, great care is taken to avoid contact between the steel and the food (through the use of soldered joints and lacquering). Therefore, the corrosion rate is only that of tin or lacquer, which is generally low.

As another example, rivets of a material different from the basic structure are often used, especially as an expedient in repair. When copper rivets are used in steel sheet, there is a *large anode area* (the steel), and the galvanic action is not serious. However, when steel rivets are used in a copper sheet, all the metal loss is concentrated in a small anodic region and there is a large area for cathode reactions to absorb electrons. The corrosion is catastrophic. Figure 12.9a shows the effect of anode-to-cathode area on the corrosion rate

Table 12.4 WEIGHT LOSS OF ZINC AND STEEL IN THE UNCOUPLED AND COUPLED CONDITIONS

	Weight Change of Each Sample, g			
	Uncoupled		*Coupled*	
Solution	*Zinc*	*Steel*	*Zinc*	*Steel*
0.05 molar Na_2SO_4	−0.17	−0.15	−0.48	+0.01
0.05 molar NaCl	−0.15	−0.15	−0.44	+0.01
0.005 molar NaCl	−0.06	−0.10	−0.13	+0.02

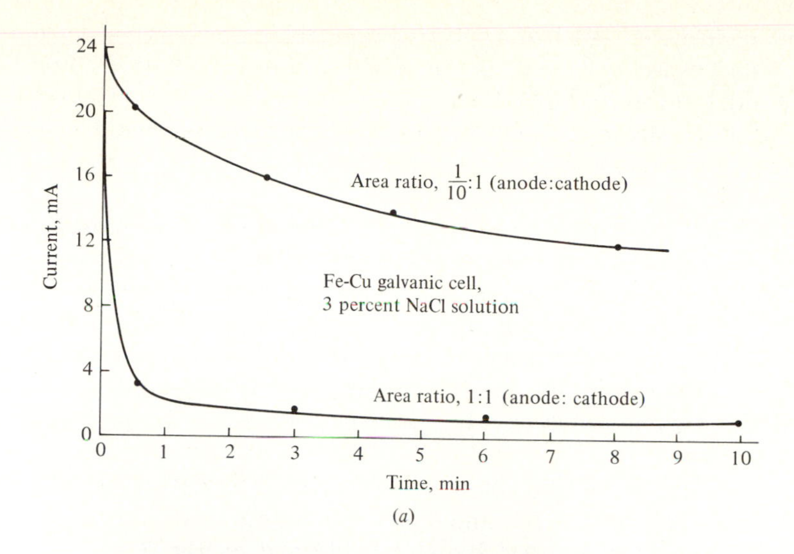

(*a*)

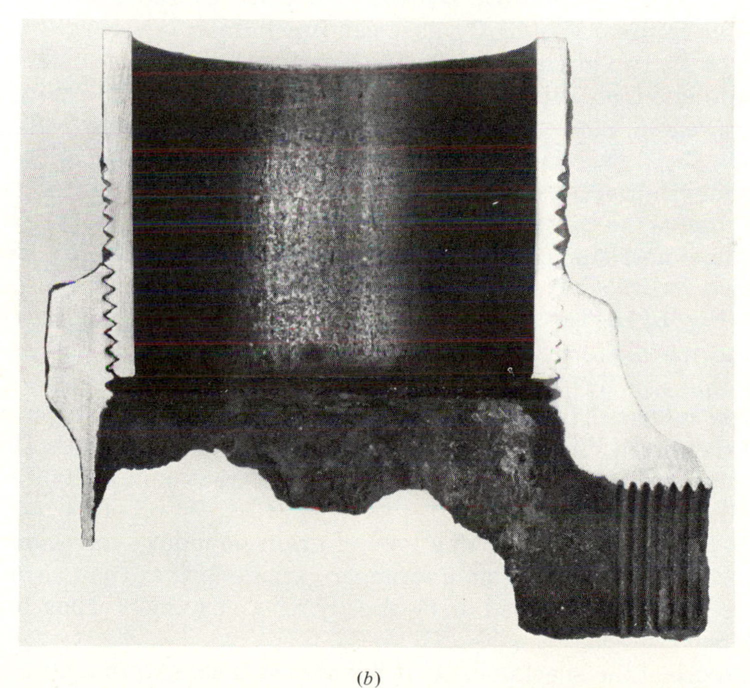

(*b*)

Fig. 12.9 *(a) Rate of corrosion (current) as related to anode-to-cathode area (25°C, 77°F). (Upper curve shows higher corrosion with large cathode and small anode.) (b) Condensate trap of iron fitting in brass pipe (cross section), with iron fitting badly corroded.*

(current) of just such an iron-copper system. Therefore, galvanic coupling is to be avoided, especially in systems exposed to aqueous solutions. Figure 12.9*b* shows another example, in which a brass pipe and a cast-iron fitting were used as part of a steam condensate trap. Since the iron is anodic, it has almost completely corroded away.

Microscopic Cases of Galvanic Corrosion The question may be asked: Suppose that we have a sample of an alloy with several phases present. Won't these act like small galvanic couples and corrode? As a matter of fact, this effect is the reason we can distinguish different phases in the microstructure. Usually there is not much to be seen in the as-polished condition; it is only after etching that the details of structure are visible. There are many commercial cases in which this behavior is important. We shall now discuss some examples.

WELD DECAY OF 18% CR, 8% NI STAINLESS STEEL. Type 304 stainless steel contains 18% Cr, 8% Ni, and 0.08% C maximum. It is usually delivered in the single-phase austenitic structure obtained by rapid cooling from elevated temperatures. The Cr_4C phase can precipitate, however, if the steel is reheated in the two-phase field. During welding the weld zone is heated to the liquid state and cools rapidly enough to avoid carbide precipitation. However, there is a region adjacent to the weld that has been heated just enough to precipitate Cr_4C. This usually takes place at grain boundaries. Because of the high chromium content of the carbide, the nearby regions are impoverished in chromium as the carbide is formed. The chromium level falls below 10% in these regions near the grain boundary. Hence the low-chromium regions are not passive (<10% Cr), whereas the remainder of the matrix is passive, as shown in Fig. 12.10. The result is galvanic action between the grain boundary region and the higher-chromium regions within the grain. It should be emphasized that it is not the Cr_4C particles that are corroded, but the low-chromium–iron matrix. The Cr_4C particles have been recovered after corrosion. Under the electron microscope they appear as platelike crystals. This type of corrosion can be avoided by using a solution heat treatment at 2000°F (1093°C) after welding.

We can also minimize this type of grain boundary corrosion by using 18% Cr, 8% Ni steels to which a stronger carbide-former than chromium (titanium or niobium) is added or by specifying a very low-carbon level, as in Table 12.5.

The use of the special steels is not always a safeguard against galvanic action because chromium carbide may still form in preference to other carbides in a certain temperature range. The special case called "knife-line attack" can be encountered in type 347. All carbides are dissolved in the weld, and niobium carbide then precipitates in the range 2250 to 1450°F (1232 to 788°C) on cooling. However, if the material at the edge of the weld is cooled rapidly in the range for niobium carbide formation but more slowly from 1450

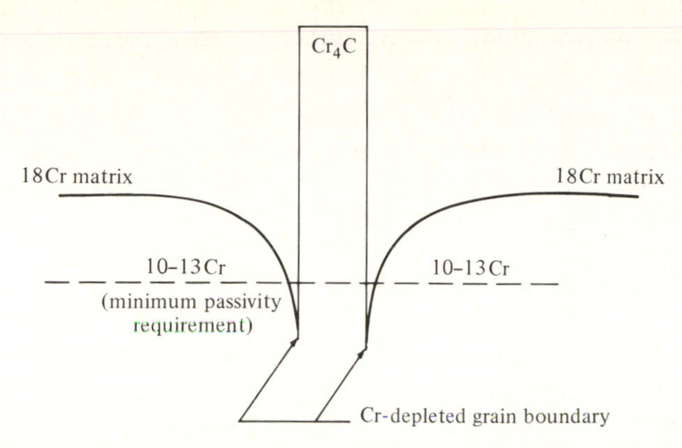

Fig. 12.10 *The lowering of the amount of chromium in a stainless steel below the critical amount required for passivity. Corrosion of grain boundary results when chromium carbides precipitate.*

to 950°F (788 to 510°C), *chromium* carbide can precipitate, again forming an active-passive condition. The weld will be corroded in a narrow region; hence the name "knife line."

Cases of galvanic action are not limited to the high-alloy steels. An interesting case in plain-carbon steel piping is called *ringworm corrosion*. This selective attack takes place near the end of the pipe that has been *especially heated* to forge the flange portion. This treatment results in a spheroidized iron carbide structure with a different solution potential than the untreated balance of the pipe. A circular form of attack occurs near the junction of the two regions.

In aluminum alloys there is a marked difference between pure aluminum and some of the age-hardened alloys, particularly those containing copper, in corrosive liquids such as salt water. In a 4% copper alloy, for example, the potential after solution heat treatment is uniform (−0.69 volt). After aging, however, there is a change in potential near the $CuAl_2$ particles that precipitate. The matrix near the grain boundaries when the precipitation is heavy is −0.78 volt, whereas the potential of the balance of the structure is

Table 12.5 CHEMICAL COMPOSITIONS OF SEVERAL STAINLESS STEELS

	Chemical Analysis, percent			
Stainless Steel	C	Cr	Ni	*Other*
Type 304	0.08 max.	18	8	
321	0.08 max.	18	8	Ti = 5 × percent C
347	0.08 max.	18	8	Nb = 10 × percent C
304L	0.03 max.	18	8	

still −0.69 volt. In this condition the alloy is subject to *intergranular corrosion*. For protection a pure aluminum coating is used.

Even in single-phase alloys galvanic effects will develop. A common source of potential difference is segregation during solidification (Chapter 4). Figure 12.11 shows an as-cast cupronickel alloy. By reheating the alloy to below the solidus and allowing diffusion to take place, one homogenizes the material.

In a homogeneous alloy the material at the grain boundaries is at a high solution potential because the dislocation concentrations lead to poorer bonding and higher strain energy. After cold working, the material tends to corrode more than annealed material. Figure 12.12 shows an interesting illustration of the use of this effect in criminal investigation. A series of identification numbers was stamped on an engine part (1). Next the identification was ground away (2). However, upon etching the ground surface, the numbers reappeared (3). The reason is that the metal is cold-worked below the visible bottom of the impression. It therefore still etches differentially after the impression is ground off. To avoid this action in the corrosion of cold-worked parts, at least partial annealing is necessary (recovery).

12.14 Selective leaching (dezincification, etc.)

Selective leaching is a term that covers a number of mouth-filling classifications, such as "dezincification" and "dealuminization." The case of brass illustrates this type of failure.

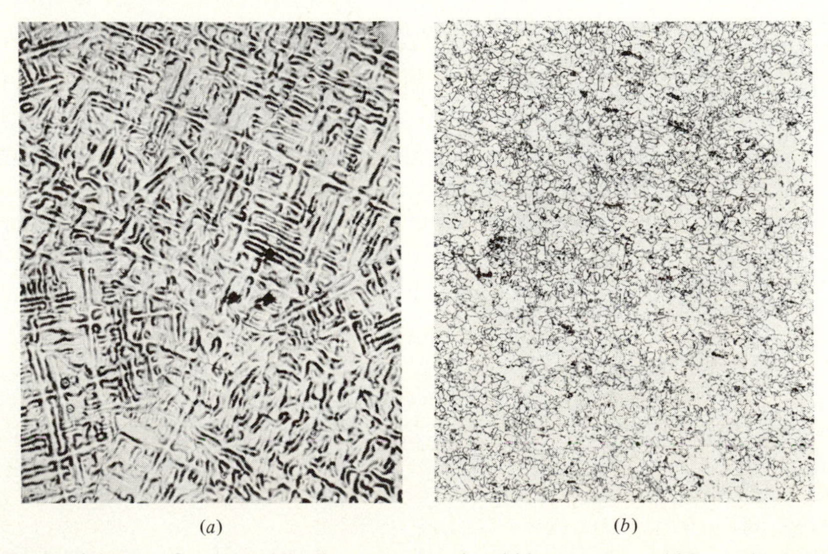

Fig. 12.11 *(a)* 70% Cu, 30% Ni *alloy as-cast, dendritic structure; 100×, chromate etch. (b) Same as (a), but after rolling and homogenization at* 1700°F (927°C); *100×, chromate etch.*

Fig. 12.12 *Process by which one can reveal stamped numbers after they have been "ground away."* Top, *numbers are stamped on the bar.* Middle, *numbers are ground away.* Bottom, *after 10 min in boiling 50 percent sulfuric acid, the numbers reappear because the severely cold-worked region beneath the stamped number etches more rapidly. This illustrates the higher solution potential of cold-worked metal.*

After exposure to fresh or salt water, brass may develop copper-colored layers or plugs of material (Fig. 12.13*a* and *b*, page 504). One finds that these are in fact spongy low-strength regions of copper from which the zinc has leached out. The usually accepted mechanism is that the brass dissolves slightly, then the copper ions are displaced by more zinc going into solution, and the copper plates out. This is obviously a dangerous phenomenon when spongy plugs of copper form in a pipe under pressure. In these cases a copper-nickel alloy is used.

To avoid the problem, the zinc content of the pipe should be lowered. The most sensitive alloys contain 40% zinc and have a second phase (β) that aggravates the problem. In severe cases, in which the 30% zinc alloys are troublesome, the addition of tin and arsenic (1% Sn, 0.04% As) or reduction of zinc to below 20% alleviates the problem.

In gray cast-iron pipe the iron matrix may be slowly leached away, leaving the insoluble graphite behind. This is called *graphitic corrosion* or, often improperly, *graphitization*. This is usually a very slow effect, and pipe showing this condition has been in service for centuries. On the other hand, it can occur in a few years under extreme soil conditions, such as where cinders have been used for backfill. Figure 12.13*c* shows the microappearance of this type of corrosion.

Selective leaching of aluminum from copper alloys and of cobalt from a cobalt-tungsten-chromium alloy has also been encountered.

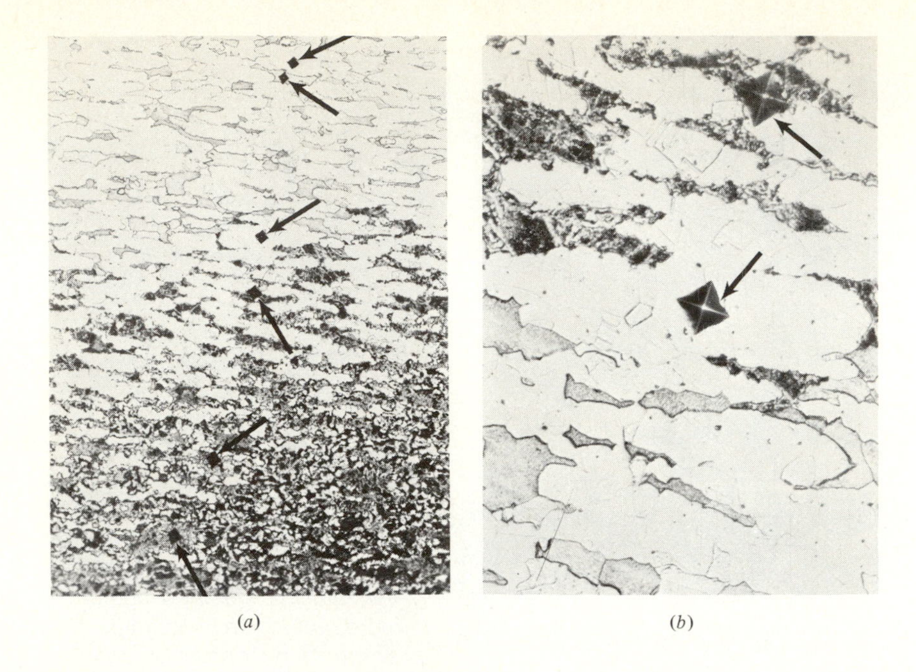

(*a*) (*b*)

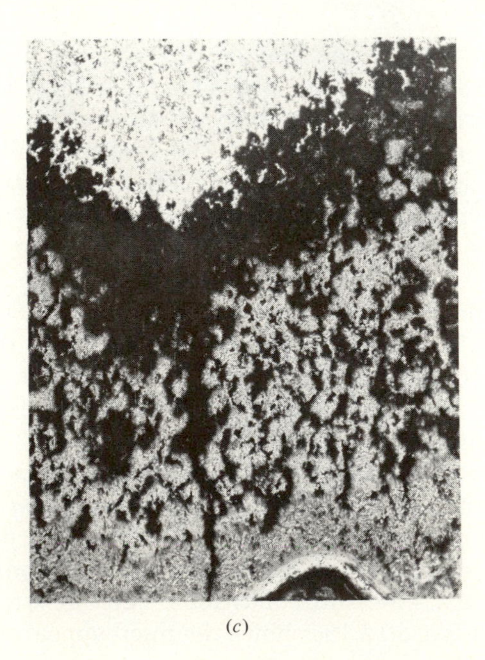

(*c*)

Fig. 12.13 *(a)* 60% Cu, 40% Zn *brass.* FCC α *phase, light,* VHN *100.* BCC β *phase, gray,* VHN 130. *Spongy copper, dark,* VHN 52; *100×, chromate etch. (b)* 60% Cu, 40% Zn *brass.* FCC α, *light.* BCC β *phase, gray. Spongy copper, dark. Note that corrosion takes place in the* β *regions (higher zinc content and more active). The selective leaching is called dezincification; 500×, chromate etch. (c) Gray cast-iron pipe showing graphitic corrosion. Original diameter of pipe included all the black and gray corrosion product in lower two-thirds of photomicrograph. Only a spongy mass of graphite and some silicate remain. White area at top is unattacked gray iron, in which graphite flakes can be seen. 50×, unetched.*

12.15 Hydrogen damage

In discussing the basic atom movements in corrosion, we said that hydrogen discharged at the cathode could form bubbles *or* dissolve in the metal and diffuse through it. There are cases involving either hydrogen evolution from applied currents, as in electroplating, or hydrogen from corrosion in acid solution. In *hydrogen blistering* the atomic hydrogen (H) diffuses through the metal and, finding a void, diffuses into it. It then forms molecular hydrogen (H_2) and exerts high pressure, tending to spread out the void. The pressure of molecular hydrogen in equilibrium with atomic hydrogen is more than 10^5 atm (1.033×10^6 kg/m^2). In *hydrogen embrittlement* the dissolved interstitial atoms lead to low ductility and low impact strength, perhaps because of interaction with microcracks. These effects can be minimized by baking for a long time to remove hydrogen and by avoiding couples that produce it.

12.16 Oxygen-concentration cells; water-line corrosion

In contrast to the galvanic effects just discussed, corrosion can appear to occur without any starting potential. A drop of water corrodes a polished iron surface; a homogenized steel tank corrodes at the water line. To explain these cases, let us refer to the equations shown in the emf series:

$$O_2 + 2H_2O + 4e \longrightarrow 4OH^- \qquad +0.401 \text{ volt}$$

$$Fe \longrightarrow Fe^{2+} + 2e \qquad -0.440 \text{ volt}$$

We see, therefore, that in an oxygen-concentration cell involving solution of iron at the anode and discharge of electrons by forming OH^- at the cathode, a potential of 0.841 volt exists. In the apparently innocuous case of the drop of water we can predict the following sequence of events.

Some iron dissolves in the water, and oxygen and water react to absorb the electrons. The oxygen in the water can be replaced rapidly only at the edge of the drop. This region of high oxygen concentration develops as the cathode (Fig. 12.14). Iron dissolves inside the drop, but more goes into solution closer to the edge regions than at the center. This is because there is a higher resistance to electron travel from the center; i.e., anodes and cathodes are favored when they are close together. A ring of rust is formed at the edge of

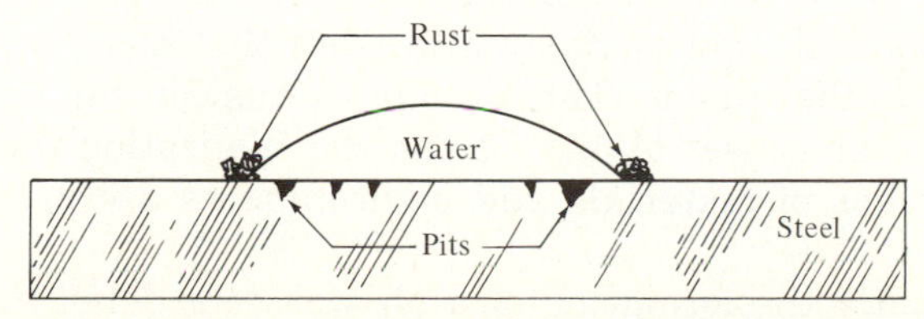

Fig. 12.14 *Corrosion of steel by a drop of water*

the drop because the iron ions migrate faster than the OH^- ions. The ferrous hydroxide is oxidized to hydrated ferric hydroxide (rust) in a secondary reaction.

The case of water-line corrosion is similar. We find that a tank that is kept only partly filled with water corrodes more rapidly than a completely filled one. Corrosion takes place at the water line because of the gradient in oxygen concentration that develops between the solution at the water surface and that at depth. The greatest attack is *just below* the water line, where there is a pronounced dropoff in oxygen concentration plus a short electron path to the water line, where hydroxyl ions are formed.

There are, of course, other examples of oxygen affecting the corrosion rate. Home hot-water-heating systems are usually "closed systems." The initial oxygen is very rapidly used up, and the corrosion rate becomes very low because of the low amount of oxygen. On the other hand, if an "open system" is used (fresh water is continuously added), the high amount of dissolved oxygen results in a continued high corrosion rate.

Stains on the bottom of cooking utensils from "cooked-on food" are usually the result of concentration cells. The liquid under barnacles on a ship's hull develops a different concentration of ions than normal salt water. Hence a concentration corrosion cell forms.

EXAMPLE 12.3 Water allowed to dry on stainless steel usually results in "water spots." How do these spots occur?

ANSWER This is an oxygen-concentration-cell effect that can be increased by other ions. If a large area of water, say several inches in diameter, is followed during its process of evaporation, the spot occurs in the very last portion to evaporate. This may be explained by the higher concentration of oxygen and other ions in the last liquid. (Dissolved salts and gases remain in the liquid. Since the volume of liquid becomes smaller, these concentrations become greater.)

12.17 Pit and crevice corrosion

In many applications pit and crevice corrosion, rather than the overall corrosion rate, dictates the choice of materials. There is little comfort in owning a gasoline storage tank that is 99.9% intact but that has numerous pits that have penetrated to the outside. This unhappy situation can be avoided by an understanding of the causes of pitting and the realization that certain well-known combinations of materials and environments are prone to this phenomenon.

Until recently the formation of a pit was considered merely a special situation of an oxygen-concentration cell. However, this did not explain the important role of ions such as chlorides. The most recent concept is that a pit begins at a surface discontinuity such as an inclusion or grinding mark. An oxygen-concentration cell develops between the discontinuity and the surrounding material. The chloride ions are involved as follows: Within the incipient pit positive metal ions dissolve and accumulate. These attract chloride ions. The metal chloride concentration begins to build up in the pit. If the chloride is iron chloride, for example, this hydrolyzes to give HCl: $M^+Cl^- + H_2O = MOH + H^+Cl^-$.

The combination of chloride and hydrogen ions then accelerates the attack. In proof of this mechanism, it has been found that the fluid within crevices exposed to an overall neutral dilute sodium chloride solution contains 3 to 10 times as much chloride ion as the bulk solution and has a pH of 4 rather than 7.

Crevice corrosion has the same mechanism as pit corrosion, since the crevice serves as a ready-made pit in which the oxygen concentration is low.

In combating pitting, the most important point is to avoid combinations of materials and environments known to be susceptible. For example, many materials with passive surfaces exhibit pitting because of the large potential difference between the passive and active regions of the pit. Furthermore, chloride ions are known to destroy passive layers locally. Among the stainless materials, the following sequence, from the most to the least susceptible, is found for salt water.

Type 304 stainless steel (18% Cr, 8% Ni)
Type 316 stainless steel (18% Cr, 8% Ni, 2% Mo)
Titanium

It is also interesting that titanium forms a very stable passive layer under oxidizing conditions. Therefore, it has excellent corrosion resistance and is extensively used in the chemical industry.

12.18 Combined mechanical–corrosion effects

In many cases a component fails because of the combined effect of mechanical or hydraulic factors and corrosion. These cases are of three types: stress corrosion, corrosion fatigue, and liquid velocity effects (corrosion, erosion, cavitation).

The typical test for *stress corrosion* is to take a sample of metal in the form of a small beam, apply a permanent bending load, and place the beam in a corrosive liquid. Failure occurs far more rapidly than in an unstressed part. Another method is to take a deformed sample such as a stamping and observe the onset of cracking. Classic cases are the *season cracking* of drawn-brass cartridge cases and the *caustic embrittlement* of steel in boilers. In both these cases we have a highly deformed structure with high residual stress and a suitable environment. In the case of the brass cartridge cases, the cracking is associated with the presence of ammonia from decaying organic matter plus the high humidity of the tropics. In the boiler tubes the elastic strain is produced by cold-rolling the ends, and caustic is present in the environment.

Stress corrosion is usually accompanied by an intergranular fracture in which the material shows no ductility, even though tensile test results may indicate a high plastic elongation capability. Although many examples are due to residual stresses from processing, the stresses can also result from the service conditions. Thermally induced stresses resulting from different coefficients of expansion have caused stress-corrosion failure in some bimetal plumbing fixtures.

EXAMPLE 12.4 Where does corrosion normally occur first in a car bumper?

ANSWER The portions that have been most cold-worked are usually the first to corrode. Just as on a microscopic scale grain boundaries are *anodic* to the surrounding grain, the macroscopic area of high residual stress is anodic to the nonworked portion. Of course, a corrosive medium such as salt spray from roads is necessary to cause the corrosion. Chromium (cathodic to steel) does not offer any galvanic protection for the steel. Once ruptured, it makes the problem worse because of the cathode-to-anode area relationship.

Chapter 3 discussed fatigue failures or cyclic stress failures. It pointed out that such failures are very sensitive to surface conditions. Pits from corrosion cause a surface stress concentration that easily propagates as a crack under a cyclic stress; hence the name *corrosion fatigue.*

Normally an increase in liquid velocity is accompanied by an increase in corrosion rate. Therefore a spinning disk in water would have a higher corrosion rate at its periphery. (Recall that high velocities decrease concentration polarization.) A normal manifestation is a gouging out or *erosion.* However, not all erosion results from high liquid velocities. A dripping action has also been known to "wear a hole" in materials such as copper sewer pipe, as shown in Fig. 12.15. *Cavitation,* on the other hand, results from bubbles of vapor popping against a surface such as a ship propeller.

Because of the importance of stress-corrosion cracking, we shall treat it more extensively in Chapter 13.

Corrosive Environments

12.19 General

The most important corrosion problems occur in three types of environments:

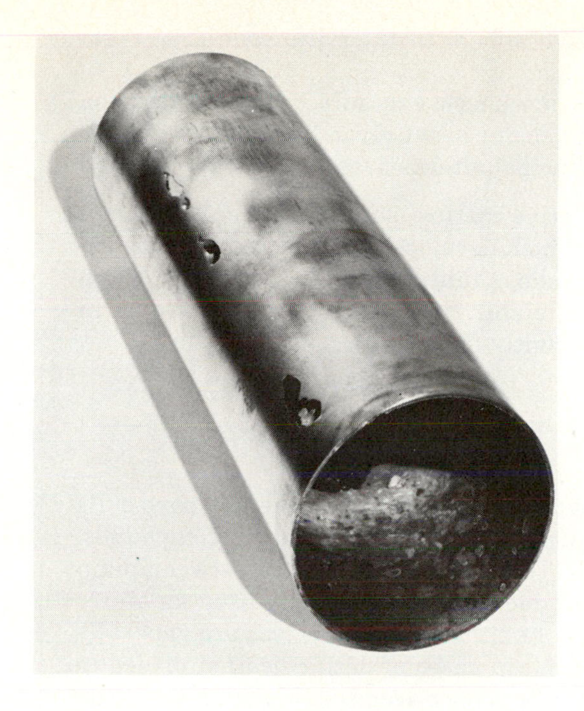

Fig. 12.15 *Copper sewer pipe with holes eroded by dripping waste fluids. Although copper has excellent resistance to corrosion, waste fluids that dripped and concentrated on standing finally penetrated the pipe.*

1. The atmosphere
2. Water, fresh and salt
3. Chemicals: acids, alkalis, salts

We shall discuss the types of reactions encountered in each case.

12.20 Atmospheric corrosion

Although atmospheric corrosion is not spectacular, the cost of it is. The annual bill in the United States is more than $2 billion. After much experimentation, corrosion engineers have found three distinctly different corrosion rates in industrial, marine, and rural environments.

The problems in the industrial environment arise from SO_2, which leads to H_2SO_4 and H_2SO_3. Salt and other contaminants from roads also lead to accelerated corrosion rates.

In marine atmospheres the chief problem is salt spray.

In rural atmospheres rain and dust cause the principal problems. By contrast, in the Atacama Desert in Chile, automobiles wrecked 50 years ago appear just as they did at the time they were discarded.

Recently a change in atmosphere has changed the life of galvanized eave troughs. In the past, in highly industrialized areas it was not uncommon to have to replace such items in as little as 5 years. The present concern with air pollution has resulted in an extension of their life to more than 20 years, which compares favorably with their life span in a rural atmosphere.

In selecting a material for atmospheric exposure, your first decision is whether a shiny metallic surface such as that given by stainless steel is necessary. If not, there are two principal alternatives.

1. Use an alloy that forms a protective coating, such as a steel with small amounts of copper and nickel, which develops a brown surface. The extreme case is the use of copper alloys that develop an attractive green patina. Styles change, and while shiny metal surfaces used to be in fashion, muted surfaces have recently found favor.
2. Apply paint or plastic coatings.

12.21 Water

The attack from fresh water varies widely, depending on the dissolved salts and gases. The principal contaminants are chloride ions, sulfur compounds, iron compounds, and calcium salts. There is little difference between plain and low-alloy steels. Cast iron and ductile iron are widely used for water pipe. At critical junctions such as valves the mating surfaces are generally specified as copper alloys. In general, in cases of dezincification or dealuminization, alloys with more than 80% copper are used. Monel, aluminum, some stainless steels, and cupronickel are also employed, depending on the application.

Sea water attacks ordinary steel and cast iron fairly rapidly, and protection by painting or a sacrificial anode is used. For example, oceangoing vessels have zinc sacrificial anodes bolted at intervals to the hull. Pitting is encountered in stainless steel, and brass with less than 80% copper may dezincify. Titanium has excellent resistance.

12.22 Chemical corrosion

The petroleum and chemical industries have the most severe chemical corrosion problems. In the petroleum industry, salt water, sulfide, organic acids, and other contaminants accelerate corrosion. Stainless steel, Stellite (a cobalt-base alloy), and Monel are used. There are a few specific cases of successful combating of chemical corrosion.

Corrosive Chemical	*Resistant Material*
Nitric acid	Stainless steels
Hot oxidizing solutions	Titanium
Caustic solutions	Nickel alloys
Concentrated sulfuric acid	Steel
Dilute sulfuric acid	Lead
Pure distilled water	Tin

To show the effects of concentration and temperature, Fig. 12.16, pages 512–513, lists materials used to resist sulfuric acid. Note that the temperature-concentration graph is divided into 10 zones of increasing severity of attack. As the attack becomes more severe, the number of materials that can be used to give a corrosion rate of less than 0.020 in./yr dwindles rapidly until only gold, glass, and platinum are left. The behavior of 316 stainless steel is interesting: It can resist higher acid concentrations better than intermediate concentrations. This is related to passivation, whereby a high concentration of oxidizing acid is required to form a *stable* passive layer. Passivity is not achieved at low concentrations, and therefore the corrosion rate is higher.

Corrosion in Gas

12.23 General

At first a topic such as oxidation may seem out of place in the same chapter with corrosion. However, as we delve into the actual mechanisms it will be evident that oxidation is closely related to other forms of corrosion, since in both processes ions and electrons must be transferred. We shall consider first the types of scales or oxides encountered, then how they form, next the rate of oxidation, and finally the scale-resistant alloys and special cases such as catastrophic oxidation and internal oxidation.

12.24 Types of scales formed

When a metal is exposed to air, for example, it is important to determine first whether the scale occupies a smaller or larger volume than the metal it came from. For instance, if we oxidize the outer layer of a piece of magnesium, the volume of the scale formed will be less than the volume of metal from which it was formed. In the case of iron the volume of the scale will be greater than that of the parent metal. Therefore, the scale is protective. Magnesium oxidizes rapidly because cracks appear in the scale, giving ready access to the metal beneath.

EXAMPLE 12.5 What is the ratio of oxide volume to metal volume for the oxidation of magnesium? (The specific gravity of magnesium is 1.74, and that of MgO is 3.58.) *NOTE:* The answer to this example continues on page 514.

ANSWER Assume that 100 g of Mg is oxidized to MgO:

$$Mg + \tfrac{1}{2}O_2 \rightarrow MgO$$

$$\text{Volume of Mg} = \frac{100\text{ g}}{\text{density of Mg}} = \frac{100\text{ g}}{1.74\text{ g/cm}^3} = 57.5\text{ cm}^3$$

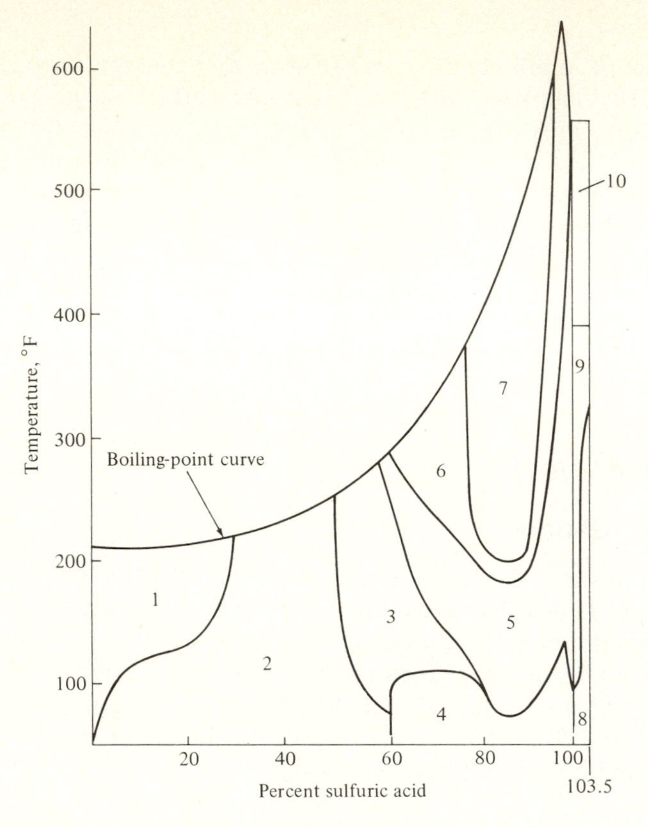

MATERIALS IN DESIGNATED ZONES HAVING REPORTED CORROSION RATES LESS THAN 20 MPY

Zone 1	
10 percent aluminum bronze (air-free)	Gold
Glass	Platinum
Lead	Silver
Copper (air-free)	Zirconium
Monel (air-free)	Tungsten
Rubber (up to 170°F)	Molybdenum
Impervious graphite	Type 316 stainless (up to 10% H_2SO_4, aerated)
Tantalum	
Zone 2	
Glass	Tantalum
Silicon iron	Gold
Lead	Platinum
Copper (air-free)	Silver
Monel (air-free)	Zirconium
Rubber (up to 170°F)	Tungsten
10 percent aluminum bronze (air-free)	Molybdenum
Austenitic ductile iron	Type 316 stainless (up to 25% H_2SO_4 at 75°F, aerated)
Impervious graphite	

MATERIALS IN DESIGNATED ZONES HAVING REPORTED CORROSION RATES LESS THAN 20 MPY (*Continued*)

	Zone 3
Glass	Tantalum
Silicon iron	Gold
Lead	Platinum
Monel (air-free)	Zirconium
Impervious graphite	Molybdenum
	Zone 4
Steel	Impervious graphite (up to 96% H_2SO_4)
Glass	Tantalum
Silicon iron	Gold
Lead (up to 96% H_2SO_4)	Platinum
Austenitic ductile iron	Zirconium
Type 316 stainless (above 80% H_2SO_4)	
	Zone 5
Glass	Tantalum
Silicon iron	Gold
Lead (up to 175°F and 96% H_2SO_4)	Platinum
Impervious graphite (up to 175°F and 96% H_2SO_4)	
	Zone 6
Glass	Gold
Silicon iron	Platinum
Tantalum	
	Zone 7
Glass	Gold
Silicon iron	Platinum
Tantalum	
	Zone 8
Glass	Gold
Steel	Platinum
18% Cr, 8% Ni stainless steel	
	Zone 9
Glass	Gold
18% Cr, 8% Ni stainless steel	Platinum
	Zone 10
Glass	Platinum
Gold	

Fig. 12.16 *Resistance to corrosion of materials exposed to sulfuric acid.* (M. B. Fontana and N. D. Green, *Corrosion Engineering*, McGraw-Hill Book Company, New York, 1967)

Letting x = MgO produced, we have

$$\frac{100 \text{ g}}{\text{mol. wt. Mg}} = \frac{x \text{ g}}{\text{mol. wt. MgO}}$$

$$x = \frac{100 \times 40.32 \text{ g}}{24.32} = 167 \text{ g}$$

$$\text{Volume of MgO} = \frac{167 \text{ g}}{\text{density of MgO}} = \frac{167 \text{ g}}{3.58 \text{ g/cm}^3} = 46.6 \text{ cm}^3$$

or 57.5 cm^3 Mg → 46.6 cm^3 MgO. The ratio is then 46.6/57.5 = 0.810.

The ratio of the volume of the scale to the volume of the parent metal is called the *Pilling-Bedworth ratio.* It may be calculated from the formula

$$\text{P-B ratio} = \frac{Wd}{Dw}$$

where W = molecular weight of the oxide, d = density of the metal, w = atomic weight of the metal, and D = density of the oxide.

When this ratio is much less than 1, the scale is nonprotective because it will crack. On the other hand, if the ratio is much greater than 1, the scale may crack off because the volume difference (P-B ratios greater than 2.3) is so great. This ratio, therefore, serves as a rough screening test. However, we must realize that other factors such as coefficient of expansion, melting point, vapor pressure, and high-temperature plasticity are also important characteristics.

The P-B ratio is not just an experimental observation, but is based on the concept that ratios less than 1 result in tensile strains in the oxide scale. As pointed out in Chapter 8, oxides and ceramics, in general, do not possess high tensile strengths. Therefore scale cracking can occur. However, their compressive strengths are also not infinite, and large P-B ratios can result in spalling off of the protective scale.

12.25 Mechanism of scale formation

Assuming that we have a relatively adherent scale, how does it grow? There are two possibilities: The oxygen or oxygen ion can diffuse through the scale to the metal. *Or* the metal or metal ion can diffuse through the scale to the surface and react with oxygen. This is a rather important distinction; in the first case the scale would grow at the metal-oxide interface, whereas in the second it would grow at the oxide-air interface. Wagner suspected that because metal ions are generally smaller than oxygen ions the growth occurred at the outer surface, and he performed the experiment illustrated in Fig. 12.17. Instead of oxidation with air he chose the reaction of silver

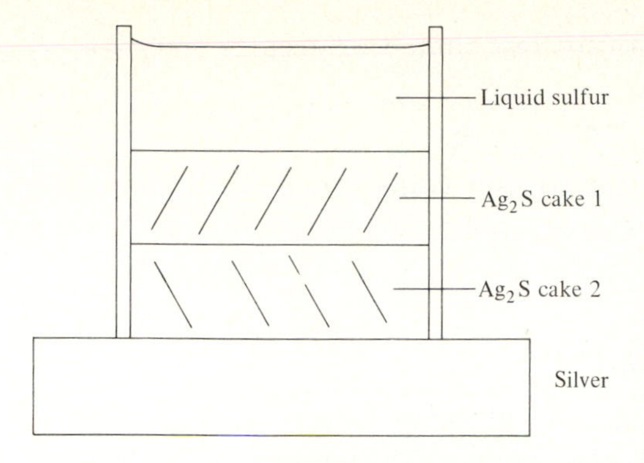

Fig. 12.17 *Diffusion experiment to show whether metal ions (Ag^+) or anions (S^{2-}) diffuse more rapidly. The results can be compared to formation of oxide scale in metal systems.*

with liquid sulfur in a cell. Two weighed cakes of silver sulfide were placed on a specimen of silver, and then liquid sulfur was placed over the cakes. At a given temperature two reactions were possible. Ag^+ ions could travel through the sulfide and react at the upper surface with sulfur, or sulfide ions could migrate to the metal-sulfide interface. After the cell was disassembled, he found that the upper cake had gained much more weight than the lower cell, indicating that the metal ion migration was most important. The usual oxidation mechanism, therefore, is the migration of metal ions and electrons to the outer surface to react with oxygen. Since the mobility of electrons is usually an order of magnitude higher than that of the ions, this is usually not a limiting factor except under very special circumstances.

Oxide Defect Structures As discussed later in the section on semiconductors, many oxide structures do not follow the exact chemical formula but may have an excess of one ion or the other. This leads to vacancies at certain lattice points in order to attain electroneutrality. The presence of vacancies increases the diffusion rate. However, if an alloy element that reduces the vacancies is added, the oxidation rate is lowered. A rather startling case is that the addition of a small amount of lithium, an easily oxidized element, lowers the oxidation rate of nickel by occupying vacancies in the nickel oxide lattice.

12.26 Oxidation rates

When a scale is adherent, build-up usually follows the *parabolic law*. If the rate of oxidation is proportional to the thickness of the scale (see the problems), then we obtain the relation

$$W^2 = k_p t + C$$

where W = weight of scale, t = time of exposure, and k_p and C = constants for the particular reaction.

If the scale cracks, as at P-B ratios less than 1 or greater than 2.3, the rate is *linear:*

$$W = kt$$

In cases of fairly thick scale containing voids or when a limiting thickness is approached (such as aluminum in air), an empiricial *logarithmic* relation is obeyed: $W = k \log (Ct + A)$, where k, C, and A are constants.

12.27 Scale-resistant materials

Scale-resistant materials are most needed at temperatures from 800 to 2000°F (427 to 1093°C). Iron, nickel, and cobalt oxidize appreciably above 1000°F (538°C). The addition of chromium, silicon, and aluminum leads to the formation of protective, adherent scale. Although these elements oxidize more readily than iron, they form new structures with iron, such as the spinel types discussed in Chapter 7. The importance of forming the proper scale is especially apparent when one tries to use pure molybdenum at elevated temperatures. Despite the excellent creep resistance, the volatility of MoO_3 leads to reduction of section thickness at a rapid rate. Similar problems are encountered with tungsten and niobium.

12.28 Special cases

Certain alloys oxidize very rapidly under special conditions. This phenomenon is called *catastrophic oxidation*. The most insidious type is the effect of small amounts of vanadium pentoxide or lead oxide in the gas phase. These oxides can combine with the normal metal scale to provide a low-melting-point phase in the scale. This provides a liquid with rapid transport of oxygen to the metal, so that attack is accelerated.

Another interesting case is called *internal oxidation*. In this case the scale is more permeable to oxygen than to the metal ion. The oxygen dissolves in the metal, migrates, and forms oxide particles in the metal itself. Although this causes some hardening, the embrittling effect is usually undesirable in high-temperature alloys.

EXAMPLE 12.6 Determine by calculation whether aluminum would be resistant to oxygen or chlorine gas. The density of Al_2O_3 is 3.8 g/cm^3 and that of $AlCl_3$ is 2.44 g/cm^3.

$$2Al + \tfrac{3}{2}O_2 \rightarrow Al_2O_3 \qquad Al + \tfrac{3}{2}Cl_2 \rightarrow AlCl_3$$

ANSWER The P-B ratio is *Wd/Dw*.

$$\text{The P-B ratio of } Al_2O_3 = \frac{(2 \times 26.98 + 3 \times 16)(2.699)}{(3.8)(2 \times 26.98)} = 1.34$$

We would therefore expect aluminum in oxygen to show a parabolic rate of scale formation.

$$\text{The P-B ratio of } AlCl_3 = \frac{(26.98 + 3 \times 35.46)(2.699)}{(2.44)(26.98)} = 5.47$$

Although compressive stresses result, they are too high to maintain without fracture (the P-B ratio is greater than 2.3). Hence we would expect a linear rate of scale formation with chlorine.

We might also anticipate, in the absence of reaction between oxygen and chlorine or the formation of complexes with aluminum, that pre-oxidation of aluminum might form a scale that would be adherent and protective against chlorine and the formation of $AlCl_3$. In practice this might not offer complete protection, since some aluminum ions might migrate through the Al_2O_3 scale. However, the chlorination rate would be lower than it would if the Al_2O_3 scale were not present.

Corrosion of Ceramics and Plastics

12.29 General

Ceramics and plastics, often in conjunction with metals, offer alternative solutions to corrosion problems. The main drawback of ceramics is their lack of ductility. For plastics, it is the effects of low and elevated temperature and in some cases water absorption and flammability. On the other hand, none of the effects associated with metal corrosion, such as pitting and galvanic action, is encountered. In many cases plastic parts can be used to insulate metal parts from interaction as long as oxygen-concentration cells do not form under the plastic. We shall consider the resistance of ceramics and plastics to the same group of corrosive environments that attack metals: atmosphere, water, and chemicals.

An important general rule in predicting the performance of an organic material in a solvent is that *like dissolves like*. Structures that contain polar groups such as OH, C≡N, COOH will be attacked or swelled by polar solvents such as acetone, alcohol, and water. (Remember that a polar group is one in which the electrons are shifted to one of the elements or are unbalanced around the carbon atom because of the stronger attractive forces.) These polar groups resist solution by the balanced nonpolar solvents, such as carbon tetrachloride, gasoline, and benzene.

On the other hand, polymers with nonpolar groups such as methyl (CH_3) and phenyl (C_6H_5) are resistant to the polar solvents, but they are swollen or dissolved by the nonpolar solvents.

Furthermore, straight-chain (aliphatic) polymers tend to dissolve in straight-chain solvents such as ethyl alcohol, whereas those with benzene rings (aromatic polymers) tend to dissolve in aromatic solvents such as benzene. As molecular weight increases, solubility decreases. Also, with greater crystallinity and consequently denser packing and stronger intermolecular forces, solubility decreases.

12.30 Atmosphere

Ceramics are only slowly affected by the atmosphere, as shown by the many structures made of brick and cement that have stood for centuries. The principal dangers in weathering are the effects of water entering cracks or joints and expanding on freezing. Salt in water aggravates the problem.

Plastics are affected only slowly by the atmosphere, particularly by sunlight. There are wide differences in this effect on various plastics, as shown in Table 12.6.

12.31 Water

Nonporous ceramics are widely used for containers and piping. Glass-lined and enameled-steel tanks have been used for many years with great success. Plastics are also generally resistant to water; they make excellent protective coatings. However, there is a small percentage of water absorption in all cases except Teflon and a very slight amount for polyethylene and polypropylene. This absorption can aggravate problems such as creep in fibers that are used for clothing; thus we get wrinkles in summer weather.

12.32 Chemicals

There are wide differences in the resistance of ceramics to chemicals. In the glasses pure silica and borosilicate are very resistant, but the soda-lime glasses are slowly attacked by alkalis. Basic refractories such as magnesia are attacked by acids. Organic solvents have no effect on the typical ceramic.

Plastics also show a great deal of variation in resistance to chemicals. Most are resistant to weak acids and alkalis. However, strong acids decompose cellulose acetate; some oxidizing acids decompose melamines and phenol-formaldehyde. Strong alkalis and organic solvents also attack certain plastics. The most resistant materials are Teflon, polyethylene, and vinyl.

Sometimes the corrosive media come from unexpected sources. As an example, we often have considerable difficulty trying to maintain eyeglass frames. A combination of body fluids, exposure to airborne hydrocarbons, and stress often results in failure in plastic eyeglass frames in 12 to 24 months, as shown in Fig. 12.18.

12.33 Summary of steps used to prevent corrosion

In summary, the methods commonly used to inhibit corrosion are:

1. Careful selection of materials, to prevent galvanic cells from forming.

Table 12.6 CORROSION RESISTANCE OF PLASTICS*

Material	Acids		Alkalis		Organic Solvents	Water Absorption, percent/24 hr	Oxygen and Ozone	High Vacuum	Ionizing Radiation	Temperature Resistance, °F	
	Weak	Strong	Weak	Strong						High	Low
Thermoplastics											
Fluorocarbons	Inert	Inert	Inert	Inert	Inert	0.0	Inert		P	550	G, 275
Polymethylmethacrylate	R	A-O	R	A	A	0.2	R	Decomp.	P	180	
Nylon	G	A	R	R	R	1.5	SA		F	300	G, 70
Polyether (chlorinated)	R	A-O	R	R	G	0.01	R			280	G
Polyethylene (low density)	R	A-O	R	R	G	0.15	A	F	F	140	G, 80
Polyethylene (high density)	R	A-O	R	R	G	0.1	A	F	G	160	G, 100
Polypropylene	R	A-O	R	R	R	<0.01	A	F	G	300	P
Polystyrene	R	A-O	R	R	A	0.04	SA	P	G	160	P
Rigid polyvinylchloride	R	R	R	R	A	0.10	R		P	150	P
Vinyls (chloride)	R	R	R	R	A	0.45	R	P	P	160	
Thermosetting Plastics											
Epoxy (cast)	R	SA	R	R	G	0.1	SA		G	400	L
Phenolics	SA	A	SA	A	SA	0.6			G	400	L
Polyesters	SA	A	A	A	SA	0.2	A		G	350	L
Silicones	SA	SA	SA	SA	A	0.15	R		F	550	L
Ureas	A	A	A	A	R	0.6	A		P	170	L

*Abbreviations: R = resistant, A = attacked, SA = slight attack, A-O = attacked by oxidizing acids, G = good, F = fair, P = poor, L = little change, decomp. = decomposes.
Source: M. B. Fontana and N. D. Green, *Corrosion Engineering,* McGraw-Hill Book Company, New York, 1967.

Fig. 12.18 *Plastic eyeglass frames that failed because of applied stress and corrosive media. Although thermoplastics generally have good corrosion resistance, the combination of stress and body fluids caused this failure at normal temperature.*

2. Alteration of environment—temperature, velocity, ion concentration—and the use of inhibitors
3. Design changes, such as removing pockets that may hold corrosive fluids
4. Cathodic protection, which can be applied to almost any metal
5. Anodic protection, which can be used only for materials that passivate
6. Application of coatings—metallic, ceramic, or organic—including metallic coatings as sacrificial anodes

Finally, as we pointed out at the beginning of the chapter, corrosion is seldom totally prevented. Economics plays a most important role, and an optimization of both costs and required life span is the normal ultimate solution.

EXAMPLE 12.7 For the cases below, indicate the type of corrosion cells and state how the corrosion might be minimized.

a. Bolts and nails stored in the basement corrode.
b. A painted trash barrel corrodes on the inside.
c. A so-called "magwheel" for an automobile corrodes between the aluminum hub and the steel rim.
d. Paint blisters from an automobile fender.

ANSWER

a. The corrosion cells are localized galvanic cells accentuated by residual stress in the bolts and nails resulting from cold forming and machining. The basement suggests dampness, and storage in a drier atmosphere would reduce corrosion.

The components are also available in zinc- or cadmium-plated forms to increase corrosion resistance. In hardware stores prepackaging in plastic containers excludes high-humidity atmospheres.

b. Again localized galvanic action. Fluids should be excluded from the barrel, by covering it to keep out rain and by placing small holes in the bottom to drain off fluids. A corrosion-inhibiting paint would also decrease the corrosion rate.

c. The design ignores a fundamental principle in corrosion control: Do not join dissimilar materials. The result is a galvanic cell made worse by the environment of the salt used on roads for deicing. The remedy is to use only one material, to use a coating over the materials that will maintain its integrity over many years, or to electrically insulate the two materials.

d. Although the corrosion may have been initiated by a scratch or stone bruise, the continuation is the result of a concentration cell. The obvious solution is sanding and repainting. To prevent recurrence, one must completely remove the rust and the paint must have no pinhole porosity.

SUMMARY

Corrosion takes place because of the solution or other reaction of the surface of a component with its environment. In the case of metals each instance of corrosion can be studied as an electrochemical cell in which the metal goes into solution, i.e., corrodes at the anode. For the process to continue, the electrons left in the solid at the point of solution must migrate through the structure and be consumed at the cathode. By far the most common reaction, that seen in the rusting of iron, is the reaction with water and dissolved oxygen to produce OH^- ions: $2H_2O + O_2 + 4e \rightarrow 4OH^-$.

A starting point in estimating the tendency of a metal or alloy to corrode is its half-cell potential, the voltage it develops when connected to a standard electrode. However, although this shows that elements such as aluminum and magnesium have a high tendency to corrode, it does not tell us anything about the nature of the corrosion product, which is important. If a passive or unreactive film forms, corrosion stops. A great many specialized terms, defined below, have developed, and practically all involve applications of simple electrochemical cell theory.

Oxidation of metals in gases, which is a specialized case of corrosion, is governed by a combination of the tendency of a metal to react and by whether a protective scale is formed. Finally, corrosion can occur in both ceramic and polymeric materials, although usually it is not as prevalent as it is in metals.

DEFINITIONS

Activation polarization A decrease in cell potential due to intermediate steps necessary to complete the reaction.

Anode The electrode of a cell at which metal goes into solution; the negative pole in the external circuit.

Cathode The electrode of a cell at which metal ions are plated out or negative ions are created, as in the reactions $2H_2O + O_2 + 4e \rightarrow 4OH^-$ and $Cu^{2+} + 2e \rightarrow Cu^0$.

Cathodic protection Plating or attachment of a more active metal that dissolves, preventing solution of the component to be protected.

Cell potential The electromotive force (emf) developed by a cell.

Concentration polarization A decrease in the emf of a cell due to build-up of ions around the electrodes.

Corrosion The solution or harmful reaction of the structure of a material with its environment.

Corrosion fatigue Fatigue failure aggravated by corrosion.

Erosion The wearing away of a surface because of combined mechanical-corrosion effects.

Galvanic corrosion Corrosion that is accelerated by the presence of electrically connected dissimilar metals.

Graphitic corrosion Selective corrosion of gray cast iron in which only a structure of graphite and some oxides remain.

Half-cell potential of an element The emf developed when an element is coupled with a hydrogen half-cell.

Hydrogen blistering The development of blisters as a result of the diffusion of hydrogen into voids.

Hydrogen half-cell A reference half-cell involving hydrogen adsorbed on a platinum wire in a standard solution of hydrogen ions.

Inhibitor A chemical added to produce a film or coating that slows down the corrosion reaction.

Intergranular corrosion Preferential corrosion at grain boundaries resulting from the formation of an anode because of precipitation of a second phase at the grain boundaries.

Oxygen-concentration cell A galvanic cell caused by differences in oxygen concentrations. An example is the water line in a partially filled tank.

Passivating The formation of a film of reaction product that inhibits further reaction.

Pilling-Bedworth ratio The ratio of volume of scale to volume of parent metal. Values much different from 1 may lead to high scaling rates.

Pit corrosion Corrosion resulting from differences in oxygen and ion concentrations at the base of a pit compared with the surface.

Sacrificial anode An active metal that, when attached to the object to be protected, makes the object cathodic and therefore causes solution to take place only on the active metal.

Scale A reaction product formed on a surface that is attacked by gas or liquid.

Season cracking The acceleration of corrosion owing to high residual stresses. *Caustic embrittlement* is a similar effect in an alkaline environment.

Selective leaching Preferential solution of one element in an alloy. In brass the zinc dissolves and a spongy copper deposit remains (dezincification).

Weld decay Corrosion at or adjacent to a weld as a result of galvanic action resulting from differences in structure produced by the welding.

PROBLEMS

Let us be perfectly candid about the problems in this section. We could put together a group of electrochemical calculations that would be a great exercise in calculator manipulation. But this might give the erroneous impression that we can go out and calculate such things as the corrosion rate of a car fender in the spring mush of Michigan roads. M. Pourbaix has done some excellent work in the application of thermodynamics to corrosion, but this cannot yet be applied directly to the average complex situation. Also, quantitative calculations of potentials necessary for protection of pipes have been made.

Therefore, in this problem section, we shall illustrate the application of the general principles set forth in the sections on chemical principles of metallic corrosion, corrosion phenomena, and gas corrosion.

12.1 Write the ion-electron equations for the anode and cathode reactions (half-cell reactions) in the following cases. (If you believe corrosion will not occur, write "no reaction.") (Sections 12.1 through 12.7)

a. An opened (punctured) tin can at the bottom of a fresh-water lake
b. An opened (punctured) tin can in the ocean. Why would the rate be more rapid here than in part *a*?
c. A stainless steel (302) piece of trim on a car that is exposed to intermittent splashing with salt solutions
d. A car fender (1010 steel) covered on the inside with a coat of asphalt with a few holes through the asphalt
e. A copper heating coil brazed to a steel pipe with a 70% Cu, 30% Zn braze. The junction is exposed to dripping from a hot-water tank.
f. The effect of attaching a "copper ground" from a "live" electric stove to a steel water pipe that sweats in the summer
g. A punctured Alclad aircraft wing (pure aluminum covering alloy 2014) exposed to salt water

12.2 Large amounts of copper are obtained from copper mine water by immersing iron scrap in the solution and later collecting fine copper powder (cement copper). Write the ion-electron equations and discuss the economics of this technique compared with other possible methods, such as electrolysis, evaporation, and reduction of the salt. (Sections 12.1 through 12.7)

12.3 Write the *ion-electron* equations for the reactions taking place at the anode and cathode, respectively, for the following conditions, describing the cathode and anode materials adequately to show the difference. (Sections 12.1 through 12.7)

a. A piece of impure zinc containing a second-phase β dissolves in acid, giving hydrogen bubbles. β is nobler than the zinc-rich phase α.

b. It is found that to protect a steel boat hull against corrosion in salt water, pure zinc should be bolted to the hull. The zinc gradually corrodes. Why is *pure* zinc specified?

c. A piece of 18% Ni, 8% Cr, 0.08% C stainless steel is held at 1800°F to develop a grain boundary precipitate of Cr_4C. The steel then corrodes in the grain-boundary regions.

12.4 If a mercury-sodium alloy is used, the sodium can be effectively oxidized in a battery without detrimental side reactions. (Sections 12.1 through 12.7)

a. For a battery containing an electrolyte with a sodium ion concentration of 1 g-mol/liter (one molar) and a standard hydrogen electrode, what should be the battery voltage?

b. If the sodium ion concentration in the electrolyte were only 0.1 g-mol/liter, what should be the battery potential?

c. If the reference hydrogen electrode in (*a*) is replaced with a silver electrode, what should be the potential?

d. Indicate the correct answers for case (*a*) above in the following table by filling in the blanks correctly.

Electrode	*Anode*	*Cathode*	*Positive*	*Negative*
Sodium				
Hydrogen				

12.5 Assume that concentration polarization is present in a certain cell. Modify Fig. 12.9 to show (schematically) the effect of (*a*) increased temperature, (*b*) increased electrolyte concentration. (Sections 12.8 through 12.12)

12.6 A battery is nothing more than controlled corrosion. What is the purpose of a battery charger or an alternator in your car? (Sections 12.8 through 12.12)

12.7 Indicate whether the corrosion rate of a piece of iron placed in tap water is increased or decreased by doing the following. (Sections 12.8 through 12.12)

a. Adding NaCl to the water
b. Imposing electron flow *into* the iron (by a battery)
c. Placing nickel in contact with the iron
d. Adding chromate ion to the water
e. Freezing the water

12.8 Explain *in one sentence* why the following statements are either correct or incorrect. (Sections 12.8 through 12.12)

a. A high concentration of oxidizing acid may render a metal more corrosion-resistant.
b. When we see an etched microstructure under the microscope, the grain boundaries were anodic during the etching process.
c. Oxygen dissolved in water has no effect on the corrosion rate of iron that is exposed to the water.
d. Aluminum rivets in a steel structure should offer longer life against corrosion than steel rivets in an aluminum structure.

12.9 Compare the merits of using tin and zinc: (*a*) As a protective coating for cans for food, (*b*) for outdoor fencing. (Sections 12.13 through 12.18)

12.10 A student argues that dead spots in the circulation of a tank should not be harmful because corrosion products could build up and protect the region from corrosion. Discuss. (Sections 12.13 through 12.18)

12.11 With most alloys, reducing the concentration of oxygen in the solution retards corrosion. For what group of alloys is this not true? (Sections 12.13 through 12.18)

12.12 Why are zinc sacrificial anodes preferred to magnesium sacrificial anodes for protecting steel ship hulls? A number of years ago it was found that the zinc had to be quite pure in order to be effective and, in particular, had to be free from second phases. Why? (Sections 12.13 through 12.18)

12.13 Describe two ways in which buried ductile-iron pipelines can be protected from corrosion. (Sections 12.13 through 12.18)

12.14 One device for protecting boat hulls uses a platinum electrode protruding from the steel hull. How does this work? To which electrode of a battery should it be connected, and to which should the steel be connected? (Sections 12.13 through 12.18)

12.15 Antique brass pots and kettles may not be great "finds," since they often leak because of a corrosion phenomenon. What would the corrosion be called if it had the following appearances? (Sections 12.13 through 12.18)

a. Fine cracks that appear to follow grain boundaries
b. Copper rather than the normal yellow brass color

12.16 How would Fig. 12.9 appear if current density rather than current was used? (Sections 12.13 through 12.18)

12.17 *a.* In recent years automobile manufacturers have used galvanized steel sheet in body parts to combat corrosion. Considering the principal corrosive to be a dilute NaCl solution, write the ion-electron equations for the corrosion taking place before the changeover (in ordinary steel) and after (in galvanized steel) (Sections 12.13 through 12.18)

b. A sign manufacturer makes small signs of 18% Cr, 8% Ni, 0.08% C stainless steel by welding letters to a plate of the same material using a weld rod of the same material. Corrosion occurs $\frac{1}{4}$ in. from the weld. (Sections 12.13 through 12.18)

1. Write the ion-electron equations.
2. Why is the stainless steel not "stainless" (in 10 words)?
3. What could be done to prevent the corrosion *without* changing the composition of the parts? (No painting.)

12.18 The adherence of Al_2O_3 to aluminum involves a close match of (*a*) the aluminum-to-aluminum distances in the (111) planes of the metal with (*b*) the close-packed planes of Al_2O_3 (Fig. 7.19). What is the percentage difference in spacing? (Sections 12.23 and 12.24)

12.19 Magnesium parts are heat-treated in an atmosphere containing 1% SO_2, forming a sulfate coating. Why is this preferable to air? (The density of $MgSO_4$ is 2.66 g/cm^3 and that of MgO is 3.58 g/cm^3.) (Sections 12.23 and 12.24)

12.20 Why does the addition of chromium, a rapidly oxidized element, improve the oxidation resistance of austenitic nickel steels? (Sections 12.23 and 12.24)

12.21 A heat-treat furnace is to operate at 1800°F (1032°C). Parts are to be placed on molybdenum trays. There is some question about the oxidation resistance of molybdenum. Show by calculation whether molybdenum will stand up under these conditions. (Sections 12.23 and 12.24)

$$Mo + \tfrac{3}{2} O_2 \rightarrow MoO_3$$

Densities (g/cm^3):		
	Mo	10.22
	MoO_3	4.50
	O_2	1.43×10^{-3}

12.22 Show, by calculation, the anticipated oxidation resistance of titanium. The specific gravity of TiO_2 = 4.26 g/cm^3. (Sections 12.23 and 12.24)

12.23 When we see galvanic cells in corrosion, it means that electrons are flowing. Show by a suitable sketch how electrons may also flow in oxidation (therefore oxidation is also corrosion). (Sections 12.23 and 12.24)

12.24 Under what conditions might you expect concrete to corrode? (Recall the type of bonding in concrete.) (Sections 12.29 through 12.32)

12.25 Medicine has advanced to the point at which more and more synthetic components are considered as replacements for natural bone and tissue. What are some of the requirements of the material if it is to resist corrosion? (Sections 12.29 through 12.32)

12.26 There is current interest in the development of biodegradable plastics so that plastic trash thrown along the roadside will decay. What would be some of the important characteristics of these corrodible plastics? (Sections 12.29 through 12.32)

PROBLEMS COVERING OPTIONAL SECTIONS

12.27 A nail is immersed completely in oxygenated water. At which locations would corrosion occur? Write the anode and cathode reactions. (Sections 12.19 through 12.22)

12.28 Home and industrial water systems often use dissimilar piping materials. Local codes suggest using "dielectric unions," which are commonly made of nonmetallic materials. What is the advantage of using them? (Sections 12.19 through 12.22)

12.29 Derive the equation for the parabolic oxidation rate. (Let the rate of oxidation dW/dt be proportional to $1/x$, where x is the thickness of the scale.) (Sections 12.25 through 12.28)

12.30 Show by calculation whether the nitrification of titanium might be expected to be linear or parabolic (related to oxidation). The reaction is

$$2Ti + N_2 \longrightarrow 2TiN$$

The specific gravity of titanium is 4.50 and that of TiN is 5.43. (Sections 12.25 through 12.28)

12.31 Your company is going to buy a large quantity of cerium metal and hopes to store it in air. On the basis of the following data, you are asked to decide whether it can form a protective film. (Sections 12.25 through 12.28)

	Ce	*CeO_2*
Density (g/cm^3)	6.77	7.30
Atomic or molecular wt. (g/mole)	140	172
Crystal structure	HCP	cubic

Atomic radius of Ce = 1.82 Å. Ionic radius of Ce^{+4} = 1.02 Å. Make an appropriate calculation to determine whether it is possible for a protective film of CeO_2 to form.

12.32 Copper is known to form two oxides, Cu_2O and CuO. Indicate by calculation whether or not scales of the two oxides would have comparable stability (protective qualities). (Sections 12.25 through 12.28)

Density: Cu: 8.96 g/cm^3 Cu_2O: 6.00 g/cm^3 CuO: 6.40 g/cm^3

12.33 Cesium metal is known to form an oxide, Cs_2O. From the following data, show by calculation whether you might expect cesium to exhibit a linear or parabolic oxidation law (that is, give a porous or nonporous oxide scale). (Sections 12.25 through 12.28)

	Ce	*O_2*	*Cs_2O*	
Crystal structure	BCC	Gas		–
Atomic radius (Å)	2.62	0.62	Cs^+ = 1.65	O^{2-} = 1.32
Molecular weight	132.9	32	281.8	
Specific gravity	1.87	Gas	4.36	
Melting point	28°C	−218.4°C	400°C	

12.34 A common method of removing corroded bolts and nuts is to heat them with a torch. Why should this aid disassembly? (Sections 12.25 through 12.28 and 12.1 through 12.7)

12.35 Below are several phenomena associated with corrosion. Indicate by letter(s) which of the following corrosion terms would apply to each observation. (Summary of Chapter 12)

a. Cavitation
b. Passivation
c. Stress corrosion
d. Selective leaching
e. Galvanic action
f. Concentration cell
g. Corrosion fatigue

1. A cadmium-plated steel bolt seems to be losing its cadmium coating.
2. A knife blade turns black after cutting grapefruit.
3. Corrosion of a metal component results in an intergranular failure.
4. A brass kettle appears to be copper-colored rather than yellow in certain areas.
5. A golf club head shows corrosion on areas left covered with mud.
6. A titanium valve corrodes in concentrated HCl acid but not in an oxidizing acid such as HNO_3.

12.36 Here are different appearances of five corrosion cells. Give the name associated with each description. (Summary of Chapter 12)

a. A stainless steel in a corrosive medium ultimately fractures along the grain boundaries.
b. A 90% Cu, 10% Al alloy exposed to sea water for an extended period forms a high-copper spongy layer.
c. An aluminum pulley on an automatic clothes washer breaks away from its steel shaft.
d. A high-speed mixer pits at its outer periphery.
e. Steel rivet in a steel fender corrodes to the point where rivet falls out.

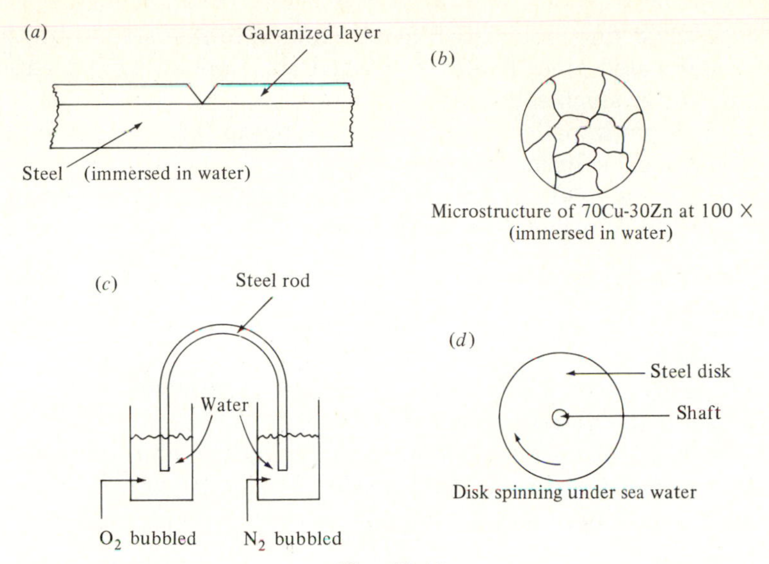

Fig. 12.19

12.37 For the corrosion cells pictured in Fig. 12.19, indicate the anode and cathode. (Summary of Chapter 12)

12.38 Figure 12.20 shows a number of corrosion cells made up of components immersed in sea water. Label the anodes and cathodes. Indicate the cathode reactions: (1) in the sea water, (2) if HCl is added. (Summary of Chapter 12)

12.39 For the following corrosion conditions, indicate the anode reaction, cathode reaction, and type of corrosion. (Summary of Chapter 12)

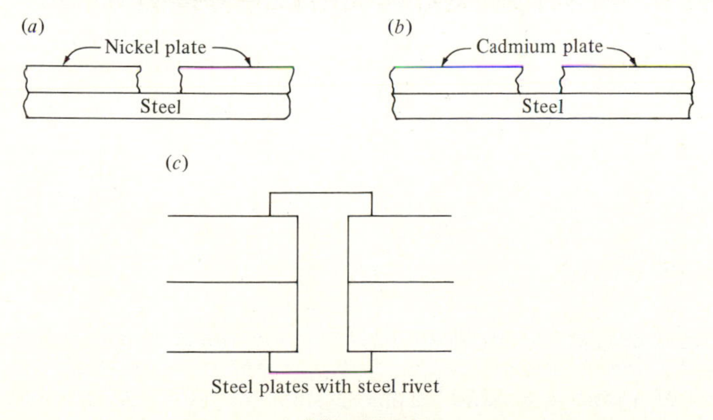

Fig. 12.20

a. Scratched zinc-plated steel fence exposed to industrial atmosphere
b. Undercoated steel fender scratched through undercoat exposed to salt-water slush solution
c. Aluminum impeller in an aluminum pump corrodes while pumping a mild acid solution
d. A stainless steel (18% Cr-8% Ni) corrodes at grain boundaries in a mild acid solution after being welded

12.40 For each of the corrosion-related situations shown in Fig. 12.21:

1. Draw and label arrows touching the cathode and anode region(s), respectively.
2. Write balanced ion-electron equations, giving the reactions taking place at the anode and cathode.
3. Label the corrosion phenomenon, using the one of the following terms that describes it most closely. *Do not use any given term more than once.* (Summary of Chapter 12)

Galvanic corrosion
Intergranular corrosion
Caustic embrittlement
Oxygen-concentration cell corrosion
Graphitic corrosion
Selective leaching
Sacrificial corrosion

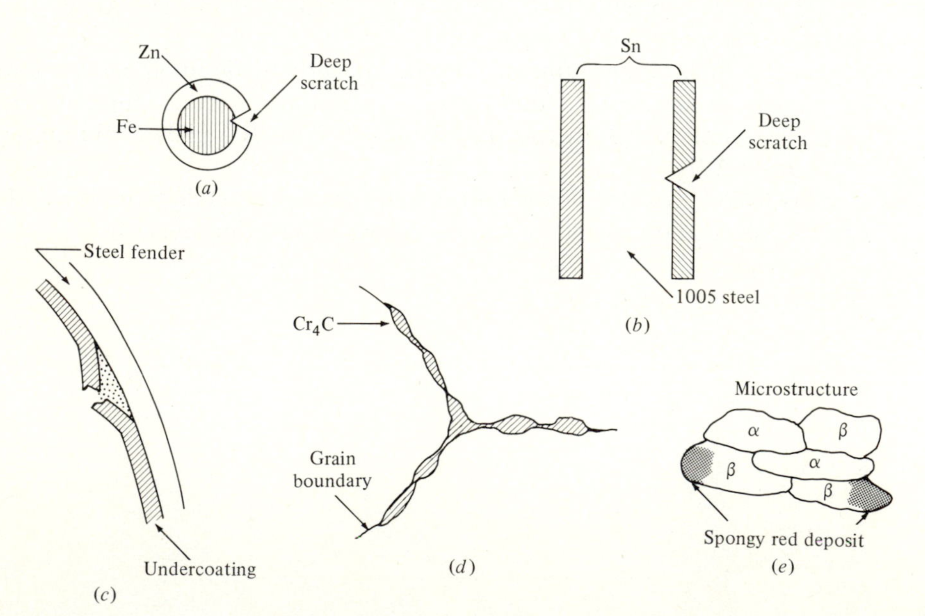

Fig. 12.21 *(a) Cross section of galvanized-iron fence wire in moist air. (b) Cross section of tin can in the rain. (c) Hole in undercoating of fender of car driven in the rain. (d) Microstructure of a weld in type 304 stainless steel (18% Cr, 8% Ni, 0.08% C, bal. Fe) showing particles of carbide precipitate at grain boundaries; immersed in sea water. (e) 60Cu-40Zn brass propeller corroding in sea water.*

4. How could corrosion be avoided in situation *d*? Select the *one best* remedy from the following group. (Indicate choice by letter.)
5. How could corrosion be avoided in situation *e*? Select the *one best* remedy from the following group. (Indicate choice by letter.)

Possible remedies. (Select only one for questions 4 and 5.)

a. Heat the metal to 750°F (448°C), hold for 5 hr, then water-quench.
b. Use a similar alloy containing 1% aluminum.
c. Heat the metal to 2000°F (1143°C), hold for 5 hr, then water-quench.
d. Attach a platinum plate.
e. Increase the copper content of the metal to 80%.
f. Increase the zinc content of the metal to 80%.
g. Use a similar alloy containing 0.4% niobium.

12.41 In each of the following cases, write balanced ion-electron equations for anode and cathode reactions and calculate (where requested) the voltage developed at startup. Assume 1-molar ion concentrations. (Summary of Chapter 12)

a. A tin can
Anode:
Cathode:
Voltage if a 1-molar solution of each ion is present at the respective electrode:

b. A galvanized steel fence (zinc on steel)
Anode:
Cathode:
Voltage if a 1-molar solution of each ion is present at the respective electrode:

c. Weld decay in stainless steel
Anode:
Cathode:

d. Dezincification of brass
Anode:
Cathode:

e. Graphitic corrosion
Anode:
Cathode:

f. Pit corrosion of steel
Anode:
Cathode:

g. Liquid sulfur in contact with silver
Anode:
Cathode:

13

ANALYSIS AND PREVENTION OF FAILURE

CO-AUTHORED WITH J. W. JONES†

THE catastrophic failure of certain steel ships, often while moored quietly at dockside, has led to extensive investigation of the "fracture toughness" of materials. A tensile specimen cut from the steel plate of these ships invariably exhibits good strength and ductility. The brittle fracture of the ship results from a combination of constraints and lowered temperature, as discussed in the text.

---

†Assistant Professor, Materials and Metallurgical Engineering, University of Michigan, Ann Arbor

13.1 General

When an engine tore loose from a DC-10 aircraft during takeoff at O'Hare Airport in 1979, millions of people asked why. In many other failures, such as railroad train derailments, trucks jackknifing, private cars going out of control, or tools failing, people are also asking why. These are important questions because it is possible to act on the evidence and prevent further failures if the selection of material, design of the part, or manufacturing process is at fault.

Many engineers are finding an exciting and rewarding career in failure analysis and prevention. The engineer with management responsibility must also understand the nature of these problems. Many lives as well as billions of dollars in litigation are involved each year, and a fair settlement of the claim depends in large measure on the engineer's ability to find the answers to two simple questions:

1. Why did the part fail?
2. Whose fault was it?

Because of the rapidly increasing importance of this field, we have included this chapter to show how the topics we have investigated can be applied not only to the analysis of failure but to the prevention of failure. Of course we cannot cover the entire subject in one chapter, but this overview will give examples of the interrelation of structure, properties, and service performance that we have presented in the rest of the book.

We shall follow the following outline:

1. *Fracture mechanics* (How do materials fail?) We have already discussed tensile, creep, impact, and fatigue testing. However, the concept of *fracture toughness* is essential if we are to understand fractures in service, especially those in which the stress is below the yield strength of the material. Further discussion of fatigue, stress corrosion, wear, and abrasion is added for background.
2. *Experimental methods for investigating failures* In previous chapters we have given examples of macro- and microstructures and their interpretation, but here we shall present an orderly progression of test procedures for examining and interpreting failures. In addition to macro- and microexamination, we shall briefly discuss nondestructive testing and the determination of residual stress.
3. *Defects and structural anisotropy caused by manufacturing processes* Often a failure is due not to chemical makeup or general microstructure but to flaws such as gas holes, shrinkage, inclusions, quenching cracks, rolling defects, or welding practice. We shall illustrate some of the typical structures of the most serious of these defects and suggest solutions. The designer must realize that it is not the *average* structure and properties that are important. Rather, it is the *minimum* atypical conditions that can be encountered in a portion of production that cause failures.

13.2 Fracture toughness and fracture mechanics

For many years engineers have been pressed to develop strong, light-weight alloys for aircraft. Recently, with the need to conserve energy, people have been seeking higher-strength alloys for cars, railroad equipment, and moving parts of machinery. As long as designers worked at ambient temperatures with the lower-strength, high-ductility materials, they could generally avoid failures by designing with stresses below the yield strength. However, when they used the same methods with the new high-strength, low-ductility materials, there were many catastrophic failures. In addition, the fractures were brittle and did not exhibit even the lower levels of ductility of the tensile test bar. New design criteria have been developed for the safe use of these high-strength alloys on the basis of the concept of *fracture toughness* and equations developed from fracture mechanics.

13.3 What is fracture toughness?

A simple experiment illustrates the physical significance of fracture toughness, the behavior of material with a crack present. If we try to bend and break a glass rod, we find that it takes considerable force. However, if we place a small notch on the surface of the glass rod, we find that the force needed to cause fracture is greatly reduced. When we repeat the experiment on a copper rod, we find that the small notch has no effect on the force required to bend the rod. In fact, the notched copper rod can be bent into a U-shape without fracturing. Figure 13.1 shows specimens on which such tests were

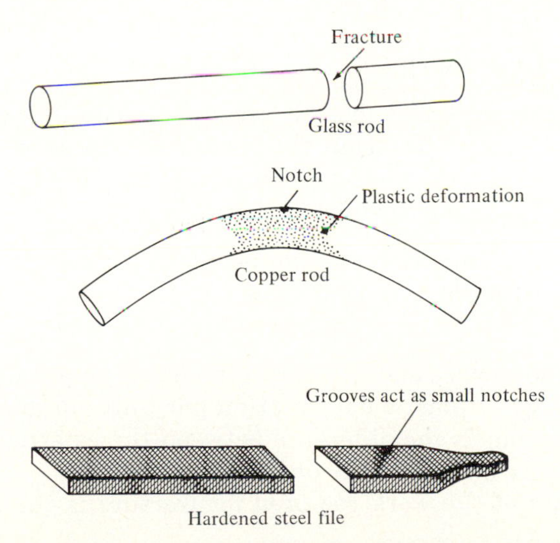

Fig. 13.1 *Results of bending tests of glass and copper rods, of equal diameter, each containing a notch of the same depth. Note that a file makes a poor lever because—even though it is very high in strength—it is low in fracture toughness.*

performed. Note that, in the case of the glass rod, no detectable local plastic deformation accompanied fracture. However, considerable plastic deformation occurred in the copper rod, and yet no fracture occurred. From the results of this simple experiment, we may conclude that glass has a very low fracture toughness and that copper has a much higher fracture toughness. We may generalize these observations and say that when a material fractures without gross local yielding, it is brittle, whereas if considerable plastic deformation occurs before fracture, the material is ductile.

Copper and glass represent two extremes in fracture behavior: extremely ductile and extremely brittle. Common sense tells us to avoid extremely brittle materials in structural applications and to use ductile materials instead. Unfortunately, the more ductile materials are not strong enough for many applications. For example, high-strength alloys must be used where minimizing the weight of the component is critical. As we have mentioned, the aircraft and automotive industries are prime examples of areas where the strength-to-weight ratio of materials is important. Unfortunately, as the strength of alloys increases, the ductility and fracture toughness generally decrease and the susceptibility to brittle fracture increases.

We can understand why some materials have low fracture toughness and some have high fracture toughness if we consider two important aspects of fracture.

1. The response of materials to high local stresses (which can be many times greater than the average stress of the cross section).
2. The role of notches, holes, cracks, and other defects in producing very high local stresses in a part.

13.4 Fracture energy

To fracture a material, work must be performed. This work is required to supply the energy needed to create the fracture surfaces and to plastically deform the material if local yielding occurs prior to fracture. This "energy-balance" approach to fracture can be summarized as:

$$\begin{array}{lcl} \text{Energy input (work)} & & \text{surface energy } (\gamma_s) & & \text{energy of plastic} \\ \text{to produce fracture} & \geqslant & \text{of fracture} & + & \text{deformation } (\gamma_p) \\ & & \text{surfaces} & & \end{array}$$

Here γ_s is the surface energy† per unit surface area (in.-lb/in.^{2}) or J/m^2, and γ_p is the energy of plastic deformation per unit volume (in.-lb/in.^{3}) or J/m^3. The energy input is the difference between the external work supplied and the stored elastic energy at the onset of fracture.

One measure of the work required for fracture is the area under the

---

†γ_s may be thought of as the energy required to break the atomic bonds and create the new surfaces.

true stress–true strain curve. To illustrate this, let us reconsider the fractures of the glass and copper rods. If we performed a tensile test on both types of rods, the result would be curves such as those shown in Fig. 13.2(a) and (b).

For glass, no plastic deformation occurs prior to fracture, and the energy required for fracture is quite small. In fact, a portion of the elastic deformation (the stored elastic energy term) is recovered during fracture.† The total energy consumed, which is considerably less than the area under the true stress–true strain curve, is equal to the increase in surface energy (γ_s × fracture surface area); see Fig. 13.3 on page 538.

For copper, considerable plastic deformation occurs before fracture. The area under the true stress–true strain curve is quite large, and the energy required for fracture is correspondingly large (γ_s × fracture surface area + γ_p × volume of plastic deformation). Generally in metals, $\gamma_p \approx 10^4 \gamma_s$, which explains why the fracture toughness of copper is much greater than the fracture toughness of glass.

13.5 Stress concentration

If no flaws are present, all important structural alloys deform, at least locally, before failure. Yet every year catastrophic failures occur in components in which the operating stress was well below the yield stress. Often such failures occur because the stresses in particular regions of the components have been amplified by the presence of notches, holes, cracks, and other geometrical discontinuities. The following example illustrates the local increase in stress at a defect.

If we drill a hole of radius r through a bar, we reduce the cross section of the bar (Fig. 13.4, page 539), and calculate that the effective stress in the bar at the hole would be

$$S_{\text{nom}} = \frac{P}{(w - 2r)t} \tag{1}$$

†In springback of the fractured parts.

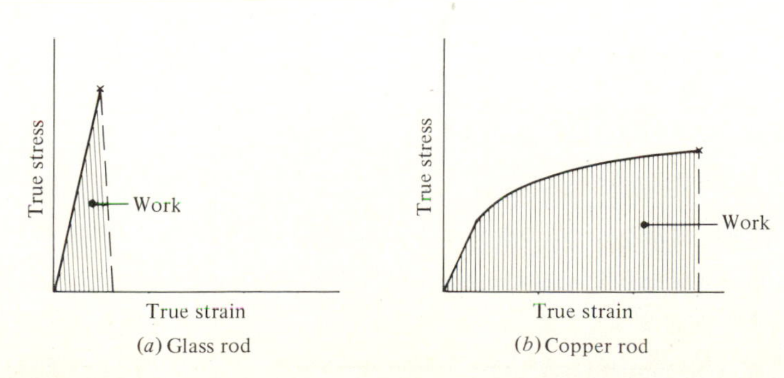

Fig. 13.2 *True-stress–true-strain curves for glass and copper*

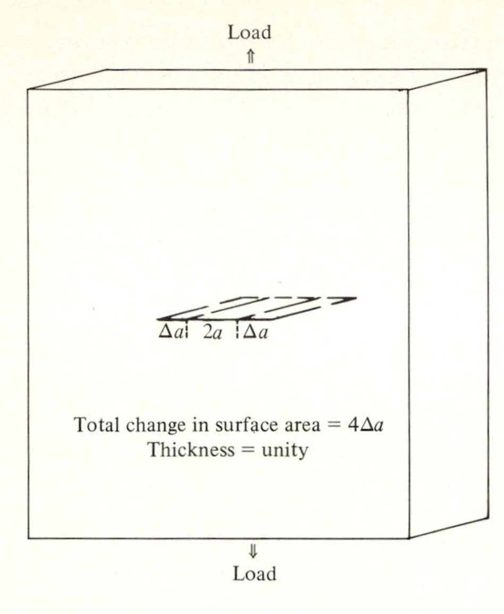

Fig. 13.3 *Extension of a center crack in a large sheet. Extension of the crack by Δa in both directions produces an increase in surface area of 4Δa (two new surfaces, top and bottom, are produced on each end).*

where S_{nom} = nominal stress, P = load, w = width, and t = thickness. (The normal area of the bar would be wt, and we reduce this by $2rt$ to take care of the area of the cross section lost by the hole.)

Calculations such as this have led to the disastrous failure of many highly stressed components, especially in the presence of cyclic stresses. In fact, the stress measured at the sides of the hole is not this nominal stress, but is much higher. This is because the stresses that would normally be supported by the material where the hole was drilled are *concentrated* at the edge of the hole (Fig. 13.4). Furthermore, the smaller the ratio of the radius of the hold to the width of the bar, the *higher* the *stress-concentration factor* K_σ reaching a value of about 2.8 at a radius-to-width ratio of 0.05. Measurements beyond this point are difficult to make.

EXAMPLE 13.1 Given P = 20,000 lb_f, r = 0.1 in., w = 1 in., t = 1 in., calculate the maximum stress in a bar similar to that in Fig. 13.4.

ANSWER

$$S_{nom} = \frac{20{,}000}{(1 - 0.2)1} = 25{,}000 \text{ psi (172.5 MPa)}$$

but since $K_\sigma = 2.7$ $(r/w = 0.1)$,

$$\text{Stress at edges of hole} = 2.7 \times 25{,}000 = 67{,}500 \text{ psi (466 MPa)}$$

(This is why so many fatigue failures start at holes and other "stress raisers" such as fillets, notches, etc.)

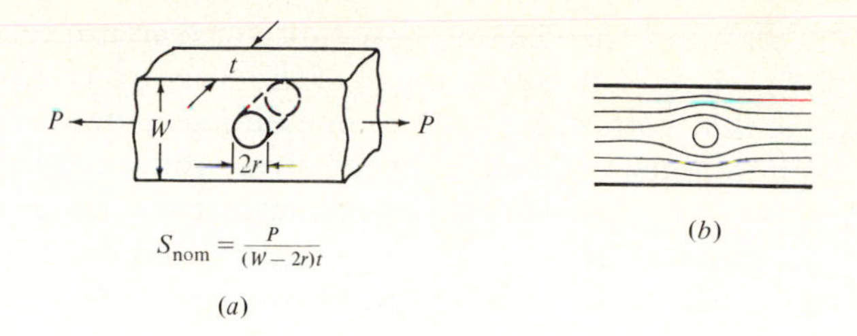

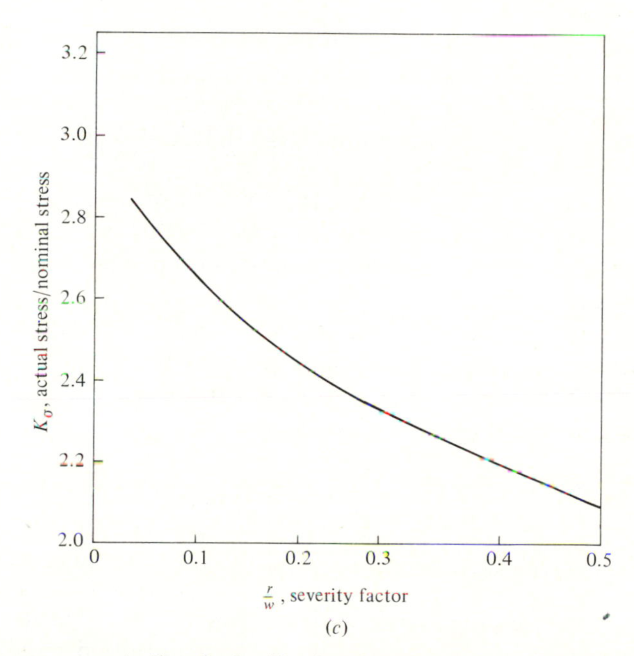

Fig. 13.4 *Stress concentration factor* K_σ *due to a transverse hole in the tension plate member.* (C. Lipson, G.C. Noll, and L.S. Clock, *Stress and Strength of Manufactured Parts*, McGraw-Hill Book Company, New York, 1950)

When a hole or a notch approaches the geometry of a sharp crack, we can no longer readily measure the stress-concentration factor K_σ. (Note that the curve in Fig. 13.4 is not continued for low values of r/w.) For very sharp cracks, we must resort to the methods of fracture mechanics, described in the following section.

13.6 Need for fracture mechanics

We have discussed how stress concentration can lead to very high stresses locally, and how we may expect materials to respond to this region of locally high stresses. It is the response of the material to these high stresses that in large measure determines the fracture resistance or fracture toughness of the material. Let us again refer to our example of the fracture of copper and glass rods. If we assume that both rods are the same diameter and that the notches are identical, then we must conclude that the stress concentration is the same in each rod. For equal applied loads, the stresses at the tip of the notch are the same in both materials. Yet only in the glass rod did the notch cause catastrophic brittle fracture! We may draw two important conclusions from these observations.

1. The ability of a particular *flaw* or stress concentrator to cause catastrophic failure depends on the *fracture toughness*, a material property.
2. The *stress concentration* depends on the *geometry* of the flaw and the *geometry* of the component, but not on the properties of the material.

Therefore, to predict the fracture strength of a component, we must know both the severity of the stress concentration *and* the fracture toughness of the material. This theme will recur throughout our discussions of the analysis of fracture and failure.

With this general background, let us investigate the quantitative evaluation of fracture toughness. We shall begin with the simplest case, glass, in which no plastic deformation is involved, then advance to the metallic materials in which plastic deformation is dominant.

13.7 Griffith theory: fracture mechanics of glass

In the 1920s, A. A. Griffith conducted a series of experiments to determine the fracture strength of glass components that contained small flaws. From his experiments, he concluded that the stress required to cause failure decreased as the size of the flaw increased, even after correcting for the reduced cross section of good material. He also showed that very thin fibers of glass had much higher tensile strength than coarse fibers because the thin fibers were essentially flaw-free. Griffith then developed an expression relating the fracture stress to the flaw size:

$$\sigma_f = \sqrt{\frac{2E\gamma_s}{\pi a}} \tag{2}$$

where σ_f = fracture stress, psi (MPa), $a = \frac{1}{2}$ crack length (this was chosen to simplify the derivation), in. (m), E = Young's modulus of elasticity, psi (MPa), and γ_s = energy required to extend the crack by a unit area, in.-lb/in.^{2} (J/m^2).

For glass, γ_s is simply equal to the surface-energy term we discussed earlier. For a given glass, E and γ_s are constants and Eq. (2) can be simplified to

$$\sigma_f = \frac{C}{\sqrt{\pi a}} \tag{3}$$

Note that the constant C, which has units of psi $\sqrt{\text{in.}}$ (MPa$\sqrt{\text{m}}$), is proportional to the energy required for fracture.

EXAMPLE 13.2 Fused silica has a surface energy of ~ 17.1×10^{-5} in.-lb/in.^{2} (4.32 J/m^2) and an elastic modulus of 10×10^6 psi (70,000 MPa). A large plate of this material is to withstand a nominal stress of 5000 psi (35 MPa). What is the largest flaw that can be tolerated without causing fracture?

ANSWER Rearrange the Griffith equation as:

$$a = \frac{2E\gamma_s}{\pi\sigma_f^2}$$

Substitute the appropriate values and solve for a.

$$a = \frac{2 \times (10 \times 10^6 \text{psi})(17.1 \times 10^{-5} \text{ in.-lb}_f/\text{in}^2)}{3.14(5000 \text{ psi})^2}$$

$$a = 4.4 \times 10^{-5} \text{ in. } (1.1 \times 10^{-4} \text{ cm})$$

Since a is half the crack length, the actual crack length is 8.8×10^{-5} in. (2.2×10^{-4} cm.)

Note: In this example the critical crack length is only 88 μinches (2.2 microns) in length. Such a small flaw is invisible to the naked eye, yet is common in commercial glass. This explains why the actual strength of glass is far below the theoretical strength calculated from the breaking of atom–atom bonds.

13.8 Application of fracture mechanics to metals

With slight modification, the Griffith approach can be applied to the fracture of metals. As we mentioned earlier, the difference in the fracture of metals is the occurrence of plastic deformation at the tip of the propagating crack. The fracture toughness is proportional to the energy consumed in the plastic deformation. Unfortunately, it is hard to measure accurately the energy required for this plastic deformation.

We use a parameter called the *stress-intensity factor,* K_I, to determine the fracture toughness of most materials. The stress-intensity factor, as the name suggests, is a measure of the concentration of stresses at the tip of a sharp crack. It is similar to the stress-concentration factor K_σ, but the two are not equivalent. For a given flawed material, catastrophic fracture occurs when the stress-intensity factor reaches a critical value, denoted as K_{IC}. This critical value is called the fracture toughness of the material.

An expression relating the fracture stress to the fracture toughness and the flaw size is:

$$\sigma_f = \frac{K_{\mathrm{IC}}}{Y\sqrt{\pi a}} \tag{4}$$

where K_{IC} = fracture toughness, psi $\sqrt{\text{in.}}$ (MPa$\sqrt{\text{m}}$), σ_f = nominal stress at fracture, psi (MPa), a = crack length (or one-half crack length, depending on geometry), in. (m), and Y = a dimensionless correction factor that accounts for the geometry of the component containing the flaw. Note that when $Y = 1$, K_{IC} is equivalent to the constant C in Eq. 2. The actual derivation of Eq. (4) is rather complex.†

Confusion sometimes arises as to how the crack length is measured. For an *edge crack,* the crack length is a. For a *center crack,* the crack length is $2a$. Similarly in Eq. (4), for an edge crack the a value is the crack length, and for a center crack the a value is the crack length divided by 2. One accounts for the two cases by using the correction factor Y.

The relationship between stress intensity K_{I} and fracture toughness K_{IC} is similar to the relationship between stress and tensile strength. The stress intensity K_{I} represents the level of "stress" at the tip of a crack in a test specimen or component containing a crack (stress dependent), and the fracture toughness K_{IC} is the highest value of stress intensity that the specimen can withstand without fracturing (material dependent). The units of stress intensity and fracture toughness, psi$\sqrt{\text{in.}}$ (MPa$\sqrt{\text{m}}$), may seem strange. They can best be thought of as a combination of the units of stress (psi) and crack length (in.). When we note the general expression for stress intensity K_{I} as

$$K_{\mathrm{I}} = \sigma\sqrt{\pi a}\, Y \tag{5}$$

where σ is the nominal applied stress and a is the crack length, we see that, for a given crack length, the stress intensity is zero when $\sigma = 0$, and increases linearly with applied stress and the square root of the crack length. As we shall see later, a knowledge of the stress intensities developed at flaws during service is important in predicting failure due to fatigue and stress corrosion cracking.

---

†Discussed further in John Knott, *Fundamentals of Fracture Mechanics,* Halsted Press, New York, 1973.

EXAMPLE 13.3 Consider a plate containing a crack of length $2a$ extending through the thickness. Assume that the length of the crack is negligible compared with the width of the plate. For a fracture toughness of 25 ksi$\sqrt{\text{in.}}$ (27.5 MPa$\sqrt{\text{m}}$) and a yield strength of 65 ksi (448.5 MPa), calculate the fracture stress, σ_f, and the ratio, $\sigma_f/\sigma_{Y.S.}$. Assume $Y = 1$.

ANSWER

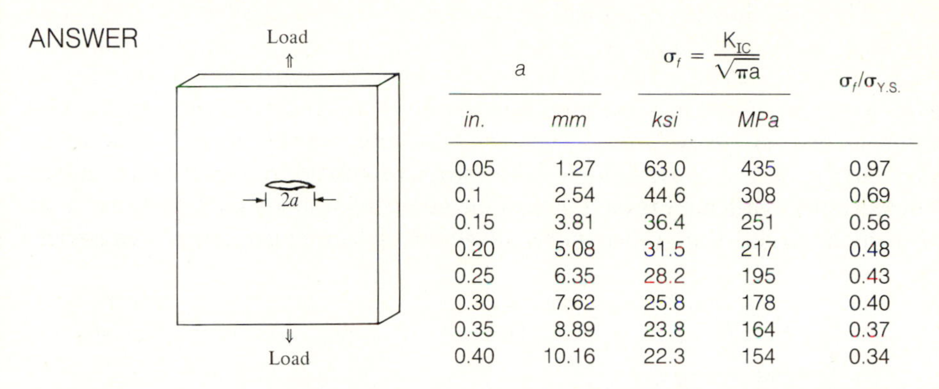

a		$\sigma_f = \frac{K_{IC}}{\sqrt{\pi a}}$		$\sigma_f/\sigma_{Y.S.}$
in.	*mm*	*ksi*	*MPa*	
0.05	1.27	63.0	435	0.97
0.1	2.54	44.6	308	0.69
0.15	3.81	36.4	251	0.56
0.20	5.08	31.5	217	0.48
0.25	6.35	28.2	195	0.43
0.30	7.62	25.8	178	0.40
0.35	8.89	23.8	164	0.37
0.40	10.16	22.3	154	0.34

Note: With extremely small cracks, the general yield strength is reached before catastrophic failure occurs. However, with longer cracks and/or a low-toughness material, catastrophic failure occurs long before the yield strength is reached.

13.9 The role of specimen thickness in fracture toughness

In presenting Eq. (4), we introduced a geometric correction factor Y. One must consider width and thickness of specimen, in addition to the structural behavior and the simple stress-concentration effects. The factor Y includes these effects, except thickness, which is important because a material may show ductile behavior in a thin sheet but may fracture in brittle fashion in a thick plate. We shall see from an example later that this ductile-to-brittle transition takes place only at very thick sections for low-strength, very ductile materials, but at thinner sections for high-strength, low-ductility alloys. To explain this effect, we need to define conditions of plane stress and plane strain.

Plane Stress and Plane Strain

Let us consider the conditions leading to plane stress and plane strain in a given material.

Plane Stress Consider a notch in a thin plate loaded in simple tension in the *Y* direction. Since there is a volume of material missing in the notch, the tensile stress normally borne by this volume is transferred to the notch region. Figure 13.5 shows the higher level of stress in the *Y* direction near the notch, which finally falls off to the average value away from the notch. This is a typical stress-concentration effect.

There is, however, a *local stress* in the *X* direction even though the specimen is stressed only in the *Y* direction overall. To understand this, consider the material in the region of the notch as small tensile specimens [Fig. 13.5(*b*)]. If these specimens were separate and free to deform, the diameter of each would contract as the specimen was stretched. (The ratio of the strain in the *X* direction to the strain produced by the tension in the *Y* direction is called *Poisson's ratio;* for steel it is about 0.3.) However, in the real material the small tensile specimens cannot contract away from

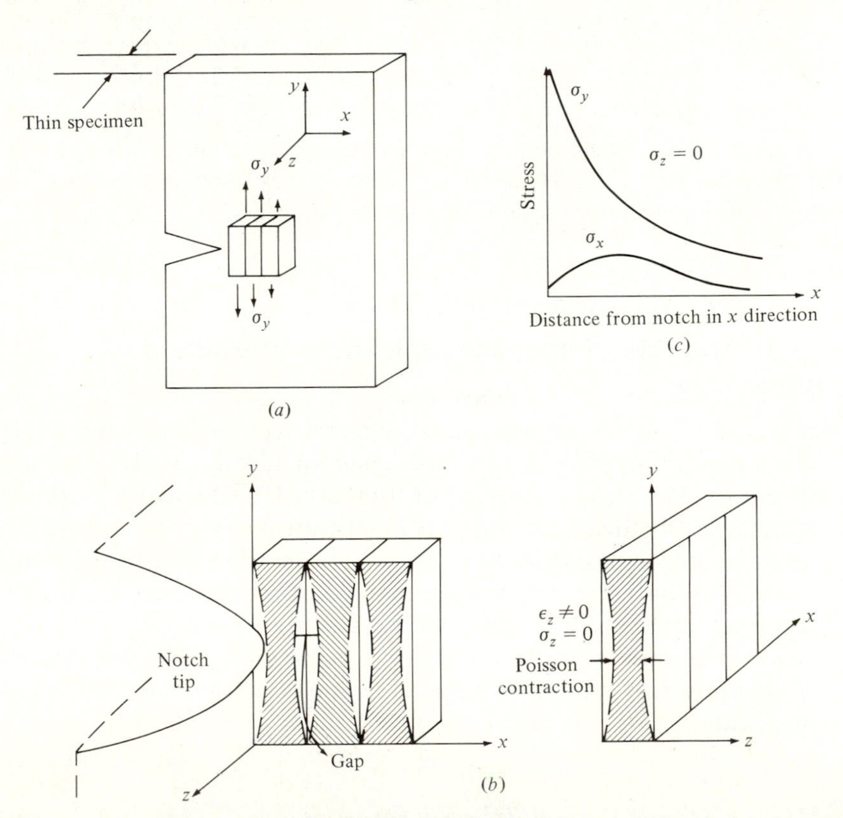

Fig. 13.5 *Plane stress in a thin sheet. (a) Region near notch is divided into small tensile bars. (b) If each bar were free to contract in the x direction, gaps would form between bars. Tensile stresses develop in the x direction to keep such gaps from forming. (c) In the z direction (thickness), the specimen is only one bar thick. Each bar is free to contract in this direction, and thus* $\sigma_z = 0$ *while* $\epsilon_z \neq 0$.

each other because this would leave voids in the material. A local stress, σ_x, results, preventing the contraction in the X direction. This stress is zero at the notch because there is no material in the empty space to react with. It reaches a maximum at a short distance from the notch and then falls off [Fig. 13.5*c*)].

Note particularly that since the specimen is thin (only one bar thick in our example), a similar condition does not develop across the thickness that we call the Z direction in this case. We therefore have finite stresses only in the X and Y directions, and these vectors are in a plane. This is called a condition of *plane stress,* and exists for thin plates. Also, since the thin specimen can contract in the Z direction all across the thickness, the strain in the Z direction is not zero.

As a further explanation, we should point out that a stress arises as a result of a *strain gradient.* In the X direction [Fig. 13.5(*b*)], the gap between our small ligaments becomes large near the crack tip, reaches a maximum, then becomes smaller. In fact, if we move far enough away from the crack tip, there is no tendency for a gap to form. However, the gap represents a strain. Since the width of the gap changes, there is a strain gradient and hence a stress. This stress reaches a maximum and then falls, since the strain gradient does the same [Fig. 13.5(*c*)].

In an analogous fashion [Fig. 13.5(*b*)], the ligament or bar in the Z direction has a width equal to the sheet thickness. A gap does not exist, and there is no stress because there is no strain gradient. In other words, ϵ_Z is constant but uniform across the thickness.

Plane Strain Let us now extend the same reasoning to a thick plate with a notch. We see that the tensile cylinders are restrained in the Z direction as well, a stress develops in the Z direction, and we no longer have a condition of plane stress. However, the material at the center of the notch (thickness) is no longer free to neck inward from the sides, as in the thin specimen, because it is restrained by the mass of material at the sides in the Z direction. Therefore the *strain* in the Z direction is approximately zero. Since the strain is finite in the X and Y directions, these vectors are in a plane. Figure 13.6 (p. 546) sums up this condition, called *plane strain.*

In summary, for this material we have a condition of plane stress in thin specimens and plane strain in thick specimens. The condition of plane strain or plane stress influences the development of plastic deformation at the notch tip. A large plastic zone develops in plane stress and consequently thin specimens have a higher value of fracture toughness. As shown in Fig. 13.7, the minimum fracture toughness at greater thicknesses is called the plane-strain fracture toughness and is denoted as K_{IC}. In thinner sections, the fracture toughness is denoted as K_C†. For plane-stress conditions, the

†This is also called K_Q.

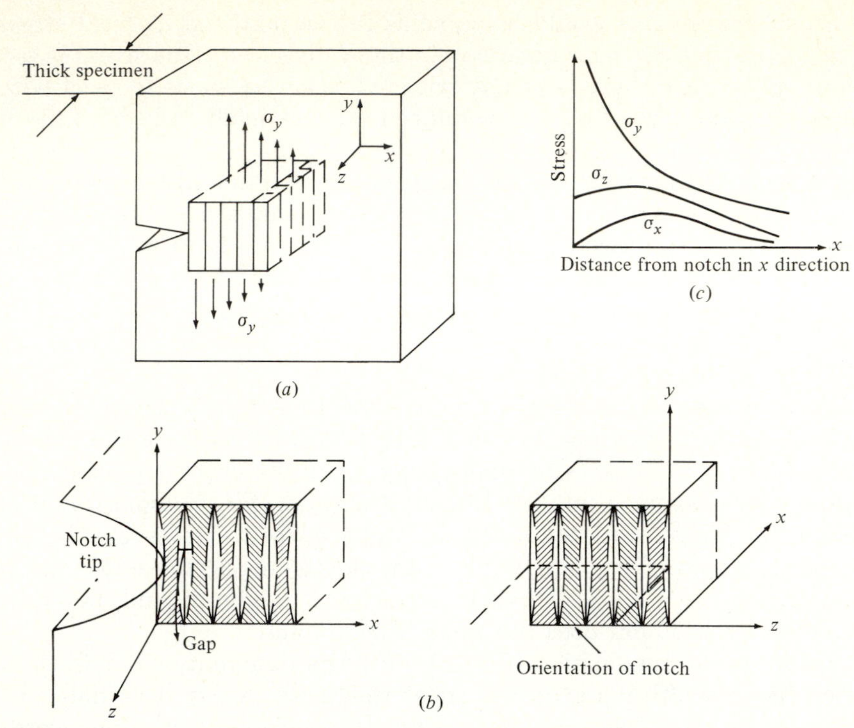

Fig. 13.6 *Plane strain in a thick sheet. (a) Region near notch is divided into small tensile bars, as in Fig. 13.5. However, the sheet is several bars thick. (b) As with plane stress, contraction of the bars in the x direction is prevented by the development of stresses* σ_x. *In the z direction, stresses now develop because the greater number of bars (thickness) prevents the bars from contracting. (c) Schematic of stresses developed in plane strain.*

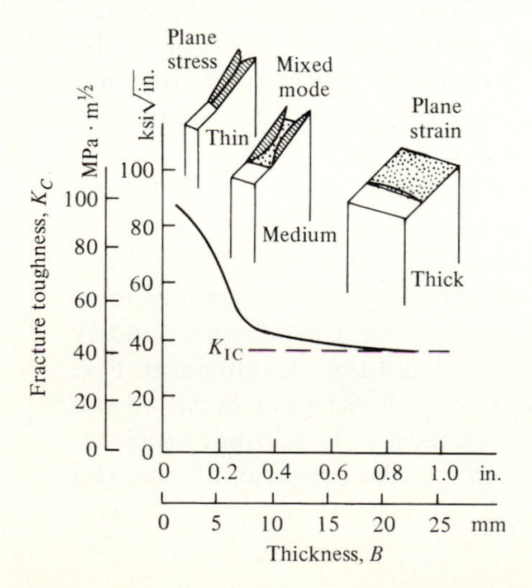

Fig. 13.7 *Variation in fracture toughness with thickness. The appearance of the fracture varies with thickness.* (From *Welding Handbook*, Vol. 1, 7th ed. American Welding Society, 1976)

thickness must be reported with fracture toughness, since K_C depends strongly on thickness. We shall see later that the transition from plane-stress to plane-strain fracture toughness depends on the yield stress, and varies for different materials.

Years of testing have shown that plane-strain conditions generally prevail when

$$\text{Thickness} \geqslant 2.5 \left(\frac{K_{IC}}{\sigma_{Y.S.}}\right)^2 \tag{6}$$

This relation is incorporated in the ASTM standards for fracture toughness testing. If a material of unknown fracture toughness is tested and the resulting fracture toughness does not satisfy the requirements of Eq. (6), the thickness of the test sample must be increased and the test performed again.

EXAMPLE 13.4 What are the minimum thicknesses of specimens from which valid plane-strain fracture-toughness values can be determined for two structural alloys, one with $\sigma_{Y.S.} = 70$ ksi and $K_{IC} = 150$ ksi$\sqrt{\text{in.}}$ and the other with $\sigma_{Y.S.} = 77$ ksi and $K_{IC} = 27$ ksi$\sqrt{\text{in.}}$?

ANSWER For the high-toughness alloy, Eq. (6) becomes

$$\text{Minimum thickness} = 2.5 \left(\frac{K_{IC}}{\sigma_{Y.S.}}\right)^2 = 2.5 \left(\frac{150 \text{ ksi}\sqrt{\text{in.}}}{70 \text{ ksi}}\right)^2 = 11.5 \text{ in. } (292 \text{ mm})$$

For the low-toughness alloy, Eq. (6) becomes

$$\text{Minimum thickness} = 2.5 \left(\frac{K_{IC}}{\sigma_{Y.S.}}\right)^2 = 2.5 \left(\frac{27 \text{ ksi}\sqrt{\text{in.}}}{77 \text{ ksi}}\right)^2 = 0.3 \text{ in. } (7.6 \text{ mm})$$

Note that, with the exception of thick-walled pressure vessels, structural alloys are not commonly used in thicknesses approaching 12 in. The use of K_{IC} for thinner sections would be overly conservative. It would be more appropriate to use K_C for the thickness. However, for the high-strength alloy with relatively low toughness, thicknesses of 0.3 in. and greater are common, and K_{IC} would be the appropriate parameter to use.

13.10 Fracture mechanics and design

The application of fracture mechanics to the design of fracture-resistant structures and to the prediction of catastrophic failure in existing structures depends on the determination of three important parameters.

1. Fracture toughness, K_{IC} or K_C. This parameter is determined from precracked test coupons using procedures set forth by the American Society for Testing and Materials (ASTM E399).
2. Existing crack length, a. This can be determined from rigorous inspection, proof testing, or a conservative estimate if inspection or testing is not possible.
3. Operating stress, σ. The distribution of nominal stress can be determined by stress analysis. This is usually, although not always, a design variable.

The magnitude of any two of these parameters determines the third, through their interdependence as given by Eq. (4). For example, if the design stress is fixed and the choice of materials depends on properties other than toughness and is also fixed, then the size of the critical flaw is automatically established. Proper inspection must ensure that cracks of this size or larger do not exist prior to service, and periodic inspection must assure that such flaws do not develop during service. Actually, as an additional safety factor, one should allow a crack length only a fraction of the critical size.

13.11 Fracture toughness vs. yield strength

In a manner similar to the variation of ductility with yield strength, the fracture toughness of metals and alloys generally decreases as the yield strength increases. Thus, if we use very high-strength materials to minimize the size or weight of components, the size of flaws that can be tolerated becomes increasingly smaller. Figure 13.8 illustrates the variation of toughness with yield strength.

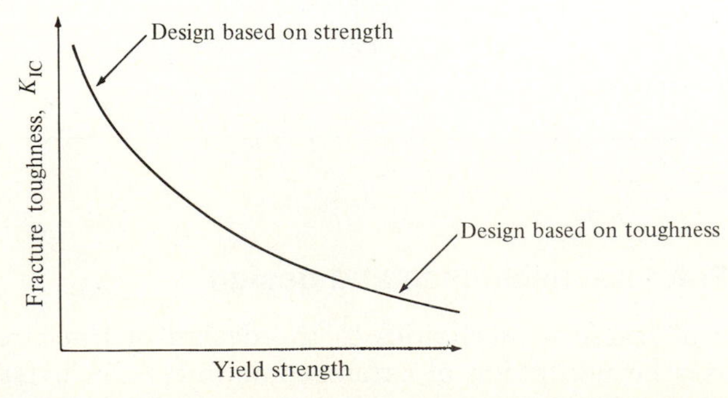

Fig. 13.8 *Variation of fracture toughness with yield strength*

EXAMPLE 13.5 Two rectangular supports carry equal tensile loads. Support A is fabricated from 4340V (Ni, Cr, Mo + Vanadium steel) given an 800°F (427°C) temper, and support B is fabricated from 4340V steel given a 500°F (260°C) temper. The thickness of each member has been adjusted so that each supports a stress σ_D equal to 60% of the respective yield strength. For each support, what is the longest edge crack (see figure) that can be tolerated without causing catastrophic failure?

ANSWER

Step I. Determine the appropriate expression for the stress-intensity factor for the geometry in question.

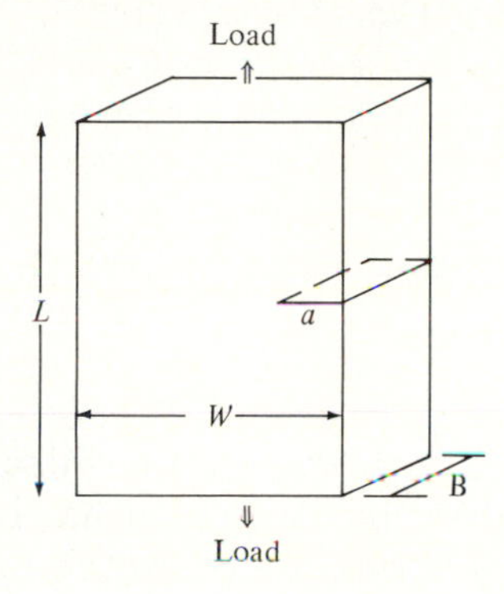

The stress intensity for the configuration of the support is given by†

$$K_I = \sigma\sqrt{\pi a}\, Y\left(\frac{a}{w}\right) \qquad \text{where } Y\left(\frac{a}{w}\right)$$

is the geometric correction factor that takes into account the finite width of the plate. It is calculated as

$$Y\left(\frac{a}{w}\right) = 0.265\left(1 - \frac{a}{w}\right)^4 + \frac{0.857 + 0.265\left(\frac{a}{w}\right)}{\left(1 - \frac{a}{w}\right)^{3/2}}$$

[*NOTE:* In this discussion, $Y\,(a/w)$ is not intended to represent an algebraic form, but rather Y is a function of a/w.]

Step II. Determine the relevant dimensions of the supports and the appropriate material properties for use in the calculations.

Support Dimensions	*Support A*	*Support B*
L	16 in. (.406 m)	16 in. (.406 m)
W	4 in. (.102 m)	4 in. (.102 m)
B	0.7 in. (.018 m)	0.59 in. (.015 m)

Material Properties	*4340V 800°F (427°C)*	*4340V 500°F (260°C)*
$\sigma_{Y.S.}$	191 ksi	228 ksi
σ_D	115 ksi	137 ksi
K_{IC}	97 ksi$\sqrt{\text{in.}}$	51 ksi$\sqrt{\text{in.}}$

(ksi × 6.9 = MPa; ksi$\sqrt{\text{in.}}$ × 1.1 = MPa$\sqrt{\text{m}}$)

†Taken from *The Stress Analysis of Cracks Handbook,* Hiroshi Tada, Del Research Corporation, Hellertown, Pa. 1973.

Step III. Calculate the critical flaw sizes.
At fracture the governing equation is

$$\frac{K_{IC}}{\sigma_D} = \sqrt{\pi a_c}\left[0.265\left(1 - \frac{a_c}{w}\right)^4 + \frac{0.857 + 0.265\frac{a_c}{w}}{\left(1 - \frac{a_c}{w}\right)^{3/2}}\right]$$

where a_c is the critical flaw size and K_{IC} is the critical stress intensity. Substitute the appropriate values of K_{IC} and σ_D for each support and solve for a_c by iteration (trial and error). Calculated critical flaw sizes:

Support A: a_c = 0.171 in. (4.34 mm) Support B: a_c = 0.035 in. (0.89 mm)

Note: Even though both members are designed with the same safety factor in terms of the yield stress, the size of the critical flaw is less than one-fifth as large, and hence much harder to detect in the support fabricated from the higher-strength steel.

13.12 Delayed failure: fatigue and stress corrosion cracking

Even if we design for stress levels in accordance with the relationships we have discussed, catastrophic failure may occur after the component has been in service for some time. Cyclic stresses or corrosive conditions are the cause of failure in many cases. In both these situations small flaws are initiated and grow until the critical crack size is reached; then rapid failure transpires. These delayed failures are more dangerous than rapid failures because consumers have false confidence in a component that has been in service for months or years.

13.13 Initiation of fatigue cracks

In Chapter 3 we discussed fatigue in terms of the *S-N* curve, which is the overall summary of the cycles for failure at different stresses for a given material. The new insight into the relation between flaws and fracture toughness has made possible analysis of fatigue failure in more detail.

We may begin by dividing the region of the *S-N* curve prior to failure into two zones that have been found by experimental observations with the scanning electron microscope—initiation of cracks and growth of cracks (Fig. 13.9). Cracks may initiate from essentially flaw-free regions or from existing defects such as inclusions. This variation in initiation is one of the factors leading to the scatter in the data for failure.

In the case of a defect-free region, cracks initiate from minute stress concentrators caused by localized plastic deformation on particular slip bands.

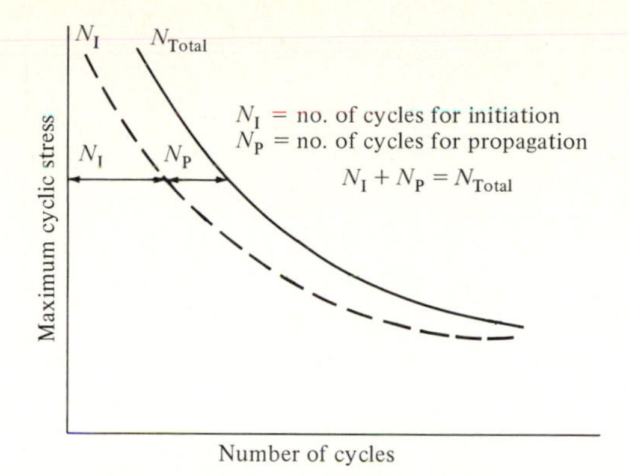

Fig. 13.9 *Schematic of* S-N *curve, showing that the total fatigue life consists of two phases: crack initiation and crack propagation. Note that, at high stresses,* $N_I < N_P$; *at low stresses,* $N_P < N_I$.

Figure 13.10*a* shows an example of such localized plastic flow. After repeated cycling, a small crack develops at the intersections of these localized deformation bands with the surface of the specimen. Figure 13.10*b* shows the early stages of such crack development. Even minute surface scratches may eliminate or greatly shorten the crack-initiation phase of fatigue failure. This explains the strong dependence of total fatigue life on surface finish.

The presence of internal defects shortens the time required for initiation of cracks. Figure 13.11 (p. 552) shows fatigue cracks arising from a nonmetallic inclusion and from a gas pore in a nickel-base alloy tested in fatigue.

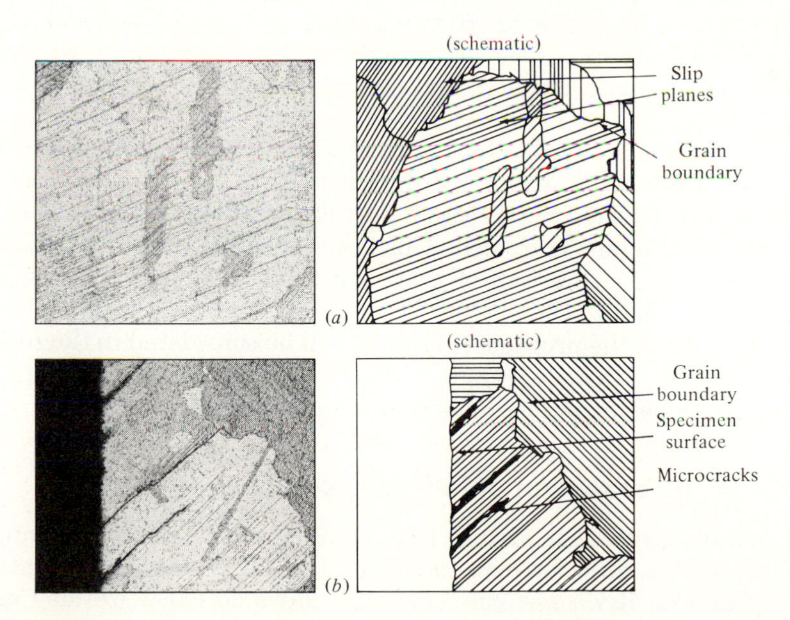

Fig. 13.10 *(a) Regions of intense slip that developed during fatigue of a nickel alloy. (b) Crack forming at intersection of surface with slip bands.* (Courtesy of J.M. Hyzak, Sandia National Laboratory)

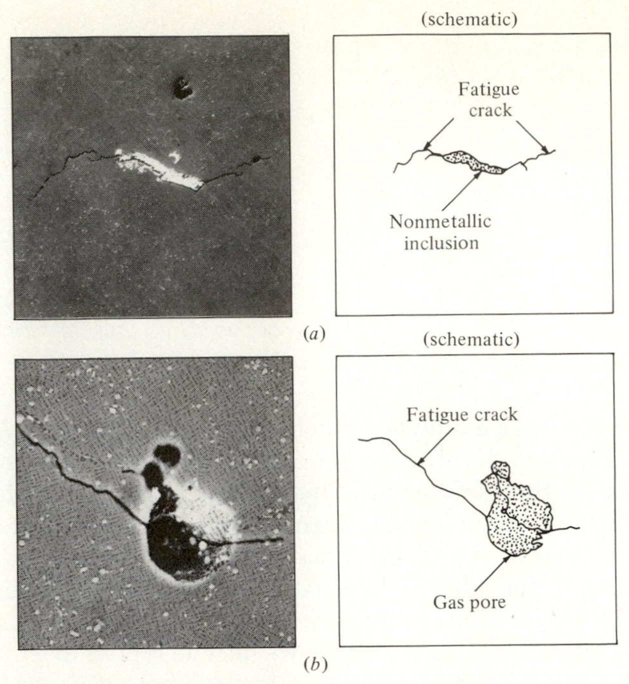

Fig. 13.11 *(a) Fatigue crack initiating from a nonmetallic inclusion in a nickel alloy. (b) Fatigue crack initiating from a gas pore in a nickel alloy.* (Courtesy of J.M. Hyzak, Sandia National Laboratory)

In the early stages of crack propagation, the direction of propagation depends on orientation of the grain. Cracks proceed on specific crystallographic planes and thus the direction may change abruptly, as in Fig. 13.12, when a grain boundary is crossed. This is called stage I crack growth. As the crack lengthens, the direction of growth becomes independent of grain orientation and normal to the applied stress. This phase of growth, called stage II, involves most of the propagation life. It will be considered in the remainder of our discussion of fatigue. Figure 13.13 is a schematic illustrating the various phenomena of initiation and growth of fatigue cracks.

13.14 Propagation of fatigue cracks

It is often advantageous to assume that small flaws do exist in the component. This bypasses the need for information on the early stages of initiation and growth of cracks. If we assume that small flaws do exist, we may estimate fatigue life on the basis of the measured rates of crack propagation. Such an approach may be conservative if in fact no flaws exist, but it eliminates the danger of premature failure if flaws are present initially and are not detected.

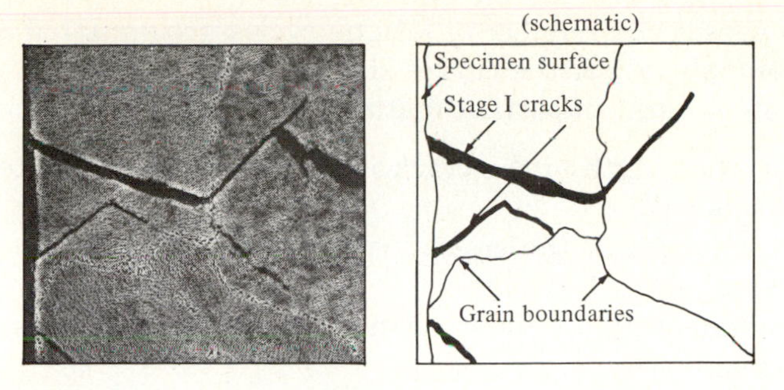

Fig. 13.12 *Propagation of a stage I fatigue crack.* (Courtesy of J.M. Hyzak, Sandia National Laboratory)

This method, as we shall see, also has the advantage that fracture mechanics can be used in a quantitative manner in predicting the life of the component.

Let us consider the growth of fatigue cracks in two identical specimens subjected to identical fatigue loading. However, let us assume that the initial length of the crack in specimen I, a_I, is greater than the initial length of the crack in specimen II, a_{II}. During the course of fatigue tests, we may stop the tests periodically and measure the lengths of the cracks. If we plot the lengths

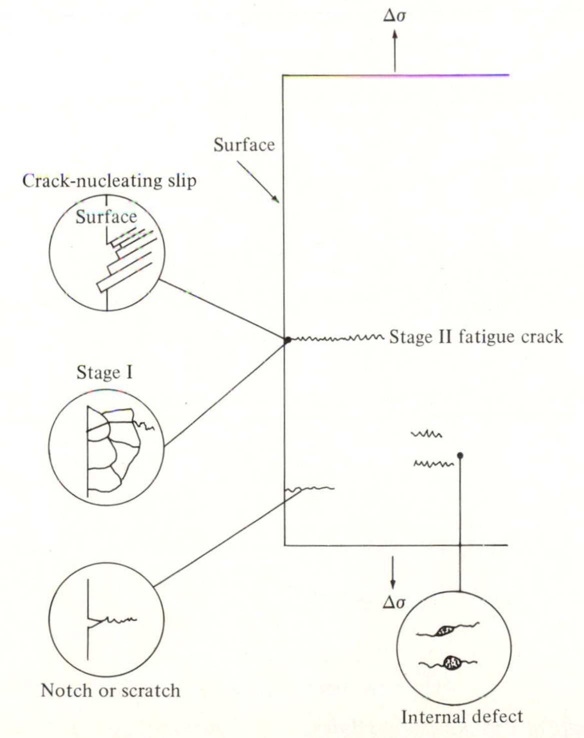

Fig. 13.13 *Schematic of the various ways that fatigue cracks initiate and grow*

of the cracks versus the number of fatigue cycles accumulated at the time of each measurement, we get the curves shown in Fig. 13.14.

Also note several important features of these curves.

- When the length of the crack is small, the growth rate of the crack $\Delta a/\Delta N$ is also small.
- As the length of the crack increases, the growth rate of the crack also increases.
- Under identical cyclic stressing, larger initial cracks propagate to failure in fewer cycles.
- If the cyclic stressing is the same and if the geometry of the specimen and the crack are the same, the length of the crack at failure will be the same, regardless of the starting length of the crack and the number of cycles to failure.

For many years it was assumed that some function of the stress controlled the rate of propagation of cracks. It is now generally accepted that the rate of propagation of cracks is a function of the *stress intensity factor*, K_I. For many engineering alloys, the rate of propagation of cracks da/dN can be expressed as a function of the *range of the stress intensity* ΔK_I that the crack experiences during the stress cycle.

$$\frac{da}{dN} = C\,\Delta K_I^{\,m} \tag{7}$$

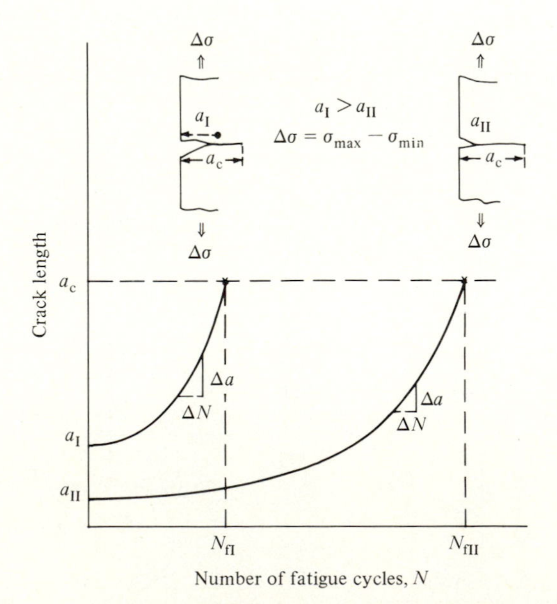

Fig. 13.14 *Growth of cracks during fatigue for two identical specimens that are under identical cyclic loading, but with cracks that were initially of different lengths*

C and m are constants that depend only on the material. We can calculate ΔK_I as follows: At the maximum stress, σ_{max}, of the fatigue cycle, we set up the equation for K_I using the formula

$$K_{I\,max} = \sigma_{max} \sqrt{\pi a}\, Y \qquad \text{(See Eq. 4)}$$

and at the minimum stress

$$K_{I\,min} = \sigma_{min} \sqrt{\pi a}\, Y$$

Then

$$\Delta K_I = K_{I\,max} - K_{I\,min} = (\sigma_{max} - \sigma_{min}) \sqrt{\pi a}\, Y \qquad (8)$$

Note that ΔK_I is not a constant, but varies as the crack length a changes.

Figure 13.15 shows a schematic representation of rate of crack growth versus stress intensity range. There are three regions on this curve that generally appear in actual data. In region I, at low values of ΔK_I, the rate of growth increases rapidly with a small increase in ΔK_I. In region II, at intermediate values of ΔK_I, Eq. (7) is followed. The curve is linear when plotted on log-log scales, and has a slope of m. In region III, where ΔK_I approaches K_{IC}, da/dN increases sharply with increasing ΔK_I. In many materials the majority of life is spent in region II, and Eq. (7) can be used to predict fatigue life.

In region I a limiting value of ΔK_I is reached below which no measurable growth of cracks takes place. This limiting value ΔK_{th} is called the *threshold stress intensity range*. Its significance is similar to that of the fatigue limit determined from an S-N curve. If the combination of starting

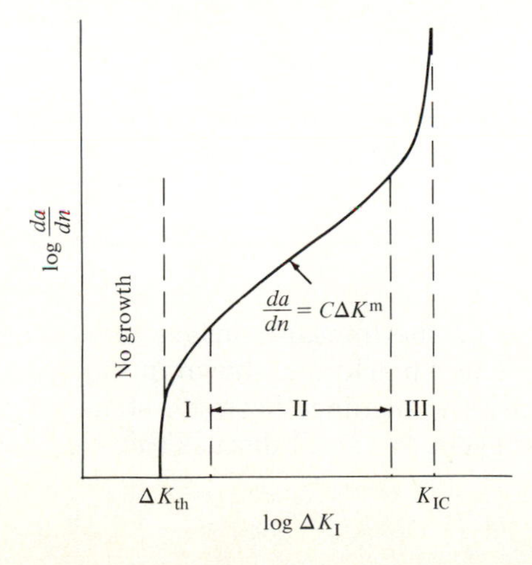

Fig. 13.15 *Growth rate of fatigue cracks as a function of the range of stress intensity,* ΔK_I

size of the crack and applied cyclic stress is such that the stress intensity range is below ΔK_{th}, then no growth of cracks should occur and failure of the component by fatigue will be not a problem. This assumes, of course, that no larger flaws are introduced during service.

In Example 13.7 we consider an example of how we can use data on crack growth to predict the safe lifetime of a component operating under cyclic stress. First, however, let us see how ΔK_I is determined.

EXAMPLE 13.6 A metal strip (4 in. wide and 0.2 in. thick) is subjected to a cyclic load ranging from 6000 to 43,000 lb. There is a crack in the center of the strip that extends through the thickness. For half-crack lengths of $a = 0.1$ and 0.4 inch, calculate ΔK_I. (Recall that a is one-half of the crack length for a center crack.)

ANSWER For this geometry, $K_I = \sigma\sqrt{\pi a}\, Y(a/w)$, and in this example we assume that $Y(a/w) = 1$.

Compute stresses:

$$\sigma_{max} = \frac{43{,}000 \text{ lb}}{0.2 \times 4} = 53{,}750 \text{ psi (371 MPa)}$$

$$\sigma_{min} = \frac{6000 \text{ lb}}{0.2 \times 4} = 7500 \text{ psi (51.8 MPa)}$$

Compute stress intensities:

$$K_{I\,max} = \sigma_{max}\sqrt{\pi a} = 53{,}750\sqrt{\pi(0.1)} \approx 30.1 \text{ ksi}\sqrt{\text{in.}}\ (33.1 \text{ MPa}\sqrt{\text{m}})$$

$$K_{I\,min} = \sigma_{min}\sqrt{\pi a} = 7500\sqrt{\pi(0.1)} = 4.2 \text{ ksi}\sqrt{\text{in.}}\ (4.6 \text{ MPa}\sqrt{\text{m}})$$

$$\Delta K_{I\,max} = 30.1 \text{ ksi}\sqrt{\text{in.}} - 4.2 \text{ ksi}\sqrt{\text{in.}} = 25.9 \text{ ksi}\sqrt{\text{in.}}\ (28.5 \text{ MPa}\sqrt{\text{m}})$$

Repeat the calculations for $a = 0.4$:

$$K_{I\,max} = 53{,}750\sqrt{\pi(0.4)} = 60.3 \text{ ksi}\sqrt{\text{in.}}\ (66.3 \text{ MPa}\sqrt{\text{m}})$$

$$K_{I\,min} = 7500\sqrt{\pi(0.4)} = 8.4 \text{ ksi}\sqrt{\text{in.}}\ (9.2 \text{ MPa}\sqrt{\text{m}})$$

$$\Delta K_I = 60.3 \text{ ksi}\sqrt{\text{in.}} - 8.4 \text{ ksi}\sqrt{\text{in.}} = 51.9 \text{ ksi}\sqrt{\text{in.}}\ (57.0 \text{ MPa}\sqrt{\text{m}})$$

Note that for a constant load range the stress intensity range increases as the crack length increases. Thus the crack extension per cycle increases as the crack grows.

For many materials distinct markings on the fracture surface indicate the occurrence of stage II crack growth. These markings, shown in Fig. 13.16, are called *fatigue striations*. Generally the distance between striations represents the crack growth per stress cycle. We shall discuss this in greater detail later.

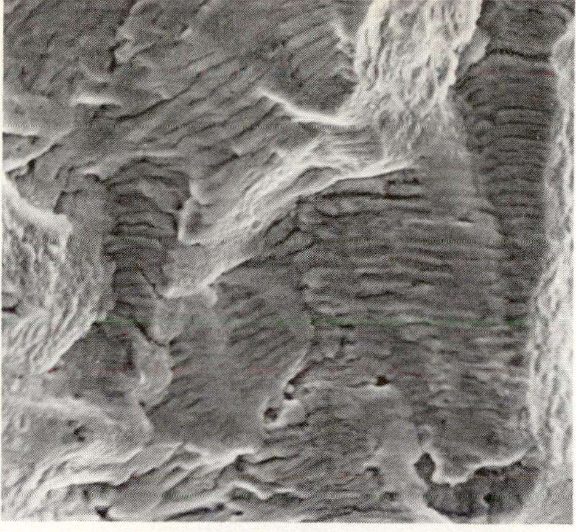

Fig. 13.16 *Fatigue striations observed on the fatigue fracture surface of an iron-nickel-chromium alloy; 1200×*

EXAMPLE 13.7 A large panel with a central crack through the thickness with half length a = 0.10 in. is cyclically stressed between 14,000 psi and zero stress. The panel is 30 in. wide and 0.5 in. thick and is fabricated from a material with a fracture toughness of 25,000 psi$\sqrt{\text{in.}}$ Growth of fatigue cracks follows the law [Eq. (7)]:

$$\frac{da}{dN} = C\,\Delta K_{\mathrm{I}}^{m}$$

with $C = 1.8 \times 10^{-18}$ in./(cycle · psi$\sqrt{\text{in.}}$) and $m = 3.0$. Estimate the number of fatigue cycles it takes to cause failure of the panel.

ANSWER Calculate the length of the crack at which failure occurs. For the "center-cracked" panel,

$$K_{\mathrm{I}} = \sigma\sqrt{\pi a}\,Y(a/w)$$

Since $a << w$, we shall set $Y(a/w) = 1$.

$$a_c = \left(\frac{K_{\mathrm{IC}}}{\sigma_f}\right)^2 \frac{1}{\pi} = \left(\frac{25{,}000 \text{ psi}\sqrt{\text{in.}}}{14{,}000 \text{ psi}}\right)^2 \frac{1}{\pi} = 1.0 \text{ inch (25.4 mm)}$$

Note: Keep in mind that the values for the crack lengths used here are one-half the total length in the panel. At failure we would see a fatigue crack of 2 in. total length.

We shall now estimate the number of cycles it takes to extend the crack by a small amount, then repeat this operation until the crack reaches the critical length and sum the cycles to failure.

We choose an increment of growth of 0.05 in. For extension from 0.10 in. to 0.15 in., we use the average crack length of 0.125 in. to compute an average growth rate over this interval. From this we compute the number of cycles required. A sample calculation follows.†

$$a_0 = 0.10 \text{ in.} \qquad a_1 = 0.15 \text{ in.} \qquad a_{av} = 0.125 \text{ in.}$$

$$\Delta K_I = \Delta\sigma\sqrt{\pi a_{av}} = 14{,}000 \text{ psi}\sqrt{\pi(0.125)} = 8770 \text{ psi}\sqrt{\text{in.}}\ (9.65 \text{ MPa}\sqrt{\text{m}})$$

Next, compute the growth rate for this ΔK_I:

$$\frac{da}{dN} = C\,\Delta K_I^m = (1.8 \times 10^{-18})(8770)^3$$

$$= 1.21 \times 10^{-6} \text{ in./cycle } (30.7 \times 10^{-6} \text{ mm/cycle})$$

Compute the number of cycles needed to extend the crack by $\Delta a = 0.05$ in.

$$N = \frac{\Delta a}{da/dN} = \frac{0.05 \text{ in.}}{1.21 \times 10^{-6} \text{ in./cycle}} \qquad \text{(First interval) } N = 41{,}100$$

Repeat the above procedure for each interval of growth. The results are:

a *(in.)*	a_{av} *(in.)*	*N (cycles)*	ΣN *(cycles)*
0.10	0	0	0
0.15	0.125	41,100	41,100
0.20	0.175	24,500	65,600
0.25	0.225	16,900	82,500
0.30	0.275	12,600	95,100
0.35	0.325	9,800	104,900
0.40	0.375	7,900	112,800
0.45	0.425	6,500	119,300
0.50	0.475	5,500	124,800
0.55	0.525	4,800	129,600
0.60	0.575	4,200	133,800
0.65	0.625	3,700	137,500
0.70	0.675	3,300	140,800
0.75	0.725	3,000	143,800
0.80	0.775	2,700	146,500
0.85	0.825	2,400	148,900
0.90	0.875	2,200	151,100
0.95	0.925	2,000	153,100
1.0 (a_c)	0.975	1,800	154,900

†Since $Y = 1$, we may use calculus to solve this problem in closed form.

$$\frac{da}{dN} = C\,\Delta K^m \qquad \frac{da}{C\,\Delta K^m} = dN$$

$$N = \int_{a_0}^{a_f} \frac{1}{C}(\Delta\sigma\sqrt{\pi a})^{-3}\,da = \frac{1}{C(\Delta\sigma)^3\pi^{3/2}}\int_{a0}^{af} a^{-3/2}\,da$$

$$N = \left(\frac{-2}{C(\Delta\sigma)^3\pi^{3/2}}\right)\left(a^{-1/2}\Big|_{a_0}^{a_f}\right) = \frac{2}{C(\Delta\sigma)^3\pi^{3/2}}\left[\frac{1}{\sqrt{a_0}} - \frac{1}{\sqrt{a_f}}\right]$$

$$N = \frac{2}{(1.8 \times 10^{-18})(14{,}000)^3 \pi^{3/2}} \left(\frac{1}{\sqrt{0.1}} - \frac{1}{\sqrt{1}} \right) = 1.57 \times 10^5 \text{ cycles}$$

Note: The actual and approximated answers differ. If, however, we use smaller increments of crack growth in our calculations, the answers agree more closely. For a more complete description of fatigue, consult *Deformation and Fracture Mechanics of Engineering Materials,* by R.W. Hertzberg, published by John Wiley in New York in 1976; also *Fracture and Fatigue Control in Structures,* by S.T. Rolfe and J.M. Barson, published by Prentice-Hall, Englewood Cliffs, N.J., 1977.

13.15 Stress corrosion cracking

As we saw in Chapter 12, the environments in which most engineering materials are used are seldom inert. In many cases environmental attack can limit the useful lifetime of components. Corrosion is the most common result of environmental attack. But there are more subtle interactions that may limit service life even if general corrosion problems have been eliminated. One such particularly insidious interaction is stress corrosion cracking (SCC).

Stress corrosion cracking, as the name suggests, is the advance of a crack in a material subjected to stress in the presence of a gaseous or liquid environment. SCC may be particularly difficult to detect because:

1. Environments that are only mildly corrosive to the material may cause severe SCC.
2. The required concentration of the harmful component in the environment may be extremely small and its presence difficult to detect.
3. The attack may be highly localized as one or a number of small cracks that may propagate undetected to failure.
4. Residual stresses in components are often great enough to cause stress corrosion cracking even in the absence of applied stresses.

As with fatigue cracking, there are two approaches to the measurement of lifetimes in components experiencing SCC. One approach involves determining the time required to cause failure of smooth uncracked specimens subjected to stresses in the environment in question. Figure 13.17 shows that, as the applied stress level decreases, the time to failure from SCC increases. If the stress is low enough, time to failure becomes excessively long and an apparent threshold stress below which SCC does not occur can be defined.

In such a test the time to failure necessarily involves the time required to initiate the crack by localized chemical attack and the time required to propagate the crack to failure. As in fatigue, if small cracks are already present, the time to failure may be much shorter than that predicted from tests on smooth specimens.

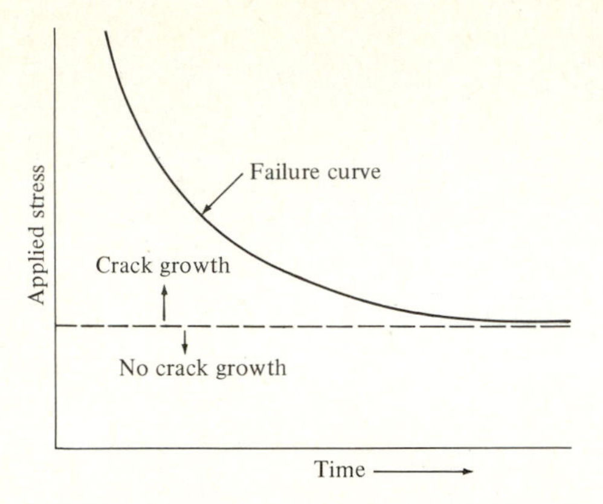

Fig. 13.17 *Variation of time to failure by stress corrosion cracking as a function of applied stress*

If we measure the rate of growth of stress corrosion cracks for a particular alloy–environment combination, we find, just as in fatigue, that the rate is determined by the stress intensity factor K_I. Figure 13.18 illustrates schematically how the growth rate of stress corrosion cracks varies with K_I. Note that, since loading is static rather than cyclic, K_I is used rather than ΔK_I.

This figure has three regions of interest. In region I the rate of crack growth increases with increasing K_I. In region II the rate of crack advance is independent of the intensity of the stress. In region III, further increases in K_I cause a rapid increase in rate of growth of the crack.

As in fatigue, a threshold value of stress intensity exists. Below this threshold value—designated K_{ISCC}—no growth of cracks from stress corrosion occurs. Values of K_{ISCC}, which can be determined from laboratory testing, are an indication of the relative susceptibility of a material to SCC. High values of K_{ISCC} are desired.

Figure 13.19 shows an example of cracking due to the combination of corrosive environment and high residual stresses. The cup was made by

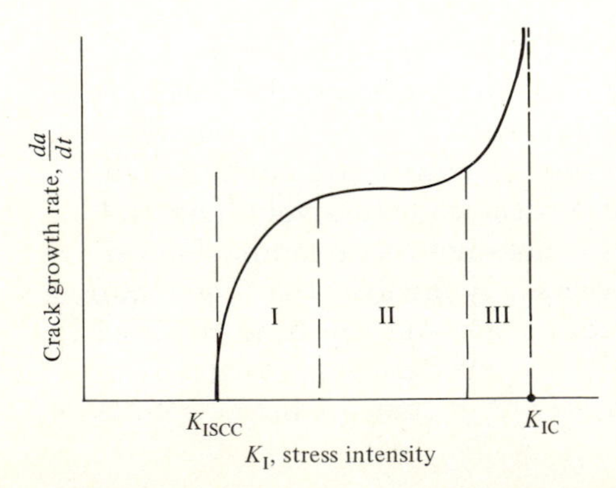

Fig. 13.18 *Variation of rate of growth of stress corrosion crack with applied stress intensity, K_I*

Fig. 13.19 *Deep-drawn cup that cracked while sitting unused on a shelf. Atmospheric contaminants, combined with high residual tensile stress, can cause failure by stress corrosion cracking.* (Photo courtesy D. Meuleman)

deep-drawing stainless steel. After drawing, the cup, which contained no cracks, was placed on a laboratory shelf. When it was inspected the following day, the cracks shown in Fig. 13.19 were observed. Apparently stress corrosion resulted from the laboratory air or from contaminants in the lubricating oil. In many cases drawn parts must be annealed immediately after drawing to remove residual stresses and to avoid the start of stress-corrosion cracking by immediate reaction with the atmosphere.

As an example of resistance to stress corrosion, Table 13.1 shows the considerable variation in crack susceptibility among copper alloys exposed to an ammonia-bearing atmosphere.

Table 13.1 RELATIVE SUSCEPTIBILITY OF COPPER ALLOYS TO AMMONIACAL SCC*

Very low susceptibility	Cupronickels, tough pitch copper, silicon bronze
Low susceptibility	Phosphorized copper
Intermediate susceptibility	Brasses containing less than 20% Zn, such as red brass, commercial bronze, aluminum bronze, nickel silver, phosphor bronze, and gilding metal
High susceptibility	Brass containing over 20% Zn, with or without small amounts of Pb, Sn, Mn, or Al (such as leaded brass, Naval brass, Admiralty brass, manganese bronze, or aluminum brass). The higher the Zn, the higher the susceptibility.

*From B.F. Brown, *Stress Corrosion Cracking Control Measures,* National Bureau of Standards Monograph No. 156, 1977.

Experimental Methods for Examining Failed Parts

13.16 General approach, nondestructive methods

It is essential to divide experimental methods into two groups—nondestructive and destructive. In many cases the investigator is limited to nondestructive methods, although it can be reasoned that the careful cutting of a specimen for microexamination and for exhibit is not really destructive, compared with grinding up a sample for chemical analysis.

Nondestructive Methods The budding metallurgist is often too eager to cut up the part and etch a microspecimen, as illustrated by the following case. Our laboratory was notified that a failed gear was being sent in for examination. After waiting several months, we checked with a colleague who was in charge of the engineering design laboratory next door to see if the part had gone astray. We received a pretty blunt answer: "Sure, the gear came in a month ago. We checked the dimensions against the blueprint and found some bad conflicts with the mating gear. It's a good thing it didn't go to you or you would have cut it up right away and we never would have found the trouble!"

This story illustrates that, before proceeding with the actual examination, one should obtain all relevant background material, such as drawings and material specifications, conditions of service, history of similar parts (failures and successes), performance data, and design calculations. Also, one should complete all possible nondestructive examinations. An equally important precaution is to repress the urge that possesses laypeople and engineers alike—to force the two pieces of the fracture together and exclaim, "They fit!" Imagine the effect on the clear scanning electron photomicrographs shown throughout the text of gashes looking like Alpine landslides caused by mashing the fractures together. Some investigators have even interpreted these gashes as fatigue striations!

The first step in nondestructive examination is careful viewing in a good light with the unaided eye, then with a 10× hand lens or low-power binocular microscope. This examination is often adequate to distinguish between different types of fracture, such as between a ductile tensile failure and fatigue failure. This is also an essential step in beginning to locate the point of initiation of the fracture and in selecting specimens for fractographic examination with the scanning electron microscope.

The SEM† (scanning electron microscope) is of great value for investigating the fracture in detail. If the part cannot be placed in the vacuum chamber of the SEM or a specimen cut, another method called the *replica technique* may be used. One places a few drops of acetone (generally noncor-

---

†The principles of the SEM and the TEM are described in standard texts on physics.

rosive) on the surface and presses a strip of cellular acetate plastic against the surface. After a few seconds, one strips off the film. It bears an accurate replica of the fracture. For examination in the transmission electron microscope (TEM),† a thin layer of metal such as gold is vapor-deposited on the replica. This does not alter the detail, and excellent photomicrographs are obtained.

Other nondestructive techniques include magnaflux, use of penetrating dye, electrical tests, and radiography. For magnaflux examination, the component must be ferromagnetic. The part is magnetized and a colored magnetic powder is shaken on the surface. High concentrations of powder are attracted to cracks, just as iron filings are attracted to the gap in a horseshoe magnet. The part should be magnetized in several directions to provide favorable conditions for revealing all cracks. This method is called *dry magnafluxing*. Some people use a more sensitive method employing a fluorescent magnetic powder contained in a low-viscosity oil. In this case, when the part is viewed under ultraviolet light, the fluorescent powder delineates the defects.

Inspection with a penetrating dye is used for nonmagnetic materials. First the part is sprayed with a red-colored oil that penetrates defects. Next the oil is washed from the surface, leaving a residue in the cracks. Finally, a white coating is sprayed over the surface. The oil in the cracks gradually oozes through the white surface, delineating the defects.

Electrical and sonic methods include a variety of techniques. One technique is ultrasonic testing. Sound waves are transmitted into the part from a vibrating crystal pressed against the part. These sound waves are then received and displayed on an oscilloscope. One can determine discontinuities and variations in structure from the oscilloscope trace or evaluate them quantitatively and display them as numerical values. Other methods involve evaluating hysteresis and eddy-current losses as well as acoustic emission from the part.

Radiography by x ray, cobalt 60, or other high-energy sources is well known from its uses in medical diagnosis; an extensive description is not needed here. We should point out that excellent ASTM standard radiographs are available for specification of quality.

13.17 Summary of fracture appearance

Before we can determine the cause of failure, we must first determine how the component failed. For example, components may fail because the stresses imposed during service exceeded the ultimate strength of the material. If this is the case, we say that the part failed because of a tensile overload. If fatigue is responsible, we say that the component failed due to fatigue. The fracture surfaces produced by a tensile overload appear quite different from those produced by fatigue failure. In fact, it is often possible to classify the type of failure by macroscopic observation of the fracture surfaces themselves.† In this section we shall briefly discuss features of the more common types of failure. Such types of failure include:

1. Tensile overload
2. Propagation of brittle cracks
3. Fatigue
4. Stress corrosion

Tensile Overload Such failure results when the applied stress exceeds the ultimate tensile strength of the material. The appearance of the fracture depends strongly on the material's ductility. Figure 13.20*a* shows the general types for the ductile and brittle cases. The appearance of the fracture is generally dull. In low-strength, high-ductility materials, evidence of plastic deformation can be found.

Propagation of Brittle Cracks We may think of brittle cracks as initiating at a pre-existing defect and rapidly propagating through the component to cause failure. In many materials this rapid propagation leaves characteristic marks on the surfaces of the fracture. We can use these marks to locate the origin of fracture. Figure 13.20*b* shows two examples.

Fatigue Often we can easily note the presence of fatigue by observing the surface of the fracture. Fatigue cracks generally propagate normal to the stress axis. No shear lips are present. (A *shear lip* is a slanting ridge at the edge of a fracture.) Many times a series of concentric rings called clamshell or beach marks are visible on the surface (see Fig. 3.17). These marks represent abrupt changes in fatigue loading during service and help pinpoint the origin of the fatigue crack. Figure 13.20*c* shows two cases of fatigue failure.

Stress Corrosion Familiarity with each material under study is important in assessing stress-corrosion failure. Some materials may show branching or multiple cracking. The fracture path may be transgranular or intergranular (Fig. 13.20*d*). Again, familiarity with the specific material is of utmost importance.

In some instances a number of different events combine to produce the final failure. Such a case is shown schematically in Fig. 13.20*e*. Here a fatigue crack initiated from a corrosion pit. The crack propagated, and final failure resulted from rapid propagation of the crack. Each area shows different features. The failure analyst must rely on experience and a knowledge of the

---

†Although there are some general characteristics of fracture surfaces that we shall summarize in this section, you should be aware that in many cases a combination of fracture types may be present. Also the characteristics of one type of failure may differ from material to material.

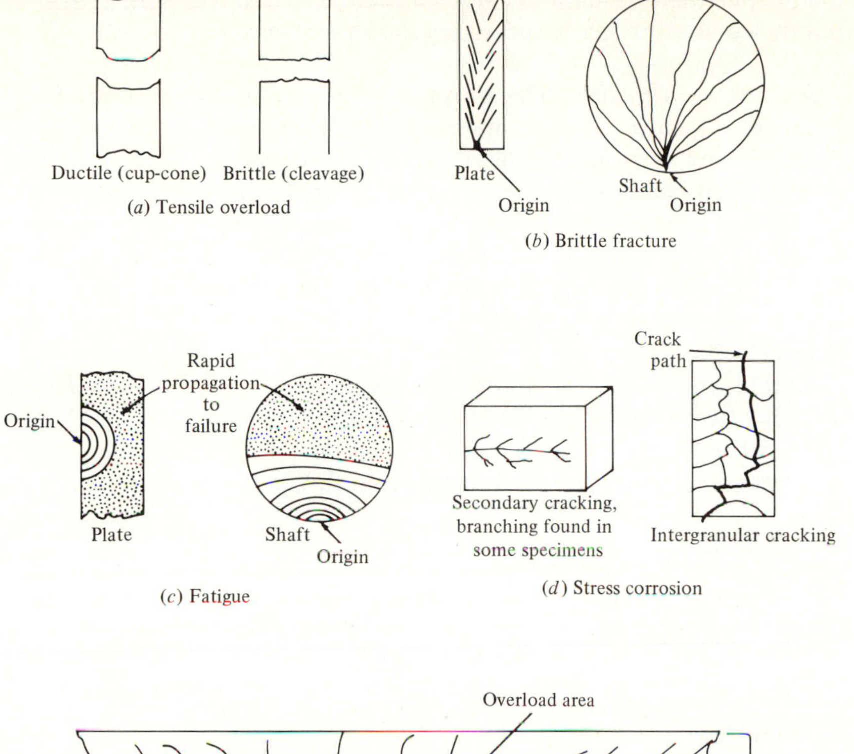

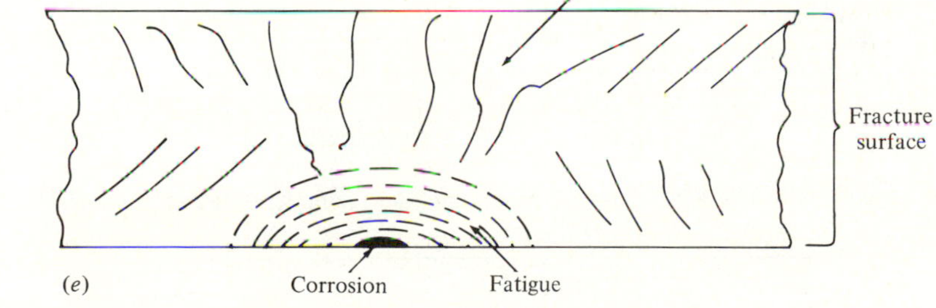

Fig. 13.20 *Features common to various types of fractures. (a) Tensile overload fracture. (b) Brittle fracture. (c) Fatigue fracture. (d) Stress-corrosion fracture. (e) A combination of types of fracture which results in failure of a plate.*

properties of the failed material to sort out the multiple types of degradation. We suggest that you consult *The Metals Handbook,* Vols. 7, 8, and 9, 8th ed. 1972, 73, 74, for a detailed coverage of the macroscopic and microscopic aspects of fracture surfaces.

13.18 Destructive methods of analysis

Although these methods are not literally destructive, they do involve sectioning and some etching of the component. Indeed, a well-prepared, etched cross section is often the best permanent exhibit of all. We shall not discuss preparation of specimens in detail (see *ASM Metals Handbook,* Vol. 8, 8th edition, 1973), but we shall indicate some important precautions.

Macroexamination The purpose of this step is to disclose two types of structure: (1) flaws, and (2) macroscopic inhomogeneity. In many cases it is helpful to cut sections of the specimen in two perpendicular directions, polish and etch. This brings to light larger inclusions, such as slag, sand, entrapped metal oxides (also called *dross*), and shrinkage cavities (see later illustrations). Also, if the structure near the surface is different as a result of conditions of casting or heat treatment, this is evident from differences in etching. Finally, the etch discloses the direction of hot and cold working, which affects the properties.

Extensive tables of etchants are given in the *ASM Metals Handbook,* Vol. 8, 8th edition, 1973, but we shall give a few favorites for most materials here. (In all cases you should consult a standard chemistry safety manual before proceeding, because the reagents and fumes are dangerous.)

Material	*Method*
Steel or cast iron	Add 1 part conc. HCl to 1 part water. Heat in Pyrex or other suitable container to 160°F (71°C). Immerse sample for 10 to 40 minutes to desired depth of etch. Use well-ventilated hood. Handle sample with tongs and rubber gloves and rinse carefully with water.
Aluminum alloys	Dissolve 10 g sodium hydroxide in 90 ml water; use at 150°F (66°C), 5 to 15 min. Remove etching products with concentrated nitric acid, then wash with water.
Magnesium alloys	Mix ethylene glycol (75 ml), distilled water (24 ml), and concentrated nitric acid (1 ml). Etch sample with mixture at room temperature; wash with water.
Copper alloys	Etch sample with concentrated nitric acid, under hood. Remove acid with cold water.

In all cases avoid spurious indications caused by excessive heating of the cut surface during cutting or pronounced cold working by machining. For example, it is extremely difficult to section a hard material such as white cast iron or hardened steel with an abrasive cutoff wheel without altering the structure near the cut surface. Use the softest cutoff wheel, with a copious flow of coolant followed by extensive wet grinding after sectioning.

Microexamination The microspecimen should be cut only after careful macroexamination to indicate critical areas. Observe the same precautions in cutting micro- as macro-specimens. Polishing and etching procedures are given in detail in the *ASM Metals Handbook,* Vol. 8, 8th edition, 1973, and in earlier editions. A few useful etches for microspecimens of different alloys that we recommend are the following.

Material	*Method*
Iron and steel (less than 8% alloy)	Use 2% nitric acid in ethanol.
Stainless and high-alloy steels	Use a mixture of nitric acid (10 ml), hydrochloric acid (25 ml), and glycerol (25 ml).
Aluminum alloys	Use a mixture of hydrofluoric acid (conc.) (0.5 ml) and water (99.5 ml). Hydrofluoric acid is particularly dangerous. It penetrates the skin even in low concentration.
Magnesium alloys	Use a mixture of ethylene glycol (60 ml), glacial acetic acid (20 ml), concentrated nitric acid (1 ml), and distilled water (19 ml).
Copper alloys	1. Use ammonium hydroxide (5 ml), hydrogen peroxide 3% (5 ml), and water (5 ml). Use fresh for brass.
	2. Use a mixture of 2 g of $K_2Cr_2O_7$, 1.5 g of NaCl, 8 ml of H_2SO_4 conc., and 100 ml of H_2O. (For more resistant alloys, such as silicon brass)

When you conduct a metallographic examination of the structure, you should examine the entire polished surface, especially the areas adjacent to the fracture, at low magnification (50 to 100×) before proceeding to higher magnification. You can miss the cause of failure if you do not follow this procedure. In many cases the cause of failure has been improperly assessed because a nonrepresentative field was selected for study and photographing at high power.

Mechanical Testing Hardness tests may be made on carefully prepared surfaces of the component or on microspecimens. Microhardness tests, as illustrated in Chapter 5 and elsewhere, are particularly valuable in determining the properties of thin layers or of individual phases. Tensile, compressive, torsion, and impact test blanks may be cut carefully from selected areas to check the mechanical properties of the material in the component. If the results are slightly at variance with the specified values, the manufacturer is not necessarily at fault. Only if the specifications specifically require minimum values *in specimens cut from specified regions of components* can the cut specimen be used in place of special test bars.

13.19 Determining residual stresses

We shall discuss this large and important field only briefly. To illustrate its importance, consider the following case. A large casting (an experimental railroad car wheel) was subjected to a new heat treatment in which the entire part was heated to 1600°F (871°C). Next the rim was subjected to a water spray to form a hard martensitic surface. After this the wheel was laid aside to cool. In the middle of the night, the wheel exploded into several pieces, one traveling several hundred feet. Examination of the piece showed that the metal was perfectly sound, of the desired structure, and with a tensile strength of 60,000 psi (414 MPa) in the fracture area.

The reason for this catastrophic failure was the presence of residual stresses resulting from high elastic strain that developed in the part. We shall discuss the generation of these strains under processing and confine ourselves here to the methods of measurement.

Let us recall the simple tensile test of Chapter 3. Suppose that a student stressed a tensile specimen of steel to 30,000 psi (207 MPa), causing an elastic strain of 0.001 in. (0.0025 cm). If the student locked the machine with the load intact, we would have a system in which one member (the specimen) has a residual stress of 30,000 psi (207 MPa). If the dials on the machine were covered, how could we determine the residual stress? We could inscribe two marks very accurately, 1 in. apart on the gage length, then make a saw cut above the gage length across the bar. We could then unload the specimen and relieve the elastic strain. When we remeasured the distance between the scribe marks, we would find that it had contracted to 0.999 inch, indicating 0.001 in./in. elastic strain. We could then calculate residual stress (from $\sigma = \epsilon E$) as $0.001 \times 30 \times 10^6$ psi = 30,000 psi (207 MPa).

In practice, instead of using scribe marks (which are relatively inaccurate), we use elastic strain gages. These are essentially fine wires held in a paper matrix that are cemented to the specimen in the stressed condition. We take an initial reading of the electrical resistance of the gage with a precision of 1 microhm. We then relieve the elastic strain by sectioning and take a second resistance reading. If the gage length shortens, the resistance decreases, indicating residual *tension*; the resistance increases for residual compression. The precision of the method is excellent, since one microhm is equivalent to about 0.000001 strain, or a residual stress of 30 psi (0.21 MPa) for steel.

When residual stresses are present in several directions, we use strain gage rosettes with three strain gages. A Mohr circle analysis is used to interpret the results. This is discussed in standard texts on mechanics.

Although the fracture usually relieves the residual stresses in a failed part, it is often possible to obtain a similar unfractured part for stress analysis. In many cases a project engineer has felt secure with a steel of 60,000 psi (414 MPa) tensile strength operating in a given service, only to find that the processing operations led to a residual stress of 50,000 psi (345 MPa), so that a service stress of only 10,000 psi (69 MPa) was required for failure.

Defects and Anisotropy Caused by Manufacturing Process

13.20 Processing defects

This is another very large field requiring a study of the individual processes to fully appreciate the problems. However, we shall give examples of some of the severe problems in casting, working, and heat treatment of metals.

Castings The major defects are caused by gas porosity, shrinkage porosity, nonmetallic inclusions, and hot tears. Let us look at Fig. 13.21, which examines these in some detail.

1. *Cut and wash* Metal erodes sharp corners of ingate and protruding core. Letters A and B denote sequence of steps.
2. *Rat tail, buckle, scab* Heat from stream of metal, either radiant heat or heat caused by conduction, locally and non-uniformly expands sand, producing either a minor extension (rat tail), a buckle, or spalling away (scab). The material from the mold forms a sand inclusion at another spot.
3. *Fusion, penetration* Sand can fuse by action of hot metal, producing a mass that may adhere to casting. Metal can penetrate the sand grains after the surface of the grains has fused.
4. *Crush* When the mating halves of a mold do not fit, the protruding portion is crushed.
5. *Sticker* Sand sticks to the pattern, producing projection of ragged edges of metal after casting.
6. *Swell* Hydrostatic pressure of metal (plus expansion on solidification in some cases) pushes mold walls outward.
7. *Shrinkage* When the density of the solid metal is greater than that of the liquid (the usual case), porosity is produced.
8. *Shift* Mismatch of halves of the mold

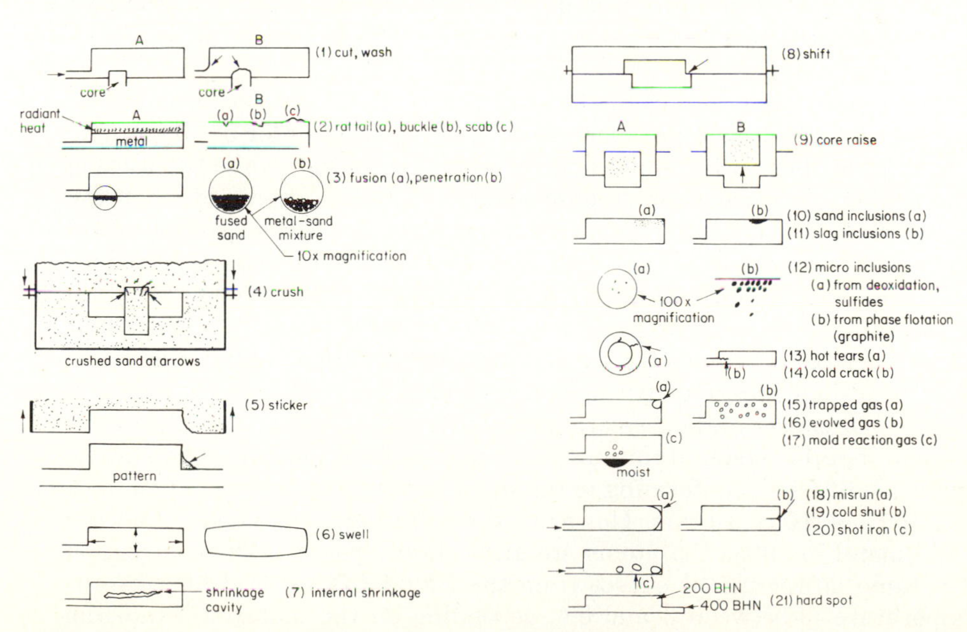

Fig. 13.21 *Typical defects found in casting*

9. *Core raise* Flotation of the core due to buoyancy of liquid metal
10. *Sand inclusions* Sand that is entrapped in the mold as a result of crushing or spalling, or as a result of loose sand being in the mold
11. *Slag inclusions* Result from the presence of slag in the metal.
12. *Microinclusions* Sulfides and oxides result because of separation from the melt. Graphite flotation may occur in hypereutectic ductile iron.
13. *Hot tears* Interdendritic ruptures produced by tensile stresses near the solidus temperature
14. *Cold cracks* Produced by residual stresses or by improper removal of gate and riser
15. *Trapped gas* Gas that is mechanically entrapped in the mold
16. *Evolved gas* Gas that evolves from precipitation of dissolved gas
17. *Mold reaction gas* Gas evolved when the liquid metal reacts with sand that is damp or that produces hydrocarbons on the mold. *Note:* Pores due to gas are often spherical in shape, while pores due to internal shrinkage show a ragged surface.
18. *Misrun* Incomplete filling of the mold due to metal that is of low fluidity (cold metal)
19. *Cold shut* A section of mold that is imperfectly bonded due to low pouring temperature and poor flow
20. *Shot iron* Particles of frozen metal caught in the stream of molten metal and improperly bonded; often caused by interrupted flow
21. *Hard spot* Occurs especially in gray and ductile iron. More rapid rate of solidification of light sections and corners may result in very hard carbidic structure.

Gas porosity is caused by the evolution of gas as the casting solidifies. Gases such as hydrogen are more soluble, often by an order of magnitude, in the liquid than in the solid, and therefore precipitate when the casting solidifies. Other more complex gases, such as CO, are produced when the oxygen dissolved in liquid steel precipitates when the casting solidifies and reacts with the carbon of the steel. The remedy in both cases is to reduce the amount of the offending gas in the liquid to a level below its solid solubility. One way of doing this is to add an active element, such as aluminum, to steel to remove the oxygen in the form of inert aluminum oxide.

You should learn to distinguish shrinkage porosity from gas porosity, because the causes and cures are quite different. If we pour a cube of the common metals—steel, aluminum, or copper—of 10 mm edge, the surface layers solidify first, conforming to the shape of the mold. We therefore have a shell $10 \times 10 \times 10$ mm. However, the solid metal is denser than the liquid metal, because the atoms are more closely packed. The contraction in volume as the metal passes from the liquid to the solid at constant temperature is between 3 and 6%, depending on the material. Since the outer walls are fixed, the difference in density results in a shrinkage cavity inside the cube. There is, of course, further contraction as the solid cools, but it is the contraction during the liquid-to-solid phase change that is responsible for shrinkage porosity. The remedy is to provide a reservoir of liquid metal, called a *riser*. The riser is attached to the casting, and freezes afterward. The shrinkage of both casting and riser is contained in the riser, which is later cut off and remelted.

Gray cast iron is an exception because graphite, the less-dense phase, precipitates during freezing, so that there is little or no shrinkage porosity in the casting. For this reason, very complex castings that would be hard to riser, such as automotive engine blocks, are made of gray cast iron.

Nonmetallic inclusions are of two types. One is a gross type made up of foreign materials such as sand, slag, and oxide dross. These can be controlled by care in the pouring operation and in the design of the gating system for the casting. The second type of inclusion, such as sulfides or complex silicates, forms on solidification. Controlling these inclusions is more difficult. We control them either by reducing the offending element to very low levels or by making ladle additions that change the shape of the inclusions to a less harmful geometry—for example, that would change their shape from a continuous-grain boundary type to a spherical shape.

Hot tears are fractures that develop in the casting when solidification is almost complete. Although we generally associate high ductility with a metal part at elevated temperatures, the casting exhibits very little plastic elongation while a small amount of liquid is present. This liquid is generally located between grains or dendrites. When a relatively small amount of strain is called for by interaction between the casting and the mold or between sections of the casting, tearing takes place. These tears exhibit oxidized dendritic fractures. If the part is put in service, severe stress concentrations that can lead to failure are present at the root of the tear.

13.21 Defects produced during metal working

Many components are produced by casting a metal ingot, then working the ingot by a combination of rolling, forging, extrusion, piercing, swaging, stamping, and more complex processes. Since all these processes involve hot or cold deformation, we can discuss the defects in one group. Figure 13.22(*a*) on page 572 illustrates each of the defects that can occur during forging, rolling, and forming.

1. *Segregation* Inclusions and alloying elements segregate during solidification and are rolled into bands.
2. *Pipe* (shrinkage cavity) If oxidized, it does not weld. If forged, it produces a fissure in the rolled product.
3. *Inclusions* In the ingot, inclusions are uniformly distributed; after forming, they elongate in the forming direction.

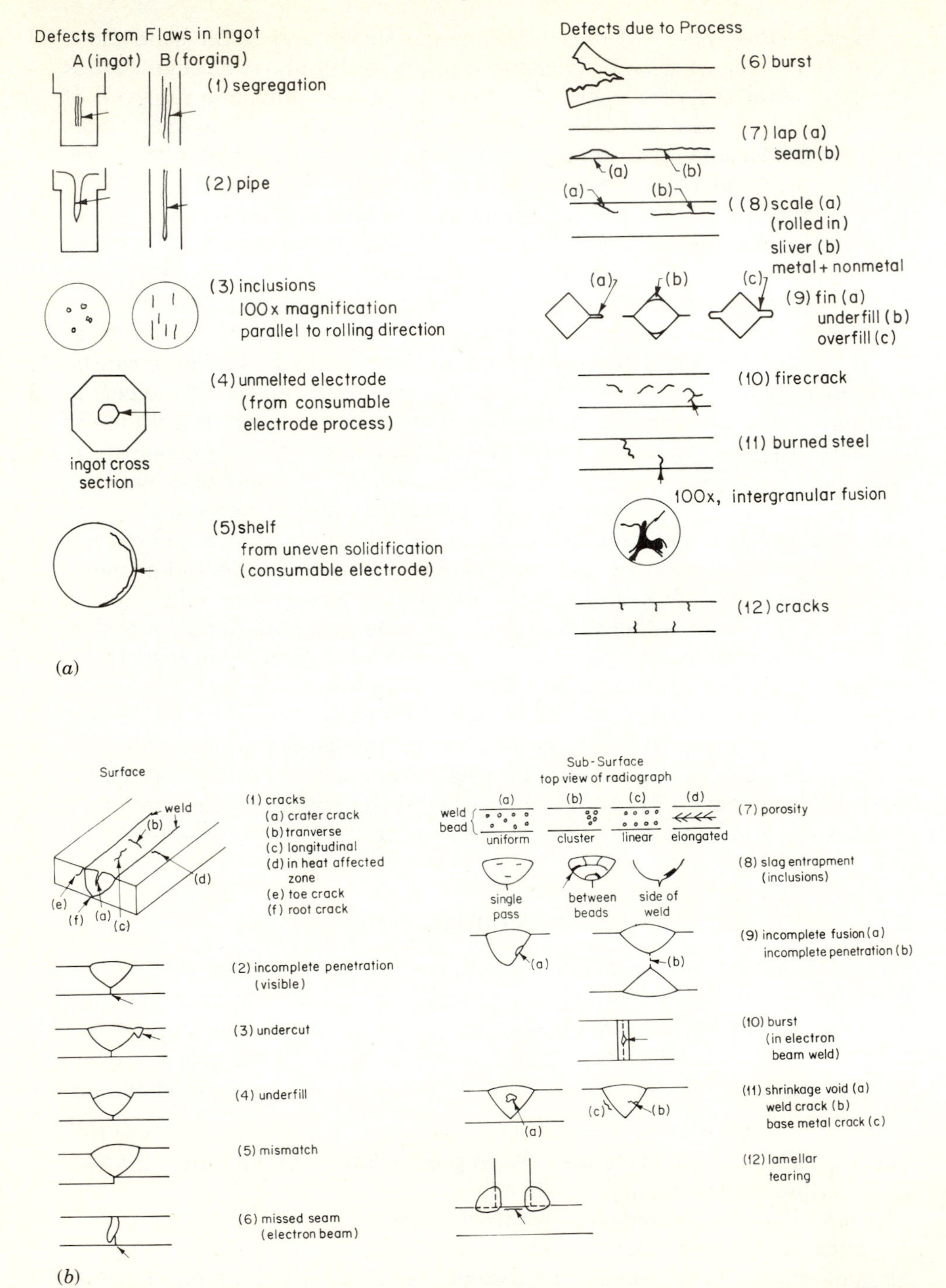

Fig. 13.22 *(a) Typical defects found in forming. (b) Typical defects found in welding*

4. *Unmelted electrodes* During the process of melting metal using electrodes, pieces of the electrode fall into the ingot.
5. *Shelf* When solidification occurs in an uneven wave, there are lap-like defects.
6. *Burst* (also called alligatoring) When surface stresses at rolled edges are higher than in the interior, bursting occurs.
7. *Lap* Folding over of rolled surface at one region causes lap.
   Seam Continuous folding over of rolled surface
8. *Scale* Oxidized scale rolled into surface, or internal deoxidation products strung out in the direction of rolling (in the form of slivers)
9. *Fin* An extrusion of metal, excessive when the fit of the die is poor. *Underfill* Insufficient metal between dies. *Overfill* Excess metal between dies.
10. *Fire cracks* Caused by cracking of rolls and imprinting of striations on work piece
11. *Burned steel* Steel that is heated to temperatures above the melting point of one of the phases (normally the sulfide phase); also called hot shortness.
12. *Cracks* Caused by improper forging temperature or not enough force behind the hammer

Hot Working Defects encountered during hot working (above the recrystallization temperature) can be caused by the following factors:

1. Working temperature too high
2. Working temperature too low
3. Pre-existing defects in the ingot
4. Residual stresses
5. Defective working procedure

EXCESSIVE WORKING TEMPERATURE, BURNING If the steel is heated at or near the solidus, the material at grain boundaries melts and forms voids. These are not sealed by subsequent working. It is not necessary to reach the solidus of the pure alloy if other elements, such as sulfur in the form of metal sulfides, form a liquid with a melting point below the solidus of the pure alloy. Segregation of some of the desired elements may lead to melting below the estimated equilibrium temperature. Overheating is related to burning, occurs at slightly lower temperatures, and is not really a well-defined condition. It may be evidenced by loss of ductility and cracking at high temperatures without definite evidence of fusion, as in the case of burning. Heating at temperatures that are too high for the particular composition may also lead to excessive grain growth.

LOW WORKING TEMPERATURES It is generally advisable to work at as high a temperature as possible because of the greater ductility, lower energy requirements, and ease of flow of the metal, provided that excessive temperatures are avoided. Many hot processes are designed with this high ductility in mind. When the temperature standards are not met, *cracking* takes place. There are several special cases. In forging and rolling, corner and edge cracks develop. Another rolling defect, called a *center burst,* is quite striking because it takes place inside the bar being rolled. It arises from the fact that the material next to the rolls is being rapidly deformed relative to the undeformed metal at the center.

PRE-EXISTING DEFECTS IN THE INGOT An ingot is, after all, a casting and is subject to the same defects, such as gas holes, shrinkage porosity, and inclusions. Fortunately in most cases the gas holes are welded shut by rolling. One grade of steel, called *rimming steel,* is made deliberately with gas holes that are not open to the atmosphere and that prevent shrinkage cavities from developing. Shrinkage porosity, however, results in a *pipe* in the ingot that is compressed but not welded shut during rolling because the internal surfaces are oxidized. If this portion of the ingot is not cut off, piping defects occur. *Inclusions* are either elongated, as in the case of manganese sulfide, or fractured, as in the alumina type. When the metal is tested in the rolling direction, the effect of inclusions is minimal, but across the rolling direction, ductility is lowered. Actually the inclusions may be treated as cracks from the point of view of fracture toughness. *Segregation* in the ingot may lead to *banding*; for example, in hypoeutectoid steel, there may be alternate layers of ferrite and pearlite. *Seams* are regions of inhomogeneity that may open up during forging. They can be caused by segregation.

RESIDUAL STRESSES These develop when a part is cooled from the hot-working range with a severe temperature difference (thermal gradient) existing between different sections. When it reaches the elastic range, hotter sections contract a greater amount than cooler sections, resulting in residual stresses. These stresses may crack the part during the cooling cycle, or, if unrelieved, may lead to premature failure. When this condition is encountered, the part should be reheated within a reasonable time to a temperature at which creep occurs, to relieve the elastic strain. Generally this temperature is near the recrystallization temperature. The part should be slowly cooled to prevent redevelopment of the thermal gradient.

DEFECTIVE WORKING PROCEDURE Improper manipulation of the piece during forging or poor die design leads to the development of laps, also called a *cold shut.* This defect is caused when one surface region is forced over an adjacent region. The overlap traps a portion of the usually oxidized and scaled surface beneath the surface, producing a plane of weakness.

Cold Working During cold working, which is generally conducted at about room temperature, the ductility of the metal is much lower and recrystallization does not take place to remove work hardening (except in low-melting-point metals such as lead). As a result, the part is more subject to cracking if defects such as inclusions are present, especially if they have been exaggerated by previous hot working. If harmful residual stresses are present, the condition may be aggravated by cold working and the part may actually break. The same terms are used to describe typical defects, except that the phrase "cold short" denotes a material that cracks particularly easily during operations at room temperature, most often because of the presence of brittle phases.

13.22 Welding defects

One way to look at welding is to consider it as a "fast casting" process. One group of defects—such as gas porosity, shrinkage, hot tearing, and inclusions—are similar to those discussed in casting and need not be considered further. However, there are other defects peculiar to the welding process that should be described.

Figure 13.22(b) illustrates defects that can be introduced during welding (see page 572).

1. *Cracks* (a) crater, (b) transverse, (c) longitudinal, (d) in heat-affected zone, (e) toe crack, (f) root crack (due to stresses on heating and cooling)
2. *Incomplete penetration* Due to too little heat
3. *Undercut* Due to poor control of weld rod
4. *Underfill* Due to poor deposition of weld rod
5. *Mismatch* Shift of mating parts in fixture.
6. *Missed seam* In an electron-beam weld, due to poor alignment of parts to be welded
7. *Porosity* Due to evolution of gas from weld metal; also due to shrinkage
8. *Slag entrapment* Due to improper control and poor removal of slag between passes
9. *Incomplete fusion* Due to poor manipulation of weld rod or dirty joint
10. *Burst void* Due to shrinkage and/or gas in an electron-beam weld
11. *Shrinkage void* Internal defects caused by shrinkage during solidification, and by weld stresses
12. *Lamellar tearing* Cracks due to weld stresses

Lack of Fusion and Penetration This occurs when the welding conditions do not produce a hot enough weld bead to dissolve (penetrate) the base metal. This leads to a weak joint. It is quite evident compared with a normal weld.

Undercutting This takes place when the weld melts the base metal at the side wall, but does not fill in the groove. This gives a longitudinal notch that can lead to failure.

Hydrogen Cracking If hydrogen from the materials of the weld-rod coating is dissolved in the bead, it may precipitate later at inclusion sites at very high pressures and generate cracks.

Chapter 12 discusses certain types of corrosion peculiar to welds.

13.23 Defects due to heat treatment

There are three sources of defects due to heat treatment.

1. Effect of thermal gradient, especially in quenching
2. Structural changes resulting in brittle phases and/or high stresses due to transformation
3. Reactions with the furnace atmosphere

Chapter 6 discussed the role of the thermal gradient produced by quenching a component from a high temperature. The portions of the component that cool most rapidly form a rigid, high-strength structure. The other regions, still at higher temperature, later contract and develop residual stresses or cracks on cooling. The remedy is to design equal sections, use a quenchant that develops a slower cooling rate, or use high-velocity quenching jets at the heavy sections.

A brittle microstructure may result from improper heat treatment. For example, a slow-cooled hypereutectoid steel develops a brittle carbide network. Also, quenched steel develops a hard, brittle martensitic structure that must be tempered to avoid cracking.

The reaction of the part with the furnace atmosphere that often leads to fatigue failures is called *decarburization*. If the furnace atmosphere is oxidizing while a steel part is being austenitized prior to quenching, considerable loss of carbon can occur in the surface layers. These layers do not harden properly and their fatigue strength is much less than that of the nonaffected deeper regions. Other surface reactions such as excessive carburization and absorption of hydrogen, may lead to undesirable structures.

13.24 Wear and abrasion

We shall now complete our discussion of failure analysis by considering wear and abrasion.

The most important characteristic of wear is its unpredictability. For example, a gray cast iron is an excellent engine-block material because it retains lubricant and avoids seizing of the piston. On the other hand, the graphitic areas crumble under high stresses and cannot be present in a component such as a railroad car wheel or grinding ball. In another case a general "rule" is to avoid contact between similar metals. However, gray iron piston rings perform well against a gray iron engine block and gray iron tappets† do well against a gray iron camshaft.†

In spite of this apparent confusion, problems of wear can be approached reasonably if the mechanism of wear is known. Often a careful observation of worn parts discloses which structures failed and what conditions developed at the interface. Before we discuss specific cases, let us consider the fundamentals of wear.

1. The nature of wearing surfaces
2. Effects of pressing surfaces together (static contact)
3. Interaction between sliding surfaces
4. Effects of lubrication
5. Resistance to wear of different material combinations (Sec. 13.25)

The Nature of Wearing Surfaces Even the best polished surface has two types of inhomogeneity. By using an accurate diamond stylus we can find pits and scratches of the order of 1000 Å, or hundreds of times greater than the size of the unit cell. In addition to these physical differences, we know that variations in grain orientation, in the nature of phases and inclusions, can occur at the wearing surface. In other words, even with the best machining, we have two relatively rough inhomogeneous mating surfaces. Furthermore, we know that it is possible to have two very different microstructures of the same overall macrohardness. For example, one structure might contain martensite at BHN 600 microhardness plus graphite flakes that would lower the overall hardness to BHN 500. Another structure might be homogeneous martensite at BHN 500. Under one type of wear, involving high loads and good lubrication, the second structure would probably perform better, particularly if the load crushed the iron at the edges of the graphite. However, with moderate loads and intermittent lubrication the first structure would resist galling (sticking) better than the second.

Static Contact Consider the simple case of a ball bearing on a flat plate. This is incorrectly called *point contact* because the ball and plate both deform, giving a circular area of contact. It can be shown that the shear stress is a maximum not at the surface, but beneath it. If a is the radius of the circle of contact, the maximum shear stress is at $0.6a$ beneath the surface.

We can calculate the maximum stress at this level as a function of the radius of the ball or other "point" contact. From this value we can calculate the load required to produce flow for different radii of the ball and different plate materials (Table 13.2 on page 578).

In conventional terms, a high load is required to deform tool steel with

---

†These contain controlled amounts of massive iron carbide.

Table 13.2 LOAD REQUIRED TO PRODUCE FLOW IN DIFFERENT MATERIALS USING INDENTERS (STEEL BALLS) OF DIFFERENT RADII

	Load, g, for Ball Radius Indicated		
Plate Material	*10^{-4} cm*	*10^{-2} cm*	*1 cm*
Copper	2.5×10^{-6}	2.5×10^{-2}	250
Mild steel	4.7×10^{-5}	0.47	4,700
Tool Steel	1.4×10^{-3}	14	140,000

a ball. However, when we consider that a large ball has tiny rough points of radii as small as 10^{-4} cm, then we must expect some deformation even with very light loading (1.4×10^{-3} g).

Interaction between Sliding Surfaces When surfaces slide over each other, the projections produce very high stresses and flow in the areas of contact. If we accept this, we can understand the inevitable breaking-in period better. In many cases the high stresses result in high friction, and welding followed by shearing takes place between the parts.

1. If the weld is weaker than either metal, little wear takes place. This is the case, for example, when a tin-base-alloy bearing wears against steel.
2. If the junction is stronger than one of the metals, then shear takes place in the weaker metal. For example, when steel wears on lead, the fracture takes place in the lead.
3. If the junction is stronger than both metals, for example, when martensite forms in the junction, tearing occurs in both metals and wear is rapid.

The effects of surface temperature are also important. Under many simple conditions of sliding wear, lead actually melts and covers the surface, whereas steel is hardened by local heating above the austenite transformation temperature. The presence of freshly formed martensite is often noted in steel or iron parts subjected to wear, such as brake drums.

Effects of Lubrication When a lubricant is present, there is either hydrodynamic lubrication or boundary lubrication. In hydrodynamic lubrication one tries to preserve a fluid film so that the surfaces do not touch. Many factors operate against this, including intermittent operation, breakdown of the lubricant molecule, and heavy loads with slow speeds. Under these conditions welding or galling can take place.

With boundary lubrication one tries to coat the surface with lubricant molecules. The metal reacts with the lubricant to form a metal soap. In other cases extreme pressure lubricants containing sulfur, phosphorus, or chlorine react with the metals to reduce seizing. Molybdenum disulfide and graphite are also used where extreme pressure exists, to serve as solid fragments in the bearing area with easy basal cleavage.

13.25 Wear-resistant combinations

It is possible to dissolve lead in liquid copper alloys. On solidification, the lead precipitates in fine globules. These alloys are often used against steel in slow-speed bearings with intermittent lubrication. The copper is hardened by the addition of tin and zinc, as in the alloy 85% Cu, 5% Sn, 5% Zn, 5% Pb. For some applications a higher-strength alloy is needed and aluminum bronze (89% Cu, 11% Al) is used. For gears a favorite combination is hardened steel wearing against a nickel-tin bronze (11% Sn, 87% Cu, 2% Ni). The tin provides a hard δ phase, and the nickel gives solid-solution hardening of the copper.

Many applications involve ceramic and plastic bearing materials. Sapphire bearings have long been used in precision equipment and watches, and it is possible to obtain 1-in.- (2.54-cm-) diameter sapphire balls. Nylon composites are used in applications in which metal was required previously, such as fishing reel gears.

13.26 Resistance to abrasion

In the usual wear application one avoids introducing gritty foreign material. However, in some cases, such as grinding ore and conveying sand, the abrasive is part of the system. The chief variables in abrasion are the hardness of the abrasive and the type of loading of the equipment, whether impact or steady compression.

Unfortunately, the hardest and most abrasion-resistant alloys, such as tungsten carbide, can be used in only a few installations because the carbides crack out. The most useful abrasion-resistant alloys in mining and other grinding are martensitic white cast iron, hardened steel, and austenitic manganese steel. The white irons are the most wear-resistant, whereas manganese steel is the toughest. In some instances, in which the part is not heated or cut by the abrasive, rubber-coated parts are very successful. Conveyor belts and gloves for sandblast operators are examples.

SUMMARY

To provide a basis for understanding the analysis and prevention of failure, the discussion is divided into three parts.

1. *Fracture toughness and fracture mechanics,* plus additional material on fatigue, stress corrosion cracking, wear, and abrasion. To avoid catastrophic failures in new high-strength, low-ductility alloys, one must consider

fracture toughness as well as yield strength. In cases in which the plane strain fracture toughness is the criterion, the important formula is:

$$\sigma_f = \frac{K_{IC}}{Y\sqrt{\pi a}}$$

where σ_f is the fracture stress, Y is a factor calculated from the crack length and the geometry of the part, K_{IC} is the plane strain fracture toughness, and a is the crack length.

Fatigue fracture depends on the initiation of a crack, its growth to a critical length for the geometry of the part, and the stress involved. Parts can be designed to avoid fatigue failure using the same concepts of fracture toughness as for static stresses. Stress corrosion cracking also depends on the concept of development of a crack under the combined effects of stress and corrosion.

Resistance to wear is not a function of hardness alone, but involves the interaction of the mating structures and the lubrication. Abrasion is an extreme case of wear. It is a vital problem in the mining and materials-handling industries.

2. *Examination of failed parts* Once we have obtained all data relevant to the failure, we should subject the part first to nondestructive testing. This includes macroexamination, with photography and, where indicated, radiography, dye penetrant, magnaflux, and ultrasonic inspection. Small pieces may be viewed directly in the scanning electron microscope. If metallography is permitted, microspecimens should be cut from relevant areas, polished, etched, and examined. One should consider hardness and mechanical tests to determine how the properties of the material compare with specifications.

3. *Defects due to processing* Failure may result if defective regions are produced in the component during the processes of casting, working, welding, and heat treatment. Typical defects are discussed and defined.

DEFINITIONS

Boundary lubrication In this condition the lubricant film is discontinuous in places, and special lubricants are added to prevent galling.

Burning Heating of steel or other alloy above the solidus during heat treatment, producing intergranular voids.

Center burst Rupture along the centerline of a billet during rolling, caused by greater flow at the edges than at the center.

Dye-penetrant inspection Use of a penetrating oil to penetrate cracks, followed by background spray to delineate oil-saturated cracks.

Fracture stress, σ_f The nominal stress at fracture. When computing the nominal stress, one generally ignores the presence of a flaw.

Galling This occurs when two metal surfaces are in contact and sticking or welding takes place, resulting in surface roughening and increased wear.

Gas porosity Rounded voids in a casting or ingot caused by evolution of dissolved gases during solidification.

Hot tears Fissures in a casting caused by restraint during attempted contraction near the solidus.

Hydrodynamic lubrication In this type of lubrication the motion of the bearing surfaces produces a continuous lubricant film between the surfaces.

Hydrogen cracking In welding, cracking produced by dissolving of excessive amounts of hydrogen during welding, as a result of high hydrogen content of the weld rod.

K_{ISCC} The value of stress intensity above which cracks grow because of stress corrosion. Below this value, crack growth due to stress corrosion does not occur.

Macroscopic examination Examination with the unaided eye or low magnification (usually up to $50\times$).

Magnaflux inspection Use of a magnetic field in conjunction with magnetic powder to find surface flaws or cracks.

Microscopic examination Examination at greater magnification than $50\times$, usually using a light microscope with a polished and etched section.

Nondestructive methods, NDT Nondestructive testing is the examination of a part by methods such as x ray or magnaflux that do not change the surface or interior of a part, so that it can be used afterward in service.

Nonmetallic inclusions Brittle oxides, sulfides, or silicates that result from the entrapment of foreign matter during pouring or from reactions within the liquid metal.

Plane strain fracture toughness, K_{IC} The minimum stress intensity required to cause catastrophic failure. K_{IC} can be measured only above a particular thickness, which depends on the material, and is constant for all thicknesses above this value.

Plane stress fracture toughness, K_C The value of stress intensity required to cause catastrophic failure in components with thicknesses below that required for plane strain. K_C varies with thickness.

Residual stresses Stresses that result from the presence of elastic strains remaining in a component as a result of prior working or treatment.

Scanning electron microscope (SEM) examination Use of an electron beam with scanning equipment that magnifies 50 to $50{,}000\times$. A fractured surface can be used without sectioning and polishing.

Shrinkage porosity Dendritic cavities in a casting or ingot caused by shrinkage in the inner region (often near the centerline) during solidification.

Stress concentration, K_σ The ratio of the stress near a notch or hole to the nominal stress.

Stress intensity factor, K_I A measure of the magnitude of the stresses near the tip of a sharp crack. K_I is a function of nominal stress, crack length, and component geometry.

Stress intensity range, ΔK_I When cyclic stressing occurs between K_{max} and K_{min}, $\Delta K_I = K_{max} - K_{min}$.

Surface energy, γ_s The increase in energy of a system per unit area of new surface created. Related to the number of atomic or molecular bonds broken to create the new surface.

Threshold stress intensity range, ΔK_{th} The value of ΔK_I below which crack propagation due to fatigue does not occur.

Ultrasonic testing Use of ultrasonic impulses, coupled with a crystal receiver, to monitor reflections and speed of sound waves in a component. Reflections by internal surfaces indicate defects.

Undercutting In welding, undercutting is melting away of the base metal at the side of the mold without filling in of the welding groove.

PROBLEMS

13.1 An air-operated hoist had a hook one inch in diameter at its end. The shank portion of the hook was too long, so the engineer shortened the shank by flame-cutting out a portion and rewelding the end sections. Although the shank of the hook was one inch in diameter, the diameter at the weld was excessive ($1\frac{3}{4}$ in. in diameter). Subsequently the rewelded shank failed in service at the junction of the weld and the shank. It was found that grinding back the weld eliminated the service failures (see sketch on page 583). Explain why removal of the excess material seemed to strengthen the component. (Sections 13.1 through 13.8)

13.2 Considerable scatter is generally found in *S-N* fatigue data. What are the possible causes of this scatter? [*Hint:* Review the section on initiation of cracks.] (Sections 13.12 and 13.13)

13.3 A thin section of a fan blade has failed in service. You believe that the mode of failure is fatigue. Close examination of the fracture surfaces does not reveal the characteristic clamshell marks (as shown in Figure 13.20), although a portion does show a shear lip and rapid propagation of cracks. In fact, a large portion of the surface of the fracture appears worn and shiny. Why might this still represent a fatigue failure? (Sections 13.12, 13.13, and 13.16 through 13.19)

13.4 Stress corrosion cracking (SCC) occurs in stainless steels, and is often intergranular (along the grain boundaries). As a result, a common conclusion is that SCC is *always* intergranular. Explain why intergranular

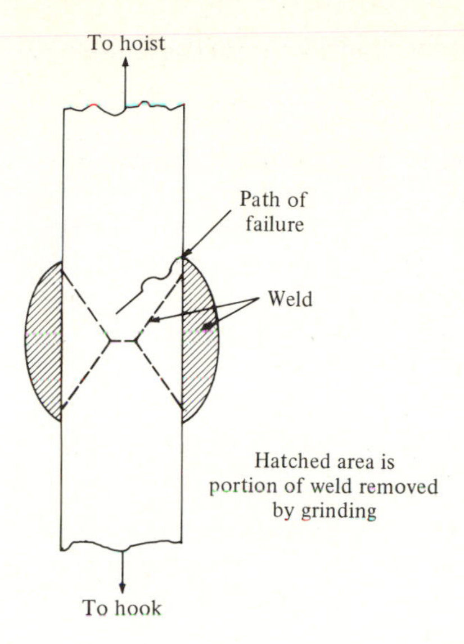

SCC may occur in stainless steel, and under what circumstances it may also be transgranular. (Section 13.15)

13.5 A quenched and lightly tempered 0.6C alloy steel is specified to have a surface hardness of 57 R_C. When it is placed in a wear environment, the wear is found to be excessive. The surface hardness is determined to be only 40 R_C. Suggest what might have happened to cause this situation. (Sections 13.24 through 13.26)

13.6 Discuss, in terms of microstructure, some exceptions you could take to the widely published generality, "Wear is approximately proportional to hardness." (Sections 13.24 through 13.26)

13.7 A common rule to follow when choosing materials for applications involving wear is to avoid having both wearing surfaces made of the same material. Explain why this is true for single-phase materials. Also explain why it is possible to use an automotive piston ring made of pearlitic gray iron that wears against a cylinder wall of the same structure—a pearlitic gray iron. (Sections 13.24 through 13.26)

PROBLEMS COVERING OPTIONAL SECTIONS AND PREVIOUS CHAPTERS

13.8 The fracture toughness of a 0.6-inch- (1.5 cm-) thick section of an alloy is 25,000 psi$\sqrt{\text{in.}}$ (27.5 MPa$\sqrt{\text{m}}$). The yield strength of this alloy is 35,000 psi (241.5 MPa). We wish to use this fracture-toughness value in designing a component from the same material, but with a thickness of 1.6 inches (0.041 m). Is this approach valid? (Sections 13.9 through 13.11)

13.9 Assume that the tension support member described in Example 13.5 must support a load of 320,000 lb (145,280 kg). Because of inspection difficulties, the smallest edge flaw that can be detected with confidence is 0.15 in. (3.8 mm) long. Which material—4340V given a 500°F (260°C) temper or 4340V given an 800°F (427°C) temper—should you choose to minimize the weight of the support member? Assume that the width remains constant at 4 in. (Sections 13.9 through 13.11)

13.10 The stress distribution in plane stress near the tip of a crack is as follows. (Sections 13.9 through 13.11)

$$\sigma_y = \frac{K_I}{\sqrt{2\pi r}} \cos\frac{\theta}{2}\left(1 + \sin\frac{\theta}{2}\sin\frac{3\theta}{2}\right)$$

$$\sigma_x = \frac{K_I}{\sqrt{2\pi r}} \cos\frac{\theta}{2}\left(1 - \sin\frac{\theta}{2}\sin\frac{3\theta}{2}\right)$$

$$\tau_{xy} = \frac{K_I}{\sqrt{2\pi r}} \sin\frac{\theta}{2}\cos\frac{\theta}{2}\cos\frac{\theta}{2}$$

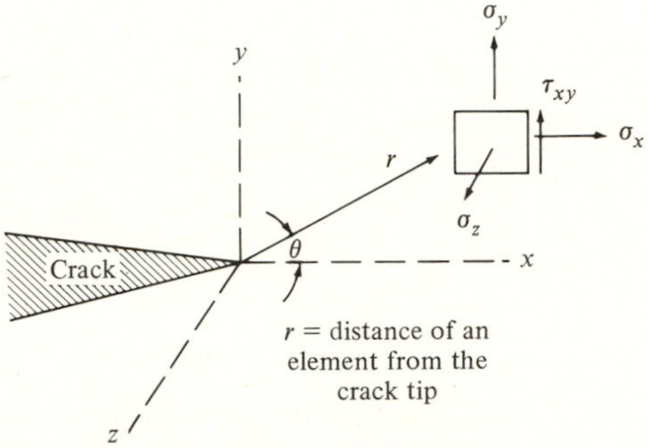

a. For a stress intensity of 20,000 psi$\sqrt{\text{in.}}$ (22 MPa$\sqrt{\text{m}}$), show the variation of σ_y along the projected path of the crack (x direction, $\theta = 0$) from $r = 0.05$ to $r = 1.0$ in. (1.3 to 25.4 mm).

b. Repeat (*a*) for $K_I = 40{,}000$ psi$\sqrt{\text{in.}}$ (44 MPa$\sqrt{\text{m}}$)

c. Given that the yield stress is 35,000 psi (241.5 MPa), approximate the size of the plastic deformation field at the crack tip for both K_I values.

13.11 Refer to Example 13.7.

a. For the initial crack length given, calculate the cycles to failure if the fracture toughness of the material is doubled.

b. Using the original toughness given in Example 13.7, calculate the cycles to failure if the initial size of the flaw is decreased to $a = 0.05$ in. (1.3 mm).

c. Explain the reasons for the differences between your answers in (*a*) and (*b*). (Sections 13.14 and 13.15—math portion)

13.12 The $K_{I\,SCC}$ for a particular alloy is 15,000 psi$\sqrt{\text{in.}}$ (16.5 MPa$\sqrt{\text{m}}$). What is the largest flaw that will not be subject to growth by stress corrosion, given that this flaw is located in the central portion of a large plate that is subjected to a static stress of 25,000 psi (172.5 MPa)? (Assume $Y = 1$.) (Sections 13.14 and 13.15—math portion)

13.13 Hardness values, especially of cast irons, are often used to predict the tensile strength. Explain why this might be a useful technique for quality control. What might the limitations be? (Sections 13.1 through 13.8 and 6.32 through 6.36)

13.14 Some components are sold as being of "x-ray quality." That is, radiographs are used to determine the existence of flaws. Does this necessarily mean that, in high-strength materials, we need not worry about brittle fracture if they have passed an x-ray examination? Explain. (Sections 13.1 through 13.8 and 13.16)

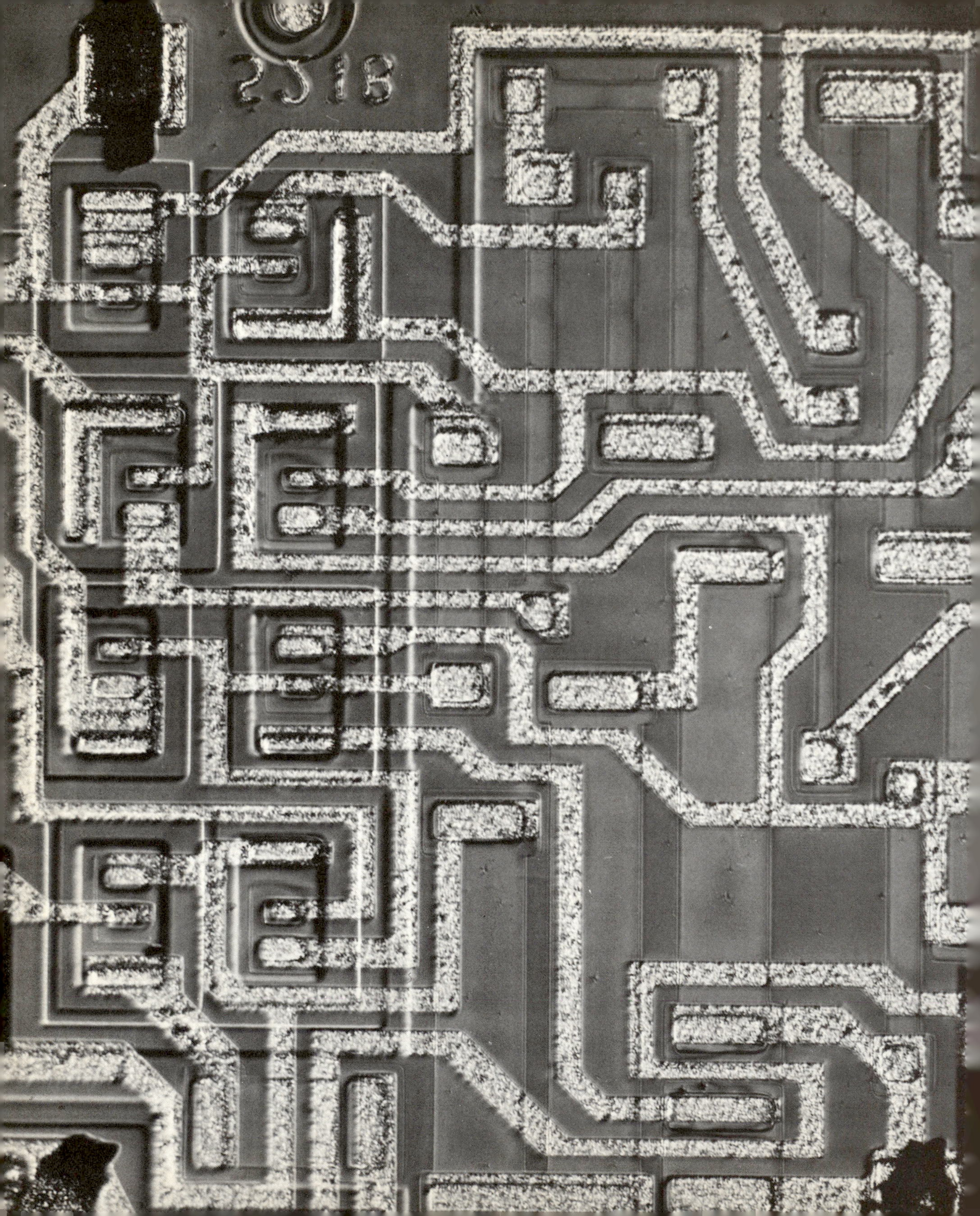
2318

14

ELECTRICAL PROPERTIES OF MATERIALS

CO-AUTHORED WITH PAUL A. FLINN*

THIS illustration shows a linear integrated circuit used in a power amplifier. The grayish background is a "chip" of silicon 0.1 mm across alloyed with a small amount of material to make it an n-type material, as described in the text. By a combination of diffusion, etching, insulation with SiO_2*, and further diffusion, a system of transistors, capacitors, and resistors is built up into a complete circuit. There are eight transistors in a group in the left-hand side of the photograph.*

This is a medium-scale integrated circuit as used in a radio. In the past decade, large-scale integrated circuits with up to ten layers have become very important as memory units.

In this chapter we shall consider conduction first in the high-conductivity materials, such as metals, then in the insulators and semiconductors. We shall then take up other important materials such as dielectrics and piezoelectrics and their uses.

---

*Formerly Professor of Physics and Materials Science, Carnegie-Mellon Institute. At present Senior Research Scientist, Intel Corporation, Santa Clara, California

14.1 General

We shall now take up some new and exciting fields for materials—their use in devices for transmitting electrons, heat, and light. These are the rapidly growing fields of the semiconductors, magnetic tapes, computer memory units and switches, television tubes, and lasers. Even in these phenomenal developments we shall see that the properties depend on the structure. Therefore a knowledge of the structure helps us to apply the materials to maximum advantage.

In this chapter we shall discuss the theory of electrical conduction in metals, semiconductors, and insulators. The applications are to be found in metallic conductors, superconductors, resistors, and semiconductor devices. Then we shall take up other electrical effects—dielectrics, thermionic emission, and piezoelectricity. Here the applications include capacitors, photocells, thermocouples, and crystal oscillators.

In Chapter 15 we shall study magnetic properties, then in Chapter 16 optical and thermal properties.

Electrical Conductivity

14.2 Conduction and carriers

Let us first consider conductivity from the large-scale engineering point of view, then see how we can develop the same relations from a knowledge of elementary particles. If we have a wire and apply a potential E, the current I that flows will depend on the circuit resistance R, as given by the well-known Ohm's law: $I = E/R$. The resistance depends on the nature of the wire itself: A copper wire has a lower resistance than an iron wire of the same size (length and cross section). We use the term *resistivity* (ρ) to characterize the inherent ability of the wire to affect current flow, and multiply this by l/A to give the resistance:

$$R = \rho\frac{l}{A} \quad \text{or} \quad \rho = R\frac{A}{l} \qquad \rho = \text{ohm}\,\frac{\text{m}^2}{\text{m}} = \text{ohm-m}$$

where R = resistance, ρ = resistivity, l = length of wire, and A = cross-sectional area of wire.

EXAMPLE 14.1 A student wants to build a dc heating coil rated at 110 volts and 660 watts. She has some Chromel wire 0.1 in. (2.54 mm) in diameter with a resistivity of 1.079 microhms-m. What length wire should she use?

ANSWER

$$\text{Power} = EI \qquad I = \frac{660}{110} = 6\ \text{A} \qquad R = \frac{E}{I} = \frac{110}{6} = 18.3\ \text{ohms}$$

$$R = \rho\frac{l}{A} \quad \text{or} \quad l = \frac{RA}{\rho}$$

$$l = \frac{18.3\ \text{ohms} \times (0.1\ \text{in.} \times 0.0254\ \text{m/in.})^2 \times (\pi/4)}{1.079 \times 10^{-6}\ \text{ohm-m}} = 86\ \text{m}$$

It is more positive and simpler in the text ahead to think of the material as conducting rather than resisting the passage of current, so we use the well-known parameter *conductivity* instead of resistivity. This is simply the reciprocal of resistivity:

$$\sigma = \frac{1}{\rho} = (\text{ohm-m})^{-1} = \frac{\text{mho}}{\text{m}}$$

Now let us examine the structural factors that go into conductivity. Given a cube of material, 1 m on the edge, the conductivity between opposite faces depends directly on three factors:

1. The number of charge carriers, n (carriers/m^3)
2. The charge per carrier, q (coulombs/carrier)
3. The mobility of each carrier, μ (m/sec)/(volt/m)

The conductivity depends on the product of all three of these factors:

$$\sigma = nq\mu$$

or, checking the units,

$$\sigma = \frac{\text{carriers}}{\text{m}^3} \times \frac{\text{coulombs}}{\text{carrier}} \times \frac{\text{m}}{\text{sec}} \times \frac{\text{m}}{\text{volt}}$$

Since coulombs = A-sec and volts = A-ohms,

$$\sigma = \frac{1}{\text{ohm-m}} = (\text{ohm-m})^{-1} = \frac{\text{mhos}}{\text{m}}$$

as in the large-scale example. This is important because we shall be studying how large-scale electrical properties are due to the movement of the elementary carriers.

14.3 Types of carriers

There are four different types of carriers that give the phenomenon of "current flow."

1. *The electron* (charge = 1.6×10^{-19} coul). We recall that an ampere is a coulomb per second. Therefore, a movement of 6.25×10^{18} electrons is the

motion of a coulomb of charge. If it occurs across our cell each second, we have an ampere of current flowing.

2. *The electron hole* (charge = 1.6×10^{-19} coul). In Chapter 7, we discussed briefly the concept of an electron hole in the $(Fe^{2+}, Fe^{3+})O^{2-}$ lattice. Associated with each Fe^{3+} ion there is an electron hole, a place to which a traveling electron is attracted because the electrical field of O^{2-} ions around the Fe^{3+} is adjusted for an Fe^{2+}. We can visualize the motion of three electrons, each moving one unit distance to the left (Fig. 14.1). As a result of the movement of the first electron, the Fe^{2+} marked "A" becomes Fe^{3+} in Step 1, then the Fe^{2+} marked "B," and so forth. Since we have electron motion, we have conductivity. However, instead of keeping track of small movements of many electrons, it is simpler to focus on the movement of the hole. Instead of saying that three electrons (negative charges) moved three unit distances to the left, we say that one positive charge moved the same total distance to the right (A to D). Thus an electron hole is a charge carrier with the same magnitude of charge as an electron but opposite in sign.

3 and 4. *Positive and negative ions* (charge = $n \times 1.6 \times 10^{-19}$ coul, where n = valence). Since the current is the net charge transferred per second, if a positive ion such as Ca^{2+} moves from left to right, the electrical effect is the same as if two electrons moved from right to left. In this way ion movement can contribute to conductivity. Of course, the movement of an O^{2-} from right to left would be equivalent (in terms of charge) to the movement of two electrons to the left. In a perfect crystal the movement of ions would be difficult, but real crystals have vacancies that make ion movement possible. It is important to distinguish between electron hole movement that takes place by *electron* jumps from ion to ion and ion movement that takes place by *ion* jumps from lattice position to position. Nat-

	electron hole	(A)			
Original	Fe^{3+}	Fe^{2+}	Fe^{2+}	Fe^{2+}	Fe^{2+}
	← e		(B)		
Step 1	Fe^{2+}	Fe^{3+}	Fe^{2+}	Fe^{2+}	Fe^{2+}
		← e		(C)	
Step 2	Fe^{2+}	Fe^{2+}	Fe^{3+}	Fe^{2+}	Fe^{2+}
			← e		(D)
Step 3	Fe^{2+}	Fe^{2+}	Fe^{2+}	Fe^{3+}	Fe^{2+}

Fig. 14.1 *Schematic representation of countercurrent flow of electrons and vacancies*

urally, we expect to find ions much less mobile than electrons because of their larger size.

14.4 Conductivity in metals, semiconductors, and insulators

Let us now discuss why there are vast differences in conductivity in different materials (Table 14.1).

In the metals we already have a partial picture of the metallic bond, in which the valence electrons are contributed by the atom as it becomes an ion in the unit cell. We therefore expect to find a high mobility of these charge carriers. We have to refine this picture a little to explain why a metal with two electrons such as magnesium (2.5×10^7 mho/m) does not have better conductivity than copper and silver with one valence electron and why silicon with four electrons has low conductivity (4×10^{-4} mho/m). To explain this and other concepts, we have to use the "band model," which runs contrary to classical thinking. Let us study the model and its explanation of conductivity.

First let us review the structure of sodium and the energy levels of its electrons (Fig. 14.2*a*, page 592). The electrons closer to the nucleus in the inner shells, such as 1*s*, 2*s*, and 2*p*, need not concern us here. It is the 3*s* electron, which is furthest from the nucleus and therefore at a higher energy level, that leaves the atom to form the electron gas. It is important to examine this electron gas more closely. When we discussed the structure of an atom, we gave the Pauli exclusion principle, which holds that only two electrons in an atom can have the same energy level (and these have opposite spins). A group of atoms in a metal block have the same exclusion tendency. The electrons that form the electron gas in sodium, for example, must have energies slightly different from one another. Therefore, instead of having the sharp energy level of the single 3*s* electron in an isolated sodium atom, a band of

Table 14.1 ELECTRICAL CONDUCTIVITY OF SELECTED ENGINEERING MATERIALS

Material	*Conductivity, mho/m*
Silver (commercial purity)	6.30×10^7
Copper (high conductivity)	5.85×10^7
Aluminum (commercial purity)	3.50×10^7
Ingot iron (commercial purity)	1.07×10^7
Stainless steel (301)	0.14×10^7
Graphite	1×10^5
Window glass	2×10^{-5}
Lucite	10^{-12} to 10^{-14}
Borosilicate glass	10^{-10} to 10^{-15}
Mica	10^{-11} to 10^{-15}
Polyethylene	10^{-15} to 10^{-17}

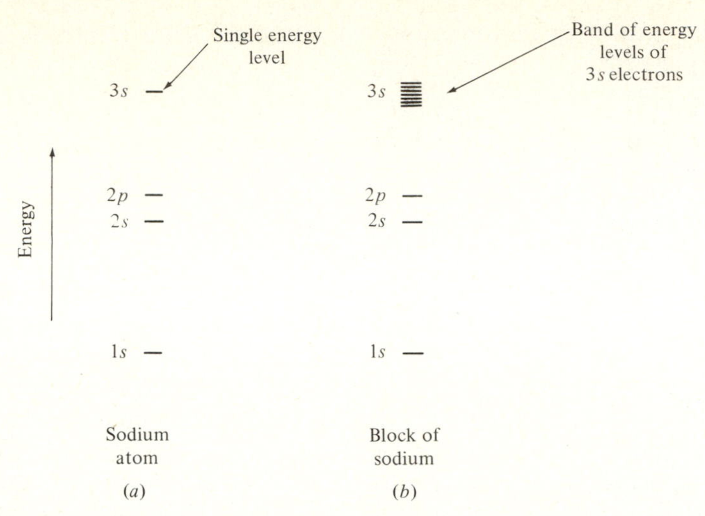

Fig. 14.2 *Development of a band of energy levels for (a) a single sodium atom, and (b) a block of sodium. The inner electrons (1s, 2p) are sufficiently shielded so that they form no bands.*

energy levels exists; hence the name *band theory* (Fig. 14.2*b*). It is postulated further that the number of energy levels in a band is equal to the number of electrons that can occupy the energy level times the number of atoms present in the block. This leads to the model of the conduction band of sodium having the number of "states" in the 3*s* band equal to twice the number of 3*s* electrons.

The energy levels or states at the top of the band are higher in energy. Also, since there is only one 3*s* electron per atom of sodium and the number of states equals twice the number of atoms, half of the states are unoccupied. Since the higher states have higher energy, we would expect that at absolute zero only the lower half of the states would be occupied, and this is indeed the case (Fig. 14.3*a*). As we warm up the metal, some electrons have higher energy and leave lower states unoccupied (Fig. 14.3*b*). It is useful for later discussion to define the *Fermi energy level* (E_F), where the probability of occupancy of a state by a conduction electron is 0.5. In other words, at the Fermi level half the states are occupied, and there are as many occupied states above this level as there are unoccupied states below it.

Note that the number of available states at different energy levels is a complex function. This is important in analyzing the action of a thermocouple.

Returning now to sodium, we explain conductivity by saying that the electrons in the 3*s* band can move readily to empty states in the band. Therefore, when an electrical potential is applied, the electron is accelerated and acts as a charge carrier.

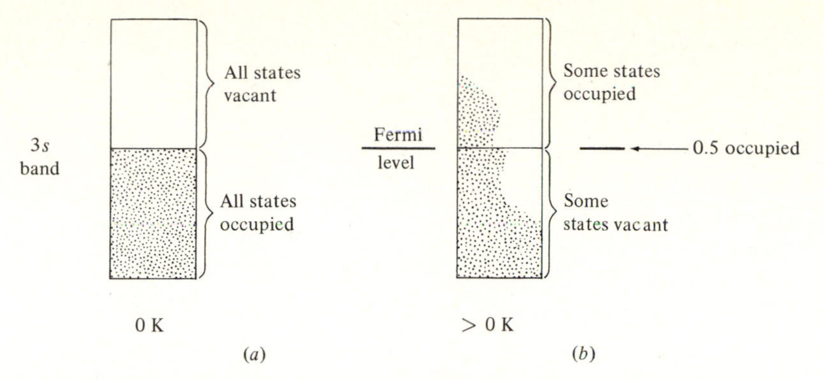

Fig. 14.3 *Enlarged (schematic) view of the 3s band of sodium at (a)* 0 K *and (b) higher temperatures*

The important point is that in a monovalent metal the energy band is half-filled, and the electrons in the upper part of the band can be energized and accelerated easily because there are open higher-level states nearby.

How do we explain the conductivity of magnesium with two valence electrons? Using the Pauli exclusion principle, we could calculate that all the spots in the 3*s* band, for example, would be filled because we have two valence electrons per atom. Since there are no empty adjacent spots, the element would be an insulator.

However, the energy levels of the 3*p* band overlap the 3*s* band, and there is a continuous series of possible states (Fig. 14.4, page 594). Electrons can be accelerated into the upper levels of the combined band and serve as conductors. However, magnesium is not as good a conductor as sodium because of the complexity of electron motion accompanying the overlap of the 3*s* and 3*p* bands.

14.5 Applications

Let us review the conductivity of a few alloys (Fig. 14.5, page 595). We see that the monovalent elements silver, copper, and gold are best. Pure aluminum also has good conductivity because there are many unfilled *p* levels above its higher-energy electrons (i.e., the *p* level is only half-filled for a trivalent metal). Iron and the other transition metals have lower conductivity because of the complex energy levels in the region where *s* and *p* or *d* bands overlap.

The wide range of conductivity in any family of alloys deserves careful analysis. The conductivity of all elements decreases as a second element is added in solid solution. Figure 14.6 (p. 595) shows data for copper. The more dissimilar an element is to copper, the greater the change in resistivity. An example is the effect of phosphorus vs. the effect of silver. When two phases are present, the conductivity is determined by the volume fraction of each.

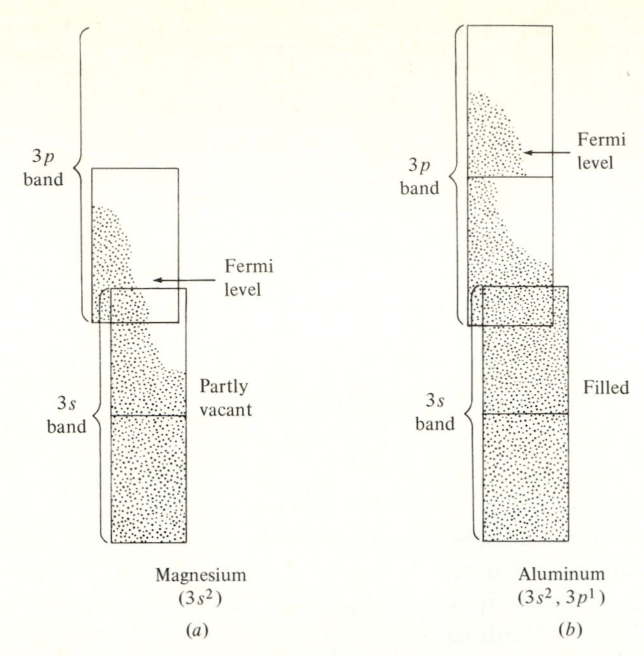

Fig. 14.4 *The 3s and 3p bands in divalent and trivalent metals*

Because of the severe increase in resistivity that accompanies alloying, it is often useful to improve the mechanical properties of a part by cold-working pure metal rather than by adding an alloy. Cold working increases the resistivity only a few percent because there are large blocks of pure undistorted metal available. On the other hand, even a small percentage of another element in solid solution produces irregularities in the lattice every few atoms. This has a pronounced effect in impeding electron motion, thereby raising the resistivity.

As an example of conductor selection, let us review the competition between copper and aluminum. The conductivity of copper is higher, but the *conductance* of a wire of the same weight per foot is lower than for aluminum. Therefore aluminum is used for high-tension wires. However, the distance between towers is an important cost factor, so a steel core is used to strengthen the cable. Once the power is close to the consumer, it is more convenient to use less bulky, more fatigue-resistant copper wires that have the added advantage of being easier to solder. For high frequencies the current flows only in the outer regions near the surface. Therefore copper tubing is used extensively in this application.

Resistors are of great importance in circuitry and in heating. In industrial furnaces resistivity is not important compared with oxidation resistance. For this reason the heat-resistant nickel-chromium alloys such as 80% Ni,

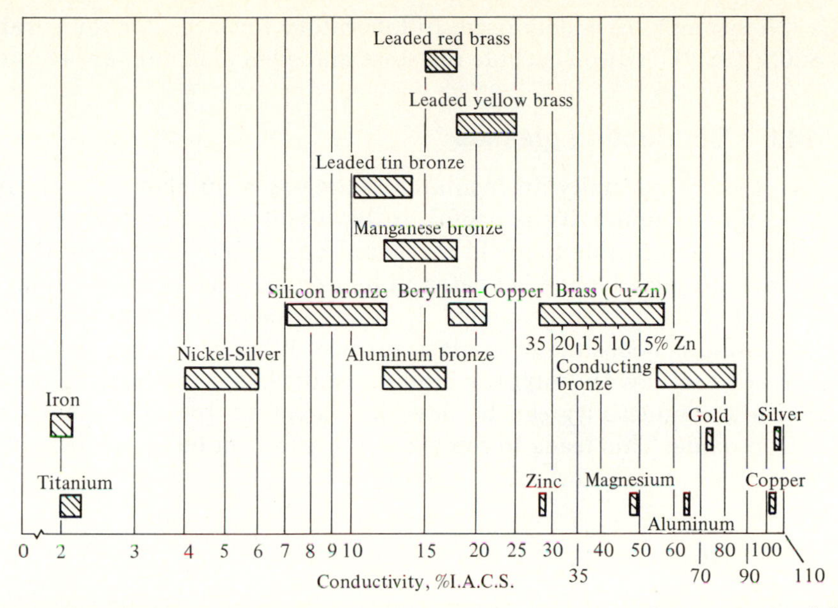

Fig. 14.5 *Conductivity of common metals and alloys in terms of percentage of the international annealed copper standard (which is not of optimum purity)*

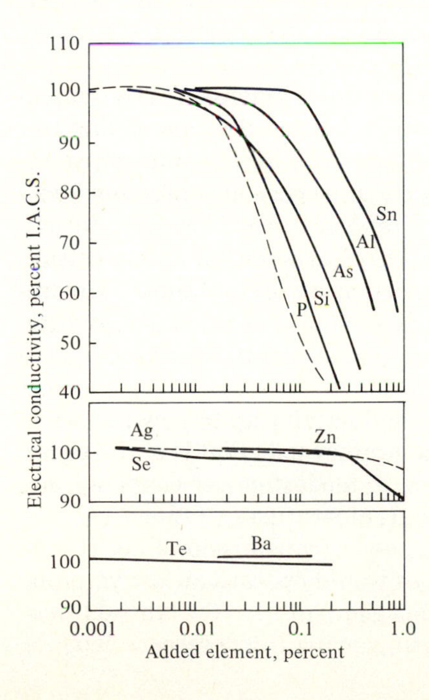

Fig. 14.6 *Effect of various elements on the conductivity of annealed copper*

20% Cr (Nichrome) are widely used. In laboratory furnaces, for temperatures over 1800°F (982°C) silicon carbide resistors and even platinum are employed.

14.6 Conducting glasses

Although glass is normally an insulator, there are a number of applications in which slight conductivity is useful. In high-voltage glass devices such as x-ray tubes, it is desirable to avoid a gradual build-up of charge. In this case a glass containing lead oxide is heated in hydrogen, producing a thin layer of metallic lead. This is then grounded. In another case a transparent layer of tin oxide, SnO_2, is deposited on a glass surface. When a voltage is applied, a small current can flow through the high resistance, and the heating prevents fogging. Bulk conductivity can be developed in glass by adding substances such as iron oxide. This leads to conduction by electron holes, as described in Chapter 7.

14.7 Superconductivity

Many years ago Kamerlingh Onnes found that when mercury was cooled below a critical temperature, the electrical resistance fell to zero. In one demonstration he showed that a current flowed indefinitely in a ring of mercury, and the phenomenon was called *superconductivity*. This effect has since been discovered in a number of other elements and even in alloys.

To point out some of the features of this effect, we shall review first normal conductivity and then some of the special characteristics of superconductors.

From elementary physics we learn that, as the temperature decreases, the vibration frequency of the positive ions of the lattice decreases. The resistance decreases because there is less conflict with electron motion (less scattering). Therefore, in normal metals, as we approach absolute zero, resistance gradually decreases with decreasing temperature. However, to explain the *sudden* appearance of superconductivity at a *critical* temperature above zero, Bardeen used the concept of *pair formation*. Below the superconducting critical temperature, electrons of opposite spin and the same energy form pairs. In this condition the electrons are exempt from scattering and therefore exhibit perfect conductivity.

The main practical application of superconductivity up to now has been in constructing large high-field magnets for laboratory use. Fields of the order of 10 teslas can be produced in large volumes with modest power consumption (primarily from refrigeration to liquid helium temperatures). Other applications are under development; superconducting power transmission lines transmitting power over long distances with low loss would operate at low voltage, eliminating the difficulty of providing high-voltage insulation for buried lines. New computer circuitry based on superconducting switching devices is capable of operating at switching speeds of 50 to 100 picoseconds, which is 5 to 10 times faster than can be obtained with semiconductor devices. Superconducting devices can also be used to measure extremely weak magnetic fields, such as those produced by electric currents in living organisms. Magnetic measurements of brain waves in humans show promise of providing considerably more detailed information than is available from conventional electroencephalograms.

14.8 Semiconductors: general

We come now to a fascinating group of materials which, although they are lower in electrical conductivity than the metals, are essential in a number of newly developed devices, from Dick Tracy wrist-size radios to tiny remote communication units in distant space satellites. The transistor, the solar battery, and the integrated miniature circuit in TV games and auto engine controls all depend on the properties of semiconductors.

The heart of all these appliances is not complex circuitry but materials that can marshal and direct electron motion in a tiny space with a precision never before attained. The key to these devices is a very small crystal of a semiconductor such as silicon, with controlled amounts of impurities in solid solution.

To understand the operation and vast potential of these materials, we shall first observe the differences in the band structures of these elements compared with the metals. Then we shall see the effect of adding different types of impurities (called *dopants*), leading to n- and p-type semiconductors. Finally, we shall take up a few applications of these materials.

14.9 Semiconductors and insulators

Semiconductors have a filled band (Fig. 14.7*b*, page 598); the next energy level into which we can accelerate electrons is *separated* by a small energy gap E_g in contrast to the metals, in which open states exist. For example, in the series carbon, silicon, germanium, tin (four valence electrons), the next bands are separated by the following values.

Material	*Energy Gap*, E_g (at 20°C), eV
Diamond	5.30 (insulator)
Silicon	1.06 (semiconductor)
Germanium	0.67 (semiconductor)
Tin	0.08 (conductor)

We express the width of the energy gap in electron volts. The *electron volt* (eV) is the energy required to move an electron through a field of 1 volt. To give a feel for the magnitude of this quantity, the thermal energy of an elec-

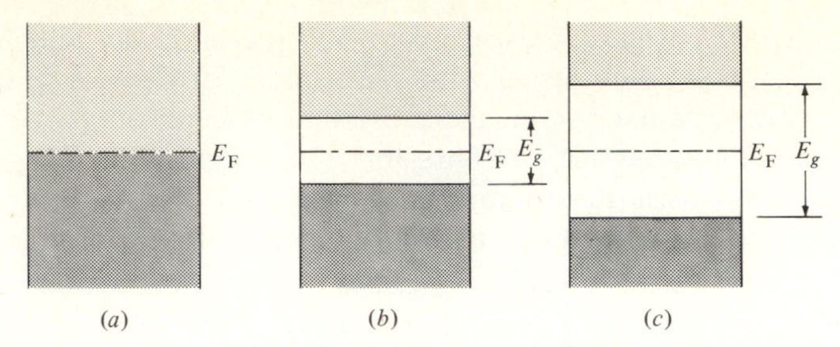

Fig. 14.7 *Models of energy bands of (a) a metal, (b) a semiconductor, and (c) an insulator.* E_F *is the Fermi level (0.5 of possible states are occupied).*

tron at room temperature is only about 0.03 eV on the average, so we would not expect many carriers to be produced from this source except in tin. Diamond, therefore, is a good insulator, whereas the small number of conduction-band electrons in silicon and germanium leads us to call them semiconductors.

At this point we face a difficult question as to the existence of an energy gap, since there are four electrons in the outer shell in each of these cases. Why should the gap exist if we have filled only two of the six 2*p* states in carbon, for example? The explanation is rather complex and depends principally on the fact that, although eight electrons can be accommodated in the two 2*s* plus six 2*p* states, an effect called *hybridization* takes place in a diamond. Instead of the four valence electrons falling into two 2*s* and two 2*p* levels, a *hybrid* group of four electrons is formed, giving rise to four equal tetrahedral covalent bonds. This acts like a filled band, and there is an energy gap between this band and the conduction band above it.

Now we have the problem of why in silicon, for example, we encounter any conductivity *at all* if the band is filled and a gap exists to the next higher band. The answer lies in the fact that there is a variation in the energies of the outer-shell electrons. As we raise the temperature above 0 K, an increasing, though small, number of electrons have enough energy to reach the upper band (Fig. 14.8*a*). Furthermore, shifting these electrons to the upper band creates other carriers by forming electrons holes in the lower band (Fig. 14.8*b*).

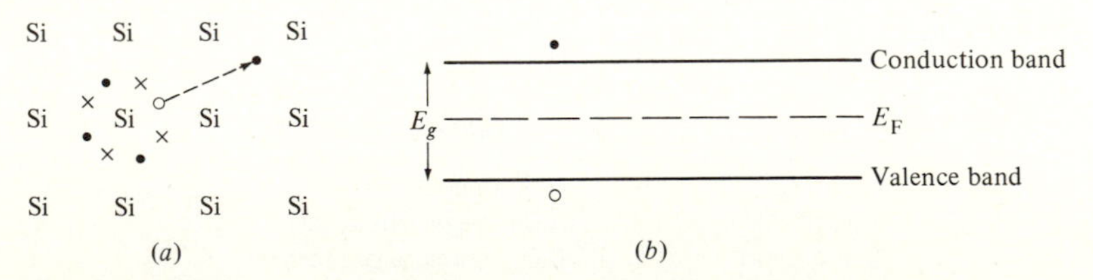

Fig. 14.8 *(a) Formation of a conduction electron and an electron hole in pure silicon (intrinsic semiconductor). Through the acquisition of thermal energy, an electron attains a high enough energy to leave its valence position. (b) The process shown in (a) has led to the presence of an electron in the conduction band and a hole in the valence band. The Fermi level is halfway between the bands (50 percent occupation).*

The band above the valence band is called the *conduction band,* although conduction also occurs in the valence band because of the electron holes.

In addition to the group IV elements, semiconductors have been produced by combining elements with three outer-shell electrons with elements with five outer-shell electrons to give an average of four, as for example in AlP and InAs. These are called III-V compounds. The band structure is similar to that of group IV, and a range of energy gaps is obtained for different combinations.

14.10 Extrinsic vs. intrinsic semiconductors

Up to now we have discussed only semiconductors with an average of four electrons per atom, or *intrinsic* semiconductors. If we vary this balance by *doping* with impurities, we obtain *extrinsic* semiconductors of two types, *n* and *p*. To understand these important materials, let us consider the effect of adding impurities or dopants to an intrinsic semiconductor such as silicon. If we add a small quantity of phosphorus, these atoms form a substitutional solid solution with silicon, giving the structure shown in Fig. 14.9.

We have shown the valence electrons contributed by the four silicon atoms adjacent to a phosphorus atom by the symbol ×. However, phosphorus has five outer-shell electrons, shown by •, so we have an extra electron. The phosphorus, therefore, is an electron donor, and the structure is called an "*n*-type extrinsic semiconductor" (*n* refers to the *negative* extra electron). The electrons from the phosphorus are not in exactly the same positions as those that would be produced by an electron in the valence-band, but they lie just below it in energy.† This is because the electrical field of the phosphorus is

†However, at room temperature, there is enough excitation by thermal energy to raise the electron to the conduction band.

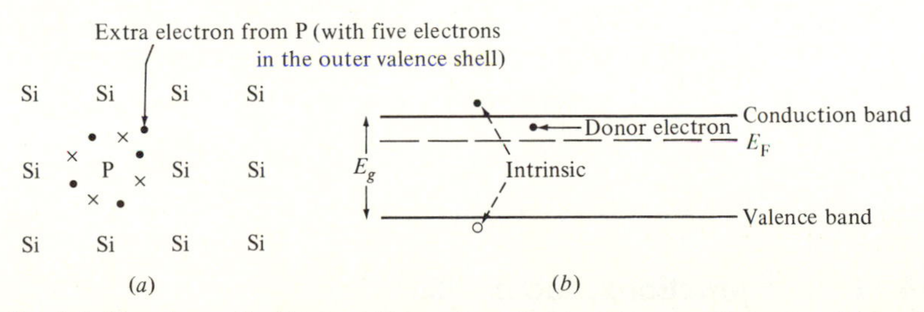

Fig. 14.9 *(a) Structure of silicon with a phosphorus atom as dopant, giving rise to an extra electron, thus making an n-type semiconductor (extrinsic). (b) The electron from the phosphorus shown in (a) is found just below the conduction band. The Fermi level E_F is higher than in Fig. 14.8 because of the addition of an electron above the old Fermi level and no increase in the electron holes below.*

not identical with that of a silicon atom. The Fermi level E_F is close to the conduction band. We have added electrons to the structure without adding holes, and so E_F shifts upward. By doping silicon in this way, we can increase its conductivity 10,000 times!

Now let us consider the effect of adding an element with less than enough electrons to satisfy the covalent bonds of silicon. Aluminum is such an atom (Fig. 14.10). Again the silicon electrons are shown by × and the aluminum electrons by •. We see that there is an electron missing. Aluminum is an acceptor element, and the structure is called a "*p*-type semiconductor" (electron holes are *positive*). The band structure is also shown in Fig. 14.10, with the holes just adjacent to the valence band. In this case the conduction is mainly from electron holes, and the 50% occupied level (E_F) is closer to the valence band, since we have added holes without adding electrons.

EXAMPLE 14.2 Calculate the conductivity of the *intrinsic* semiconductor germanium from the following characteristics (300 K):

Hole density = electron density = 2.4×10^{19} carriers/m^3
Electron mobility = 0.39 m^2/volt-sec
Hole mobility = 0.19 m^2/volt-sec
Charge/electron = charge/hole = 1.6×10^{-19} coulomb/carrier

ANSWER

$$\sigma = nq\mu \qquad \text{(from Sec. 14.2)}$$
$$= nq_n\mu_n + pq_p\mu_p$$

where n = number of electrons/m^3, q_n = charge/electron, μ_n = mobility of electron carrier, p = number of holes/m^3, q_p = charge/hole, and μ_p = mobility of hole carrier. Since, in an intrinsic semiconductor, $n = p$ and $q_n = q_p$, we have

$$\sigma = nq(\mu_n + \mu_p)$$
$$= (2.4 \times 10^{19}\ \text{carriers/m}^3) \times (1.6 \times 10^{-19}\ \text{coulomb/carrier}) \times [(0.39 + 0.19)\ \text{m}^2\text{/volt-sec}]$$
$$= 2.23\ \frac{\text{coul}}{\text{volt-sec-m}} = 2.23\ \frac{\text{A}}{\text{volt-m}}$$
$$= 2.23\ \frac{1}{\text{ohm-m}} = 2.23\ \frac{\text{mho}}{\text{m}}$$

14.11 *P-n* junctions, rectification

One of the important uses of semiconductors is in rectifying alternating to direct current. For this a junction is made between *n*- and *p*-type material. We can do this by changing the impurity (dopant) from a *p* type such as aluminum to an *n* type such as phosphorus during crystallization. Let us examine the operation of this device.

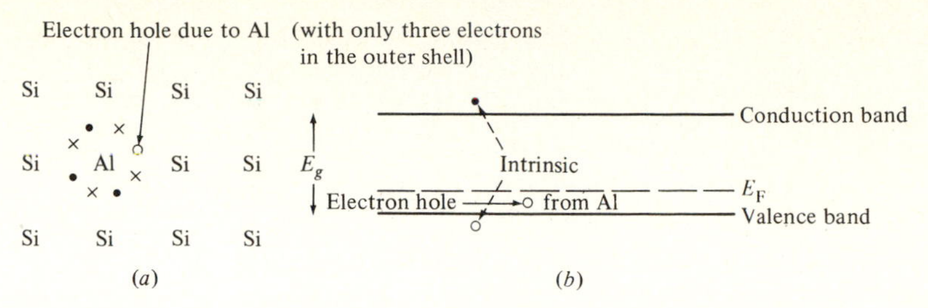

Fig. 14.10 *(a) Structure of silicon with an aluminum atom added as a dopant, giving rise to an electron hole, thus making a p-type semiconductor (extrinsic). (b) The electron hole produced by the aluminum shown in (a) is found just above the valence band. The Fermi level* E_F *is lower than in Fig. 14.8 because of the addition of a hole below the old Fermi level and no increase in the number of electrons above.*

Assume that electrically neutral blocks of n-type and p-type material are brought into contact (Fig. 14.11*a*, page 602). Note that when one matches the band diagrams, the two Fermi levels must be equal, indicating no overall flow of current in either direction. (Some electrons, attracted to the holes of the p-type material, flow from the region of the junction. Similarly, some holes move from the junction region to the n-type material.) Although the potentials of the two conduction and valence bands are different, the Fermi levels are the same. Note that the dopant atoms do not move, only the carriers. The migration of carriers away from the junction creates a *depleted* region that is higher in resistance than the material on either side. The Fermi level is constant throughout the assembly. Even though there are fewer carriers in the depleted region, the potential at which the probability of occupancy by an electron is 0.5 is still the same. The constant Fermi level also indicates no current flow. Also, the depleted region has length l in the x direction.

Now let us apply an external voltage, ΔV, across the junction. This builds up electrons on the n side of the junction, raising the Fermi level (Fig. 14.11*b*). This is called a *forward bias.* This pile-up raises the Fermi level on the n side. The block of n material is therefore raised to fit more closely with the p-type material (Fig. 14.11*b*). In turn, this better fit reduces the length of the depleted region of high resistance, and current flows relatively easily. Another way to look at this is that an electron at the bottom of the conduction band has less of a potential hill to climb in flowing toward the p material. Within the depleted region, there is *recombination* of n and p conductors.

On the other hand, if the potential is reversed (reverse bias), electrons are added to the p side and the Fermi level is raised. This increases the mismatch with the n-type material (Fig. 14.11*c*). The length of the depleted high-resistance zone increases, and the flow of current is impeded. The electrons in the conduction band have a greater potential hill to climb in passing from the n to the p block. Note that the displacement of the Fermi level and of the level of the conduction and valence bands corresponds to the applied potential ΔV.

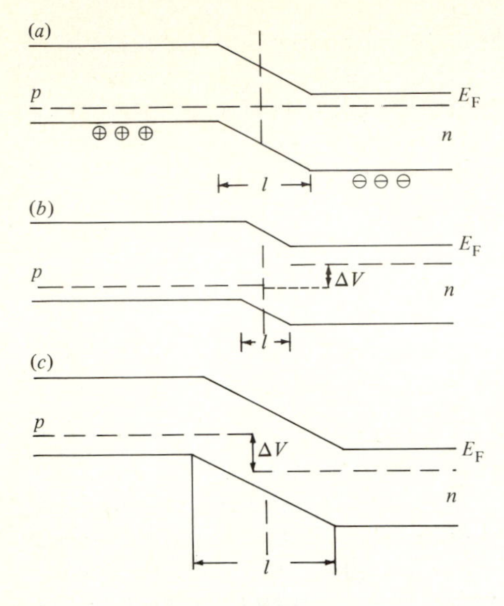

Fig. 14.11 *A* p-n *junction device. (a) No bias (current flow). (b) Forward bias (electrons are pumped into the* n *material). The depleted region becomes smaller, allowing current to flow. (c) Reverse bias (electrons are pumped into the* p *material). The depleted region becomes larger, inhibiting flow of current.*

Figure 14.12 shows the net result of these effects. If a small voltage is applied in the forward direction, a large current is obtained, whereas very little current results with a normal range of voltages in the reverse direction. Therefore, if we impose the output of an alternating current transformer, it is rectified to a series of dc pulses.

Now let us analyze these effects quantitatively.

The current across the *p-n* junction is given by the equation:†

$$I = I_s \exp(qV/kT - 1)$$

where V = applied voltage and I_s = maximum current obtained with reverse bias (a small value).

The maximum or limiting current I_s obtained with reverse bias may be expressed as:

$$I_s = Aqn_i^2\left(\frac{D_e}{L_eN_d} + \frac{D_p}{L_pN_s}\right)^{\dagger}$$

†See D. A. Fraser, *The Physics of Semiconductor Devices,* Clarendon Press, Oxford, 1977.

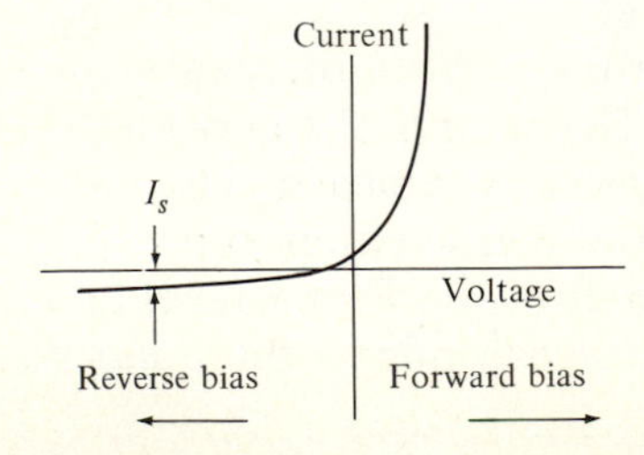

Fig. 14.12 *Amplification of current, obtained by applying a current with a forward bias.* I_s *is maximum current obtained with reverse bias.*

where A = area of junction, n_i = number of intrinsic carriers (a function of material and temperature, as discussed in Example 14.5), D_e = diffusion coefficient for electrons, D_p = diffusion coefficient for holes, L_e, L_p = constants (diffusion lengths related to the distance carriers travel before recombination), and N_d, N_s = concentrations of donor and acceptor elements.

EXAMPLE 14.3 Calculate the ratio of the currents with forward and reverse bias, respectively, with imposed voltages of +1 volt and −1 volt at room temperature (300 K).

ANSWER The current in the forward direction, I_F:

$$I_F = I_s \exp\left(\frac{qV}{kT} - 1\right)$$

in the reverse direction, I_R:

$$I_R = I_s \exp\left(\frac{qV}{kT} - 1\right)$$

$$I_F = I_s \exp\left[\frac{(1.602 \times 10^{-19}\ \text{coul})(1\ \text{volt})}{(1.381 \times 10^{-23}\ \text{J-K}^{-1})(300\ \text{K})} - 1\right]$$

$$= I_s\,(2.28 \times 10^{16})\ \frac{\text{coul-volt}}{\text{A-volt-sec}^{-1}}$$

$$I_R = I_s \exp\left[\frac{(1.602 \times 10^{-19}\ \text{coul})(-1\ \text{volt})}{(1.381 \times 10^{-23}\ \text{J-K}^{-1})(300\ \text{K})} - 1\right] = I_s\,(5.92 \times 10^{-18})$$

$$\text{Ratio}\ \frac{I_F}{I_R} = 3.85 \times 10^{33}$$

EXAMPLE 14.4 Calculate the change in I_s caused by doubling the concentration of the dopants in the p and n sides of a semiconductor.

ANSWER

$$I_{s,\,\text{orig}} = \text{const}\left(\frac{D_e}{L_e N_d} + \frac{D_p}{L_p N_s}\right)$$

$$I_{s,\,\text{new}} = \text{const}\left(\frac{D_e}{2L_e N_d} + \frac{D_p}{2L_p N_s}\right) = \tfrac{1}{2} I_{s,\,\text{orig}}$$

In other words, although the conductivity of both the n and p legs is increased, the limiting current is decreased and rectification is improved. However, in silicon p-n junctions the reverse current is already very low—approximately 0.001 microampere. This is why a p-n junction may also function as a capacitor.

14.12 Solar cells

The solar cell (Fig. 14.13) is another interesting use of *p-n* junctions. In the preparation of a solar cell, an *n*-type layer is formed on the surface of a relatively thick *p*-type silicon substrate by diffusion of an element such as phosphorus. Contact strips are imprinted on the *n* and *p* layers. The *n* layer is exposed to sunlight or other radiation. Electrons receiving the radiation are raised to higher energy levels and move to the conduction band, and so holes are formed. This is called *electron–hole pair production*. It raises the Fermi level of the *n* layer. It is equivalent to attaching an electric generator across the diode.

The holes flow to the *p* region, raising the Fermi level, which forward-biases it, giving a current flow in the forward direction. An open-circuit voltage develops. This is essentially an equilibrium between the formation of hole pairs, the light current, and the migration of carriers across the *p-n* junction.

The equation for the migration across the interface, as before, is

$$I = I_s \exp\left(\frac{qV}{kT} - 1\right)$$

The V is the open-circuit voltage of the junction. At equilibrium, $I = I_L$, where I_L is the current produced by the radiation, and

$$I_L = I_s \exp\left(\frac{qV}{RT} - 1\right)$$

Solving for V, we obtain

$$V = \frac{RT}{q}\left(\ln\frac{I_L}{I_s} + 1\right)$$

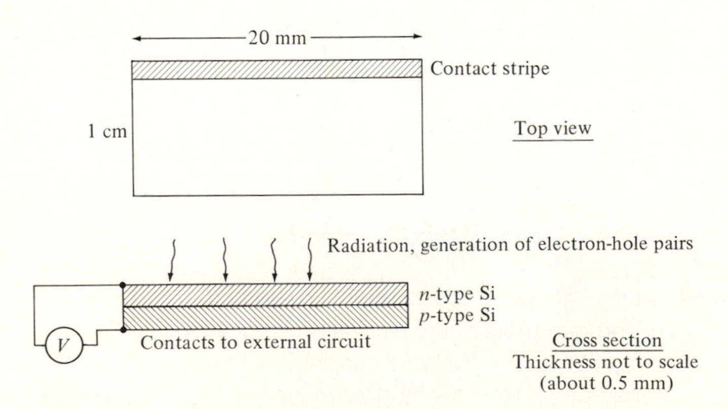

Fig. 14.13 *Schematic representation of a solar cell*

14.13 Amplifier circuits

To illustrate the use of p-n junctions in more complex circuits, let us consider two different devices in amplifier circuits, the bimodal transistor and the MOSFET. The principle of the amplifier circuit is that the magnitude of a strong current passing between a point A at ground potential and a point B at higher potential is generally greatly modified by a relatively weak signal voltage. The signal is said to be amplified because the strong current varies by a greater amount than the pulses of the signal. The current also has the strength to drive a radio speaker, for example.

In the bimodal transistor two p-n junctions are assembled with the n portions in contact (Fig. 14.14). A potential is applied between the p section (grounded) at the left of the figure and the p section at the right in such a way that the left section is forward-biased, and therefore the right section is reverse-biased. As a result, the energy bands are aligned as shown. This is a composite of the situation for a single p-n junction, discussed earlier. Let us simply consider the movement of holes from the left-hand p region. There is a potential against the hole movement from the first p to n material. To explain this we must consider that the effect of the slopes of the lines on holes is the reverse of the effect on electrons. Therefore, since the valence-band line drops, this indicates resistance to flow of holes.

However, once the holes reach the n-n junction, they are accelerated through the next n-p junction because the lines rise sharply. If we now superimpose the voltage from a weak signal at the n-n junction, this has a pronounced effect on the strong current, and hence the weak current is amplified.

In the MOSFET (*m*etal-*o*xide-*s*emiconductor *f*ield-*e*ffect *t*ransistor) (Fig. 14.15, p. 606), p-type silicon is altered to n-type in small patches. A thin silicon dioxide layer is deposited, followed by an aluminum conducting layer called a *gate*. If we apply a potential between the two n spots, no appreciable current flows because one spot is reverse-biased and we can

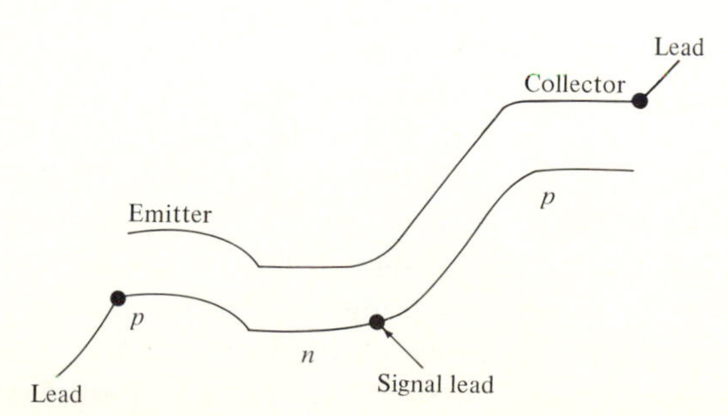

Fig. 14.14 *Schematic representation of a bimodal transistor (two* p-n *junctions)*

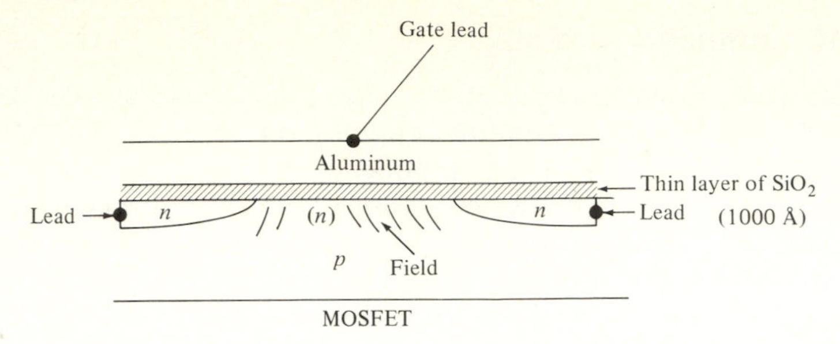

Fig. 14.15 *Schematic representation of a MOSFET (metal oxide silicon field effect transistor)*

obtain only I_s. If we now apply a signal flowing through the aluminum gate, the field affects the p layer below. If the signal is in the "right" direction, the *field effect* distorts the layer of p material close to the cap, converting it to an n-type conduction band. This permits current to pass easily. On the other hand, if the signal is in the opposite direction, only a small limiting current can flow. The signal is therefore amplified.

14.14 The Hall effect

We have spoken about the motion of holes and electrons as carriers without describing how we know which major carriers are present. This identification is accomplished using the *Hall effect*. It is more important because we can turn the principle around to measure magnetic fields in the analog multiplier.

To explain the Hall effect, recall that when a wire carrying a current is placed in a magnetic field, there is a force on the wire—or, more correctly, on the carriers in the wire. This is of course the principle of the electric motor. Now consider the case shown in Fig. 14.16, where a current is flowing in a block of metal because of the motion of carriers. We want to find out whether the carriers are holes or electrons. If we superimpose a magnetic field B, this exerts a force on electrons or holes acting at right angles to both the direction of current flow and the magnetic field, that is, toward the vertical side walls of the block. If the majority of the carriers are electrons, the motion of the side current is toward one wall, whereas if they are holes, the motion is opposite. These differences can be transmitted as voltages of opposite sign, therefore, to the voltage-measuring probes on the side walls. The magnitude of the voltage measured is also a measure of the number of carriers.

In actual experimentation we measure the voltage developed by a given magnetic field. The Hall constant is derived from the data. It is large for semiconductors and varies inversely with the number of carriers. The Hall voltage is a function of the applied magnetic field and the normal current flowing through the conductor.

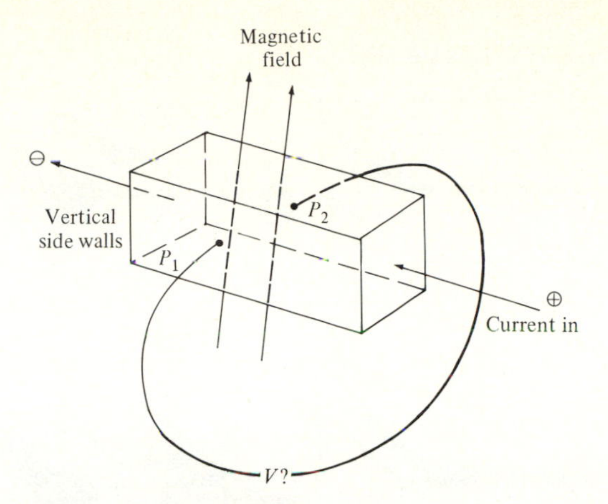

Fig. 14.16 *A block of metal containing a flow of current and exposed to a magnetic field exhibits Hall voltage, V. This is called the Hall Effect.*

We may also use the Hall effect to measure the intensity of a magnetic field if we use a conductor with known carriers.

14.15 Production of transistors

To illustrate what is involved in selecting and processing materials for semiconductors, let us review some of the operations involved in producing a two-level n-channel polysilicon gate MOS (metal-oxide-semiconductor) circuit element† (Fig. 14.17, page 608).

1. *Preparation of silicon-chip base.* First we must prepare high-purity silicon because of the effects of traces of n- and p-type elements on conductivity. We do this by a process called *zone melting,* whose principle depends on the phase equilibria discussed in Chapter 4. In most cases, if we have a given impurity B dissolved in solid solution in the silicon, the solubility of B is higher in liquid silicon than in the solid. We can express this by the partition coefficient discussed earlier. Now let us consider a mass of solid silicon contained in a long refractory boat. If we melt one end by placing an induction coil around and across the boat, then move the boat slowly so that the liquid zone sweeps from one end to the other, the impurity concentrates in the liquid and is swept to the far end of the boat. If we make a number of passes in this manner, we can reduce the impurities in the main body of the boat to one part per billion. We discard the end portion containing the impurities.

†Reference: W. G. Oldham, "The Fabrication of Microelectronic Circuits," *Scientific American,* September 1977.

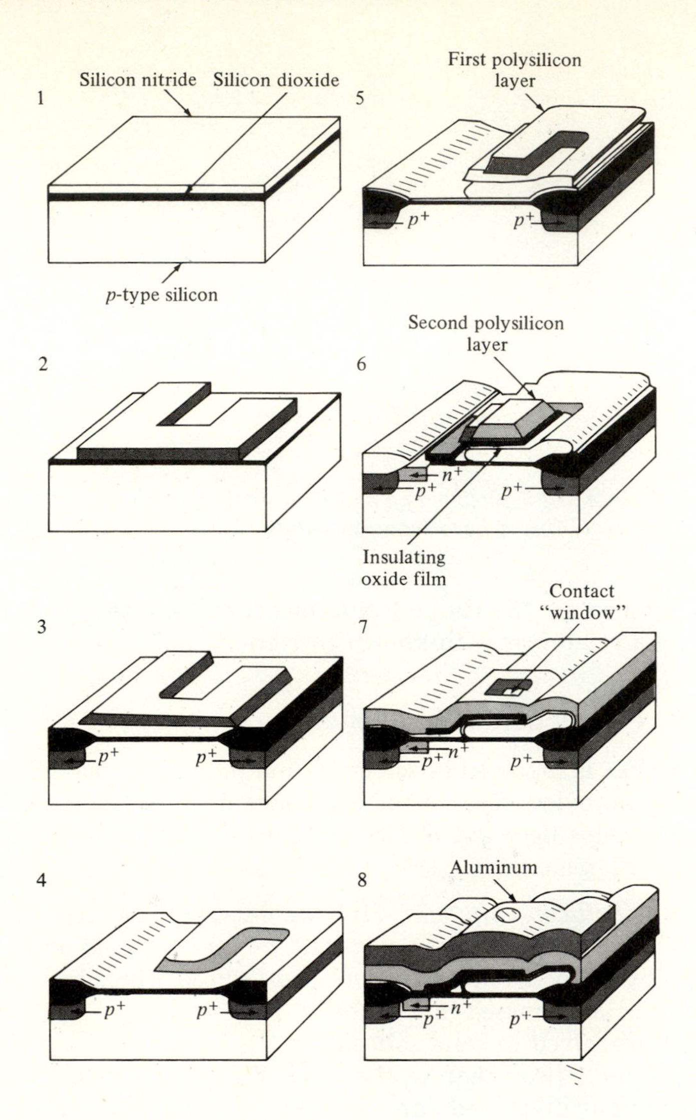

Fig. 14.17 *Various steps required in the fabrication of a two-level* n*-channel polysilicon-gate metal-oxide-semiconductor (MOS) circuit element. (1) Selective oxidation of silicon with the aid of vapor-deposited silicon nitride film. (2) Selective removal of layer of silicon nitride by photolithography (first masking). (3)* p*-type dopant is added, followed by oxidation of silicon; silicon nitride acts as a mask. (4) Silicon nitride is removed by chemical etching, leaving the unattached silicon and silicon dioxide. (5) First polysilicon layer is deposited (second masking). (6) Silicon is oxidized to produce an insulating layer; then a second polysilicon layer is deposited (third masking). (7) Hydrofluoric acid etch attacks silicon dioxide to expose some regions to diffusion of* n*-type dopant. More silicon dioxide is deposited and contact windows are opened in this fourth mask. (8) Deposition of aluminum provides the fifth mask.*

A single crystal of silicon is required to avoid the effects of grain boundaries and change in grain orientation. To get this, we remelt the silicon in a crucible under a protective gas atmosphere to avoid oxidation. If we want *p*-type silicon, we add a small amount of an acceptor element to the melt. A small crystal of silicon of the desired orientation is placed at the end of a rod that is lowered to touch the melt, which is close to the freezing point. As the silicon freezes onto the nucleus, the crystal is pulled upward and rotated, and as material continues to freeze on the rod, the crystal is withdrawn. The long, rodlike crystal, about 70 mm in diameter, is then carefully sectioned into $\frac{1}{2}$-mm slices with a diamond-tipped saw wheel, and the slices are polished. These form the silicon wafers.

The key to the economics of the process from this point onward is to produce many identical circuits on adjacent squares of the chips *before* the wafer is broken up into "dice" that must be handled separately. Although the prices of silicon and of processing have risen moderately in recent years, the price of the circuits has fallen drastically. This is because the size of the dice has been constantly decreased, allowing many dice to be processed for the same overall cost.

2. *Growing a film of silicon dioxide.* While it is still in the gassy state, a film of SiO_2 is then grown on the surface of the wafer by exposing the disk to an atmosphere of pure oxygen at 1000 to 1200°C. For example, a layer of oxide a tenth of a millimeter thick can be produced at 1050°C in 1 hour. The ease of production and the insulating qualities of this SiO_2 layer have resulted in the dominance of silicon as a semiconductor material over germanium or the compound materials such as gallium arsenide. The silica can also be used as a dielectric in a capacitor formed between layers. Several hundred wafers can be processed simultaneously with computer control of the operation.
3. *Etching the circuit design.* We now wish to etch away the oxide to expose the silicon for a selected circuit design. To do this, we want to deposit a protective film in the desired pattern. A special organic lacquer called a *photoresist* is dropped on the wafer. When the wafer is spun, a uniform film is formed and dried. The organic film has the special property that, if it is exposed to ultraviolet light, the polymer cross-links and is insoluble in organic solvents, whereas the original material is soluble. Therefore light is passed through a carefully produced mask that defines the desired circuit on the wafer surface. After exposure, the unexposed lacquer is washed off, leaving a bare silica surface in the desired circuit pattern. The wafer is then immersed in a solution of hydrofluoric acid that cuts through the silica layer to the silicon. The photoresist protection layer covering the SiO_2 is removed by another chemical treatment.

We can now produce *p-n* junctions at exposed areas in the silicon chip by diffusion. A boat of silica wafers is placed at 1100°C in an atmosphere containing phosphorus. A layer of phosphorus-rich material builds up in the surface, and the reaction is governed by relations similar to those studied in the carburization of steel. For example, a layer one micrometer deep is attained in 1 hour. A second heat treatment in a neutral atmosphere called a *drive-in* treatment is used to develop a deeper layer and reduce the concentration gradient. Ion bombardment may also be used to implant dopant atoms. This has the advantage that it can be performed *through* a thin layer of silica.

When circuits are to be formed above the lower layers, thin films may be evaporated in predetermined patterns, using the same photoresist techniques just discussed. Polycrystalline silicon is laid down by deposition of chemical vapor by heating silane, SiH_4, which decomposes into silicon and hydrogen. The polycrystalline silicon is used as a resistor, not as a *p-n* junction. In these steps silicon nitride is used in place of silicon oxide to produce a thinner, flatter film.

After production, the individual circuits on each wafer are tested with a computer-controlled probe. Unsatisfactory circuits are automatically marked with a tiny paint dot and later discarded.

The wafers are then notched and broken into dice and assembled into packages. This is a crucial step to protect against corrosion. Carefully formulated ceramic materials are generally used.

14.16 Effects of temperature on electrical conductivity

One of the salient features of a metal is the decrease in conductivity with increasing temperature. We discussed the reasons for this briefly in the section on superconductivity. In contrast, the conductivity of semiconductors and insulators increases with temperature. When an intrinsic semiconductor is heated, more electrons are pumped up to energies where they can enter the conduction band, leaving electron holes. This effect is disadvantageous in a circuit that depends on a *p-n* junction because the number of natural carriers can obscure the effects of the carriers added by the dopant.

On the other hand, because of the sensitivity of the semiconductor to temperature, the change in resistance can be used to indicate temperature accurately from 1 to 723 K (−273 to 450°C). These devices are called *thermistors* (Fig. 14.18).

EXAMPLE 14.5 The general formula for the number of electrons and holes in an *intrinsic* semiconductor is

$$n \cdot p = AT^3 e^{-11{,}600 E_g / T}$$

where $n \cdot p$ = number of carriers/m^3, A = constant, T = absolute temperature in K (K = 273 + °C), E_g = energy gap (eV), and e = 2.718.

We wish to construct a thermistor from silicon. What is the change in the number of carriers per cubic centimeter in going from 27 to 47°C? For silicon we are given that $n = p = 1.5 \times 10^{16}$ carriers/m³ at 300 K (27°C). Also, $E_g = 1.06$ eV and remains constant over the temperature range in question.

ANSWER First we must arrive at the general equation, which requires calculation of A from the known conditions:

$$2.25 \times 10^{32} \text{ carriers/m}^3 = A(300 \text{ K})^3 e^{(-11,600 \times 1.06)/300} \qquad A = 5.26 \times 10^{42}$$

$$\text{At } 47°\text{C}, \quad n \cdot p = n^2 = 5.26 \times 10^{42}(320)^3 e^{(-11,600 \times 1.06)/320} = 3.54 \times 10^{33}$$

$$n = 5.95 \times 10^{16} \qquad \text{or} \qquad \frac{5.95 \times 10^{16}}{1.5 \times 10^{16}} = 3.97$$

Therefore, for the 20°C change, the number of carriers increases by 3.97 times, which of course increases the conductivity by this factor plus a factor for the corresponding increase in mobility of both holes and electrons.

The conductivity of ionic materials also increases with temperature. This is shown quantitatively by the effect on mobility:

$$\mu = \frac{qD}{kT}$$

where μ = mobility (m²/volt-sec), q = charge on carrier (coulomb/carrier), D = diffusion coefficient (m²/sec), T = absolute temperature (K), and k = Boltzmann's constant (1.38×10^{-16} erg/K).

Because of their large size, the mobility of the ions in a solid is orders of magnitude lower than that of electrons or holes. Ions move by jumps, using

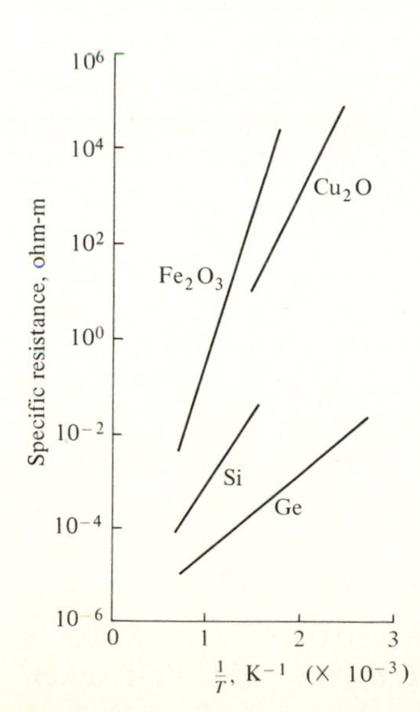

Fig. 14.18 *Variation in resistivity of semiconductors with temperature*

vacancies. In liquid salts the conductivity is higher because ion movement in the liquid state is easier.

Other Electrical Properties

14.17 Dielectric properties

A dielectric generally has two functions: (1) as an *insulator* and (2) to add to the total *capacitance* of a condenser compared with an air gap between the plates. Use as an insulator involves two principal factors: the breakdown strength and the useful temperature range. Cotton, silk, and many plastics are used below 90°C. Inorganic fillers such as mica and asbestos are used with plastics up to 130°C. The range can be extended with silicone to 180°C. Above this range inorganic materials such as mica, porcelain, and glass are necessary.

Typical values of breakdown voltage [the voltage across 1 mil (0.001 in. = 0.00254 cm) which cause considerable discharge], called the *dielectric strength,* are given in Table 14.2, on pages 614–615. It is evident that for ordinary operating temperatures only mica exceeds plastics. However, plastics can absorb water, as explained earlier. This lowers the breakdown voltage and should be taken into account. Therefore the breakdown voltage is governed by a complex number of internal and environmental factors.

14.18 Dielectric constant

The relative permittivity (dielectric constant) ϵ_r is the quantity used to evaluate the charge-storing capacity of a dielectric in a capacitor:

$$\epsilon_r = \frac{\epsilon}{\epsilon_0}$$

where ϵ is the permittivity of the dielectric and ϵ_0 that of a vacuum. Typical values of the dielectric constant are given in Table 14.2.

In absolute values, for a parallel-plate capacitor, the capacitance C is

$$C = \frac{0.224\epsilon_r A}{10^6 d}(n - 1)$$

where C = microfarads (μF), A = area in square inches of one plate, d = distance between plates in inches, n = number of plates, and ϵ_r = dielectric constant.

The value of the dielectric constant depends on the ability of the material to react and orient itself to the field. The greater the reaction, the greater the energy stored, and hence the higher the dielectric constant. The behavior of the dielectric can be made up of the following effects.

1. *Electronic polarization.* This is present in all dielectrics. The positions of outer-shell electrons around atoms are affected by the field. This takes place very rapidly.

2. *Ionic polarization.* Ions of opposite sign move elastically because of the effect of the field. This is also rapid and takes place only in ionic solids.
3. *Orientation of molecules.* When asymmetric (polar) molecules are present, their orientation is changed by the field.
4. *Space charge.* This is the development of charge at the interface of phases.

Of these effects, the orientation of molecules contributes most heavily to the differences in the dielectric constant. As an example, a greater charge can be built up on the plates of a capacitor if a material with a strong dipole is placed between the plates. In other words, the material with larger dipoles has a higher dielectric constant. Similar reasoning suggests that liquids have higher dielectric constants than solids, since polarization or dipole orientation is easier. This effect is shown in Fig. 14.19 on page 616. After the change due to the melting, the gradual fall-off with increasing temperature of the liquid reflects a higher atomic or molecular mobility, which decreases polarization.

Polymers have the same general characteristics. The amorphous polymers are not as tightly bonded as the crystalline polymers. Therefore, amorphous polymer dipoles are more easily aligned, giving a higher dielectric constant. The crystalline polymers show a large increase in the dielectric constant at their melting point, as shown in Fig. 14.19.

The effect of frequency is different for the various materials. The small physical movements encountered with electronic and ionic polarization suggest that these changes take place over a broad range of frequencies. This is not the case, however, for dipole motion of molecules. Generally, the dielectric constant decreases as the frequency increases, since it becomes difficult for the dipole to shift at high frequencies.

Figure 14.20 (p. 616) shows that the dipoles can rotate when the polarity of the capacitor changes. But it takes time to shift from (*a*) to (*c*), and as the complexity of the molecule increases, it takes a longer time to shift the dipole.

If we impose a frequency that causes a slight shift each time, we can have dielectric heating because of the energy loss in each cycle. This phenomenon has been used in industry in such applications as the setting of glues in furniture manufacture, and now it has also found use in the home. Since most of our foods are made up of organic molecules with dipoles (e.g., proteins and water), it is possible to rapidly "cook them from within" by exposing them to suitable frequencies, delivered by the microwave oven.

14.19 Barium titanate–type dielectrics (ferroelectrics)

As manufacturers try to reach objectives such as pocket-size television sets, they demand materials with higher dielectric constants to reduce the size of the capacitors. Since the mineral rutile, TiO_2, was known to have a value of ϵ_r about 10, the search led to the titanates. Finally, it was found that barium

Table 14.2 DIELECTRIC STRENGTH AND DIELECTRIC CONSTANT FOR A NUMBER OF ENGINEERING MATERIALS

Dielectric Strength of Nonmetallics,* volts/mil								
Material	*High*	*Low*	*Material*	*High*	*Low*	*Material*	*High*	*Low*
Micas, natural and synthetic	2,000	1,000	Nylon, glass-filled	500	400	Silicones (molded)	400	250
Polymethylstyrene	1,950	890	Nylons 6 and 11	500	420	Ureas	400	300
Polyvinylchloride	1,400	24	Polyesters (cast), allyl type	500	330	Melamines, shock-resistant	370	130
Acetal copolymer	1,200		TFE fluorocarbons	500	400	Phenolics (molded), very high shock resistance	370	200
Polyvinyl formal	1,000	860	Polyethylenes	480		Alkyds	350	300
Plastic laminates, high pressure	1,000	70	Polycarbonate, filled	475		Polycrystalline glass	350	250
Polypropylene	800	520	Nylons 6/6 and 610	470	385	Phenolics, heat-resistant	350	100
Plastic laminates, low pressure	800	100	Epoxies (molded)	468	334	Melamines, GP	330	310
Modified polystyrenes	650	300	Cellulose propionate	450	300	ABS resins, extra high impact	312	
Polyallomer	650		Diallyl phthalate	450	275	Beryllia	300	250
Cellulose acetate	600	250	Phenolics (cast), GP	450	300	Alumina ceramics	300	200
Cellulose nitrate	600	300	Melamines, electrical	430	350	Standard electrical ceramics	300	55
CFE fluorocarbons	600	530	Phenolics (molded), GP	425	200	Zircon	290	60
Hard rubber	600	344	Polystyrenes, glass-filled	425	340	Steatite	280	145
Mica, glass-bonded	600	270	ABS resins, high impact	416	350	Forsterite	250	
Polyesters (cast), rigid	570	340	Cellulose acetate butyrate	400	250	Phenolics (cast), GP transparent	250	75
Epoxies (cast)	550	350	Chlorinated polyether	400		Cordierite	230	140
Acrylics	530	400	Melamines, cellulose, electrical grades	400	350	Polyethylene foam, flexible	220	
Polystyrenes, GP†	>500		Phenolics (cast), mechanical and chemical grades	400	350			
Acetal	500		Polycarbonate	400				
Ethyl cellulose	500	350	Polyesters (cast), nonrigid	400	220			
			Polyvinyl butyral	400				

Dielectric Constant of Nonmetallics‡

Material	*High*	*Low*	*Material*	*High*	*Low*	*Material*	*High*	*Low*
Mica, glass-bonded	40.0	6.9	Alkyds, GP†	5.0	4.8	Acrylics	3.2	2.7
Phenolics (cast)	11.0	4.0	Rubber phenolics	5.0		Polyvinyl formal	3.0	
Alumina ceramics	9.6	8.2	Vinylidene chloride	5.0	3.0	Polycarbonate	2.96	
Lead silicate glass	9.5	6.6	Hard rubber	4.95	2.90	Chlorinated polyether	2.92	
Zircon	9.2	5.3	Polyesters (cast), allyl type	4.8	3.3	Methylstyrene-acrylonitrile	2.81	
Polyvinylchloride	9.1	2.3	Alkyds, electrical and impact	4.5	4.2	Epoxies, resilient	2.8	2.6
Micas, natural and synthetic	8.7	5.4	Diallyl phthalate	4.5	3.3	Polystyrenes, GP	2.65	2.45
Phenolics (molded)	8.0	4.0	Nylons 6 and 11	4.5	3.5	Polymethylstyrene	2.48	
Soda-lime glass	7.4	7.2	Epoxies (cast)	4.4	2.6	CFE fluorocarbons	2.37	2.30
Melamines	7.2	4.7	Epoxies, GP	4.4	3.4	Polyethylenes	2.3	
Beryllia	7.0	6.4	Boron nitride	4.2		Propylene-ethylene polyallomer	2.29	
Cellulose acetate	7.0	3.2	ABS resins	4.1	2.8	Polypropylene	2.1	2.0
Standard electrical ceramics	7.0	5.4	Epoxies, heat-resistant	4.0	3.5	TFE fluorocarbons	2.0	
Ureas	6.9	6.4	Modified polystyrenes	4.0	2.5	Prefoamed epoxy, rigid	1.55	1.19
Plastic laminates, high pressure	6.8	3.3	Polyesters (cast), rigid	4.0	2.8	Polyethylene foam, flexible	1.49	
Forsterite	6.5	6.2	Nylon, glass-filled	3.9	3.4	Urethane rubber foamed-in-place, rigid	1.40	1.05
Steatite	6.5	5.5	Phenoxy	3.8	3.7	Silicone foams, rigid	1.26	1.23
Aluminum silicate glass	6.3		Silica glass	3.8		Polystyrene foamed-in-place, rigid	1.19	
Cellulose acetate butyrate	6.2	3.2	Acetal	3.7		Prefoamed cellulose acetate, rigid	1.12	1.10
Cordierite	6.2	4.0	Cellulose propionate	3.6	3.4	Prefoamed polystyrene, rigid	<1.07	
Polyesters (cast), nonrigid	6.1	3.7	Ethyl cellulose	3.6	2.8			
Plastic laminates, low pressure	5.6	3.4	Nylons 6/6 and 610	3.6	3.4			
Polycrystalline glass	5.6		Polycarbonate, filled	3.50				
Borosilicate glass	5.1	4.0	Polystyrenes, glass-filled	3.41	2.74			
Silicones (molded)	5.1	3.6	Modified polystyrenes, extra high impact	3.3	1.9			
			Polyvinyl butyral	3.3				

* Values represent high and low sides of a range of *typical* values. To obtain volts/cm, multiply by 393.7.

† GP = general-purpose grades.

‡ Values represent high and low sides of a range of *typical* values at 10^6 cycles.

Source: "Materials Selector Guide," *Materials and Methods,* Reinhold Publishing Corp., New York, 1973.

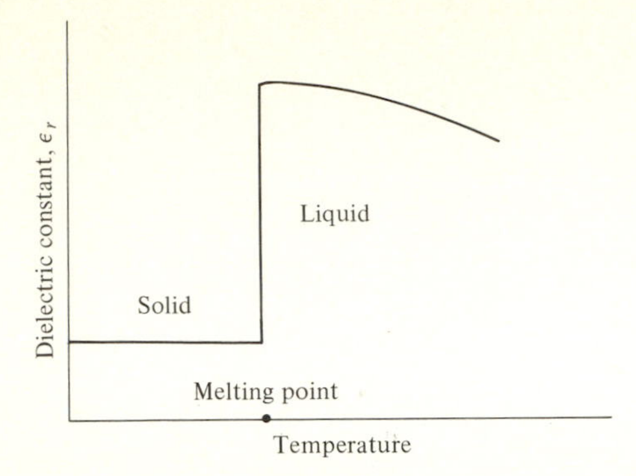

Fig. 14.19 *Variation in dielectric constant with temperature (polarizable crystalline polymer)*

titanate showed values of ϵ_r over 1000, several orders of magnitude better than any known material. In addition, it was noticed that a permanent charge was developed. This is called *ferroelectric behavior.*

It is interesting to examine how these effects result from the structure of $BaTiO_3$. In Chapter 7 we discussed the fact that the Ti^{4+} ion was larger than the octahedral site formed by the six oxygen ions (radius Ti^{4+} = 0.64 Å vs. 0.625 Å for the site) (Fig. 14.21). As a result, the titanium ion is located to one side of the center. Thus in each unit cell one side of the center is positive and the other negative, or, in other words, a dipole develops.

When an electric potential is applied across the capacitor plates, the Ti^{4+} ions are attracted to the negative side. This leads to a high charge storage in the plates, and therefore a higher dielectric constant.

Let us now examine the hysteresis effect (Fig. 14.22). By placing a solution of magnetized powder over an etched sample of $BaTiO_3$, we can show that groups of unit cells within a grain of the ceramic are organized into *domains* even before the electrical field is applied. In the unit cells of a given domain, the Ti^{4+} ions are oriented in the same direction. When the field is applied, domains that have Ti^{4+} ions situated in the *right* direction (i.e., toward the negative plate) grow at the expense of those with the *wrong* orientation by a change in orientation of the adjacent unit cells. This domain growth finally slows down, leading to a smaller increase in charge (polarization) as applied fields are increased. If the electric field is cut to zero, a charge

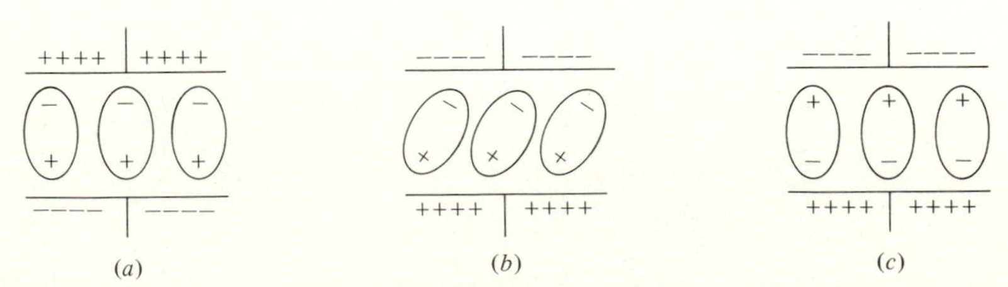

Fig. 14.20 *Dipole motion caused by a change in polarity. When the polarity reverses, (a) → (b), it takes a finite time to make the complete shift to (c).*

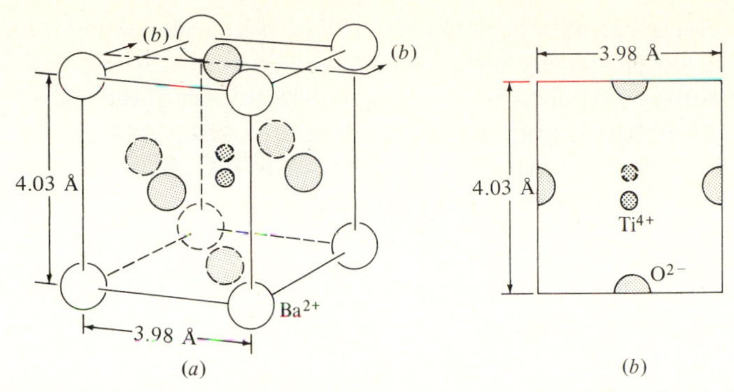

Fig. 14.21 *(a) Structure of barium titanate. (b) The source of the dipole is due to two possible positions for the* Ti^{4+} *ion.* (L.H. Van Vlack, *Elements of Materials Science*, 2d ed., Addison-Wesley Publishing Company, Reading, Mass., 1964, Fig. 8-26, p. 222)

remains because of the domain alignment. This is called the *residual polarization* $\mathcal{P}_r$. To remove this, one must reverse the electric field from the original condition to $-\mathcal{E}_c$.

There is an important effect of temperature on this "ferroelectric" behavior of barium titanate. At 120°C the structure changes from tetragonal to cubic and the effect disappears. This is called the *Curie temperature,* in honor of Pierre Curie. This general term is used to specify the temperature at which

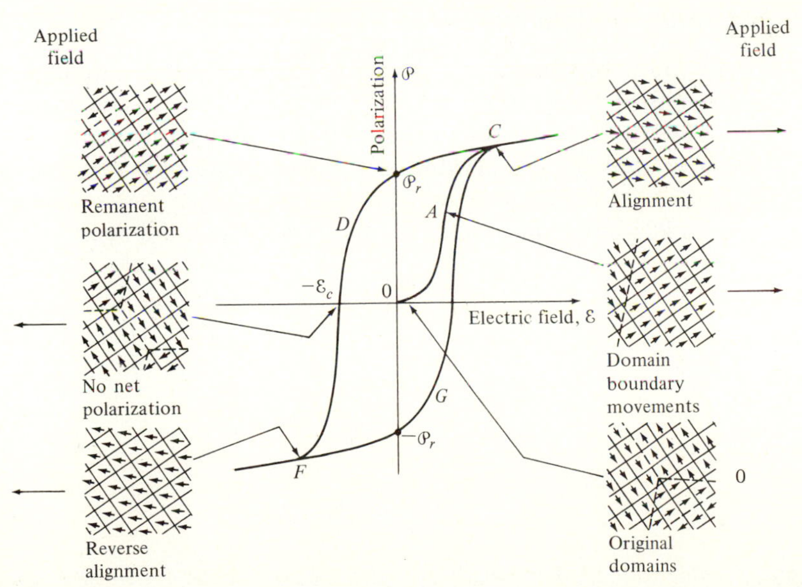

Fig. 14.22 *Ferroelectric hysteresis. The spontaneous polarization can be reversed; however, energy is consumed with each cycle.* (L.H. Van Vlack, *Materials Science for Engineers*, Addison-Wesley Publishing Company, Reading, Mass., 1970, Fig. 13.24, p. 273)

a change in electrical or magnetic properties takes place. The change may or may not be accompanied by a phase change. To obtain ferroelectric materials for use at higher temperatures, researchers investigated other compounds with similar structures, and found lead titanate, with a Curie point of 480°C, and lead meta-niobate, with a Curie point of 570°C. The material most commonly used at present is PZT [$Pb(Zr,Ti)O_3$].

EXAMPLE 14.6 Show that lead titanate ($PbTiO_3$) would be expected to have a structure equivalent to $BaTiO_3$.

ANSWER

	Ba^{2+}	Ti^{4+}	O^{2-}	
Ionic radius, Å:	1.43	0.64	1.32	(From Table 7.2)
	Pb^{2+}	Ti^{4+}	O^{2-}	
Ionic radius, Å:	1.32	0.64	1.32	

In general, in the perovskite structure (Figs. 7.20 and 14.21), the ions in the unit cell corners will have an average valence of 2 and radii close to Ba^{2+}. To obtain extensive solid solution, the divalent radii should be $\pm$15% or 1.21 to 1.55 Å. Ions in the $\frac{1}{2}, \frac{1}{2}, \frac{1}{2}$ sites must average 4+ (3+ in conjunction with Nb^{5+} in niobates, for example) and be about the radius of Ti^{4+}. See the problems for further data.

14.20 Interrelated electrical–mechanical effects (electromechanical coupling)

Every radio amateur has heard of the piezoelectric effect of a quartz crystal. If pressure is applied to a quartz crystal, the ends become charged, or conversely, if an electric field is applied, the crystal changes length. If an alternating voltage is applied, the crystal oscillates, giving out a sound wave of constant frequency.

Figure 14.23 shows the reason for the generation of a sound wave. Let

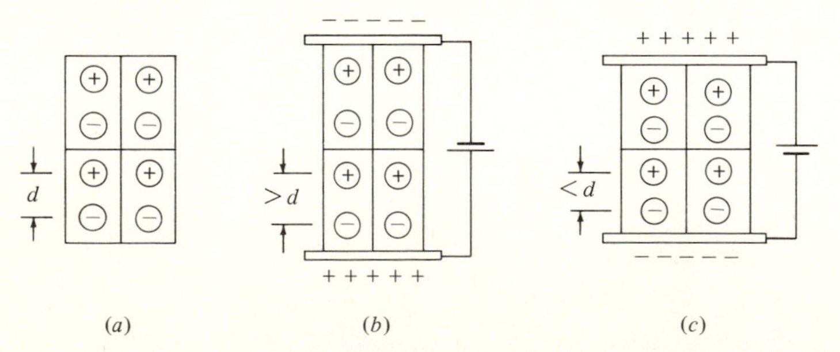

Fig. 14.23 *(a) Piezoelectric material. (b) An electric field induces dimensional expansion. (c) Reverse polarity causes a corresponding contraction. The procedure can be inverted by applying a pressure and obtaining a change in the voltage.* (L.H. Van Vlack, *Elements of Materials Science*, 2d ed., Addison-Wesley Publishing Company, Reading, Mass., 1964)

us assume that we have a single crystal with the dipoles of the unit cells aligned as shown. If we apply an electric potential as shown, the crystal lengthens because the ions are attracted to the pole plates. If we use an ac voltage, the crystal alternately expands and contracts, sending out a wave into the surrounding medium, whether air or water. On the other hand, if we supply the mechanical motion, we receive an electric charge by the reverse procedure. This phenomenon is called *piezoelectricity,* from the Greek word meaning "to press."

We can obtain effects similar to the piezoelectric effect, but stronger—with ferroelectric materials. We must distinguish between piezoelectric and ferroelectric effects. In piezoelectricity, as in quartz, a single crystal is required if the phenomenon is to be used in a device. However, ferroelectric materials contain domains that grow in response to the applied field. This has two important consequences. First, the charge storage is greater because of the greater structural change. Second, a single crystal is not required. This ability to use a polycrystalline material, rather than a single crystal as in purely piezoelectric materials, is naturally a great commercial advantage.

(In a problem at the end of the chapter, we shall take up the use of ferroelectrics in such applications as sonar and phonograph pickups.)

14.21 Thermocouples; thermoelectric power

If we produce a temperature difference in a rod, the energy levels of the electrons will be different, giving more high-energy electrons in the hotter region (Fig. 14.24). High-energy electrons flow toward the cold end, producing a difference in charge.

If we now attach a voltmeter with contact wires of the same material, there is no difference at the meter. If the wires are of a different material than the rod, a different voltage is induced and the net voltage appears on

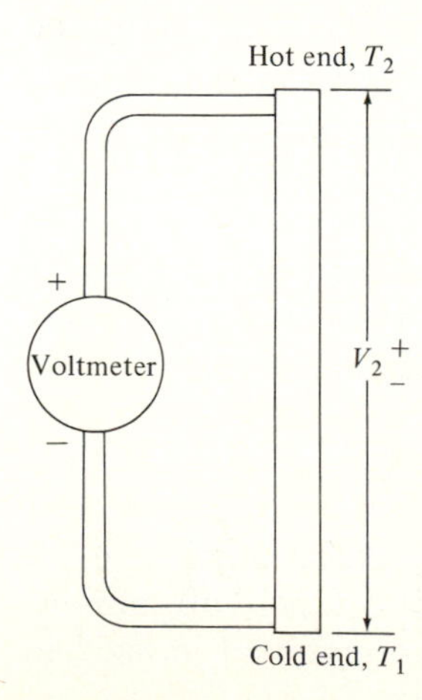

Fig. 14.24 *Development of a voltage in a thermocouple.* (R.M. Rose, L.A. Shepard, and J. Wulff, *The Structure and Properties of Materials,* Vol. 4: Electronic Properties, John Wiley & Sons, Inc., New York, 1965)

the meter. This voltage is the *Seebeck potential S,* used in thermocouples. Examples are:

Thermocouple	*Maximum Useful Temperature, °C*
Copper vs. constantan (60% Cu, 40% Ni)	315
Iron vs. constantan	950
Chromel (90% Ni, 10% Cr) vs. Alumel (94% Ni, 2% Al, 3% Mn, 1% Si)	1200
Platinum vs. platinum–10 to 13% rhodium	1500
Tungsten vs. tungsten-rhenium, also tungsten vs. molybdenum	>1500

There is a basic reason for the potential difference of a thermocouple that goes back to the concept of the Fermi level. In different metals the Fermi level changes at a different rate with increased temperature. Therefore, in a pair of metals the element with the greatest temperature coefficient of change has more energetic electrons, which flow into the element with the lower coefficient.

When we survey the metals used for thermocouples, we see that many are transition metals. This is because the distribution of electrons at different energy levels shows the greatest difference with temperature in metals with unfilled $3d$ and $4f$ shells.

SUMMARY

The electrical properties of materials depend mainly on the number and mobility of the charge carriers in the structure: electrons, electron holes, and ions.

The flow of current is related to the conductivity, σ = (number of carriers/m^3) $\times$ (charge/carrier) $\times$ mobility. The number of carriers per cubic meter is relatively great in a metal because the electrons donated by the atoms exist in a band of energies that has many vacant states to which the electrons can be accelerated and thereby carry current. By contrast, in an insulator the valence electrons are bound in a filled energy band. To raise an electron to an energy level at which it can conduct, the conduction band, requires considerable energy. The intrinsic semiconductors occupy an intermediate position. The number and type of carriers in a semiconductor can be increased by "doping" with elements that are electron donors or acceptors. By setting up junctions of these "extrinsic" semiconductors, we can form *p-n* junctions. Certain metals and alloys exhibit superconductivity below critical temperatures.

Dielectrics with greatly improved dielectric constants can be made from ferroelectric materials such as barium titanate. In these materials there is a permanent dipole or charge imbalance in the unit cell. These dipoles result in the formation of electrical domains or regions of similar dipole alignment. This configuration leads to a high capacitance when the material is used in a capacitor. Electromechanical coupling from these materials or materials with fixed dipoles leads to valuable devices for conversion of electrical impulses to sound. Thermocouples are important for measuring temperature. They operate because of the development of an emf by pairs of metals subjected to two different temperatures.

DEFINITIONS

Band model A model in which the outer electrons of the atoms are combined in a band of energy levels. If there are empty levels within a band, the material is a good conductor, because upper-level electrons can easily be raised to a higher level of energy and therefore move to conduct current.

Capacitance The charge-storing ability of a capacitor. Capacitance is a function of the geometry of the device and the dielectric constant of the material between the plates. The unit of capacitance is the farad (F).

Charge carrier A tiny particle with an electric charge, specifically electrons, holes, and ions.

Conductivity, σ A measure of the ease of passage of an electric current. $\sigma = 1/\text{resistivity}$.

Dielectric constant The ratio of the capacitance obtained using the material between the plates of a capacitor compared with that obtained when the material is replaced by a vacuum.

Dipole An atomic or ionic grouping such as a molecule or unit cell in which the positive and negative centers of charge do not coincide. The *dipole moment* is the product of the charge and the distance between the charges.

Electromechanical coupling A material that develops a charge at its ends when compressed or, conversely, changes dimensions when an electric potential is applied.

Electron A negative charge carrier with a charge of 1.6×10^{-19} coulomb.

Electron hole A positive charge carrier with a charge of 1.6×10^{-19} coulomb.

Electronic polarization The movement of the electrons in a dielectric toward the positively charged plate.

Energy gap, E_g The energy required to move an electron from the valence band to the conduction band; usually expressed in electron volts.

Extrinsic semiconductor A semiconductor material that has been doped with an *n-type element* (such as phosphorus) that donates electrons that have energies close to the conduction band, or with a *p-type element* (such as aluminum) that provides electron holes close to the valence-band level.

Fermi level, E_F The energy level at which the probability of occupancy of an energy state is 0.5. In other words, at this level the possible states that an electron or electron hole can occupy are half-filled.

Ferroelectric A material such as barium titanate in which the permanent dipoles are aligned in domains.

Forward bias Electron flow into the *n*-material of a *p-n* semiconductor.

Hall effect The development of a voltage by applying a magnetic field perpendicular to the direction in which a current is flowing. Electron carriers are forced to one side and electron holes to the other, giving a voltage that is perpendicular to both the field and the current.

Insulator A material with a filled valence band and a large energy gap between the valence band and the conduction band.

Ionic polarization The displacement of ions in the material toward oppositely charged plates.

Intrinsic semiconductor A semiconductor material that is essentially pure and for which conductivity is a function of the temperature and the energy gap of the material.

Metal A material with an unfilled valence band, leading to a very high concentration of carriers.

Mobility A measure of the ease of motion of the carrier.

MOSFET Metal-oxide-semiconductor-field-effect-transistor.

p-n junction The boundary between *n*- and *p*-type materials.

Pauli exclusion principle No more than two electrons in an atom or block of material can occupy the same energy level.

Piezoelectric material A material that shows electromechanical coupling. All ferroelectric materials are piezoelectric, but only materials with electrical domain structures are ferroelectric.

Rectification The conversion of alternating to direct current.

Resistance $R = E/I$, where E is the voltage and I the current. The resistance of a conductor to the passage of a current increases with length and decreases with increasing cross section. Note that this applies to direct current, not to high-frequency alternating current, in which case most of the current flows in the surface layers.

Resistivity $\rho = R(A/l)$, where A is the area of a conductor, l is its length, and R is its resistance.

Resistor An electrical component designed to provide a desired resistance. Examples include a tiny component in a radio or the heating element in a toaster.

Reverse bias Electron flow into the *p*-material of a *p-n* semiconductor.

Semiconductor A material with a filled valence band and a small energy gap between the valence band and the conduction band.

Solar cell Forward biasing of an *n-p* junction by exposure of the *n*-material to radiant energy.

Superconductivity The phenomenon of zero resistivity that occurs in some metals and alloys below a critical temperature that is different for each material.

Thermistor A device using the change in resistivity with temperature to measure temperature.

Thermocouple A pair of materials that, when joined, develop an emf when one junction is at a different temperature from the other.

Transistor A three-element, two-junction, semiconductor device (*n-p-n* or *p-n-p*) composed of an emitter, base and collector. Through forward bias of the emitter and reverse bias of the collector the transistor becomes a current amplification device.

PROBLEMS

14.1 Suppose that in Example 14.1 the student considers the length of the wire too awkward, and decides to use a 10-ft (3.05-m) length. What diameter Chromel wire should she buy? Suppose that she decides to use only a 1-in. (0.0254-m) length. What diameter should she use? What practical difficulties might she encounter in using such short wires? (Sections 14.1 through 14.7)

14.2 Sketch the electron occupation of the valence and conduction bands for the following materials at 0 and 273 K: potassium, magnesium, aluminum, pure silicon, silicon doped with aluminum, germanium doped with phosphorus. Indicate the approximate Fermi level E_F in each case. (Sections 14.1 through 14.7)

14.3 Chromium is insoluble in copper at room temperature, yet dissolves to over 1% at elevated temperatures. It can form a precipitation-hardening system. Compare the relative usefulness of this alloy with a solid solution of copper or work-hardened copper in improving the yield strength when maximum electrical conductivity is desired. (Sections 14.1 through 14.7)

14.4 From the calculations of Problem 14.1, explain why the manufacturer's catalog of Chromel wire gives recommendations for the wire diameters for different currents. Why is the allowable current lower for wires embedded in refractory than for wires in air when the thermal conductivity of air is lower than that of the refractory? (Sections 14.1 through 14.7)

14.5 The resistance of a one-foot (0.305-m) length of metallic conductor was increased 20% by cold-working the wire and decreasing its cross-sectional area by 10%. By what percentage was the mobility of the electrons changed? (Sections 14.1 through 14.7)

14.6 A well-insulated wire-wound small-tube furnace requires 900 watts to achieve a temperature of 2000°F (1093°C). To prevent melting the wire, the maximum current draw is limited to 6 A. (Sections 14.1 through 14.7)

a. Calculate the voltage requirement and the wire resistance.
b. The diameter of the wire is 0.050 in. (0.0013 m) and a 1-foot- (0.305-m) long piece has a resistance of 0.205 ohm. Calculate the resistivity of the wire and the required length of wire to consume the 900 watts at 6 amps.
c. The outside diameter of the tube furnace is 2.00 in. (0.051 m). How many turns of wire are required for the length in part (b)?

14.7 Assume that copper has 2 electrons per atom available for electrical conduction. With a resistivity of 1.7×10^{-8} ohm · m, calculate the number of carrier electrons per m^3 of copper. Then calculate the electron mobility. (Sections 14.1 through 14.7)

14.8 The following table summarizes the characteristics of the two intrinsic semiconductors, silicon and germanium, at 300 K.

Characteristic	*Silicon*	*Germanium*
Electrical resistivity, microhm-m	2.3×10^9	4.5×10^5
Energy gap, eV	1.06	0.67
Density of carriers, (electrons plus holes)/m^3	1.5×10^{16}	2.4×10^{19}
Lattice drift mobility, m^2/volt-sec		
Electrons	0.135	0.390
Electron holes	0.048	0.190

The conductivity of germanium was calculated in Example 14.2. Calculate the conductivity of silicon and check the way it corresponds to the resistivity given in the table. (Sections 14.8 through 14.11)

14.9 In a doped or extrinsic semiconductor, the carriers produced by the dopant usually overpower the intrinsic carriers. Assume that we add phosphorus to silicon and find a conductivity of 100 mho/m and a mobility of 0.17 m^2/volt-sec. What is the concentration of conduction electrons per cubic meter? [*Hint:* Determine what type of carriers the phosphorus contributes.] (Sections 14.8 through 14.11)

14.10 An extrinsic *n*-type semiconductor (phosphorus in silicon) has a resistivity of 0.005 ohm-m. Conduction is primarily by the excess electrons, with a mobility of 0.120 m^2/volt · sec. (Sections 14.8 through 14.11)

a. What is the density of the carriers, per cubic meter?
b. Calculate the phosphorus atoms per cubic meter of silicon. How many grams of phosphorus would there be per cubic meter of silicon?

14.11 A *p*-type semiconductor is produced by adding aluminum to silicon. The resistivity is 5×10^{-4} ohm-m, with an extrinsic carrier mobility of 0.1625 m^2/volt · sec. (Sections 14.8 through 14.11) Calculate:

a. the density of the carriers

b. the carriers per atom of added aluminum
c. the carriers per kilogram of silicon (density of silicon is 2.33×10^3 kg/m^3)
d. the kilograms of aluminum per kilogram of silicon

14.12 An intrinsic semiconductor is produced from a compound of Group III and Group V compounds. The density of the electrons is 2×10^{15} electrons/cm^3, with an electron mobility of 0.30 m^2/volt · sec and a hole mobility of 0.10 m^2/volt · sec. (Sections 14.8 through 14.11)

a. Calculate the conductivity due to the combined movement of holes and electrons.
b. Suppose that antimony is added to produce an extrinsic semiconductor with a conductivity of 200 mho/m. Calculate the antimony atoms per cubic meter of semiconductor. Assume the same hole and electron mobilities.

14.13 Assume that we were to use germanium rather than silicon, as proposed in Example 14.5. Show by calculations whether the germanium thermistor would be more or less sensitive within this temperature range (27 to 47°C). (Sections 14.14 through 14.16)

14.14 Indicate how we might conduct an experiment on a crystal of LiCl to determine the diffusion coefficient of Li^+ from a measurement of the electrical conductivity. Be as specific as possible as to what measurements must be made and what assumptions would be necessary. In most cases the major portion of the ionic conductivity is due to the cation. For this problem, assume that this portion is 100%. (Sections 14.14 through 14.16)

14.15 Several years ago there was a rash of failures due to brittle failure at contacts of transistors. The fractures showed a purple color, and the effect was promptly called the "purple plague." From the following facts, deduce the cause and suggest a remedy: (1) The transistor is made of silicon doped with aluminum. (2) The contact wire is gold. (3) Gold and aluminum form a compound, $AuAl_2$, which is purple. (Sections 14.14 through 14.16)

14.16 Silicon is fast eclipsing germanium as a semiconductor material. The main reason is not cost, purity, band gap, or properties of the silicon itself. What *is* the main reason? (Sections 14.14 through 14.16)

14.17 One of your associates says that the following equation relates the electrical resistance of a conductor to its temperature.

$$R = \frac{lkT}{nq^2DA}$$

a. List the fundamental equations necessary to the derivation of the relationship.
b. What limitations, if any, should be included with the relationship? (Sections 14.14 through 14.16)

14.18 The electrical conductivity of a compound AX is controlled by the mobility of A^+ ions (4.8×10^{-14} m^2/volt · sec). Suppose that AX is a NaCl structure with radii of A^+ = 0.88 Å and X^- = 1.61 Å. (Sections 14.14 through 14.16) Calculate: (*a*) the volume of the AX unit cell, (*b*) the electrical conductivity of AX.

14.19 Assume that we have an intrinsic silicon semiconductor wire 1 cm long and 0.01 cm^2 in cross-sectional area. Using the data in Example 14.5, calculate the temperature-measuring capability of the semiconductor, given that resistance can be measured to the nearest 0.001 ohm. Compare this result to a thermocouple which has an accuracy of 0.1°C within the same temperature range. (Sections 14.14 through 14.16)

14.20 A capacitor consists of two plates 2 by 3 in. (50.8 by 76.2 mm), spaced 0.01 in. (0.254 mm) apart and filled with mica. Calculate the capacitance. (Sections 14.17 through 14.21)

14.21 Which dielectric would you use for the smallest two-plate capacitors for the following application, assuming constant plate thickness?

Capacitance: 0.001 μF; Operating temperature: 300°F (149°C); Voltage: 1000 volts

Specify plate size and the distance between plates, using a safety factor of 5 against puncturing due to breakdown. [*Hint:* Compare several of the materials in Table 14.2.] (Sections 14.17 through 14.21)

14.22 Indicate which material of each pair would have the higher dielectric constant and explain why. (Sections 14.17 through 14.21)

a. Polyvinylchloride (PVC) and polytetrafluoroethylene (PTF)
b. PVC at 25°C and 100°C
c. PVC at 10^2 Hz (hertz) and PVC at 10^{10} Hz

14.23 Why should the cooling of barium titanate ceramics through the Curie temperature in the presence of a strong electric field improve their piezoelectric properties? (Sections 14.17 through 14.21)

14.24 How do the characteristics of the barium titanate structure lead to its use in the following devices? (*a*) Ultrasonic applications such as emulsification of liquids, mixing of powders and paints, homogenization of milk (*b*) Microphones and phonograph pickups (*c*) Accelerometers (*d*) Strain gages (*e*) Sonar devices (Sections 14.17 through 14.21)

14.25 As mentioned in the text, a variety of alloys are used for thermocouples. Suppose that you can read a given potentiometer (a device for measuring emf) to 0.01 millivolt at 400°F (204°C). What would the precision of your temperature measurement be using a Chromel vs. Alumel, iron vs. constantan, copper vs. constantan, or a platinum vs. platinum-10% rhodium thermocouple? There are many places to find charts, such as the *Handbook of Chemistry and Physics,* and instruction manuals accompanying the potentiometer. (Sections 14.17 through 14.21)

PROBLEMS COVERING OPTIONAL SECTIONS AND OTHER CHAPTERS

14.26 Calculate the ratio of I_s current in silicon vs. germanium, considering all variables the same except for the value of the band gap. [*Hint:* n_i is related to the band gap. What does this tell you about the relative usefulness of silicon vs. germanium *p-n* junctions as rectifiers?] (Math portion of Section 14.11)

14.27 Arsenic is preferred to phosphorus as a dopant in certain applications, especially those in which extensive processing at high temperatures (over 1000°C) is involved. The reason is not cost or effect of electrical properties, but involves the ability to produce a stable circuit. What is the property involved? (Sections 14.10, 14.11, and 14.15)

14.28 Explain why an *n-p* junction can be used as a capacitor. (Sections 14.11, 14.15, and 14.18)

14.29 Iron oxide is made up with a ratio of Fe^{3+} to Fe^{2+} of 0.1. What is the mobility of the electron holes if the oxide has a conductivity of 100 mho/m and 99% of the charge is carried by electron holes? $a_0 = 4.3$ Å [*Hint:* FeO has an NaCl-type lattice; for every two Fe^{3+} there must be one vacancy at an Fe^{2+} position to balance the charge. If we take 100 Fe^{2+}, there are 10 Fe^{3+} and 5 vacancies. Accompanying these atoms will be 115 oxygen atoms. Since the oxygen ion lattice is perfect, calculate the number of unit cells formed by 115 oxygen atoms at 4 oxygen atoms per unit cell. Since each Fe^{3+} has one electron hole, we can calculate the density of carriers, then the mobility.] (Sections 14.16 and 7.8)

14.30 Show that the following complex materials meet the criteria for the perovskite structure, and therefore might react in a way similar to barium titanate: (*a*) $KLaTi_2O_6$ (*b*) Sr_2CrTaO_6 (*c*) $BaKNbTiO_6$ (*d*) $Pb(Zr, Ti)O_3$ (Sections 14.19 and 7.6, 7.10)

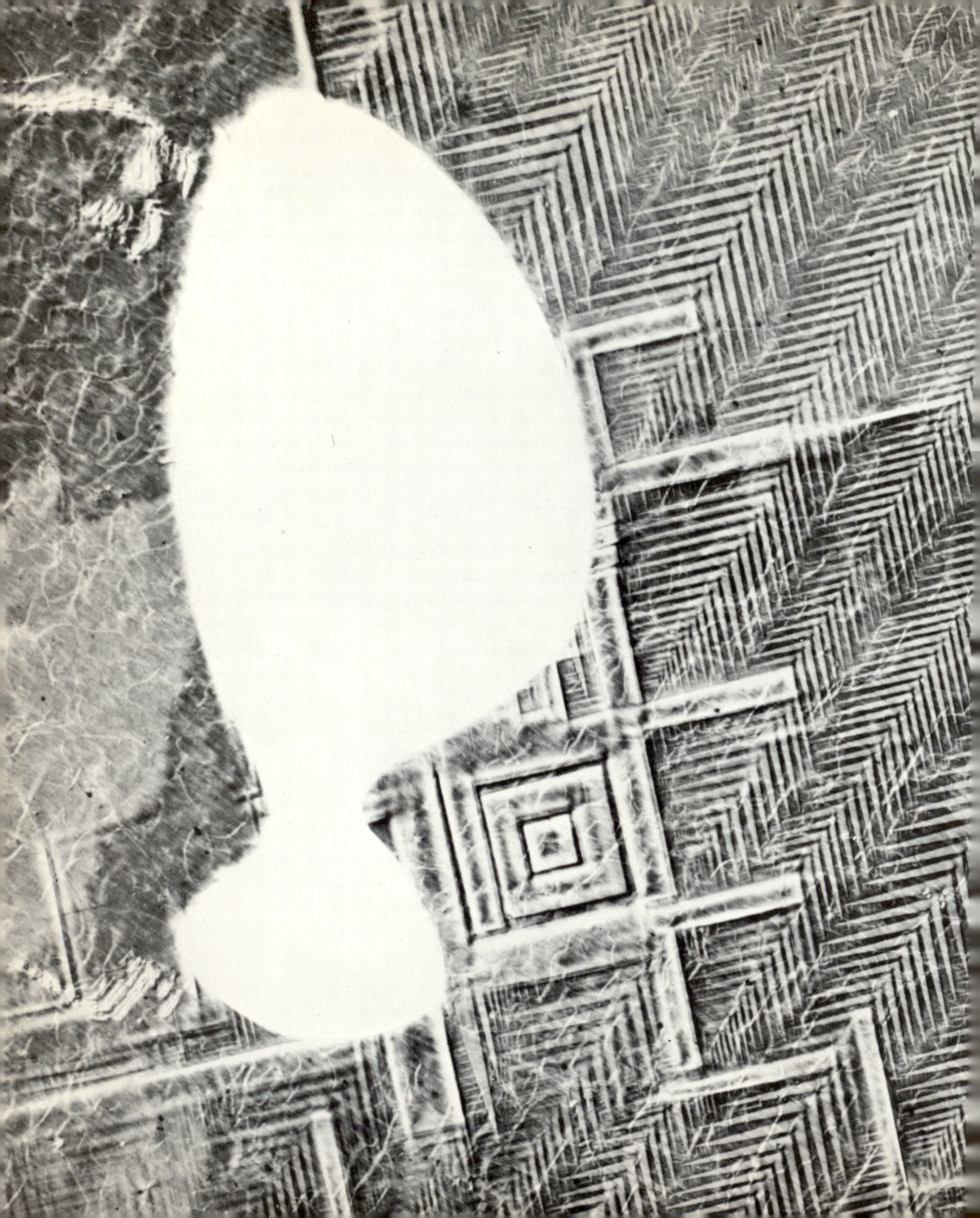

15

MAGNETIC PROPERTIES OF MATERIALS

THIS is an x-ray diffraction pattern from an iron–3% silicon single crystal. The interaction of the x rays with the magnetic domains in the iron caused the features shown. The "magnetic fish" shows no structure because the orientation of this region does not satisfy the Bragg law, discussed in Chapter 2, for diffraction. The square region to the right of the tail of the "fish" is a ferromagnetic domain. The chevron markings show other domains of different orientation.

In this chapter we shall take up the relation of magnetic properties to structure in two groups of materials—the metallic and the ceramic. We shall see that in each group important recent growth has occurred to meet the demands of the electronics industry.

15.1 General

It is not generally recognized that the development of materials with new high levels of magnetic properties is responsible for the growth of many new devices, such as the tape recorder, ferrite magnets in television sets, memory cores in computers, permanent magnets in motor controls, and particle accelerators used in basic research. We shall see how these new developments involve two important groups of materials.

1. The metallic materials in which the advances have taken place in recent years, from the conventional iron-base alloys to Alnico and, more recently, to samarium-cobalt magnets.
2. The ceramic materials called *ferrites,* which are spinels and related structures. The growth of this field has been comparable to that of the semiconductors in the electrical materials.

To understand the specification of magnetic materials and how their structure is related to magnetic properties, we must first review a few definitions of the most important properties. These are relatively simple and are all related to the magnetic flux that is produced in a material as a result of a magnetic field. These properties, such as permeability and remanent induction, are given quite simply by the graph of the flux density B as a function of the field strength H, called a *hysteresis loop* or *magnetic hysteresigraph.*

Following this review we shall take up important metallic magnetic materials, encountering magnetic domains and domain movement, just as we found domains in ferroelectric materials. We shall see that the structure required for soft or low-hysteresis magnets is quite different from that required for the hard or permanent magnets. Finally, we shall discuss the magnetic properties available in the ceramic magnets or ferrites and their relation to structure.

15.2 The magnetic circuit and important magnetic properties

First recall that, in an electric circuit (Fig. 15.1), if we apply a voltage E, the current I that flows in the conductor is related to the conductivity σ of the material. Mathematically,

$$E = IR \text{ and } R = \frac{\rho l}{A} = \frac{l}{\sigma A} \quad \text{or} \quad E = \left(\frac{I}{A}\right)\left(\frac{l}{\sigma}\right)$$

Therefore

$$\sigma = \left(\frac{I}{A}\right)\left(\frac{1}{E/l}\right)$$

where I/A is the current density with units of A/m^2 and E/l is a voltage gradient with units of volts/m. Then

$$\sigma \text{ (electrical conductivity)} = \frac{\text{current density}}{\text{voltage gradient}}$$

In an analogous way (Fig. 15.1), if we apply a magnetic field across a gap, the flux in the gap is proportional to the permeability μ of the material in the gap. If we represent the field strength or gradient by H and the flux density by B, then

$$\mu = \frac{B}{H} = \frac{\text{flux density}}{\text{field strength or gradient}}$$

The usual meter-kilogram-second (mks) units for the quantities in this equation are derived in physics as

$$H = 1 \text{ A/m} \qquad B = 1 \text{ weber/m}^2 = 1 \text{ tesla} = 1 \text{ volt-sec/m}^2$$

For a vacuum,

$$\mu_0 = 4\pi \times 10^{-7} \text{ henry/m}$$

where 1 henry = 1 ohm-sec. Often centimeter-gram-second (cgs) units are still used, and the following conversions apply:

$$H\text{: } 1 \text{ A/m} = 4\pi \times 10^{-3} \text{ oersted}$$
$$B\text{: } 1 \text{ tesla} = 10^4 \text{ gauss}$$
$$\mu_0\text{: } 4\pi \times 10^{-7} \text{ henry/m} = 1 \text{ gauss/oersted}$$

The important difference between the electric and magnetic circuits is that the conductivity in an electric circuit is constant (independent of the values of E and I), whereas the permeability μ of a magnetic material changes with the applied field strength H. The ratio B/H is not constant. Also, the flux can persist after the field is removed. This is called *remanent magnetization,* as in a permanent magnet. Clearly, we need to examine these important effects in a magnetic circuit and to relate them to the structure of the material if we are to understand and construct magnetic devices.

Let us first discuss the B-H curve, which is as important in expressing magnetic properties as the stress-strain curve is for understanding mechanical properties. We place a sample in a gap in which we provide a controlled magnetic field strength H (as, for example, by changing the current in an

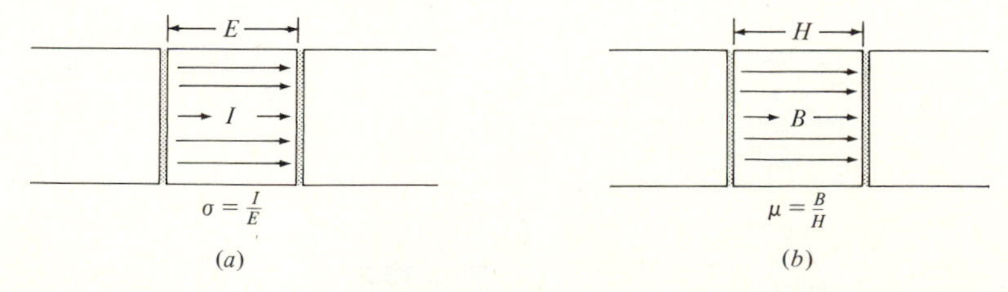

Fig. 15.1 *Analogy between (a) electric and (b) magnetic circuits. (Assume unit length and cross-sectional area.)*

electromagnet with its poles at the sides of the gap), and we measure the magnetic flux density B and plot the B-H curve (Fig. 15.2). In a typical virgin curve the value of B rises rapidly, then practically levels off at B_s, called the *saturation induction*. (The curve obtained the first time the material is magnetized—the virgin curve—is the dashed curve.) As the field strength is reduced to zero, there is still a magnetic flux density B_r, called the *remanent induction* or *residual induction*. To lower the flux to zero, we need a definite field strength H_c in the opposite direction to the original, and this is called the *coercive magnetic force* (or the *coercive field*). Since the graph never retraces the virgin curve but traces out a loop, it is called a *hysteresis loop* and is an index of the energy lost in a complete cycle of magnetization.

The shape of the B-H curve varies greatly with different magnetic materials, and various types of curves have different uses. To illustrate this, consider the B-H curves for a transformer core, a permanent magnet, and a computer memory unit, as shown in Fig. 15.3. In the curve for a transformer core the hysteresis loop encloses only a small area (Fig. 15.3*a*). This represents work done per cycle. It is related to power loss and is kept as small as possible. By contrast, in a memory unit we wish to magnetize easily to a sizable value of B and then retain most of this flux density when the power is off, so that it can be used as "memory" to activate the controls when called on. For this we want a "square loop," so that even with a reversed field we still obtain a strong flux for "readout" (Fig. 15.3*b*). In the permanent magnet the "power" of the magnet is related to the BH product.† We note that in this graph the x axis, the field strength, is many times more compressed than in the other two graphs, which reflects the large value of H needed to produce a permanent magnet with a large B_r. Note this large value of remanent magnetization (Fig. 15.3*c*).

†The maximum BH product in the second quadrant of the B-H curve is used.

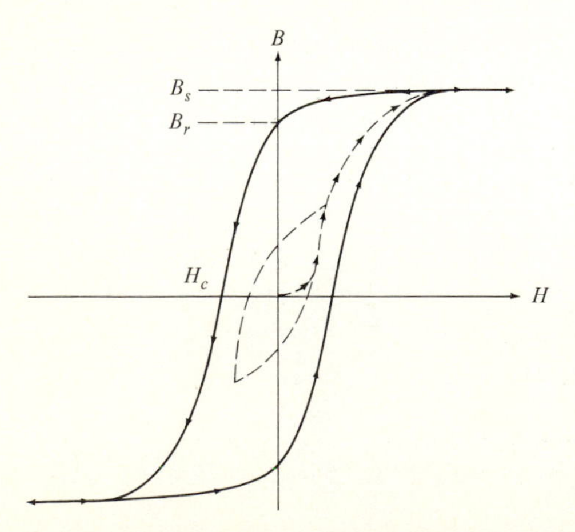

Fig. 15.2 *Hysteresis loop for a ferromagnetic material, showing* B *(flux density) versus* H *(field strength)*

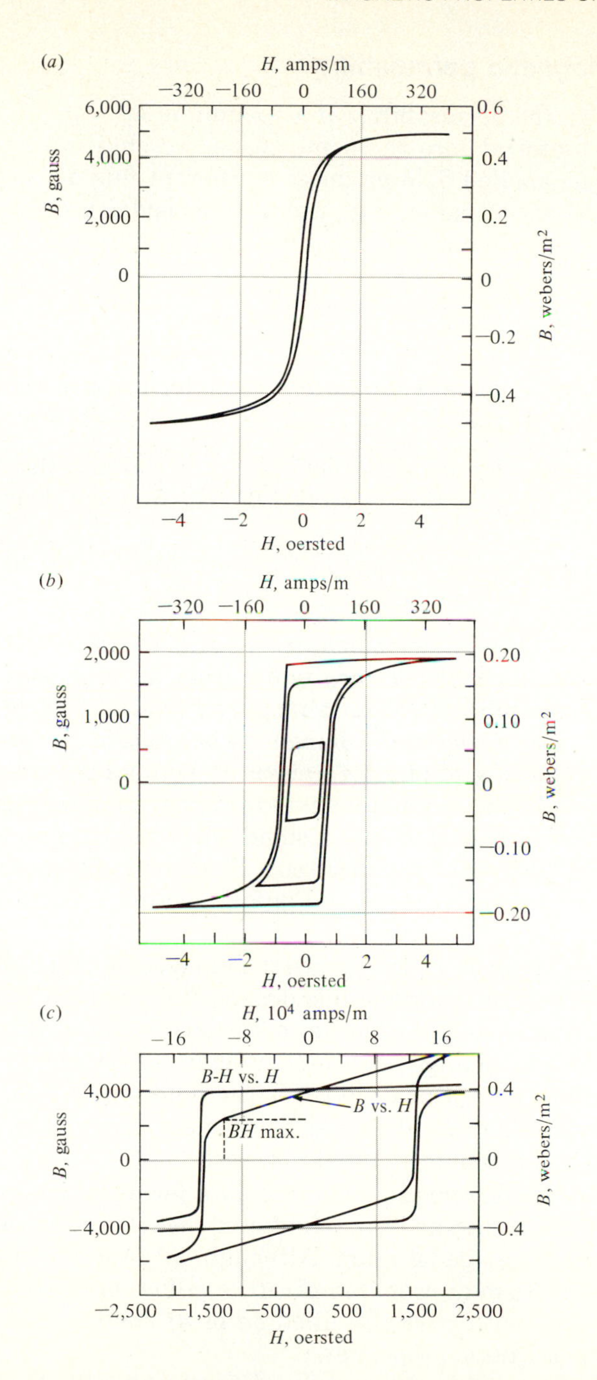

Fig. 15.3 *(a) Hysteresis curve of a soft ferrite. (b) Hysteresis curves of a square-loop ferrite. (c) Hysteresis curve of a permanent-magnet ferrite.* (E.C. Henry, *Electronic Ceramics*, Doubleday & Company, Inc., Garden City, N.Y., 1969)

15.3 Magnetic permeability

As stated earlier, the permeability of a vacuum is $\mu_0 = 4\pi \times 10^{-7}$ henry/m. If we insert a material into the same space, we obtain a new value for μ because the same applied field produces a different flux density B. Instead of discussing the absolute value, it is useful to consider the relative value vs. that of the vacuum. We define the *relative permeability* as

$$\mu_r = \frac{\mu}{\mu_0} \qquad \text{and for a vacuum:} \qquad \mu = \mu_0 \text{ and } \mu_r = 1.0$$

After determining μ_r for a wide variety of materials, we find that they fall into three general classes:

1. If μ_r is less than 1 by about 0.00005, the material is called *diamagnetic*.
2. If μ_r is slightly greater than 1 up to 1.01, the material is called *paramagnetic*.
3. If μ_r is much greater than 1 (up to 10^6), the material is called *ferromagnetic*, or in the case of ceramic magnets, *ferrimagnetic*.

All three of these phenomena are related to the structure of the material. Diamagnetism, which is negligible from an engineering standpoint, can be explained as follows: When a magnetic field is applied to a material, the motions of all the electrons change. This produces a local field around each electron. By a law of physics known as Lenz's law, this field is in the opposite direction to the applied field and gives a small decrease in the total magnetism. Paramagnetism is also a small effect noted in all atoms and molecules having an odd number of electrons. Here the electrons line up with the applied field, reinforcing the field and giving a value of permeability slightly greater than 1.

Ferromagnetism is of the greatest engineering importance because a high permeability leads to a great flux density B for a given magnetizing force H compared with a vacuum or normal materials. Without this high flux density and the ability to change or reverse it (as in soft magnetic materials such as electromagnets) or maintain it after the field is removed (as in hard or permanent magnets), most electrical devices such as transformers, motors, tape recorders, and computers would not work.

We encounter ferromagnetism in only a few elements: those with partially filled $3d$ shells, such as iron, cobalt, and nickel, and those with partially filled $4f$ shells, such as gadolinium. Although only the four elements named exhibit ferromagnetism at room temperature, alloys or ceramic compounds of the other elements with partially filled $3d$ or $4f$ shells, such as manganese, have important magnetic properties.

Now let us examine the structural factors responsible for these magnetic properties. We shall see that, as with mechanical properties, both the atomic structure and the microstructure govern the magnetic properties. We shall first discuss the effects of atomic structure on permeability and remanent induction in general terms, then take up metallic and ceramic magnets.

15.4 Atomic structure of ferromagnetic metals

Recall from Chapter 2 that, in addition to its other quantum numbers, each electron revolving about the nucleus has the quantum number m_s, which describes the way an electron is spinning on its axis. Also, when $m_s = +\frac{1}{2}$, we think of the electron as spinning in one direction, and when $m_s = -\frac{1}{2}$, in the other. A spinning charge generates a magnetic field, and we can say by definition that $+\frac{1}{2}$ means that the axis of the field is up and $-\frac{1}{2}$ means it is down. Furthermore, we find that in an atom of manganese, for example, the five $3d$ electrons all have their spins in the same direction rather than alternating in direction. This is an example of Hund's rule, which states that as electrons are added in progressing to higher elements in the Periodic Table, there is a strong tendency to attain the maximum number of spins in one direction, such as five for the $3d$ level, before electrons of opposite spin are added. This leads to a magnetic moment for the atom as a whole. Each unpaired electron, i.e., an electron not balanced by one of opposite spin, contributes a magnetic moment that is given the special name of the *Bohr magneton*. This has the value of 9.27×10^{-24} A-m^2 in mks units or 0.927 erg/gauss in cgs units. The isolated manganese atom, therefore, has a magnetic moment of 5 Bohr magnetons. It is important to establish the magnetic moments of some other nearby elements (Fig. 15.4).

We note that iron has one less Bohr magneton than manganese because there are only four unpaired electrons and that the moments for cobalt and nickel are correspondingly reduced. The magnetic moment of copper is zero because the tendency to attain a second completed group of electrons of the

Magnetic moment	Element	Number of electrons	Electronic structure $3d$ shell					$4s$ electrons
1	Sc	21	↑					2
2	Ti	22	↑	↑				2
3	V	23	↑	↑	↑			2
5	Cr	24	↑	↑	↑	↑	↑	1
5	Mn	25	↑	↑	↑	↑	↑	2
4	Fe	26	↑↓	↑	↑	↑	↑	2
3	Co	27	↑↓	↑↓	↑	↑	↑	2
2	Ni	28	↑↓	↑↓	↑↓	↑	↑	2
0	Cu	29	↑↓	↑↓	↑↓	↑↓	↑↓	1

Fig. 15.4 *Magnetic moments (un-ionized) of the transition elements*

same spin is stronger than that to add a second 4*s* electron. The 4*s* electrons do not contribute to the magnetic moment.

From this table we would predict that a block of manganese metal would have the greatest magnetic moment because its atoms have the highest moment. This is not the case because when the atoms are placed together in a block, complex alignment called *exchange interaction* occurs. In the case of *metallic* manganese the atoms line up so that the directions of the magnetic fields in the individual atoms are antiparallel and cancel out each other. This is called *antiferromagnetism*. (We shall find later that this is not the case for manganese in ceramic magnets, where the ions are isolated.) However, in the case of iron, cobalt, and nickel a fraction of the atoms do line up to produce ferromagnetism.

To get a feel for the relation between the atomic-scale property of magnetic moment and the engineering-scale measurement of the gross effect called *saturation induction,* discussed earlier, let us make a calculation to compare theory with experiment, just as we related atomic radius and density in Chapter 2.

First, we write again for a vacuum with a given field strength H

$$B = \mu_0 H$$

If we place a ferromagnetic material in the circuit, B will rise greatly, and we can satisfy the equation by changing μ_0 to a new μ to show the new permeability. However, it is simpler to express the increase in B over the original value by adding a term $\mu_0 M$, where M is called the *magnetization*:

$$B = \mu_0 H + \mu_0 M$$

The term $\mu_0 M$, therefore, is the increase in field strength, which is the same as the increase in magnetic moment caused by the new material.

As a final step, for ferromagnetic materials, we can usually omit $\mu_0 H$, since it is so small compared with $\mu_0 M$. Let us now consider an example.

EXAMPLE 15.1 Calculate the saturation induction B_s you would expect from iron in webers per square meter (teslas).

ANSWER $B = \mu_0 H + \mu_0 M$, but since $\mu_0 M$ is always many times greater than $\mu_0 H$ for a ferromagnetic material,

$$B \simeq \mu_0 M$$

The magnetic moment of an iron atom is 4 Bohr magnetons. Let us assume that when $B = B_s$, all magnetic moments are aligned. Converting to mks units, we have

$$\frac{\text{Magnetic moment}}{\text{Atom}} = \frac{4 \text{ Bohr magnetons}}{\text{Fe atom}} \left(9.27 \times 10^{-24} \frac{\text{A-m}^2}{\text{Bohr magneton}} \right)$$

Since the density of iron is 7.87 g/cm³, there are n atoms per cubic meter:

$$n = \frac{(7.87 \times 10^6 \text{ g/m}^3)(0.602 \times 10^{24} \text{ atoms/at. wt.})}{55.85 \text{ g/at. wt.}}$$
$$= 8.5 \times 10^{28} \text{ Fe atoms/m}^3$$

The magnetization will be the magnetic moment per atom times n, or

$$M = \frac{4 \text{ Bohr magnetons}}{\text{Fe atoms}} \left(9.27 \times 10^{-24} \frac{\text{A-m}^2}{\text{Bohr magneton}}\right)\left(8.5 \times 10^{28} \frac{\text{Fe atoms}}{\text{m}^3}\right)$$
$$= 3.15 \times 10^6 \frac{\text{A}}{\text{m}}$$
$$B_s = \left(4\pi \times 10^{-7} \frac{\text{volt-sec}}{\text{A-m}}\right)\left(3.15 \times 10^6 \frac{\text{A}}{\text{m}}\right) = 3.96 \frac{\text{volt-sec}}{\text{m}^2} \text{ (3.96 teslas)}$$

We find experimentally that the actual B_s is less, 2.1 teslas. This is because the actual interaction between iron atoms in the metallic state does not lead to the sum of the magnetic moments. In other words, not all the magnetic moments are aligned. Only when the atoms are isolated and ionized, as in a ceramic, will the full magnetic moment be realized.

At this point the question arises: If the atoms (and therefore their electrons) are lined up in a block of iron, why is the block not spontaneously a magnet without the necessity of applying a magnetic field? This leads us to the topics of magnetic domains and Bloch walls.

15.5 Magnetic domains

For many years physicists speculated that a block of iron contained many tiny magnets that would line up under the influence of a magnetic field. In 1912 Weiss went further and postulated that there were *domains* (regions) where the atoms were aligned to produce a magnetic field in one direction and that these were balanced by domains of opposite alignment, so that no magnetic flux was present outside the specimen. Finally, in 1931 Francis Bitter decided to look for these domains under the microscope. He sprinkled magnetic powder on the polished sample and found something like grain boundaries but within the individual grains of the material. These were indeed the domains of Weiss, and the structure is shown in Fig. 15.5 on page 638.

In the unmagnetized sample the domains are unaligned and assume balanced orientation, with a resultant magnetic moment of zero. However, if a magnetic field is applied, those domains with magnetic directions parallel to the field will grow at the expense of those opposed. If the microstructure is such that the new domain structure is retained, we have a permanent magnet; if the material returns to the original state, we have a soft magnet.

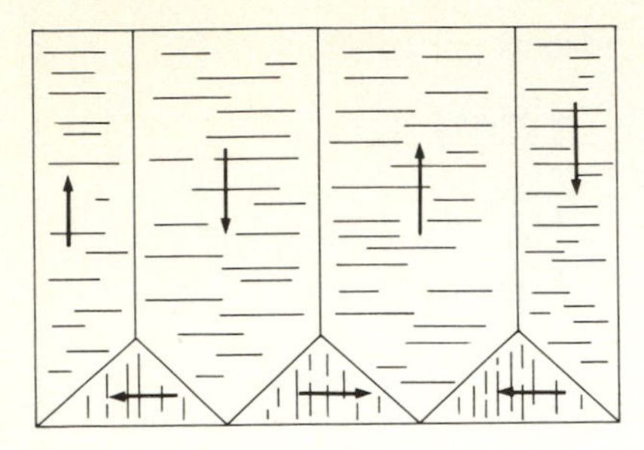

Fig. 15.5 *Domains shown by a magnetic powder etch in a single crystal of iron*

As one domain grows at the expense of another (Fig. 15.6), a change in alignment of the individual atomic magnets takes place. The domain boundary is called a *Bloch wall.* It is highly important because it helps us to understand the reasons for the magnitude of the magnetic field required to magnetize a material. First, the ease of magnetization varies with direction in a crystal, just as the ease of straining elastically varies in a single crystal, as shown by the modulus of elasticity (Chapter 3). Figure 15.7 shows typical magnetization data for iron and nickel. Note that the preferred orientation for iron is [100] (higher induced magnetization for low magnetic field strengths). We would expect to encounter a domain polarized in the [001]

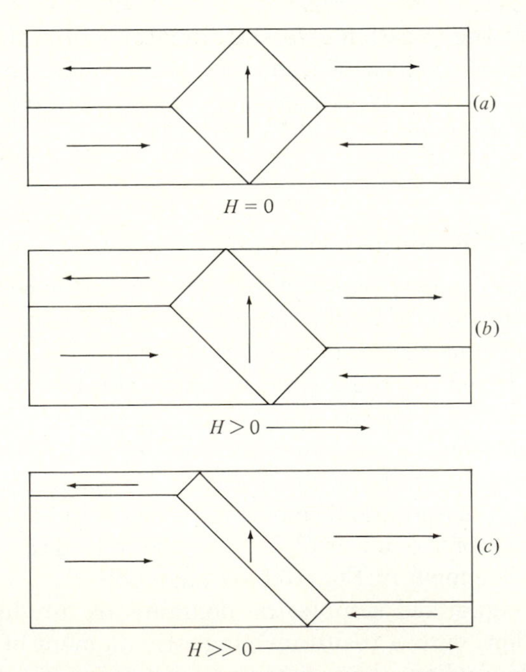

Fig. 15.6 *Two-dimensional representation of growth of domains. (a) No magnetic field (orientation of domains shown by arrows). (b) Application of a field causes growth of those domains with correct orientation. (c) Applying fields with still greater strengths causes further selective growth of correctly oriented domains at the expense of those with incorrect orientation.*

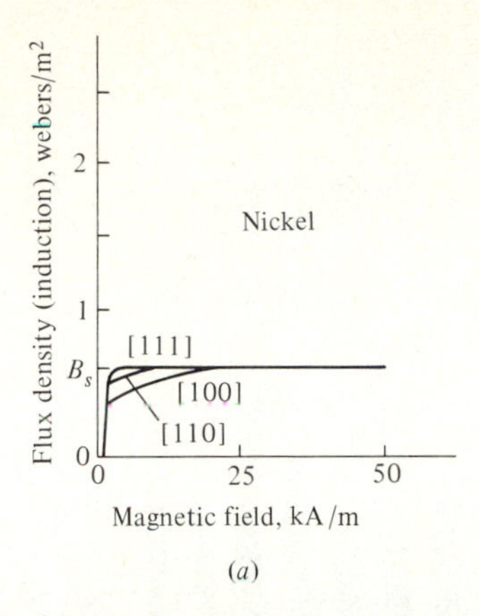

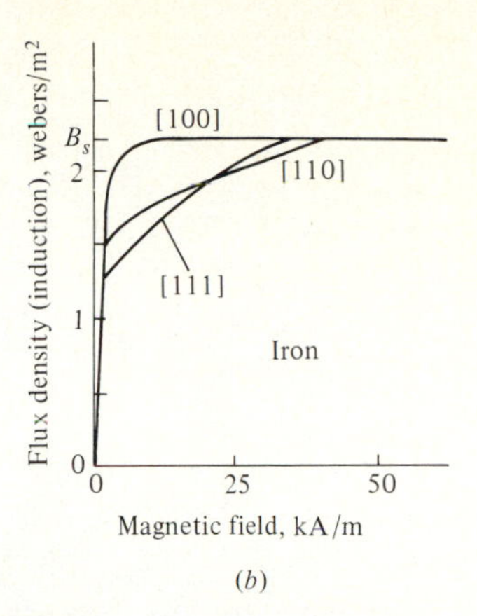

Fig. 15.7 *Magnetic anisotropy: (a) nickel, (b) iron.* [After S. Kaya et al., *Sci. Reports* (Tohoku Imp. Univ.). (1)**15**: 721 (1926); (1)**17**: 1,157 (1928)]

direction on one side of a Bloch wall and another in the $[00\bar{1}]$ direction on the other side. In the wall itself (about 1000 Å wide) we find intermediate polarizations and pass through the directions of difficult magnetization (Fig. 15.8). These domains are shown clearly in Fig. 15.9 (page 640) by another technique that uses the interaction of the domains with polarized light. Figure 5.9 shows the effect of a magnetic probe in growing domains.

EXAMPLE 15.2 A physics experiment suggests that a magnetic compass can be made using a glass of water, a small cork, a sewing needle, and a horseshoe magnet. How can we accomplish this?

ANSWER The needle is ferrous and hence ferromagnetic. By rubbing the magnet along the length of the needle in *one direction,* we may partially align the domains in the needle. If we place the needle on the small cork floating on the water surface, this allows the needle to align with the earth's magnetic field. Since the domain alignment in the needle is not permanent, the effect will be lost with time, depending on chemistry and on the heat treatment of the needle and the ambient temperature.

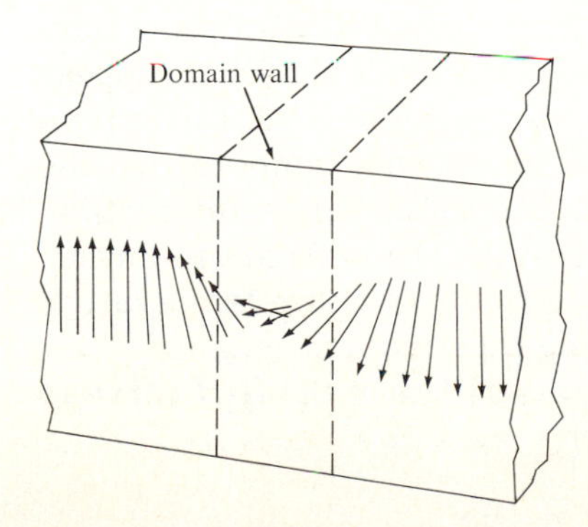

Fig. 15.8 *Domain wall, showing gradual change in direction of field between adjacent domains*

Fig. 15.9 *Magnetic domains in a thin (0.1-mm) single crystal of a magnetic ceramic, samarium terbium orthoferrite (perovskite structure). A beam of polarized light passes through the crystal, then through an analyzer. The domains then show on the screen because of the effect of their magnetic field on the light beam (Faraday effect). The magnetic needle (white finger) has caused growth of the domains in the crystal (black areas) which have the same magnetic direction as the domains in the needle.* (Courtesy of P.C. Michaelis, Bell Telephone Laboratories)

15.6 Effect of temperature on magnetization

Every ferromagnetic material will lose its magnetic properties in some temperature range upon heating, and the ferromagnetism will finally disappear at a temperature called the *Curie temperature*. For example, the Curie temperature for iron is 770°C, similar to Fig. 15.10. At this temperature the thermal oscillations overcome the orientation due to exchange interaction, and a random grouping of the atomic magnets results. Domains are re-formed on cooling, so the application of a magnetic field from the Curie temperature to room temperature can be helpful in producing a permanent magnet.

Let us now, for review, consider the characteristics of the B-H curve in terms of domain theory.

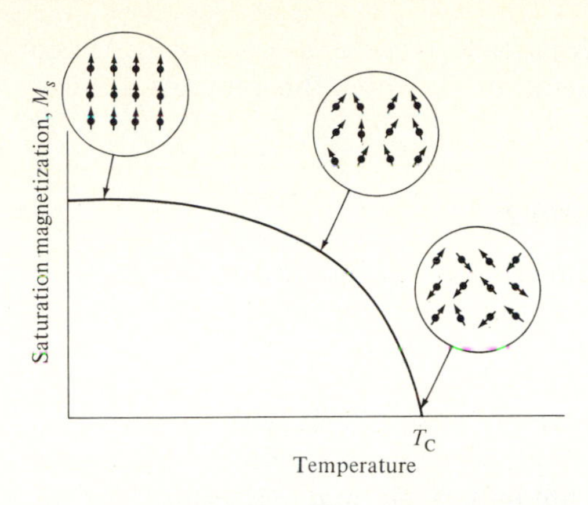

Fig. 15.10 *Effect of temperature on the saturation magnetization* M_s *of a ferromagnetic material below the Curie temperature* T_C.

15.7 Magnetic saturation

If we start with a small field strength H and then increase it, the flux density B will rise rapidly and attain *saturation induction* (Fig. 15.2, p. 632). After this point the flux will only increase slightly with H. This phenomenon can be readily explained by the effect of increasing field strength on domains, as discussed earlier. To begin with, as the field is increased, the domains in line with the field grow and the permeability is high. As the flux approaches saturation, the atomic magnets in the domains are practically all aligned and contributing to the high flux density. Beyond this point of saturation induction, further increase in field strength produces only the small increases in flux density that would be obtained without the core of magnetic material.

15.8 Remanent induction

When the field strength H that was applied originally is decreased to zero, a magnetic flux B_r will persist in a magnetic material (Fig. 15.2). In a permanent magnet this will have a high value, whereas in a soft magnetic material the value will be low. The reverse field H_c needed to demagnetize the bar and produce zero flux is called the *coercive field*. Figure 15.3 (page 633) shows the extreme differences between soft and hard magnets.

To explain this phenomenon structurally, we must observe the effects of the metal structure on retaining the domain structure that is present at high imposed field strengths. In a bar of pure iron, a soft magnetic material, the domains revert to the original balance when the field is removed. However, in a magnetic material with obstacles to uniform domain movement, such as inclusions, highly cold-worked material, or strained regions, a considerable portion of the magnetism is retained (B_r). For example, a quenched or cold-worked bar of high-carbon steel makes a good permanent magnet. Other methods of observing these effects include isolating domains by using powders suspended in an insulator so that each particle is a domain or, as in Alnico magnets, the development of a fine precipitate in which the domains are isolated from each other.

15.9 Metallic magnetic materials

Let us now compare the properties of some typical engineering materials with their structures. There are two principal groups: soft magnetic materials and hard or permanent magnetic materials.

Soft Magnetic Materials In these materials, such as those used in a transformer core, the most sought-after properties are minimum hysteresis loss (area of the *B-H* loop) and maximum saturation magnetization. The attainment of these properties depends on the ease of domain movement, and for this reason cold work and the presence of second phases are avoided. In the silicon-iron alloys, the grain orientation in the sheet is also controlled. The ease of magnetization varies with direction in a crystal, being easiest in the ⟨100⟩ directions. The sheet for transformers is prepared with the (100) plane parallel to the plane of the sheet and the [001] direction parallel to the rolling direction. This is the so-called "cubic" texture. The advantage of this is that the greater the ease of magnetization, the smaller the hysteresis loop and, consequently, the lower the power loss in the transformer core.

We obtain the special properties of "permalloy" (very low hysteresis) by controlled cooling to avoid the ordered structure that occurs in these iron-nickel alloys. This alloy is especially sensitive to work-hardening, and for this reason the shape or wire is formed and then annealed.

Hard Magnetic Materials As we would expect, the characteristics we want in hard or permanent magnetic materials are quite the opposite of those we want in soft materials. When the field is removed, we want a high value of remanent magnetization B_r and of coercive field H_c. It has been found, furthermore, that the lifting force of a magnet is related to the area of the largest rectangle that can be drawn in the second quadrant of the hysteresis curve, called the "*BH* product," and this energy product is taken as an index of the "power" of a magnet (Fig. 15.3*c*, page 633).

As discussed earlier, the key to retaining magnetism is preventing the domains from relapsing into a balanced orientation, giving no external magnetic moment. There are two principal methods for this:

1. Use of a strained metal structure
2. Division of the magnetic part of the structure into small volumes that reorient with difficulty

In the first method high-carbon or alloy steels are quenched to produce martensite, which has a high internal stress.

The second method uses alloys such as Alnico (Al, Ni, Co, and Fe) and

Cunife (Cu, Ni, Fe). The key to the Alnico alloys is that a single-phase BCC structure, stable at 1300°C, decomposes into two different BCC phases, α and α', at 800°C. The α', rich in iron and cobalt, has a higher magnetization and precipitates as fine particles. It takes a substantial field to magnetize these particles because each is essentially a single domain and must be rotated. However, once this is done, the B_r is high and the coercive force to demagnetize is also high. Many interesting techniques to improve the characteristics of hard magnets are used, such as controlled solidification to develop preferred orientation and heat treatment in a magnetic field.

More recently, samarium-cobalt and platinum-cobalt permanent magnets have been developed with very high maximum energy (BH) products, as shown in Fig. 15.3 and Table 15.1.

In the platinum-cobalt magnets a phase transformation (order to disorder) results in isolated domains, as in Alnico. This material is machinable but expensive—about $2000/lb (1973 prices).

The samarium-cobalt magnet and magnets involving rare earths such as praseodymium and lanthanum are prepared as intermetallic compounds such as Co_5Sm (37% by weight samarium). Then the material is ground to fine powder, so that each particle is domain size. These tiny grains (of hexagonal structure) have great differences in ease of magnetization, depending on the direction in the crystal—easy along the c axis and difficult in the a direction. The powder particles are aligned by a strong magnetic field in the die, then pressed together and sintered. Care is taken to avoid grain growth during sintering. One use of these Co_5Sm magnets is in step motors in electronic wristwatches. The name of these motors arises from the fact that each time the motor receives a pulse from the vibration of a quartz crystal, it moves the hands a step. The present cost of the magnet is $200/lb (1973 prices), but important decreases are expected.

15.10 Ceramic magnetic materials

For many years there was a search for magnetic materials with high electrical resistance. For example, we know that a transformer heats up because the alternating electrical field generates a stray current in the core, which leads

Table 15.1 MAXIMUM *BH* PRODUCTS FOR SEVERAL MAGNETIC MATERIALS

Material	*Maximum BH Product, amp-weber/m³*
Samarium-cobalt	120,000
Platinum-cobalt	70,000
Alnico	36,000
Ferroxdur*	12,000
Iron-cobalt sinter	7,000
Carbon steel	1,450

*A synthetic ceramic magnet ($BaFe_{12}O_{19}$) described in the following section.

to heating (eddy currents). In high-frequency equipment this heating is very severe, and so a high-resistance magnetic material was sought. It was recognized that the mineral magnetite, Fe_3O_4, was a natural high-resistance ceramic magnet, so the search led to synthetic ceramic magnets of similar structure, generally called *ferrites* although many do not contain iron. (It is important not to confuse the use of the word "ferrite" for these magnetic oxides and the same name for α iron!)

As a starting point, we should look at the unit cell structure of magnetite itself. If we write the formula $Fe^{2+}Fe_2{}^{3+}O_4{}^{2-}$, we see that it is analogous to the spinel $Mg^{2+}Al_2{}^{3+}O_4{}^{2-}$ discussed in Chapter 7 as a refractory, and we find that it has a similar unit cell. We have waited until this point to describe the unit cell (Fig. 15.11), because it is most important here to understand the magnetic properties. To explain the important ion positions, it is necessary to take a multiple of 8 times the chemical formula written in general terms, with the divalent and trivalent metal ions written as M^{2+} and M^{3+}: $[M^{2+}M_2{}^{3+}O_4{}^{2-}]_8$.

The 32 O^{2-} ions are arranged to form eight FCC cells. We can determine from inspection that there are twice as many tetrahedral sites as FCC atoms in a unit cell, whereas the number of octahedral sites is equal to the number of FCC atoms (i.e., four per unit cell). Figure 15.11 shows examples of these sites. In the model there are 64 tetrahedral and 32 octahedral sites, but only a portion of them are occupied in spinel. In the ferrites the iron ions are distributed as follows in the so-called "inverse" spinal structures:

Iron Ions	*Tetrahedral Sites*†	*Octahedral Sites*†
16 Fe^{3+}	8	8
8 Fe^{2+}	0	8

†A *site* is a position that can be occupied by an atom. A tetrahedral site is at the center of a tetrahedron formed by other atoms. Six atoms form an octahedron and surround an octahedral site.

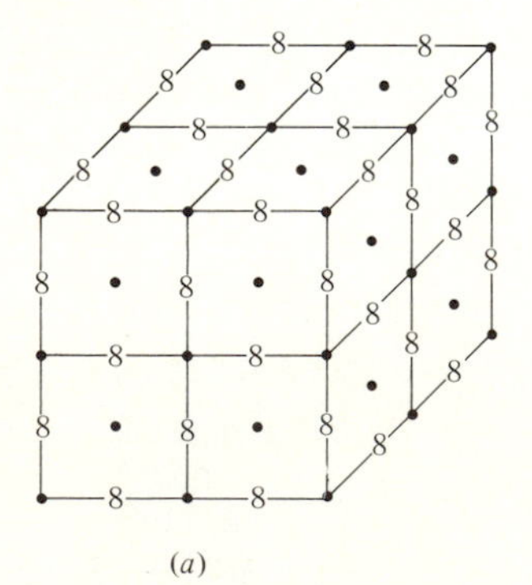

(*a*)

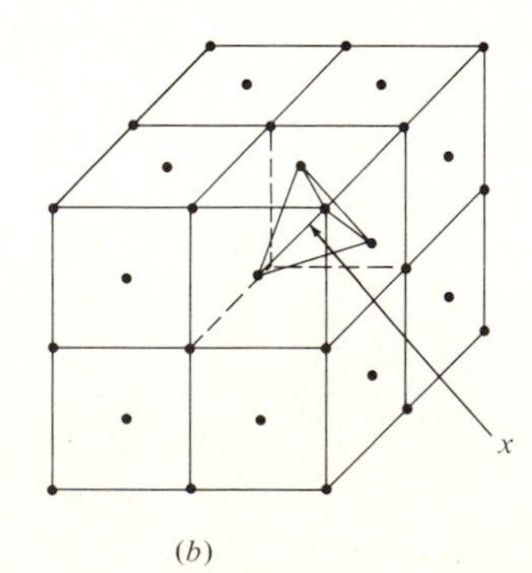

(*b*)

Fig. 15.11 *Octahedral and tetrahedral sites in a spinel. (See footnote above.) Oxygen atoms are indicated by •; (a) possible octahedral sites are indicated by 8; and (b) location of the tetrahedral site is indicated by x.*

Table 15.2 MAGNETIC MOMENTS FOR IONS FOUND IN SPINELS

Ion	*Total Electrons*	*3d Electrons*	*Bohr Magnetons*
Fe^{3+}	23	$3d^5$	5
Mn^{2+}	23	$3d^5$	5
Fe^{2+}	24	$3d^6$	4
Co^{2+}	25	$3d^7$	3
Ni^{2+}	26	$3d^8$	2
Cu^{2+}	27	$3d^9$	1
Zn^{2+}	28	$3d^{10}$	0

This is called *inverse* because in *normal* spinel, $MgAl_2O_4$ for example, the M^{2+} ions are in the tetrahedral sites and the M^{3+} are all in the octahedral sites.

Now, just as in ferromagnetic iron, the magnetism of a ceramic magnet is due to the magnetic moments of the ions. These are calculated in the same way as the magnetic moments of the atoms, but they are not the same if $3d$ electrons are removed in ionizing. For example, Fe^{2+} is formed by removing the two $4s$ electrons, so that it has the same moment as Fe^0. However, in forming Fe^{3+} we develop a magnetic moment of 5 Bohr magnetons because one electron of opposite spin is removed. Table 15.2 gives common magnetic moments.

The eight Fe^{3+} ions in the tetrahedral sites are always opposite in magnetic alignment to the eight Fe^{3+} ions in the octahedral sites, and therefore no net magnetic moment results from the Fe^{3+} ions. This is called *antiferromagnetism,* as in metallic manganese. However, the moments of the eight Fe^{2+} ions that also occupy octahedral sites are all aligned in the same direction in the unit cell and are responsible for the net magnetic moment (ferrimagnetic condition) (Fig. 15.12). Let us check this arrangement with a calculation.

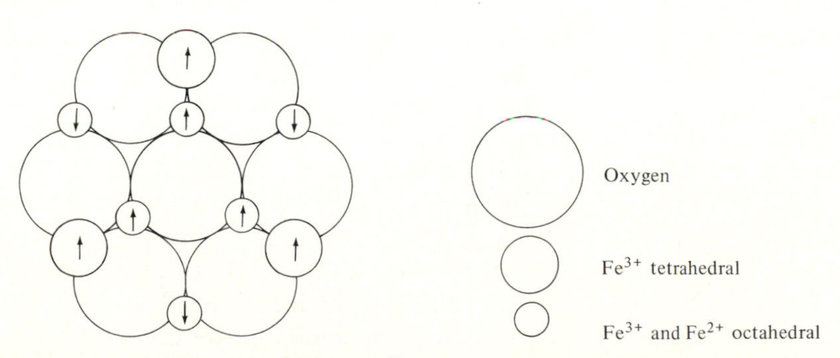

Fig. 15.12 *Spin directions in* Fe_3O_4. Fe^{3+} *ions in tetrahedral sites have opposite spin directions from* Fe^{3+} *ions in octahedral sites. The net magnetization is due to the* Fe^{2+} *ions in the octahedral sites.*

EXAMPLE 15.3 Calculate the saturation induction of $[NiFe_2O_4]_8$. The unit cell is cubic, $a_0 = 8.34$ Å.

ANSWER The magnetic moment is due entirely to nickel ions, since the spins for the 16 Fe^{3+} ions cancel each other out. Therefore, we have

$$8 \times 2 = 16 \text{ Bohr magnetons/unit cell contributed by the } Ni^{2+} \text{ ions}$$

$$M = \left(\frac{16 \text{ Bohr magnetons/unit cell}}{(8.34 \times 10^{-10})^3 \text{m}^3\text{/unit cell}}\right)\left(9.27 \times 10^{-24} \frac{\text{A-m}^2}{\text{Bohr magneton}}\right)$$

$$= 2.56 \times 10^5 \frac{\text{A}}{\text{m}}$$

$$B_s = \mu_0 M \qquad (\text{neglecting the } \mu_0 H \text{ term})$$

Assuming that all magnetic moments are aligned at B_s,

$$B_s = \left(4\pi \times 10^{-7} \frac{\text{volt-sec}}{\text{A-m}}\right)\left(2.56 \times 10^5 \frac{\text{A}}{\text{m}}\right) = 0.32 \frac{\text{volt-sec}}{\text{m}^2} \text{ (0.32 tesla)}$$

The experimental value 0.37 tesla is even greater than the calculated value because, as a result of cation vacancies in the crystal structure, some Fe^{3+} is present with Ni^{2+} and contributes a higher magnetic moment per atom (see Table 15.2).

However, note that although the saturation induction is greater than the theoretical value, it is less than that for pure iron in Example 15.1. There are two reasons for this: the lower magnetic moment per atom and the lower density of atoms in the ceramic that contribute to the magnetic moment.

Returning to the development of ceramic magnets in general, we find that in most cases the Fe^{2+} ions are replaced partly or completely with other ions. Since the ions in these positions are responsible for the magnetization, a partial replacement with an ion with lower moment reduces the overall magnetization. Another technique is to add nonmagnetic ions that will replace some of the Fe^{3+} ions and force them into the Fe^{2+} sites.

For *soft* magnets for electronic and radio communications the following substitutions for Fe^{2+} ions are used: $Mn^{2+} + Zn^{2+}$, $Ni^{2+} + Zn^{2+}$, $Ni^{2+} + Cu^{2+} + Zn^{2+}$, giving a hysteresis loop similar to that in Figure 15.3*a* (p. 633). The Mn^{2+}, Ni^{2+}, and Cu^{2+} ions have magnetic moments, but the role of the Zn^{2+} ions is interesting in that they have no magnetic moment themselves but raise the overall moment in the following way. We mentioned that in the normal spinel structure the divalent ions occupy the tetrahedral sites. The Zn^{2+} ions continue to follow this tendency when they are added to the inverse spinel. They thereby force some Fe^{3+} from their "unproductive" tetrahedral sites to the octahedral sites, where they align to increase the magnetic moment. Just as we added alloying elements in solid solution to change mechanical properties, in this case we are adding ions in solid solution to modify magnetic properties.

To provide magnets with square loops (Fig. 15.3*b*), the substitutional ions are Mn^{2+} + Mg^{2+} and Co^{2+}. These are important for computer operations.

The field of ceramic *permanent* or *hard* magnets is especially exciting. Already there are a variety of uses for these materials, such as the magnetic inserts in refrigerator doors and in small motors. Formerly an electromagnet was needed in these motors, but the use of small, powerful permanent magnets to provide a field greatly simplifies the design. These operate windshield wipers, heater blowers, air conditioners, and window raisers in modern automobiles. The 30-billion-eV accelerator at the Brookhaven National Laboratory uses over 7 tons of ferrites. As an illustration of possible future use, Figure 15.13 shows a drawing of an elevated magnetic roadway in which the car would be floated by magnetic repulsion.

The crystal structure of these permanent ceramic magnets is slightly different from the spinel structure (see the problems), but since it follows the same principles, the materials are still called "ferrites." Just as the ferrites we have discussed have the structure of a naturally occurring mineral (spinel), so these hard ferrites have the hexagonal structure of a mineral called *magnetoplumbite,* $Pb(Fe,Mn)_{12}O_{19}$. The source of the magnetism is similar: the alignment of ions such as Fe^{2+} and Mn^{2+} with magnetic moments and the formation of domains. The synthetic ceramic magnets use simpler formulas, such as $PbFe_{12}O_{19}$ or $BaFe_{12}O_{19}$. One of the ways of obtaining a strong initial alignment of domains is to heat the material above its Curie temperature, then cool it under the influence of an externally imposed magnetic field. This

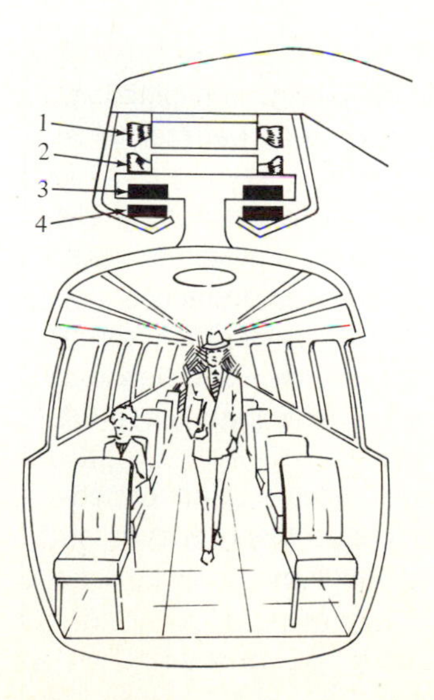

Fig. 15.13 *An artist's conception of a magnetic roadway.* Electric drive: *(1) Portion of linear motor attached to roadway and serving the same function as the stator of a conventional electric motor. (2) Portion of linear motor attached to the car, functioning as a rotor.* Magnetic suspension: *(3) Ceramic (ferrite) permanent magnets attached to the car. (4) Ferrite magnets attached to the roadway.* (Courtesy of Westinghouse Corporation)

is analogous to polarizing a ferroelectric material by cooling it under an electric field.

In addition to the soft and hard types, there are two subtypes of great importance.

The square-loop materials retain their full remanent magnetization as the field is reversed until a sharp cutoff is reached (Fig. 15.3*b*). Therefore, in using the magnet as an information-giving device, we can get full strength without weakening until the cutoff is reached. If we want to clear out the information, we can do this cleanly and the device is ready to remagnetize. For these magnets manganese-magnesium-iron oxides and cobalt-iron oxides are used.

For microwave communication it is important to have low losses per cycle because of the high frequencies. In iron-core transformers we can tolerate the hysteresis losses at 60 Hz, but at 10^6 Hz they would be prohibitive. Here the high resistance and very narrow loops of certain ferrites are important. Typical oxides are of the (Al,Ni,Zn)O type.

Another family of magnetic ceramics is the iron garnet group. The general formula is $3M_2O_3 \cdot 5Fe_2O_3$. The unit cell is more complex than that of spinel. The chief advantage is that elements with high magnetic moment such as samarium and yttrium can be accommodated as M in the formula.

EXAMPLE 15.4 Explain the following observations using magnetic domains:

a. A tie bar of soft iron is placed on a horseshoe magnet when it is not being used.

b. An induction furnace operating at 3000 Hz is used to melt steel.

ANSWER

a. A more random orientation of domains is the lowest energy configuration. Therefore the residual magnetism decreases with time. Using a tie bar maintains the domain alignment between the magnet ends.

b. The heating of steel in an alternating magnetic field *is not* completely explained by magnetic domains. It is true that one circuit around a hysteresis loop can be treated as an energy loss that will cause heating. However, the amount of energy loss is insufficient for melting. A second and more important source of energy loss is due to eddy currents. In an alternating field a voltage and hence a current (eddy current) is induced in the core (the steel in our example). These I^2R losses can be large enough to cause melting.

In induction hardening, we use a higher frequency with lower depth of penetration, since melting is not desired. The total heat in the core then depends on frequency, time, heat transfer to the surroundings, and conductivity.

SUMMARY

In the electrical materials the most important concept was the effect of structure on the ease of movement of charge carriers. In an analogous way, in magnetic materials the important concept is the effect of structure on the value of magnetic flux. There are two important differences to be established immediately.

The value of magnetic flux density B obtained by the application of a magnetic field of strength H is a function of the permeability μ, but at some value of H there is very little increase in B. This is the saturation magnetization B_s.

Second, the flux density does not fall to zero when the field is removed, but only drops to the residual magnetization B_r, which is low in a soft magnetic material and high in a permanent or hard magnetic material.

The features of hard and soft metallic magnetic materials are reviewed first. It is established that the key feature is the magnetic domain, a region of structure in which the magnetic moments of the atoms are all aligned in one direction. When a field is applied, the domains with moments aligned parallel to the field grow at the expense of others. In a soft magnetic material, domain wall movement is made easy by simple grain structure, avoidance of cold work, etc. Conversely, in hard magnetic materials the desired structure consists of tiny (300 Å) particles, each with only one domain, aligned to provide a strong magnetic field.

In ceramic magnets the inverse spinel structure is used to produce soft magnets of the desired characteristics. For hard magnets fine domain-sized particles of the magnetoplumbite structure are used.

DEFINITIONS

Antiferromagnetic Material with ferromagnetic atoms that are aligned with magnetic moments in opposing directions, leading to no net magnetic moment.

Bohr magneton The magnetic moment produced in a ferromagnetic atom by one unpaired electron. 1 Bohr magneton $= 9.27 \times 10^{-24}$ A/m^2 (0.927 erg/gauss).

Coercive field, H_c The field needed to reduce the flux density to zero after initial application of a magnetic field to obtain B_s.

Diamagnetism, paramagnetism Very small deviations from $\mu_r = 1$.

Domain A region of a metal or ceramic structure in which the magnetic moments due to the individual atoms are aligned in the same direction.

Domain wall, Bloch wall The boundary between two domains. On either side of the wall the magnetic moments are oriented at 90 or 180° to each other.

Ferrimagnetic Material with ferromagnetic atoms that are aligned in opposing directions in the unit cell but with a net moment.

Ferrite A magnetic inverse spinel. Do not confuse this term with the same word used earlier to mean α iron.

Ferromagnetism Large values of μ_r produced by alignment of certain atoms such as iron, cobalt, and nickel into domains.

Hard magnetic material Material that is magnetized and demagnetized with difficulty and that has a large area in the B-H loop.

Inverse spinel Material with the same general formula as a normal spinel but with the trivalent ions occupying tetrahedral and octahedral sites and the divalent ions occupying only octahedral sites.

Magnetization, M The magnetization times the permeability of a vacuum gives the increase in flux due to insertion of a given material into a field of strength H. Since μ_0 is a constant ($4\pi \times 10^{-4}$ volt-sec/A-m), the magnetization is a measure of the increased flux density produced by the material.

Normal spinel A ceramic of the general formula $M^{2+}M_2^{3+}O_4$. There are divalent ions in tetrahedral sites and trivalent ions in octahedral sites.

Permeability, μ The ratio of the flux density B developed when a magnetic field of strength H is applied; $\mu = B/H$. This is analogous to the concept that the conductivity σ is equal to the ratio of the current I produced by a voltage E in a wire (for unit length and area of the conductor).

Relative permeability, μ_r The ratio of permeability of a material to the permeability of a vacuum; $\mu_r = \mu/\mu_0$.

Remanent induction, B_r The flux density remaining when the magnetic field is removed.

Saturation induction, B_s The maximum magnetic flux obtainable in a material. For a ferromagnetic material, B_s may be taken as $\mu_0 M$.

Soft magnetic material An easily magnetized and demagnetized material, with a small area in the B-H loop.

PROBLEMS

15.1 An important characteristic in a power-transformer core is the power loss, which leads to heating and lowered efficiency. The power loss is made up of stray currents induced in the core and the hysteresis loss, which is related to the area inside the B-H loop for a complete cycle. (Sections 15.1 through 15.4)

a. Describe the type of metallic material you would select for a transformer and explain how it would be fabricated (laminations, preferred orientation, etc.).
b. Describe the type of ceramic material you would select and the ideal *B*-*H* curve you would specify.

15.2 Explain how the following characteristics of a ferromagnetic material could affect its *B*-*H* curve: (*a*) Grain size, (*b*) Inclusions, (*c*) Residual stresses, and (*d*) Heat treatment (specify) (Sections 15.1 through 15.4)

15.3 Why does a hysteresis loop occur in most magnetic materials? (Sections 15.1 through 15.4)

15.4 After an external magnetic field is removed, the domains tend to revert to their original size and distribution. Explain. What structural conditions inhibit this motion, i.e., tend to retain the induced magnetism? (Sections 15.5 through 15.9)

15.5 Estimate the saturization magnetization of gadolinium (atomic weight = 157.26 g/mole, density = 7.8 g/cm^3), assuming that the magnetic moment per atom is equal to the number of unpaired 4*f* electrons and that the atom follows Hund's rule. In this case Hund's rule states that the first seven 4*f* electrons will have the same spin direction and the next seven the opposite spin direction. (Sections 15.5 through 15.9)

15.6 In the manufacture of Alnico magnets the alloy is solidified and cooled in the magnetic field. What is the advantage? (Sections 15.5 through 15.9)

15.7 One can buy a high-purity vacuum-melted iron for use as a solid-core material in an electromagnet. The recommended processing procedure is to (*a*) machine it to the desired shape, (*b*) anneal in an inert atmosphere, and (*c*) furnace (slow) cool the component. Explain why each step in this sequence is desirable. (Sections 15.5 through 15.9)

15.8 Compare the magnetization of pure iron in the [100] and [111] directions. Which direction would result in the smallest hysteresis loop? In what application is this information useful? (Sections 15.5 through 15.9)

15.9 You wish to obtain maximum residual magnetization B_r in a component. It has been suggested that a compaction and sintering process be used. Explain why each of the following observations may be advantageous to maximum B_r. (Sections 15.5 through 15.9)

a. Powder size should be equal to domain size.
b. Compaction, sintering, and cooling should be done in the presence of a magnetic field.
c. Incomplete sintering or the presence of inclusions at the interface between particles may be an advantage.

15.10 Ceramic materials with square *B*-*H* curves are used for memory units in computers. Explain the differences in B-H curves you would specify compared with ceramic *transformer* materials. (Section 15.10)

15.11 Estimate the magnetic moment per unit cell of the following substances. (Section 15.10) (*a*) $[Fe_3O_4]_8$, considered as $[Fe^{2+}Fe_2{}^{3+}O_4]_8$ (*b*) $[MgAl_2O_4]_8$ (*c*) $[CoFe_2O_4]_8$

15.12 Calculate the saturation induction B_s of $[Li_{0.5}Fe_{2.5}O_4]_8$, assuming that a_0 = 8.37 Å in an inverse spinel. (Section 15.10)

15.13 In making magnetic sealer sticks for refrigerators, the magnetic material is placed in a thermoplastic in the liquid state, and a magnetic field is maintained while the plastic sets. Why? (Section 15.10)

15.14 Show how the net magnetic moment of $[(Zn_{0.5}Ni_{0.5})Fe_2O_4]_8$ is greater than that of $[NiFe_2O_4]_8$. [*Hint:* The zinc ions displace Fe^{3+} ions from tetrahedral sites.] (Section 15.10)

15.15 Ferrite $(MnFe_2O_4)_8$ is experimentally measured and shows a magnetization of 3.30 × 10^5 A/m. (Section 15.10)

a. How many Bohr magnetons are there per unit cell? Show calculations.

b. If the unit cell is cubic, what is its size?

15.16 An investigator proposes making a new ferrite in which half the nickel atoms in $[NiFe_2O_4]_8$ would be replaced by manganese atoms. What would be the change in the magnetic moment in Bohr magnetons/unit cell? Assume that all the nickel and manganese atoms are in sixfold sites in all cases, and have their spins aligned in the same direction. The Fe^{3+} ions are half in sixfold sites, half in fourfold sites, and the spins in the sixfold sites are opposite those in the fourfold sites. (Section 15.10)

15.17 It has been suggested that we synthesize a ferrite to achieve higher saturation induction than is obtained in $[NiFe_2O_4]_8$, which we found to be 0.32 tesla in Example 15.3. In the new synthesized ferrite, manganese ions would be substituted for one-half the nickel ions to produce $[Ni_{0.5}Mn_{0.5}Fe_3O_4]_8$. The size of the unit cell increases to 9.03 Å because manganese is larger than nickel. (*a*) What is the magnetic moment contribution from each ion? (*b*) What is the magnetism M? (*c*) What is the saturation magnetization B_s? (Section 15.10)

15.18 In ferrites, Fe_3O_4 would be written $[Fe^{2+}Fe_2{}^{3+}O_4]_8$. (*a*) The material $[Li_{0.5}Fe_{2.5}O_4]_8$, which is also an inverse spinel, would be written $[Li_{0.5}(?)_{0.5}Fe_2O_4]_8$. What ion is present at (?), and what sites in the inverse spinel does it occupy? (*b*) Calculate the net magnetic moment. (Section 15.10)

15.19 The inverse spinel structure consists of a FCC lattice of oxygen ions (ionic radius = 1.32 Å) with certain M^{3+} ions at both octahedral and tetrahedral sites. (1) Consider the points in the inverse spinel unit cell with the following coordinates: (*a*) 0, 0, 0; (*b*) $\frac{1}{2}$, $\frac{1}{2}$, $\frac{1}{2}$; (*c*) $\frac{1}{4}$, $\frac{1}{4}$, $\frac{1}{4}$; (*d*) $\frac{3}{4}$, $\frac{1}{4}$, $\frac{1}{4}$; (*e*) 0, 0, $\frac{1}{2}$; (*f*) 0, 0, $\frac{1}{4}$. List by code letter (a, b, c, etc.) all the points for which coordinates are given that are tetrahedral sites for M^{3+}. (2) Calculate the smallest ionic radius in angstroms (Å) that an ion may have to occupy (in stable condition) an octahedral site in the inverse spinel structure. Show calculations. (Section 15.10)

PROBLEMS COVERING ALL SECTIONS IN CHAPTER 15 AND OTHER CHAPTERS

15.20 Consider that you have three magnet cores, A for a 60-cycle transformer, B for a memory core, and C for a permanent magnet. Select from the following vocabulary the *magnetic characteristic and material* you consider most important for each application. (Summary of Chapter 15)

A: transformer
B: memory core
C: permanent magnet

Characteristics (*a*) *BH* product in second quadrant, (*b*) square *B-H* curve, (*c*) low coercive field, (*d*) maximum B_s (saturization magnetization)

Materials (*a*) Alnico, (*b*) 18-4-1 tool steel, (*c*) 3% Si steel, (*d*) alumina dispersed in a polymer, (*e*) iron oxide dispersed in a polymer

15.21 The figure shows magnetization curves for three different ferromagnetic alloys. On the basis of these curves, answer the following questions. (Summary of Chapter 15)

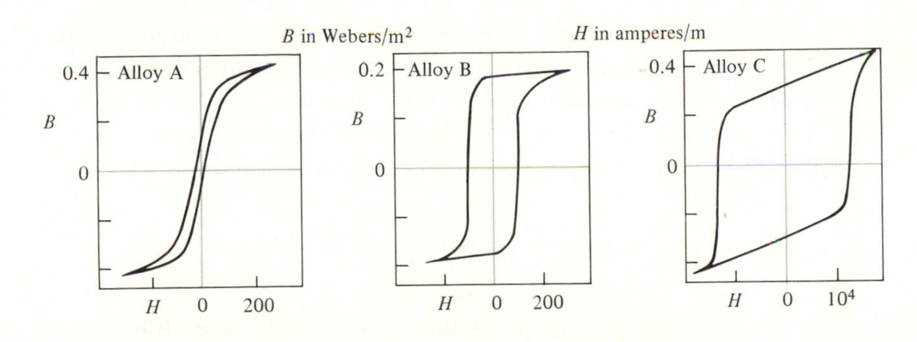

1. Which alloy has the highest coercive field?
2. Which alloy has the highest hysteresis loss?
3. Which alloy has the highest residual magnetization?

(*Continued on page* 654)

4. Which alloy would be preferred for a permanent magnet?
5. Which alloy would be best for a computer memory?
6. Which alloy would be best for a transformer core?
7. Which alloy probably has the highest fraction of domains oriented in parallel when $H = 0$ (after first having been magnetized to saturation)?
8. Which alloy would have the highest mechanical hardness?
9. Ferrite magnets are generally preferred to metallic magnets in applications involving (*a*. high, or *b*. low) frequency electric fields.
10. Ferromagnetic materials lose their ferromagnetic properties above the (   ) temperature.

15.22 The terms "ferromagnetic," "antiferromagnetic," and "ferrimagnetic" are used to describe the different types of alignment of electron spins in iron, manganese, and magnetite (Fe_3O_4, spinel), respectively. For each case, sketch a group of atoms illustrating the difference in spin alignment. In the ferrimagnetic structure, mark which group of ions is antiferromagnetic and which group is aligned to produce magnetization. (Summary of Chapter 15)

15.23 Alnico magnets and ferrites are produced by different processing techniques. Explain what techniques are used (e.g., forming, casting, sintering, machining, etc.) and why the methods must be different. (Summary of Chapter 15)

15.24 *a*. Assume an FCC unit cell of element A. Check whether the following sites are classified as *octahedral, tetrahedral,* or *neither* when an atom B is added. (1) $\frac{1}{2}, \frac{1}{2}, \frac{1}{2}$; (2) $\frac{1}{2}$, 1, 1; (3) $\frac{3}{4}, \frac{3}{4}, \frac{1}{4}$; (4) 1, 1, 1

b. Referring to an FCC lattice of atom A *with no B present,* extending through the space: (1) What is the coordination number of the atom at 1, 1, 1? (2) List the coordinates of all its nearest neighbors in the unit cell. (3) If B atoms are added to occupy all nearby tetrahedral sites, what will the coordination number of atom A become?

c. What is the significance of tetrahedral and octahedral sites in an inverse spinel structure? (Sections 15.10 and 7.6)

15.25 In the magnetoplumbite structure of hard magnets, the general formula is $AO \cdot 6B_2O_3$, where A is barium, lead, or strontium; O is oxygen; and B is aluminum, chromium, gadolinium, or iron. (Sections 15.10, 7.6, and 7.7)

a. The valences of A and B are constant but different. What are they?

b. Using these valences, what is the percentage variation in ionic radius for A metals and B metals?

15.26 A magnetic ferrite for an oscilloscope component is to have a final rectangular shape measuring 0.621 by 0.356 by 0.109 in. (1.58 by 0.905 by

0.277 cm). An investigator finds that the specific gravity of a similar test piece changes from 3.57 green (before sintering) to 5.02 after sintering. Design and give dimensions of a die for producing green shapes that will sinter to the desired dimensions. Assume that linear shrinkage is the same percentage in all directions. Also, note that volumetric shrinkage is not 3 times linear shrinkage for appreciable amounts of shrinkage. (Sections 15.10 and 8.6)

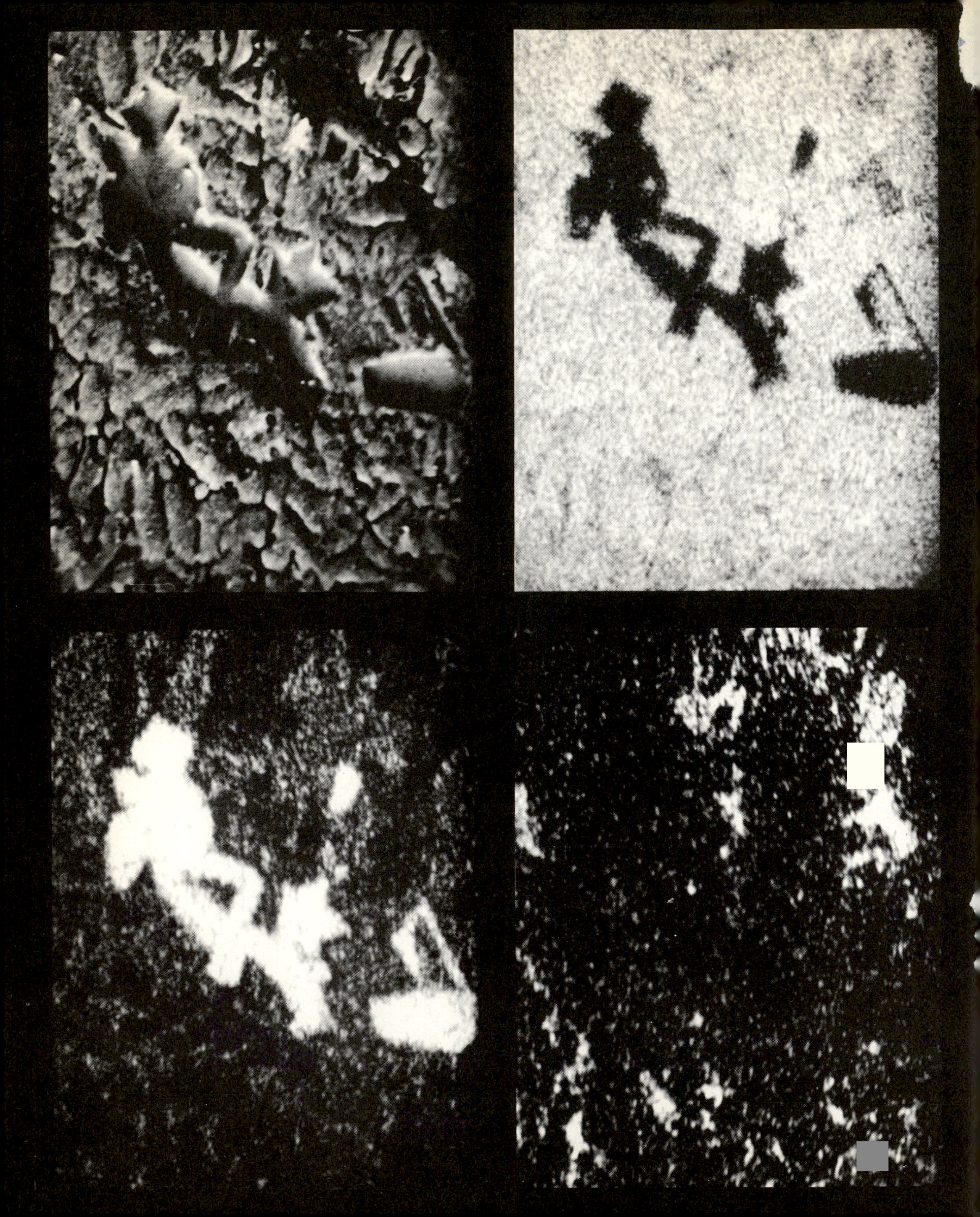

16

OPTICAL AND THERMAL PROPERTIES OF MATERIALS

THIS illustration shows the use of the microprobe to find the chemical composition of the different phases in the aluminum silicon copper engine block alloy discussed in Chapter 4. The overall microstructure is shown in the upper left. It consists of a silicon-rich phase in relief in a matrix of aluminum-rich alloy. This structure is then exposed to a sweeping beam of electrons that excites x rays from the surface. The wavelength of the x rays depends on the atoms emitting them.

To find the location of silicon atoms, for example, the x rays are filtered so that only those characteristic of silicon are passed through to the TV screen that shows the image of the structure. Therefore bright areas show the silicon-rich phase, lower left. In the upper right, the x rays from aluminum atoms were admitted, showing the aluminum-rich matrix. Finally, in the lower right, the locations of the copper-rich phase are indicated.

In this chapter we shall first take up the optical properties of materials, relating the emission, absorption, etc., of light to the structure. Then we shall consider the thermal properties, which have not yet been described.

16.1 Introduction

In this final chapter we shall take up some of the optical and thermal properties of materials. At this point we should hardly need to mention that these properties, like all the others we have studied, depend on the structures of the materials. We shall begin with the optical properties because they are closely related to the electrical properties recently discussed in Chapter 14.

Optical Properties

By *optical properties,* we mean those properties governing the emission, absorption, transmission, reflection, and refraction of light. This field covers not only well-known properties such as emission from a tungsten filament, the sources of color, transparency, opacity, and index of refraction, but also newer areas of explosive growth, such as fluorescence in lighting, phosphors for television tubes, and emissions from lasers.

16.2 Emission

There are two main types of emitted light: light with a continuous spectrum, such as "white" light from a tungsten light bulb, and light confined to definite wavelengths, such as that from a sodium-vapor lamp. We shall concentrate on the emission of definite wavelengths because of its great importance in new devices.

When light is emitted from any source, we need first to think of it as traveling in discrete units, or little energy packets called *photons* (Fig. 16.1). Second, we must recognize that light has a definite wavelength that is related to the energy of the photon. We shall need to use both these aspects in explaining the interaction of light with materials, especially with the electrons of materials.

Since visible light can originate from radiation of nonvisible wavelengths, let us review a broad portion of the spectrum. On one side of the visible spectrum we find the infrared and longer-wavelength photons, and on the other, the ultraviolet and x-ray wavelengths (Fig. 16.2).

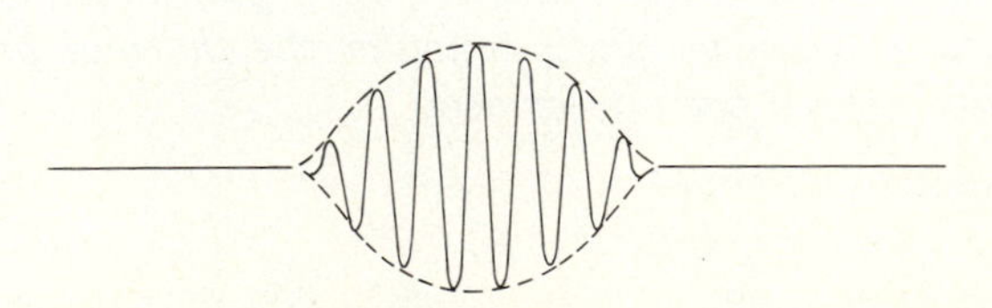

Fig. 16.1 *Schematic drawing showing both particle and wave characteristics of a photon*

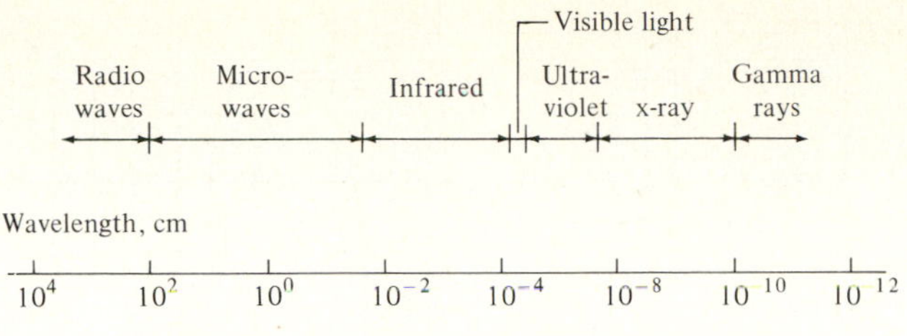

Fig. 16.2 *The electromagnetic spectrum*

From physics we have the relation between energy and the wavelength of a photon:

$$E = \frac{hc}{\lambda}$$

where E = energy, h = Planck's constant = 6.62×10^{-27} erg-sec, c = velocity of light = 3×10^{10} cm/sec, and λ = wavelength in centimeters.

If we express this relation in simple terms of electron volts, which we used in Chapter 14, we have

$$E = \frac{1.24 \times 10^{-4}}{\lambda} \text{ eV}$$

EXAMPLE 16.1 Calculate the energy of photons of infrared (λ = 11,500 Å), yellow (λ = 5800 Å), and ultraviolet (λ = 2000 Å) light.

ANSWER

Infrared $$E = \frac{1.24 \times 10^{-4}}{11{,}500 \times 10^{-8}} = 1.08 \text{ eV}$$

Yellow $$E = \frac{1.24 \times 10^{-4}}{5800 \times 10^{-8}} = 2.14 \text{ eV}$$

Ultraviolet $$E = \frac{1.24 \times 10^{-4}}{2000 \times 10^{-8}} = 6.20 \text{ eV}$$

This interrelation of energy and wavelength is important in understanding both emission and absorption of light. We shall consider emission first.

We are all familiar with the emission of yellow light that occurs when we throw salt into a fire. Let us examine the reason for this phenomenon. We recall from Chapters 2 and 14 that the electrons in an atom may occupy different energy levels. For example, Fig. 16.3 (p. 660) shows the energy levels above the 3*s* level in sodium. If an electron in the 3*s* level is raised to the 3*p* level and then falls back to the 3*s* level, a photon of about λ = 5890 Å, or

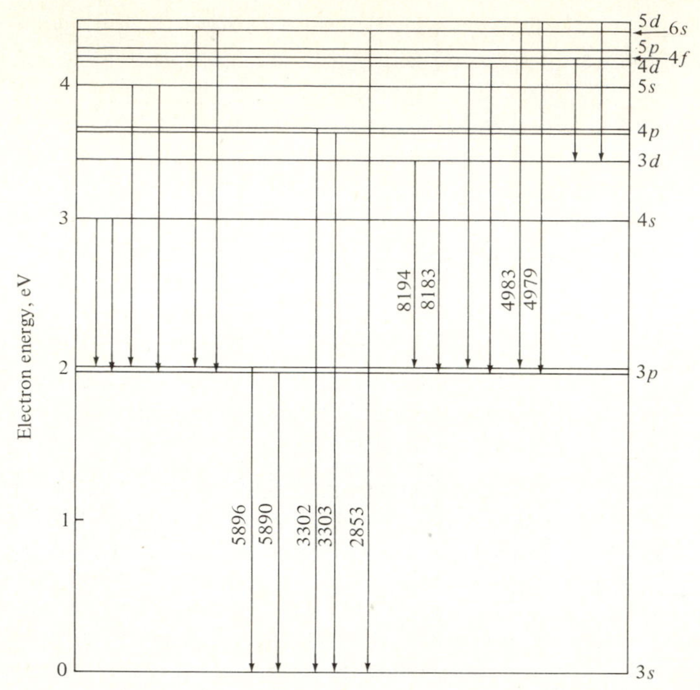

Fig. 16.3 *Electron energy level diagram for a sodium atom. The photon wavelengths associated with some of the transitions of electrons from excited states are given in units of 10^{-8} cm.*

yellow light, is emitted. The steps involving energy absorption (raising the electron to an upper level) and the fall need not be the same. For example, we could supply the energy to raise the electron to the 6*s* level, and it could jump down to the 3*p* level and then to the 3*s*. An analogy is to consider playing a pinball machine. The ball is sent to the top in one stroke but it can come down by a series of jumps. Some of these may give a blink of light, others just noise or heat.

At this point we should also consider the companion problem of the emission of shorter wavelengths (x rays) when an electron falls to a still lower level. If we bombard sodium atoms (or ions) with sufficiently energetic electrons, we can knock out an electron from the first (*K*) shell. This creates a hole into which an *L*-shell electron can fall. For example, the transition $2p \rightarrow 1s$ gives a photon of wavelength 11.91 Å or an x-ray wavelength. Just as the color of visible light emitted from a given jump depends on the particular levels involved, so the wavelength of the x ray depends on the specific case. For instance, it takes much more energy to knock out an electron from the inner ring (*K* shell) of an atom of high atomic number because of the greater charge on the nucleus, and we expect to find a shorter-wavelength x ray emitted when an L-shell electron falls into the K shell. Typical wavelengths illustrating this effect for the K_α x ray are manganese: $\lambda = 2.10578$ Å; iron: $\lambda = 1.939980$ Å; tungsten: $\lambda = 0.213828$ Å.

The scanning electron microprobe that we mentioned in Chapter 4 depends on the emission of these characteristic x rays. A scanning electron beam is swept across a given microstructure. Each tiny region of the structure produces x rays with wavelengths that depend on the elements present. For example, if we want an indication of the amount of silicon in a given region, we put in a filter that allows only the silicon wavelength rays from this region to pass through. Then the picture screen showing the microstructure is irradiated by these x rays and shows bright spots only at regions of high silicon concentration (Chapter 16 frontispiece). Similarly, we can find regions of high aluminum concentration by allowing only the x rays that are characteristic of aluminum to pass through in another picture.

We should discuss one more phenomenon, called *luminescence,* which is quite important in the new light-emitting materials. A familiar sight in natural history museums is a case of fluorescent and phosphorescent minerals. After exposure to ultraviolet light the specimens glow different colors, and some continue to glow after the ultraviolet light source is turned off. This conversion of ultraviolet light to visible light is given the general term "luminescence." If this glow ceases when the ultraviolet stimulus is removed, it is called *fluorescence.* If the glow persists for over 10^{-8} sec after the ultraviolet light is turned off, the light is called *phosphorescence.* In either case we have to explain why light of a visible wavelength results from the exposure to ultraviolet light or bombardment by electrons.

In a luminescent material a solid, such as zinc sulfide, contains a small amount (one part per million) of an impurity, such as copper. This impurity is the key to luminescence. Again the process is a little like a sophisticated pinball machine, in which an electron is raised to a high level and may fall into traps from which it is later released (Fig. 16.4). There are essentially three steps in this process.

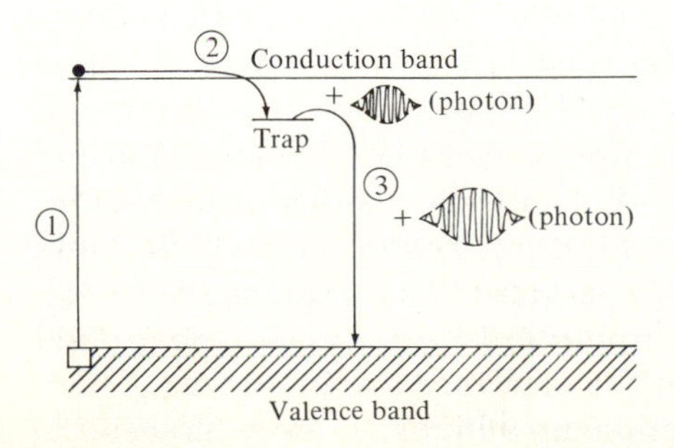

Fig. 16.4 *Development of luminescence. Step 1: The electron is excited by ultraviolet radiation into the conduction band. Step 2: The electron falls into the trap, emitting an infrared photon (the trap is an intermediate energy level due to an impurity). Step 3: The electron falls into a lower level, emitting a photon of visible light. Note: The hole is also attracted to the trap and also emits light.*

1. A photon or electron is shot at the solid, such as zinc sulfide, and its energy is absorbed. This enables an electron of the solid to rise to the conduction band, which is separated from the valence band by a definite energy gap. A hole is created in the valence band as well.
2. The electron wanders through the structure until it falls into a trap or luminescent center. These traps are produced by the impurity, in this case the copper ions. The "energy level" traps lie between the more elevated conduction band of the solid and its valence band.
3. After a time the electron acquires enough energy to leave the trap and fall to the valence band, giving up a photon of definite wavelength (in the visible spectrum) related to the ion producing the trap, in this case copper. Phosphorescence occurs when materials have deep traps. Therefore, it takes more time for the trapped electrons to receive enough energy to escape. As a consequence, they are liberated over a longer period by continued electron movement. It is possible to produce approximately "white" light by using a mixture of impurities or luminescent centers, such as silver, which gives a peak at 4300 Å, copper at 5100 Å, and manganese at 5900 Å.

Applications These variables in color and in duration of luminescence are of great practical importance. In the dots of a color television tube, for example, we want phosphors that give pure primary colors and do not have long decay times. The development of materials with these characteristics provides an example of the extremes to which manufacturers will go for color quality. It was found that europium, a rare earth, provided good red-color centers. As a result, a new mining and chemical-separation industry developed, and the other rare earths that are found with europium are now available as low-cost byproducts. Another application of these effects is in electroluminescent panels in which the initial radiation step is eliminated. Here the application of a voltage across a strip provides directly the energy required for electrons of the material to move to the phosphor centers, and the jumps that occur later give off light. The color of the panel is a function of the active ions chosen.

Lasers *L*ight *a*mplification by *s*timulated *e*mission of *r*adiation (*laser*) is an even more sophisticated effect. In the ruby laser, for example, a single crystal rod of alumina (Al_2O_3) contains a small quantity of Cr^{3+} ions as impurities (Fig. 16.5). The rod is ground with flat surfaces at each end. One end is heavily silvered to be opaque and the other is lightly silvered for partial reflection. Shining on the rod is a tube containing xenon gas, which emits constant-wavelength light (5600 Å) when energized. The sequence is as follows: At first only a few electrons in the chromium ions are above the ground state, but after the xenon lamp is flashed, the electrons are energized. These electrons either fall to the ground state or take up a higher-energy metastable

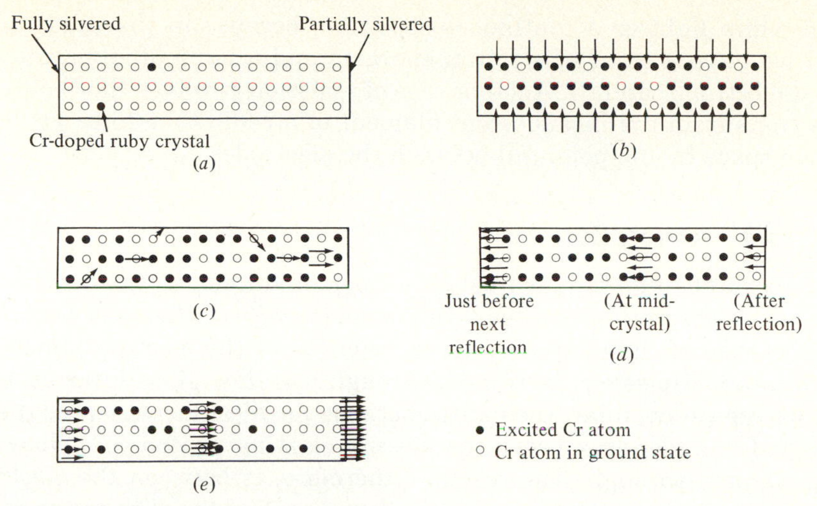

Fig. 16.5 *Schematic drawing of stimulated emission and amplification of a light beam in the ruby laser. (a) At equilibrium. (b) Excitation by xenon light flash. (c) A few spontaneously emitted photons start the stimulated emission chain reaction. (d) Reflected back, the photons continue to stimulate emission as they travel. (e) Increasing in power, the beam is finally emitted.* (R.M. Rose, L.A. Shepard, and J. Wulff, *The Structure and Properties of Materials,* Vol. 4: Electronic Properties, John Wiley & Sons, Inc., New York, 1966)

state for an average of 3 millisec. A high percentage are pumped up to this metastable state. Next a few electrons fall to the ground state, emitting photons of a characteristic wavelength. A percentage of these emitted photons are reflected from the ends and stimulate emission throughout the structure in a short period, $\frac{6}{1000}$ sec.† A pulse of coherent wavelength emission thus travels through the lightly silvered end. (By *coherent* we mean that the photons are in phase with one another.) It is a beam with very low divergence because photons emitted through the sides of the rod are lost.

Many uses have been developed for lasers, including surgical and metallurgical heating, surveying, and precision measurement. Various materials, including gas, can be used in the laser. In a semiconductor laser with a *p-n* junction, an electrical impulse energizes the carriers, so that as they fall through an energy gap, a stimulated emission of photons of longer wavelength (6500 to 8400 Å) is obtained. Also, lasers with continuous emission of light (rather than pulsed) have been developed.

Emission from Solid Masses As distinguished from the emission of light by separated ions or electrons, the radiation from heated solids such as a tungsten wire filament in a conventional light bulb is important. Here we

†There is a critical relation between the length of the rod and the wavelength of the emission, so that a "standing wave" of photons is built up within the rod.

obtain white light or a continuous spectrum, because in the solid tungsten metal we have a *band* of electron energies and a band of sites into which excited electrons can fall. Another case of emission involves the "boiling off" of electrons from the incandescent filament of a radio tube to be accelerated through space by the potential between the electrodes.

16.3 Absorption

It is important to distinguish between color produced by emission and color produced by absorption. We see a yellow color *emitted* when salt is thrown in a fire because of the characteristic wavelength of the electron jumps in the sodium atom. However, if we look through a yellow glass filter at a white light, we see yellow light transmitted because the filter has absorbed the other colors and has merely allowed the already existing photons of yellow wavelength to pass through. The reaction, therefore, is between the electrons of the material and the photons of nonyellow wavelengths. The range of colors produced this way is called the *absorption spectrum*.

For example, glass is colored only by ions of the transition elements. Copper is considered a transition element in the Cu^{2+} state because it has an unfilled 3*d* shell. Typical colors are Cr^{2+}, blue; Cr^{3+}, green; Cu^{2+}, blue-green; Cu^{+}, colorless; Mn^{2+}, orange; Fe^{2+}, blue-green; U^{6+}, yellow (unfilled 5*f* shell). These colors result from the interaction of the inner-shell electrons with the photons of the various wavelengths of white light. The characteristic color that emerges is the unabsorbed part of the spectrum. Because the ions do not act independently of the glass matrix, the color is not a sharp wavelength, as it is in the emission spectrum, and the ions in the vapor have no such interaction with a matrix. The rare earths also lead to colored glasses because of the electron interaction in the 4*f* shell.

Another example of absorption is the photoelectric effect, in which the absorption of a photon results in the emission of an electron.

When a photon of light of energy hv (v = frequency = c/λ) strikes an electron in a block of metal and the electron is sufficiently energized to leave the block, we say that the energy of the electron is made up of two components:

1. The energy required to leave the block ($e\phi$)
2. The kinetic energy (K.E.) of the electron outside the block

Therefore, $hv = e\phi + \text{K.E.}$

The value $e\phi$ is a constant for the material at a given temperature. It is called the *work function*. Multiplying the work function by the charge on the electron gives the energy barrier that is surmounted as the electron leaves the material.

We would expect that the alkali metals would make good emitters because of the ease with which their electrons leave. This is the case. The yield of electrons varies with the wavelength of the light and the element. As the atomic number increases from sodium to cesium, the maximum yield occurs at larger wavelengths. In other words, it takes a photon of higher energy to knock out an outer-shell electron in sodium than in cesium because the outer electrons in cesium are further from the nucleus and are more loosely held.

EXAMPLE 16.2 The work function $e\phi$ for tungsten is 4.52 *eV*. What is the maximum wavelength of light that will give photoemission of electrons from this element?

ANSWER The energy of the photons will just equal the value they require to escape; in other words, kinetic energy is equal to zero. Therefore,

$$h\nu = e\phi = 4.52 \text{ eV}$$

$$(6.62 \times 10^{-27} \text{ erg-sec}) \left(10^{-7} \frac{\text{J}}{\text{erg}}\right) \nu = 4.52 \text{ eV} \left(1.6 \times 10^{-19} \frac{\text{J}}{\text{eV}}\right)$$

or

$$\nu = 1.09 \times 10^{15} \text{ sec}^{-1}$$

and

$$\lambda = \frac{c}{\nu} = \frac{(3 \times 10^{10} \text{ cm/sec})(10^{8} \text{ Å/cm})}{1.09 \times 10^{15} \text{ sec}^{-1}} = 2750 \text{ Å}$$

16.4 Reflection

Reflection can be an apparently simple phenomenon or a rather complex one. In the simple case we merely shine a beam of white light at a surface and receive a reflected beam of almost the same intensity and "spectrum," with the angle of incidence equaling the angle of reflection. On the other hand, we find a red color in the reflected beam from copper, a yellow color from gold, and similar but less noticeable variations from other metals. In all cases the photons interact with the electrons, since a photon can be considered an electromagnetic wave. However, the interaction is not the same for all wavelengths of light and depends on the energies of the different electron levels of the material. In metallic reflection, significant interaction can occur between the photons and the outer-shell electrons of the metal. Thus, in the case of copper, the photons corresponding to blue wavelengths are absorbed by $3d$ electrons, leaving a reddish color to be reflected.

16.5 Transmission

Now let us consider why metals are opaque in all but very thin sections and why ionic and covalent solids are transparent to visible light. At the same time we can discuss the more interesting subject of why materials we consider opaque, such as silicon and germanium, are transparent to infrared.

Metals We recall that we can think of the valence electrons of a metal as having a band of energies with many unfilled energy states at the upper energy levels. When a photon of light strikes an electron, the electron can be raised to an unoccupied higher level, and the radiation is absorbed rather than transmitted. We can transmit visible light in metals only in very special cases, such as in a very thin sheet of gold, in which the probability of collision during transmission is reduced.

Ionic and Covalent Solids In this case the electrons are bound to the atoms. It takes a high-energy photon to break an electron bond and accelerate the electron to the conduction band. (Recall that there is a gap between the valence and conduction bands in these materials.) For this reason photons of visible and infrared wavelengths pass through ionic solids with ease. However, very high-energy radiation such as ultraviolet is absorbed because the energy of these photons can be used to accelerate the electrons to the conduction band. If atoms of an impurity are present, absorption can occur. Such is the case with chromium ions in ruby, which absorb the photons of blue and green wavelengths. The remaining photons (or other wavelengths) in white light pass through, giving an overall red spectrum.

Semiconductors These materials provide a very interesting special case of transmission. Only relatively little energy is required to raise the carriers to the conduction band and therefore absorb radiation. It happens that the energy of the photons in infrared is not sufficient (long wavelength and, therefore, low energy), so that semiconductors are transparent to infrared radiation and can be used as windows or lenses for efficient transmission. However, the energy of visible light is sufficient to raise the carriers to the conduction band; the semiconductors appear opaque in visible light.

EXAMPLE 16.3 Explain the following transmission effects of white light on the given materials.

Material	*Energy gap, eV*	*Color*
Diamond	5.6	Colorless
Sulfur	2.2	Yellow
Silicon	1.1	Opaque

ANSWER Referring to Example 16.1, we see that none of the photons of the visible spectrum is energetic enough to react with the electrons of diamond and raise them to the conduction band. Therefore, the light is transmitted unchanged. In sulfur the green and blue wavelengths are absorbed, leaving yellow (and red). In silicon all the visible light is absorbed because only 1.1 eV is required for photon–electron interaction, and this is available from the light.

The use of light to activate electrons of a semiconductor to the conduction band is called *photoconductivity*. It is possible to make a semiconductor with an energy gap that corresponds to the energy of photons of light of a certain wavelength. If the wavelength is longer than the given wavelength, the photons are not energetic enough to activate electrons to the conduction band. If the wavelength is much shorter, the light is so heavily absorbed that only the surface is affected; since most of the crystal is inactive, the overall photoconductivity is small. At just the critical wavelength the response is high, as shown in Fig. 16.6. By this technique one can detect a missile or aircraft, both rich infrared emitters, from hundreds of miles away.

Color Centers The characteristics of transmission may be altered by impurities. This is especially evident in the colors developed in transparent crystals such as ruby and sapphire. This is due to the development of color centers. You can find the most striking evidence of this effect by heating a transparent uncolored crystal of sodium chloride in sodium vapor and observing the development of a yellow color in the material. The mechanism is as follows: Sodium atoms condense on the surface, and chloride ions migrate from the inner regions to combine with the sodium. This leaves ionic vacancies. To balance the charges in the crystal, these vacancies attract electrons. These electrons are not held as rigidly as the normal valence electrons and can therefore be raised to the conduction band by lower-energy photons, such as those in the visible spectrum. Thus, when white light passes through the

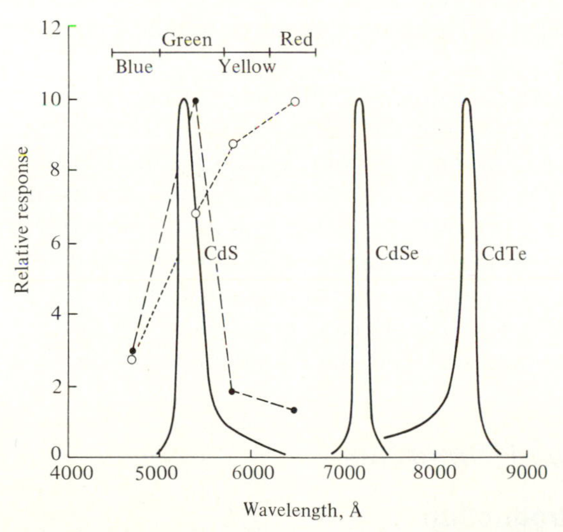

Fig. 16.6 *Photoconductivity of* CdS, CdSe, *and* CdTe, *showing a maximum response at wavelengths corresponding to the width of the forbidden energy gap. The data circles show the response of commercial* CdS *and* CdSe *photocells to colored light bulbs.* (J.J. Brophy, *Semiconductor Devices*, McGraw-Hill Book Company, New York, 1964)

crystal, certain wavelengths are absorbed, and the remaining beam is colored. Ions of impurities can cause the same effect by providing electrons that are easily raised to the conduction band. Therefore photons of certain wavelengths are selectively absorbed.

A number of important new glasses depend on the interaction of light with impurities in the glass. For years people noticed that the windowpanes in old buildings became purple with age. It was found that this was due to the reaction of ultraviolet light with trace amounts of manganese. The Mn^{2+} ion is colorless, but reaction with a photon causes it to lose another electron and become Mn^{3+}, which absorbs most of the visible spectrum except violet, thereby giving the glass a purple hue. In this case the electron from the Mn^{2+} is probably trapped by iron impurities, so that the process is not reversible. However, the popular photochromic eyeglasses and windows containing silver ions use a reversible process. When the light is bright, the photons reduce Ag^{+} ions to metallic silver, causing a darkening. When the light source is removed, the process reverses and colorless Ag^{+} ions are regenerated.

16.6 Refraction

Everyone has seen that a stick immersed in water appears to be bent, starting at the water–air interface. A useful concept is that the velocity of light appears to be lower in water. The index of refraction n is defined as

$$n = \frac{c}{v}$$

where c = velocity of light in a vacuum and v = velocity of light in the material.

Liquids or solids with dense atomic packing or elements of high atomic number generally have higher indices of refraction (i.e., lower velocities of light). In some crystals the index of refraction is not the same in all directions, just as there are differences in magnetic properties in different directions. Crystals of lower symmetry, such as calcite and quartz, split the incident beam into polarized beams. This is called *birefringence* or *double refraction*. Not only is this characteristic useful in identifying minerals, but it is used in the material Polaroid. In this case crystals of iodoquinone are encased in plastic. These filters absorb the glare component of the light that is polarized.

Thermal Properties

16.7 Introduction

When a solid is heated, there are three important engineering effects. The solid itself absorbs heat, transmits heat, and expands. The first effect is described by the heat capacity C_p, which is the energy required to raise the temperature of a mole of the solid 1 K, or cal/mole-K. The second is defined by the thermal conductivity or the energy (calories) flowing per second from one face to another of a cube of 1-cm dimensions with 1 K temperature difference between the faces, that is, $(cal/cm^2)/(cm/sec\text{-}K)$ or cal/cm-sec-K. The third is the change in length per unit of length per K, that is, cm/cm-K or simply $(K)^{-1}$. [The interrelation of these units with British thermal units (BTU) and with the joule—the preferred energy unit—is given in the problems.] The interesting problem is to relate these characteristics to structure.

We can start by considering what happens as we heat a simple metal such as silver from 0 K. The energy is taken up by the increasing amplitude of vibration of silver ions, and the electrons at the upper level of the conduction band are accelerated. From this simple model we can explain all three thermal properties.

16.8 Heat capacity

For our purposes it is sufficient to discuss C_p, the heat capacity at constant pressure. It is simpler to begin with the heat capacity at say 20°C. We find that most simple solids show a value of approximately 6 cal/mole-K (25.1 J/mole-K). Boltzmann showed that this could be rationalized by considering the energy of the vibrating atoms. From elementary chemistry we know that the internal energy of a mole of perfect gas is $\frac{3}{2}RT$, where R is the gas constant of 1.987 cal/mole-K (8.3 J/mole-K) and T is the absolute temperature (K). For bound atoms vibrating in a solid there is an added potential-energy term of the same $\frac{3}{2}RT$, giving total energy of $3RT$ or approximately $6T$. Since the energy equals C_pT (from 0 to TK), then $C_p = 6$ cal/mole-K (25.1 J/mole-K).

This simple relation does not apply at low temperatures because of effects discussed in quantum mechanics, nor does it apply to complex solids in which the rotational energy of molecules is involved. The contribution of the electrons to specific heat is also low, except at low temperatures.

We should note in passing that, although the heat capacity per *mole* is constant for many simple substances, in engineering calculations we must realize that there is a large difference in heat capacity per *gram*. For example, the value for iron is 0.107 cal/g (0.448 J/g), whereas the value for lead is 0.031 cal/g (0.1296 J/g). Also, for light elements the heat capacity per mole is lower until elevated temperatures are reached.

16.9 Thermal conductivity

The carriers of thermal energy are electrons in metals and phonons (quanta of energy) in other solids. The velocity of a phonon is the speed of sound, but since many collisions occur, the net thermal conductivity is low compared with that of metals (Table 16.1, page 670).

Table 16.1 THERMAL CONDUCTIVITY OF VARIOUS MATERIALS AT 300 K

Material	*Conductivity, cal/cm-sec-K**
Aluminum	0.53
Copper	0.94
Iron	0.18
Silver	1.00
Carbon (diamond)	1.50
Germanium	0.14
70% Ni, 30% Cr solid solution	0.034
70% Cu, 30% Zn solid solution	0.24
Steel	0.12
Sodium chloride	0.017
Potassium chloride	0.017
Silver chloride	0.0026
Glass	0.0019

*Multiply by 4.1868 to obtain J/cm-sec-K.
Source: R. M. Rose, L. A. Shepard, and J. Wulff, *The Structure and Properties of Materials,* Vol. 4: Electronic Properties, John Wiley & Sons, Inc., New York, 1966.

Since electrons serve as carriers in metals, we would expect good electrical conductivity to be an indication of good thermal conductivity. This is shown by the Wiedemann-Franz relationship:

$$\frac{\text{Thermal conductivity}}{\text{Electrical conductivity} \times T} = L \text{ (practically a constant)}$$

where the Lorenz number $L = 1.6$ to $2.5\ (\text{volts/K})^2 \times 10^{-8}$ for most metals at $T = 20°\text{C}$ (293 K).

As alloys are added, the thermal conductivity, like the electrical conductivity, falls off rapidly, often to a value only one-tenth that of the pure metal. An example is in the construction of stainless-steel cooking utensils. For better conductivity a sandwich of stainless steel with a low-carbon iron center is used so that the influence of the practically pure iron will level out temperature gradients. An alternative is the use of a copper base bonded to the stainless steel.

16.10 Thermal expansion

If we use a classical model of atom or ion vibrations back and forth about fixed centers as a sample is heated, we cannot explain expansion. However, let us review the attractive and repulsive forces between two ions that determine the interionic distance (Fig. 16.7).

As we bring two ions together, we see that the attractive forces dominate until we reach the bottom of the trough at radius r_0. To bring the atoms

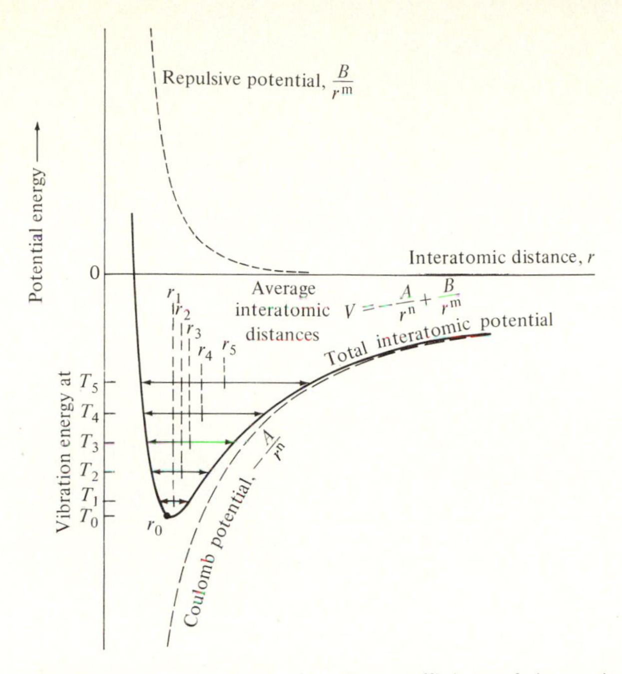

Fig. 16.7 *Basic relationships that determine the coefficient of thermal expansion. One can find the change in interatomic distance with temperature as follows. At* 0 K, *as the electron clouds overlap, the balance between the attractive forces of oppositely charged ions (the Coulomb potential) and the repulsive forces due to the overlap leads to the interatomic distance* r_0. *As the material is heated, the average distance between atoms becomes* r_1, r_2, *and so forth. The change in distance per* K *is the coefficient of expansion. Values of exponents* n *and* m *depend on the type of bonding; that is, metallic, ionic, or covalent.* (R.M. Rose, L.A. Shepard, and J. Wulff, *The Structure and Properties of Materials*, Vol. 4: Electronic Properties. John Wiley & Sons, Inc., New York, 1966)

closer requires additional energy. Now the atoms show radial spacing r_0 only at 0 K. At higher temperatures the atoms vibrate with increasing amplitude, as at T_1 and T_2. Note that the average distance of separation is r_1 and r_2, which do not lie above r_0 because of the different effects of temperature on the attractive and repulsive forces. The quantity $\Delta r/\Delta T$ is the familiar coefficient of thermal expansion, which is characteristically different for different structures.

Ceramics have low coefficients because the attractive forces are almost as great as the repulsive forces as the temperature increases. Values range from 1.2×10^{-6} (K)$^{-1}$ for diamond to 6.7×10^{-6} (K)$^{-1}$ for alumina (Table 16.2, page 672). There is great variation in glasses. In pure fused silica the coefficient is low, 0.05×10^{-6} (K)$^{-1}$, while in window glass the bonding is weaker and the coefficient is 7.2×10^{-6} (K)$^{-1}$. Correspondingly the melting and softening points of fused silica are higher than those of window glass.

Table 16.2 THERMAL PROPERTIES OF SELECTED MATERIALS

Material	Specific Heat C_p(at 300 K), cal/g-K*	Atomic or Molecular Weight, g/mole	Linear Coefficient of Expansion [at 300 $(K)^{-1} \times 10^{-6}$]	Lorenz Number† $L = \sigma_T/\sigma_e T$, $(volts/K)^2 \times 10^{-8}$	Melting Point T_m, K
Aluminum	0.22	27	24.1	2.2	933
Carbon					
Graphite	0.18	12	2.3 to 2.8		3760
Diamond	0.12		1.2		
Copper	0.092	63.5	17.6	2.23	1356
Gold	0.031	197	13.8	2.35	1336
Iron	0.11	55.9	14.0	2.47	1810
Lead	0.32	207.2	28	2.47	600
Molybdenum	0.065	96	5.55	2.61	2880
Nickel	0.105	58.7	13.3	2.2	1725
Niobium	0.074	92.9	7.4		2740
Platinum	0.031	195.1	8.8	2.51	2040
Silver	0.056	107.9	19.5	2.31	1333
Tantalum	0.036	181	6.7		3270
Tin	0.54	118.7	23.5	2.52	505
Tungsten	0.034	183.9	3.95	3.04	3680
Type 304 stainless steel (austenitic)	0.14		17.3		~1690
Invar	0.12		1.26		~1700
Alumina ceramic	0.18	101.9	6.7		2470
Boron nitride		24.8	7.72		1950
Magnesia (MgO)	0.21	40.3	14		3020
Fused silica (or fused quartz)	0.19	60.1	0.05		1950
Glass	0.2		7.2		
Mica	0.12 to 0.25		40		
Phenolic resins	~0.4		15 to 45		
Polymethylmethacrylate	0.35		81		
Polytetrafluoroethylene	0.25		100		
Low-density polyethylene	0.5		180		
Natural rubber			670		
Silicone rubber			1,200		

*Multiply by 4.186 to obtain J/g-K. †At 20°C (239K).
Source: R. M. Rose, L. A. Shepard, and J. Wulff, *The Structure and Properties of Materials*, Vol. 4: Electronic Properties. John Wiley & Sons, Inc., New York, 1966.

In the metals there is a close correlation between the melting point, which is an index of bonding, and this coefficient. Aluminum, for example, has a coefficient of 24.1×10^{-6} $(K)^{-1}$, whereas that of tungsten is 3.95×10^{-6} $(K)^{-1}$. Of special interest is the difference between α iron, with a coefficient of about 14×10^{-6} $(K)^{-1}$, and γ iron, with 18×10^{-6} $(K)^{-1}$. As illustrated in the problems, the austenitic stainless steels exhibit the same level of expansion as γ iron, whereas other steels resemble α iron. This difference can cause severe stresses when structures containing different types of steel are heated. In gas turbines, interference between mating parts made of α and γ structures has been encountered.

A bimetallic strip is an example of how differential thermal expansion can be used in temperature control relays. If we bond two metals together, as by hot rolling, we form a bimetallic strip. If a piece is held at one end and heated, it will bend if the metals have different coefficients of expansion. The amount the strip will bend depends on the temperature.

EXAMPLE 16.4 (This example will help you to understand the action of a thermostat.)

Given strips of iron and copper 0.010 in. wide, 0.005 in. thick, and 1 in. long ($0.254 \times 0.127 \times 25.4$ mm), bond them together lengthwise along the flat faces at 70°F (21°C). Then heat the bimetal to 170°F (77°C).

a. Which way will the strip curl?
b. If the bimetal is restrained from curling (as in a die), what will be the stress developed in the iron and in the copper portions?

Assume the following data:

	Linear Coefficient of Expansion, in./in.-°F (m/m-K)	*Modulus of Elasticity, psi (MPa)*
Iron	7.8×10^{-6} (14×10^{-6})	30×10^{6} (2.07×10^{5})
Copper	10.0×10^{-6} (18×10^{-6})	16×10^{6} (1.10×10^{5})

ANSWER

a. If unbonded, the copper strip would expand more than the iron strip. The bonded strip, therefore, curls into an arc with the copper on the outside and the iron on the inside.
b. If the bonded strip is restrained between flat plates, the copper is prevented from attaining the length it would have reached if it were not bonded to the iron. It is under compressive stress. The iron, on the other hand, is stretched by the tendency of the copper to expand a greater amount, and so the iron is under tension. Since the cross-sectional areas are equal, the compressive stress in

the copper is the same magnitude as the tensile stress in the iron (or movement would occur due to the unbalanced force):

$$\sigma_{Cu} = \sigma_{Fe}$$

Now for a moment consider the expansion that would have occurred in each metal if it had not been bonded to the other during heating from 70 to 170°F. The change in length in the iron:

$$\Delta l = (170 - 70)°F \times 7.8 \times 10^{-6} \text{ in./in.-°F}$$
$$= 780 \text{ microinches (in an inch length) or 780 microns/m}$$

For the copper

$$\Delta l = (170 - 70)°F \times 10.0 \times 10^{-6} \text{ in./in.-°F}$$
$$= 1000 \text{ microinches (in an inch length) or 1000 microns/m}$$

Therefore the free iron specimen would be 1.000780 in. long and the free copper 1.001000 in.

Returning to the bonded case, we find that the specimen would have a length between these values. Furthermore, the difference in elastic strain between the copper and the iron would equal 1000 − 780 microinches/in., or 220 microinches/in. In other words, the copper attempts to expand 1000 microinches/in. and the iron only 780. An accommodation is reached in which the copper side is under compression and the iron side in tension. The important point is that the elastic strain in the copper plus that in the iron will equal 220 microinches/in. or 220 microns/m:

$$\epsilon_{Cu} + \epsilon_{Fe} = 220 \times 10^{-6}$$

Substituting the relation $\epsilon = \sigma/E$ in each case,

$$\frac{\sigma_{Cu}}{E_{Cu}} + \frac{\sigma_{Fe}}{E_{Fe}} = 220 \times 10^{-6}$$

But we know that

$$\sigma_{Cu} = \sigma_{Fe} \qquad \text{or} \qquad E_{Cu}\,\epsilon_{Cu} = E_{Fe}\,(220 \times 10^{-6} - \epsilon_{Cu})$$

Substituting the numerical values given for E_{Cu} and E_{Fe}, we obtain:

$\epsilon_{Cu} = 143 \times 10^{-6}$ in./in. or 143 microns/m (σ_{Cu} = 2300 psi or 15.9 MPa compression)

$\epsilon_{Fe} = 77 \times 10^{-6}$ in./in. or 77 microns/m (σ_{Fe} = 2300 psi or 15.9 MPa tension)

In the polymers the coefficients of expansion are orders of magnitude greater as a result of the low intermolecular bonding forces. However, where bonding is weakest, in the elastomers, the coefficients are 670 to 1200 × 10^{-6} $(K)^{-1}$, and in Teflon the coefficient is 100 × 10^{-6} $(K)^{-1}$. These values must be kept in mind if cracking resulting from differences in expansion on heating is to be avoided in metal–brittle plastic assemblies.

The low coefficient of expansion of the iron-nickel alloy known as Invar *below* 200°C is an interesting example of the interaction between thermal and magnetic effects. Increasing the temperature tends to produce expansion, but because of the special strong magnetic attraction between atoms in this alloy, only a small net expansion takes place. However, above the Curie temperature (where ferromagnetism disappears) the alloy expands normally.

Other anomalous effects are the differences in coefficients of expansion in different crystal directions in cubic as well as other systems. These exist in both metals and ceramics. Finally, a word of caution in design: The coefficient of expansion should never be taken as an average over a temperature range where phase transformation takes place, or catastrophic failure can occur as a result of the great dilation due to transformation alone.

SUMMARY

The optical properties of major importance are emission, absorption, reflection, transmission, and refraction. The basis for understanding each of these phenomena is the relation between electrons and photons, which may be thought of as little energy packets with definite energy and wavelength.

Emission of a photon occurs when an electron falls from a higher to a lower state or quantum level in an atom. For example, sodium atoms emit yellow light because the jump of the predominant electron is accompanied by the emission of a photon of about $\lambda = 5890$ Å. In fluorescent and phosphorescent materials, electrons absorb photons of higher energy, such as x rays, then fall back through a series of electron traps formed by atoms of impurities. The fall from an intermediate trap to the normal level is accompanied by the emission of visible light.

Absorption is the reverse of emission. A photon strikes an electron that can be raised to an existing higher state by absorbing energy. This explains why metals are opaque. There is such a selection of empty energy states for electrons that are continuous with valence bands that photons of practically any wavelength can be absorbed.

Reflection is the re-emission of light from a surface. If there is no selective absorption, i.e., absorption of photons of certain energy levels by electrons, the light is re-emitted with the color unchanged. In some cases, such as copper and gold, photons in the blue and green portions of the spectrum are absorbed.

Transmission is closely related to absorption. A diamond, for instance, is transparent to visible light because photons of these wavelengths do not have enough energy to raise the electrons through the energy gap to the conduction band. Therefore, the light passes through unaffected. Color centers may lead to a change from white to colored light. These centers result from impurities or lattice defects that lead to regions in the structure with electrons that can absorb certain photons of the visible spectrum. The remaining transmitted light is colored.

Refraction is due to the slower speed of light in a material compared with a vacuum, and the index of refraction increases with density.

The thermal properties of major importance are heat capacity, thermal conductivity, and thermal expansion.

Heat capacity is the energy required to raise the temperature of a system 1 K. It may be thought of as increasing the kinetic and potential energies of the atoms. Since a mole contains a definite number of atoms, for most solids the heat capacity per mole is constant [6 cal/mole-K (25.1 J/mole-K)].

Thermal conductivity is due to the motion of carriers, such as free electrons and photons. Because of the relatively great quantity of free electrons in metals, the conductivity is great. Thermal expansion results from the fact that although atoms vibrate about an average position, the distances between average positions change with temperature. The greater the bonding forces, the lower the change in position. For this reason the covalent and some of the ionic solids have the lowest coefficients. The metals are intermediate, and finally the polymers with van der Waals bonds have the highest coefficients.

DEFINITIONS

Absorption The preferential absorption of light of certain wavelengths

Color centers The preferential absorption and transmission of photons of certain (colored) wavelengths because certain ions (transition elements) have inner-shell electrons that interact with photons of certain wavelengths and do not interact with others

Fluorescence The emission of light in the visible spectrum by a material exposed to ultraviolet light

Heat capacity, C_p The heat required to raise the temperature of a system 1 K, measured in cal/mole-K

Light emission The emission of photons of wavelengths in the visible spectrum, caused by electron jumps from higher to lower energy levels

Opacity The state in which photons interact with electrons and raise electrons to higher levels (no light is transmitted). Opacity occurs, for example, in metals.

Phosphorescence The continued emission of light in the visible spectrum after the source of the ultraviolet light has been removed.

Photoelectric effect The absorption of photons of definite wavelengths, leading to emission of electrons, i.e., photoelectric current

Thermal conductivity The quantity of heat transmitted through a unit volume in a unit time. Thermal conductivity = cal/cm-sec-K.

Thermal expansion The change in unit length per unit of temperature change; $\Delta l/\Delta T$ = cm/cm-K = $(\mathrm{K})^{-1}$

Transparency The state in which photons of transmitted light do not interact with electrons of the structure. The photon energy is insufficient to liberate bound electrons. An example of a transparent material is the diamond.

PROBLEMS

16.1 The most valuable diamonds have a bluish tint. Given that the wavelength of blue light is 4500 Å and the energy gap of pure diamond is 6 eV, explain the source of the color. (Sections 16.1 through 16.6)

16.2 An absorption-vs.-wavelength curve for a material shows a strong maximum at a wavelength equal to 1500 Å, but little absorption at longer wavelengths. (*a*) Calculate the energy gap. (*b*) Would the material be transparent, colored, or opaque in transmitting white light? (*c*) Would you expect the material to be metallic or nonmetallic? (Sections 16.1 through 16.6)

16.3 Calculate whether barium, with a work function of 2.50 eV, would be more suitable than the tungsten in Example 16.2 for use as a photocell with visible light. (Sections 16.1 through 16.6)

16.4 In expressing the transmission of visible light, one often uses the following relationship:

$$I = I_0 e^{-\alpha x}$$

where I = intensity of transmitted light, I_0 = intensity of incident light, α = absorption coefficient, and x = thickness of the material. The value of α is a material constant dependent on the wavelength of the photons. Qualitatively, how does α and hence I vary with the atomic number of the transmitting media and the wavelength of the radiation? (Sections 16.1 through 16.6)

16.5 A further refinement in the description of the transmission of light uses the relationship

$$I/I_0 = (1 - R)^2 e^{-\alpha x} \qquad \text{(see Problem 16.4)}$$

where

$$R = \left(\frac{n-1}{n+1}\right)^2 = \text{reflectivity} \qquad \text{and} \qquad n = \text{index of refraction}$$

Qualitatively explain why lenses are often coated, and why crystalline polymers are most often translucent rather than transparent. (Sections 16.1 through 16.6)

16.6 Gamma rays have higher energy than x rays. Why is lead shielding used more often than iron shielding? When one is radiographing a metal component to find internal imperfections, why does a pore appear darker on an x ray than solid material? [In Problem 16.4, for high-energy radiation, $\alpha = (\mu/\rho)\rho$, where (μ/ρ) is the mass absorption coefficient and ρ is the density.] (Sections 16.1 through 16.6)

16.7 The steady-state heat flow in a material is directly proportional to the thermal conductivity. Why are bricks for furnaces deliberately made with voids? What are the characteristics of home insulation? (Sections 16.7 through 16.10)

16.8 Calculate how close the heat-capacity values C_p of the following materials come to the ideal value of 6 cal/mole-K (25.1 J/mole-K) at 298 K (room temperature). (*a*) Copper: 0.092 cal/g-K (0.324 J/g-K) (*b*) Nickel: 0.105 cal/g-K (0.44 J/g-K) (*c*) N_2: 0.24 cal/g-K (1.04 J/g-K) (*d*) Aluminum: 0.22 cal/g-K (0.92 J/g-K) (Sections 16.7 through 16.10)

16.9 In the production of aluminum cylinder heads, wear-resistant valve guides are needed because the Inconel "X" valve stems wear the aluminum block too rapidly. After much experimentation austenitic gray cast iron (Ni-resist) valve-guide inserts were found to be successful. Hardened ferritic steel inserts also wore satisfactorily, but loosened up from a shrink fit in the aluminum when placed in service. (A shrink fit is obtained by machining the outside dimension of the valve guide slightly oversize, cooling the guide in dry ice to below the outside dimension of the hole, and inserting it while it is cold.) The coefficients of expansion of the materials are:

Aluminum block: $12 \times 10^{-6}\ (°F)^{-1}$ [$21.6 \times 10^{-6}\ (°C)^{-1}$]
Austenitic cast iron: $11 \times 10^{-6}\ (°F)^{-1}$ [$19.8 \times 10^{-6}\ (°C)^{-1}$]
Hardened steel: $7 \times 10^{-6}\ (°F)^{-1}$ [$12.6 \times 10^{-6}\ (°C)^{-1}$]

Why do the austenitic cast iron inserts maintain the shrink fit better? (Sections 16.7 through 16.10)

16.10 Calculate the stresses in the bimetallic strip of Example 16.4, given that the iron strip is 0.002 in. (0.00508 cm) thick and the copper strip is 0.005 in. (0.0127 cm) thick. (Sections 16.7 through 16.10)

16.11 Calculate the stresses per 100°F (55.6°C) temperature change in a bimetallic strip made of austenitic stainless steel and SAE 1020 steel. The coefficient of expansion of austenitic stainless steel is 9.6×10^{-6} in./in.-°F (17.3×10^{-6} m/m-°C), and that of SAE 1020 steel is 7.8×10^{-6} in./in.-°F (14.0×10^{-6} m/m-°C). Assume that the individual metals have equal cross-sectional areas. (Sections 16.7 through 16.10)

16.12 Figure 16.7 schematically shows the energy trough and equilibrium spacing of atoms for a material x. (*a*) Schematically show the trough for a more strongly bonded material. (*b*) Repeat part (*a*) for a more weakly bonded material. (*c*) Indicate the qualitative relationship between bond strength, coefficient of expansion, melting point, and modulus of elasticity. (Sections 16.7 through 16.10)

16.13 When there is a phase change, there is a discontinuity in the heat capacity of a material. The amount of energy absorbed in the phase change is referred to as the heat of transformation; examples are heat of fusion and heat of vaporization. Why is a water quench more severe than an oil quench for steel, and why does stirring provide faster cooling rates? (Sections 16.7 through 16.10)

PROBLEMS COVERING MATERIAL FROM OTHER CHAPTERS

16.14 Explain why the electrical conductivity of a semiconductor can be increased by shining a light of a given wavelength on it. This is called *photoconductivity*. How could you use this effect to activate the beam at a remote lighthouse? (Sections 16.1 through 16.6 and 14.1 through 14.10)

16.15 Glass is more easily fabricated into laser rods than single crystals of ruby. Early attempts to make lasers of glass with Cr^{3+} ions as impurity failed because the Cr^{3+} interacted with the ions of the glass rather than acting independently, as when substituted for Al^{3+} in alumina. However, satisfactory glass lasers have been made using neodymium in silicate glass. Look up the electronic structure of neodymium and explain what characteristic would lead to less interaction between the adjacent ions of the glass compared with Cr^{3+}. [*Hint:* Determine which electrons are the color centers in the neodymium.] (Sections 16.2 and 7.6, 7.7)

16.16 A thin piece of aluminum is partially oxidized at 1000°F (538°C) to Al_2O_3. It is feared that the oxide scale may crack off because of differential contraction between the two materials. Given that the tensile strength of Al_2O_3 is 25,000 psi (172.5 MPa) and the stress in both the Al_2O_3 and Al is the same (same cross-sectional area), calculate the temperature to which the composite can be cooled to just obtain the 25,000-psi (172.5 MPa) stress. [*Hint:* Treat this as a bimetallic strip.]

Thermal coefficient: Al, 12.5×10^{-6} in./in.-°F (22.5×10^{-6} m/m-°C)
Al_2O_3, 5×10^{-6} in./in.-°F (9.0×10^{-6} m/m-°C)

Moduli: Al, 10×10^{6} psi (6.9×10^{4} MPa); Al_2O_3, 50×10^{6} psi (3.45×10^{5} MPa)

(Sections 16.7 through 16.10 and 12.23, 12.24)

REFERENCES

Without trying to be comprehensive, we here list a few references for the convenience of students and instructor. They are divided into four groups: (1) Metals, (2) Ceramics, (3) Polymers, and (4) General. In each group, the first references are those that can be used readily by students who have completed this course, in order to find specifications for materials in their professional work. Then follows a list of texts for study beyond the scope of this book.

Metals

American Society of Testing Materials Specifications (*see General*)

Metals Handbook, 8th ed., American Society for Metals, Metals Park, Ohio

This is not one handbook, but eleven large volumes, beginning with volume 1, *Properties and Selection of Metals,* and advancing through topics such as heat treatment, casting, welding, phase diagrams, microstructures, failure analysis, and nondestructive testing. An authoritative source, written under the supervision of committees of technical experts. The 9th edition is composed of two volumes treating ferrous and nonferrous metals separately.

Metals Reference Book, 4th ed., C. J. Smithells, Plenum Publishing Corp., New York, 1967

Three volumes; has many tables of properties: diffusivity, thermochemical data, etc.

The Structure of Metals, 3d ed., C. S. Barrett and T. B. Massalski, McGraw-Hill, New York, 1966

This text features: (1) a thorough discussion of crystallographic methods such as x-ray diffraction techniques, and (2) excellent discussions of metallic structures and transformation mechanisms. Much of the material is at senior or graduate level.

Textbooks on Metallurgy

Atomic Theory for Students of Metallurgy, W. Hume-Rothery, Institute of Metals, London, 1955

Elements of Mechanical Metallurgy, W. J. M. Tegart, Macmillan, New York, 1966

Elements of Physical Metallurgy, A. G. Guy, Addison-Wesley, Reading, Mass., 1959

Elements of X-Ray Diffraction, B. D. Cullity, Addison-Wesley, Reading, Mass., 1956

High Strength Materials, V. F. Zackay, Wiley, New York, 1965

Mechanical Metallurgy, G. E. Dieter, McGraw-Hill, New York, 1961

The Mechanical Properties of Matter, A. H. Cottrell, Wiley, New York, 1964

Phase Diagrams in Metallurgy, F. N. Rhines, McGraw-Hill, New York, 1956

Physical Metallurgy, E. Birchenall, McGraw-Hill, New York, 1959

Physical Metallurgy, Bruce Chalmers, Wiley, New York, 1959

Physical Metallurgy Principles, 2d ed., R. E. Reed-Hill, Van Nostrand Reinhold, New York, 1973

Science of Metals, N. H. Richman, Blaisdell, Waltham, Mass, 1967

Structure and Properties of Alloys, 3d ed., R. M. Brick, R. B. Gordon, and A. Phillips, McGraw-Hill, New York, 1965

Ceramics

American Society of Testing Materials Specifications (*see General*)

Phase Diagrams for Ceramics, E. M. Levin, C. R. Robbins, and H. F. McMurdie, American Ceramic Society, Columbus, Ohio, 1964

Concrete-Making Materials, S. Popovics, Hemisphere Publishing Corp; McGraw-Hill, New York, 1979

Design and Control of Concrete Mixtures, 11th ed., Portland Cement Association, Skokie, Ill., 1968

Textbooks on Ceramics

Electronic Ceramics, E. C. Henry, Doubleday, Garden City, N.Y., 1969

Elements of Ceramics, F. H. Norton, Addison-Wesley, Reading, Mass., 1952
Excellent discussions of technique, as well as basic material

Glass-Ceramics, P. W. McMillan, Academic, London, 1964

Glass Science, R. H. Doremus, Wiley, New York, 1973

Introduction to Ceramics, W. D. Kingery, Wiley, New York, 1960
General coverage of ceramic structures and properties

Physical Ceramics for Engineers, L. H. Van Vlack, Addison-Wesley, Reading, Mass., 1964
Basic coverage with many illustrative problems

Polymers

American Society for Testing Materials Specifications (*see General*)

"Materials Selector Guide," *Materials and Methods*, Reinhold, Stanford, Conn., published yearly

Modern Plastics Encyclopedia, McGraw-Hill, New York, published yearly

Textbooks on Polymers

Fundamental Principles of Polymeric Materials for Practicing Engineers, S. L. Rosen, Barnes and Noble, New York, 1971

Mechanical Properties of Polymers, L. E. Nielsen, Van Nostrand Reinhold, New York, 1962

Organic Polymers, T. Alfrey and E. F. Gurnee, Prentice-Hall, Englewood Cliffs, N.J., 1967

Properties and Structure of Polymers, A. V. Tobolsky, Wiley, New York, 1960

Textbook of Polymer Science, 2d ed., F. W. Billmeyer, Wiley, New York, 1971

General

American Society for Testing Materials Specifications, American Society for Testing Materials, Philadelphia. A number of separate volumes issued triennially that cover metals, ceramics, and polymers; the most commonly used standard for specification.

Crystal Structures, R. W. G. Wykoff, Wiley, New York, 1963. Several volumes giving details of crystal structures

Materials Handbook, G. S. Brady, McGraw-Hill, New York, 1951

"Materials Selector Guide," *Materials and Methods*, Reinhold, Stanford, Conn., published yearly

Textbooks on Materials in General

Corrosion Control, H. H. Uhlig, Wiley, New York, 1963

Corrosion Engineering, M. B. Fontana and N. D. Green, McGraw-Hill, New York, 1967

Electronic and Magnetic Properties of Materials, A. Nussbaum, Prentice-Hall, Englewood Cliffs, N.J., 1967

Electronic Processes in Materials, L. V. Azaroff and J. J. Brophy, McGraw-Hill, New York, 1963

Introduction to Properties of Materials, D. Rosenthal, D. Van Nostrand, Princeton, N.J., 1964

Introduction to Solids, S. V. Azaroff, McGraw-Hill, New York, 1960

Mechanical Behavior of Engineering Materials, J. Marin, Prentice-Hall, Englewood Cliffs, N.J., 1962

Modern Composite Materials, L. J. Broutman and R. H. Krock, eds., Addison-Wesley, Reading, Mass., 1967

Physics of Solids, 2d ed., C. A. Wert and R. M. Thomson, McGraw-Hill, New York, 1970

Principles of Engineering Materials, C. R. Barrett, W. D. Nix, and A. S. Tetelman, Prentice-Hall, Englewood Cliffs, N.J., 1973

The Structure and Properties of Materials, A. T. Di Benedetto, McGraw-Hill, New York, 1967

The Structure and Properties of Materials, 4 vols., J. Wulff et al., Wiley, New York, 1965

An Introduction to Materials Science and Engineering, K. M. Ralls, T. H. Courtney, and J. Wulff, Wiley, New York, 1976

Elements of Materials Science and Engineering, 4th ed., L. H. Van Vlack, Addison-Wesley, Reading, Mass., 1980

The Nature and Properties of Engineering Materials, 2d ed., Z. D. Jastrzebski, Wiley, New York, 1977

Materials Science in Engineering, 3d ed., C. A. Keyser, Merrill, Columbus, Ohio, 1980

INDEX

Italicized numbers indicate pages on which terms are defined

FGHIJ-D-89876543

PHYSICAL PROPERTIES OF SELECTED ELEMENTS

Element	*Symbol*	*Atomic Number*	*Atomic Weight*	*MP (°C)*	*Density (g/cm^3)*	*Crystal Structure*	*Atomic Radius (Å)*	*Ionic Radius (Å)*	*Most Common Valence*
Aluminum	Al	13	26.98	660	2.699	FCC	1.43	0.57	+3
Argon	A	18	39.99	−189	1.78×10^{-3}	FCC	1.92	—	—
Barium	Ba	56	137.36	714	3.5	BCC	2.17	1.43	+2
Beryllium	Be	4	9.01	1277	1.85	HCP	1.14	0.54	+2
Boron	B	5	10.82	2030	2.34	Ortho.	0.97	0.2	+3
Bromine	Br	35	79.92	−7.2	3.12	Ortho.	1.19	1.96	−1
Cadmium	Cd	48	112.41	321	8.65	HCP	1.50	1.03	+2
Calcium	Ca	20	40.08	838	1.55	FCC	1.97	1.06	+2
Carbon[1]	C	6	12.01	3727	2.25	Hex.	0.71	<0.20	+4
Cerium	Ce	58	140.13	804	6.77	HCP	1.82	1.18	+3
Cesium	Cs	55	132.91	28.7	1.90	BCC	2.65	1.65	+1
Chlorine	Cl	17	35.46	−101	3.21×10^{-3}	Ortho.	1.07	1.81	−1
Chromium	Cr	24	52.01	1875	7.19	BCC	1.25	0.64	+3
Cobalt	Co	27	58.94	1495	8.85	HCP	1.25	0.82	+2
Copper	Cu	29	63.54	1083	8.96	FCC	1.28	0.96	+1
Fluorine	F	9	19.00	−220	1.70×10^{-3}	—	—	1.33	−1
Germanium	Ge	32	72.60	937	5.32	Dia.	1.22	0.44	+4
Gold	Au	79	197.00	1063	19.32	FCC	1.44	1.37	+1
Helium	He	2	4.00	−270	0.18×10^{-3}	HCP	1.79	—	—
Hydrogen	H	1	1.01	−259	0.09×10^{-3}	HCP	0.46	1.54	−1
Iodine	I	53	126.91	114	4.94	Ortho.	1.36	2.20	−1
Iron	Fe	26	55.85	1536	7.87	BCC	1.24	0.87	+2
Lead	Pb	82	207.21	327	11.36	FCC	1.75	1.32	+2
Lithium	Li	3	6.94	180	0.534	BCC	1.52	0.78	+1
Magnesium	Mg	12	24.32	650	1.74	HCP	1.60	0.78	+2
Manganese	Mn	25	54.94	1245	7.43	Cubic	1.12	0.91	+2
Mercury	Hg	80	200.61	−38.4	13.55	Rhomb.	1.50	1.12	+2
Molybdenum	Mo	42	95.95	2610	10.22	BCC	1.36	0.68	+4
Neon	Ne	10	20.18	−249	0.90×10^{-3}	FCC	1.60	—	—
Nickel	Ni	28	58.71	1453	8.90	FCC	1.25	0.78	+2
Niobium	Nb	41	92.91	2468	8.57	BCC	1.43	0.74	+4
Nitrogen	N	7	14.01	−210	1.25×10^{-3}	Cubic	0.71	0.1 to 0.2	+5
Oxygen	O	8	16.00	−219	1.43×10^{-3}	Ortho.	0.60	1.32	−2
Phosphorus[2]	P	15	30.98	44.3	1.83	Ortho.	1.09	0.3 to 0.4	+5
Platinum	Pt	78	195.09	1769	21.45	FCC	1.38	0.52	+2
Potassium	K	19	39.10	63.7	0.86	BCC	2.31	1.33	+1
Scandium	Sc	21	44.96	1539	2.99	FCC	1.60	0.83	+2
Silicon	Si	14	28.09	1410	2.33	Dia.	1.17	0.39	+4
Silver	Ag	47	107.88	961	10.49	FCC	1.44	1.13	+1
Sodium	Na	11	22.99	97.8	0.971	BCC	1.86	0.98	+1
Strontium	Sr	38	87.63	768	2.60	FCC	2.15	1.27	+2
Sulfur[3]	S	16	32.07	119	2.07	Ortho.	1.06	1.74	−2
Tin	Sn	50	118.70	232	7.30	Tetra.	—	0.74	+4
Titanium	Ti	22	47.90	1668	4.51	HCP	1.47	0.64	+4
Tungsten	W	74	183.86	3410	19.3	BCC	1.37	0.68	+4
Uranium	U	92	238.07	1132	19.07	Ortho.	1.38	1.05	+4
Vanadium	V	23	50.95	1900	6.1	BCC	1.32	0.61	+4
Zinc	Zn	30	65.38	419	7.13	HCP	1.33	0.83	+2
Zirconium	Zr	40	91.22	1852	6.49	HCP	1.58	0.87	+4

[1] Present as graphite—sublimes rather than melts.
[2] White phosphorus.
[3] Yellow sulfur.